T0221297

Handbook of
Small Animal Imaging

IMAGING IN MEDICAL DIAGNOSIS AND THERAPY

Series Editors: Andrew Karellas and Bruce R. Thomadsen

Published titles

IMAGING IN MEDICAL DIAGNOSIS AND THERAPY

Series Editors: Andrew Karellas and Bruce R. Thomadsen

Published titles

Ultrasound Imaging and Therapy
Aaron Fenster and James C. Lacefield, Editors
ISBN: 978-1-4398-6628-3

Cardiovascular and Neurovascular Imaging: Physics and Technology
Carlo Cavedon and Stephen Rudin, Editors
ISBN: 978-1-4398-9056-1

Handbook of Small Animal Imaging: Preclinical Imaging, Therapy, and Applications
George Kagadis, Nancy L. Ford, Dimitrios N. Karnabatidis, and George K. Loudos, Editors
ISBN: 978-1-4665-5568-6

Forthcoming titles

Physics of PET and SPECT Imaging
Magnus Dahlbom, Editor

Hybrid Imaging in Cardiovascular Medicine
Yi-Hwa Liu and Albert Sinusas, Editors

Scintillation Dosimetry
Sam Beddar and Luc Beaulieu, Editors

Handbook of
Small Animal Imaging

Preclinical Imaging, Therapy, and Applications

Edited by
George C. Kagadis
Nancy L. Ford
Dimitrios N. Karnabatidis
George K. Loudos

CRC Press
Taylor & Francis Group
Boca Raton London New York

CRC Press is an imprint of the
Taylor & Francis Group, an **informa** business

Cover Image: Courtesy of Nikolaos Karnabatidis

CRC Press
Taylor & Francis Group
6000 Broken Sound Parkway NW, Suite 300
Boca Raton, FL 33487-2742

First issued in paperback 2019

© 2016 by Taylor & Francis Group, LLC
CRC Press is an imprint of Taylor & Francis Group, an Informa business

No claim to original U.S. Government works

ISBN-13: 978-1-4665-5568-6 (hbk)
ISBN-13: 978-0-367-86735-5 (pbk)

This book contains information obtained from authentic and highly regarded sources. Reasonable efforts have been made to publish reliable data and information, but the author and publisher cannot assume responsibility for the validity of all materials or the consequences of their use. The authors and publishers have attempted to trace the copyright holders of all material reproduced in this publication and apologize to copyright holders if permission to publish in this form has not been obtained. If any copyright material has not been acknowledged please write and let us know so we may rectify in any future reprint.

Except as permitted under U.S. Copyright Law, no part of this book may be reprinted, reproduced, transmitted, or utilized in any form by any electronic, mechanical, or other means, now known or hereafter invented, including photocopying, microfilming, and recording, or in any information storage or retrieval system, without written permission from the publishers.

For permission to photocopy or use material electronically from this work, please access www.copyright.com (http://www.copyright.com/) or contact the Copyright Clearance Center, Inc. (CCC), 222 Rosewood Drive, Danvers, MA 01923, 978-750-8400. CCC is a not-for-profit organization that provides licenses and registration for a variety of users. For organizations that have been granted a photocopy license by the CCC, a separate system of payment has been arranged.

Trademark Notice: Product or corporate names may be trademarks or registered trademarks, and are used only for identification and explanation without intent to infringe.

Library of Congress Cataloging-in-Publication Data

Names: Kagadis, George C., editor. | Ford, Nancy L., editor. | Karnabatidis, Dimitrios N., editor. | Loudos, George K., editor.
Title: Handbook of small animal imaging : preclinical imaging, therapy, and applications / editors, George C. Kagadis, Nancy L. Ford, Dimitrios N. Karnabatidis, and George K. Loudos.
Other titles: Small animal imaging | Imaging in medical diagnosis and therapy.
Description: Boca Raton : Taylor & Francis, 2016. | Series: Imaging in medical diagnosis and therapy | Includes bibliographical references and index.
Identifiers: LCCN 2015024670 | ISBN 9781466555686 (hardcover : alk. paper)
Subjects: | MESH: Diagnostic Imaging--methods. | Diagnostic Imaging--veterinary. | Animal Diseases--diagnosis.
Classification: LCC SF757.8 | NLM SF 757.8 | DDC 636.089/60754--dc23
LC record available at http://lccn.loc.gov/2015024670

Visit the Taylor & Francis Web site at
http://www.taylorandfrancis.com

and the CRC Press Web site at
http://www.crcpress.com

To my wife, Voula, and our children, Orestis and Filippos,
who have supported me in so many ways during yet another project.
To my very good friend Bill Hendee, for his endless advice and for
our discussions at the West Branch Mill Pond, Wisconsin.

George C. Kagadis

To my parents, Alan and Sharon, for teaching me
the value of education and learning.

Nancy L. Ford

To my beloved family.

Dimitrios N. Karnabatidis

To my parents, Konstantinos and Kyriakoula, whose support
has given me the means to face all difficulties in life till now.
To my wife, Maria, and my daughter, Aggeliki, who show me
every day that there is more in life to learn and explore.

George K. Loudos

Contents

SECTION I—Introduction to Small Animal Imaging, Therapy, and Applications

SECTION II—Small Animal Imaging: Ionizing Radiation

SECTION VII—Image Quantification

SECTION VIII—Applications

Series Preface

Advances in the science and technology of medical imaging and radiation therapy are more profound and rapid than ever before, since their inception over a century ago. Further, the disciplines are increasingly cross-linked as imaging methods become more widely used to plan, guide, monitor, and assess treatments in radiation therapy. Today, the technologies of medical imaging and radiation therapy are so complex and so computer-driven that it is difficult for the persons (physicians and technologists) responsible for their clinical use to know exactly what is happening at the point of care when a patient is being examined or treated. The persons best equipped to understand the technologies and their applications are medical physicists, and these individuals are assuming greater responsibilities in the clinical arena to ensure that what is intended for the patient is actually delivered in a safe and effective manner.

The growing responsibilities of medical physicists in the clinical arenas of medical imaging and radiation therapy are not without challenges, however. Most medical physicists are knowledgeable in either radiation therapy or medical imaging, and expert in one or a small number of areas within their discipline. They sustain their expertise in these areas by reading scientific articles and attending scientific meetings. In contrast, their responsibilities increasingly extend beyond their specific areas of expertise. To meet these responsibilities, medical physicists periodically must refresh their knowledge of advances in medical imaging or radiation therapy, and they must be prepared to function at the intersection of these two fields. How to accomplish these objectives is a challenge.

At the 2007 annual meeting of the American Association of Physicists in Medicine in Minneapolis, this challenge was the topic of conversation during a lunch hosted by Taylor & Francis Group involving a group of senior medical physicists (Arthur L. Boyer, Joseph O. Deasy, C.-M. Charlie Ma, Todd A. Pawlicki, Ervin B. Podgorsak, Elke Reitzel, Anthony B. Wolbarst, and Ellen D. Yorke). The conclusion of this discussion was that a book series should be launched under the Taylor & Francis Group banner, with each book in the series addressing a rapidly advancing area of medical imaging or radiation therapy of importance to medical physicists. The aim for each book would be to provide medical physicists with the information needed to understand technologies driving a rapid advance and their applications to safe and effective delivery of patient care.

Each book in the series is edited by one or more individuals with recognized expertise in the technological area encompassed by the book. The editors are responsible for selecting the authors of individual chapters and ensuring that the chapters are comprehensive and intelligible to someone without such expertise. The enthusiasm of book editors and chapter authors has been gratifying and reinforces the conclusion of the Minneapolis luncheon that this series of books addresses a major need of medical physicists.

Imaging in Medical Diagnosis and Therapy would not have been possible without the encouragement and support of the series manager, Luna Han, of Taylor & Francis Group. The editors and authors, and most of all I, are indebted to her steady guidance of the entire project.

William R. Hendee
Founding Series Editor
Rochester, Minnesota

Preface

In the healthcare system and in preclinical health research, the important goal for imaging is to detect and identify disease (if possible at an early stage) and monitor response to treatment. Increasingly, multiple imaging modalities are being used in combination to give the physician complementary information and a more complete insight into the disease. In preclinical research, the need for multimodality imaging may be important not only to detect disease, but also to understand its underlying mechanisms and the role these molecular mechanisms play in its progression and severity.

Applications of small animal imaging and therapy are an important component of current medicine and deal with the development of new imaging and therapeutic approaches that need to be first tested and validated in animals before application to humans. Due to continuous progress in image acquisition and therapy technology, the field of small animal imaging and therapy is rapidly changing and there are many books written every year to reflect this evolution. Our goal in creating this book was to collate information about various imaging and therapeutic techniques used in preclinical research into a single book. We hope this book will be a reference useful for established researchers already employing some of these techniques, and also accessible to those entering the field for the first time.

The material in the book is grouped into eight sections. Section I deals with introductory material to small animal imaging, therapy, and research ethics. Section II is concerned with ionizing radiation, while Section III covers nonionizing radiation methods in small animal imaging. In Section IV, hybrid imaging is discussed, and in Section V, special focus is given to imaging agents. Section VI addresses therapeutic research platforms, and Section VII imaging quantification. Finally, Section VIII gives an overview of various small animal imaging and therapy applications to demonstrate the potential of the these techniques. We hope that this book will inspire readers with new ideas to integrate and exploit some of these technologies in their research.

The editors thank all of the authors for the outstanding contributions to this book, thank the publishing staff for their guidance and advice throughout the project, and acknowledge the COST Action TD1007: "Bimodal PET-MRI molecular imaging technologies and applications for in vivo monitoring of disease and biological processes," and the members of the Action, who have actively contributed several chapters. Finally, they express their gratitude to the artist, Mr. Nikolaos Karnabatidis, for preparing the front cover and end cover images of this book.

George C. Kagadis
Nancy L. Ford
Dimitrios N. Karnabatidis
George K. Loudos

Editors

George C. Kagadis, PhD, FAAPM, is an associate professor of medical physics and medical informatics at the University of Patras, Greece. He received his diploma in physics from the University of Athens, Greece in 1996 and both his MSc and PhD degrees in medical physics from the University of Patras, Greece, in 1998 and 2002, respectively. He is a Greek State Scholarship Foundation grantee, a Fulbright research scholar, and a fellow of the American Association of Physicists in Medicine (AAPM). From January 8, 2014 to July 14, 2015, he was a visiting assistant professor at The University of Texas, MD Anderson Cancer Center, where he currently holds an adjunct assistant professor appointment. He has authored about 80 journal papers and presented at more than 20 conferences. He has been involved in European and national projects, including e-health. His current research interests focus on medical image processing and analysis, studies in molecular imaging, IHE, and CAD applications. He is a *Medical Physics* editorial board member, AAPM website editor, chair of the AAPM website editorial board, chair of the European Affairs Subcommittee, member of the AAPM Molecular Imaging in Radiation Oncology Work Group, Electronic Media Coordinating, Equipment Donation Subcommittee, History, International Affairs, International Scientific Exchange Subcommittee, Newsletter Editorial Board, Online Learning Services Subcommittee, Placement Service Subcommittee, and Advertising Revenue and Policy.

Nancy L. Ford, PhD, was recruited to the University of British Columbia in May 2011 and is currently an assistant professor in the Department of Oral, Biological and Medical Sciences and the director of the UBC Centre for High-Throughput Phenogenomics, a core imaging facility. Previously, she was an assistant professor in the Department of Physics at Ryerson University in Toronto, Ontario (2006–2011). She holds a PhD in medical biophysics from the University of Western Ontario (2005) and a BSc (Hons) in chemical physics from the University of Waterloo (1997). Dr. Ford's research focuses on preclinical micro-computed tomography imaging and image-based analysis, primarily studying models of respiratory diseases, along with CBCT and MSCT for medical and dental applications. She is a full member of the American Association of Physicists in Medicine, serves as an associate editor of *Medical Physics*, and publishes in journals on medical physics, radiology, and dental science.

Dimitrios N. Karnabatidis, PhD, EBIR, is currently an associate professor of interventional radiology at the University of Patras, Greece. He graduated from the School of Medicine, University of Cluj-Napoca, Romania in 1988 and completed his PhD thesis from the School of Medicine, University of Patras, Greece in 2001. He holds the certificate of expertise of the European Board of Interventional Radiology, and in 2008 started his academic carrier in the School of Medicine at the University of Patras, Greece. Over the past 20 years, he has focused his scientific interests in angiogenesis in malignant and benign diseases; the augmentation of arteriogenesis in critical limb ischemia; endothelial hyperplasia inhibition after endovascular procedures; and ureteral obstructive diseases. He has been involved as a participant/principal investigator in several national/international research projects and has 122 publications in peer-reviewed journals. He is editor/reviewer in international academic journals and has received national/international awards, including the CVIR Editor's Medal for 2008 and 2011. He is a fellow of the Cardiovascular and Interventional Radiological Society of Europe and had been a member of the CIRSE Standards of Practice Committee. He is also a member of the Hellenic Radiology Society, the Hellenic Society of Interventional Radiology, and the Western Greece Radiology Society.

George K. Loudos, PhD, is an assistant professor in the Department of Biomedical Engineering at the Technological Educational Institute of Athens. He earned his diploma in electrical engineering in 1998 and PhD in biomedical engineering in 2003 from the National Technical University of Athens, Greece. Currently, he coordinates three EU projects and a National Excellence grant, and participates in four additional EU projects and several smaller projects. He has published 89 articles in international journals and has more than 200 publications in conference proceedings and around 600 citations. He has given 35 invited lectures and holds one patent. He has been an organizer of many international conferences, workshops, and training schools. His research interests are focused on molecular imaging using nuclear medicine techniques and medical instrumentation. He strongly supports interdisciplinary cooperation and education in the field of nanomedicine and molecular imaging.

Contributors

Dmitri Artemov
Division of Cancer Imaging Research
Department of Radiology School of Medicine
Johns Hopkins University
Baltimore, Maryland

Mohammad Reza Ay
Department of Medical Physics and Biomedical
   Engineering
and
Research Center for Molecular and Cellular
   Imaging
Institute for Advanced Medical Technologies
Tehran University of Medical Sciences
Tehran, Iran

Cristian Badea
Department of Radiology
Center for In Vivo Microscopy
Duke University
Durham, North Carolina

Marcel B. Bally
Department of Experimental Therapeutics
British Columbia Cancer Agency
and
Division of Pharmacology and Toxicology
The Centre for Drug Research and Development
and
Department of Pathology and Laboratory
   Medicine
University of British Columbia
Vancouver, British Columbia, Canada

Yannick Berker
Physics of Molecular Imaging Systems
Institute of Experimental Molecular Imaging
RWTH Aachen University
Aachen, Germany

Dongmei Chen
School of Life Science and Technology
Xidian University
Xi'an, Shaanxi, People's Republic of China

Xueli Chen
School of Life Science and Technology
Xidian University
Xi'an, Shaanxi, People's Republic of China

Wenxiang Cong
Biomedical Imaging Cluster
Rensselaer Polytechnic Institute
Troy, New York

Irene Cuadrado
Cardiovascular Joint Research Unit
Department of Cardiology
University Hospital Ramon y Cajal
University Francisco de Vitoria School of
   Medicine
Madrid, Spain

Athanasios Diamantopoulos
Department of Interventional Radiology
Guy's and St Thomas' NHS Foundation Trust
King's Health Partners
London, United Kingdom

Lawrence W. Dobrucki
Department of Bioengineering
Beckman Institute for Advanced Science and
    Technology
University of Illinois at Urbana–Champaign
Urbana, Illinois

Eleni K. Efthimiadou
Sol–Gel Laboratory
Institute of Nanoscience and Nanotechnology
NCSR D
Attiki, Greece

Jesús Egido
Renal, Vascular and Diabetes Research
    Laboratory, IIS-Fundación Jiménez Díaz
Autonoma University
Madrid, Spain

Nancy L. Ford
Centre for High-Throughput Phenogenomics
and
Department of Oral Biological and Medical
    Sciences
University of British Columbia
Vancouver, British Columbia, Canada

Eirini A. Fragogeorgi
Department of Biomedical Engineering
Technological Educational Institute of Athens
Athens, Greece

Nafiseh Ghazanfari
University of Groningen
Groningen, The Netherlands

and

Research Center for Molecular and Cellular
    Imaging
Institute for Advanced Medical Technologies
Tehran University of Medical Sciences
Tehran, Iran

Matthias Glaser
Institute of Nuclear Medicine
University College London
London, United Kingdom

Tamara Godbey
Animal Care Services
University of British Columbia
Vancouver, British Columbia, Canada

William R. Hendee
Department of Radiology
and
Department of Radiation Oncology
and
Department of Biophysics
and
Department of Population Health
Medical College of Wisconsin
Milwaukee, Wisconsin

David Jaffray
Princess Margaret Cancer Centre
and
Department of Radiation Oncology
and
Department of Medical Biophysics
and
Institute of Biomaterials and Biomedical
    Engineering
University of Toronto
and
The Techna Institute for the Advancement of
    Technology for Health
Toronto, Ontario, Canada

G. Allan Johnson
Department of Radiology
Center for In Vivo Microscopy
Duke University
Durham, North Carolina

George C. Kagadis
Department of Medical Physics
University of Patras
Rion, Greece

Aleksandra Kalinowska
Department of Mechanical Engineering
Massachusetts Institute of Technology
Cambridge, Massachusetts

Jessica Kalra
Department of Experimental Therapeutics
British Columbia Cancer Agency Cancer
and
Department of Biology and Health Sciences
Langara College
Vancouver, British Columbia, Canada

Irene S. Karanasiou
Institute of Communication and Computer
   Systems
School of Electrical and Computer Engineering
National Technical University of Athens
Athens, Greece

Dimitrios N. Karnabatidis
Department of Radiology
School of Medicine
Patras University Hospital
Patras, Greece

Veerle Kersemans
CRUK/MRC Oxford Institute for Radiation
   Oncology
Department of Oncology
University of Oxford
Oxford, United Kingdom

Michael V. Knopp
The Wright Center for Innovation in
   Biomolecular Imaging
Department of Radiology
The Ohio State University College of Medicine
Columbus, Ohio

George Kordas
Sol–Gel Laboratory
Institute of Nanoscience and Nanotechnology
NCSR D
Attiki, Greece

Maria Koutsoupidou
Institute of Communication and Computer
   Systems
School of Electrical and Computer Engineering
National Technical University of Athens
Athens, Greece

Krishan Kumar
The Wright Center for Innovation in
   Biomolecular Imaging
Department of Radiology
The Ohio State University College of Medicine
Columbus, Ohio

Matthew A. Lewis
Department of Radiology
University of Texas Southwestern Medical Center
   at Dallas
Dallas, Texas

Jimin Liang
School of Life Science and Technology
Xidian University
Xi'an, Shaanxi, People's Republic of China

FengLin Liu
ICT Research Center
Chongqing University
Shapingba, Chongqing, People's Republic of China

George K. Loudos
Department of Biomedical Engineering
Technological Educational Institute of Athens
Athens, Greece

Sven Macholl
inviCRO Ltd
and
Barts Cancer Institute
Queen Mary University of London
London, United Kingdom

Panagiotis Papadimitroulas
Department of Medical Physics
University of Patras
Rion, Greece

Dimitrios Psimadas
Department of Biomedical Engineering
Technological Educational Institute of Athens
Athens, Greece

Volkmar Schulz
Physics of Molecular Imaging Systems
Institute of Experimental Molecular Imaging
RWTH Aachen University
and
Philips Research Europe
Aachen, Germany

Vesna Sossi
Department of Physics and Astronomy
University of British Columbia
Vancouver, British Columbia, Canada

Stavros Spiliopoulos
Department of Radiology
Patras University Hospital
School of Medicine
Patras, Greece

Michael G. Stabin
Department of Radiology and Radiological
    Sciences
Vanderbilt University
Nashville, Tennessee

James Stewart
Princess Margaret Cancer Centre
and
Institute of Biomaterials and Biomedical
    Engineering
University of Toronto
Toronto, Ontario, Canada

Istvan Szanda
Cancer Research UK Cambridge Institute
University of Cambridge
Cambridge, United Kingdom

and

Division of Imaging Sciences & Biomedical
    Engineering
King's College London
London, United Kingdom

Michael F. Tweedle
The Wright Center for Innovation in
    Biomolecular Imaging
Department of Radiology
The Ohio State University College of Medicine
Columbus, Ohio

Frank Verhaegen
Department of Radiation Oncology (MAASTRO)
Maastricht University Medical Center (MUMC)
Maastricht, The Netherlands

Matthew Walker
Department of Physics and Astronomy
University of British Columbia
Vancouver, British Columbia, Canada

Ge Wang
Biomedical Imaging Cluster
Rensselaer Polytechnic Institute
Troy, New York

Murray Webb
Division of Pharmacology and Toxicology
The Centre for Drug Research and Development
Vancouver, British Columbia, Canada

Jakob Wehner
Physics of Molecular Imaging Systems
Institute of Experimental Molecular Imaging
RWTH Aachen University
Aachen, Germany

Donald T. Yapp
Department of Experimental Therapeutics
British Columbia Cancer Agency Cancer
Vancouver, British Columbia, Canada

Jose Luis Zamorano
Cardiovascular Joint Research Unit
Department of Cardiology
University Hospital Ramon y Cajal
University Francisco de Vitoria School of
    Medicine
Madrid, Spain

Carlos Zaragoza
Cardiovascular Joint Research Unit
Department of Cardiology
University Hospital Ramon y Cajal
University Francisco de Vitoria School of
    Medicine
Madrid, Spain

Section I
Introduction to Small Animal Imaging, Therapy, and Applications

1

Defining Small Animal Imaging, Therapy, and Applications

Nancy L. Ford, George K. Loudos, Dimitrios N. Karnabatidis, and George C. Kagadis

1.1 DEFINITIONS

The use of small animal models in basic and preclinical sciences constitutes an integral part of testing new pharmaceutical agents prior to commercial translation to clinical practice. Whole-body small animal imaging is a particularly elegant and cost-effective experimental platform for the timely validation and commercialization of novel agents from the bench to the bedside. Biomedical imaging is now listed along with genomics, proteomics, and metabolomics as an integral part of biological and medical sciences. Miniaturized versions of clinical diagnostic modalities, including but not limited to microcomputed tomography, micromagnetic resonance imaging, microsingle photon emission tomography, micropositron emission tomography, optical imaging, digital angiography, and ultrasound, have all greatly improved our investigative abilities to longitudinally study various experimental models of human diseases in rodents.

On the other hand, advances in conformal radiation therapy and advancements in preclinical radiotherapy research have recently stimulated the development of precise micro-irradiators for small animals such as mice and rats. These devices are often kilovolt x-ray radiation sources combined with high-resolution computed tomography imaging equipment for image guidance as the latter allows precise and accurate beam positioning. This is similar to modern clinical radiotherapy practice.

These devices are considered a major step forward compared to the current standard of animal experimentation in cancer radiobiology research.

One of the major advantages of small animal imaging, which is steadily being appreciated, is its compliance with the "Three Rs":

1. *Replacement*: While small animal imaging does not fully replace the use of animals, it facilitates performing studies in fewer animal species before clinical translation.
2. *Reduction*: There is definitely a dramatic reduction of the number of animals required when *in vivo* imaging studies replace complex *ex vivo* biodistribution studies.
3. *Refinement*: *In vivo* imaging is mainly performed in anesthetized animals, replacing blood sampling and other painful interventional techniques since anatomical or functional information can be obtained noninvasively.

1.2   BOOK CONTENT

The field of preclinical imaging, therapy and its applications is a rapidly growing area of active research. To date, no comprehensive collection of imaging and treatment modalities, techniques, and applications has been produced. Our goal was to fill this niche with a textbook designed to introduce the reader to all aspects of preclinical imaging, from initiating a research program using animals, through imaging and therapeutic techniques, with additional chapters on the analysis of the imaging data and methods of ensuring that data obtained are of the highest quality. Finally, we include a variety of applications to show the strengths of small animal imaging and therapy in a variety of biomedical fields. This book is designed to be accessible for beginning researchers, including graduate students and postdoctoral fellows, and as a reference book for researchers in medical physics, biomedical engineering, or radiology.

This book begins with an overview of the practicalities of using small animals in research programs and the motivation for including imaging in medical research. In this section, the requirements for ethics approval, appropriate animal handling and care during experimentation, and selecting an animal model are discussed.

In the following sections, imaging modalities are discussed individually, with information on how the images are formed and optimization of the techniques to exploit the contrast and resolution available in each modality. The modalities are separated into sections on imaging techniques that use ionizing radiation and that do not use ionizing radiation.

Hybrid imaging is discussed in detail in the next section. In hybrid imaging, two or more modalities are combined to compile a more complete dataset. Generally, the imaging modalities are chosen to complement each other; the weaknesses of one modality are the strengths of the other and vice versa. Recent advances in preclinical imaging have introduced multimodality imaging systems with two or three modalities included on a single gantry system, and in addition to sequential imaging, tools for simultaneous multimodal imaging are now available. Imaging agents are discussed in three chapters, including contrast agents, radiochemistry, and molecular targets and probes.

Therapeutic research platforms comprise four chapters. To ensure that the results of preclinical studies are relevant to the clinical practice of medicine, it has become increasingly important to use clinically acceptable treatment modalities in the preclinical testing phase of drug or device development. In this section, the use of ionizing radiation for therapeutic purposes is discussed.

Image quantification is the next topic, which is important for both preclinical imaging and planning therapeutic interventions. Measurements and sources of error in those measurements are discussed, along with ways to minimize them through calibrating equipment, the use of quality assurance programs, and simulation using Monte Carlo techniques.

Finally, the book concludes with a number of chapters dedicated to small animal imaging and therapy for specific disease models. These chapters are included to give an overview of the current status of preclinical techniques over a range of applications. However, there are many other applications and disease models that utilize small animal imaging and/or therapeutic techniques in addition to those listed here, with new applications being developed constantly.

1.3   FUTURE DIRECTIONS OF SMALL ANIMAL IMAGING AND THERAPY IN TRANSLATIONAL MEDICINE

Imaging and therapy in preclinical medical research has rapidly developed into necessary experimental tools for medical research. The ability to identify organs and detect anomalies has been demonstrated with numerous imaging modalities that are based on radiological devices commonly found in patient care. More recently, therapeutic devices specifically designed for small animals have been developed to enable researchers to subject their experimental animals to the same types of treatment options available clinically.

In the future, small animal imaging and therapy will continue to develop new techniques, multimodality equipment, and automated image processing to ensure accurate detection of disease. As imaging times are reduced, and image quality and resolution improve, more researchers will employ noninvasive imaging and therapies in their research programs. Imaging and therapy will be combined, potentially into the same instrument, to enable accurate monitoring of the disease response to interventions, including therapies using ionizing radiation, laser, sound, or thermal waves in addition to novel drug formulations. The ability to monitor the same population over the course of a treatment will be especially powerful in drug development and in ensuring the efficacy and safety of new medicines prior to clinical trials.

For translational medicine, and drug development in particular, imaging only provides a portion of the information required. The more important data to obtain are the specific mechanisms that the disease is utilizing and how the drug can impact these mechanisms. For small animal imaging to continue to grow and develop, targeted imaging will become necessary. In this book, the idea of delivering imaging agents is discussed, but this is an important area for future development. Developing imaging agents that can be targeted to specific metabolic processes or receptors will enable more accurate imaging of the disease site. Developing targeted therapeutic agents will enable targeted destruction of diseased tissues and cells, leaving surrounding tissue and cells untouched. The combination of these technologies together would enable accurate detection and destruction of diseased regions. Once verified in preclinical experiments, these targeted agents could then be applied in clinical trials to improve diagnosis and treatment for a variety of diseases.

The role of small animal imaging is expected to be further extended as the sensitivity on bioethics in preclinical research continues to increase, leading to more strict, yet necessary rules and procedures. To this direction, modeling tools, which now are limited only to a research level, are expected to play an important role, if the tools and the procedures that allow the transfer of experiments at the *in silico* level are well developed and validated. Currently, *in vivo* imaging, using different modalities, plays an important role toward the establishment of those platforms.

2

Ethics and Regulations for Research with Animals

William R. Hendee

2.1   INTRODUCTION

Most societies with a well-developed research infrastructure have regulations of some form to permit the use but prevent the abuse of animals as experimental subjects in biomedical research. In a particular society, the nature and extent of the regulations reflect the philosophical foundation that underlies the perceived roles and rights of animals in the society. Consequently, to understand and appreciate the intent and limitations of the regulations, one must have some knowledge of the philosophical framework that defines the roles and rights of animals in the society. In some societies, this framework is challenged by groups who have differing views on the roles and rights of animals. Occasionally, these challenges result in dramatic demonstrations and, less frequently, violent actions. Such actions reflect different ethical points of view among the groups, and these points of view must be understood in order to put the demonstrations and actions into a societal context.

2.2   ETHICAL POINTS OF VIEW

There are four different points of view concerning the roles and rights of animals in a society. These are as follows: (1) Animals have roles but no rights in a society. (2) Animals have roles and rights, but the rights are less than those of humans, thereby permitting humans to exercise dominion over animals in ways that may benefit humans. (3) Animals have roles and equal rights in a society, and humans must not infringe on these rights in ways that might harm animals in any way. (4) The issue is not one of roles and rights of animals, but instead is the responsibility of humans to ensure the well-being of animals and to protect them from abuse and exploitation.

2.2.1   ANIMALS HAVE NO RIGHTS

One philosophical position is that animals have no rights and no moral status, and they can be used in ways to benefit humans without consideration of the consequences to the animals. The French philosopher Rene Descartes, for example, proposed that animals are simply automata and not conscious beings, and therefore can be used to improve the human condition without ethical repercussions (Cottingham 1991). According to Descartes, animals react mechanistically and do not feel pain even though they may react spontaneously and automatically to what would be a painful process for humans. The German philosopher Immanuel Kant proposed that humans have the ability to select a rational course of action rather than simply to submit to immediate desires, something that animals are unable to do (Branham 2005). Consequently, animals should not be accorded moral status in a society, but instead should be viewed as a resource for human use. The Italian Dominican priest Thomas Aquinas believed that animals have no rights, but humans should not be unkind to animals because the unkind behavior might carry over to unkind acts against other humans. More recently, the British philosopher Scruton suggested that the concept of rights cannot apply to animals, because rights are accompanied by obligations, and obviously animals are not imbued with a sense of obligation (Scruton 2000). The American philosopher Raymond Frey has argued that animals do not have rights and interests because they do not have desires, beliefs, or a language. Another American philosopher, Carl Cohen, has proposed that animals have no rights because the attribution of rights implies the capacity of the holder to distinguish between what is right and what serves his or her self-interest (Cohen 1986). Since animals do not possess the capacity to exercise moral judgment between what is right and what is in their self-interest (i.e., animals have no "sense of duty"), animals should not be considered to have rights. The idea that animals have no rights is consistent with early Greek thought (e.g., Aristotle and the Stoics) that the world is populated by many life-forms arranged in a hierarchy of sentience, with humans at the top because they exercise rational thought. This hierarchy was known as the Great Chain of Being, with lower life-forms (e.g., plants) serving higher life-forms (e.g., animals), and with all lower life-forms subservient to humans at the top.

2.2.2 ANIMALS HAVE LESSER RIGHTS

A second philosophical position is that animals have rights, but these rights are considerably less than those possessed by humans. Persons advocating this position argue that humans have capacities such as rational thought, a sense of self, and a desire for autonomy, which are nonexistent, or at least underdeveloped, in animals. This position implies that animals can be used to benefit humans (e.g., in biomedical research), provided that the benefit is judged to outweigh the pain and suffering of the animals. Jean-Jacques Rousseau espoused the view that humans should never hurt any other human, nor any sentient being, except in exceptional circumstances where the human's life is at stake (Rousseau). He adopted this position because he felt an obligation to animals not because they are rational but because they are sentient beings (i.e., conscious of and responsive to impressions on the senses). Some observers have noted that nature is cruel, with lions killing zebras and anacondas preying on rabbits. This "survival of the fittest" Darwinian outlook has been countered by what is referred to as the naturalistic fallacy, explained by the Scottish philosopher David Hume as essentially "what is may not necessarily be what should be" (Hume 1969).

2.2.3 ANIMALS HAVE EQUAL RIGHTS

A third philosophical position is that animals have equal rights to humans. Many recent philosophers have taken the view that the rights of animals should not be distinguished from the rights of humans. In his book *The Case for Animal Rights*, the American philosopher Tom Regan argues that animals have rights in the same manner as humans, and that there is no distinction in the moral status of animals and humans (Regan 2004). He says that every being has values and deserves respect, and no being should suffer in order to benefit another. Regan proposes that equal rights must be attributed to animals in the same manner as equal rights are accorded to humans of marginal functionality in a society. All humans and all animals possess a life, and life imposes a responsibility to attribute an equal moral status, and therefore equal rights, to all.

The concept that humans are inherently superior to animals has been defined as "specieism" by a group of philosophers in Oxford, England, included Richard Ryder (1985). Specieism is defined as "discrimination against or exploitation of certain animal species by human beings, based on an assumption of mankind's superiority." Ryder also developed the concept of "painism," which implies that any life-form (human or animal) that can experience pain has moral standing and therefore rights that must be protected.

The Australian philosopher Peter Singer published a seminal work in 1975 entitled *Animal Liberation* (Singer 2002). This book, which has become a pivotal work in support of the Animal Liberation Movement, suggested that the utilitarian principle of (the greatest good for the greatest number) should be applied not to just human behavior but should be extended to animals as well. Once this expansion is accomplished, then the utilitarian principle becomes the only valid criterion of ethical behavior. Singer suggests that all beings capable of suffering should be given equal consideration, and that reduced consideration based on species is no more justifiable than discrimination based on skin color.

In a short time following the publication of Singer's book, ethnologists such as Dian Fossey and Jane Goodall described emotions such as affection, jealousy, and deceit among primates, which were interpreted as almost-human attributes by many followers. These anthropomorphic views of primates contributed to the passage of legislation in several countries that reduced the exploitation and abuse of animals in general.

2.2.4 RESPONSIBILITY OF HUMANS

Many individuals do not see the issue of experimentation with animals to be framed in terms of the rights (or lack thereof) of animals. Instead, they accept the belief that humans have the right of dominion over other beings, often as a result of religious teachings in Christianity, Islam, Judaism, and other belief systems. However, that belief is accompanied by a responsibility of humans to protect other beings and prevent their abuse and exploitation. As the English philosopher Jeremy Bentham asked in 1780, What "insuperable line" prevents humans from extending moral regard to animals? Benthan stated: "The question is not, Can they reason? Nor, Can they talk? But, Can they suffer?" (Bentham). Most individuals who use animals for experimentation in biomedical research willingly abide by rules and regulations to diminish pain and suffering to the extent possible in animal experimentation. It was this sense of responsibility, together with reaction of the public to the justification for animal experimentation arising from Darwin's theory of evolution, that gave rise in 1876 to the British Cruelty to Animals Act that regulated experimentation with animals. Today, the Animal Welfare Act carries a similar purpose in the United States.

2.3 REGULATORY OVERSIGHT OF ANIMAL EXPERIMENTATION IN THE UNITED STATES

2.3.1 ANIMAL WELFARE ACT (PUBLIC LAW 89-544)

The Animal Welfare Act (AWA) was enacted in the United States in 1966 to establish standards for the treatment of animals in research, exhibition, transport, and marketing (1966). The AWA establishes the baseline for what is acceptable concerning the use of animals bred for commercial purposes, used in research, transported commercially, or exhibited to the public. The law has been amended several times since its original signing, and is enforced by the United States Department of Agriculture (USDA; Animal and Plant Health Inspection Service [APHIS]). Requirements under the AWA concerning the welfare of animals used in research include the following:

- A veterinarian must be available and responsible for the health and welfare of the animals.
- An Institutional Animal Care and Use Committee (IACUC) must oversee the research use of animals, and a veterinarian must be a member of the committee.
- All investigators must accept the responsibility to minimize pain and distress of research animals.
- All investigators and others involved in animal research must be educated and qualified to protect the welfare of animals to the extent possible.
- Every research facility must ensure that pain and suffering of experimental animals are minimized and that the facility abides by the regulations of the AWA.
- Housing for all experimental animals must be reasonably comfortable and the animals must be fed and watered appropriately and regularly.
- Dogs must be exercised, and the psychological well-being of nonhuman primates must be considered and they must be protected from conditions such as anxiety, boredom, separation, and isolation.
- Research facilities must maintain reports and records consistent with AWA requirements.

2.3.2 INSTITUTIONAL ANIMAL CARE AND USE COMMITTEE

Every institution in which animals are used in research must establish an IACUC to oversee and monitor the use of the animals and to ensure that the institution remains in compliance with AWA requirements. The IACUC is appointed by the Chief Executive Officer (CEO) of the institution and is required to report to the CEO any substantial noncompliance with accepted animal use protocols, along with actions taken to correct the noncompliance. The IACUC is required also to report any substantial noncompliance to the Office of Laboratory Animal Welfare (OLAW) of the

National Institutes of Health (NIH) if any research in the institution is funded by the NIH. Further, the IACUC is required to file an annual report with the OLAW. The authority of OLAW to oversee federally sponsored research with animals is provided in the Health Research Extension Act of 1985 (Public Law 99-158). The IACUC must consist of at least five members, including a veterinarian, a research scientist, a nonscientist, and a member of the public. Responsibilities of the IACUC include the following:

- At 6-month intervals or more frequently, review the institution's program for the humane care and use of animals.
- At 6-month intervals or more frequently, inspect all of the institution's animal care facilities.
- Submit reports of all reviews and inspections to the institution's administration.
- Review and take appropriate action with regard to any and all concerns that arise concerning the care and welfare of research animals.
- Review and approve (when appropriate) applications from investigators wishing to use animals for research purposes.
- Suspend research projects when issues arise concerning the care of animals in the projects.

The activities of the IACUC are subject to inspection by the NIH staff (for projects supported by NIH funds), APHIS (for compliance with the AWA), and AAALAC (a nongovernmental accrediting agency discussed in the following section).

2.4 RESOURCES IN THE USE OF ANIMALS IN RESEARCH

2.4.1 ANIMAL WELFARE INFORMATION CENTER (AWIC)

The AWIC was established in 1986 and is a segment of the National Agricultural Library operated by the USDA in Beltsville, MD. The purposes of the AWIC are to prevent unintended duplication of animal experimentation, improve methods of research with animals, reduce or replace animal use where possible, and minimize pain and distress of research animals by developing better anesthetic and analgesic techniques. The AWIC is primarily an educational resource that promotes training of investigators and provides informational material consistent with the humane use of animals in research.

2.4.2 ASSOCIATION FOR ASSESSMENT AND ACCREDITATION OF LABORATORY ANIMAL CARE (AAALAC) INTERNATIONAL

AAALAC International is a private, not-for-profit organization that offers a voluntary accreditation program for institutions conducting research with animals. Its purpose is to promote the humane use of animals in research through accreditation of facilities and provision of educational programs and information. AAALAC is not a governmental regulatory body, and has no enforcement authority other than the withdrawal of accreditation from noncompliant institutions. AAALAC relies on accepted guidelines for animal use such as the National Research Council's publication *Guide for the Care and Use of Laboratory Animals*, and supplements these guidelines with its own position statements and Rules of Accreditation.

2.5 ORGANIZATIONS PROTECTING THE RIGHTS AND WELFARE OF ANIMALS

Two types of organizations exist to protect the rights and welfare of animals. One type, such as the Society for the Prevention of Cruelty to Animals (SPCA) discussed in the following, serves to protect the welfare of animals without claiming that animals have rights equal to or less than those of humans. Other organizations such as People for the Ethical Treatment of Animals (PETA; also described in the following) state that animals have the same rights as humans and must not be exploited or abused to benefit humans. It is interesting to note that many, if not most, animal rightists live their beliefs; they are vegetarian and many are vegans, eschewing the use of any animal product, including eggs, milk, leather, and fur.

2.5.1   SOCIETY FOR THE PREVENTION OF CRUELTY TO ANIMALS

The SPCA was founded in England in 1824 primarily for the purpose of preventing the abuse of carriage horses. These animals were often abused by their owners who worked them in all weather conditions, often without adequate food, water, and rest. The SPCA labored to pass legislation to regulate the use and abuse of carriage horses, and ultimately expanded its purpose to encompass cruelty to other animals, especially dogs. The first American SPCA was established in 1866 in New York City, and today there are SPCA offices around the world dedicated in part to providing shelters for homeless and abused animals. These offices tend to be independent and not affiliated with a national organization. SPCA International was founded in 2006 with a mission to advance the safety and well-being of all animals and to provide an information and networking resource for independent SPCA offices.

2.5.2   PEOPLE FOR THE ETHICAL TREATMENT OF ANIMALS (PETA)

PETA was founded in 1980 as an international not-for-profit organization based in Norfolk, VA. It is dedicated to acknowledging and defending the rights and humane treatment of all animals. PETA's underlying principle is that animals are not subservient to humans and are not intended to be eaten, worn, experimented with, or used for entertainment. PETA's mission statement says in part:

> PETA focuses its attention on the four areas in which the largest numbers of animals suffer the most intensely for the longest periods of time: on factory farms, in the clothing trade, in laboratories, and in the entertainment industry. We also work on a variety of other issues, including the cruel killing of beavers, birds and other "pests" as well as cruelty to domesticated animals.

PETA chooses to use provocative tactics that engender media attention because it claims to not have access to financial or other resources to advertise its mission and purposes. PETA opposes all forms of research with animals with the following justifications:

✦ Human clinical and epidemiological studies, cadavers, and computer simulators are faster, more reliable, less expensive, and more humane than animal tests.

✦ Most people will agree that it is wrong to sacrifice one human for the "greater good" of others because it would violate that individual's rights. There is no logical reason to deny animals the same rights that protect individual humans from being sacrificed for the common good.

✦ Animal care and use committees were meant to stop pointless and redundant studies, but most merely rubber-stamp animal experiments without question. They are composed of animal experimenters with a vested interest in the continuation of animal experimentation, with only one person to represent the community and the interests of the animals.

The most well-known PETA activity to date is the 1981 Silver Springs Monkey Case in which a PETA member worked undercover in a laboratory in Silver Springs, MD, that was conducting research on macaque monkeys to improve the treatment of brain-damaged humans. Alleged abuse of the animals prompted a police raid on the laboratory during which the monkeys were removed, leading to a 9-year battle over custody of the monkeys. This battle culminated in a ruling of the U.S. Supreme Court rejecting PETA's appeal for custody. Autopsies of the monkeys revealed new knowledge about the brain's neuroplasticity that led to a new form of therapy that is being used to help stroke victims regain use of their extremities. The original conviction of the principal investigator on six misdemeanor charges was overturned on appeal in 1983.

2.5.3   ANIMAL LIBERATION FRONT (ALF)

The ALF is not an organized national or international movement. Rather, it consists of small autonomous groups of individuals who conduct direct acts to liberate and protect animals according to loosely structured ALF guidelines. These guidelines promote the following objectives:

✦ To liberate animals from places of abuse, i.e., laboratories, factory farms, fur farms, etc., and place them in good homes where they may live out their natural lives free from suffering
✦ To inflict economic damage to those who profit from the misery and exploitation of animals
✦ To reveal the horror and atrocities committed against animals behind locked doors by performing nonviolent direct actions and liberations
✦ To take all necessary precautions against harming any animal, human or nonhuman
✦ To analyze the ramifications of all proposed actions, and never apply generalizations when specific information is available

Although the ALF claims to be nonviolent, some of its actions have vandalized investigators' homes and laboratories, and may have resulted in threats of assault and death of the investigators. In 1991, the U.S. Federal Bureau of Investigation identified ALF as a domestic terrorist threat, and the U.S. Department of Homeland Security did so as well in 2005.

2.6   CURRENT STATUS OF RESEARCH WITH ANIMALS

Without question, the use of animals in biomedical research is diminishing, particularly larger animals and nonhuman primates. This diminution is due to several reasons, including the following:

✦ The cost of using animals in research continues to grow at a time when financial resources in support of research are increasingly constrained. Often, investigators are forced to limit the number of animals in an experiment because of a shortage of funds.
✦ Surrogate methods for conducting research, particularly mathematical algorithms and computer models, are becoming increasingly powerful tools in research, or at least decrease, the need for experimentation on animals. These methods are less expensive and, because they can be run an unlimited number of times, often yield more reliable and reproducible data compared with the animals they replace.
✦ Research with animals is subject to an array of rules and regulations that require a complex institutional infrastructure and considerable time and effort on the part of investigators to ensure compliance.
✦ Public sentiment against the use of animals in research sometimes complicates the social networking of scientists in their communities, and occasionally makes them the target of animal activists.

Public sentiment is also slowly evolving against the use of animals for research purposes. In 1985, 63% of American respondents agreed that "scientists should be allowed to do research that causes pain and injury to animals like dogs and chimpanzees if it produces new information about human health problems"; in 1995, the percentage of respondents agreeing with the statement dropped to 53% (Mukerjee 1997). In attitudinal surveys in several countries, women tend to be more antivivisectionist and protective of animals compared with men. These findings may reflect in part the growing separation of most individuals and families from farms and the rural environment in which animals are raised for purposes of food production. When animals are increasingly viewed as pets rather than as food sources for humans, a shift in sentiment might be expected towards the protection of animals and away from their use in support of human health and well-being. Also, surveys have shown that older and less educated people are less troubled by the concept of animals as resources than are those who are younger and better educated.

Nevertheless, animals are still being used, and this use is supported by several groups that believe the benefits of research with animals greatly outweigh the detriments. Among these groups is the organization known as Americans for Medical Progress (AMP).

2.6.1 AMERICANS FOR MEDICAL PROGRESS

The mission of the AMP is described on the organization's website as follows:

> Americans for Medical Progress (AMP) protects society's investment in research by nurturing public understanding of and support for the humane, necessary, and valuable use of animals in medicine.

> Threats by animal rights extremists hurt medical progress. AMP provides accurate and incisive information to foster a balanced public debate on the animal research issue, ensuring that among the voices heard are those whose lives have been touched by research and those who work in the field. Through various specialty publication, outreach initiatives, and the media, AMP informs the public of the facts of animal-based research. AMP also distributes timely and relevant news, information, and analysis about animal rights extremism to the research community through its news service.

> AMP's Board of Directors is composed of physicians, researchers, veterinarians, university officials, and two Nobel laureates in medicine.

2.7 THE THREE Rs

Many scientists are advocates for reducing the number of animals used in research and for using only the essential number of animals always under humane conditions. In 1959, Russell and Burch published The Principles of Humane Experimental Technique in which they proposed the "three Rs" as goals for the conscientious researcher (Russell and Burch 2012). The three Rs are as follows:

1. *Replacement* of animals by in vitro or test tube methods, or replacement of higher order by lower order animals
2. *Reduction* of the number of animals by means of statistical techniques
3. *Refinement* of the experiment so that suffering is reduced

Ways to reduce the number of animals in a study include the following:

- Perform a pilot study to validate the experimental design.
- Design the study so that animals serve as their own controls.
- Gather the maximum information attainable from each animal.
- Consult a statistician to use only the number of animals required for statistical significance.
- Minimize variables between animals.
- Perform an extensive literature search before embarking on the research.
- Use the appropriate species of animal.
- Replace with lower order animals when possible.

Refinement of a study may include the following objectives:

- Identify pain and distress and plan to alleviate or reduce it.
- Set the earliest possible endpoint for the study.
- Ensure adequate training of all involved in the study.
- Use proper handling methods for animals.
- Ensure that drug doses are appropriate and that drugs are not expired.
- Verify that procedures to be performed are appropriate.
- Use analgesics and anesthetics for potentially painful procedures.
- Perform surgeries and procedures aseptically.
- Perform only a single major surgery on any one animal whenever possible.
- Provide appropriate postsurgical care.

In 1993, the U.S. Congress directed the National Institutes of Health (NIH) to develop a plan for implementing the three Rs. The NIH responded by producing a document entitled "Plan for the Use of Animals in Research," which provides an overview of biomedical models for research, including animals, and charges investigators with the responsibility for selecting the optimum approach to their research.

Today, these goals underlie the search for alternatives to animals in biomedical research. Several European countries, the United States, and Canada have funded explorations of alternatives to the use of animals in research. In 1981, Johns Hopkins University established the Center for Alternatives to Animal Testing (CAAT), which exists today with the following three goals:

1. Providing funding for new research
2. Disseminating information about alternative methods
3. Creating a forum for industry, regulatory agencies, and the animal welfare community to work together for progress

As one example, CAAT provides an online course entitled Enhancing Science/Improving Animal Research that is available on the CAAT website (http://caat.jhsph.edu/programs/).

2.8 CONCLUSIONS

Animals have been used for centuries as research objects in order to gain insight into physiological phenomena. Greek writings as early as 500 BCE describe the dissection of living animals, and the Roman physician Galen used vivisection as a tool to understand biological processes. The use of animals in research by later physicians such as Vesalius and Harvey were not challenged on philosophical or moral grounds. These challenges began to surface only in the mid-nineteenth century. Since these early challenges, arguments against experimentation with animals have become increasingly sophisticated so that today there is a schism in developed cultures between those who support animal experimentation because of its contributions to human health and well-being and those who oppose animal experimentation on the grounds that it is unethical. This schism waxes and wanes, but its overall impact has been to reduce animal experimentation in most developed countries, and to ensure that when it is done that the animals are treated as humanely as possible. The humane use of experimental animals is fostered in the United States and other developed countries by regulations and by institutional rules and procedures that scientist are obliged to follow. At the present time, research with animals is diminishing slowly as algorithms and computer models serve increasingly as surrogates for animal experimentation. Whether these approaches will ever completely replace animals as research objects is impossible to foresee.

ACKNOWLEDGMENTS

Many resources were accessed in preparing this chapter. One in particular has been particularly helpful—the educational web module *Research with Vertebrate Animals* prepared by Daniel Eisenberg, Jed Peterson, and Kenneth Zeitzer, which is part of the Ethics and Professionalism Modules series prepared by the Radiological Society of North America in cooperation with several other professional organizations. This module may be accessed at http://ep.rsna.org/section/default.asp?id=EP1107.

REFERENCES

American Society for the Prevention of Cruelty to Animals. Retrieved October 10, 2012, from http://www.aspca.org.

Americans for Medical Progress. Retrieved October 10, 2012, from http://www.amprogress.org.

Animal Liberation Front. Retrieved October 10, 2012, from http://www.Animalliberationfront.com.

Association for the Assessment and Accreditation of Laboratory Animal Care International. Retrieved October 10, 2012, from http://www.aaalac.org/.

Bentham, J. On the suffering of non-human animals. Retrieved October 10, 2012, from http://www.utilitarianism.com/jeremybentham.html.

Branham, A. (2005). Quick Summary of philosophy and animals. Retrieved October 10, 2012, from http://www.animallaw.info/topics/tabbed%20topic%20page/spusphilosophy_animals.htm.

Center for Alternatives to Animal Testing Programs. Retrieved October 10, 2012, from http://caat.jhsph.edu/programs/.

Cohen, C. (1986). The case for the use of animals in biomedical research. *N. Engl. J. Med.* **315**(14): 865–870.

Cottingham, J., R. Soothoff et al. (1991). *The Correspondence of the Philosophical Writings of Descartes.* Cambridge University Press, Cambridge, U.K.

Hume, D. (1969). *A Treatise of Human Nature.* Penguin Books, Middlesex, England.

Mukerjee, M. (1997). Trends in animal research. *Sci. Am.* **276**: 86–93.

People for the Ethical Treatment of Animals. Retrieved October 10, 2012, from http://www.peta.org.

Public Law 89-544. (1966). Retrieved October 10, 2012, from http://awic.nal.usda.gov/public-law-89-544-act-august-24-1966.

Regan, T. (2004). *The Case for Animal Rights.* University of California Press, Berkeley, CA.

Rousseau, J.-J. Animal rights: A history. Retrieved October 10, 2012, from http://www.think-differently-about-sheep.com/Animal_Rights_A_History_Jean_Jacques_Rousseau.htm.

Russell, W. and R. Burch. (2012). *The Principles of Humane Experimental Technique.* Johns Hopkins School of Public Health, Baltimore, MD.

Ryder, R. (1985). Speciesism in the laboratory. In Singer, P. (ed.), *In Defense of Animals.* Blackwell Publishing, New York, pp. 77–88.

Scruton, R. (2000). *Animal Rights and Wrongs.* Metro Books, London, U.K.

Singer, P. (2002). *Animal Liberation.* Harper-Collins Publishers, New York.

St. Thomas Aquinas. Retrieved October 10, 2012, from http://www.newadvent.org/cathen/14663b.htm.

The Animal Welfare Information Center. Retrieved October 10, 2012, from http://awic.nal.usda.gov/.

3

Small Animal Handling, Care, and Anesthesia

Tamara Godbey

3.1   ANIMAL CARE: INTRODUCTION

The use of animals in any type of biomedical research brings with it implicit ethical considerations that should be thoroughly reviewed prior to performing the research. At most institutions, these ethical reviews will be conducted by an institutional animal care and use committee (IACUC) or an animal care committee (ACC) after a research group has accepted the need to use animals and defined their use. This review is mostly based on accepted national standards such as the *Guide for the Care and Use of Laboratory Animals* (Institute of Laboratory Animal Resources et al. 1996) in the United States or the guidelines and policies defined by the Canadian Council on Animal Care in Canada (Olfert et al. 1993). These guidelines, standards, and policies attempt to ensure improvement in animal welfare: reduction in animal numbers, animal pain, and stress, while at the same time not compromising the results of animal-based research.

In the book, *The Principles of Humane Experimental Technique* (Russell and Burch 1959), William Russell and Rex Burch defined their "3Rs" concept—replacement, reduction, and refinement—in relation to the use of animals in research. In this fundamental publication, they were ahead of their time, defining accepted practices in the design and implementation of animal-based research today.

"Replacement" refers to using a nonanimal model or excised animal tissue to replace the use of live animals, eliminating stress and pain for the animal. It may also refer to using a "less sentient" species such as replacing the use of a vertebrate animal with an invertebrate. The use of cell cultures and animal and human tissues would also be a "replacement." In small animal imaging, there may be limited opportunities for "replacement" available in a study as most imaging equipment is specialized for small rodents. The use of models and postmortem carcasses or tissues may be considered a viable replacement in some studies.

"Reduction" refers to minimizing the number of animals used in a study by appropriate experimental design and improved data analysis. This can be further enhanced by using genetically homogeneous animals and controlling for other research variables. Small animal imaging often allows for substantial "reduction" as living animals are often followed longitudinally, allowing for fewer animals within a study, and for noninvasive research. Imaging-based research also allows each animal to serve as its own control, thereby increasing statistical power.

"Refinement" is concerned with improvements in an experiment that reduces stress, distress, and pain in the animal. Refinement can also apply to improving an animal's environment allowing for decreased stress and normal species-specific behaviors. Examples would include providing appropriate analgesia and postoperative care to an animal, establishing and adhering to early experimental endpoints, or group housing social animals allowing for positive conspecific behaviors such as grooming. Imaging is inherently nonpainful and may allow for the use of animals without clinical disease. Unlike human imaging, animals often require anesthetics during imaging procedures and this may be done repeatedly. Repeated recovery anesthesia should be considered a negative aspect of imaging and attempts to limit the frequency and duration of anesthetic sessions should be sought.

As mice, and secondly rats, are the mostly widely used rodent species in biomedical research worldwide, this chapter deals with only these two species.

3.1.1 HOUSING

Rodents can be housed and maintained in many types of environments within the laboratory setting. In conventional housing, rodents are often housed in open-topped cages or cages with a filter lid. They are often handled in open air at the room or procedure room level. Negative aspects of conventional housing systems are that they may allow for human exposure to allergens and rodent disease transmission within the colony. Alternatively, modified barrier or barrier housing may have rodents contained within individually ventilated cages or filtered cages and animals are only handled within a biological safety cabinet or other equipment that protects them from pathogens that may be transferred by contact or via room air. Depending on the type of containment used in housing, the requirements for moving animals to imaging equipment may differ among research facilities.

The standard housing environments for rodents have inherent stressors but it is most important to ensure proper groupings of rodents as this may have research and welfare implications. It is typical to keep rodents in groups of three to five animals per cage. Not only is this useful to maximize holding space but it has a strong welfare implication. Rodents are social species (Smith and Hargaden 2001), and it has been shown that even male rodents will work to be in contact with another male rodent (van den Broek et al. 1993; Van Loo et al. 2001). However, caution should be taken when group housing certain strains of male mice as they are known to fight which can result in injury or even death. Studies should be designed to allow group housing when possible, and stable groups and group sizes should be maintained throughout a study so as to decrease potential variability.

Enrichment or environmental enrichment is the addition of objects or alterations in an animal's environment that allows an animal to display species-specific behaviors. For example, group housing rodents would be considered an environmental enrichment as it allows for conspecific grooming. Other standard enrichments in rodent environments may include nesting, group housing, hiding spaces/devices, climbing devices, foraging devices, or food treats. Enrichment, while often not standardized or well defined, does have the potential to produce positive or negative effects on many types of research (Bayne 2005). Little is known about the effects of standard enrichments such as nesting on imaging outcomes. However, for example, nesting has a significant impact on body temperature (Gaskill et al. 2012), which may be critical for physiological homeostasis and survival post-anesthesia/imaging. Enrichment should always be a study component and should be evaluated prior to its use.

3.1.2   HANDLING

3.1.2.1   Basics: Mouse and Rat

In addition to general handling required for husbandry, animal handling is a standard practice for preparing animals for imaging, monitoring animals post-imaging, and providing treatments. It is generally an accepted practice to handle mice by first gently grasping them by the base of the tail and then lifting them and supporting their body weight in the same or other hand (Figure 3.1a). In rats, it is general practice to avoid handling them by the tail, as there is potential for damage. Instead, "scooping" a rat into an open hand is preferred (Figure 3.1b).

Handling can induce unwanted stress or distress that has the potential to complicate research. Certain types of handling may mitigate some of this stress. It appears that handling mice by the tail induces aversion to the handling process and induces anxiety, whereas using tunnels to handle mice or using an open hand improves these unwanted conditions (Hurst and West 2010) (Figure 3.1c). Rodents may not habituate to repeated handling, so all handling should be thought of as a potential stressor (Balcombe et al. 2004; Longordo et al. 2011).

3.1.3   ADMINISTRATION OF SUBSTANCES

In all cases, it is preferable to get specific training on handling and injections long before a research project is to take place. At most institutions, training is available from veterinary staff or facility staff. As consistency and technique are critical to both animal welfare and animal research, training should continue until an individual is proficient at all the techniques required for a study.

(a)

(b)

(c)

FIGURE 3.1   (a) Open hand technique for mouse handling. The tail is gently held to ensure the mouse does not fall or jump. (b) "Scooping" a rat while holding the tail gently to control the animal. (c) Mouse being handled in a red plastic tube.

3.1.3.1 Subcutaneous Routes of Administration

The subcutaneous route is frequently used for the administration of substances and fluids to prevent dehydration that can ensue following anesthesia and/or surgery, but rarely used for the administration of anesthetics. The rate of absorption will depend on the formulation of the substance. In general, the smallest needle should be used taking into account the viscosity of the substance. The recommended volumes are 10 mL/kg (mouse) and 5 mL/kg (rat) (Diehl et al. 2001). This would be approximately a 0.25 mL injection for a 25 g mouse. However, for subcutaneous fluid administration, up to 1 mL is normally given for a mouse. Normally the scapular region is the preferred site for subcutaneous injections; however, due to rodent's loose skin, these injections can be given along the back and along the abdomen. A "tent" of skin should be created by lifting the skin in the area, and the needle should be inserted at an angle into the tented area. The entire needle should be placed into the subcutaneous space to ensure that the injected substance does not leak out of the puncture site. Before injecting, aspiration should occur to ensure that the needle has not penetrated a blood vessel or exited the skin, in which case air will be aspirated.

3.1.3.2 Intramuscular Routes of Administration

In almost all instances, intramuscular injections are not recommended for rodents as their muscle bellies are quite small, leading to pain and potential muscle and nerve damage on injection. Intramuscular injections should only be given when no other alternatives exist, and well-trained personnel should perform these injections.

Intraperitoneal and intravenous injections are the most common routes for injecting substances in rodents, and they are discussed in detail in the following sections.

3.1.4 TRANSPORTATION

In most instances, animals will need to be moved or relocated to a facility that houses imaging equipment. Transport, even from the simplest move of animals from one room to another, can alter basic physiological functions, such as elevated heart rate and blood pressure. Consideration of potential stress induced by transport between facilities, between rooms within a facility, and between holding areas and imaging areas should be given. Short-distance transport may result in stress due to changes in the environment including diurnal light/dark cycles, temperature, and relative humidity (Tuli et al. 1995). Rats transported between holding room and testing room showed a significant elevation in body temperature that lasted for 2 h (Dallmann et al. 2006). Long distance transport also causes alterations in physiological function including behavioral changes, coticosterone levels, food and water consumption, and body weight (van Ruiven et al. 1998). Extreme temperature variations during air shipment have been documented and this should be expected to cause changes in the animals. Other factors to consider in movement of animals are the change in the microenvironment (cage type, bedding type, animal density, temperature, diet, etc.) and/or the change in personnel that can cause a stress response.

While it may be difficult to quantify the stress imposed by transporting animals to and from imaging equipment, it is still critical to establish standards of acclimatization to attempt to mitigate these mostly unforeseen effects. In general, a minimum period of 2–3 days of acclimatization is recommended when transporting animals between facilities (Conour et al. 2006).

3.1.5 DISEASE TRANSMISSION CONCERNS

One of the most important considerations in small animal imaging is that of disease transmission or biosecurity at the level of multiuser equipment. As imaging equipment is expensive and often scarce within an institution, it is likely that many different research groups will use the equipment. If mice from one facility that allows certain opportunistic or pathogenic organisms in their rodent colonies are scanned the day before mice from a barrier facility that excludes all known pathogenic organisms, what are the risks? There are very few publications on this topic yet it is likely the most difficult-to-manage issue in small animal imaging. Methods and standard operating procedures to exclude pathogens from imaging equipment and imaging facilities should be well established.

Regardless of the movement of animals within an imaging facility or suite, it is best practice to disinfect equipment either between groups of same health status animals or between individual animals. In most cases, equipment such as the imaging bed, the induction chamber, and the anesthetic circuits can be disinfected with an appropriate solution. In some instances, equipment that cannot be appropriately disinfected can be covered with a disposable material to help reduce contamination. Note that certain pathogens of rodents are highly resistant to disinfection and so there are inherent risks to sharing equipment between animals of a different health status.

Some strategies that are commonly used to maintain rodent facility health status is to move rodents to imaging equipment and then not allow those rodents to return to their facility of origin. In some cases, rodents are held and euthanized within the imaging facility. Another option would be to provide a short-term holding facility that accepts rodents after imaging and allows them to be transported back and forth to the equipment as needed, without returning to and possibly contaminating the facility of origin. Another strategy is to have the biosecurity level at the individual animal, where each animal is in a mobile self-contained container that can be used for imaging.

In cases where animals are held post-imaging in groups that arrived from different sources or facilities, sentinel testing, or health monitoring should be considered. Sentinel rodents are used to monitor the pathogen status of a group or colony of rodents. Sentinels are often exposed to the air that the colony is exposed to, and to the dirty bedding of colony rodents. In this way, sentinels are exposed to whatever pathogens may be present in the colony. At the end of a designated monitoring period, the sentinels are sampled and tested for a specific list of infectious agents. Another option would be to sample directly from the animals that are used for research purposes. There are commercially available diagnostic tests, including serological and polymerase chain reaction based tests, for determining the status of either sentinel or research animals. Monitoring strategies should be developed in conjunction with experienced laboratory animal veterinarians who are familiar with the risks of rodent pathogens to research.

3.2 ANIMAL PREPARATION AND SUPPORTIVE CARE FOR ANESTHESIA

3.2.1 FASTING

In imaging studies, there is sometimes a need to fast rodents; however, it is not typical to fast rodents for anesthesia or manipulations as in other larger species of animals. Vomition during anesthetic induction and anesthesia is not a concern in rodents due to the limiting ridge of the forestomach (Luciano and Reale 1992). It is still possible to have regurgitation though, so appropriate animal positioning without weight on the abdomen is important. Fasting can be an effective way to ensure uniformity in PET imaging especially in fluorodeoxyglucose (FDG) imaging to decrease blood glucose levels (Hildebrandt et al. 2008). However, fasting in rodents should be done with care as rodents have high metabolic rates causing quicker physiologic alterations compared to larger animals. Food should not be withheld from small rodents for more than 2–3 h as hypoglycemia and dehydration may occur, as these species typically will not drink water if they do not eat (Harkness et al. 2010). Regardless, water should be available at all times during the fasting period.

Fasting may have significant research effects. When food was withheld from the beginning of the dark period, even short periods of food deprivation will affect variations in metabolism. A 3 h fasting period during the dark phase caused a substantial decrease in liver weight and glycogen content in rats (Palou et al. 1981). Fasting rats resulted in depleted glycogen stores and a significantly lower liver attenuation observed in computed tomography (CT) images compared to normal rats (Leander et al. 2000). The fact that rodents reingest feces as part of their normal behavior also means that the ability to actually fast these animals is limited.

3.2.2 INJECTIONS OF ANESTHETICS AND CONTRAST AGENTS

Many guidelines are available on the administration of substances to animals regarding volumes and routes of administration. Properties of substances to be injected including pH, tonicity, sterility,

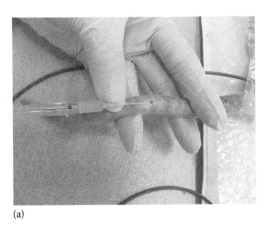
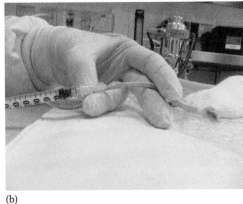

(a) (b)

FIGURE 3.2 (a) Rat tail vein catheterization with a 24-gauge over-the-needle catheter while the rat is anesthetized. (b) Mouse tail vein intravenous injection being performed with a 27-gauge needle and a 1 mL syringe.

viscosity, and rate of absorption are all important to consider. In imaging, contrast agents are usually given intravenously, although in very small animals, such as a 20 g mouse, this can be difficult. An alternative route in rodents is to give contrast agents via the intraperitoneal route that is most similar to the uptake rates of intravenous administration. However, substances given by the intraperitoneal route are possibly first absorbed into the portal circulation, resulting in potential biotransformation of the substance before it reaches circulation (Nebendahl 2000).

Intravenous injections, while more complicated in rodents than larger species, have the advantage of rapid absorption and allow the injection of some substances that are too irritating to be given by other routes. Intravenous access can be obtained in both mice and rats. In rats, the use of 24-gauge over-the-needle catheter in the lateral tail vein can be accomplished relatively easily (Figure 3.2a). In mice, there are no commercially available catheters for tail vein catheterization but catheters can be self-made with a small-gauge needle and polyethylene tubing. Alternatively, injections can be given directly with a needle but there is increased chance of injecting the substance outside of the vessel (Figure 3.2b). In mice, a 25- to 27-gauge needle is typically used, and in rats, a 25-gauge needle is standard. In both species, the typical maximum volume for bolus administration would be 5 mL/kg (Diehl et al. 2001). For rapid IV administration, a substance must be compatible with blood and not be too viscous. If large volumes are given, the substance should be warmed to body temperature. For injections in a conscious animal, many commercially made restrainers are available, and in some cases, these restrainers have built-in heating systems to ensure vasodilation.

While over-the-needle catheters may be useful for a single imaging session, longer-term catheterization of many vessels can be achieved surgically. Frequently used vessels for chronic catheterization are the internal or external jugular veins, the femoral vein, or the carotid artery. This allows for multiple injections or blood samples to be performed over time. Several commercial companies can provide catheterized rats or mice for studies. If catheters are not properly maintained, clotting or infection may occur. It may be necessary to house rodents with external catheters separately.

Intraperitoneal injections are common in rodents but may cause complications including injecting into the intestinal tract, laceration of organs/vessels, and peritonitis caused by irritating substances or bacteria. These injections are generally done without anesthesia and proper restraint and injection techniques are important to avoid complications. Mice and rats can be properly restrained and injected by one person, but in the case of rats, it is preferable to have two people for an IP injection, with one person responsible for restraint. In mice, the scruff hold works best for these injections, and in rats, the animal can be restrained with a "v" hold, and the hind legs restrained by one person and the injection by another person (Figure 3.3a). If only one person is available to hold and inject the rat, the authors preference is to gently towel wrap the rat and cradle it in dorsoventral hold

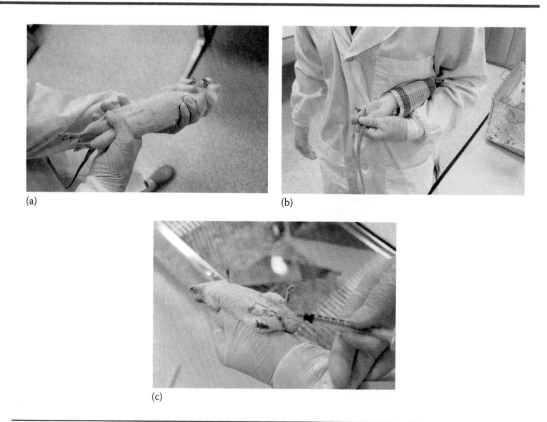

FIGURE 3.3 (a) Rat being held for a two-person intraperitoneal injection. (b) Rat restained for a single-person intraperitoneal injection. The rat is gently restrained in a small hand towel ensuring that there is no respiratory compromise. (c) Intraperitoneal injection in the mouse. The landmarks for injection are shown here with pen marks on the fur. Injection is being performed using a 5/8″ 25-gauge needle and a 1 mL syringe.

in the forearm while retracting the hind legs (Figure 3.3b). The abdomen can be thought of as four quadrants separated by midline and a perpendicular line through the umbilical area (Figure 3.3c). Injections should be given into one of the lower quadrants of the abdomen to avoid damage to internal organs such as the liver or the spleen. To avoid accidental injection into the cecum, a large organ of the abdominal cavity, injections in mice should be given in the lower left quadrant, and in rats, the lower right quadrant (Suckow 2001). Typically, the animal is held in a slight head-down position and injections are given with a quick motion at a 20°–30° angle. Only the tip of the needle needs to enter the peritoneal space. Aspiration should occur before injection to ensure that no blood, urine, or other fluid is aspirated. Failure of intraperitoneal injections is common and injections by this route should be justified prior to the study (Gaines Das and North 2007).

3.2.3 Supportive Care

Basic supportive care, regardless of the rodent species, is critical for ensuring that physiological parameters are maintained during and after anesthesia. Supportive care also has implications for imaging. The five essential components of supportive care are discussed next.

3.2.3.1 Heat

Rodents are prone to hypothermia due to their large surface-area-to-body-mass ratio and their rapid metabolism (Balaban and Hampshire 2001). Anesthesia lasting longer than 10–15 min may result in hypothermia in rodents. Therefore, monitoring and maintaining body temperature is imperative. The normal body temperature of a mouse is 37.4°C and that of a rat is 38°C (Flecknell 1996). With select types of imaging equipment, heating devices are built in (Figure 3.4). In some cases, a heated imaging bed is provided with a rectal probe that provides feedback to the bed and adjusts the temperature appropriately. In the case of optical imaging equipment, the temperature within

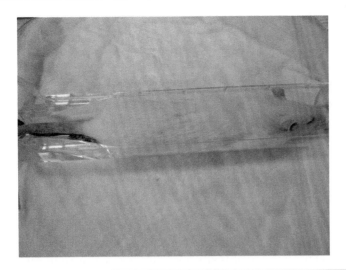

FIGURE 3.4   Rat anesthetized in an imaging bed with built-in heating device and rectal probe. A drape or bubble wrap could be used to insulate the animal and further maintain body temperature.

the chamber can be regulated. If a heating device is not part of the imaging equipment, a commercial heating device should be obtained as well as a way to continuously monitor body temperature. Circulating warm water blankets, warm air devices, or thermal pads with rectal temperature probes are recommended. Electrical heating pads, microwavable heating pads, water bags or bottles, and lamps should be used with extreme caution as overheating and thermal burns are common. In addition to heating devices, small rodents can be covered or wrapped in a lightweight drape or bubble wrap, taking care not to impede respiration. Heat is important in the anesthetic recovery period as well. Rodents should be maintained in environmental temperatures of 30°C–35°C (86°F–95°F) during recovery (Harkness et al. 2010).

3.2.3.2   Fluids
Preoperative or preanesthesia administration of fluids given subcutaneously may be beneficial for long-duration anesthesia or debilitated small rodents. Fluids may also be provided during the imaging session, especially if intravenous access has been obtained with a catheter. Lactated ringers solution or 0.9% sodium chloride can be given intravenously, subcutaneously, or intraperitoneally at a rate of 5–10 mL/kg/h (Harkness et al. 2010). Fluids should be warmed as cold fluids can cause hypothermia. Syringe drivers or syringe infusion pumps may be very useful at delivering these small volumes. Rodents will require 40–80 mL/kg/day of fluids in the postimaging period until it is certain that the animal is eating and drinking sufficiently.

3.2.3.3   Oxygen
While oxygen is typically delivered with inhalant anesthetics, it is a critical additional supportive care when using injectable anesthesia and may hasten recovery from anesthesia. Cyanosis and pulse oximeter readings of less than 70% are often observed in rodents given injectable anesthetics, especially ketamine and xylazine. In order to counteract hypoxemia, oxygen can be provided via facemask for the duration of the anesthesia. Corneal hypoxia resulting in damage to the cornea may be a consequence of low oxygen saturation (Harkness et al. 2010). Rodents that are debilitated or having subclinical disease may benefit from pre-oxygenation with 100% oxygen for 3–5 min prior to anesthetic induction.

3.2.3.4   Eye Lubrication
Due to their large size and somewhat exophthalmic eyes, rodents may not be able to close their eyelids completely during anesthesia. Certain injectable drugs may also cause further exophthalmia of the eyes. Sterile, nonmedicated ophthalmic ointments or drops must be used to minimize corneal drying and corneal damage. This should be placed in the eyes immediately after anesthetic induction.

Animals should be positioned to avoid eye contact with heated or other surfaces and inhalant anesthetics to avoid corneal damage and desiccation. Eye lubrication should be repeated as needed especially for long-term imaging.

3.2.3.5 Analgesia

While pain is not expected during imaging sessions, if rodents have undergone painful procedures such as surgery prior to imaging, appropriate analgesics should be used. Some analgesics, such as opioids, may be used in conjunction with anesthesia to lower the required amount of anesthetics needed.

3.3 ANESTHESIA

Unlike imaging in humans, small animal imaging almost always requires anesthesia in order to maintain the animal in one position without movement. Repeated anesthesia and recovery brings challenges to small animal imaging. It is also a variable that may have research implications when comparing results in nonanesthetized animals or correlating to human research.

3.3.1 STRAIN, GENDER, CIRCADIAN RHYTHM EFFECTS

The first parameter to consider in mouse or rat anesthesia is rodent strain. There are a vast number of strains, particularly of mice, many of which have been genetically altered. Strain differences alone are known to influence the sleep time of anesthetics (Kohn et al. 1997). Genetically modified rodents may have unexpected variability in their response to anesthetics since the location of insertion and number of copies of genetic material will vary (Gaertner et al. 2008). Anesthetics doses cannot easily be extrapolated from one strain to another and mice often require much higher doses of anesthesia than rats. Pilot studies should be undertaken when changing to a new anesthetic regimen in research models (Flecknell 1993). Gender also has significant effects on pharmacokinetics, metabolism, and other physiological parameters (Curry 2001). This effect must be considered when anesthetizing rodents. Rodents are nocturnal and most of their activity including food consumption happens in the dark cycle. Because of this, the time of day plays an important role in anesthesia of rodents and it has been documented that the same doses of anesthetics produce different levels of anesthesia depending on the time of day that they are given (Challet et al. 2007). There is less concern for these variables when inhalant anesthetics are used as the depth of anesthesia can be easily altered.

3.3.2 OPTIONS FOR ANESTHESIA: INHALANTS VERSUS INJECTABLES

It is important to define the use of anesthetics in imaging studies. In almost all cases, the goal is immobilization and decreased stress (light anesthesia) compared to anesthetizing an animal for surgery that would also require loss of consciousness and analgesia (deep anesthesia). However, due to the size of rodents and difficulty with vascular access, there are very few anesthetics or combinations of anesthetics commonly used. Drugs given to induce and maintain general anesthesia are either inhalant anesthetics or injectable anesthetics. There are many online resources for determining the doses of anesthetics in rodent species and, therefore, doses will not be discussed here. In some cases, it may be beneficial to consider the use of injectable and inhalant anesthetics together, often allowing the dose of each anesthetic to be lowered. This may result in fewer unwanted side effects of each anesthetic.

3.3.2.1 Inhalant anesthetics versus inhalants

Inhalant anesthetics are considered the gold standard for rodents as they are easy to administer, easy to remotely administer via a breathing system, safe for many ages and strains, and allow for control of the depth of anesthesia. Inhalant anesthetics include halothane, isoflurane, sevoflurane, and desflurane. Minimum alveolar concentration (MAC) is the alveolar concentration of an anesthetic required to block the response to a specific stimulus in 50% of animals, so the lower the MAC value, the more potent the anesthetic. The most commonly used inhalant anesthetic in veterinary medicine and rodent research is isoflurane. The MAC for isoflurane in rats is 1.38% (Flecknell 1987). It is readily available and relatively inexpensive. Isoflurane is usually delivered at 3.5%–4.5% gas in oxygen to induce anesthesia, which is then maintained with a concentration of 1.5%–3% (Flecknell 1996). For longer-term anesthesia, Constantinides et al. (2011) showed isoflurane at 1.5% to provide the most stable heart rate and mean arterial pressure in mice anesthetized for 90 min. Isoflurane maintains better cardiac function than most injectable anesthetics but is a respiratory depressant.

Due to the respiratory depressant concerns with the use of inhalant anesthetics, it has become common to intubate rodents for prolonged imaging sessions. Intubation allows for positive pressure ventilation and gives a direct connection from the rodent to the anesthetic machine and a secure airway. Rodent intubation kits are available commercially and there are many choices of techniques available for rodent intubation. The most common method is to use intravenous catheters or laboratory tubing of appropriate size to intubate. Artificial ventilation can be done via commercially available rodent ventilators and may reduce variability in studies by decreasing hypoxemia and hypercapnia.

Induction of inhalant anesthetics is safely done in an induction chamber, with appropriate scavenging systems to prevent human exposure to waste gases. The induction chamber should allow for the visualization of the animal and should be sized as small as possible to accommodate the animal and to ensure a rapid induction phase of anesthesia (Figure 3.5).

Once induced, the animal should be quickly removed from the induction chamber and either intubated or attached to a nonrebreathing anesthetic circuit with a tight-fitting nose cone with minimal dead space. Nose cones are an ideal way to deliver inhalant anesthetics to rodents, as rodents are obligate nasal breathers; however, anesthetic-induced respiratory depression is common and rodents may require positive pressure ventilation. As rodents have a compliant pulmonary system, it is possible to ventilate a rodent with a nose cone (Gaertner et al. 2008). Appropriate scavenging should be used to ensure human safety, especially when a facemask is used. Because rodents have a relatively large abdomen, which could lead to thoracic compression, they should be positioned at a slight incline with the head slightly above the tail (Balaban and Hampshire 2001).

For inhalant anesthetics, an anesthetic machine consisting of a calibrated vaporizer, a flow meter, and a delivery circuit is necessary for proper administration. The carrier gas is normally oxygen, allowing for a built-in supportive care mechanism. Recovery from inhalant anesthetics is rapid, which for a nonpainful procedure such as imaging can be an advantage.

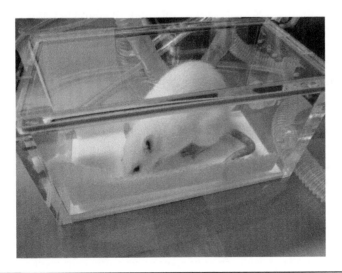

FIGURE 3.5 Rat in an appropriately sized, clear, induction chamber.

3.3.2.2 Injectable anesthetics versus injectables

Injectable anesthetics come with inherent risks in small animals, as once drugs are given by injection, the dosage cannot be reduced. Therefore, injectable drugs should be used at low doses and drugs with a wide safety margin should be used. It is also advisable to do a small pilot study on the particular species and strain to be anesthetized as there are significant strain effects on anesthesia. There are occasions where the use of injectable anesthetics are warranted such as when inhalant anesthetics may interfere with the research in question or when appropriate anesthetic equipment and scavenging are not available. Injectable anesthesia requires only a needle, syringe, and appropriate training to give the drug in the appropriate location (Flecknell 1993).

As close as possible to the timing of anesthetic induction, animals should be evaluated for health status and weighed to ensure that accurate dosing is used. Because of the small body size of rodents and the fact that many drugs are designed for larger species and must be diluted, it is easy to inadvertently overdose or underdose an individual rodent. In almost all instances, injectable drugs are given intraperitoneally. Obese mice may have an altered biodistribution of lipophilic agents and a high incidence of liver dysfunctions and are, therefore, at high anesthetic risk because of hypoventilation and hypoxia. However, cachectic mice present low plasma protein binding and might hide renal, hepatic, or cardiac deficiencies (Gargiulo et al. 2012).

Injectable anesthetics may be a single drug (e.g., pentobarbital) or a combination of drugs (e.g., ketamine/xylazine). In general, combinations of drugs are preferred allowing for lower doses of individual drugs and the potential to reduce side effects. The most common anesthetic drugs and combinations are outlined in the following sections including their advantages and disadvantages. In the author's experience, published doses of injectable anesthetics for rodents can be used at lower doses when a surgical plane of anesthesia is not being sought, such as for most imaging studies.

3.3.2.2.1 Alpha-2 Agonists (Xylazine or Dexmedetomidine)

Alpha-2 adrenergic agonists xylazine (Rompun®) and dexmedetomidine (Dexdomitor®) are commonly used with ketamine to provide surgical anesthesia in rodents. Used as sole agents, these drugs are muscle relaxants and sedatives often used for chemical restraint. In some species, dexmedetomidine appears to lead to greater anesthetic depth than xylazine, and it is more reliably antagonized by a reversal agent.

> *Advantages*: Advantages of alpha-2 agonists are that they provide analgesia and can be combined with ketamine ± acepromazine to produce surgical anesthesia in some animals. These drugs are not controlled substances and they are reversible with intravenous or subcutaneous reversal agents.
>
> *Disadvantages*: Disadvantages in most species include cardiovascular depression (decreased heart rate, decreased cardiac output, and hypotension) and, therefore, these drugs are not recommended for the studies of cardiac function. In addition, these drugs can cause more profound hypothermia and irreversible corneal opacities. They also act as diuretics making fluid replacement important. Oxygen supplementation is helpful to counteract the effects on the cardiovascular and respiratory systems. These anesthetics cause a transient hyperglycemia, which may have research implications.

3.3.2.2.2 Ketamine

Ketamine is a dissociative anesthetic used in a wide variety of rodent species. At low doses, ketamine provides chemical restraint with minimal analgesia. In most instances, ketamine is used in combination with other injectable agents, as it does not lead to muscle relaxation or a surgical plane of anesthesia when used alone. Incremental additional doses of ketamine can be given to extend the period of anesthesia but can cause severe respiratory depression (Flecknell 1987).

> *Advantages*: Advantages of ketamine are its wide margin of safety in most species, its *N*-methyl-D-aspartic acid receptor blocking action providing special pain control, and its limited effects on the cardiovascular system. In combination with other drugs, it can provide a surgical plane of anesthesia for about 30 min when given intraperitoneal.

Disadvantages: Disadvantages of ketamine include a mild irritancy on injection due to low pH and insufficient anesthesia in some species and strains (especially mice) for some procedures. Ketamine is a controlled substance and requires a license for use.

3.3.2.2.3 *Ketamine Combinations*

3.3.2.2.3.1 Ketamine and Alpha-2 Agonists (Xylazine or Dexmedetomidine) Ketamine may be combined with the alpha-2 agonists, xylazine or dexmedetomidine, and is normally the preferred anesthetic combination when the equipment for inhalant anesthesia is not available (Gaertner et al. 2008).

> *Advantages*: Advantages of ketamine/alpha-2 agonist combinations are that they may be combined for injection and that they may produce short-term surgical anesthesia with good analgesia in some animals. This combination can be partially reversed by reversing the alpha-2 agonist with reversal agent.
>
> *Disadvantages*: Disadvantages of ketamine/alpha-2 agonist combinations are that rodents will not reliably reach a surgical plane of anesthesia in all cases in addition to the disadvantages of both drugs listed earlier. The reversal of the alpha-2 agonist results in the reversal of the analgesic component of the combination and should only be done when there is no need for pain control. If a ketamine/alpha-2 agonist combination is used for surgery longer than 20–30 min, animals will likely require additional anesthetic. It is likely to get a much longer duration of sedation. Redosing with a lower dose of ketamine rather than the combination is usually safer, as the cardiovascular depression of alpha-2 agonists is often longer lasting than the sedation or analgesia produced. However, repeated redosing with ketamine alone will not produce a surgical plane of anesthesia.

Adding acepromazine to the ketamine–alpha-2 agonist combination may result in deeper and/or longer plane of anesthesia in small rodents, especially rats, and possibly some strains of mice as well (Arras et al. 2001). These combinations should be mixed immediately prior to use since the drugs are incompatible and their efficacy, once mixed, will decrease over time.

3.3.2.2.4 *Ketamine and Benzodiazepines (Midazolam or Diazepam)*

Benzodiazepines such as diazepam (valium) are used in combination with ketamine as anesthesia or as a premedication combination for inhalant anesthesia. Diazepam alone is a muscle relaxant and works on centers in the brain to cause a calming effect. In combination with ketamine, it counteracts the muscle rigidity of ketamine. This sedative combination will require an inhalant agent or other anesthetic to achieve surgical anesthesia.

> *Advantages*: Advantages of ketamine/benzodiazepine combinations are that they may be combined in one syringe and can produce profound sedation without the negative aspects of ketamine/xylazine or dexmedetomidine.
>
> *Disadvantages*: Disadvantages of ketamine/benzodiazepine combinations are that rodents will not reach surgical anesthesia. This combination, however, is preferred for imaging and other nonpainful procedures as it is safer than the ketamine/alpha-2 agonist combinations. Diazepam should be restricted to intravenous or intraperitoneal use.

3.3.3 BARBITURATES

Although the other injectable anesthetics addressed have many advantages over barbiturate drugs, barbiturates are still frequently used for rodent anesthesia. They are most frequently used in terminal or acute studies, as recovery can be prolonged and unpleasant. Concurrent use of an analgesic (opioid or nonsteroidal anti-inflammatory drug) is encouraged as it may improve pain relief with barbiturate use and lower the required dose of barbiturate. The most commonly used barbiturate in rodents is sodium pentobarbital given intraperitoneally. For long-term nonrecovery imaging, urethane is a commonly used barbiturate anesthetic in rodents. Urethane is likely to cause minimal respiratory and cardiovascular depression compared to other injectable anesthetics (Flecknell 1987). However, urethane is considered a mutagen and a carcinogen and its use should be well justified and precautions taken to protect personnel.

Advantages: Barbiturates may provide longer sleeping times than other commonly used anesthetics or anesthetic combinations. They are often preferred for physiological recordings and appear to provide more stable anesthesia in rats than in mice.

Disadvantages: Disadvantages of barbiturates include a narrow margin of safety, primarily associated with respiratory depression. Pain sensation is only decreased at surgical planes of unconsciousness and may even be heightened (hyperalgesia) at subanesthetic doses. Barbiturates are controlled substances and require a license for use.

3.3.3.1 Equipment Requirements for Anesthesia

With the number of rodents used in research, there are many commercially available products specifically for mice and rat anesthesia. This includes rodent ventilators and kits for endotracheal intubation, heating devices, induction chambers, nose cones, anesthetic circuits, and a variety of monitoring equipment. It is necessary to ensure that appropriate monitoring equipment is available that will work with the species being used (e.g., a pulse oximeter that can register heart rates above 400 bpm for mice under isoflurane anesthesia).

While injectable anesthetic regimes for short-duration imaging may not require much equipment, where possible a heating device with a rectal feedback probe and a source of oxygen should be used.

3.3.3.2 Monitoring Anesthesia and Monitoring Equipment

Monitoring of rodents under anesthesia can be difficult due to their small size and their rapid heart and respiratory rates, which can exceed 350 beats/min and 90 respirations/min. Monitoring will also depend on the imaging modality, the type and length of anesthesia, and the access to the animal. The goal of monitoring is to recognize expected changes in physiological mechanisms and to adjust the anesthesia or supportive care to prevent short- or long-term adverse effects on the animals. Many types of monitoring devices for large animals have been adapted for use with rodents and are commercially available.

Anesthetic monitoring of small rodents, without the use of monitoring devices, typically includes testing of rear foot reflexes before any painful procedure is started, and continuous observation of respiratory pattern, mucous membrane color, and responsiveness to manipulations and rear foot reflexes throughout the procedure. This, however, may be difficult with many types of imaging equipment and often a surgical plane of anesthesia is not warranted. Some types of imaging machines have built in monitoring devices such as beds with heating devices and rectal probes that continuously record body temperature and ECG leads that record heart rate and function. Without these devices, other than the monitoring noted earlier, it is critical to have a reliable way to monitor body temperature for longer-term imaging. Any device purchased must work with the particular imaging modality (i.e., nonferrous metal for MRIs).

As anesthetics are known to cause respiratory depression, and in some cases cardiovascular depression, monitoring of these vital signs should occur. Monitoring respirations and circulation in rodents can be difficult due to rodent heart rates exceeding 250 beats/min. To monitor these parameters, it is necessary to use specialized respiratory monitors and electrocardiograms specifically designed for rodents (Hildebrandt et al. 2008). Many commercially available monitoring devices such as pulse oximeters are available that can register heart rates over 400 beats/min. Because magnetic resonance imaging can produce artifacts in electrocardiograms, it is necessary to synchronize the heart and respiratory data with the imaging data, which is known as "gating."

3.3.3.3 Anesthetic Recovery

Rodents should be removed from imaging equipment prior to recovery. It is beneficial to provide oxygen until the animal starts purposeful movement. A recovery cage should be ready prior to recovery; it should be prewarmed and ideally not contain bedding or other devices that could harm the animal during recovery. The author prefers a standard rodent cage lined with paper towel and supplied with prepared nesting material allowing the animal to seek shelter from light. The rodent should be continuously monitored until it is conscious and moves around the cage easily. The time for recovery is normally directly related to the time under anesthesia for inhalant anesthetics. Rodents receiving injectable anesthetics may take over 30 min to begin to recover. Heat is critical during recovery and the animal's cage may be placed on a surface with supplemental heat (e.g., water circulating heating pad). Be cautious with supplemental heat sources; hyperthermia can be as detrimental as hypothermia. Provide replacement fluid therapy as needed.

Animals should not be returned to the animal room until they can stand and move about their cage. Once returned to the holding room, animals should be monitored sufficiently to ensure their complete recovery. Animals that are not fully awake may injure cage mates, so group housing of animals at recovery should be avoided.

3.3.3.4 Influence of Anesthesia on Imaging

Anesthetics themselves and their complications including hypothermia can have profound effects on imaging. It is critical to understand the potential physiological effects of a particular anesthetic to be used, to carefully consider the time period between imaging sessions and to standardize anesthetic techniques. In the case of specific types of anesthetics affecting specific imaging modalities, some general anesthetics are known to inhibit the luciferase enzyme reaction. Isoflurane, sevoflurane, desflurane all inhibited luciferase activity, which was thought to be due to their hemodynamic effects (Keyaerts et al. 2012). Other effects such as immunomodulatory effects have been documented in mice for up to 9 days following three 40 min weekly exposures to sevoflurane (Elena et al. 2003).

Hypothermia can negatively affect the quality of some imaging procedures such as PET imaging with FDG, increasing FDG uptake by interscapular brown fat, and meddling the visualization of nearby structures (Fueger et al. 2006). It was also shown that isoflurane impeded FDG uptake in mouse heart and brain (Toyama et al. 2004). Respiratory acidosis, a complication of anesthesia with increased pCO_2 and reduced pH, has been shown to affect the uptake of tracers (Fuchs et al. 2012).

3.4 ANIMAL WELFARE CONCERNS

While imaging generally improves upon the 3Rs in research, including the longitudinal analysis of often fewer animals, the potential elimination for the need for surgery and serial sacrifice, and the ability to use earlier endpoints, there are still some concerns faced by ethics committees, veterinarians, and researchers. The decision to use live animal imaging should be well justified, as the possibility of postmortem imaging exists. In the case of postmortem imaging, it can likely provide morphological data and biochemical data without motion and without conventional gross pathology procedures.

3.4.1 NUMBER AND DURATION OF ANESTHETIC SESSIONS

The duration and number of imaging sessions and the intervals between them will depend on factors such as the duration of time it takes to acquire the images, an animal's tolerance to anesthesia, and the half-life of the contrast agents. Anesthesia, depending on the type, can cause profound physiological changes and these need to be accounted for in imaging studies.

Knowledge of the negative physiological effects and pharmacological effects of repeated imaging and anesthesia, as well as the animal welfare implications, is still limited. Isoflurane used at six time points within a 24 h period for short-duration anesthesia caused an elevation in corticosterone, which was highest during the initial anesthesia with further elevation at the second anesthesia

(Altholtz et al. 2006). Suggestions for limiting anesthetic sessions to 2–3 h in a 24 h period and when performing repeated anesthesia no more than five sessions in a 1–2 week period with no more than one anesthetic session per day have been made; however, further research in this area is needed (Workman et al. 2010).

3.4.2 STRESS

Longitudinal imaging studies create numerous occasions for rodents to be stressed including repeated injections, repeated handling, transportation, experimental conditions including tumor burden, anesthesia effects including hypothermia, and fasting (Hildebrandt et al. 2008). These acute or chronic stressful conditions need to be understood and studies should be planned to reduce these where possible.

3.4.3 IMAGING OF CONSCIOUS ANIMALS

Because of the effects of anesthesia on many physiological functions, some imaging research has incorporated imaging of conscious animals, mostly rats. Restraint devices have been developed including stereotaxic devices for MRI and PET imaging. These devices require training and conditioning of the animals for up to 2 weeks in order for them to remain visually calm during imaging (Hildebrandt et al. 2008). However, there is a major welfare and physiological concern as restraint stress is an established model for stress induction in rodents causing elevated heart rate and blood pressure, increased cortisol levels, and gastric and duodenal ulceration (Glavin et al. 1994). Some optical imaging systems are designed to image freely moving conscious animals negating the need to restrain them.

3.4.4 REPEATED INJECTIONS/RADIATION EXPOSURE

Radiation exposure during single or serial imaging depends on the tracer dosages and the imaging parameters (e.g., micro-CT). In CT, the radiation dose experienced by an animal is related to the desired resolution of the scan, the number and duration of scans, and the energy of the X-ray used (Glavin et al. 1994). Small animal CT causes radiation doses ranging from 70 to 400 mGy. However, different scan protocols can be used to reduce this exposure level. A longitudinal imaging study in mice consisting of five FDG–PET and CT scans resulted in up to 1 Gy of exposure in a single animal. While 6.5–7 Gy is a lethal exposure dose for mice, these lower doses can show biological effects (Taschereau et al. 2006). Common radiotracers used in micro-PET are relatively short-lived but do leave the animal briefly radioactive. Whole body radiation doses of 6–90 cGy for mice and 1–27 cGy for rats are typical with common PET and SPECT radioisotopes (Funk et al. 2004). This whole body dose does not take into account the biodistribution of the radionuclide and, therefore, may underestimate the radiation risk to the animal (Hildebrandt et al. 2008). Care must be taken to ensure that animals are kept physically separate from other animals until background radiation levels are measured. It is important that all persons working with these animals are trained appropriately to handle radioactive material.

3.4.5 EXPERIMENTAL CONDITIONS

Imaging modalities and longitudinal studies will benefit research related to aging, and the progression of diseases such as cancer, heart disease, and neurological conditions. However, the consequences of an aged rodent population with inherent spontaneous diseases should be considered as well as the lifetime degree of stressors to the animal.

Typically, study endpoints should be defined such as tumor volume or the progression of disease. With imaging, it should be possible to refine these endpoints as more detailed information about disease state/stage can often be obtained.

3.5 FUTURE OF SMALL ANIMAL IMAGING

It is clear that rodents as models of human disease will continue to be the most widely used animals in research in the near future. As imaging equipment adapted for these small animals becomes more available and accessible, it appears that imaging as an adjunct to biomedical research will increase.

The intent of this chapter is to introduce the reader to the animal care and use aspects of rodent-based research and imaging. They are many resources, many online, in which to find details relating to the topics discussed and further information should be sought prior to imaging-based studies. In addition, a laboratory animal veterinarian is a great resource in helping to plan studies using best practices.

REFERENCES

Altholtz, L. Y., K. A. Fowler et al. (2006). Comparison of the stress response in rats to repeated isoflurane or $CO_2:O_2$ anesthesia used for restraint during serial blood collection via the jugular vein. *J. Am. Assoc. Lab. Anim. Sci.* **45**(3): 17–22.

Arras, M., P. Autenried et al. (2001). Optimization of intraperitoneal injection anesthesia in mice: Drugs, dosages, adverse effects, and anesthesia depth. *Comp. Med.* **51**(5): 443–456.

Balaban, R. S. and V. A. Hampshire. (2001). Challenges in small animal noninvasive imaging. *ILAR J.* **42**(3): 248–262.

Balcombe, J. P., N. D. Barnard et al. (2004). Laboratory routines cause animal stress. *Contemp. Top. Lab. Anim. Sci.* **43**(6): 42–51.

Bayne, K. (2005). Potential for unintended consequences of environmental enrichment for laboratory animals and research results. *ILAR J.* **46**(2): 129–139.

Challet, E., S. Gourmelen et al. (2007). Reciprocal relationships between general (Propofol) anesthesia and circadian time in rats. *Neuropsychopharmacology* **32**(3): 728–735.

Conour, L. A., K. A. Murray et al. (2006). Preparation of animals for research—Issues to consider for rodents and rabbits. *ILAR J.* **47**(4): 283–293.

Constantinides, C., R. Mean et al. (2011). Effects of isoflurane anesthesia on the cardiovascular function of the C57BL/6 mouse. *ILAR J.* **52**: e21–e31.

Curry, B. B., 3rd. (2001). Animal models used in identifying gender-related differences. *Int. J. Toxicol.* **20**(3): 153–160.

Dallmann, R., S. Steinlechner et al. (2006). Stress-induced hyperthermia in the rat: Comparison of classical and novel recording methods. *Lab. Anim.* **40**(2): 186–193.

Diehl, K. H., R. Hull et al. (2001). A good practice guide to the administration of substances and removal of blood, including routes and volumes. *J. Appl. Toxicol.* **21**(1): 15–23.

Elena, G., N. Amerio et al. (2003). Effects of repetitive sevoflurane anaesthesia on immune response, select biochemical parameters and organ histology in mice. *Lab. Anim.* **37**(3): 193–203.

Flecknell, P. (1987). *Laboratory Animal Anaesthesia: A Practical Introduction for Research Workers and Technicians*. Academic Press, San Diego, CA.

Flecknell, P. (1996). *Laboratory Animal Anaesthesia: A Practical Introduction for Research Workers and Technicians*, 2nd edn. Academic Press, London, U.K.

Flecknell, P. A. (1993). Anaesthesia of animals for biomedical research. *Br. J. Anaesth.* **71**(6): 885–894.

Fuchs, K., D. Kukuk et al. (2012). Oxygen breathing affects 3′-deoxy-3′-18F-fluorothymidine uptake in mouse models of arthritis and cancer. *J. Nucl. Med.* **53**(5): 823–830.

Fueger, B. J., J. Czernin et al. (2006). Impact of animal handling on the results of 18F-FDG PET studies in mice. *J. Nucl. Med.* **47**(6): 999–1006.

Funk, T., M. Sun et al. (2004). Radiation dose estimate in small animal SPECT and PET. *Med. Phys.* **31**(9): 2680–2686.

Gaertner, D., T. Hallman et al. (2008). Anesthesia and analgesia for laboratory rodents. In R. Fish, M. Brown, P. Danneman, and A. Karas (eds.), *Anesthesia and Analgesia in Laboratory Animals*, 2nd edn. Elsevier, Inc., London, U.K., pp. 239–297.

Gaines Das, R. and D. North. (2007). Implications of experimental technique for analysis and interpretation of data from animal experiments: Outliers and increased variability resulting from failure of intraperitoneal injection procedures. *Lab. Anim.* **41**(3): 312–320.

Gargiulo, S., A. Greco et al. (2012). Mice anesthesia, analgesia, and care, Part I: Anesthetic considerations in preclinical research. *ILAR J.* **53**(1): E55–E69.

Gaskill, B. N., C. J. Gordon et al. (2012). Heat or insulation: Behavioral titration of mouse preference for warmth or access to a nest. *PLoS One* **7**(3): e32799.

Glavin, G. B., W. P. Pare et al. (1994). Restraint stress in biomedical research: An update. *Neurosci. Biobehav. Rev.* **18**(2): 223–249.

Harkness, J. E., P. V. Turner et al. (2010). *Harkness and Wagner's Biology and Medicine of Rabbits and Rodents*, 5th edn. Wiley-Blackwell, Ames, IA.

Hildebrandt, I. J., H. Su et al. (2008). Anesthesia and other considerations for in vivo imaging of small animals. *ILAR J.* **49**(1): 17–26.

Hurst, J. L. and R. S. West (2010). Taming anxiety in laboratory mice. *Nat. Methods* **7**(10): 825–826.

Institute of Laboratory Animal Resources, Commission on Life Sciences et al. (1996). *Guide for the Care and Use of Laboratory Animals.* National Academy Press, Washington, DC.

Keyaerts, M., I. Remory et al. (2012). Inhibition of firefly luciferase by general anesthetics: Effect on in vitro and in vivo bioluminescence imaging. *PLoS One* **7**(1): e30061.

Kohn, D. R., S. K. Wixson et al. (1997). *Anesthesia and Analgesia in Laboratory Animals.* Academic Press, New York.

Leander, P., S. Mansson et al. (2000). Glycogen content in rat liver: Importance for CT and MR imaging. *Acta Radiol.* **41**(1): 92–96.

Longordo, F., J. Fan et al. (2011). Do mice habituate to "gentle handling?" A comparison of resting behavior, corticosterone levels and synaptic function in handled and undisturbed C57BL/6J mice. *Sleep* **34**(5): 679–681.

Luciano, L. and E. Reale. (1992). The "limiting ridge" of the rat stomach. *Arch. Histol. Cytol.* **55**(Suppl.): 131–138.

Nebendahl, K. (2000). Routes of administration. In G. J. Krinke (ed.), *The Laboratory Rat.* Academic Press, San Diego, CA, pp. 463–483.

Olfert, E. D., B. M. Cross et al. (1993). *Guide to the Care and Use of Experimental Animals*, Vol. 1. Canadian Council on Animal Care, Ottawa, Ontario, Canada.

Palou, A., X. Remesar et al. (1981). Metabolic effects of short term food deprivation in the rat. *Horm. Metab. Res.* **13**(6): 326–330.

Russell, W. M. S. and R. L. Burch. (1959). *The Principles of Humane Experimental Technique.* Methuen, London, U.K.

Smith, M. M. and M. Hargaden. (2001). Developing a rodent enrichment program. *Lab. Anim. (NY)* **30**(8): 36–41.

Suckow, M. A. (2001). Experimental methodology. In M. A. Suckow, P. Danneman, and C. Brayton (eds.), *The Laboratory Mouse.* CRC Press, Boca Raton, FL, pp. 113–134.

Taschereau, R., P. L. Chow et al. (2006). Monte carlo simulations of dose from microCT imaging procedures in a realistic mouse phantom. *Med. Phys.* **33**(1): 216–224.

Toyama, H., M. Ichise et al. (2004). Evaluation of anesthesia effects on [18F]FDG uptake in mouse brain and heart using small animal PET. *Nucl. Med. Biol.* **31**(2): 251–256.

Tuli, J. S., J. A. Smith et al. (1995). Stress measurements in mice after transportation. *Lab. Anim.* **29**(2): 132–138.

van den Broek, F. A., C. M. Omtzigt et al. (1993). Whisker trimming behaviour in A2G mice is not prevented by offering means of withdrawal from it. *Lab. Anim.* **27**(3): 270–272.

Van Loo, P. L., J. A. Mol et al. (2001). Modulation of aggression in male mice: Influence of group size and cage size. *Physiol. Behav.* **72**(5): 675–683.

van Ruiven, R., G. W. Meijer et al. (1998). The influence of transportation stress on selected nutritional parameters to establish the necessary minimum period for adaptation in rat feeding studies. *Lab. Anim.* **32**(4): 446–456.

Workman, P., E. O. Aboagye et al. (2010). Guidelines for the welfare and use of animals in cancer research. *Br. J. Cancer* **102**(11): 1555–1577.

4

Preclinical Models

Irene Cuadrado, Jesús Egido, Jose Luis Zamorano, and Carlos Zaragoza

4.1   PRECLINICAL MODELS IN BIOMEDICAL RESEARCH

The use of preclinical animal models in biomedical research is a key step in elucidating the plethora of host signals triggered in response to pathogenesis and implies a significant advancement in the evaluation and improvement of procedures focused on early diagnosis (a critical step for patient survival in certain types of pathologies), progression of disease, and the evaluation of novel therapeutic approaches.

Recent progress in biological disciplines including animal genetics and physiology has today made possible the generation of almost *a la carte* animal models for every type of pathology reported today. It is common to find different models used to recreate the same pathology, including small and big animal models of the same disease. However, genetic and environmental factors play a significant role in pathophysiology, and each model also exhibits strengths but at the same time weaknesses, making difficult to match a particular disease, with a single experimental model, leading to translational limitations. On the other hand, and given the wide range of models, it is now easier to devise the best strategy to find the more efficient and reliable solution against a particular pathology. Animal models should be first selected in terms of recapitulation of human disease, but additional considerations based on infrastructure, the requirement for additional personnel (which is crucial for the development of the model and for screening), and available budgets are key steps to consider in advance.

Here, we will describe small and large preclinical animal models of disease, focusing on cardiovascular, cancer, and neurodegenerative pathologies, since they represent the most important leading causes of prevalence and death; therefore, they constitute a very significant piece of extensive investigation from the bench to bedside.

4.2   ANIMAL MODELS OF CARDIOVASCULAR DISEASE

Cardiovascular diseases (CVDs) are the first leading cause of death in developed countries. Cardiac and vascular complications are multifactorial pathologies, thus making them very complicate to prevent. The development of animal models for research on CVD, including cardiac and atherothrombotic diseases, has provided us with important insights into the pathophysiology, and they were provided as essential tools to evaluate new therapeutic strategies and to predict and prevent complications. In this section, we summarize the latest advances in the use of animal models of atherothrombosis, including models of expanding vascular disease like abdominal aortic aneurysms (AAA), occlusive atherosclerotic diseases, and models of heart failure.

4.2.1   Animal Models of Atherosclerosis

4.2.1.1   Mouse Models of Atherosclerosis

Mouse models of atherosclerosis have been very useful in unveiling the importance of inflammatory and immunological mechanisms in the formation and progression of atheroma plaques. Recently, an enormous interest for the use of noninvasive magnetic resonance imaging (MRI) in mouse models of atherosclerosis has arisen (Weinreb et al. 2007) since MRI accurately characterizes the location, the size, and the shape of lesions. In addition, MRI allows the differentiation between fibrous and lipid components of regression in mouse plaques. Therefore, and in combination with noninvasive imaging technologies, mouse models of atherosclerosis are one of the best examples for testing novel contrast agents and to target specific molecules involved for full characterization of atheroma plaques at high risk of rupture.

Current mouse models for atherosclerosis are based on genetic modifications of lipoprotein metabolism with additional dietary changes. Among them, low-density lipoprotein (LDR) receptor–deficient mice (LDLR–/– mice) and apolipoprotein E–deficient mice (apoE–/– mice) are the most widely used models. Atherosclerotic lesions seen in these animals can be exacerbated by the addition of risk factors such as hypertension or diabetes.

4.2.1.1.1   LDLR–/– Mice

The LDLR–/– mouse represents a model of familial hypercholesterolemia due to one of the mutations affecting the LDLR, and the plasma lipoprotein profile resembles that of humans. Mice that are genetically deficient in LDLR manifested delayed clearance of very low density lipoprotein (VLDL) and LDL from plasma. As a result, LDLR–/– mice exhibit a moderate increase of plasma cholesterol level and develop atherosclerosis slowly on normal chow diet (Ishibashi et al. 1993; Bentzon and Falk 2010). Interestingly, the severity of the hypercholesterolemia and atherosclerotic lesions in LDLR–/– mice can be accelerated by feeding a high-fat, high-cholesterol diet (Knowles and Maeda 2000), by mutating the apoB gene into an uneditable version (Veniant et al. 1998), and by crossing with either leptin deficient mice (Hasty et al. 2001) or apoB100 transgenic mice (Sanan et al. 1998). Under these conditions, the lesions in the aorta can progress beyond the foam-cell fatty-streak stage to the fibro-proliferative intermediate stage.

4.2.1.1.2   ApoE–/– Mice

Homozygous deficiency in apoE gene results in a marked increase in the plasma levels of LDL and VLDL due to a failure in their clearance through the LDLR and LDLR-related proteins. The apoE–/– mouse contains the entire spectrum of lesions observed during atherogenesis and was the first mouse model described to develop lesions similar to those of human (Plump et al. 1992; Zhang et al. 1992).

Under normal dietary conditions, apoE–/– mice have dramatically elevated plasma levels of cholesterol, and they develop extensive atherosclerotic lesions widely distributed throughout the aorta (Plump et al. 1992; Nakashima et al. 1994; Zhang et al. 1994). This process can be exacerbated on a high-fat diet, the female mice being more susceptible than male mice (Zhang et al. 1994).

4.2.1.1.3   Diabetic Mice

Diabetes is considered a high-risk factor of CVD, including atherosclerosis and cardiomyopathy. Therefore, several models are available to study atherosclerosis and cardiomyopathy associated with diabetes, including apoE–/– and LDLR–/– mice in which type 1 diabetes is induced by streptozotocin or viral injection (Reaven et al. 1997; Shen and Bornfeldt 2007). In both mice, diabetes induction did not markedly change plasma lipid levels, thereby mimicking the accelerated atherosclerosis seen in patients with type 1 diabetes.

4.2.1.2   Rabbit Models of Atherosclerosis

Rabbit models of atherosclerosis have been used for the study of human atherosclerosis, including noninvasive MRI quantification of the fibrotic and lipid components of atheroma plaques overtime, evaluating therapeutic strategies focused on atherosclerotic plaque stabilization (Helft et al. 2001).

4.2.1.2.1   High-Cholesterol Diet

The high-cholesterol diet rabbit model has been widely used for experimental atherosclerosis. Rabbit model has largely been used to study the influence of lipid lowering (induced by diet or by pharmacological treatment with statins) on plaque formation and stabilization and also have contributed to shed light into the mechanisms by which lipid lowering reduces macrophages accumulation during the early steps of inflammation (Bustos et al. 1998; Aikawa et al. 2002; Hernandez-Presa et al. 2003).

4.2.1.2.2 Combination of Hyperlipidemia and Vascular Injury

In order to examine the relevance of inflammation on atherosclerotic plaque, recently a rabbit model was set up, by inducing femoral endothelial denudation in high-fat diet hyperlipidemic rabbits. The animals exhibit more intensive vascular lesions and represent a novel approach to study inflammation-related atherosclerosis (Largo et al. 2008).

4.2.1.2.3 Vulnerable Plaque model

Shimizu et al. (2009) have developed a simple rabbit model of vulnerable atherosclerotic plaque, with the combination of aggressive vascular injury and hyperlipidemic diet, which resembles three human features or atherosclerosis, including a "vulnerable plaque" lipid core, macrophage infiltration, and thin fibrous cap. In addition, a LDL receptor–deficient animal model (the WHHL rabbit) has been developed. This model resembles human familial hypercholesterolemia.

4.2.1.3 Porcine Models of Atherosclerosis

Currently, there is no single and golden standard animal model of vulnerable plaque, but pig models are probably the best way to recreate human plaque instability. The combination of diabetes and hypercholesterolemia constitute a good model of accelerated atherosclerosis (Gerrity et al. 2001) and it was relevant to study the role of certain biomarkers, such as the Lp-PLA2, since these animals share a similar plasma lipoprotein profile to humans. In this regard, the selective inhibition of Lp-PLA2 by darapladib decreased progression to advanced coronary atherosclerotic lesions and confirmed a crucial role of vascular inflammation not associated to hypercholesterolemia, in the development of lesions implicated in the pathogenesis of myocardial infarction (MI) and stroke (Wilensky et al. 2008).

Swine models of atherosclerosis, including LDL receptor–deficient pigs, are also used to develop and validate new imagining methods for the study and design of devices of aortic and coronary atherosclerosis, including drug-eluting stents (Schinkel et al. 2010; Tellez et al. 2010).

We cannot forget to mention that a new swine model based on intramural coronary lipid injection was found to recapitulate early human inflammation during the onset of coronary plaque formation with very promising results (Tellez et al. 2011).

4.2.2 ANIMAL MODELS OF ATHEROTHROMBOTIC ANEURYSMS

Preclinical evaluation of atherothrombotic AAA in animal models has provided essential tools to evaluate new therapeutic strategies to suppress aneurysmal degeneration. As detailed next, models of AAA diseases include rat, mice, rabbit, and pig species.

4.2.2.1 Rat Models of AAA

Localized aortic perfusion of elastase. This model consists in the isolation of a segment of the abdominal aorta, in which elastase is transmurally perfused. The model is characterized by a secondary invasion of the vascular wall by inflammatory cells leading to elastic fiber degradation and aneurysm formation, which is potentially increased by the addition of plasmin (Anidjar et al. 1990).

Decellularized xenografts. This model is based on the observation that the extracellular matrix is immunogenic when grafted into a different species. The model consists of preparing a tube of extracellular matrix by detersion of all the cellular components (SDS elution) of a segment of abdominal aorta in one species (guinea pig, for example) and grafting it into another morphologically compatible species (usually rat) (Allaire et al. 1994). The xenogenic grafted extracellular matrix then becomes the site of immune injury, involving both cellular and humoral immune responses, leading to matrix destruction and aneurysm formation. This model was used to demonstrate that SMC seeding could prevent aneurysm development and rupture (Allaire et al. 1998).

4.2.2.2 Mouse Models of AAA

With regard to AAA, most of the mechanistic insights of human AAA based on animal models came from the study of different AAA procedures implemented in mice.

Calcium chloride–induced AAA. This method was initially tested in rabbits (Gertz et al. 1988), and it consist in the periaortic application of calcium chloride between the renal arteries and the iliac bifurcation, leading to a significant dilatation of the aorta 14 days after the procedure, and showed much more significant results when high cholesterol diet and thioglycolate application were combined (Freestone et al. 1997). In comparison with other models, calcium chloride application leads to the development of luminal dilatation in the absence of mechanistic effects.

Elastase-induced AAAs. Pyo et al. modified the elastase-induced model that was originally characterized in the rat (Pyo et al. 2000). This model involves transient perfusion of the isolated abdominal aorta with dilute pancreatic elastase. Elastase injury results in only mild-moderate immediate aortic dilatation, with subsequent development of aneurysmal dilatation (>100% increase in aortic diameter) within 14 days. The development of AAAs in this model is temporally associated with delayed degradation of medial elastin and aortic wall inflammation that consists of mononuclear phagocytes throughout the adventitial and medial layers, with relatively few polymorphonuclear cells localized to the adventitial aspect of the aortic wall.

Angiotensin II (AII)-induced AAAs. The first description of this model was provided by Daugherty et al. (2000) who showed that AII infusion into female hyperlipidemic apoE–/– mice induced the development of supra-renal AAA in ~25% of mice. Although the reasons behind the preferential localization of AAA to the supra-renal mouse abdominal aorta are still unknown, current knowledge indicates that the mechanistic pathways responsible for AII-induced AAA are most probably related to inflammatory activation of the vessel wall through AT1a receptors (Cassis et al. 2007), with activation of the NADPH oxidase p47phox (Thomas et al. 2006), c-JUN N-terminal (Yoshimura et al. 2005), and Rho kinases (Wang et al. 2005); enhanced recruitment of monocytes/macrophages (Saraff et al. 2003) through CCL2/CCR2 chemokine pathway (Ishibashi et al. 2004); and enhanced production and activation of several proteases (Manning et al. 2003; Deguchi et al. 2009).

Spontaneous mutated mice. The blotchy mouse is a mouse strain containing a spontaneous mutation on the X chromosome, which leads to abnormal intestinal copper absorption. Interestingly, these animals do not crosslink elastin and collagen, showing weaker elastic tissue and developing aortic aneurysms (Brophy et al. 1988) mainly in the aortic arch, the thoracic aorta, and occasionally in the abdominal aorta. The use of this mouse model, however, is difficult, since many side effects associated with the mutation including emphysema condition the results obtained with these mice.

4.2.2.3 Rabbit Models of AAA

The use of rabbit models over other species offers a more comprehensive translation to humans, since rabbit aneurysms hemodynamically and histologically resemble human aneurysms more precisely than other species including mice, thus providing an excellent model for testing endovascular therapies (Dai et al. 2006, 2008). As in mice, rabbit models of AAA include porcine elastase infusion and calcium chloride application (Freestone et al. 1997) in the abdominal aorta, as well as elastase infusion in the right carotid artery (Fujiwara et al. 2001).

4.2.2.4 Porcine Models of AAA

Porcine models of AAA have provided reliable information about the changes that occur after AAA induction and about the responses to stent deployment. A recently developed porcine model combines mechanical dilatation by balloon angioplasty with enzymatic degradation by infusion of a collagenase/elastase solution. The model is characterized by gradual AAA expansion associated with degradation of aortic wall elastic fibers, an inflammatory cell infiltrate, and persistent smooth muscle cell loss (Molacek et al. 2009). A broad number of similarities were found between this model and human AAA, and the procedure may also represent an excellent method to evaluate endovascular-related procedures. Despite the benefits, however, pigs have significant disadvantages, including complex animal handling, the requirement of special housing and surgical room facilities, the elevated cost of the animals, and the reduced sample sizes per assay.

Thoracic aortic aneurysm. Mouse models of disease contributed to significant progress in the understanding of thoracic aortic aneurysm (TAA). Marfan syndrome (MFS) is a common disorder of the connective tissue that involves the cardiovascular system and it is caused by mutations that affect the structure or expression of the extracellular matrix protein fibrillin-1, a glycoprotein that associates with extracellular proteins, including integrin receptors and insoluble elastin (Ramirez and Dietz 2007). Fibrillin-1 mutations in MFS decrease extracellular matrix sequestration of latent transforming growth factor-beta (TGFβ), thus rendering it more prone to or accessible for activation (Neptune et al. 2003; Habashi et al. 2006). TAA progression in MFS is driven by elastic fiber calcification, improper ECM proteins synthesis and matrix-degrading enzymes (matrix metalloproteinases), vascular wall inflammation, intimal hyperplasia, structural collapse of the vessel wall, as well as improper activation of MAP kinase signaling (Carta et al. 2009). In view of these results, systemic TGFβ antagonism is applied to mitigate vascular disease in mouse models of MFS and in children with severe and rapidly progressive MFS (Ramirez and Dietz 2007). In addition, murine models have recently shown that fibulin-4 and LRP1 are also associated to TAA (Boucher et al. 2007; Hanada et al. 2007).

4.2.3 Animal Models of Heart Failure

4.2.3.1 Rat Models of Heart Failure

Rat models have dominated research in heart damage since they share many of the benefits of mice, but in addition, the size of the rat greatly facilitates surgical and postsurgical procedures. A first method to induce heart failure in rat consists in the subcutaneous administration of isoproterenol, which causes myocardial necrosis in the heart, finding extensive left ventricle dilatation and hypertrophy 2 weeks after the procedure (Zbinden and Bagdon 1963). Alternatively, an electrical procedure that applies a 2 mm tipped soldering iron to the epicardium of the left ventricle, is causing a similar effect (Adler et al. 1976).

The third method today by far is the surgical procedure most commonly used in rodents; this consists in the ligation (permanent or transient) of the left coronary artery, between the pulmonary artery outflow tract and the left atrium (Pfeffer et al. 1979).

4.2.3.2 Mouse Models of Heart Failure

Mouse models of permanent or partial occlusion of the left coronary artery, efficiently reproduce human ischemia/reperfusion injury (Michael et al. 1999). The method consists in the occlusion of the left anterior descending coronary artery followed by reperfusion, thus allowing flow through the previously occluded coronary artery bed.

The method has been further modified to analyze ischemic preconditioning of the heart. In this method, the left coronary artery is repeatedly occluded to subject the heart to several rounds of brief ischemia and reperfusion, followed by permanent occlusion. This approach has identified several ischemia-induced genes that confer tolerance to a subsequent ischemic event (Xuan et al. 2007).

Another model of MI was implemented in mice and rats, by creating series of cryo injuries on the epicardial size, results (Li et al. 1999).

4.2.3.3 Large Animal Models of Heart Failure

The first large animal models used to study heart failure were dogs, in which MI and serial microembolization of the coronary artery models were performed (Adamson and Vanoli 2001). However, the preferred large species is the pig, in which different models of MI were implemented, including balloon occlusion of the left anterior descending coronary artery, by inserting a guide catheter in the femoral position and positioning a angioplasty balloon over a guide wire in the artery in a distal position to the second largest diagonal branch, inducing the infarction by inflating the balloon (Suzuki et al. 2008). The similar size and cardiac physiology of pigs and humans means that this model offers major advantages over the models in other species. However, the method requires specialized equipment, dedicated surgical facilities, and skilled personnel, thus limiting the number of laboratories able to conduct these studies.

Rabbit models of heart failure, including coronary artery occlusion models, have major advantages over other species (Gonzalez et al. 2009). Interestingly, a spontaneous model of MI also exist in rabbits: the WHHLMI rabbit strain, developed by selective breeding of coronary atherosclerosis-prone WHHL rabbits without any interventions such as ligation (Shiomi et al. 2003). A limitation of this model is the absence of coronary plaque rupture, which is a critical step and a major cause that leads to acute MI in humans. However, the model is still valid for the study of atherosclerotic-related heart complications (Kuge et al. 2008; Shiomi et al. 2008).

Additional models of heart failure in large and small animals include pressure overload models of the left ventricle by transverse aortic constriction in mice (Rockman et al. 1994), aortic banding in rats and dogs (Schunkert et al. 1995; Nagatomo et al. 2000), and left ventricle hypertrophy models of renal artery constriction or aortic stenosis in rats, hamsters, mice, and dogs (Koide et al. 2000; Shimizu et al. 2006; Henderson et al. 2007).

4.3 ANIMAL MODELS IN CANCER RESEARCH

In cancer research, rodents, and in particular mice, are the species that lead to extensive development in human cancer research. As in cardiovascular research, mice are more convenient for reasons that include small size, easy handling and breeding, and large size populations, and more important, extensive research in mouse genomics made very accessible to induce quick and reliable genetic modifications at low cost, when compared to other species of interest. Currently, mouse models of human cancer contain specific mutations that include the expression of selected oncogenes, inhibition of specific genes, or inactivation of tumor suppressor genes, together with the use of tumor cell and/or metastatic tissue transplantation. However, mouse research limitations include species-specific differences and the failure to recapitulate human tumor progression for certain types of cancer.

4.3.1 LOSS OF GENE FUNCTION MOUSE MODELS OF CANCER RESEARCH

Targeted disruption of selected genes is a very powerful tool in therapy research on specific types of cancer as it provides with substantial information about the signaling pathways in which the gene or genes of interest participate as tumor suppressor or tumor formation genes. However, conventional knockout mouse models have sometimes significant limitations based on the pleiotropic effects of the targeted gene of interest. Such is the case of genes that lead to embryonic or infant lethality, or infertility. Breast cancer *BRCA* genes are a clear example, since single knockout mice die before birth or adulthood (Evers and Jonkers 2006).

To overcome these genetic limitations, conditional mouse models are designed to shut down gene expression based under experimental conditions in two different ways: temporary or a combination of temporary and spatial gene deletion, in which a tissue-specific promoter allows tissue-specific deletion of the gene of interest whenever it is best convenient for the investigation. This technology allows targeted gene disruption site and is time specific (Sykes and Kamps 2003; Borowsky 2011; Hlady et al. 2012).

Alternatively, a more recent approach of gene targeting is the use of small interfering RNA (siRNA) and short hairpin RNA (shRNA), causing long-term in vivo silencing expression of the gene of interest by specific binding to messenger RNA sequences (Dillon et al. 2005; Dohmen et al. 2012).

4.3.2 GAIN-OF-FUNCTION MOUSE MODELS OF CANCER RESEARCH

Gain-of-function mouse models are based on the overexpression of the gene or genes of interest in the whole body, or by using conditional overexpression models, to track time and tissue specificity of gene expression.

Good examples of this approach are the knocking models of p53 and the variety of distinct phenotypes reported in the literature. More recently, a new study was conducted to investigate the specific roles of mutant p53 dominant negative forms or different gain of function (p53 knock-in) mutants on tumorigenesis, concluding that mutant p53 dominant negative affects acute response, whereas gain of function is not universal, and it is mutation type specific (Lee et al. 2012).

4.3.3 ALLOGRAFT AND XENOGRAFT TRANSPLANTATION MODELS

Mouse models of gain or loss of gene function allow the study of almost all types of cancer forms alone and/or in combination with procedures where tumor cells are injected from their own species (syngeneic models) (Waldmeier et al. 2012) or from different species (xenogeneic models) (Huszthy et al. 2012). In the syngeneic model, tumor cells are inserted into naïve members of the same mouse strain, whereas in the xenogeneic model, xenotransplantation is mostly performed by subcutaneous injection of human tumor cells in mice. The selection of one particular model depends on different factors, but it should be noticed that syngeneic models are more reproducible and easy to work with, but they often fail to reproduce human cancerous conditions. To this regard, a significant increase in specificity came from an alternative method, which consists of xenograft injection at the same location as the human tumor source (orthoptic models). Xenotransplantation models are more difficult to implement but show significant predictor value of human tumors, which explain that anticancer therapies are mostly based on this approach in combination with gain- and loss-of-function genetic mouse models. It is reported that patient-derived human tumor tissue xenografts models implanted subcutaneously or in subrenal capsule in immunodeficient mice (athymic nude mice or severe combined immunodeficient mice) are increasingly used, providing better preclinical testing of new therapies for the treatment and better outcome for cancer (Jin et al. 2010). Interestingly, a clear example of xenotransplantation was recently reported in which zebrafish embryos or adult fish are used as a recipient model of human tumor cell xenotransplantation. However, despite the high conservation of gene function between fish and humans, a significant concern remains that potential differences in zebrafish tissue niches and/or missing microenvironmental cues could limit the relevance and translational utility of data obtained from zebrafish human cancer cell xenograft models (Konantz et al. 2012).

4.4 ANIMAL MODELS OF NEURODEGENERATIVE DISEASE

Age-related neurodegenerative disease models are almost exclusively developed in genetically modified mice, although some nonhuman primates and other big mammalian species show phenotypes similar to those found in human brain aging. Such are the cases of superoxide dismutase–linked canine degenerative myelopathy (Grabitz et al. 1990), cerebral beta-amyloidosis in nonhuman primates and dogs (Wei et al. 1996; Chambers et al. 2011), or neurofibrillary tangles in nonhuman primates, bears, and sheep (Cork et al. 1988; Nelson and Saper 1995; Rosen et al. 2008). Besides, genetically engineered animals have provided with substantial information about the onset and treatment of a vast number of age-related neurodegenerative human pathological conditions.

As in other pathologies and for the reasons described earlier, mice are also considered as the number one species of interest in age-related neurodegenerative research.

4.4.1 MOUSE MODELS OF ALZHEIMER'S DISEASE

Mouse models of Alzheimer's disease are genetically modified animals lacking the amyloid-beta precursor protein (Guo et al. 2012) or the presenilins1 or 2 (*PSEN1* or *PSEN2*) genes (Price et al. 1998; Beglopoulos and Shen 2006) since mutations of these genes cause autosomal dominant

Alzheimer's disease. Animal models carrying single, double, and even triple targeted mutations are consistent in terms of Alzheimer's event recapitulation overtime. However, significant discrepancies arise between mouse models and human trials and hence require additional effort and intellectual revisiting some critical concepts of disease.

4.4.2 PARKINSON'S DISEASE

Models of Parkinson's disease are based on single point mutations of the gene encoding for alpha-synuclein (Kahle et al. 2000), whereas overexpression of this protein in mice (Masliah and Hashimoto 2002) strongly resembles the Lewy bodies and neuritis of Parkinson's and other neurodegenerative pathologies (Kirik and Bjorklund 2003). Alpha-synuclein models, as well as mutated forms of the LRRK2 gene (Lee et al. 2012), recapitulate most of the symptoms of human Parkinson's disease, but the early dopaminergic neuronal loss in the substantia nigra, as early cause of disease, is still under investigation, and yet no models are available to reproduce this pathological condition.

A new platform for the study of Parkinson's disease came from the recent advancements on induced pluripotent stem cell (iPSC) research (Cundiff and Anderson 2011). At least two iPSCs containing mutations at the alpha-synuclein and *LRRK2* genes were generated (Soldner et al. 2011), and these were found able to differentiate into dopaminergic neurons, resulting in extensive apoptosis and expression of oxidative stress and ubiquitin–proteasome responsive genes, suggesting that these iPSC-derived neurons exhibit early phenotypic signs shown in vivo during the onset of disease.

4.4.3 MODELS OF DEMENTIA

Dementia is a pathological age-related condition, progressive and yet incurable. Common forms of dementia include Alzheimer's disease, semantic dementia, or frontotemporal dementia, featuring a frontotemporal lobar degeneration (FTLD) of the brain. FTLD-tau is a form of FTLD in which the microtubule-associated protein tau (MAPT) does not correctly polymerize (Goedert et al. 1996). Transgenic mice overexpressing different mutated forms of FTLD-tau were found useful at the molecular level to characterize intracellular signals triggered during the progression of disease (Goedert et al. 1998; Probst et al. 2000).

At postmortem, another form of FTLD subtype described is FTLD-TDP43 (Colombrita et al. 2009). TDP-43 is encoded by the TARDBP gene. Mutated forms of TARDBP are associated with FTLD (Yin et al. 2010) and amyotrophic lateral sclerosis (Herman et al. 2011; Liscic and Breljak 2012). Other mutations associated with familial FTLD include mutations at the superoxide dismutase (SOD-1) gene (Maekawa et al. 2009), 14 mutations at the fused sarcoma (FUS) gene (Lanson and Pandey 2012), and the progranulin (GRN) gene (Cruts and Van Broeckhoven 2008).

It is assumed that no single mouse model of neurodegenerative disease fully recapitulates human onset and the progression of pathology for several reasons, including differences in brain complexity, size, and architecture. Considering this major limitation, investigation now is focusing much on creating specific models toward early pathogenic events and new animal platforms for the development of more efficient screening compounds against disease.

4.5 CONCLUSIONS

Animal models serve as valuable and powerful tools in preclinical investigation toward a common horizon focused on translational medicine. Thanks to the latest technological advances, today it is feasible to recreate any human pathological condition in different animal templates. Depending on the nature of the studies, some models would be more accurate than others, but it is important to realize that none of them perfectly match the human disease by which they were originally developed. Having this concern in mind, animal models facilitated a much better knowledge of disease and significantly contribute with formulation of more effective diagnostic, treatment, and preventive strategies against disease. In combination with the newest advances in noninvasive molecular imaging technologies, fascinating upcoming events are yet to come.

REFERENCES

Adamson, P. B. and E. Vanoli. (2001). Early autonomic and repolarization abnormalities contribute to lethal arrhythmias in chronic ischemic heart failure: Characteristics of a novel heart failure model in dogs with postmyocardial infarction left ventricular dysfunction. *J. Am. Coll. Cardiol.* **37**(6): 1741–1748.

Adler, N., L. L. Camin et al. (1976). Rat model for acute myocardial infarction: Application to technetium-labeled glucoheptonate, tetracycline, and polyphosphate. *J. Nucl. Med.* **17**(3): 203–207.

Aikawa, M., S. Sugiyama et al. (2002). Lipid lowering reduces oxidative stress and endothelial cell activation in rabbit atheroma. *Circulation* **106**(11): 1390–1396.

Allaire, E., C. Guettier et al. (1994). Cell-free arterial grafts: Morphologic characteristics of aortic isografts, allografts, and xenografts in rats. *J. Vasc. Surg.* **19**(3): 446–456.

Allaire, E., D. Hasenstab et al. (1998). Prevention of aneurysm development and rupture by local overexpression of plasminogen activator inhibitor-1. *Circulation* **98**(3): 249–255.

Anidjar, S., J. L. Salzmann et al. (1990). Elastase-induced experimental aneurysms in rats. *Circulation* **82**(3): 973–981.

Beglopoulos, V. and J. Shen. (2006). Regulation of CRE-dependent transcription by presenilins: Prospects for therapy of Alzheimer's disease. *Trends Pharmacol. Sci.* **27**(1): 33–40.

Bentzon, J. F. and E. Falk. (2010) Atherosclerotic lesions in mouse and man: Is it the same disease? *Curr. Opin. Lipidol.* **21**(5): 434–440.

Borowsky, A. D. (2011). Choosing a mouse model: Experimental biology in context--the utility and limitations of mouse models of breast cancer. *Cold Spring Harb. Perspect. Biol.* **3**(9): a009670.

Boucher, P., W. P. Li et al. (2007). LRP1 functions as an atheroprotective integrator of TGFbeta and PDFG signals in the vascular wall: Implications for Marfan syndrome. *PloS One* **2**(5): e448.

Brophy, C. M., J. E. Tilson et al. (1988). Age of onset, pattern of distribution, and histology of aneurysm development in a genetically predisposed mouse model. *J. Vasc. Surg.* **8**(1): 45–48.

Bustos, C., M. A. Hernandez-Presa et al. (1998). HMG-CoA reductase inhibition by atorvastatin reduces neo-intimal inflammation in a rabbit model of atherosclerosis. *J. Am. Coll. Cardiol.* **32**(7): 2057–2064.

Carta, L., S. Smaldone et al. (2009). p38 MAPK is an early determinant of promiscuous Smad2/3 signaling in the aortas of fibrillin-1 (Fbn1)-null mice. *J. Biol. Chem.* **284**(9): 5630–5636.

Cassis, L. A., D. L. Rateri et al. (2007). Bone marrow transplantation reveals that recipient AT1a receptors are required to initiate angiotensin II-induced atherosclerosis and aneurysms. *Arterioscler. Thromb. Vasc. Biol.* **27**(2): 380–386.

Chambers, J. K., M. Mutsuga et al. (2011). Characterization of AbetapN3 deposition in the brains of dogs of various ages and other animal species. *Amyloid* **18**(2): 63–71.

Colombrita, C., E. Zennaro et al. (2009). TDP-43 is recruited to stress granules in conditions of oxidative insult. *J. Neurochem.* **111**(4): 1051–1061.

Cork, L. C., R. E. Powers et al. (1988). Neurofibrillary tangles and senile plaques in aged bears. *J. Neuropathol. Exp. Neurol.* **47**(6): 629–641.

Cruts, M. and C. Van Broeckhoven. (2008). Loss of progranulin function in frontotemporal lobar degeneration. *Trends Genet.* **24**(4): 186–194.

Cundiff, P. E. and S. A. Anderson. (2011). Impact of induced pluripotent stem cells on the study of central nervous system disease. *Curr. Opin. Genet. Dev.* **21**(3): 354–361.

Dai, D., Y. H. Ding et al. (2006). A longitudinal immunohistochemical study of the healing of experimental aneurysms after embolization with platinum coils. *AJNR* **27**(4): 736–741.

Dai, D., Y. H. Ding et al. (2008). Endovascular treatment of experimental aneurysms with use of fibroblast transfected with replication-deficient adenovirus containing bone morphogenetic protein-13 gene. *AJNR* **29**(4): 739–744.

Daugherty, A., M. W. Manning et al. (2000). Angiotensin II promotes atherosclerotic lesions and aneurysms in apolipoprotein E-deficient mice. *J. Clin. Invest.* **105**(11): 1605–1612.

Deguchi, J. O., H. Huang et al. (2009). Genetically engineered resistance for MMP collagenases promotes abdominal aortic aneurysm formation in mice infused with angiotensin II. *Lab. Invest.* **89**(3): 315–326.

Dillon, C. P., P. Sandy et al. (2005). Rnai as an experimental and therapeutic tool to study and regulate physiological and disease processes. *Annu. Rev. Physiol.* **67**: 147–173.

Dohmen, C., T. Frohlich et al. (2012). Defined folate-PEG-siRNA conjugates for receptor-specific gene silencing. *Mol. Ther.* **1**: e7.

Evers, B. and J. Jonkers. (2006). Mouse models of BRCA1 and BRCA2 deficiency: Past lessons, current understanding and future prospects. *Oncogene* **25**(43): 5885–5897.

Freestone, T., R. J. Turner et al. (1997). Influence of hypercholesterolemia and adventitial inflammation on the development of aortic aneurysm in rabbits. *Arterioscler. Thromb. Vasc. Biol.* **17**(1): 10–17.

Fujiwara, N. H., H. J. Cloft et al. (2001). Serial angiography in an elastase-induced aneurysm model in rabbits: Evidence for progressive aneurysm enlargement after creation. *AJNR* **22**(4): 698–703.

Gerrity, R. G., R. Natarajan et al. (2001). Diabetes-induced accelerated atherosclerosis in swine. *Diabetes* **50**(7): 1654–1665.

Gertz, S. D., A. Kurgan et al. (1988). Aneurysm of the rabbit common carotid artery induced by periarterial application of calcium chloride in vivo. *J. Clin. Invest.* **81**(3): 649–656.

Goedert, M., R. A. Crowther et al. (1998). Tau mutations cause frontotemporal dementias. *Neuron* **21**(5): 955–958.

Goedert, M., R. Jakes et al. (1996). Assembly of microtubule-associated protein tau into Alzheimer-like filaments induced by sulphated glycosaminoglycans. *Nature* **383**(6600): 550–553.

Gonzalez, G. E., I. M. Seropian et al. (2009). Effect of early versus late AT(1) receptor blockade with losartan on postmyocardial infarction ventricular remodeling in rabbits. *Am. J. Physiol.* **297**(1): H375–H386.

Grabitz, K., E. Freye et al. (1990). The role of superoxide dismutase (SOD) in preventing postischemic spinal cord injury. *Adv. Exp. Med. Biol.* **264**: 13–16.

Guo, Q., Z. Wang et al. (2012). APP physiological and pathophysiological functions: Insights from animal models. *Cell Res.* **22**(1): 78–89.

Habashi, J. P., D. P. Judge et al. (2006). Losartan, an AT1 antagonist, prevents aortic aneurysm in a mouse model of Marfan syndrome. *Science (New York, N.Y)* **312**(5770): 117–121.

Hanada, K., M. Vermeij et al. (2007). Perturbations of vascular homeostasis and aortic valve abnormalities in fibulin-4 deficient mice. *Circ. Res.* **100**(5): 738–746.

Hasty, A. H., H. Shimano et al. (2001). Severe hypercholesterolemia, hypertriglyceridemia, and atherosclerosis in mice lacking both leptin and the low density lipoprotein receptor. *J. Biol. Chem.* **276**(40): 37402–37408.

Helft, G., S. G. Worthley et al. (2001). Atherosclerotic aortic component quantification by noninvasive magnetic resonance imaging: An in vivo study in rabbits. *J. Am. Coll. Cardiol.* **37**(4): 1149–1154.

Henderson, B. C., N. Tyagi et al. (2007). Oxidative remodeling in pressure overload induced chronic heart failure. *Eur. J. Heart Fail.* **9**(5): 450–457.

Herman, A. M., P. J. Khandelwal et al. (2011). beta-amyloid triggers ALS-associated TDP-43 pathology in AD models. *Brain Res.* **1386**: 191–199.

Hernandez-Presa, M. A., M. Ortego et al. (2003). Simvastatin reduces NF-kappaB activity in peripheral mononuclear and in plaque cells of rabbit atheroma more markedly than lipid lowering diet. *Cardiovasc. Res.* **57**(1): 168–177.

Hlady, R. A., S. Novakova et al. (2012). Loss of Dnmt3b function upregulates the tumor modifier Ment and accelerates mouse lymphomagenesis. *J. Clin. Invest.* **122**(1): 163–177.

Huszthy, P. C., I. Daphu et al. (2012). In vivo models of primary brain tumors: Pitfalls and perspectives. *Neuro Oncol.* **14**(8): 979–993.

Ishibashi, M., K. Egashira et al. (2004). Bone marrow-derived monocyte chemoattractant protein-1 receptor CCR2 is critical in angiotensin II-induced acceleration of atherosclerosis and aneurysm formation in hypercholesterolemic mice. *Arterioscler. Thromb. Vasc. Biol.* **24**(11): e174–e178.

Ishibashi, S., M. S. Brown et al. (1993). Hypercholesterolemia in low density lipoprotein receptor knockout mice and its reversal by adenovirus-mediated gene delivery. *J. Clin. Invest.* **92**(2): 883–893.

Jin, K., L. Teng et al. (2010). Patient-derived human tumour tissue xenografts in immunodeficient mice: A systematic review. *Clin. Transl. Oncol.* **12**(7): 473–480.

Kahle, P. J., M. Neumann et al. (2000). Physiology and pathophysiology of alpha-synuclein. Cell culture and transgenic animal models based on a Parkinson's disease-associated protein. *Ann. N. Y. Acad. Sci.* **920**: 33–41.

Kirik, D. and A. Bjorklund. (2003). Modeling CNS neurodegeneration by overexpression of disease-causing proteins using viral vectors. *Trends Neurosci.* **26**(7): 386–392.

Knowles, J. W. and N. Maeda. (2000). Genetic modifiers of atherosclerosis in mice. *Arterioscler. Thromb. Vasc. Biol.* **20**(11): 2336–2345.

Koide, M., M. Hamawaki et al. (2000). Microtubule depolymerization normalizes in vivo myocardial contractile function in dogs with pressure-overload left ventricular hypertrophy. *Circulation* **102**(9): 1045–1052.

Konantz, M., T. B. Balci et al. (2012). Zebrafish xenografts as a tool for in vivo studies on human cancer. *Ann. N. Y. Acad. Sci.* **1266**: 124–137.

Kuge, Y., N. Kume et al. (2008). Prominent lectin-like oxidized low density lipoprotein (LDL) receptor-1 (LOX-1) expression in atherosclerotic lesions is associated with tissue factor expression and apoptosis in hypercholesterolemic rabbits. *Biol. Pharmaceut. Bull.* **31**(8): 1475–1482.

Lanson, N. A., Jr. and U. B. Pandey. (2012). FUS-related proteinopathies: Lessons from animal models. *Brain Res.* **1462**: 44–60.

Largo, R., O. Sanchez-Pernaute et al. (2008). Chronic arthritis aggravates vascular lesions in rabbits with atherosclerosis: A novel model of atherosclerosis associated with chronic inflammation. *Arthritis Rheum.* **58**(9): 2723–2734.

Lee, B. D., V. L. Dawson et al. (2012). Leucine-rich repeat kinase 2 (LRRK2) as a potential therapeutic target in Parkinson's disease. *Trends Pharmacol. Sci.* **33**(7): 365–373.

Lee, M. K., W. W. Teoh et al. (2012). Cell-type, dose, and mutation-type specificity dictate mutant p53 functions in vivo. *Cancer Cell* **22**(6): 751–764.

Li, R. K., Z. Q. Jia et al. (1999). Smooth muscle cell transplantation into myocardial scar tissue improves heart function. *J. Mol. Cell. Cardiol.* **31**(3): 513–522.

Liscic, R. M. and D. Breljak. (2012). Molecular basis of amyotrophic lateral sclerosis. *Prog. Neuropsychopharmacol. Biol. Psychiatry* **35**(2): 370–372.

Maekawa, S., P. N. Leigh et al. (2009). TDP-43 is consistently co-localized with ubiquitinated inclusions in sporadic and Guam amyotrophic lateral sclerosis but not in familial amyotrophic lateral sclerosis with and without SOD1 mutations. *Neuropathology* **29**(6): 672–683.

Manning, M. W., L. A. Cassis et al. (2003). Differential effects of doxycycline, a broad-spectrum matrix metalloproteinase inhibitor, on angiotensin II-induced atherosclerosis and abdominal aortic aneurysms. *Arterioscler. Thromb. Vasc. Biol.* **23**(3): 483–488.

Masliah, E. and M. Hashimoto. (2002). Development of new treatments for Parkinson's disease in transgenic animal models: A role for beta-synuclein. *Neurotoxicology* **23**(4–5): 461–468.

Michael, L. H., C. M. Ballantyne et al. (1999). Myocardial infarction and remodeling in mice: Effect of reperfusion. *Am. J. Physiol.* **277**(2 Pt 2): H660–H668.

Molacek, J., V. Treska et al. (2009). Optimization of the model of abdominal aortic aneurysm—Experiment in an animal model. *J. Vasc. Res.* **46**(1): 1–5.

Nagatomo, Y., B. A. Carabello et al. (2000). Differential effects of pressure or volume overload on myocardial MMP levels and inhibitory control. *Am. J. Physiol.* **278**(1): H151–H161.

Nakashima, Y., A. S. Plump et al. (1994). ApoE-deficient mice develop lesions of all phases of atherosclerosis throughout the arterial tree. *Arterioscler. Thromb.* **14**(1): 133–140.

Nelson, P. T. and C. B. Saper. (1995). Ultrastructure of neurofibrillary tangles in the cerebral cortex of sheep. *Neurobiol. Aging* **16**(3): 315–323.

Neptune, E. R., P. A. Frischmeyer et al. (2003). Dysregulation of TGF-beta activation contributes to pathogenesis in Marfan syndrome. *Nat. Genet.* **33**(3): 407–411.

Pfeffer, M. A., J. M. Pfeffer et al. (1979). Myocardial infarct size and ventricular function in rats. *Circ. Res.* **44**(4): 503–512.

Plump, A. S., J. D. Smith et al. (1992). Severe hypercholesterolemia and atherosclerosis in apolipoprotein E-deficient mice created by homologous recombination in ES cells. *Cell* **71**(2): 343–353.

Price, D. L., R. E. Tanzi et al. (1998). Alzheimer's disease: Genetic studies and transgenic models. *Annu. Rev. Genet.* **32**: 461–493.

Probst, A., J. Gotz et al. (2000). Axonopathy and amyotrophy in mice transgenic for human four-repeat tau protein. *Acta Neuropathol.* **99**(5): 469–481.

Pyo, R., J. K. Lee et al. (2000). Targeted gene disruption of matrix metalloproteinase-9 (gelatinase B) suppresses development of experimental abdominal aortic aneurysms. *J. Clin. Invest.* **105**(11): 1641–1649.

Ramirez, F. and H. C. Dietz. (2007). Marfan syndrome: From molecular pathogenesis to clinical treatment. *Curr. Opin. Genet. Dev.* **17**(3): 252–258.

Reaven, P., S. Merat et al. (1997). Effect of streptozotocin-induced hyperglycemia on lipid profiles, formation of advanced glycation endproducts in lesions, and extent of atherosclerosis in LDL receptor-deficient mice. *Arterioscler. Thromb. Vasc. Biol.* **17**(10): 2250–2256.

Rockman, H. A., S. P. Wachhorst et al. (1994). ANG II receptor blockade prevents ventricular hypertrophy and ANF gene expression with pressure overload in mice. *Am. J. Physiol.* **266**(6 Pt 2): H2468–H2475.

Rosen, R. F., A. S. Farberg et al. (2008). Tauopathy with paired helical filaments in an aged chimpanzee. *J. Comp. Neurol.* **509**(3): 259–270.

Sanan, D. A., D. L. Newland et al. (1998). Low density lipoprotein receptor-negative mice expressing human apolipoprotein B-100 develop complex atherosclerotic lesions on a chow diet: No accentuation by apolipoprotein(a). *Proc. Natl. Acad. Sci. USA* **95**(8): 4544–4549.

Saraff, K., F. Babamusta et al. (2003). Aortic dissection precedes formation of aneurysms and atherosclerosis in angiotensin II-infused, apolipoprotein E-deficient mice. *Arterioscler. Thromb. Vasc. Biol.* **23**(9): 1621–1626.

Schinkel, A. F., C. G. Krueger et al. (2010). Contrast-enhanced ultrasound for imaging vasa vasorum: Comparison with histopathology in a swine model of atherosclerosis. *Eur. J. Echocardiogr.* **11**(8): 659–664.

Schunkert, H., E. O. Weinberg et al. (1995). Alteration of growth responses in established cardiac pressure overload hypertrophy in rats with aortic banding. *J. Clin. Invest.* **96**(6): 2768–2774.

Shen, X. and K. E. Bornfeldt. (2007). Mouse models for studies of cardiovascular complications of type 1 diabetes. *Ann. N. Y. Acad. Sci.* **1103**: 202–217.

Shimizu, M., R. Tanaka et al. (2006). Cardiac remodeling and angiotensin II-forming enzyme activity of the left ventricle in hamsters with chronic pressure overload induced by ascending aortic stenosis. *J. Vet. Med. Sci.* **68**(3): 271–276.

Shimizu, T., K. Nakai et al. (2009). Simple rabbit model of vulnerable atherosclerotic plaque. *Neurol. Med. Chir.* **49**(8): 327–332; discussion 332.

Shiomi, M., T. Ito et al. (2003). Development of an animal model for spontaneous myocardial infarction (WHHLMI rabbit). *Arterioscler. Thromb. Vasc. Biol.* **23**(7): 1239–1244.

Shiomi, M., S. Yamada et al. (2008). Lapaquistat acetate, a squalene synthase inhibitor, changes macrophage/lipid-rich coronary plaques of hypercholesterolaemic rabbits into fibrous lesions. *Br. J. Pharmacol.* **154**(5): 949–957.

Soldner, F., J. Laganiere et al. (2011). Generation of isogenic pluripotent stem cells differing exclusively at two early onset Parkinson point mutations. *Cell* **146**(2): 318–331.

Suzuki, Y., J. K. Lyons et al. (2008). In vivo porcine model of reperfused myocardial infarction: In situ double staining to measure precise infarct area/area at risk. *Catheter. Cardiovasc. Interv.* **71**(1): 100–107.

Sykes, D. B. and M. P. Kamps. (2003). Estrogen-regulated conditional oncoproteins: Tools to address open questions in normal myeloid cell function, normal myeloid differentiation, and the genetic basis of differentiation arrest in myeloid leukemia. *Leuk. Lymphoma* **44**(7): 1131–1139.

Tellez, A., C. G. Krueger et al. (2010). Coronary bare metal stent implantation in homozygous LDL receptor deficient swine induces a neointimal formation pattern similar to humans. *Atherosclerosis* **213**(2): 518–524.

Tellez, A., D. S. Schuster et al. (2011). Intramural coronary lipid injection induces atheromatous lesions expressing proinflammatory chemokines: Implications for the development of a porcine model of atherosclerosis. *Cardiovasc. Revasc. Med.* **12**(5): 304–311.

Thomas, M., D. Gavrila et al. (2006). Deletion of p47phox attenuates angiotensin II-induced abdominal aortic aneurysm formation in apolipoprotein E-deficient mice. *Circulation* **114**(5): 404–413.

Veniant, M. M., C. H. Zlot et al. (1998). Lipoprotein clearance mechanisms in LDL receptor-deficient "Apo-B48-only" and "Apo-B100-only" mice. *J. Clin. Invest.* **102**(8): 1559–1568.

Waldmeier, L., N. Meyer-Schaller et al. (2012). Py2T murine breast cancer cells, a versatile model of TGFbeta-induced EMT in vitro and in vivo. *PLoS One* **7**(11): e48651.

Wang, Y. X., B. Martin-McNulty et al. (2005). Fasudil, a Rho-kinase inhibitor, attenuates angiotensin II-induced abdominal aortic aneurysm in apolipoprotein E-deficient mice by inhibiting apoptosis and proteolysis. *Circulation* **111**(17): 2219–2226.

Wei, L. H., L. C. Walker et al. (1996). Cystatin C. Icelandic-like mutation in an animal model of cerebrovascular beta-amyloidosis. *Stroke* **27**(11): 2080–2085.

Weinreb, D. B., J. G. Aguinaldo et al. (2007). Non-invasive MRI of mouse models of atherosclerosis. *NMR Biomed.* **20**(3): 256–264.

Wilensky, R. L., Y. Shi et al. (2008). Inhibition of lipoprotein-associated phospholipase A2 reduces complex coronary atherosclerotic plaque development. *Nat. Med.* **14**(10): 1059–1066.

Xuan, Y. T., Y. Guo et al. (2007). Endothelial nitric oxide synthase plays an obligatory role in the late phase of ischemic preconditioning by activating the protein kinase C epsilon p44/42 mitogen-activated protein kinase pSer-signal transducers and activators of transcription1/3 pathway. *Circulation* **116**(5): 535–544.

Yin, F., M. Dumont et al. (2010). Behavioral deficits and progressive neuropathology in progranulin-deficient mice: A mouse model of frontotemporal dementia. *FASEB J.* **24**(12): 4639–4647.

Yoshimura, K., H. Aoki et al. (2005). Regression of abdominal aortic aneurysm by inhibition of c-Jun N-terminal kinase. *Nat. Med.* **11**(12): 1330–1338.

Zbinden, G. and R. E. Bagdon. (1963). Isoproterenol-induced heart necrosis, an experimental model for the study of angina pectoris and myocardial infarct. *Revue canadienne de biologie* **22**: 257–263.

Zhang, S. H., R. L. Reddick et al. (1992). Spontaneous hypercholesterolemia and arterial lesions in mice lacking apolipoprotein E. *Science* **258**(5081): 468–471.

Zhang, S. H., R. L. Reddick et al. (1994). Diet-induced atherosclerosis in mice heterozygous and homozygous for apolipoprotein E gene disruption. *J. Clin. Invest.* **94**(3): 937–945.

Section II
Small Animal Imaging
Ionizing Radiation

5

Microcomputed Tomography

Nancy L. Ford

5.1 BASIC PRINCIPLES OF MICROCOMPUTED TOMOGRAPHY

Microcomputed tomography (CT) is a 3D x-ray imaging technique based upon the principles of clinical CT. Specialized equipment has become available that tailors the micro-CT scanner to high-resolution imaging of small objects or small animals. These specialized devices can be classified into two types: specimen scanners that provide very high-resolution images of small objects, excised tissues, or postmortem scans of intact rodents and specimens, and *in vivo* scanners that are capable of producing images of live rodents.

5.1.1 SCANNER COMPONENTS

Micro-CT scanners all consist of an x-ray source, an area detector, and a means of producing images at regularly spaced angular intervals around the object being imaged. There are many different options for each of these components, combinations of which are selected by the vendor, and the following should be viewed as a set of general guidelines.

5.1.1.1 X-Ray Tube

Typical commercial machines include a tungsten target tube operating with a small focal spot size. As a result of the small focal spot size, all of the heat is deposited in a very small volume of the anode, which limits the power output of the scanner. For the highest resolution images, micro-focus tubes with focal spot size of 10 μm and total power output of up to 8 W are often used, with voltage settings of 45–90 kVp and up to ~0.2 mA tube current.

To increase the tube output, some micro-CT machines employ a clinical x-ray tube with a rotating anode to dissipate heat more efficiently during the scan. Using a clinical tube allows tube settings of up to 120 kVp and 50 mA, with a peak power of ~5 kW. The drawback of the clinical tube is that the focal spot size is usually 1–3 mm, which will degrade the image resolution compared to the micro-focus tube.

Prototype micro-CT scanners built in individual research labs can be more flexible in adopting new technologies, such as carbon nanotube field emission micro-focus x-ray sources. As the technology has matured, the tube output power and lifetime have improved substantially to provide high-resolution images suitable for micro-CT applications (Lee et al. 2011).

5.1.1.2 X-Ray Detector

X-rays emitted from the source pass through the object and strike the detector, which is located directly across from the x-ray tube at a known source-to-detector distance. The detector is pixelated or divided into a regular array of square detector elements. Theoretically, each detector element is electrically isolated from its neighbors, meaning that each ionization event is only registered by a single detector element. The detectors operate in integration mode where charges accumulate over the entire duration of the x-ray exposure, which ranges from tens to hundreds of milliseconds in length per projection image. At the end of the exposure for each projection, the detector is read out and the signal is stored as a 2D projection image.

There are two categories of detectors employed in micro-CT imaging: indirect and direct conversion detectors. Indirect detectors have some similarities with film-screen imaging by employing a phosphorescent material as a conversion layer that is often made from the same material as x-ray screen. One of the common indirect detectors in micro-CT imaging is a phosphor-coated fiber optic taper that is optically coupled to a charge-coupled device (CCD). In this scenario, the incident x-rays are absorbed in the phosphor causing light photons to be released. The CCD camera will detect the light photons and convert the signal to charge. The CCD consists of regularly spaced "wells" that collect charge during the scan. The number of x-rays that strike the detector is proportional to the number of charges accumulated in each well.

Direct conversion detectors absorb the incident x-rays, creating charge pairs that are separated to opposing sides of the detector using a bias voltage. The charge that is read out is a direct measure of the ionization in the detector. Flat panel detectors are usually made with amorphous silicon (a-Si) or selenium (Se) as the absorbing material and are commonly used in micro-CT.

5.1.1.3 Gantry Rotation and Geometry

To achieve the 3D imaging aspects of micro-CT, both the x-ray source and the detector must be able to "view" the object from all angular positions. Early micro-CT scanners operated in step-and-shoot mode. Step-and-shoot imaging begins with the acquisition of the first projection image. While the detector is read out and the image is written to disk, the rotation to the next angular position occurs. In this mode, the scanner alternates between image acquisition and rotation until all of the projection views have been acquired. More recently, continuous rotation systems have become commercially available. In continuous rotation mode, the projection images are acquired during the rotation similar to clinical CT scanning. In order to achieve sufficient temporal resolution, the detector must have very fast readout times to reduce the amount of blurring in the images due to the rotation. Some scanners have the option of operating in either continuous rotation or step-and-shoot mode to allow the operator the flexibility to tailor the imaging protocol to the specific requirements of the imaging task.

There are two methods of achieving the desired angular positions around the object: gantry rotation or object rotation. In the rotating gantry system, both the x-ray source and the detector are mounted on a gantry with the object positioned at the center of rotation at a precisely known source-to-object distance (SOD). Generally, the rotating gantry system is used for *in vivo* imaging, with the animal lying on a bed similar to a clinical CT unit and can be implemented as step-and-shoot and/or continuous rotation. Rotating the object is generally easier to achieve, as the x-ray tube and detector are fixed in place, with the specimen mounted on a rotating stage. Often, the stage can be positioned at various distances from the x-ray source with very accurate motors, but the SOD is always very accurately known. The majority of the early designs of micro-CT scanners and virtually all of the commercially available specimen scanners operate by rotating the object in step-and-shoot mode.

5.1.2 Image Formation

At each projection angle, the detector reads out a 2D projection image representing a map of the attenuation of the object. Sample projection images of the thorax of a rat are given in Figure 5.1, representing two angular positions separated by approximately 90°. Note the differences in attenuation in the lung region compared with the spine. The attenuation is dependent upon the x-ray energies of the beam and the chemical composition of the object. For the spectral energies used in micro-CT, attenuation curves are plotted in Figure 5.2 for soft tissue, bone, and iodine (ICRU 1989), which is a common active ingredient in CT contrast agents. The attenuation differences provide contrast between different materials—the larger the separation between the curves for a particular energy, the larger the difference in the measured signals representing the different compositions and the more visible the different materials will be in the final reconstructed volume.

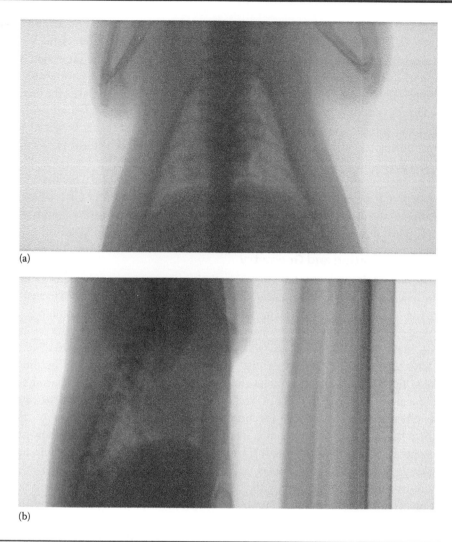

(a)

(b)

FIGURE 5.1   2D projection views of a rat thorax, where (a) shows a view with X-rays traveling along the dorsoventral axis, whereas (b) is a view rotated by approximately 90° with the X-rays traveling along the mediolateral axis. In (b), the gap between the rat (left) and the bed (right) is due to the animal being positioned on low-density foam to ensure the center of rotation of the scan aligned well with the midline of the rodent.

5.1.3   IMAGE RECONSTRUCTION

5.1.3.1   Preprocessing the Projection Images

Before the projection images can be reconstructed, they are preprocessed. An unwarping algorithm is used to remove geometric distortions and magnification. Bad pixels, such as pixels that do not respond to incident x-rays or pixels that saturate and read out the maximum signal regardless of the incident x-ray fluence at that location, can be mapped out for the detector and all of the affected pixels can be corrected in the projection by replacing the bad pixels with an average value of its neighbors. Flat field corrections are applied to equalize the response of the individual pixels across the detector, using a bright field image acquired with no object in the x-ray beam and a dark field image of the detector with no x-ray signal. From these two images, a pixel-by-pixel correction map can be developed and applied to each projection image. Finally, since the intensity of the beam is reduced exponentially as it passes through the object, the logarithm is taken of each projection image (Smith 1997). Corrected projection images ensure that the reconstructed image will have similar signal levels for regions of the object with similar attenuation properties.

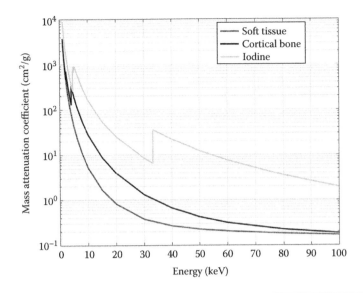

FIGURE 5.2   Attenuation curves for soft tissue, bone, and iodine (the active ingredient in x-ray contrast media) are shown for the range of energies typical in micro-CT imaging. Data for this plot were obtained from the NIST standards website (http://www.nist.gov/pml/data/xraycoef/).

5.1.3.2 Reconstruction Algorithms

Commercially available micro-CT equipment mainly utilize filtered back projection reconstruction algorithms that are well described elsewhere. Briefly, each of the corrected projection views is convolved with a ramp filter to reduce blurring and intensity gradations in the final 3D image. Each filtered projection view is then back projected through the image volume at the angle of acquisition. Back projection is a "smearing" of the pixel intensities of the filtered projection images along the path covering the entire image space. The result is a 3D volumetric image that is built up from summing the intensities from all of the projection views that "smeared" through each voxel. A sample image of a mouse is given in Figure 5.3, showing the 3D volume and the individual slices

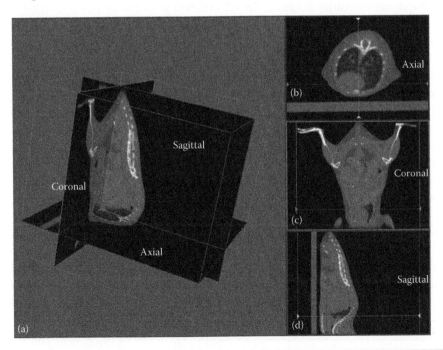

FIGURE 5.3   Reconstructed image of a mouse showing the 3D volume (a) and selected slices in the axial (b), coronal (c), and sagittal (d) planes. These images were reconstructed with 0.15 mm isotropic voxel spacing and show the inherent contrast between different tissue types (e.g., bone, soft tissue, fat, and lung).

through the axial, coronal, and sagittal planes. Alternate approaches to reconstructing micro-CT images are an active area of research (Ertel et al. 2009; Badea et al. 2011; Jin et al. 2012) and algorithms designed for cone beam clinical CT scanners will also be applicable to micro-CT imaging. Images can be scaled into Hounsfield units (HU), linear attenuation coefficients (cm^{-1}), or density of hydroxyapatite (mg HA/cm^3).

5.1.4 Image-Based Analysis

Quantitative assessment of the 3D images allows one to obtain noninvasive measurements of dimensions, volumes, attenuation properties, distribution of structures within the object or animal, and how structures are connected. Semi-automated and automated algorithms have been developed to quantitatively assess the disease model and to enable comparisons between diseased and control animals, monitor the progression of disease, or assess response to treatment over time. Analysis of bone morphometrics from the micro-CT images, such as bone volume fraction, trabecular thickness, or trabecular spacing, has been developed (Uchiyama et al. 1997; Chappard et al. 2008). Measurements of lung metrics include airway diameter, lung volume, functional residual capacity, tidal volume, CT density, and airway branching structure (Chaturvedi and Lee 2005; Ford et al. 2007a; Artaechevarria et al. 2009). Automated segmentation of the vasculature from contrast-enhanced micro-CT images has also been described (Dorr et al. 2007; Yang et al. 2010).

5.2 OPTIMIZATION OF IMAGE ACQUISITION

5.2.1 Image Quality

Image quality refers to the appearance of the 3D volumetric image. Image noise describes the variation in signal between different voxels in the image that represent the same material. Theoretically, a single material, like tissue or water or air, should have a uniform appearance in the image; every voxel representing that material should have the same signal intensity. However, due to the random nature of x-ray photon generation, electronic noise in the detector, and the presence of scattered radiation, the signal intensities fluctuate. This fluctuation gives the images a mottled appearance and can be measured quantitatively as the standard deviation of the signal intensities measured in a region of interest containing a uniform material. One can reduce the image noise by increasing the number of x-ray photons that form the image, although the x-ray dose and image acquisition time will increase.

Contrast refers to the ability of an observer to visualize the desired object in the image. Contrast arises from differences in the x-ray attenuation properties of the different materials in the sample and is related to chemical composition. Materials with a high effective atomic number (Z_{eff}) will absorb x-rays more strongly. For biomedical imaging, this implies that denser materials like bone or metal will attenuate the x-ray beam more effectively than soft tissues like muscle or fat, as shown by the x-ray attenuation curves in Figure 5.2. The larger the separation between the curves at a given energy, the larger the difference in x-ray attenuation and, therefore, the image contrast. In the 3D volume, contrast is the difference in the signal intensity between the desired object and the background. However all micro-CT images have some inherent image noise, which degrades the contrast or the conspicuity of the desired object. The contrast-to-noise ratio (CNR) is a commonly used metric to describe the ability to visualize objects within the image, taking into account the differences in signal intensities and the image noise. The easiest method to increase the CNR is to decrease the image noise, as described earlier. Altering the contrast itself is more difficult, as the Z_{eff} for the target object must be altered to artificially increase the difference in the signal intensity compared to the background material. For biomedical studies, a contrast agent can be introduced that will increase the attenuation by the target organ, while the surrounding tissues remain unaltered. Contrast agents have high atomic numbers, and with a sufficient concentration, serve to increase the Z_{eff} of the mixture (contrast media plus target object).

5.2.2 RESOLUTION

Resolution describes the ability to distinguish between two closely spaced objects. The resolution achieved depends upon many factors along the imaging chain, which are mostly fixed by the specifications of the scanner itself (focal spot size, geometry, quantum efficiency of the detector, etc.). Changing the voxel size can alter the resolution of the image, which is achieved by binning adjacent detector elements together during the image acquisition or in the reconstruction. Even if the two objects are separated by the limiting resolution of the system, they may not appear as separate objects in the image depending on how they align with the reconstruction grid, and on the image noise and object contrast. High-contrast objects are generally easier to distinguish than low-contrast objects. To maintain the image noise, an increase in the x-ray dose is required as the voxel spacing is reduced, which also implies an increased scan time for higher resolution imaging (Ford et al. 2003).

In the scientific literature describing micro-CT procedures, the term resolution generally refers to the reconstructed voxel spacing in the 3D image. The voxel spacing is determined by the pixel size on the detector scaled by a magnification factor to determine what the element size would be at the isocenter, or center of rotation, of the scanner. The smallest achievable voxel spacing varies depending on the technical specifications of individual scanners, with specimen scanners achieving the best resolution but at a reduced field of view. Specimen scanners that have been developed for biomedical purposes can produce images with an isotropic voxel spacing ranging from 5 to 100 μm. For *in vivo* scanners, the voxel spacing is larger, between 20 and 200 μm, to enable larger fields of view, faster scan times, and reduced radiation dose to the animal.

5.2.3 X-RAY DOSE

For micro-CT imaging of specimens, the x-ray dose is of little concern. However, for *in vivo* studies, the x-ray dose received by the animal may lead to unwanted effects that interfere with the measurements of the disease model. For all *in vivo* imaging, but particularly for disease models of oncology, bone formation, and respiratory disease, the acquisition protocols should be customized to minimize the x-ray dose. Minimizing the dose comes at a penalty for the image quality and a compromise must be achieved to ensure that the details of the image are retained for low-dose imaging protocols.

5.2.4 SCAN TIME

Micro-CT scan times can range from less than a minute to hours or even days. For specimen imaging, scan time is not a concern, as the samples can sit in the scanner over many hours with no loss of integrity. For biological specimens, care must be taken to ensure that the samples are well preserved and prepared in a sealed container so they do not dry out during the scan. Specimens can be submerged in water, alcohol, or other solutions that do not significantly attenuate the x-ray beam. For *in vivo* studies, the maximum scan time is guided by the duration of the anesthesia; this duration is related to the animal welfare regulations that prescribe how long each animal may be anesthetized in a 24 h period. For inhaled agents, the animals are continually breathing the anesthetic and can remain sedated up to the maximum duration of anesthesia that the animal can tolerate. For injectable anesthesia, the duration is more variable depending on the drug dose delivered and the individual metabolic rate of the animal. Care must be taken to ensure that the scan time is shorter than the expected duration of the anesthesia to avoid motion artifacts in the images and potentially unusable data, keeping in mind that there will be some additional time for setting up the scanning protocol and positioning the animal in the cradle.

5.3 SPECIMEN IMAGING

5.3.1 SCANNER FEATURES

Specimen micro-CT scanners are generally cabinet style machines, with all of the components and the sample contained within a lead-shielded compartment during image acquisition. The components include a micro-focus x-ray source and high-resolution detector mounted directly opposite

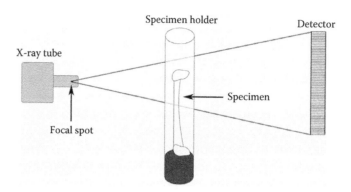

FIGURE 5.4 Schematic diagram of a specimen micro-CT scanner. The micro-focus x-ray tube and detector are typically fixed in place, with the specimen located on a moveable stage. The specimen is shown in a sample holder, which rotates during the scan to enable the acquisition of projection views from all angular positions.

to each other and operate in step-and-shoot mode. The specimen is loaded into a sample chamber that is fixed at a known position on a rotation table that rotates to a new angular position for each projection view. Usually, the specimen holder can be positioned at a number of locations along the midline of the scanner to obtain different geometric magnifications of the object, resulting in different image resolutions. The components of a specimen scanner are depicted in Figure 5.4. To obtain high-quality, high-resolution images, specimen scanners have the ability to run continuously for many minutes, hours, or even over a few days. Numerous projection views can be acquired at each angular position and averaged together to reduce image noise. The higher the required resolution, the longer the image acquisition will be.

To enable very high spatial resolution images of larger objects, multiple image acquisitions can be performed and stitched together in software. In this mode, the complete set of projection views can be acquired for one extreme section of the sample. The specimen is then translated to a new position and the next set of projection views are acquired and saved to disk. In this way, an entire specimen can be imaged piecewise and then reassembled in the reconstructed image.

5.3.2 PROTOCOL DESIGN

Optimizing the protocol design involves *a priori* knowledge of the desired resolution in the reconstructed image and the expected contrast between different materials in the sample. Increasing the image resolution will lengthen the scan time, especially when low-contrast objects are being imaged. The problem with a longer scan time is the possibility of motion artifacts degrading the image. Although the specimens are not alive, there may still be some motion if the sample shifts at any point during the scan. Furthermore, for very long scan times, the focal spot itself may move due to thermal effects in the anode.

To ensure that the contrast is sufficient for identifying the different materials in the sample, a low-noise image is required. To achieve good contrast, the spectral energy of the x-ray beam can be adjusted by setting the kVp or by adding aluminum or copper filtration to the beam. Additional filtration may reduce the artifacts surrounding highly attenuating materials, like metal or dense bone, but will reduce the number of x-ray photons that reach the detector, resulting in more image noise. For different specimen sizes and compositions, tube settings (kVp and mA), filtration, and exposure time will have to be separately optimized.

A final consideration is the amount of data that each scan will generate. Higher resolution scans will require more disk space to store the projection views and the reconstructed 3D image. In addition, these files will take longer to transfer to remote drives and will require more powerful computer workstations to analyze and manipulate the images. Although acquiring the highest resolution possible may sound like a technological advance, there are practical considerations that need to be included at the planning stage of every study to ensure that the data are providing the benefit intended.

5.3.3   APPLICATIONS

Originally, micro-CT scanners were developed for high-resolution 3D imaging of excised bone samples from rodents to study the degeneration of bone. Applications include rodent models of osteoporosis and arthritis (Kapadia et al. 1998; Day et al. 2001; Batiste et al. 2004; Siu et al. 2004; Park et al. 2007) or biopsies from larger animals or patients (Ito et al. 1998; Siu et al. 2004; Jiang et al. 2005). Studies have expanded to include other excised tissues, organs, and tumors (Jorgensen et al. 1998; Bentley et al. 2002). Casting agents are available to provide contrast between tissues with similar attenuation properties. These agents contain heavy metals, such as lead, titanium, or barium, and can only be used for postmortem scanning (Maehara 2003; Langheinrich et al. 2004; Marxen et al. 2004; Granton et al. 2008). Other areas of research have also embraced micro-CT as a high-resolution imaging alternative to destructive laboratory tests in fields such as engineering, material science, archeology, zoology, and other fields of biology.

5.4   *IN VIVO* IMAGING

5.4.1   SCANNER FEATURES

Micro-CT scanners designed for use with live animals are similar to clinical CT scanners. The animal is positioned on a motorized cradle that moves into the x-ray beam. The x-ray tube and detector are mounted on opposing sides of a gantry ring that rotates around the cradle during the acquisition, as seen in Figure 5.5. The geometry of the system is fixed, implying only selected spatial resolutions can be achieved, unlike the specimen case. *In vivo* scanners have been developed to operate under step-and-shoot mode or continuous rotation mode, with some machines capable of delivering both modes. To ensure rapid scanning protocols, the detector readout electronics must be very fast to increase the number of projection images acquired per minute. In addition, these scanners often employ a clinical x-ray tube to ensure an adequate number of x-rays are produced during the acquisition of each projection view. Although the field of view often represents a substantial portion of the rodent anatomy, some scanners also enable a stitching mode, where the rodent can be sequentially imaged from nose to tail by translating the cradle forward between acquisitions and assembling the sections into a single reconstructed volume.

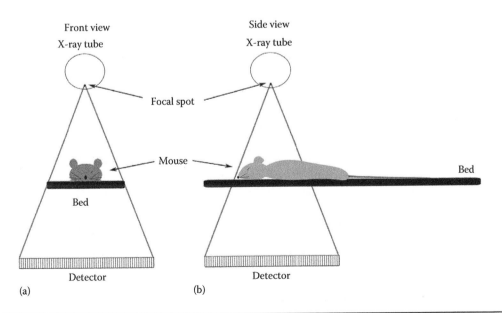

FIGURE 5.5   Schematic diagram of an *in vivo* micro-CT scanner as viewed from the front (a) and the side (b). The anesthetized animal lies on a stationary bed, while the x-ray tube and detector are mounted on a rotating gantry to enable the acquisition of projection views from all angular positions.

5.4.2 PROTOCOL DESIGN

For *in vivo* studies, protocol development is similar to that of the specimen scanner, although there is typically less flexibility in selecting resolution and added filtration. There are additional constraints as well, as the studies involve living rodents. The animals will require expert handling and care, including monitoring vital signs and maintaining body temperature during the imaging session and a safe, comfortable recovery zone following the scan that includes access to food and fluids. Scan times must be short to minimize the stress to the animals and ensure their continued well-being; published scan times range widely from under 1 min (Drangova et al. 2007; Ford et al. 2007b) to 30 min or more for a single acquisition.

To improve the contrast for imaging soft tissue and vasculature, iodinated contrast agents can be introduced to alter the x-ray attenuation characteristics of the target organ or tissue. Although there are clinically available iodinated contrast agents, they are not often used due to the relatively long scan times required for imaging compared with the metabolic rate of the rodent. To ensure constant enhancement through the imaging session, the clinical agents must be introduced via a power injection with a constant infusion rate. Alternatively, blood pool agents have been developed for preclinical applications. These blood pool agents, injected via the tail vein, recirculate in the vasculature for hours to provide exquisite contrast between the blood and surrounding tissue.

Recently, dosimetry for micro-CT studies has become more important to ensure that the radiation received during imaging does not impact the disease model under investigation (Willekens et al. 2010). Typical imaging parameters require higher x-ray doses than observed in the clinical environment; recent publications quote skin entrance doses up to 0.5 Gy per scan and absorbed doses of up to 0.28 Gy depending on the protocol used (Rodt et al. 2011). Computational models have been developed to characterize and understand the impact of the imaging doses received during micro-CT studies (Boone et al. 2004). Studies on the impact of the radiation dose received during imaging have also been described for different anatomical regions and disease models (Willekens et al. 2010; Foster and Ford 2011; Laperre et al. 2011).

To ensure the images are free of motion-induced artifacts, the rodent must be immobilized throughout the image acquisition. Motion artifacts occur when an object is located at a number of different positions in the projection views. Upon reconstruction, the 3D image will exhibit blurring around the edges of the moving object; for severe movement, there may also be streaking artifacts and the object may appear in two or more distinct locations within the 3D volume. There are a couple of approaches to eliminating motion described in the following sections: immobilization and physiological gating.

5.4.2.1   Immobilization

For *in vivo* rodent imaging, the animal should be immobilized to prevent escape of the animal and ensure that there are no gross movements during the scan. General anesthesia is used to sedate the rodent for the duration of the imaging session. Anesthesia can be administered either by injection or by inhalation. *In vivo* micro-CT systems can accommodate tubing to deliver a constant flow of gaseous anesthesia to the rodent during the scan setup and acquisition of the projection views. As the anesthesia agents have different effects on the animal, including alterations in respiratory and cardiac rates, decreased body temperature, and different duration of anesthetic effects, the agents and concentrations should be determined for each study independently.

Upon successful sedation of the rodent, positioning the animal is also important for optimized imaging. Most scanners have different sized cradles for positioning mice or for larger rodents. Animals are generally positioned lying prone (on the belly) with the limbs extended to support the body and limit lateral movement or lying supine (on the back) with additional support from the cradle or added radiolucent materials. Easily obtainable materials that are radiolucent and readily transformed into support systems include Styrofoam, foam pads, cotton, and wads of paper secured with tape or Velcro strips. For more complex imaging studies, where a specific orientation of the organ is required to study the natural function or to enable co-registration, customized support systems can be created. These customized devices, made by machining or molding plastics or hard foam, can be used to ensure identical positioning of many rodents in a single study, identical positioning of the same animal for repeated scans over weeks or months, or to hold a limb in a specific position during imaging.

5.4.2.2 Physiological Gating

During *in vivo* studies, the animal is breathing and the heart is beating. Both of these physiological functions can lead to motion artifacts in the reconstructed images because the structures in the thoracic cavity and the abdomen will be in different positions in each of the projection images. Upon reconstruction, any structures that have moved during the acquisition will have a blurry appearance. To compensate for quasi-periodic physiological motion, gating approaches have been developed to synchronize the image acquisition with the physiological motion, ensuring that all projection images are acquired during the same phase of the quasi-periodic motion. There are three types of physiological gating that can be applied to micro-CT imaging: prospective gating, retrospective gating, and image-based gating.

Prospective gating has been developed by a number of research teams for respiratory and cardiac imaging applications. Prospective gating is employed in scanners that operate under step-and-shoot conditions. The acquisition protocol is modified to ensure that after the gantry has rotated to the next projection angle, the x-ray tube will wait for a user-supplied signal before acquiring the next projection image. The user-supplied signal is a trigger based on the quasi-periodic motion of the animal—the respiratory or cardiac waveform—designed to initiate acquisition of the projection image. The respiratory trigger signal can be generated from a mechanical ventilator at a specific point in the respiratory cycle for every breath (Badea et al. 2004; Cavanaugh et al. 2004; Walters et al. 2004). The trigger will be very reproducible due to the operator control over the respiratory mechanics. Alternatively, the trigger may be based on a real-time measurement of the respiratory motion from a pneumatic cushion positioned on the diaphragm (Ford et al. 2005) or an optical measurement of the diaphragm motion (Burk et al. 2012). For both of these real-time measurements, the animal may be mechanically ventilated or free-breathing. In the free-breathing case, the respiratory mechanics are more natural, but the respiratory rate and volume of air inspired may vary from breath-to-breath. For cardiac-gated imaging, the trigger is based on the measured heart rate or ECG signal (Badea et al. 2005). Upon receiving the trigger signal, the x-ray acquisition will begin and the projection image will be captured and saved to disk, followed by rotation to the next projection angle where the system will wait for the next trigger signal. A sample of prospectively respiratory-gated images from a single free-breathing mouse is included in Figure 5.6.

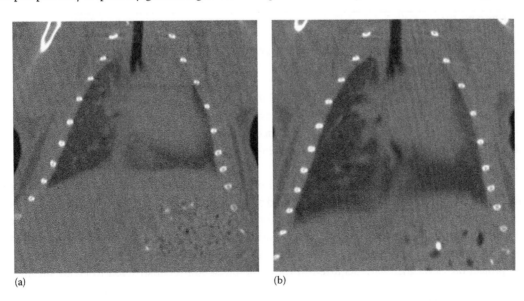

(a)   (b)

FIGURE 5.6 Example images of prospectively respiratory-gated images reconstructed with 0.087 isotropic voxel spacing. During the imaging session, the mouse was anesthetized and free breathing. Each respiratory phase was obtained in a separate 25–30 min scan, with projection images acquired during expiration (a) or inspiration (b).

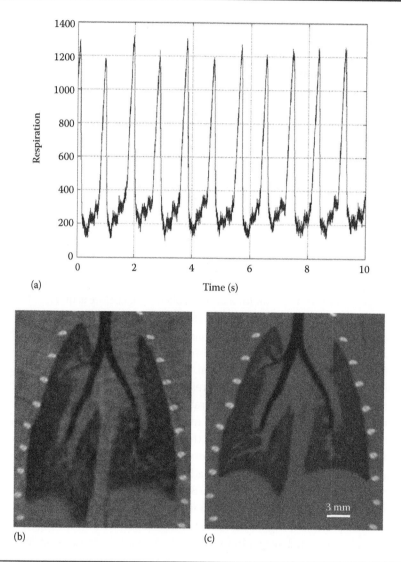

FIGURE 5.7 Retrospective respiratory gating in an anesthetized, free-breathing rat. (a) shows a segment of the respiratory trace measured during the imaging session. Projections are selected from the desired respiratory phase and reconstructed to represent (b) peak inspiration and (c) end expiration. Images are reconstructed with 0.15 mm isotropic voxel spacing.

For this sample image, two separate respiratory-gated micro-CT acquisitions were performed to capture end expiration (a) and inspiration (b) using the externally measured diaphragm motion as a surrogate for the respiratory waveform to trigger image acquisition. Images are reconstructed with 0.087 mm isotropic voxel spacing and each acquisition was 25–30 min depending on the respiratory rate during the micro-CT scan.

Retrospectively gated images are performed under continuous rotation mode or in step-and-shoot mode with multiple projection views saved at each angular position. In this case, the projection views are acquired throughout the respiratory and cardiac cycles while the respiratory and/or cardiac signals are recorded. Following acquisition, the projection views are sorted based on the recorded respiratory or cardiac phase of the animal at the time each projection was acquired. Only the projection views that were acquired in the same phase are reconstructed to create an image depicting a specific portion of the respiratory (Ford et al. 2007b) or cardiac cycle (Drangova et al. 2007). To ensure that adequate angular coverage occurs, multiple projection views are acquired at each angular position. Since all of the respiratory and cardiac phases are acquired, projection views representing different portions of the cycle can be selected for reconstruction from a single acquisition. However, there may be a number of projection views that are discarded, such as the duplicate and out of phase projection views at each angular position or the projections views acquired during portions of the cycle that are not of interest for the imaging study, leading to an unnecessary radiation exposure. In addition, some of the angular positions may not be filled if there are no projection views acquired during the desired phase, leading to increased image noise and missing view artifacts. Figure 5.7 shows a recorded respiratory trace from free-breathing rat (a), along with coronal slices representing (b) peak inspiration and (c) end expiration. The images were acquired in a single acquisition under continuous rotation mode and reconstructed with 0.150 mm isotropic voxel spacing. Figure 5.8 shows recorded cardiac and respiratory traces from a free-breathing mouse (a), along with coronal slices representing (b) systole and (c) diastole, reconstructed during end expiration to reduce respiratory motion, with 0.15 mm isotropic voxel spacing.

Image-based gating uses the projection views to determine what portion of the cycle each projection view represents. Image-based techniques can be used with step-and-shoot or continuous rotation imaging. One advantage is that no additional equipment is required for monitoring and recording the respiratory and cardiac signals. Image-based methods are inherently retrospective in nature and different algorithms have been reported (Hu et al. 2004; Farncombe 2008; Ertel et al. 2009).

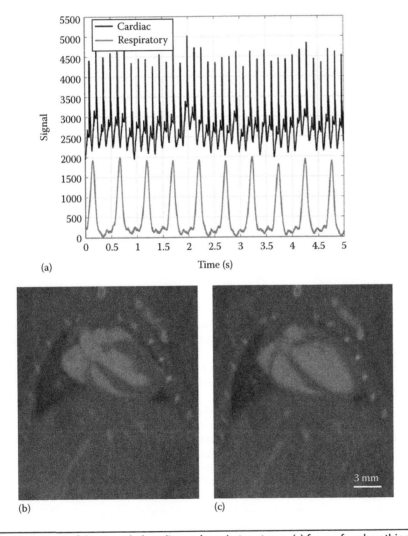

(a)

(b)   (c)

FIGURE 5.8 A segment of the recorded cardiac and respiratory traces (a) from a free-breathing anesthetized mouse obtained during the micro-CT imaging session, along with coronal slices representing (b) systole and (c) diastole. Images were reconstructed during end expiration to reduce respiratory motion, with 0.15 mm isotropic voxel spacing. (Image courtesy of Dr. S.A. Detombe and Dr. M. Drangova, University of Western Ontario, London, Ontario, Canada.)

5.4.3 APPLICATIONS

Since *in vivo* scanners typically have a larger field of view than specimen scanners, the ability to visualize tissues and organs *in situ* is possible for imaging both *in vivo* and intact specimens. The entire organ system can be imaged, often in a single scan for mice, allowing for monitoring of the same rodent over time. Contrast between different tissue types, such as blood and surrounding soft tissue, can be improved with injections of iodinated contrast agents (Weichert et al. 1998; Bakan et al. 2002; Wisner et al. 2002; Ford et al. 2006; Graham et al. 2008; Durkee et al. 2010; Boll et al. 2011; Detombe et al. 2012; Figueiredo et al. 2012). Functional information can also be obtained by taking multiple images throughout the respiratory (Ford et al. 2007b, 2009) or cardiac cycle (Drangova et al. 2007; Detombe et al. 2008). Longitudinal studies have been described for rodent models of respiratory disease (Artaechevarria et al. 2010, 2011; Namati et al. 2010), cardiac disease (Detombe et al. 2008), oncology (Namati et al. 2010; Foster and Ford 2011; Rodt et al. 2012), bone degradation (McErlain et al. 2008; Johnson et al. 2011), or regeneration (De Smet et al. 2011), and the development of imaging protocols continues for other disease models and applications.

REFERENCES

Artaechevarria, X., D. Blanco et al. (2010). Longitudinal study of a mouse model of chronic pulmonary inflammation using breath hold gated micro-CT. *Eur. Radiol.* **20**(11): 2600–2608.

Artaechevarria, X., D. Blanco et al. (2011). Evaluation of micro-CT for emphysema assessment in mice: Comparison with non-radiological techniques. *Eur. Radiol.* **21**(5): 954–962.

Artaechevarria, X., D. Perez-Martin et al. (2009). Airway segmentation and analysis for the study of mouse models of lung disease using micro-CT. *Phys. Med. Biol.* **54**(22): 7009–7024.

Badea, C., L. W. Hedlund et al. (2004). Micro-CT with respiratory and cardiac gating. *Med. Phys.* **31**(12): 3324–3329.

Badea, C. T., B. Fubara et al. (2005). 4-D micro-CT of the mouse heart. *Mol. Imaging* **4**(2): 110–116.

Badea, C. T., S. M. Johnston et al. (2011). 4D micro-CT for cardiac and perfusion applications with view under sampling. *Phys. Med. Biol.* **56**(11): 3351–3369.

Bakan, D. A., F. T. Lee, Jr. et al. (2002). Hepatobiliary imaging using a novel hepatocyte-selective CT contrast agent. *Acad Radiol.* **9**(Suppl. 1): S194–S199.

Batiste, D. L., A. Kirkley et al. (2004). Ex vivo characterization of articular cartilage and bone lesions in a rabbit ACL transection model of osteoarthritis using MRI and micro-CT. *Osteoarthritis Cartilage* **12**(12): 986–996.

Bentley, M. D., M. C. Ortiz et al. (2002). The use of microcomputed tomography to study microvasculature in small rodents. *Am. J. Physiol. Regul. Integr. Comp. Physiol.* **282**(5): R1267–R1279.

Boll, H., S. Nittka et al. (2011). Micro-CT based experimental liver imaging using a nanoparticulate contrast agent: A longitudinal study in mice. *PLoS One* **6**(9): e25692.

Boone, J. M., O. Velazquez et al. (2004). Small-animal x-ray dose from micro-CT. *Mol. Imaging* **3**(3): 149–158.

Burk, L. M., Y. Z. Lee et al. (2012). Non-contact respiration monitoring for in-vivo murine micro computed tomography: Characterization and imaging applications. *Phys. Med. Biol.* **57**(18): 5749–5763.

Cavanaugh, D., E. Johnson et al. (2004). In vivo respiratory-gated micro-CT imaging in small-animal oncology models. *Mol. Imaging* **3**(1): 55–62.

Chappard, C., A. Marchadier et al. (2008). Interindividual and intraspecimen variability of 3-D bone micro-architectural parameters in iliac crest biopsies imaged by conventional micro-computed tomography. *J. Bone Miner. Metab.* **26**(5): 506–513.

Chaturvedi, A. and Z. Lee. (2005). Three-dimensional segmentation and skeletonization to build an airway tree data structure for small animals. *Phys. Med. Biol.* **50**(7): 1405–1419.

Day, J. S., M. Ding et al. (2001). A decreased subchondral trabecular bone tissue elastic modulus is associated with pre-arthritic cartilage damage. *J. Orthop. Res.* **19**(5): 914–918.

De Smet, E., S. V. Jaecques et al. (2011). Constant strain rate and peri-implant bone modeling: An in vivo longitudinal micro-CT analysis. *Clin. Implant. Dent. Relat. Res.* 15(3): 358–366.

Detombe, S. A., J. Dunmore-Buyze et al. (2012). Evaluation of eXIA 160 cardiac-related enhancement in C57BL/6 and BALB/c mice using micro-CT. *Contrast Media Mol. Imaging* 7(2): 240–246.

Detombe, S. A., N. L. Ford et al. (2008). Longitudinal follow-up of cardiac structure and functional changes in an infarct mouse model using retrospectively gated micro-computed tomography. *Invest. Radiol.* **43**(7): 520–529.

Dorr, A., J. G. Sled et al. (2007). Three-dimensional cerebral vasculature of the CBA mouse brain: A magnetic resonance imaging and micro computed tomography study. *Neuroimage* **35**(4): 1409–1423.

Drangova, M., N. L. Ford et al. (2007). Fast retrospectively gated quantitative four-dimensional (4D) cardiac micro computed tomography imaging of free-breathing mice. *Invest. Radiol.* **42**(2): 85–94.

Durkee, B. Y., J. P. Weichert et al. (2010). Small animal micro-CT colonography. *Methods* **50**(1): 36–41.

Ertel, D., Y. Kyriakou et al. (2009). Respiratory phase-correlated micro-CT imaging of free-breathing rodents. *Phys. Med. Biol.* **54**(12): 3837–3846.

Farncombe, T. H. (2008). Software-based respiratory gating for small animal conebeam CT. *Med. Phys.* **35**(5): 1785–1792.

Figueiredo, G., H. Boll et al. (2012). In vivo x-ray digital subtraction and CT angiography of the murine cerebrovasculature using an intra-arterial route of contrast injection. *AJNR Am. J. Neuroradiol.* **33**(9): 1702–1709.

Ford, N. L., K. C. Graham et al. (2006). Time-course characterization of the computed tomography contrast enhancement of an iodinated blood-pool contrast agent in mice using a volumetric flat-panel equipped computed tomography scanner. *Invest. Radiol.* **41**(4): 384–390.

Ford, N. L., E. L. Martin et al. (2007a). In vivo characterization of lung morphology and function in anaesthetized free-breathing mice using non-invasive micro-computed tomography. *J. Appl. Physiol.* **102**(5): 2046–2055.

Ford, N. L., E. L. Martin et al. (2009). Quantifying lung morphology with respiratory-gated micro-CT in a murine model of emphysema. *Phys. Med. Biol.* **54**(7): 2121–2130.

Ford, N. L., H. N. Nikolov et al. (2005). Prospective respiratory-gated micro-CT of free breathing rodents. *Med. Phys.* **32**(9): 2888–2898.

Ford, N. L., M. M. Thornton et al. (2003). Fundamental image quality limits for microcomputed tomography in small animals. *Med. Phys.* **30**(11): 2869–2877.

Ford, N. L., A. R. Wheatley et al. (2007b). Optimization of a retrospective technique for respiratory-gated high speed micro-CT of free-breathing rodents. *Phys. Med. Biol.* **52**: 5749–5769.

Foster, W. K. and N. L. Ford. (2011). Investigating the effect of longitudinal micro-CT imaging on tumour growth in mice. *Phys. Med. Biol.* **56**(2): 315–326.

Graham, K. C., S. A. Detombe et al. (2008). Contrast-enhanced microcomputed tomomgraphy using intraperitoneal contrast injection for the assessment of tumor-burden in liver metastasis models. *Invest. Radiol.* **43**(7): 488–495.

Granton, P. V., S. I. Pollmann et al. (2008). Implementation of dual- and triple-energy cone-beam micro-CT for postreconstruction material decomposition. *Med. Phys.* **35**(11): 5030–5042.

Hu, J., S. T. Haworth et al. (2004). Dynamic small animal lung imaging via a postacquisition respiratory gating technique using micro-cone beam computed tomography. *Acad Radiol.* **11**(9): 961–970.

ICRU. (1989). Tissue substitutes in radiation dosimetry and measurements. Report 44 of the International Commission on Radiation Units and Measurements, Washington, DC.

Ito, M., T. Nakamura et al. (1998). Analysis of trabecular microarchitecture of human iliac bone using micro-computed tomography in patients with hip arthrosis with or without vertebral fracture. *Bone* **23**(2): 163–169.

Jiang, Y., J. Zhao et al. (2005). Application of micro-CT assessment of 3-D bone microstructure in preclinical and clinical studies. *J. Bone Miner. Metab.* **23**(Suppl.): 122–131.

Jin, S. O., J. G. Kim et al. (2012). Bone-induced streak artifact suppression in sparse-view CT image reconstruction. *Biomed. Eng. Online* **11**(1): 44.

Johnson, L. C., R. W. Johnson et al. (2011). Longitudinal live animal micro-CT allows for quantitative analysis of tumor-induced bone destruction. *Bone* **48**(1): 141–151.

Jorgensen, S. M., O. Demirkaya et al. (1998). Three-dimensional imaging of vasculature and parenchyma in intact rodent organs with x-ray micro-CT. *Am. J. Physiol.* **275**(3 Pt 2): H1103–H1114.

Kapadia, R. D., G. B. Stroup et al. (1998). Applications of micro-CT and MR microscopy to study pre-clinical models of osteoporosis and osteoarthritis. *Technol. Health Care* **6**(5–6): 361–372.

Langheinrich, A. C., B. Leithauser et al. (2004). Acute rat lung injury: Feasibility of assessment with micro-CT. *Radiology* **233**(1): 165–171.

Laperre, K., M. Depypere et al. (2011). Development of micro-CT protocols for in vivo follow-up of mouse bone architecture without major radiation side effects. *Bone* **49**(4): 613–622.

Lee, Y. Z., L. Burk et al. (2011). Carbon nanotube based x-ray sources: Applications in pre-clinical and medical imaging. *Nucl. Instrum. Methods Phys. Res. A* **648**(Suppl. 1): S281–S283.

Maehara, N. (2003). Experimental microcomputed tomography study of the 3D microangioarchitecture of tumors. *Eur. Radiol.* **13**(7): 1559–1565.

Marxen, M., M. M. Thornton et al. (2004). MicroCT scanner performance and considerations for vascular specimen imaging. *Med. Phys.* **31**(2): 305–313.

McErlain, D. D., C. T. Appleton et al. (2008). Study of subchondral bone adaptations in a rodent surgical model of OA using in vivo micro-computed tomography. *Osteoarthritis Cartilage* **16**(4): 458–469.

Namati, E., J. Thiesse et al. (2010). Longitudinal assessment of lung cancer progression in the mouse using in vivo micro-CT imaging. *Med. Phys.* **37**(9): 4793–4805.

Park, C. H., Z. R. Abramson et al. (2007). Three-dimensional micro-computed tomographic imaging of alveolar bone in experimental bone loss or repair. *J. Periodontol.* **78**(2): 273–281.

Rodt, T., M. Luepke et al. (2011). Phantom and cadaver measurements of dose and dose distribution in micro-CT of the chest in mice. *Acta Radiol.* **52**(1): 75–80.

Rodt, T., C. von Falck et al. (2012). Lung tumour growth kinetics in SPC-c-Raf-1-BB transgenic mice assessed by longitudinal in-vivo micro-CT quantification. *J. Exp. Clin. Cancer Res.* **31**: 15.

Siu, W. S., L. Qin et al. (2004). A study of trabecular bones in ovariectomized goats with micro-computed tomography and peripheral quantitative computed tomography. *Bone* **35**(1): 21–26.

Smith, S. W. (1997). *The Scientist and Engineer's Guide to Digital Signal Processing.* California Technical Publishing, San Diego, CA.

Uchiyama, T., T. Tanizawa et al. (1997). A morphometric comparison of trabecular structure of human ilium between microcomputed tomography and conventional histomorphometry. *Calcif. Tissue Int.* **61**(6): 493–498.

Walters, E. B., K. Panda et al. (2004). Improved method of in vivo respiratory-gated micro-CT imaging. *Phys. Med. Biol.* **49**(17): 4163–4172.

Weichert, J. P., F. T. Lee, Jr. et al. (1998). Lipid-based blood-pool CT imaging of the liver. *Acad Radiol.* **5**(Suppl. 1): S16–S19; discussion S28–S30.

Willekens, I., N. Buls et al. (2010). Evaluation of the radiation dose in micro-CT with optimization of the scan protocol. *Contrast Media Mol. Imaging* **5**(4): 201–207.

Wisner, E. R., J. P. Weichert et al. (2002). Percutaneous CT lymphography using a new polyiodinated biomimetic microemulsion. *Acad Radiol.* **9**(Suppl. 1): S191–S193.

Yang, J., L. X. Yu et al. (2010). Comparative structural and hemodynamic analysis of vascular trees. *Am. J. Physiol. Heart Circ. Physiol.* **298**(4): H1249–H1259.

6

Digital Subtracted Angiography of Small Animals

Stavros Spiliopoulos, George C. Kagadis, Dimitrios N. Karnabatidis, G. Allan Johnson, and Cristian Badea

6.1 INTRODUCTION

In vivo small animal imaging is the cornerstone of modern experimental protocols investigating disease mechanisms, novel drug development, and innovative therapies. Using previous investigational protocols, animals had to be sacrificed and prepared for pathology to obtain relevant information about the vascular system, without the possibility of longitudinal imaging. Constantly evolving imaging modalities have become indispensable for the preclinical investigation of human diseases using modern experimental animal models of both normal and genetically modified small rodents, mice, and rats, allowing in vivo follow-up of a specific disease or drug in the same animal. Currently, microimaging for small animals includes morphological, anatomical, and molecular imaging techniques such as microcomputed tomography (CT), micromagnetic resonance (MR), micropositron emission tomography (PET), microsingle-photon emission computed tomography (SPECT), micro-ultrasound, and digital subtraction angiography (DSA). The application of these imaging modalities to small animals poses several problems mainly due to the high spatial and temporal resolution required, but each method also exhibits particular advantages and limitations (Badea et al. 2008).

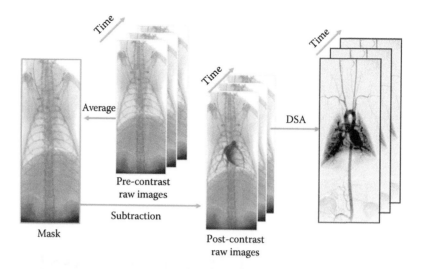

FIGURE 6.1 Principle of DSA imaging illustrated for cardiopulmonary imaging in a rat. A bolus of contrast agent is delivered via a catheter placed in the jugular vein. A few images are acquired in suspended respiration using cardiac gating prior to contrast agent injection. These images are averaged to create a mask that is subtracted from post-contrast injection images.

Imaging in small animals can be addressed particularly well using x-ray DSA, given the ease of use and its ability to capture rapid physiological changes in blood flow. DSA can be based on either temporal subtraction or K-edge subtraction. The latter technique is based on the nonlinear differences in the attenuation of contrast agents based on iodine with the x-ray beam energy. A K-edge describes a sudden increase in the attenuation coefficient of x-ray photons. K-edge DSA ideally requires imaging on both sides of the K-edge of iodine with monochromatic x-rays obtained using a synchrotron source (Schültke et al. 2010). However, the need of a synchrotron limits the availability of such a method, since the majority of x-ray imaging systems use polychromatic x-ray sources, and are therefore better suited for temporal subtraction.

We focus here on methods for in vivo DSA based on temporal subtraction. Intra-arterial DSA is performed through the acquisition of images before and following the intra-arterial injection of an extracellular contrast agent. In Figure 6.1, we illustrate the principle of DSA imaging focusing on the cardiopulmonary system of a rat. The images acquired before the injection of a bolus of contrast media are employed to generate a mask, which will be subtracted from the contrast images acquired following contrast injection. As a result, the distribution of the contrast agent is depicted in the subtracted images over time. DSA can provide real-time imaging in drug delivery with high spatial and temporal resolution and distribution studies, and is the only imaging modality that can offer the means to investigate novel endovascular devices, especially when combined with selective arterial catheterization. However, until now, only few studies have investigated DSA in small animals (Kobayashi et al. 2004; Badea et al. 2006; De Lin et al. 2006; Figueiredo et al. 2012). The aim of this chapter is to analyze the techniques, report the applications, and discuss the potential of DSA for in vivo imaging in small animals.

6.2 ANIMAL PREPARATION AND ANESTHESIA

Small animal DSA imaging is performed under anesthesia, which is of utmost importance for optimal image acquisition. Compulsory fasting is not required prior to anesthesia, as rodents do not possess a vomiting reflex. Premedication (administration of drugs prior to anesthesia) with anticholinergic drugs can be used to maintain the heart rate and decrease gut motility, while tranquilizers can relieve anxiety, create calmness, and reduce the dose of the anesthetic required. Nonsteroidal anti-inflammatory drugs (e.g., carprofen) can be given to provide longer-term pain relief (up to 24 h), especially in cases of minor surgery (Tremoleda et al. 2012). Both injectable and inhaled anesthetics can be used in rodents. Inhalation anesthesia is considered the method of choice for imaging protocols in laboratory rodents. Highly volatile agents such as halothane and isoflurane can be administered to the animals using a carrier gas (usually oxygen) through a breathing circuit with an integrated vaporizer that permits the regulation of concentration and flow. The suggested flow rate for small animals is 0.5–1.5 L/min. Isoflurane is the preferred general anesthetic agent for cardiovascular studies, because it causes less cardiac function depression than injectable agents. Nonetheless, isoflurane demonstrates an inhibitory effect on peripheral resistance, and therefore decreases blood pressure (Kersten et al. 1996). Injectable agents include fentanyl/fluanisone- and ketamine-based combinations. Usually, the intraperitoneal infusion of xylazine hydrochloride (10 mg/kg body weight) and ketamine (100 mg/kg body weight) produces safe anesthesia with good analgesic and light sedative effect. Fentanyl/fluanisone in combination with benzodiazepine (midazolam or diazepam) is also licensed for surgical anaesthesia in rodents. The combination of ketamine with midazolam can also be used to provide 20–30 min of light anaesthesia in rodents. Supplemental doses of both the aforementioned anesthetic regimens or intermittent intravenous administration of low propofol doses are usually necessary to maintain a uniform level of anesthesia in cases of prolonged imaging. Other injectable agents that that can be used are alfaxalone and tribromoethanol. Optimal imaging requires not only selecting the appropriate anaesthetic regimen, but also using monitoring systems during image acquisition. Monitoring systems are especially important for longitudinal protocols of repeated anesthesia where the exposure of ionizing radiation and the repeated use of contrast agents and anesthetic drugs can have consequences on the animal's physiology and its response to anesthesia. Pulse oximetry provides a simple method to measure oxygen saturation levels, and thus monitor anesthesia. Additionally, the investigator should maintain direct visual contact and access to the animal during imaging to assess basic parameters such as the depth and character of respiratory rate, heart rate, and body temperature. A thermoregulated surgical table is useful for maintaining steady body temperature.

6.3 IMAGING INSTRUMENTATION AND VASCULAR ACCESS

An example of DSA imaging system used for small animal imaging is shown with schematics and a photo in Figure 6.2. A clinical angiography x-ray tube with a large focal spot able to deliver sufficient flux with exposures times less than 10 ms (to limit motion blur) has been used in order to meet the requirements of fast sampling and high-resolution images necessary for optimal small animal DSA imaging (Badea et al. 2008). Moreover, an injector used for clinical angiography or a dedicated micro-injector (de Lin et al. 2008), synchronized with the cardiac or respiratory cycles, is required for optimal DSA imaging, although some investigators have used manual injections at low flow rates or remote infusion pumps (PHD 22/2000 Syringe Pumps, Harvard Apparatus) (Kobayashi et al. 2004; Badea et al. 2008). A dedicated DSA sequencer application allowing synchronization of breathing, contrast injection, radiographic exposure, and digital frame acquisition with the cardiac or respiratory cycle has been proposed (Badea et al. 2013). We present a typical DSA sampling sequence used for cardiopulmonary DSA imaging in Figure 6.3, which is implemented by a LabVIEW (National Instruments, Austin, TX, USA) application. The sequencer uses amplitude-based thresholding of electrocardiography (ECG) signal on the R peak to create TTL pulses. The injection and x-ray exposures are thus synchronized to the R wave of the ECG. A DSA imaging sequence is started by stopping the ventilator cycle during end-expiration for a few seconds (typically between 5 and 10 s) via control of the mechanical ventilator. The animal is perorally intubated and mechanically ventilated. Suspended respiration minimizes respiratory artifacts in the subtracted DSA images. A set of non-contrast images is acquired, averaged, and used to create a mask. The injection is then initiated by a command to the microinjector, which is followed immediately by the acquisition of post-contrast injection images. At the end of the DSA run, the ventilator is restarted.

A very important component of the DSA imaging setup is the microinjector. Lin et al. described a custom-built microinjector consisting of a computer-controlled solenoid valve attached to the injection catheter and a heated reservoir for the contrast agent, integrated with the imaging system

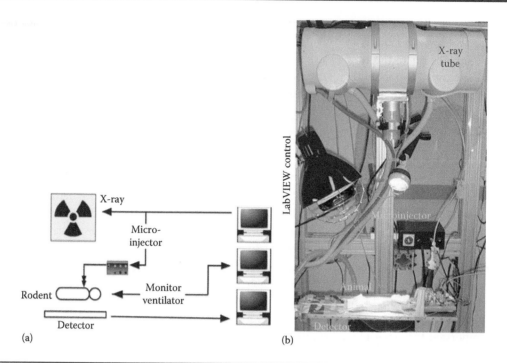

(a)   (b)

FIGURE 6.2 (a) Schematic of the DSA system and (b) a photo showing its implementation. The major components of the system are the x-ray tube, detector, and the microinjector. LabVIEW applications are used to control all components of the system and to synchronize injections to biological signals such as respiratory or ECG signals. A mechanical ventilator is also integrated and used in cardiopulmonary DSA.

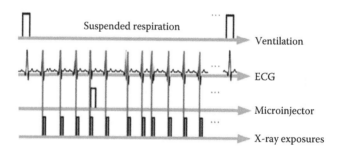

FIGURE 6.3 Sampling sequence showing the control TTL pulses delivered by a LabVIEW application used for cardiopulmonary DSA. The x-ray exposures are triggered in suspended respiration and in synchronization with the R peak of the ECG signal. The first three images are used to create a mask. DSA images are acquired before and after contrast agent injection synchronized to the R peak in the ECG signal. (Adapted from Badea, C.T. et al., *J. Med. Eng.*, 2013, 13, 2013. With permission.)

(a clinical x-ray angiography tube) and with monitoring and gating applications (Lin et al. 2009). Injection pressure was 80–90 PSI deriving from compressed N_2. Diffusible x-ray contrast agents (e.g., Isovue 370 mg I/mL, Iomeprol, Imeron 300, Bracco Diagnostics Inc., NJ, USA; Ioversol, Optiray; Tyco Healthcare, MI, USA) diluted 0.5× with saline solution have been used in DSA studies in small animals.

Placement of the catheters is critical for producing high-quality DSA images. Various access ports for contrast agent injections have been used depending on the target organ. We provide here a few examples of catheter placements. For cardiopulmonary DSA imaging, the catheter was placed in the right jugular vein at the root of the superior vena cava (De Lin et al. 2006, 2008; Lin et al. 2009). For coronary DSA imaging, the catheter was inserted into the right common carotid artery and advanced into the aortic arch. The catheter tip was positioned just cranial to the aortic valve using real-time blood pressure guidance (Badea et al. 2011). The common carotid artery and femoral artery have been used quite often for catheter placement (Kissel et al. 1987; Lin et al. 2009;

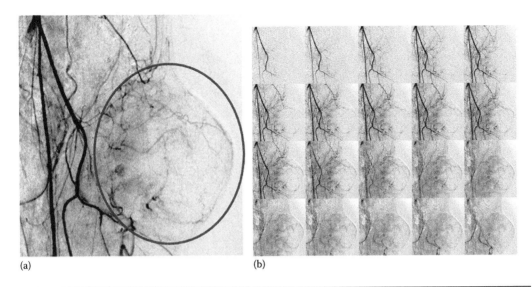

(a)   (b)

FIGURE 6.4   (a) Minimum intensity projection over a DSA sequence in a fibrosarcoma tumor implanted on the left leg of a rat. (b) DSA sequence of 20 successive images during every heartbeat. The tumor is marked by the red ellipsis.

Ortiz-Velazquez et al. 2009; Buhalog et al. 2010; Figueiredo et al. 2012). In DSA imaging of the liver, two catheter placements have been reported (Badea et al. 2013). The first catheter was used to visualize the portal vein system and was inserted through a mesenteric vein and advanced into the portal vein. To view the arterial system, a second catheter was placed into the left common carotid and into the abdominal aorta just above the celiac artery. In DSA imaging of the kidneys (see Figure 6.4), the catheter was placed in the iliac artery so that the tip was at the level just distal to the left renal artery. Facial and tail artery access points have also been described (Ortiz-Velazquez et al. 2009). Buhalog et al. developed a very interesting technique for performing sequential arterial catheterizations and DSA in rats, after the intra-arterial placement of a microcatheter through a sheath positioned using a transfemoral microsurgical approach (Buhalog et al. 2010).

A key factor in achieving a linear response in the delivery of a bolus of contrast agent for DSA imaging is the design of the catheters. The catheters are typically designed for maximum flow rate while still being small for cannulation. At constant driving pressures, a linear increase in injection volumes should occur with increasing injection times. Figueiredo et al. (2012) described a proximal external carotid artery retrograde access, which can be obtained through a small incision using a custom-made polythene catheter prepared from polyethylene (PE) tubing (Portex fine bore polythene tubing; Smiths Medical International, Ashford Kent, UK) with an external diameter of 0.38 mm, heated for a few seconds, and stretched while still hot. The stretched part of the tube was cut at the thinnest section, which resulted in a catheter with an outer diameter of less than 0.25 mm able to fit in the murine external carotid artery, and at the same time enabled the injection of contrast agent at the required flow rate (Figueiredo et al. 2012).

6.4 IMAGE ACQUISITION TECHNIQUES AND APPLICATIONS

Following vascular access, contrast injection of volumes ranging from 8 to 33 mL/s at 150 to 650 PSI has been described to achieve image acquisition with high temporal resolution (10–30 frames/s) using a tube voltage of 80 kV (160 mA and 10 ms) and high spatial resolution (100×100 μm^2 at 10 frames/s 14×14 μm at 30 frames/s) (Kissel et al. 1987; De Lin et al. 2006; Ortiz-Velazquez et al. 2009; Figueiredo et al. 2012). Lin et al. used DSA imaging to obtain cardiopulmonary vasculature visualization, quantitative blood flow in absolute metrics (mL/min), and relative blood volume dynamics from discrete regions of interest on a pixel-by-pixel basis, accomplishing repetitive concurrent in vivo visualization of the vasculature and of measurement of blood flow dynamics. Real-time anatomical

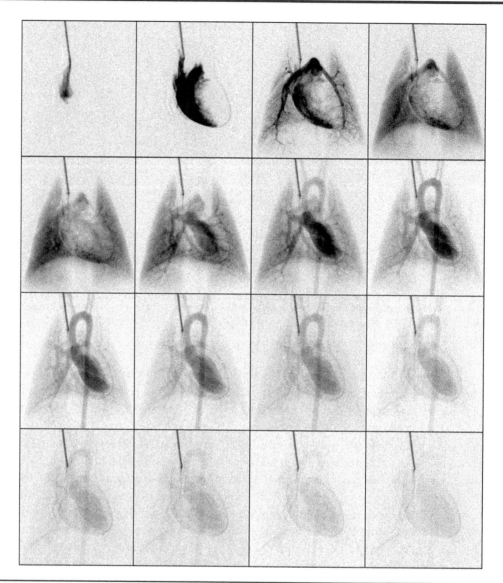

FIGURE 6.5 DSA time sequence of the cardiopulmonary system in a rat. Images are acquired with heartbeat temporal resolution and 100 μm spatial resolution. DSA images shown here correspond to 16 successive heartbeats.

and quantitative functional data acquisition can be used to evaluate drug- or disease-modulated cardiac output and pulmonary blood flow dynamics (Lin et al. 2009). Figure 6.5 shows an example of a cardiopulmonary DSA time sequence in a rat. Note how the progression of contrast agent can be followed from the right ventricle, through the lungs, and finally to the left ventricle and aortic arch. Such DSA time sequences have been used for measuring real-time physiological changes before and after administration of phenylephrine (a vasoconstrictor) in rats (De Lin et al. 2006).

Imaging coronary arteries in small animals is very challenging due to their small size and fast physiologic motion of the heart and respiration. Nevertheless, biplane coronary artery DSA has been used not only for morphological evaluation of the coronary arteries in rats, but also for assessing myocardial perfusion (Badea et al. 2011). Figure 6.6 presents DSA images of the heart in a rat treated with nitroprusside (NP) (Nitropress, Hosira, Inc., Lake Forest, IL, USA) as a model compound. NP is a smooth-muscle relaxant causing vasodilation of peripheral arteries and veins. NP is also known as a coronary artery vasodilator in humans (Yeh et al. 1977). DSA images clearly show the main branches of the coronary arteries in an intact beating heart. The drug test demonstrated that DSA can detect relative changes in coronary circulation.

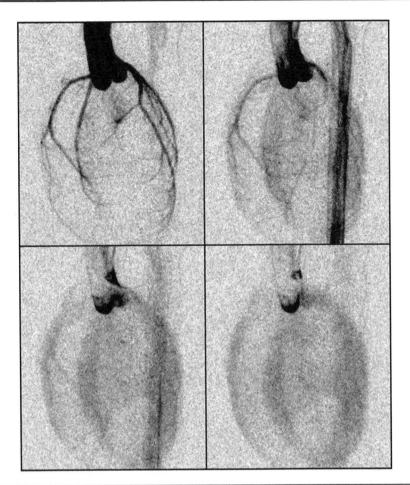

FIGURE 6.6 Coronary DSA in a rat treated with vasodilator drug (nitroprusside). Four DSA images from successive heartbeats are shown. Note that both the morphology of the coronary arteries (top row) and the myocardial perfusion (bottom row) can be assessed based on these images.

Another organ that has been imaged using DSA in rodents is the kidney. Figure 6.7 presents results from a kidney study performed in a rat (Badea et al. 2013). Figure 6.7a first shows nine DSA images post-injection cropped on the kidney level. Note in these images the arterial phase (first images) as well as the perfusion phase (last images). The selected regions of interest in different regions of the kidney (e.g., medulla and cortex) were used to measure the time density (or attenuation) curves obtained by plotting the mean enhancements over time (Figure 6.7b). Dysfunction of the kidneys in disease models can be quantified by analyzing these time attenuation curves.

A combination of micro-CT and DSA imaging has been used in cancer models to provide morphologic and functional data in a single imaging session. In this protocol, micro-CT imaging with a blood pool contrast agent assessed the 3D vascular architecture of the tumor, and DSA imaging with a conventional contrast agent assessed tumor perfusion (Badea et al. 2006). An example of DSA images of in a tumor study is shown in Figure 6.4. The DSA sequence shows both the arrival of the bolus of contrast agent through one artery followed by tumor perfusion, and the return of the bolus through the veins. The DSA sequence can serve to compute perfusion maps including the relative blood volume (rBV), relative blood flow (rBF), or relative mean transit time (rMTT) (Badea et al. 2008).

Recent technological advances in the field of medical imaging have allowed investigators to perform various other innovative angiography techniques to address the fact that DSA is a planar imaging method. Sometimes, quantitative measurements of DSA images are difficult to make because of the overlap of nonhomogeneous x-ray attenuating structures, an inherent problem with planar imaging. Ortiz-Velázquez et al. investigated 3D rotational DSA in murine experimental models. The protocol included common carotid artery surgical exposure and cannulation, followed by

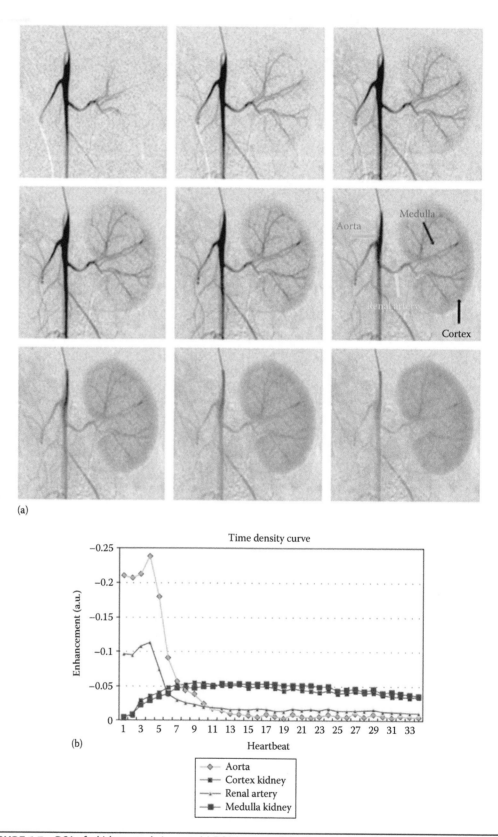

(a)

(b)

FIGURE 6.7   DSA of a kidney study in a rat. (a) DSA images of the first nine heartbeats after injection are shown. The catheter was placed in the iliac artery so that the tip was at the level just distal to the left renal artery. Note the arterial phase as well as the perfusion phase. (b) Time attenuation curves for selected regions of interest in the cortex, medulla, renal artery, and aorta. (Adapted from Badea, C.T. et al., *J. Med. Eng.*, 2013, 13, 2013. With permission.)

angiography, which was repeated up to four times to obtain high-quality images free of respiration artifacts using an automatic injection (Angiomat 6000 infusion pump, Tyco/Covidien, MA, USA) in the clinical angiography unit (Integris Allura; Philips Medical Systems, MA, USA), with volumes ranging from 8 to 16 mL at 1–2 mL stream/s and 150–650 PSI of contrast agent. Right after the standardization of the optimal imaging parameters, the authors performed digital rotational angiography (RA) (14–2 mL stream/s, 650 PSI, and 1 s of delay). Three-dimensional reconstructions from data collected using a 180° rotational arc were performed and analyzed on a dedicated 3D RA Workstation. Rotational angiography overcomes the basic methodological problems of traditional angiography by providing accurate imaging for the spatial relationship between vessels, as well as between the vasculature and the surrounding tissues, essential for computational fluid dynamics studies (Ortiz-Velazquez et al. 2009).

A similar approach to address the planar limitations of DSA is tomographic DSA (TDSA) introduced for functional lung perfusion imaging in rodents (Badea et al. 2007). The method includes the acquisition of multiple series of DSA images at different angles, resulting in the generation of 4D datasets derived using tomosynthetic reconstruction algorithms. The estimation of perfusion using TDSA offers the potential of high spatial and temporal resolution, thereby producing the best image quality while minimizing the number of contrast agent injections. Possible applications of TDSA can be found in fundamental physiology studies, tumor biology, toxicology, and drug development (Badea et al. 2008). If images are acquired over an arc of at least 180°, tomosynthesis can be replaced with cone-beam CT to provide isotropic spatial resolution (Badea et al. 2010).

To summarize, experimental studies investigating angiogenesis and tumor neovascularization can benefit from functional and anatomic imaging with DSA, as it is a reliable, low-cost in vivo vessel imaging technique with high spatial (around 90–100 µm) and temporal resolution (140 ms) (Mistretta et al. 1981). Various new techniques such as tomosynthesis, micro-CT combined imaging, 3D reconstructions, perfusion analysis, and quantitative vascular mapping can offer a variety of information essential for the preclinical investigation of novel drugs and innovative treatment modalities. Small animal DSA can be used for both anatomical and functional imaging, including lung perfusion, coronary artery angiography, tumoral neovascularization, and liver and kidney microvasculature, at a resolution of 46 µm (Mistretta et al. 1981; Kagadis et al. 2010). Combined micro-CT with iodinated blood pool and micro-DSA with conventional contrast media can be applied in animal models of cancer to provide 3D vasculature imaging and functional hemodynamics so as to analyze neo-angiogenesis and monitor antiangiogenic therapeutic protocols (Badea et al. 2006). Tomosynthetic reconstruction techniques have also been developed to provide 3D or 4D images to provide high perfusion analysis of a volume of tissue over time, such as in the lungs (Badea et al. 2008, 2010; Dobbins 2009; Ortiz-Velazquez et al. 2009). Moreover, computerized post-processing of DSA images has been used for vessel segmentation and high-resolution quantitative vascular mapping of neovascularization and collateral microcirculation down to the 50 µm scale, resulting cancer or ischemia-driven tissue reperfusion (Siablis et al. 2006; Kagadis et al. 2010). Furthermore, DSA studies can be readily translated back to the clinical setting, providing yet one more valuable imaging tool for assessing, for example, drug safety and toxicity from rodents to humans.

ACKNOWLEDGMENTS

We acknowledge the Duke Center for In Vivo Microscopy (NIH/NIBIB Biomedical Technology Resource Center, P41 EB015897). We also acknowledge the help from Dr. MingDe Lin.

REFERENCES

Badea, C. T., M. Drangova et al. (2008). In vivo small-animal imaging using micro-CT and digital subtraction angiography. *Phys. Med. Biol.* **53**(19): R319–R350.

Badea, C. T., L. Hedlund et al. (2013). A LabVIEW platform for preclinical imaging using digital subtraction angiography and micro-CT. *J. Med. Eng.* **2013**: 13.

Badea, C. T., L. W. Hedlund et al. (2006). Tumor imaging in small animals with a combined micro-CT/micro-DSA system using iodinated conventional and blood pool contrast agents. *Contrast Media Mol. Imaging* **1**(4): 153–164.

Badea, C. T., L. W. Hedlund et al. (2007). Tomographic digital subtraction angiography for lung perfusion estimation in rodents. *Med. Phys.* **34**(5): 1546–1555.

Badea, C. T., L. W. Hedlund et al. (2011). In vivo imaging of rat coronary arteries using bi-plane digital subtraction angiography. *J. Pharmacol. Toxicol. Methods* **64**(2): 151–157.

Badea, C. T., S. M. Johnston et al. (2010). Lung perfusion imaging in small animals using 4D micro-CT at heartbeat temporal resolution. *Med. Phys.* **37**(1): 54–62.

Buhalog, A., R. Yasuda et al. (2010). A method for serial selective arterial catheterization and digital subtraction angiography in rodents. *AJNR Am. J. Neuroradiol.* **31**(8): 1508–1511.

De Lin, M., L. Hedlund et al. (2006). Micro radiography imaging of the rodent with phenylephrine induced vascular hypertension. *Third IEEE International Symposium on Biomedical Imaging: Nano to Macro, 2006,* Arlington, VA.

De Lin, M., L. Ning et al. (2008). A high-precision contrast injector for small animal x-ray digital subtraction angiography. *IEEE Trans. Biomed. Eng.* **55**(3): 1082–1091.

Dobbins, J. T., 3rd (2009). Tomosynthesis imaging: At a translational crossroads. *Med. Phys.* **36**(6): 1956–1967.

Figueiredo, G., C. Brockmann et al. (2012). Comparison of digital subtraction angiography, micro-computed tomography angiography and magnetic resonance angiography in the assessment of the cerebrovascular system in live mice. *Clin. Neuroradiol.* **22**(1): 21–28.

Kagadis, G. C., G. Loudos et al. (2010). In vivo small animal imaging: Current status and future prospects. *Med. Phys.* **37**(12): 6421–6442.

Kersten, J. R., T. J. Schmeling et al. (1996). Mechanism of myocardial protection by isoflurane: Role of adenosine triphosphate-regulated potassium (KATP) channels. *Anesthesiology* **85**(4): 794–807.

Kissel, P., B. Chehrazi et al. (1987). Digital angiographic quantification of blood flow dynamics in embolic stroke treated with tissue-type plasminogen activator. *J. Neurosurg.* **67**(3): 399–405.

Kobayashi, S., M. Hori et al. (2004). In vivo real-time microangiography of the liver in mice using synchrotron radiation. *J. Hepatol.* **40**(3): 405–408.

Lin, M., C. T. Marshall et al. (2009). Quantitative blood flow measurements in the small animal cardiopulmonary system using digital subtraction angiography. *Med. Phys.* **36**(11): 5347–5358.

Mistretta, C. A., A. B. Crummy et al. (1981). Digital angiography: A perspective. *Radiology* **139**(2): 273–276.

Ortiz-Velazquez, R. I., J. G. M. P. Caldas et al. (2009). Three-dimensional rotational angiography in murine models: A technical note. *Clinics (Sao Paulo)* **64**(12): 1234–1236.

Schültke, E., S. Fiedler et al. (2010) Synchrotron-based intra-venous K-edge digital subtraction angiography in a pig model: A feasibility study. *Eur. J. Radiol.* **73**(3): 677–681.

Siablis, D., E. N. Liatsikos et al. (2006). Digital subtraction angiography and computer assisted image analysis for the evaluation of the antiangiogenetic effect of ionizing radiation on tumor angiogenesis. *Int. Urol. Nephrol.* **38**(3–4): 407–411.

Tremoleda, J. L., A. Kerton et al. (2012). Anaesthesia and physiological monitoring during in vivo imaging of laboratory rodents: Considerations on experimental outcomes and animal welfare. *EJNMMI Res.* **2**(1): 1–23.

Yeh, B. K., A. J. Gosselin et al. (1977). Sodium nitroprusside as a coronary vasodilator in man: I. Effect of intracoronary sodium nitroprusside on coronary arteries, angina pectoris, and coronary blood flow. *Am. Heart J.* **93**(5): 610–616.

7

Single-Photon Emission Computed Tomography

Matthew A. Lewis

7.1   INTRODUCTION

Single-photon emission computed tomography (SPECT) has a rich history that extends back to the post–World War II emergence of nuclear medicine imaging and the early days of computed tomography (CT) (Vaughan 2008). In recent years, SPECT has in many ways been overshadowed by positron emission tomography (PET) for technical reasons discussed elsewhere in this book. Nonetheless, SPECT benefits from both simpler instrumentation and a wider selection of radiopharmaceuticals than PET. In contrast to PET, where there is an ultimate resolution limit due to positron range, the theoretical limits on SPECT resolution are defined only by the standard exchange with sensitivity. Indeed, gamma camera microscopes have been developed (Meng et al. 2006), and discussions of *nanoSPECT* systems that can image at the resolution of nanoliter volumes have been touted.

When a radiopharmaceutical is injected into an organism, the radioisotope will be concentrated in certain tissues depending upon the biodistribution of the agent. As each radioisotope undergoes radioactive decay, one (or more) high-energy photons are emitted. The chief task in single-photon nuclear medicine imaging is to detect these emission photons in some sort of logical manner in order to image the biodistribution of the radiotracer *in vivo*. In the simplest mode of operation, one can hope to collect many gamma ray photons traveling in essentially the same direction, much as a camera lens collects optical photons traveling in the same general direction. A device that accomplishes this task is called a *gamma camera* and provides a single projection of the radiotracer distribution inside the object. As with planar x-ray imaging, this *planar scintigraphy* contains no depth information, but for many applications this single view is sufficient.

In other applications, it may be desirable to make quantitative point estimations of radiotracer concentration *in vivo*. In this case, one must turn to tomographic imaging with SPECT. In Western medicine, SPECT is now ubiquitous, with approximately 15 million patient studies being performed per year in the United States (IMV 2014). In preclinical research, however, the existing clinical infrastructure is clearly not optimal for imaging in animals such as mice and rats. With typical clinical resolutions greater than 5 mm, the cross-sectional SPECT images of a mouse would include only a few voxels. These divergent requirements have led to the research, development, and in some cases commercial availability of a variety of SPECT systems for preclinical imaging. That said, the utilization of clinical gamma cameras for small animal SPECT imaging is still commonplace and an active area of study (Aguiar et al. 2014).

7.2   PHYSICS

To develop and optimize future SPECT technologies, a firm understanding of a subset of radiation physics is required. In terms of the interaction of high-energy photons with matter, both in detectors and in tissue, one must only consider photoelectric absorption and Compton scattering for gamma rays in the range 30–400 keV. Interested readers are directed to several excellent resources (Wernick and Aarsvold 2004; Cherry et al. 2012; Flower 2012; Dahlbom 2014).

An essential tool for calculating any material properties when designing components for preclinical SPECT is a database of various photon interaction coefficients. One of the most reliable software packages for this is XMuDat (Nowotny 1998). For simulating detectors and systems, the Monte Carlo method is an established tool for predicting and optimizing resolution and sensitivity. By far the most successful open-source software package for simulating preclinical SPECT systems is GATE (Jan et al. 2004, 2011; De Beenhouwer et al. 2008). In addition to built-in support for generic SPECT geometries, GATE has been employed in the simulation of a variety of preclinical and clinical imaging systems (Momennezhad et al. 2012; Lee et al. 2013; Lin et al. 2014; Vieira et al. 2014). With the development of adaptive and optimized preclinical SPECT systems, computational optimization of design parameters for geometries and collimators is clearly playing a more important role in the development of new systems (Van Holen et al. 2013; VanAudenhaege et al. 2014).

7.2.1   RADIOISOTOPES

Preclinical SPECT benefits tremendously from the huge selection of potential radiopharmaceuticals that have been produced throughout the history of nuclear medicine imaging. This is a clear advantage over the PET radiopharmacy. The ubiquitous radioisotope for single-photon imaging is technetium-99m with a 6 h half-life and a single 141 keV photon. Furthermore, ^{99m}Tc is generator-produced and has been used to radiolabel thousands of compounds (Culter et al. 2013). Other radiometals, such as indium-111, make up the second largest group of radioisotopes proposed as SPECT tracers (Wadas et al. 2010). Unconventional radionuclides are discussed in a recent review (Holland et al. 2010).

Radiohalogenation with iodine-123 forms the last major family of SPECT radiopharmaceuticals (McConathy et al. 2012). [123]I has a 13 h half-life and emits a 159 keV gamma-ray. This photon is sufficiently separated from the 141 keV gamma ray to facilitate dual-isotope imaging in many SPECT systems. The other radioisotope of interest is iodine-125, which produces a plethora of photons below 35 keV. [125]I is a very popular radiolabel in molecular biology, and many groups have evaluated ^{125}I-labeled agents systems designed for 141 keV gamma rays. One hurdle to the adoption of a turn-key, preclinical SPECT system optimized for [125]I (Meng et al. 2009) in the molecular biology lab is the potential for significant contamination. With a 60 day half-life and excretion via urine, it is inconceivable that doses greater than 1 mCi would be routinely used when [124]I-based preclinical PET is available at higher sensitivity.

Some radioisotopes, especially those with electron capture as the primary branch, have complicated decay schemes that produce additional detectable photons from atomic-level transitions. Since these photons do not originate in the nucleus of the atom per se, they technically are not gamma rays; but for the purpose of clarity we will refer to all photons in SPECT as gamma rays.

Radioagents that target bone are often of interest to preclinical investigators since bone imaging can readily be compared to skeletal CT (Cuccurullo et al. 2013). Whether SPECT or PET radiopharmaceuticals are superior or complementary is still unresolved and may be application- or disease-specific.

In the brain, the glutamate receptor is a popular target for neuroimaging. Many radioligands have been designed and tested for crossing the blood–brain barrier, thus facilitating imaging in the living brain (Saha et al. 1994; Majo et al. 2013).

For an excellent recent review on SPECT molecular imaging agents, the interested reader is directed to Gnanasegaran and Ballinger (2014).

7.2.2 COLLIMATION

When gamma rays exit an object, they emerge in random directions. In order to make an image, one might wish to detect and then organize these random rays into a projection. In optical imaging, the lens provides the refractive power to capture and focus a subset of photons emitted from a point in space to a specific location in a focal plane. For high-energy photons, however, refraction is not viable for imaging. For gamma rays, all materials essentially appear to have a refractive index slightly less than 1, and photons travel in straight lines until they interact with electrons in atoms.

Instead, the absorptive power of dense, high-Z elements can be used to restrict the set of gamma rays that will ultimately hit the sensitive detector. Such a device is called a *collimator*, and to stop gamma rays effectively in the clinical energy range 80–200 keV, materials such as lead, tungsten, or alloys such as Rose's metal are readily utilized (Peterson et al. 2010). That said, because of the smaller size of rodent models and the applicability of radioisotopes with emissions below 80 keV, it may be possible to utilize other materials for collimation, including copper and brass. At the other extreme, several investigators have proposed the development of collimators based on gold and depleted uranium for high-energy imaging (Tenney et al. 1999, 2001; Tenney 2000, 2001; McConathy et al. 2012).

The two types of collimation most often employed in preclinical SPECT are the same as in clinical imaging: parallel hole and pinhole. However, there is an important aspect of pinhole imaging that has shown tremendous success in preclinical imaging, but with little or no implementation clinically. A parallel hole collimator consists of many identical apertures that extend through a larger block of highly attenuating material. These holes may be circular, square, triangular, hexagonal, or any other desired shape that can be easily constructed. In parallel hole collimation, the cross-sectional dimensions of the hole are invariant through the thickness T of the collimator.

The material between the parallel collimator holes is referred to as *septa*, and the septal thickness t_s is an important design criteria. But to approximate the performance characteristics of a parallel hole collimator, one can assume that the septa are very thin and made of a perfectly absorbing material. We will also assume that a perfect detector of gamma rays is positioned behind the collimator, while in practice there may be a negligible gap. Using geometric ray tracing and similar triangles, the full width at half-maximum (FWHM) resolution of a point source situated a distance F in front of the collimator can be approximated as

$$\text{FWHM}_{\text{parallel}} \approx K'_{\text{shape}}\left(\frac{D}{T}\right)(T+F) = K'_{\text{shape}} D\left(1+\frac{F}{T}\right) \tag{7.1}$$

where

D is the size of the collimator holes

K'_{shape} is a constant in the range [1, 2] that accounts for the packing of different shaped holes

From this equation, it is evident that the resolution of the parallel hole collimator degrades linearly as the source–collimator distance F is increased.

The other parameter of interest is the sensitivity, or the number of gamma rays emitted at a point source that make it through the collimator to the detector. Clearly, this can never be greater than 0.5 since half of the gamma rays are propagating away from the collimator; in practice the sensitivity will always be orders of magnitude lower, which is a major disadvantage for SPECT compared to PET or optical imaging. Again, using a geometric analysis, the collimator sensitivity g can be approximated as

$$g \approx K_{\text{shape}}^2\left(\frac{D}{T}\right)^2 \frac{D^2}{(D+t_s)^2} \tag{7.2}$$

The second shape factor $K_{\text{shape}} < 0.5$ varies only slightly with the collimator hole shape, but the sensitivity is clearly a function of the septa thickness t_s. The last fraction in this formula can be recognized as the percentage of collimator surface area taken up by the collimator holes. As the septa thickness is increased, the sensitivity will decrease.

In Equation 7.1, the resolution is proportional to D, while it is the square of this quantity that appears in Equation 7.2. Combining these approximations yields the famous resolution–sensitivity tradeoff:

$$g \propto \text{FWHM}^2 \tag{7.3}$$

and this fact is generalized and observed for all classes of collimators. If one considers the change in spatial scale going from clinical imaging (150 cm for humans) to preclinical imaging (7.5 cm for adult body length of mouse), the collimator sensitivity would decrease by a factor of 0.0025 if the required resolution is to scale similarly. One could inject 400 times the human radiotracer dose to compensate for this decreased sensitivity, but in practice the mouse typically receives the same amount of radioisotope as a human.

In reality, no septa are perfect absorbers, and septal penetration will affect both sensitivity and resolution. Gamma rays that pass through septa can be viewed as effectively shortening the overall thickness of the collimator. The choice of septal thickness t_s is therefore also related to an effective collimator thickness T_{eff}, which can be used in the approximations above. For more exact design of optimized collimators, it is possible to consider more generalized formulas and to formulate an optimization problem in three variables: the collimator thickness and two parameters related to the shape and packing of the collimator holes (Gunter 2004).

The pinhole collimator can also be analyzed using a geometric argument. Considering an extremely thin plane of material with perfect absorption, an aperture in this plane will allow gamma rays to pass through and to form an image on a detector that is now offset by a distance L from the pinhole plane. Again, the resolution for a point source a distance F away from the pinhole can be calculated using geometric ray tracing, as

$$\text{FWHM}_{\text{pinhole}} \approx D\frac{L+F}{L} = D\left(1+\frac{F}{L}\right) \tag{7.4}$$

The sensitivity can likewise be approximated by considering the ratio of the solid angle of the pinhole to the unit sphere:

$$g \approx D^2 \frac{\cos^3\theta}{16F^2} \tag{7.5}$$

Equation 7.4 illustrates the utility of pinholes in high-resolution preclinical imaging. If the gamma ray detector is sufficiently far from the collimator ($L \gg F$), then the collimator resolution is effectively the pinhole diameter D and is independent of source–collimator distance. Assuming that there are no limitations on the resolution from downstream components in the gamma camera (which is unrealistic), then resolution can be improved arbitrarily by reducing the pinhole size to microscopic dimensions. The pinhole does one more amazing feat though: the intrinsic resolution due to uncertainties in subsequent components of the gamma camera will geometrically scale down by the fact that $L > F$. Equation 7.2, however, demonstrates the pitfall though: the resolution–sensitivity tradeoff is explicit in the formula. Furthermore, in contrast to that of the parallel hole collimator, the sensitivity of the pinhole collimator diminishes precipitously as the square of the source–collimator distance F.

In practice, submillimeter resolution is feasible with pinhole collimators (Funk et al. 2006; Beekman and van der Have 2007). Surprisingly, the sensitivities of a single pinhole collimator and of a parallel hole collimator of similar resolution possess sensitivities of the same order of magnitude: 10^{-4}. This may seem counterintuitive until one realizes that the pinhole collimator allows gamma rays traveling in many directions to pass, while the parallel hole collimator restricts gamma ray to a narrow range of propagation. As with the septa in parallel hole, the knife edge or keel produced in a pinhole in a material of finite thickness is susceptible to penetration by gamma rays (van der Have and Beekman 2004). To a first-order approximation, this effect will degrade the resolution slightly and can be modeled as a pair of *effective* pinhole diameters (Cherry et al. 2012, Chapter 13). An excellent theoretical treatment of the optimal pinhole design problem has been presented by Fessler (1998).

The diverging and converging collimators possess properties of both parallel hole and pinhole collimators. Despite having resolutions and geometric efficiencies that straddle those for similar parallel hole collimators, diverging and converging collimators can perform magnification like a pinhole collimator as well as demagnification. The converging collimator actually has a sensitivity that increases with the source–collimator distance up to the point of convergence. Using geometric analysis, more complicated formulae for sensitivity and resolution can be found in standard references (Cherry et al. 2012, Chapter 13).

Initial work with dedicated high-resolution studies in small animal models of human disease used custom pinholes mounted on clinical SPECT system. Jaszczak et al. (1994) reported sensitivities on the order of 1 cps/μCi for pinhole collimators on the order of 1 mm. After obtaining sufficient counts, satisfactory proof-of-concept images were produced for micro-cold-rod and micro-Defrise phantoms, as well as two adult rats. Subsequently, Ishizu et al. (1995) demonstrated 1.65 mm FWHM resolution (deemed ultrahigh resolution at the time) using 1.0 mm pinholes, and with 4 mm pinholes they reported 4300 cps/μCi.

The typical preclinical PET scanner possesses a sensitivity on the order of $10^{-2} - 10^{-1}$ because it utilizes electronic rather than physical collimation. Improved sensitivity is one advantage that PET claims over SPECT, where a single millimeter-resolution pinhole produces 10^{-4} sensitivity. But if one considers an imaging system with 10^2 pinholes, the sensitivity of SPECT approaches that of PET with the potential benefit of better spatial resolution. Much of the research and development in preclinical SPECT after 2000 has focused on multi-pinhole collimators (Schramm et al. 2003). In a multi-pinhole design as shown in Figure 7.1, it is common for a single gamma camera to include an aperture plate with multiple pinholes. If the projections through each pinhole occupy individual regions on the scintillating crystal, then the geometry is called *nonoverlapping*. With circular

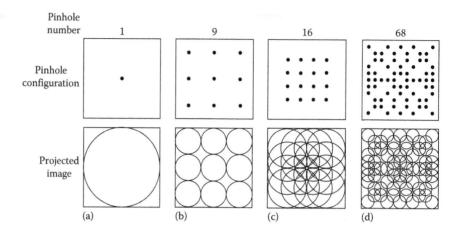

FIGURE 7.1 Classification of pinhole geometries: (a) single pinholes, (b) nonoverlapping pinholes, (c) degenerate pinholes, and (d) coded aperture. (Reprinted with permission from Soesbe, T., Lewis, M., Richer, E., Slavine, N., and Antich, P., Development and evaluation of an EMCCD based gamma camera for preclinical SPECT imaging, *IEEE Trans. Nucl. Sci.*, 54(5), 1516–1524. © 2007 IEEE.)

pinholes, this will lead to inefficient utilization of the detector face. This may be remedied by allowing the projections to slightly overlap. For even more pinholes and increased sensitivity, the projections may have significant overlap, and the system geometry may be labeled *degenerate* since a given count could have gone though one of two or more pinholes. Despite this complexity, it is possible to incorporate information on this pinhole degeneration in the image formation stage and produce a tomographic image. The optimal arrangement of multiple pinholes remains an active area of research (Rentmeester et al. 2007).

In the limit when the number of pinholes in an aperture plate becomes large, the collimator performs as a coded aperture. The arrangement of holes can be random or in a pseudo-random pattern that facilitates image reconstruction at the computational level (Chang et al. 1974; Accorsi et al. 2001a,b, 2008; Meikle et al. 2001). Schellingerhout et al. (2002) report a coded-aperture small animal imaging system with sensitivity an order of magnitude higher than parallel hole collimation. Apertures found routinely in optics have also been investigated in nuclear medicine imaging, such as the Fresnel zone plate (Barrett 1972; Itaya and Kojima 1978). Other collimators appear infrequently in the literature, including the slant-hole, rotating-slat, skew-slit (Huang and Zeng 2006; Tang et al. 2006), and the slit-slat collimator (Metzler et al. 2010), all of which are designed for improved sensitivity.

7.2.3 Scintillation

Once the gamma ray photon traverses the collimator, it must be detected and its interaction position estimated in relation to the collimator. For many decades, this process was accomplished by converting the single high-energy gamma ray photon to many lower energy optical photons, which could be counted with photomultiplier tubes. The ideal scintillator should effectively stop gamma rays through photoelectric absorption at relatively shallow depth in the scintillator, and should produce large numbers of optical photons. The statistics of scintillation light is important for both position and energy estimation.

By far the most common scintillator used in preclinical gamma cameras mirrors the technology used in clinical imaging: NaI(Tl). Sodium iodine doped with thallium produces 38 optical photons (centered at 415 nm) per keV of incident gamma ray. This blue light scintillation is well matched to bi- and multi-alkali photocathodes in many commercial photomultiplier tubes (PMTs).

Since 2000, many research groups have explored the possibility of coupling charge-coupled device (CCD) cameras to scintillating crystals (Lees et al. 2003). In theory, the CCD sensor should offer superior intrinsic spatial resolution compared to arrays of PMTs and position-sensitive PMTs (PSPMTs). The blue light of NaI(Tl) is not well-matched to the spectral response curve for almost all CCDs, so many investigators explored scintillation with CsI(Tl). CsI(Tl) has a peak emission in the green (540 nm), which is typically also the peak sensitivity for CCDs. In addition, CsI(Tl) has improved stopping power compared to NaI(Tl) due to its increased density (4.5 vs 3.7 g/cm^3), and is only slightly hydroscopic. CsI(Tl) also produces more scintillation light per event (52 per keV). The only negative aspect of CsI(Tl) is the relatively long decay time on the order of 1 μs. While PET scintillation crystals have moved to ever shorter decay times, the increased possibility of pulse pileup when using CsI(Tl) is normally not rate-limiting since the count rates with physical collimation are much lower than in PET. With overlapping pinhole and coded aperture geometries, however, this long decay time may be problematic. Preclinical PET detectors now typically utilize fast lutetium orthosilicate (LSO) scintillation crystals. Since LSO contains natural radioactivity from Lu-176 (Yamamoto et al. 2005), it is not ideal in single-photon detectors where there is no coincidence detection to remove the background counts. In general, organic scintillators are also not optimal for preclinical imaging because of their low photon yield and insufficient stopping power due to low effective Z.

7.2.4 Gamma Cameras

In gamma cameras with pinhole collimators, it is possible to minimize the intrinsic resolution of the camera by suitable magnification to the point where the total system resolution is dominated by the pinhole collimator resolution. Indeed, one might question whether a pinhole collimator has any application with a high intrinsic resolution gamma camera design. In this case, the only advantage may be the increased sensitivity through multi-pinhole collimators, but similar performance may be achieved with coded apertures.

Initial efforts in preclinical SPECT utilized traditional clinical detector technology. In many cases, clinical gamma cameras were modified for small animal imaging with pinholes. In some cases, the animal was rotated rather than the gantry (Habraken et al. 2001).

Initial desktop pinhole SPECT imaging system utilized a traditional detector approach: pixelated NaI(Tl) coupled to PSPMTs (McElroy et al. 2002; Weisenberger et al. 2003). With a single camera and a 20% energy window, sensitivities on the order of 10 cps/μCi were reported. This successful approach was duplicated in the first commercial scanner offered by Gamma Medica, Inc.

In an assembled gamma camera with an attached collimator, the total resolution is determined by both the resolution of the collimator and the intrinsic resolution of the camera, which is a function of the scintillator, optical, and electronic components.

$$\sigma_{tota}^2 = \sigma_{collimator}^2 + \sigma_{instrinsic}^2 \tag{7.6}$$

With a pinhole collimator, the intrinsic resolution will be decreased by the magnification factor M, which will of course depend upon source–collimator distance:

$$\sigma_{total}^2 = \sigma_{collimator}^2 + \frac{1}{M}\sigma_{instrinsic}^2 \tag{7.7}$$

So, there are two paths to developing high-resolution preclinical SPECT imaging system: (1) develop high intrinsic resolution detectors and match them with high-resolution collimators, or (2) use a high-resolution pinhole collimator with a normal intrinsic resolution detector.

More recently, researchers have followed the lead of PET detector developers and coupled scintillators with semiconductor-based optical detectors. For example, Funk et al. (2006) reported detector modules with CsI(Tl) and position-sensitive avalanche photodiodes for an impressive intrinsic resolution of 0.5 mm at 141 keV. With collimators, the system was designed to record 23 cps/μCi at 0.8 mm resolution, an improvement over PSPMT designs.

While PSPMT gamma cameras have been reported with intrinsic resolution on the order of 0.5 mm (Pani et al. 2004), the typical traditional gamma camera design using either PMT arrays or

PSPMTs have best intrinsic resolution on the order of 2 mm (Williams et al. 2000). To utilize these cameras, a pinhole collimator is typically used to negate the impact of the intrinsic resolution.

In the past decade, several groups have investigated high-intrinsic-resolution gamma cameras built around CCDs. While these sensors can have quantum efficiencies greater than 95% in the scintillation spectral range for CsI(Tl), they typically have small surface areas. It is well recognized that optical coupling of a mouse-sized field of view (FOV) to a small CCD is very inefficient (Liu et al. 1994; Maidment and Yaffe 1995, 1996; Yu and Boone 1997), and the detection of individual scintillation events can quickly become quantum-noise-limited.

Initial work to address this problem focused on the backend of the gamma camera. A new class of CCDs with low effective noise promised the ability to detect single photons. The electron multiplying CCDs (EMCCDs) include a special serial register where the photon-induced charge can be multiplied by impact ionization (Hynecek 2001; Hynecek and Nishiwaki 2003). In collaboration with manufacturers of columnar CsI(Tl), Beekman et al. (De Vree et al. 2004, 2005; Heemskerk et al. 2007) and Barrett et al. (Nagarkar et al. 2006; Miller et al. 2007) demonstrated intrinsic resolution below 100 μm for the first time. This approach was feasible for small animal imaging with both monolithic and columnar CsI(Tl) coupled to back-illuminated EMCCDs (Soesbe et al. 2007, 2010), but with a small FOV (Lees et al. 2011). Furthermore, the economics of commercial EMCCDs made the development of multi-camera system cost-prohibitive.

In 2008, Miller et al. (2008) reported an alternative approach to the same problem that utilized inexpensive CCD cameras. In this approach, the scintillator is immediately coupled to a multichannel plate image intensifier tube (MCP IIT) in order to amplify the weak scintillation signal. After the signal is amplified, it may be lens-coupled to a CCD camera (preferably of high frame rate)

In parallel, there is an emerging class of semiconductor-based gamma cameras. Cadmium zinc telluride (CZT) detectors have been used in both commercial preclinical and clinical nuclear medicine imaging systems. Clinically, these pixelated detectors have intrinsic resolutions on the order of 2 mm and can be coupled with pinholes for high-resolution imaging. In the research sphere, CZT detectors with small pitch (300 μm) have been explored as potential high-intrinsic-resolution gamma cameras. Because of the small pixel effect, it is unlikely that CZT gamma cameras will match the intrinsic spatial resolution of CCD-based gamma cameras.

In the photon counting mode, the position of the γ-ray interaction in the scintillating crystal must be estimated. Typically, this estimation is in two dimensions, but the depth of interaction may also be estimated when there is parallax (pinhole, converging, and diverging collimators). In addition, the energy deposited in the crystal is an important parameter to estimate because it can facilitate (1) rejection of Compton scatters in the crystal, (2) rejection of counts from gamma rays that scattered in the object before detection, and (3) separation of isotopes in dual-isotope imaging. The traditional algorithm for estimating xy count location, dating back to Anger's fundamental work, is an algebraic estimation based on the centroid of signals from the PMT array or PSPMT grid. Harrison Barrett's group at the University of Arizona has led the efforts to replace Anger logic with modern statistical estimation such as maximum-likelihood (ML) methods (Barrett et al. 2009). The ML estimation of position, energy, and time of gamma-ray interaction can incorporate *a priori* information on the noise properties of the gamma camera, although performing this task in real time requires both a lengthy calibration procedure with a collimated source and significant computational resources such as graphical processing units (GPUs).

7.3 IMAGE FORMATION

With pinhole collimation, the simplest image reconstruction model is the cone beam x-ray transform. Using backprojection algorithms such Feldkamp–David–Kress (FDK), it is possible to reconstruct a typically poor image from many pinhole projections. That said, preclinical nuclear medicine imaging has been the test bed for many advanced reconstruction methods. Algebraic reconstruction, an iterative method used early in the development of CT, has been used to exploit symmetries in a multi-pinhole and/or multi-camera system, reducing the computational cost (Israel-Jost et al. 2006).

Statistical-based iterative reconstruction has proven to be superior to analytic and algebraic reconstruction methods, and preclinical nuclear medicine imaging has been the test bed for algorithm development (Vandenberghe et al. 2001). The literature on this family of algorithms is now quite large and beyond the scope of this review, but in summary the benefits of statistical-based iterative reconstruction are numerous: (1) the incorporation of appropriate noise models for the stochastic nature of signals, (2) adoption of both frequentist and Bayesian statistical viewpoints, (3) reduction of artifacts related to sampling pattern, and (4) ready incorporation of *a priori* information. An excellent introduction to the algorithms can be obtained in the topical review by Qi and Leahy (2006).

For the multiplexing problem associated with degenerate multi-pinhole apertures, there are at least two approaches to image reconstruction. In Wilson et al. (2000), the collimator–detector distance is modulated, and from this data synthetic projections that are reconstructed are calculated using standard means. However, if the relationship between object and detector spaces is encoded in a suitably sampled system transfer matrix, then the inverse problem will take care of the apparent degeneracies. In this case, the image reconstruction task lies in the spectrum of inverse problems between wholly projection-based problems such as Radon transform-based tomography and image deblurring. Provided that the pinhole overlap is not too large, the reconstruction is feasible and realizable. The asymptotic approach is used when there are no collimators, and the reconstruction is horribly conditioned although not quite as bad as optical imaging. However, recent efforts have demonstrated that reconstruction of sparse distributions of radioisotopes is possible (Walker et al. 2015). Using a much simpler approach that avoids measuring the explicit system matrix, Vandeghinste and colleague recently demonstrated that ray-based reconstruction in preclinical SPECT exhibited in vivo quantification errors of less than 5% (Vandeghinste et al. 2014).

The noise in the reconstructed volume is dependent upon the total number of counts collected, which is directly related to the system sensitivity g. In place of the resolution–sensitivity relationship, Fessler considered spatial resolution–noise tradeoff to calculate the variance-minimizing FWHMs for two ideal pinholes: Gaussian and Laplacian (Fessler 1998). He demonstrated that the optimal pinhole for this criterion has an FWHM that is the desired spatial resolution divided by the square-root of 2. Optimal experiment design principles are potentially very important in the development of future preclinical SPECT systems.

In a related development, Barrett and colleagues have considered imaging system that can adapt to the object being imaged in order to maximize sensitivity or resolution (Barrett et al. 2008; Freed et al. 2008). In this proof-of-principle design, the collimators on multiple gamma cameras have a variable design and can be modulated depending upon the nature of the projection that each detector receives.

7.4 SYSTEMS

In this section, we highlight notable research and commercial imaging system, focusing on the design strategies outlined above. The list of instruments covered is not comprehensive, is influenced by the North American marketplace, and represents the possible instruments that be encountered in use today.*

Gamma Medica developed and marketed in the early 2000s the A-SPECT and X-SPECT imaging systems using PMT-based gamma cameras. In addition to parallel hole collimators, single pinhole collimators were initially available. In moving from their third-generation FLEX dual modality system (SPECT/CT) to the Triumph system, CZT detectors became available. Today, this line of ground-breaking preclinical imaging system continues with the Triumph II trimodality system (SPECT/PET/CT) marketed by Tri-Foil. While the current system has been upgraded to support the multi-pinhole paradigm, it has not been characterized in peer-reviewed literature.

---

* A large percentage of which are no longer commercially available.

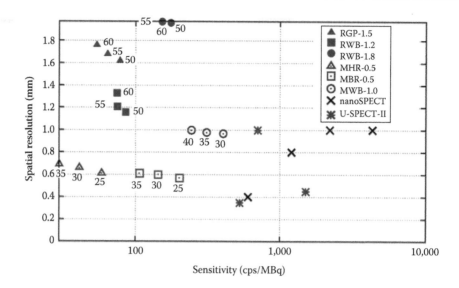

FIGURE 7.2   Sensitivity versus spatial resolution characterization of Siemens Inveon SPECT for six types of collimation. Bioscan nanoSPECT and MILabs U-SPECT-II are also shown for comparison. This was originally published in JNM. (From Boisson, F. et al., *J. Nucl. Med.*, 54(10), 1833, 2013. Figure 1. Copyright by the Society of Nuclear Medicine and Molecular Imaging, Inc. With permission.)

The Siemens Inveon SPECT (Boisson et al. 2013) uses a traditional gamma camera design of pixelated NaI(Tl) scintillator coupled to arrays of PSPMTs. The system comes with both single- and multi-pinhole collimator plates and supports reconstruction using either 3D ordered subsets-expectation maximization (OSEM) or 3D maximum *a posteriori* (MAP) algorithms. Computational model (Lee et al. 2013). As shown in Figure 7.2, multi-pinhole apertures improve system sensitivity for both rat and mouse optimized collimators.

The GE eXplore speCZT was also based on CZT detectors (Matsunari et al. 2014). The 10 CZT based detectors with 2.46 mm pixels are arranged in a ring with both 1 mm multi-pinhole collimators and multislit collimators. For image reconstruction, the system defaults to 50 iterations of maximum likelihood-expectation maximization (MLEM) with an optional recovery filter. With the recovery filter enabled and using the highest resolution mouse pinhole aperture, the system resolution has been reported to be below 0.8 mm FWHM.

The Bioscan NanoSPECT series of instruments resorted to the traditional design of a large FOV camera with pinhole collimators (Deleye et al. 2013). This system was originally based on technology developed by Mediso in Europe. Each camera in the four-head system is comprised of a 33-PMT array coupled to a 6-mm-thick NaI(Tl) crystal. With an energy resolution report below 10%, dual isotope imaging is feasible with this system. Existing Bioscan systems still in use may at this time be supported also by TriFoil.

In addition to development of EMCCD-based gamma cameras, the Beekman group at the University Medical Center, Utrecht, The Netherlands, has developed several generations of stationary detector small animal SPECT system that are marketed by MILabs. The U-SPECT-I prototype was originally designed around the triple-head clinical SPECT system (Beekman et al. 2005; Vastenhouw and Beekman 2007). In systems of this type, an important calibration step includes the spatial orientation of the pinhole insert relative to the moving detector gantry (DiFilippo 2008). U-SPECT-II also used three stationary detectors, but incorporated 75 gold pinholes for high-resolution collimation (Van Have et al. 2009; Deleye et al. 2013). For image reconstruction, the U-SPECT II utilized MLEM with image recovery using measured and interpolated PSF tables, but MLEM was not used for individual list mode event estimation. The U-SPECT-III system adopted stationary detectors with focused, multiple pinholes (Beekman and Vastenhouw 2004; Vaissier et al. 2012), mirroring contemporaneous developments in other groups. The latest generation system, called the U-SPECT⁺, is equipped with 0.6, 0.35, and 0.25 mm cylindrical multi-pinhole inserts with a nonoverlapping geometry (Ivashchenko et al. 2014). The system matrix is formulated through both robotic scanning of a point source and analytical modeling. With 3D OSEM reconstruction on an isotropic 0.125 mm voxel grid, the system is capable of evaluating molecular concentrations of volumes at the 15 nL level.

The group at the Center for Gamma Ray Imaging (CGRI) , University of Arizona, USA, directed by Harrison Barrett, has truly led the way in terms of innovations for preclinical nuclear medicine imaging. In many cases, their SPECT systems also served as test beds for the evaluation of new signal processing and task-based imaging methodologies. The FastSPECT I was originally designed for 3D clinical brain imaging and utilized 24 stationary modular cameras comprised of four PMTs coupled to NaI(Tl) (Klein et al. 1995). For collimation, the system was equipped with 24–150 pinholes and was eventually repurposed for preclinical imaging studies in oncology. FastSPECT II used 16 modular cameras of the type shown in Figure 7.3 (Furenlid et al. 2004), but

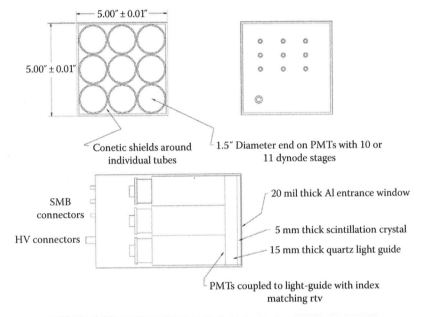

FIGURE 7.3 The modular gamma camera for FastSPECT II at the University of Arizona CGRI using a 3 × 3 array of PMTs and supported list-mode data acquisition. (Reprinted with permission from Furenlid, L., Wilson, D., Chen, Y.-C., Kim, H., Pietraski, P., Crawford, M., and Barrett, H., FastSPECT II: A second-generation high-resolution dynamic SPECT imager, *IEEE Trans. Nucl. Sci.*, 51(3 II), 631–635. © 2004 IEEE.)

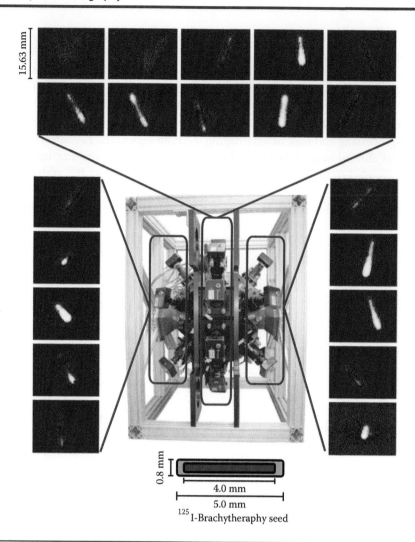

15.63 mm

0.8 mm

4.0 mm

5.0 mm

^{125}I-Brachytheraphy seed

FIGURE 7.4 The BazookaSPECT camera-based FastSPECT III at the University of Arizona CGRI demonstrating sample pinholes projection from an ^{125}I brachytherapy seed. (Reprinted with permission from Miller, B., Furenlid, L., Moore, S., Barber, H., Nagarkar, V., and Barrett, H., System integration of FastSPECT III: A dedicated SPECT rodent-brain imager based on BazookaSPECT detector technology, *IEEE Nucl. Sci. Symp. Conf. Rec.*, pp. 4004–4008. © 2009 IEEE.)

these detectors increased the PMT array to 3 × 3 and output list mode data describing the position and energy of each detected gamma ray. Rather than using the traditional Anger logic, this system provided a test vehicle for using ML methods for parameter estimation in the gamma detectors. To apply related statistical methods for 3D image reconstruction, measurement of system matrix with a robot is required. FastSPECT III shown in Figure 7.4 is the latest version of the CGRI preclinical SPECT system and is based on the successful BazookaSPECT CCD-based gamma camera (Miller et al. 2009). Each BazookaSPECT camera sends its data to a GPU in the host computer for ML processing.

The Bruker ALBIRA trimodality system is one of the most recent commercial preclinical SPECT systems to be marketed, and its performance characteristics have been documented in the recent literature (Sanchez et al. 2013; Spinks et al. 2014). Since there are no standards for the performance characterization of preclinical SPECT system, the authors adopted performance evaluation tests from the preclinical PET literature to evaluate the system's spatial resolution, sensitivity, and image uniformity. For an FOV less than 5 cm, the FWHM resolution for single and multi-pinhole collimators was superior to the associated PET system resolution. However, in terms of sensitivity, the 6% sensitivity at isocenter for the PET subsystem is superior to both the single-pinhole SPECT sensitivity (0.007%) and the nine multi-pinhole collimation (0.07%).

7.5 EXAMPLE APPLICATIONS

One advantage of preclinical SPECT over PET imaging is the ability to image simultaneously with multiple radiotracers. By setting different energy windows for the detected scintillation events, it is possible to classify data according to the radioisotope of origin, which can then be reconstructed separately. Dual-isotope SPECT imaging becomes feasible as the gamma camera energy resolution approaches 10%. Since CZT-based detectors have typically energy resolution below 10%, preclinical SPECT systems using CZT detectors exhibit excellent dual isotope performance. Waganaar and colleagues at Gamma Medica-Ideas demonstrated dual-isotope imaging with ^{111}In-DTPA and ^{99m}Tc-MDP in a mouse tumor model (Wagenaar et al. 2006). Furthermore, with a narrow window at 75 keV, they also imaged the mercury daughter x-rays from ^{201}Tl decay simultaneously with ^{99m}Tc-MDP and ^{123}I in a proof-of-principle triple-isotope SPECT imaging experiment reproduced in Figure 7.5. More recently, Hijnen and colleagues demonstrated dual-isotope imaging with ^{111}In and [177] on a Bioscan NanoSPECT/CT that uses PMT-based detectors (Hijnen et al. 2012). As shown in Figure 7.6, the crosstalk between the two energy windows, 3 h post injection, is minimal.

Iodine-125 is a radioisotope that is familiar to many molecular and cancer biologists. Although the photons that are emitted from the rather complicated decay scheme of ^{125}I are too low in energy for clinical energy, they are useful for preclinical imaging. One difficulty is that gamma detectors optimized for 141 keV photons may not have the best performance in the 25–35 keV window. Nonetheless, many groups have performed ^{125}I radiolabeling and subsequent imaging on their preclinical SPECT systems. An example of high-resolution pinhole imaging of the mouse thyroid is shown in Figure 7.7. In more recent work, Chrastina and Schnitzer used an older Gamma Medica X-SPECT system to image the in vivo biodistribution of ^{125}I-labeled silver nanoparticles injected into Balb/c mice (Chrastina and Schnitzer 2010). Accorsi et al. (2008) described a coded-aperture-based small animal imaging system for [125]I.

Imaging of the small animal skeleton with ^{99m}Tc-MDP is probably the most common preclinical imaging study. Because of the specificity of the agent and the large injected doses (1–10 mCi, which is a human dose for a mouse), the resulting SPECT images approach anatomical detail, although the quantitative value compared to lower resolution Na[18]F PET imaging is still debated. However, as shown in Figure 7.8 from (Ivashchenko et al. 2014), the details in a mouse ^{99m}Tc-MDP

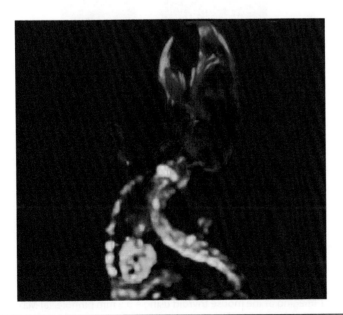

FIGURE 7.5 Tri-isotope imaging in the mouse with ^{99m}Tc-MDP (red), [123]I (blue), and [201]Tl (green) with the Gamma Medica-Ideas eV-CZT FLEX SPECT/CT. (Reprinted with permission from Wagenaar, D., Zhang, J., Kazules, T., Vandehei, T., Bolle, E., Chowdhury, S., Parnham, K., and Patt, B., In vivo dual-isotope SPECT imaging with improved energy resolution, *Nucl. Sci. Symp. Conf. Rec.*, 6, 3821–3826. © 2006 IEEE.)

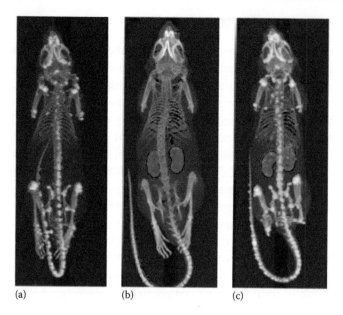

(a) (b) (c)

FIGURE 7.6 Dual-isotope imaging in the mouse with (a) ^{177}Ln (green), (b) ^{111}In (blue), and (c) fused with the Bioscan nanoSPECT/CT. (From Hijnen, N.M., de Vries, A., Nicolay, K., and Grull, H.: Dual isotope 111In/177Lu SPECT imaging as a tool in molecular imaging tracer design, *Contrast Media Mol. Imaging*, 2012, 7(2), 214–222. Copyright Wiley-VCH Verlag GmbH & Co. KGaA. Reproduced with permission.)

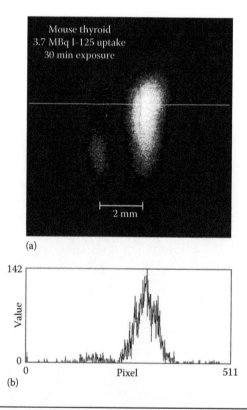

FIGURE 7.7 ^{125}I SPECT imaging of the mouse thyroid using a 0.5 mm pinhole. (a) Maximum intensity projection of MLEM reconstruction (b) profile through thyroid. (Reprinted with permission from Soesbe, T., Lewis, M., Slavine, N., Richer, E., Bonte, F., and Antich, P., High-resolution photon counting using a lens-coupled EMCCD gamma camera, *IEEE Trans. Nucl. Sci.*, 57(3), 958–963. © 2010 IEEE.)

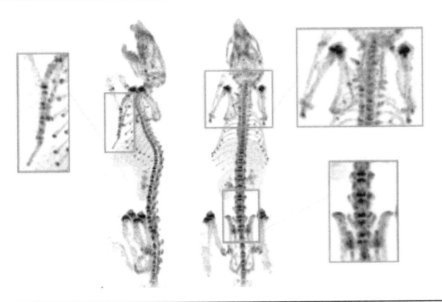

FIGURE 7.8   Extremely detailed skeletal SPECT imaging with [99mTc]-MDP on the MILabs U-SPECT+ using 0.25 mm diameter pinholes. (Reprinted with permission from Ivashchenko, O., van der Have, F., Villena, J.L., Groen, H.C., Ramakers, R.M., Weinans, H.H., and Beekman, F.J., Quarter-millimeter-resolution molecular mouse imaging with U-SPECT+, *Mol. Imaging*, 13, 1–8. © 2014 Decker Publishing.)

whole-body SPECT scan with 0.25 mm pinholes truly complement those seen in a high-resolution preclinical CT of the skeleton.

For a more detailed guide to the diverse preclinical application of molecular SPECT imaging, the interested reader is directed to the review by Khalil et al. (2011).

7.6   FUTURE PERSPECTIVES

Preclinical SPECT is especially vulnerable to competition from preclinical optical imaging, especially in studies using mouse models. In many cases, planar optical imaging is the preferred mode. For example, while much effort went into the development of bioluminescence tomography, most published studies with bioluminescence in mouse models make quantitative measures using only planar emission images. Preclinical SPECT will continue to be valued, however, in longitudinal studies where precise quantitative measurements need to be made in specific organs in animal models. While SPECT benefits from the radiotracer principle and assumes negligible biological perturbation of the systems under study, there are concerns that the agents used in optical imaging may have biological effects (Zhang et al. 2007).

With severe consolidation of the commercial preclinical imaging market, the future of preclinical SPECT is again unclear. Siemens, a late entry into the preclinical market space (Vandenberghe et al. 2001), has already announced the planned lifetime of its Inveon line of preclinical imaging instruments, and, as of September 2014, has stopped selling systems. After a period of consolidation as major vendors purchased all the smaller preclinical instrumentation vendors, it appears that the marketplace is now once again centered around multiple systems available from small vendors.

In the arena of reconstruction, a variety of problems still need to be tackled. While there is little attenuation in the mouse at 141 keV, a significant fraction of photons from iodine-125 will scatter once before leaving the mouse model. By migrating back to transport models, effective reconstruction strategies may be possible in situations where energy resolution is useless (Gallas and Barrett 1998). Another interesting area of exploration is synergistic image formation in dual-modality imaging systems. One particularly attractive approach is the utilization of *a priori* information from a high-registration MR image when performing a limited-FOV SPECT reconstruction (Lee et al. 2011). Lee and colleagues reported that the root-mean-square error due to out-of-field artifacts could be reduced by at least 50% in this approach. Other groups have also reported the development of MR-compatible SPECT systems (Hamamura et al. 2010; Meier et al. 2011). Optical imaging and SPECT are also routinely combined in the same platform (Wang et al. 2012; van Oosterom et al. 2014), particularly if some component of the optical contrast agent can be radiolabeled (Culver et al. 2008; Tian et al. 2008; Liu et al. 2011; Lu et al. 2014; Lutje et al. 2014). The U-SPECT system has successfully integrated with both bioluminescence and fluorescence imaging (van Oosterom et al. 2014), although the optical imaging modalities were not fully tomographic. As described in Chapter 10 in this handbook, Cerenkov luminescence imaging, the optical imaging of certain radiotracers, may complement limited-view single-photon nuclear medicine imaging. As with other imaging modalities, compressive sensing strategies also present opportunities for "doing more with less" (Wolf et al. 2011; Mukherjee et al. 2014).

Two recurring designs with long histories in the literature may also warrant revisiting: the Compton camera (Durkee et al. 1998a,b) and collimator-less imaging (Smith et al. 1992; Mitchell and Cherry 2009; Walker et al. 2015). Because these geometries remove physical collimation, they potentially could approach the sensitivity of PET. One should not expect, however, to see the disappearance of collimators from single-photon nuclear medicine any time soon. Recent progress of adaptive imaging with variable collimators has yet to leverage the development and optimization of multileaf collimators that are now routinely used in radiation oncology (Zhu et al. 2014). Meng et al. have even proposed combining both electronic and mechanical collimation to produce a hybrid collimator for gamma rays above 141 keV (Meng et al. 2003). Walker et al. recently reported single-photon, multi-pinhole imaging with ^{18}F with image quality superior to electronic collimation PET (Walker et al. 2014). Whether a vendor will eventually support single-photon-based imaging of PET isotopes remains to be seen, but this is an exciting development.*

One might also consider the recent revolution in optical microscopy. For over a hundred years, the optical resolution limit was taken to be the diffraction limit due to finite apertures in classical optical theory. During the past 10 years, this resolution limit has been obliterated. Through a variety of ingenious techniques involving either specialized illumination, point-spread-function engineering, and/or control of sparsely distributed fluorophores (Hell 2003; Toomre and Bewersdorf 2010; Gould et al. 2012), the limit of optical microscopy is now assumed to be limited by photon statistics. Indeed, single-particle and single-molecule tracking are also both routine. In imaging methods using sparsely distributed fluorophores that can be activated and/or deactivated (e.g., PALM, STORM), the problem of localizing a single fluorophore is analogous to the detection and estimation of a single γ-ray interaction in a gamma camera. So in case the emission source distribution is noncontinuous in space and is modulated in time, then super-resolution methods are possible. Ultimately, the distribution of the radiopharmaceutical is discrete, and one might ponder if it is feasible to map the individual locations of decay events. While it may be impossible to overcome the random nature of radioactive decay, it remains to be seen if super-resolution methods for SPECT will eventually be uncovered.

REFERENCES

Accorsi, R., L. Celentano, P. Laccetti, R. Lanza, M. Marotta, G. Mettivier, M. Montesi, G. Roberti, and P. Russo. (2008). High-resolution ^{125}I small animal imaging with a coded aperture and a hybrid pixel detector. *IEEE Trans. Nucl. Sci.* **55**(1): 481–490.

Accorsi, R., F. Gasparini, and R. C. Lanca. (2001a). Optimal coded aperture patterns for improved SNR in nuclear medicine imaging. *Nucl. Instrum. Methods Phys. Res.* **474**: 273–284.

Accorsi, R., F. Gasparini, and R. C. Lanza. (2001b). A coded aperture for high-resolution nuclear medicine planar imaging with a conventional Anger camera: Experimental results. *IEEE Trans. Nucl. Sci.* **48**(6): 2411–2417.

Aguiar, P., J. Silva-Rodriguez, M. Herranz, and A. Ruibal. (2014). Preliminary experience with small animal SPECT imaging on clinical gamma cameras. *Biomed. Res. Int.* **2014**: 369509.

---

* Actually, before the widespread use of clinical PET, low-resolution, single-photon imaging of positron emitters was utilized, so this is in fact a return to previous techniques.

Barrett, H., W. C. J. Hunter, B. Miller, S. Moore, Y. Chen, and L. Furenlid. (2009). Maximum-likelihood methods for processing signals from gamma-ray detectors. *IEEE Trans. Nucl. Sci.* **56**(3): 725–735.

Barrett, H. H. (1972). Fresnel zone plate imaging in nuclear medicine. *J. Nucl. Med.* **13**(6): 382–385.

Barrett, H. H., L. R. Furenlid, M. Freed, J. Y. Hesterman, M. A. Kupinski, E. Clarkson, and M. K. Whitaker. (2008). Adaptive SPECT. *IEEE Trans. Med. Imaging* **27**(6): 775–788.

Beekman, F. and F. van der Have. (2007). The pinhole: Gateway to ultra-high-resolution three-dimensional radionuclide imaging. *Eur. J. Nucl. Med. Mol. Imaging* **34**(2): 151–161.

Beekman, F., F. van der Have, B. Vastenhouw, A. Van Der Linden, P. Van Rijk, J. Burbach, and M. Smidt. (2005). U-SPECT-I: A novel system for submillimeter-resolution tomography with radiolabeled molecules in mice. *J. Nucl. Med.* **46**(7): 1194–1200.

Beekman, F. and B. Vastenhouw. (2004). Design and simulation of a high-resolution stationary SPECT system for small animals. *Phys. Med. Biol.* **49**(19): 4579–4592.

Boisson, F., D. Zahra, A. Parmar, M.-C. Gregoire, S. Meikle, H. Hamse, and A. Reilhac. (2013). Imaging capabilities of the Inveon SPECT system using single-and multipinhole collimators. *J. Nucl. Med.* **54**(10): 1833– 1840.

Chang, L. T., S. N. Kaplan, B. Macdonald, V. Perez-Mendez, and L. Shiraishi. (1974). A method of tomographic imaging using a multiple pinhole-coded aperture. *J. Nucl. Med.* **15**(11): 1063–1065.

Cherry, S. R., J. A. Sorenson, and M. E. Phelps. (2012). *Physics in Nuclear Medicine*, 4th edn. Elsevier Saunders, Philadelphia, PA.

Chrastina, A. and J. E. Schnitzer. (2010). Iodine-125 radiolabeling of silver nanoparticles for in vivo SPECT imaging. *Int. J. Nanomed.* **5**: 653–659.

Cuccurullo, V., G. L. Cascini, O. Tamburrini, A. Rotondo, and L. Mansi. (2013). Bone metastases radiopharmaceuticals: An overview. *Curr. Radiopharm.* **6**(1): 41–47.

Culter, C. S., H. M. Hennkens, N. Sisay, S. Huclier-Markai, and S. Jurison. (2013). Radiometals for combined imaging and therapy. *Chem. Rev.* **113**(2): 858–883.

Culver, J., W. Akers, and S. Achilefu. (2008). Multimodality molecular imaging with combined optical and SPECT/PET modalities. *J. Nucl. Med.* **49**(2): 169–172.

Dahlbom, M. (2014). *Physics of PET and SPECT Imaging*. CRC Press, Boca Raton, FL.

De Beenhouwer, J., S. Staelens, S. Vandenberghe, and I. Lemahieu. (2008). Acceleration of GATE SPECT simulations. *Med. Phys.* **35**(4): 1476–1485.

De Vree, G., F. Van Der Have, and F. Beekman. (2004). EMCCD-based photon-counting mini gamma camera with a spatial resolution <100 /spl mu/m. **5**: 2724–2728.

De Vree, G., A. Westra, I. Moody, F. Van Have, K. Ligtvoet, and F. Beekman. (2005). Photon-counting gamma camera based on an electron-multiplying CCD. *IEEE Trans. Nucl. Sci.* **52**(3 I): 580–588.

Deleye, S., R. Van Holen, J. Verhaeghe, S. Vandenberghe, S. Stroobants, and S. Staelens. (2013). Performance evaluation of small-animal multipinhole μSPECT scanners for mouse imaging. *Eur. J. Nucl. Med. Mol. Imaging* **40**(5): 744–758.

DiFilippo, F. P. (2008). Geometric characterization of multi-axis multi-pinhole SPECT. *Med. Phys.* **35**(1): 181–194.

Durkee, J. W., P. P. Antich et al. (1998a). SPECT electronic collimation resolution enhancement using chi-square minimization. *Phys. Med. Biol.* **43**(10): 2949– 2974.

Durkee, J. W., P. P. Antich et al. (1998b). Analytic treatment of resolution precision in electronically collimated SPECT imaging involving multiple-interaction gamma rays. *Phys. Med. Biol.* **43**(10): 2975–2990.

Fessler, J. (1998). Spatial resolution and noise tradeoffs in pinhole imaging system design: A density estimation approach. *Opt. Expr.* **2**(6): 237–253.

Flower, M. A. (2012). *Webb's Physics of Medical Imaging*. Taylor & Francis, Boca Raton, FL.

Freed, M., M. A. Kupinski, L. R. Furenlid, D. W. Wilson, and H. H. Barrett. (2008). A prototype instrument for single pinhole small animal adaptive SPECT imaging. *Med. Phys.* **35**(5): 1912–1925.

Funk, T., P. Despres, W. C. Barber, K. S. Shah, and B. H. Hasegawa. (2006). A multipinhole small animal SPECT system with submillimeter spatial resolution. *Med. Phys.* **33**(5): 1259–1268.

Furenlid, L., D. Wilson, Y.-C. Chen, H. Kim, P. Pietraski, M. Crawford, and H. Barrett. (2004). FastSPECT II: A second-generation high-resolution dynamic SPECT imager. *IEEE Trans. Nucl. Sci.* **51**(3 II): 631–635.

Gallas, B. and H. Barrett. (1998). Modeling all orders of scatter in nuclear medicine. *Nucl. Sci. Symp. Conf. Rec.* **3**: 1964–1968.

Gnanasegaran, G. and J. R. Ballinger. (2014). Molecular imaging agents for SPECT (and SPECT/CT). *Eur. J. Nucl. Med. Mol. Imaging* **41**(Suppl 1): 26–35.

Gould, T. J., S. T. Hess, and J. Bewersdorf. (2012). Optical nanoscopy: From acquisition to analysis. *Annu. Rev. Biomed. Eng.* **14**: 231–254.

Gunter, D. L. (2004). Collimator design for nuclear medicine. In M. N. Wernick and J. N. Aarvold (eds.), *Emission Tomography: The Fundamentals of PET and SPECT*. Elsevier Academic Press, Boston, MA, Chapter 8.

Habraken, J. B., K. de Bruin, M. Shehata, J. Booij, R. Bennink, B. L. van Eck Smit, and E. Busemann Sokole. (2001). Evaluation of high-resolution pinhole SPECT using a small rotating animal. *J. Nucl. Med.* **42**(12): 1863–1869.

Hamamura, M. J., S. Ha, W. W. Roeck, L. T. Muftuler, D. J. Wagenaar, D. Meier, B. E. Patt, and O. Nalcioglu. (2010). Development of an MR-compatible SPECT system (MRSPECT) for simultaneous data acquisition. *Phys. Med. Biol.* **55**(6): 1563–1575.

Heemskerk, J., A. Westra, P. Linotte, K. Ligtvoet, W. Zbijewski, and F. Beekman. (2007). Front-illuminated versus back-illuminated photon-counting CCD-based gamma camera: Important consequences for spatial resolution and energy resolution. *Phys. Med. Biol.* **52**(8): N149–N162.

Hell, S. W. (2003). Toward fluorescence nanoscopy. *Nat. Biotechnol.* **21**(11): 1347–1355.

Hijnen, N. M., A. de Vries, K. Nicolay, and H. Grull. (2012). Dual-isotope 111In/177Lu SPECT imaging as a tool in molecular imaging tracer design. *Contrast Media Mol. Imaging* **7**(2): 214–222.

Holland, J. P., M. J. Williamson, and J. S. Lewis. (2010). Unconventional nuclides for radiopharmaceuticals. *Mol. Imaging* **9**(1): 1–20.

Huang, Q. and G. L. Zeng. (2006). An analytical algorithm for skew-slit imaging geometry with nonuniform attenuation correction. *Med. Phys.* **33**(4): 997–1004.

Hynecek, J. (2001). Impactron: A new solid state image intensifier. *IEEE Trans. Electron Dev.* **48**(10): 2238–2241.

Hynecek, J. and T. Nishiwaki. (2003). Excess noise and other important characteristics of low light level imaging using charge multiplying CCDs. *IEEE Trans. Electron Dev.* **50**(1): 239–245.

IMV. (2014). Nm providers "consolidating" as SPECT procedures decline. *J. Nucl. Med.* **55**(2): 16N.

Ishizu, K., T. Mukai et al. (1995). Ultra-high resolution SPECT system using four pinhole collimators for small animal studies. *J. Nucl. Med.* **36**(12): 2282–2289.

Israel-Jost, V., P. Choquet, S. Salmon, C. Blondet, E. Sonnendrucker, and A. Constantinesco. (2006). Pinhole SPECT imaging: Compact projection/back projection operator for efficient algebraic reconstruction. *IEEE Trans. Med. Imaging* **25**(2): 158–167.

Itaya, G. and K. Kojima. (1978). Resolutions in imaging with Fresnel zone plate. *Radioisotopes* **27**(1): 14–19.

Ivashchenko, O., F. van der Have, J. L. Villena, H. C. Groen, R. M. Ramakers, H. H. Weinans, and F. J. Beekman. (2014). Quarter-millimeter-resolution molecular mouse imaging with U-SPECT⁺. *Mol. Imaging* **13**: 1–8.

Jan, S., D. Benoit et al. (2011). GATE V6: A major enhancement of the GATE simulation platform enabling modelling of CT and radiotherapy. *Phys. Med. Biol.* **56**(4): 881–901.

Jan, S., G. Santin et al. (2004). GATE: A simulation toolkit for PET and SPECT. *Phys. Med. Biol.* **49**(19): 4543–4561.

Jaszczak, R. J., J. Li, H. Wang, M. R. Zalutsky, and R. E. Coleman. (1994). Pinhole collimation for ultra-high-resolution, small-field-of-view SPECT. *Phys. Med. Biol.* **39**(3): 425–437.

Khalil, M. M., J. L. Tremoleda, T. B. Bayomy, and W. Gsell. (2011). Molecular SPECT imaging: An overview. *Int. J. Mol. Imaging* **2011**: 796025.

Klein, W., H. Barrett, I. Pang, D. Patton, M. Rogulski, J. Sain, and W. Smith. (1995). FastSPECT: Electrical and mechanical design of a high-resolution dynamic SPECT imager. **2**: 931–933.

Lee, K. S., W. W. Roeck, G. T. Gullberg, and O. Nalcioglu. (2011). MR-based keyhole SPECT for small animal imaging. *Phys. Med. Biol.* **56**(3): 685–702.

Lee, S., J. Gregor, and D. Osborne. (2013). Development and validation of a complete GATE model of the Siemens Inveon trimodal imaging platform. *Mol. Imaging* **12**(7): 1–13.

Lees, E., J. Bassford, E. Blake, E. Blackshaw, and C. Perkins. (2011). A high resolution small field of view (SFOV) gamma camera: A columnar scintillator coated CCD imager for medical applications. *J. Instrum.* **6**(12): C12033.

Lees, J. E., G. W. Fraser, A. Keay, D. Bassford, R. Ott, and W. Ryder. (2003). The high resolution gamma imager HRGI: A CCD based camera for medical imaging. *Nucl. Instrum. Methods Phys. Res.* **513**(1–2): 23–26.

Lin, H. H., K. S. Chuang, Y. H. Lin, Y. C. Ni, J. Wu, and M. L. Jan. (2014). Efficient simulation of voxelized phantom in GATE with embedded SimSET multiple photon history generator. *Phys. Med. Biol.* **59**(20): 6231–6250.

Liu, H., A. Karellas, L. J. Harris, and C. J. D'Orsi. (1994). Methods to calculate the lens efficiency in optically coupled CCD x-ray imaging systems. *Med. Phys.* **21**(7): 1193–1195.

Liu, Y., G. Yu, M. Tian, and H. Zhang. (2011). Optical probes and the applications in multimodality imaging. *Contrast Media Mol. Imaging* **6**(4): 169–177.

Lu, Y., K. Yang et al. (2014). An integrated quad-modality molecular imaging system for small animals. *J. Nucl. Med.* **55**(8): 1375–1379.

Lutje, S., M. Rijpkema et al. (2014). Pretargeted dual-modality immuno-SPECT and near-infrared fluorescence imaging for image-guided surgery of prostate cancer. *Cancer Res.* **74**(21): 6216–6223.

Maidment, A. D. and M. J. Yaffe. (1995). Analysis of signal propagation in optically coupled detectors for digital mammography: I. Phosphor screens. *Phys. Med. Biol.* **40**(5): 877–889.

Maidment, A. D. and M. J. Yaffe. (1996). Analysis of signal propagation in optically coupled detectors for digital mammography: II. Lens and fibre optics. *Phys. Med. Biol.* **41**(3): 475–493.

Majo, V. J., J. Prabhakaran, J. J. Mann, and J. S. Kumar. (2013). PET and SPECT tracers for glutamate receptors. *Drug Discov. Today* **18**(3–4): 173–184.

Matsunari, I., Y. Miyazaki, M. Kobayashi, K. Nishi, A. Mizutani, K. Kawai, A. Hayashi, R. Komatsu, S. Yonezawa, and S. Kinuya. (2014). Performance evaluation of the eXplore speCZT preclinical imaging system. *Ann. Nucl. Med.* **28**(5): 484–497.

McConathy, J., W. Yu, N. Jarkas, W. Seo, D. M. Schuster, and M. M. Goodman. (2012). Radiohalogenated non-natural amino acids as PET and SPECT tumor imaging agents. *Med. Res. Rev.* **32**(4): 868–905.

McElroy, D., L. MacDonald, F. Beekman, Y. Wang, B. Patt, J. Iwanczyk, B. M. W. Tsui, and E. Hoffman. (2002). Performance evaluation of aspect: A high resolution desktop pinhole SPECT system for imaging small animals. *IEEE Trans. Nucl. Sci.* **49**(5): 2139–2147.

Meier, D., D. J. Wagenaar, S. Chen, J. Xu, J. Yu, and B. M. Tsui. (2011). A SPECT camera for combined MRI and SPECT for small animals. *Nucl. Instrum. Methods Phys. Res. A* **652**(1): 731–734.

Meikle, S., R. Fulton, S. Eberl, M. Dahlbom, K.-P. Wong, and M. Fulham. (2001). An investigation of coded aperture imaging for small animal SPECT. *IEEE Trans. Nucl. Sci.* **48**(3): 816–821.

Meng, L., N. Clinthorne, S. Skinner, R. Hay, and Gross. (2006). Design and feasibility study of a single photon emission microscope system for small animal 1–125 imaging. *IEEE Trans. Nucl. Sci.* **53**(3): 1168–1178.

Meng, L., G. Fu, E. Roy, B. Suppe, and C. Chen. (2009). An ultrahigh resolution SPECT system for I-125 mouse brain imaging studies. *Nucl. Instrum. Methods Phys. Res.* **600**(2): 498–505.

Meng, L., W. Rogers, N. Clinthorne, and J. Fessler. (2003). Feasibility study of Compton scattering enhanced multiple pinhole imager for nuclear medicine. *IEEE Trans. Nucl. Sci.* **50**(5): 1609–1617.

Metzler, S. D., R. Accorsi, A. S. Ayan, and R. J. Jaszczak. (2010). Slit-slat and multislit-slat collimator design and experimentally acquired phantom images from a rotating prototype. *IEEE Trans. Nucl. Sci.* **57**(1): 125–134.

Miller, B., H. Barber, H. Barrett, L. Chen, and S. Taylor. (2007). Photon-counting gamma camera based on columnar CsI(Tl) optically coupled to a back-illuminated CCD. *Proc. SPIE Int. Soc. Opt. Eng.* **6510**: 65100N.

Miller, B., H. Barrett, L. Furenlid, H. Bradford Barber, and R. Hunter. (2008). Recent advances in BazookaSPECT: Real-time data processing and the development of a gamma-ray microscope. *Nucl. Instrum. Methods Phys. Res.* **591**(1): 272–275.

Miller, B., L. Furenlid, S. Moore, H. Barber, V. Nagarkar, and H. Barrett. (2009). System integration of FastSPECT III: A dedicated SPECT rodent-brain imager based on BazookaSPECT detector technology. *IEEE Nucl. Sci. Symp. Conf. Rec.* **2009**: 4004–4008.

Mitchell, G. S. and S. R. Cherry. (2009). A high-sensitivity small animal SPECT system. *Phys. Med. Biol.* **54**(5): 1291–1305.

Momennezhad, M., R. Sadeghi, and S. Nasseri. (2012). Development of GATE Monte Carlo simulation for a dual-head gamma camera. *Radiol. Phys. Technol.* **5**(2): 222–228.

Mukherjee, J. M., E. Sidky, and M. A. King. (2014). Estimation of sparse null space functions for compressed sensing in SPECT. *Proc. SPIE* **9033**: 9033 0X.

Nagarkar, V., I. Shestakova, V. Gaysinskiy, B. Singh, B. Miller, and H. Bradford Barber. (2006). Fast x-ray-/ray imaging using electron multiplying CCD-based detector. *Nucl. Instrum. Methods Phys. Res.* **563**(1): 45–48.

Nowotny, R. (1998). XMuDat: Photon attenuation data on PC. Tech. Rep. IAEA-NDS-95, International Atomic Energy Agency, Vienna, Austria.

Pani, R., R. Pellegrini et al. (2004). New devices for imaging in nuclear medicine. *Cancer Biother. Radiopharm.* **19**(1): 121–128.

Peterson, M., K. Ljunggren, L. Andersson-Ljus, B. Miller, and S.-E. Strand. (2010). A method for using high density fusible rose's metal with high precision machining in small animal imaging applications. *Nucl. Sci. Symp. Conf. Rec.* **2010**: 3155–3157.

Qi, J. and R. M. Leahy. (2006). Iterative reconstruction techniques in emission computed tomography. *Phys. Med. Biol.* **51**(15): R541–R578.

Rentmeester, M. C., F. van der Have, and F. J. Beekman. (2007). Optimizing multi-pinhole SPECT geometries using an analytical model. *Phys. Med. Biol.* **52**(9): 2567–2581.

Saha, G. B., W. J. MacIntyre, and R. T. Go. (1994). Radiopharmaceuticals for brain imaging. *Semin. Nucl. Med.* **24**(4): 324–349.

Sanchez, F., A. Orero et al. (2013). ALBIRA: A small animal PET-SPECTCT imaging system. *Med. Phys.* **40**(5): 051906.

Schellingerhout, D., R. Accorsi, U. Mahmood, J. Idoine, R. C. Lanza, and R. Weissleder. (2002). Coded aperture nuclear scintigraphy: A novel small animal imaging technique. *Mol. Imaging* **1**(4): 344–353.

Schramm, N., G. Ebel, U. Engeland, T. Schurrat, M. Behe, and T. M. Behr. (2003). High-resolution SPECT using multi-pinhole collimation. *IEEE Trans. Nucl. Sci.* **50**(3): 315–320.

Smith, M., C. E. J. Floyd, R. Jaszczak, and R. Coleman. (1992). Reconstruction of SPECT images using generalized matrix inverses. *IEEE Trans. Med. Imaging* **11**(2): 165–175.

Soesbe, T., M. Lewis, E. Richer, N. Slavine, and P. Antich. (2007). Development and evaluation of an EMCCD based gamma camera for preclinical SPECT imaging. *IEEE Trans. Nucl. Sci.* **54**(5): 1516–1524.

Soesbe, T., M. Lewis, N. Slavine, E. Richer, F. Bonte, and P. Antich. (2010). High-resolution photon counting using a lens-coupled EMCCD gamma camera. *IEEE Trans. Nucl. Sci.* **57**(3): 958–963.

Spinks, T. J., D. Karia, M. O. Leach, and G. Flux. (2014). Quantitative PET and SPECT performance characteristics of the Albira Trimodal pre-clinical tomograph. *Phys. Med. Biol.* **59**(3): 715–731.

Tang, Q., G. L. Zeng, and Q. Huang. (2006). An analytical algorithm for skew-slit collimator SPECT with uniform attenuation correction. *Phys. Med. Biol.* **51**(23): 6199–6211.

Tenney, C. (2000). Gold pinhole collimators for ultra-high resolution Tc-99m small volume SPECT. *Nucl. Sci. Symp. Conf. Rec.* **3**: 22/44–22/46.

Tenney, C. (2001). Optimizing gold and platinum pinhole collimators for imaging of small volumes at ultra-high resolution. *Nucl. Sci. Symp. Conf. Rec.* **3**: 1597–1599.

Tenney, C., M. Smith, K. Greer, and R. Jaszczak. (1999). Uranium pinhole collimators for I-131 SPECT imaging. *IEEE Trans. Nucl. Sci.* **46**(4): 1165–1171.

Tenney, C., M. Tornai, M. Smith, T. Turkington, and R. Jaszczak. (2001). Uranium pinhole collimators for 511-kev photon SPECT imaging of small volumes. *IEEE Trans. Nucl. Sci.* **48**(4): 1483–1489.

Tian, J., J. Bai, X. P. Yan, S. Bao, Y. Li, W. Liang, and X. Yang. (2008). Multimodality molecular imaging. *IEEE Eng. Med. Biol. Mag.* **27**(5): 48–57.

Toomre, D. and J. Bewersdorf. (2010). A new wave of cellular imaging. *Annu. Rev. Cell Dev. Biol.* **26**: 285–314.

Vaissier, P., M. Goorden, B. Vastenhouw, F. Van Der Have, R. Ramakers, and F. Beekman. (2012). Fast spiral SPECT with stationary-cameras and focusing pinholes. *J. Nucl. Med.* **53**(8): 1292–1299.

Van Audenhaege, K., C. Vanhove, S. Vandenberghe, and R. Van Holen. (2014). The evaluation of data completeness and image quality in multiplexing multi-pinhole SPECT. *IEEE Trans. Med. Imaging* **34**(2): 474–486.

van der Have, F. and F. J. Beekman. (2004). Photon penetration and scatter in micro-pinhole imaging: A Monte Carlo investigation. *Phys. Med. Biol.* **49**(8): 1369–1386.

Van Have, F., B. Vastenhouw, R. Ramakers, W. Branderhorst, J. Krah, C. Ji, S. Staelens, and F. Beekman. (2009). U-SPECT-II: An ultra-high-resolution device for molecular small-animal imaging. *J. Nucl. Med.* **50**(4): 599–605.

Van Holen, R., B. Vandeghinste, K. Deprez, and S. Vandenberghe. (2013). Design and performance of a compact and stationary microSPECT system. *Med. Phys.* **40**(11): 112501.

van Oosterom, M. N., R. Kreuger, T. Buckle, W. A. Mahn, A. Bunschoten, L. Josephson, F. W. van Leeuwen, and F. J. Beekman. (2014). U-SPECT-BioFluo: An integrated radionuclide, bioluminescence, and fluorescence imaging platform. *EJNMMI Res.* **4**: 56.

Vandeghinste, B., R. Van Holen, C. Vanhove, F. De Vos, S. Vandenberghe, and S. Staelens. (2014). Use of a ray-based reconstruction algorithm to accurately quantify preclinical microSPECT images. *Mol. Imaging* **13**: 1–13.

Vandenberghe, S., Y. D'Asseler, R. Van de Walle, T. Kauppinen, M. Koole, L. Bouwens, K. Van Laere, I. Lemahieu, and R. A. Dierckx. (2001). Iterative reconstruction algorithms in nuclear medicine. *Comput. Med. Imaging Graph* **25**(2): 105–111.

Vastenhouw, B. and F. Beekman. (2007). Submillimeter total-body murine imaging with U-SPECT-I. *J. Nucl. Med.* **48**(3): 487–493.

Vaughan, C. L. (2008). *Imagining the Elephant: A Biography of Allan MacLeod Cormack*. Imperial College Press, London, U.K.

Vieira, L., T. F. Vaz, D. C. Costa, and P. Almeida. (2014). Monte Carlo simulation of the basic features of the GE Millennium MG single photon emission computed tomography gamma camera. *Rev. Esp. Med. Nucl. Imagen Mol.* **33**(1): 6–13.

Wadas, T. J., E. H. Wong, G. R. Weisman, and C. J. Anderson. (2010). Coordinating radiometals of copper, gallium, indium, yttrium, and zirconium for PET and SPECT imaging of disease. *Chem. Rev.* **110**(5): 2858–2902.

Wagenaar, D., J. Zhang, T. Kazules, T. Vandehei, E. Bolle, S. Chowdhury, K. Parnham, and B. Patt. (2006). In vivo dual-isotope SPECT imaging with improved energy resolution. *Nucl. Sci. Symp. Conf. Rec.* **6**: 3821–3826.

Walker, K. L., M. S. Judenhofer, S. R. Cherry, and G. S. Mitchell. (2015). Un-collimated single-photon imaging system for high-sensitivity small animal and plant imaging. *Phys. Med. Biol.* **60**(1): 403–420.

Walker, M. D., M. C. Goorden, K. Dinelle, R. M. Ramakers, S. Blinder, M. Shirmohammad, F. van der Have, F. J. Beekman, and V. Sossi. (2014). Performance assessment of a preclinical PET scanner with pinhole collimation by comparison to a coincidence-based small-animal PET scanner. *J. Nucl. Med.* **55**(8): 1368–1374.

Wang, G., J. Zhang et al. (2012). Towards omni-tomography–grand fusion of multiple modalities for simultaneous interior tomography. *PLoS One* **7**(6): e39700.

Weisenberger, A., R. Wojcik et al. (2003). SPECT-CT system for small animal imaging. *IEEE Trans. Nucl. Sci.* **50**(1): 74–79.

Wernick, M. N. and J. N. Aarsvold. (2004). *Emission Tomography: The Fundamentals of PET and SPECT*. Elsevier Academic Press, Amsterdam, the Netherlands.

Williams, M. B., M. B. Williams, A. R. Goode, V. Galbis-Reig, S. Majewski, A. G. Weisenberger, and R. Wojcik. (2000). Performance of a PSPMT based detector for scintimammography. *Phys. Med. Biol.* **45**(3): 781–800.

Wilson, D., H. Barrett, and E. Clarkson. (2000). Reconstruction of two- and three-dimensional images from synthetic-collimator data. *IEEE Trans. Med. Imaging* **19**(5): 412–422.

Wolf, P., E. Sidky, and T. Schmidt. (2011). A compressed sensing algorithm for sparse-view pinhole single photon emission computed tomography. *Nucl. Sci. Symp. Med. Imaging Conf.* **2012**: 2668–2671.

Yamamoto, S.,H. Horii, M. Hurutani, K. Matsumoto, and M. Senda. (2005). Investigation of single, random, and true counts from natural radioactivity in LSO-based clinical PET. *Ann. Nucl. Med.* **19**(2): 109–114.

Yu, T. and J. M. Boone. (1997). Lens coupling efficiency: Derivation and application under differing geometrical assumptions. *Med. Phys.* **24**(4): 565–570.

Zhang, Y., J. P. Bressler, J. Neal, B. Lal, H. E. Bhang, J. Laterra, and M. G. Pomper. (2007). ABCG2/BCRP expression modulates D-Luciferin based bioluminescence imaging. *Cancer Res.* **67**(19): 9389–9397.

Zhu, X., T. Cullip, G. Tracton, X. Tang, J. Lian, J. Dooley, and S. X. Chang. (2014). Direct aperture optimization using an inverse form of back-projection. *J. Appl. Clin. Med. Phys.* **15**(2): 4545.

8

Positron Emission Tomography

Vesna Sossi and Matthew Walker

8.1   INTRODUCTION TO POSITRON EMISSION TOMOGRAPHY

A positron emission tomography (PET) scanner is designed to provide a measurement of the spatial distribution of a PET radiotracer within a subject. Measurements can be made in a time series to provide a "movie" showing quantitatively how this distribution changes over time. Such change is related to the underlying biology under investigation. PET uses the tracer principle: a compound is tagged with a radionuclide and a small (trace) amount is injected. Because it is a trace amount, it does not affect the biological system in any way (e.g., no saturation of binding sites and no clinical effects), thus allowing the measurement of the system in its natural state. Many of the PET radionuclides are isotopes of elements naturally present in the body (carbon, oxygen, etc.), thus allowing for their easy incorporation into biologically relevant compounds. The ultimate results of positron emission are two high-energy annihilation photons. These high-energy (511 keV) photons can easily penetrate through the body of the animal with minimal interactions and can be detected in coincidence. This chapter describes the physics of PET together with a short overview of data quantification and reconstruction and data analysis and interpretation.

8.2   PHYSICS PRINCIPLES OF PET

8.2.1   Coincidence Detection and Photon Interaction with Matter

PET utilizes radiotracers labeled with positron-emitting radionuclides designed to selectively target specific processes or sites of interest. The most commonly used radionuclides are cyclotron-produced ^{18}F ($t_{1/2}$ = 109.8 min), ^{11}C ($t_{1/2}$ = 20.4 min), ^{15}O ($t_{1/2}$ = 2.04 min), and ^{13}N ($t_{1/2}$ = 9.97 min). Increasing research efforts are being devoted to the development of tracers labeled with positron-emitting radiometals such as ^{64}Cu or ^{89}Zr (Zhang et al. 2011). Some of these radionuclides have relatively long half-lives that make them easily transportable and suitable for the investigation of biological processes that take place over hours or days as opposed to minutes (Wadas et al. 2010); others such as ^{82}Rb and ^{68}Ga are produced in generator systems that allow the on-site production of the tracer without the need for cyclotron-related infrastructure (Epstein et al. 2004; Ambrosini et al. 2011). A disadvantage from the imaging and dosimetry point of view is that many of these radiometals have more complex decay schemes resulting in unnecessary additional radiation dose and image background hampering accurate image quantification (Liu and Laforest 2009).

Positron emission is a nuclear decay process with three decay products (daughter nucleus, positron, and neutrino), which allows the positron (antimatter equivalent of the electron) to be emitted with a range of possible energies. After emission, the positron travels through its surrounding medium undergoing a series of collisions with nearby electrons; these collisions change the direction of the positron and decelerate it, until the positron has lost sufficient energy for a positron–electron annihilation to take place (Figure 8.1). The annihilation process produces a pair of 511 keV annihilation photons that travel in almost opposite directions. The distance (range) traveled by the positron prior to its annihilation depends on the material density and its effective atomic number and on the positron's initial kinetic energy, which is radionuclide dependent (Levin and Hoffman 1999). As an example, among all useful positron emitters, ^{18}F nuclei emit positrons with the smallest initial maximum kinetic energy of 0.63 MeV resulting into a range of less than 1 mm in tissues (Levin and Hoffman 1999). The range of positrons emitted from ^{82}Rb, on the other hand, is approximately 3.5 mm. As the PET scanner effectively measures the distribution of positron–electron annihilations

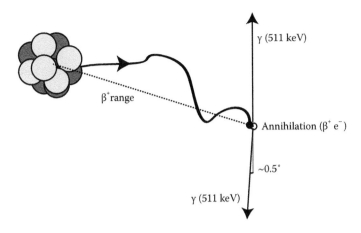

FIGURE 8.1 Depiction of positron decay, followed by positron–electron annihilation and the production of a pair of (almost) colinear annihilation photons (image not to scale).

and not positron emissions, which originate at the radiotracer location, the positron range negatively affects the PET spatial resolution in a radionuclide-dependent manner.

A valid PET event is comprised of the detection of the two annihilation photons within a temporal interval (coincidence window) by two different, opposing detectors. The definition of detector pairs that can accept a coincident event defines the imaging field of view (FOV) (Figure 8.2a). For each event, spatial information about the point of positron–electron annihilation is gained by recording the coordinates of each detector in the pair. The line connecting the two detectors is termed the line of response (LOR); it is known that the positron annihilation occurred at some point along this line (or close to it). In small animal PET, where subjects being imaged are typically less than 5 cm in diameter, the timing resolution of the detectors is insufficient to add useful information about the location of the positron–electron annihilation along the LOR by measuring the difference in the arrival times of the annihilation photons at the detectors. This is unlike human PET imaging where the so-called time-of-flight PET is now relatively common (Joshua et al. 2013).

LORs, which are at the same angle with respect to a predetermined angular offset, are grouped into angular projections, and for each detected event along an LOR, the corresponding projection bin value is incremented by one count (Figure 8.2b). An image of the radioactivity distribution can

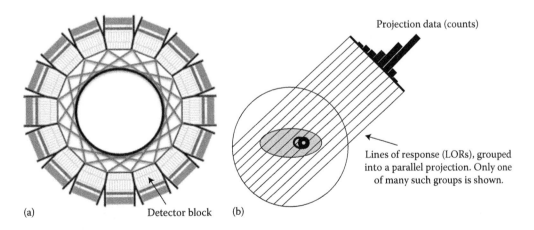

FIGURE 8.2 (a) Schematics of the field of view (FOV) definition. The edge lines of response (LORs) for sample projections (angular views) are shown and the inner black line indicates the FOV. The size of the FOV depends on the minimum distance allowed between two detector pairs that can register a coincidence event. In this figure, the ends of a valid LOR must be separated by at least four detector blocks (comprised by a crystal matrix). (b) A subset of the LORs between two edge LORs shown in Figure 8.2a are grouped to form a parallel projection through the object. In a small animal PET scanner, the LOR spacing is about 10 times finer than shown (e.g., 1 mm).

be generated from the count information simultaneously collected at several angular views, that is, from the many sets of parallel LORs contained within each projection (see Section 8.3.1).

Two physics-related factors provide an intrinsic limit to the spatial resolution. First is the afore-mentioned positron range. Methods to correct for the blurring effect of the positron range are the subject of ongoing investigation (Jødal et al. 2012). Modern small animal PET scanners have detectors that provide sufficiently high spatial resolution for the positron range–derived image degradation to be noticeable. Second, since the net momentum of the positron and electron is nonzero when they annihilate, the two emitted 511 keV photons are not exactly colinear (the deviation being 0.5° on average). The magnitude of the resulting blurring is directly proportional to the distance between detectors; fortunately, this is relatively small for small animal scanners (<0.2 mm blur for a 15 cm diameter scanner).

8.2.2   IMPACT OF PHOTON INTERACTION WITH MATTER ON DATA QUANTIFICATION AND IMAGING RESOLUTION

The two annihilation photons must exit the object and reach and interact in the detector for an event to be acquired. If the pair of annihilation photons does not interact within the subject during their passage to the detectors, and if they are both detected, a "true" coincidence event is recorded. While traversing material, 511 keV photons can, however, undergo two types of interactions: Compton scattering and photoelectric absorption (Cherry 2006). Of these, Compton scattering is the dominating effect for 511 keV photons in biological tissues. Compton scattering describes the interaction of a high-energy photon with an orbital electron, which results in a lower-energy photon emitted at a scattering angle θ. The Klein–Nishina formula provides the differential cross section (dσ/dΩ). for Compton scattering:

$$\frac{d\sigma}{d\Omega} = \alpha^2 r_c^2 P(E_\gamma, \theta)^2 [P(E_\gamma, \theta) + P(E_\gamma, \theta)^{-1} - 1 + \cos^2(\theta)]/2 \tag{8.1}$$

where

dσ is the reaction cross-section element

dΩ is the solid-angle element

α is the fine structure constant (~1/137.04)

$r_c = \dfrac{\hbar}{m_e c}$ is the "reduced" Compton wavelength of the electron

E_γ is the energy of the incoming photon

m_e is the mass of an electron (~511 keV/c^2)

$P(E_\gamma, \theta)$ is the ratio between the energy of the outgoing and incoming photons (after and before the collision):

$$P(E_\gamma, \theta) = \frac{1}{1 + (E_\gamma / m_e c^2)(1 - \cos(\theta))} \tag{8.2}$$

From consideration of Equations 8.1 and 8.2, it can be shown that for 511 keV photons, scatter through smaller angles (<45°) is favored. The photoelectric effect, in which the photon is completely absorbed, has an almost negligible probability of occurrence at this energy in tissue (Figure 8.3). In PET, the Compton interaction can give rise to two possible outcomes. First, the scattered photon can be detected by the scanner in coincidence with its partner annihilation photon, and the event is accepted and assigned to an incorrect LOR (Figure 8.4). This erroneous event is classified as a scattered coincidence. The LOR to which the event was assigned no longer passes through the location of the positron–electron annihilation; as a result of the scattering, the opportunity to measure a true coincidence is lost and in its place a scattered coincidence is recorded. Even though such events do

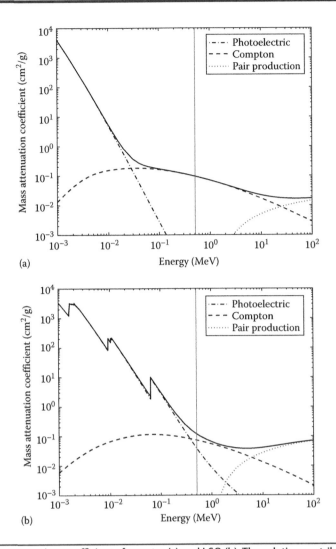

FIGURE 8.3 Mass attenuation coefficients for water (a) and LSO (b). The relative contribution of each interaction mode (photoelectric, Compton, and pair production) is shown separately; the energy of annihilation photons (511 keV) is marked with a straight vertical line.

not contribute to an overall count loss, they carry wrong information and must be subtracted or otherwise accounted for during data processing. In the second case, no coincidence event is recorded, either because the scattered photon was not detected (e.g., the new trajectory did not intersect the detectors) or because the scattered (and hence lower-than-511-keV-energy) photon was detected but failed to satisfy the constraints of the scanner's energy window settings (typically 450–650 keV). Both cases lead to a reduction in the true ("good") signal, which is termed attenuation. The magnitude of scatter and attenuation depends on the size and density of the object being imaged and on the geometry of the scanner. Increasing the solid angle subtended by the scanner, and/or increasing the width of the energy window, leads to a higher chance that a scattered photon will form a scattered coincidence. It is thus understandable why the scatter fraction (SF), typically defined as the ratio between the number of scattered events (S) and total true coincidence events (T), S/T, can be quite high in small animal imaging. Although the animal body may be small, the detectors cover a large solid angle. The SF can be as high as 30% in some scanners when imaging rats (Goertzen et al. 2012). This is comparable to the SF obtained in human brain PET imaging when using a whole body scanner (Olivier et al. 2005). A minor but interesting subtle point is the fact the detector thickness can also influence the SF. Since the scattered photons have a lower energy than the unscattered 511 keV photon, they require less detector material to interact: thinner detectors will thus preferentially detect scattered events compared to thicker detectors.

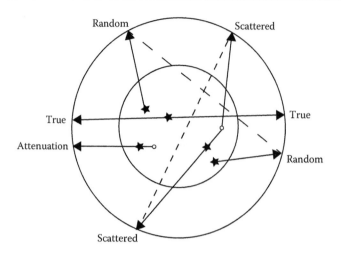

FIGURE 8.4 Different types of coincidence counts and attenuation of coincidences. A true coincidence event is shown, where the line passes through the point of positron–electron annihilation depicted by the star. A scattered coincidence event is shown, where one of the two photons underwent Compton scatter (at circle), but was still detected to form a scattered coincidence, with an incorrect LOR (short-dashed line). A random coincidence is shown, where two detected photons originated from two different annihilations that occurred within the coincidence timing window, leading to an event assigned to an LOR along which no annihilation took place (long-dashed line). Also shown is the problem of attenuation, where one of the two photons underwent Compton scatter but was not detected; no coincidence is formed.

In addition to true and scattered coincidences, random (also termed accidental) coincidence events may be measured. These "randoms" occur when two 511 keV photons originating from two separate annihilations are detected in temporal coincidence—their partner annihilation photons not being detected. The scanner, unknowingly, thus assigns the event to an LOR along which no annihilation had occurred. Random events, to be contrasted to true coincidence events, or "trues," are often expressed in terms of random fraction ($RF = R/T$, the ratio between the number of random and true events) and depend on the characteristics of the object and the scanner. Unlike the scattered and attenuated fractions, the randoms fraction has a strong dependence on the count rate, which primarily reflects the amount of radioactivity present in the FOV. The randoms rate is also linearly related to the temporal width of the coincidence window. The expected rate of random coincidences formed between two detectors is given by

$$R = 2S_1S_2t \tag{8.3}$$

where

S_1 and S_2 are the counting rates of the individual detectors prior to the application of any coincidence window (termed the singles rates)

t is the coincidence time window (Cherry 2006)

Although scatter and random events contribute to the total number of detected counts, they contain little or no information about the source location and thus form an unwanted background signal. Estimation and subtraction of this background signal amplify the level of noise, decreasing the statistical quality of the image. A semiempirical measure of the effect of such correction on the statistical quality of the images is given by the definition of the noise equivalent counts (NEC) (Strother et al. 1990). NEC is the number of true coincidences, which, if collected in the absence of scatter and randoms, would yield an image of similar statistical quality as the image formed using the corrected data. The NEC is given by

$$NEC = \frac{True^2}{True + Scatter + (2)Randoms} \tag{8.4}$$

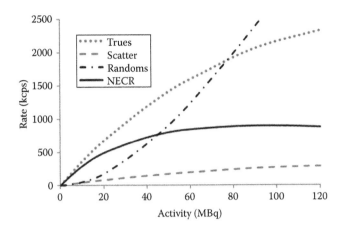

FIGURE 8.5 Representative example of a noise equivalent count rate curve together with the corresponding true event rate, scattered event rate, and random event rate obtained from a mouse size phantom. Note the strong dependence of the random rate on the amount of radioactivity.

where the factor of 2 in front of the randoms term is only required when using a delayed-coincidence window-derived method of randoms correction (Cherry 2006). Note that the conversion between NECs and statistical image quality is not trivial; while NECs have proven to be a useful surrogate measure of the image SNR in the case of analytical image reconstruction, their direct application to images formed via statistical reconstruction methods is questionable (Chang et al. 2012).

NECs are generally estimated as a function of acquired count rate (see Figure 8.5) and are often used to compare the performance of different scanners; they form part of the PET-scanner performance tests advocated by the National Electrical Manufacturers Association (NEMA) (Daube-Witherspoon et al. 2002).

8.2.3 QUANTIFICATION CORRECTION METHODS: OVERVIEW

An important characteristic of PET is its potential to provide quantitative data, that is, to provide a measure of counts/(image voxel) that is proportional to the underlying radioactivity concentration anywhere in the object. This is possible when the corrections for attenuation, scatter, and randoms are accurate and when a quantitative method of image reconstruction is used. Such images can be calibrated using an experimentally derived multiplicative factor to provide activity concentrations (e.g., kBq/mL).

The correction required for attenuation of the true signal is theoretically straightforward in PET, since for a true coincidence event, *both* of the 511 keV photons produced by the positron–electron annihilation must be detected. This necessitates that the combined paths of the photons encompass the entire LOR. The amount of attenuation is thus independent of the location of the annihilation along the LOR, and a LOR-specific attenuation correction factor (ACF) can be determined. The most direct method to obtain the ACFs is to perform two scans with an external-to-the-object positron emitting source; one scan is done with nothing in the FOV (blank scan) and one scan is performed with the object in the FOV (transmission scan). The reduced number of counts collected along each LOR during the transmission scan compared to the blank scan is due only to the attenuation caused by the object. From such a comparison, the ACFs can be easily determined empirically. While this method is potentially and theoretically the most accurate, it often suffers from significant practical drawbacks: the external source is necessarily close to the detectors, and to avoid high detection dead time (inability of a detector/electronics combination to be responsive to a new event while processing a current event), the source must not be too intense. The corresponding blank and transmission scans may hence contain relatively few coincidence counts on each LOR and yield noisy ACFs. The other family of attenuation correction methods is based on the determination of the linear attenuation coefficient (μ-values) of the object. These μ-values are then used to calculate a LOR-specific attenuation correction factor using the following relationship:

$$\text{ACF} = e^{\int_{\text{LOR}} \mu(x) x \, dx} \tag{8.5}$$

where the integral covers the specific LOR, from one detector face to the opposing one. This approach requires the estimation of μ throughout the object. Such μ-maps can be derived using various techniques, including the use of a rotating external positron-emitting source (coincidence measurements), a rotating external γ source (singles-mode transmission), a CT image (either acquired in the same, combined scanner or using a co-registered image acquired at a different time), an MRI image, or an image from an atlas.

The use of a single gamma-emitting source (deKemp and Nahmias 1994) obviates the nearby detector dead time problem. Linear attenuation coefficients depend not only on the material but also on the photon energy. μ-values measured by mono-energetic gamma rays, or by x-rays, must be appropriately scaled to an estimate of μ at 511 keV to account for this dependence. When imaging small animals, the use of low-energy single gamma emitters, such as ^{57}Co, has provided high-quality attenuation correction data, as they have a higher interaction probability in tissue compared to 511 keV photons and thus provide a better contrast between tissue types (Eric et al. 2007). While MRI provides the best contrast between different tissue types and/or organs, its use for attenuation correction is much more complicated. The MRI images depict proton density (to varying degrees depending on the acquisition settings) and are often not sensitive to the presence of bone (Bezrukov et al. 2013). The calculation of 511 keV μ-values from an MRI image is nontrivial. Often overlooked is the importance for the image of μ-values to be correctly aligned with the object; errors in attenuation correction can be introduced by the movement of the animal (e.g., between emission and transmission scanning) or by errors in the registration of CT (or MRI) data to the object.

Scatter is most often estimated from the knowledge of the radioactivity distribution estimate (emission data) and object density (obtained from the transmission data) using a single scatter approximation (Watson 2000), that is, using the assumption that only one of the two annihilation photons undergoes a single Compton scattering. Such methods simulate the scattering process to provide a relatively accurate scatter correction in small animal imaging. Typical SFs in preclinical scanners range from 5% when imaging mice to 35% when imaging rats (Goertzen et al. 2012).

8.2.4   Tracer Principle in PET Imaging: Implication for Camera and Tracer Design

PET makes use of the tracer principle, which requires that the amount of the administered marker (tracer) is sufficiently small so as not to perturb the system under observation. This requirement imposes stringent limits on the amount of administered tracer, especially when imaging saturable sites, such as receptors and transporters. Often only nanograms of a tracer substance can be used. This requires the tracers to be produced with very high specific activity (S.A. = ratio between the amount of radiolabeled and nonradiolabeled tracer) in order to be able to administer a sufficient amount of radioactivity to obtain a reasonable image (see Section 8.3.1). The relationship between the amount of radioactivity, tracer amount, and S.A. is described by

$$\text{Amount (ng)} = \frac{\text{Radioactivity injected (MBq)}}{\text{S.A. (MBq/nmol)}} \times \text{Molecular weight (ng/nmol)} \tag{8.6}$$

From Figure 8.6, it is possible to see that the amount of radioactivity that can be injected can be limited by the tracer principle, especially when the S.A. is low. This consideration further highlights the need for a high-sensitivity tomograph for these applications.

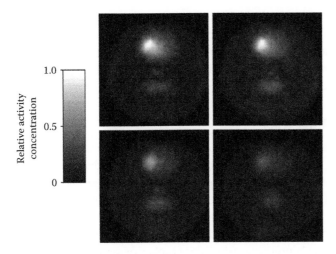

FIGURE 8.6 Image from a rat scanned with the vesicular monoamine transporter type 2 (VMAT2) marker ¹¹C-dihydrotetrabenazine (DTBZ) at an S.A. of 1445 nCi/pmol (top left), 83 nCi/pmol (top right), 35.4 nCi/pmol (bottom left), and 9.9 nCi/pmol (bottom right). The observable tracer concentration in the striata decreases since some of the transporter sites are occupied by nonradioactively labeled DTBZ as the S.A. decreases.

8.3 PET DETECTORS AND CAMERA DESIGN

High detection sensitivity and spatial resolution are two key requirements in small animal imaging. A PET camera can be roughly divided into three main components, all contributing to the overall camera performance: detector scintillating crystals, used to detect the 511 keV photons produced from positron annihilation; light detection devices, sensitive to the scintillation light produced by the interaction of an annihilation photon in the detector crystals; and electronics that process the spatial, temporal, and energy information associated with the detection of each annihilation photon.

8.3.1 SCINTILLATORS

A variety of scintillators have been used as detectors in PET. For small animal PET, the most important attribute of the scintillator is its linear attenuation coefficient (at 511 keV of course), as this primarily determines the likelihood of completely absorbing the annihilation photon in a small volume. The photoelectric fraction (the proportion of incoming annihilation photons whose first interaction with the scintillator via the photoelectric effect) is also of importance (>30% is desirable), as is the light output per unit energy deposited (scintillation photons per keV). These attributes are key in determining (1) the spatial location at which a high-energy photon enters the detector and (2) the energy resolution of the detector, and so its capability of rejecting scattered photons. In addition, the scintillator should have minimal de-excitation time (the time required for the crystal structure to decay to its ground state following the photon-induced excitation). The faster the de-excitation time, the lower the dead time associated with each photon interaction, thus minimizing system dead time. The current scintillators of choice are cerium-doped lutetium oxyorthosilicate (Lu25i05[Ce] – LSO) or cerium-doped lutetium–yttrium oxyorthosilicate (LuYSiO:Ce, LYSO), which have de-excitation times of approximately 50 ns and can yield energy resolutions of the order of 10%–20% (Melcher 2000; Pepin et al. 2004).

The geometry of the scanner must be also carefully considered. The crystals are generally made as small as possible to allow better spatial identification of the location of the photon interaction. Limitations are ability to decode the crystal position with the light detectors and electronics, and parallax errors. The parallax error occurs when a photon pair is emitted at an angle that is very oblique with respect to the crystal surface, for example, along LORs that are at the edge of the FOV. In this case, it is likely that the photons pass through the crystal they first hit but interact in the adjoining crystal. The annihilation event is thus assigned to the LOR defined by the adjoining crystal.

This effect introduces a nonuniform spatial resolution across the FOV, significantly worsening from the center to the edge of the FOV. Research on the mitigation of the parallax error while still maintaining high resolution and sensitivity is ongoing (Green et al. 2010). Improvements in performance would be expected for scintillators with higher attenuation coefficients, photoelectric fractions, and light output characteristics.

8.3.2 LIGHT DETECTORS

Upon interaction with a high-energy photon, scintillation crystals emit many photons in the visible light spectrum. The number of scintillation photons produced is proportional to the energy deposited in the scintillator; approximately, 13,000 scintillation photons (of 420 nm wavelength, violet in color) are produced when a 511 keV photon is completely absorbed in LSO. The light detection device is sensitive to these scintillation photons and should be optimized to match the characteristic wavelength (i.e., color) associated with the specific crystal. The light detector provides an output current that is proportional to the number of scintillation photons detected, which is in turn proportional to the energy deposited in the crystal by the incoming annihilation photon. Photomultiplier tubes (PMT) (Figure 8.7) are currently the most commonly used light detection and signal amplification devices. Photoelectrons resulting from the interaction of scintillation photons

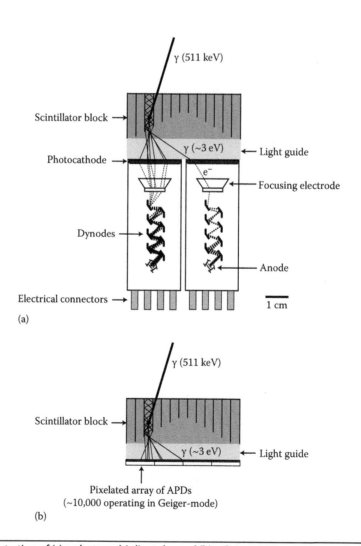

FIGURE 8.7   Illustration of (a) a photomultiplier tube and (b) a Geiger APD or SiPM coupled to a detector block comprising a crystal matrix.

in the photocathode of the PMT are amplified through a chain of dynodes (electrodes with potential differences between them) to provide a measurable current; they offer high gain (~10^6), good energy linearity, and stability.

An alternative detector for PET is the avalanche photodiode (APD) (Lecomte et al. 1990, 2001; Pichler et al. 1998), a semiconductor-based light-detecting device. The APDs are thin, compact, and resistant to the effects of a strong magnetic field since they have a high internal electric field and the charge carriers only travel for a short distance. These characteristics make them well suited for small animal cameras and for hybrid PET/MR devices, where the PET insert is placed inside a magnet bore. In contrast to PMT-based signal amplification, the amplification process in APD is affected by an excess statistical noise factor that increases with the gain (Webb et al. 1974). A multiplication gain of up to 10^3 can be obtained, but due to the excessive noise factor, they tend to be operated in a multiplication range of approximately 50–150. An interesting further improvement for gamma detection is the development of Geiger-mode diodes (G-APDs, or silicon photomultipliers [SiPMs]) (Otte et al. 2005). These devices are capable of much higher signal gains, reaching multiplicative factors up to 10^6. This is achieved by subdividing the active surface of an SiPM into very small cells (of the order of μm), which are operated in Geiger mode. While an individual cell does not provide a linear response as a function of signal energy (the output being binary), the overall linearity is maintained by ensuring that the number of cells in each SiPM is much larger than the number of expected photons (~10^6 cells/mm^2) (Pichler et al. 2008). SiPMs are predicted to become the light detector of choice in the near future for both small animal and human PET systems.

In any camera design, a crystal matrix (detector block) is coupled to a smaller number of light detection devices in such a way that a unique light distribution between the light scintillation devices serves to identify the crystal where the incoming annihilation photon interacted (Figure 8.7) (Casey and Nutt 1986). The light detectors provide an input to electronic boards where energy discrimination may be applied and where the crystal address and photon timing are determined. Coincidence electronics are connected to these boards and continually search for two detections that occurred within the specified coincidence-time window. The occurrence of a coincidence event along a specific LOR can hence be established and recorded to an acquisition computer.

8.4   IMPACT OF SOFTWARE DEVELOPMENT ON IMAGE QUALITY AND QUANTIFICATION ACCURACY

8.4.1   IMAGE RECONSTRUCTION: OVERVIEW

Image reconstruction algorithms can be roughly divided into two categories: analytical image inversion methods, such as filtered backprojection (FBP), and statistical methods, where the approach is to find an estimate of the object/image that most likely gave rise to the measured data. A good review can be found in Qi and Leahy (2006). The advantages of the analytical methods comprise speed and linearity, but no modeling of the physical processes can be included in the reconstruction process, and the corrections for randoms, scatter, attenuation, and detector efficiencies must be applied to the data prior to reconstruction. The statistical approaches, on the other hand, permit to include complex scanner geometries, the random nature of the decay process, and modeling of the scatter, attenuation, and randoms into the reconstruction process itself. They are typically iterative methods and produce images that are generally less noisy than the FBP counterparts. Disadvantages include nonlinearity with respect to the collected data and frequently no clear, uniform convergence criteria (Figure 8.8). This is of particular concern when imaging objects with a highly nonuniform distribution, where different parts of the image may converge at different rates and where local differences in image quality may result. One of the most successful statistical (and iterative) methods is based on

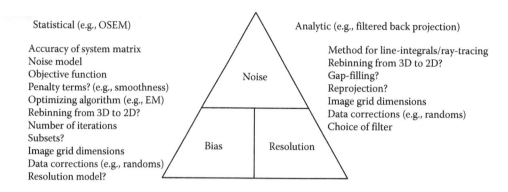

Statistical (e.g., OSEM)

Accuracy of system matrix
Noise model
Objective function
Penalty terms? (e.g., smoothness)
Optimizing algorithm (e.g., EM)
Rebinning from 3D to 2D?
Number of iterations
Subsets?
Image grid dimensions
Data corrections (e.g., randoms)
Resolution model?

Analytic (e.g., filtered back projection)

Method for line-integrals/ray-tracing
Rebinning from 3D to 2D?
Gap-filling?
Reprojection?
Image grid dimensions
Data corrections (e.g., randoms)
Choice of filter

Noise

Bias   Resolution

FIGURE 8.8   Noise, resolution, and bias are the three interrelated image reconstruction dependent image attributes. The trade-off depends on the listed parameters for each category of image reconstruction methods.

the maximum likelihood expectation maximization-ordinary Poisson (MLEM-OP) reconstruction. The image I is updated as follows:

$$\lambda_j^{m+1} = \frac{\lambda_j^m}{\sum_{i=1}^I p_{ij}} \times \sum_{i=1}^I p_{ij} \left(\frac{y_i}{\sum_{b=1}^J p_{ib} \lambda_b^m + \overline{r}_i + \overline{s}_i} \right) \tag{8.7}$$

where

j is the image voxel index

i is the LOR (or projection bin) index

λ_j^{m+1} is the new $(m + 1)$th estimate of the image intensity in image voxel j

y_i represents the number of prompt coincidences

r_i and s_i are the estimates for random and scatter coincidences for the LOR i

p_{ij} is the system matrix describing the probability that a photon pair emitted in the object (image) at voxel j is detected along the LOR i

The system matrix p_{ij} is at the heart of the statistical reconstruction methods as it is the chosen mathematical description for the imaging process. The main component of the system matrix is derived from simple ray sums (line integrals) through the image grid (as used in the analytic methods). The model can be refined, however, to include a variety of effects: attenuation, detection efficiencies and scanner response functions, positron range, etc. By modeling the imaging process more realistically (camera and physics effects), a more accurate representation of the underlying object can be reconstructed (Figure 8.9). Statistical reconstruction methods can differ in the choice of noise model for the acquired coincidence data and the use (if any) of regularization parameters; these factors determine the cost function that is to be minimized by the chosen algorithm. Various optimizing algorithms can be used to perform the iterative updates; different optimizers can take different steps (i.e., move through different images) toward the same goal of reaching the image that minimizes the cost function. Since the reconstruction is generally stopped prior to reaching convergence for all pixel values, the outputted image depends on the optimizing algorithm as well as the statistical model. Other factors that must be considered in the image reconstruction (or in the calculation of the system matrix) include the data interpolation methods, the choice of basis functions used to represent the image space (typically cubic voxels), and the amount of data used for every updating step. It is common to speed up the reconstruction by using only a subset of the acquired data to calculate the updating factors: the projection data are equally divided into a number of subsets (e.g., 16), where each successive update uses the next subset of the projection data to calculate the new image. A full iteration through the whole projection data now has a large number of sub-iterations

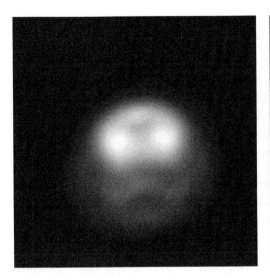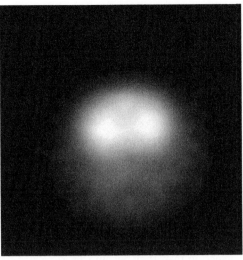

FIGURE 8.9   Data from an ^{18}F-FDG mouse brain scan reconstructed with an iterative statistical reconstruction (3DMAP) on the left and with an analytical method (2DFBP) on the right. The same slice is shown through the striatum in both cases; the slice thickness is 0.8 mm. The 3DMAP image on the left has higher spatial resolution than the 2DFBP image on the right, with less artifacts seen around the edge of the image.

(16 in this case). Ordered subsets expectation maximization-ordinary Poisson (OSEM-OP) is an example (Hong et al. 2007). The reconstruction time is decreased by a factor almost equal to the number of subsets and the result is almost identical in most situations.

8.4.2   DATA QUANTIFICATION AND PARTIAL VOLUME EFFECT

PET images are very often used to provide a quantitative estimate of the biologically relevant parameters describing the process of interest, such as uptake rate constants and tracer binding potentials (BPs) (see Section 8.4.4). A necessary prerequisite for the extraction of such parameters is for the data to accurately represent radiotracer concentration values anywhere in the image. In order to achieve this, all the corrections for the physics phenomena (scatter, attenuation, random events, and tracer physics decay) and for detector and electronics related limitations (nonuniform detection efficiency, detection sensitivity, data acquisition dead time) must be applied. However, even after proper correction and reconstruction, there is a residual limiting factor that negatively affects the quantification for small structures, called partial volume effect (PVE). When the object is smaller than approximately twice the size of the resolution volume, the number of counts originating from the objects will appear "spread" over a larger volume, thus leading to an underestimation of the object-related concentration together with an overestimation of the radioactivity concentration in the area immediately outside the object. A more subtle effect contributing to the PVE is what is called the tissue fraction effect, which arises when an image voxel includes different tissue types that have differing accumulations of the radiotracer. Most partial volume correction algorithms are applied post-reconstruction and exhibit various degrees of complexity (Soret et al. 2007). The simplest correction method is based on the use of premeasured recovery coefficients (RC); typically spherical phantoms of different size are imaged and the RC correction factors are determined by comparing the measured signal to the independently measured radioactivity concentration. Although simple, this method requires an *a priori* knowledge of the object size and suffers from the limitation that the shape is assumed to be spherical. More refined methods, which are able to correctly account for the size of the structure and the tissue fraction effect, require a very well co-registered structural image (Soret et al. 2007); correspondence between tissue structure and function is also required. Since the PVE correction factors can be as large as 10, it is very important to ensure both a correct image co-registration and tissue definition; both are aided by the availability of a CT image acquired in a multimodality environment.

8.4.3 RESOLUTION RECOVERY METHODS

Resolution (or point-spread function, PSF) recovery methods are designed to include a model of the resolution degrading effects within the reconstruction algorithm. Many commercial small animal PET scanners now include such a model in the software supplied by the manufacturer. Implementation of resolution recovery is normally done by including a model of the detector response in the system matrix (Leahy and Qi 2000). As previously discussed, the accuracy with which the system matrix is defined has a critical role in the quality of the reconstructed images. Accurate generation of the system matrix can be achieved by measuring the PSF empirically (by moving a point source through the FOV (Alessio et al. 2010)) or by estimating the PSF via Monte Carlo simulations. While these methods have shown an improvement in contrast measurements and signal detection, they change the noise structure in that data and may introduce additional variability in the biologically relevant results (Blinder et al. 2012). There are still unresolved aspects, such as presence of edge artifacts, which are currently still not fully understood (Bing and Esser 2010). While the impact of such artifacts is likely modest and outweighed by benefits when pursuing a signal detection task, they become a serious limiting factor when quantification in small structures is required as they may lead to bias and increased variability. With the present status of knowledge, it is thus important to carefully assess in which conditions the use of resolution modeling is beneficial and warranted. For a detailed review, the reader is referred to Rahmim et al. (2013).

8.4.4 IMAGE INTERPRETATION: QUANTITATIVE VERSUS SEMIQUANTITATIVE METHODS

In most applications, PET imaging is used to provide a quantitative assessment of physiological parameters. Once correct quantification of the reconstructed images is ensured, the next step consists in relating the radiotracer concentration values to underlying biochemistry using mathematical models that may rely on some prior knowledge of the tracer behavior and tracer/target interaction. Several comprehensive reviews are available in the literature (Carson et al. 1993; Laruelle et al. 2002; Cunningham et al. 2004), and only a few points will be summarized here. In each case, the radioactivity concentration observed in a structure of interest must be normalized, in some fashion, to the amount of administered radioactivity.

The simplest approach is to evaluate the standard uptake values (SUV), defined as the ratio between the tracer concentration in a given time interval post-injection and the amount of injected radioactivity normalized by body weight (Figure 8.10). This approach is technically very simple as it

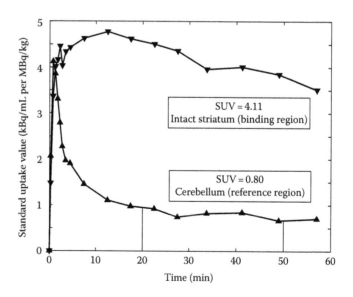

FIGURE 8.10 Example of time activity curves (TACs) obtained from ROIs placed on a rat striatum and cerebellum. The lines connect the measured points, each corresponding to a time frame, represented by the triangles. The arrows indicate a typical time interval used to estimate the SUV, that is, the average value of the radioactivity concentration over that time interval is divided by the radioactivity injected normalized by body mass.

requires a single scan (static scan) and no knowledge of the tracer concentration in the plasma, thus no need of arterial sampling. The SUVs, however, yield only a semiquantitative estimate of tracer uptake: first, the values typically vary as a function of the time interval over which they are calculated; second, they do not account for subject-specific differences in the metabolism (or clearance) of the radiotracer from the circulating plasma, which can decrease the amount of authentic tracer available for uptake; and third, they do not distinguish between nonspecific binding of the tracer and the specific binding of the tracer to its intended target (Boellaard et al. 2004).

More accurate models use the arterial tracer concentration, varying over time, as an input function to a compartmental model. In this manner, the time course of the radiotracer distribution measured by a time series of PET images (dynamic PET imaging) can be used to estimate meaningful parameters that relate to the specific processes taking place between the radiotracer and the tissue (Figure 8.10). The general approach used in modeling is the concept of compartments (Figure 8.11), where each compartment relates to a distinct state of the tracer (Laruelle et al. 2002). Note that many compartments may share the same physical space. Some of these compartments are of general nature, not related to the process of interest (such as tracer concentration in the plasma or nonspecifically bound in tissue) and some refer to the process of interest, such as tracer binding to a specific receptor site. The researcher is normally interested in determining the exchange parameters between compartments related to specific binding to the target of interest and those related to nonspecific binding. In practice, for tracers that are involved in reversible processes the tracers' BP, calculated with respect to either the tracer concentration in the plasma or the concentration

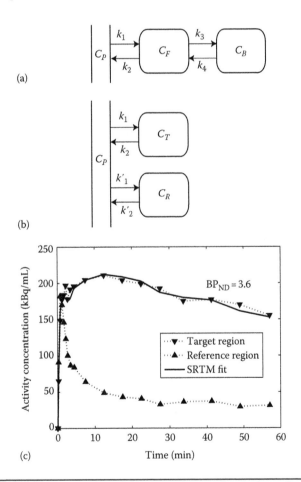

FIGURE 8.11   (a) Two tissue compartment model (tracer exists in a free (unbound) state) (C_F) and specifically bound to the target of interest (C_B). (b) Both the target and reference region are described by one tissue compartment model. (c) Example of a fit to the target (C_T) and reference (C_R) tissue data obtained with the simplified reference tissue model (SRTM) (Gunn et al. 1997), which is based on the compartmental model described in (b).

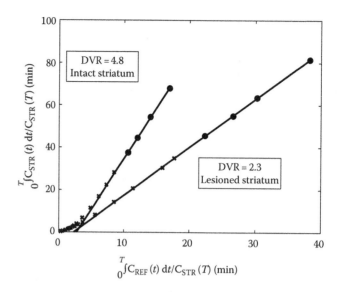

FIGURE 8.12 Example of a tissue input Logan plot. C_{REF} represents the tracer concentration in a reference region assumed to be devoid of specific binding and C_{STR} is the tracer concentration measured in a target region of interest, in this case the striatum, by PET. The x-axis represents the running integral of the reference region activity concentration (C_{REF} between time $t = 0$ and $t = T$), divided by target activity concentration C_{STR} at $t = T$. The y-axis is the running integral of the target region activity concentration (C_{STR} between time $t = 0$ and $t = T$), divided by target region activity concentration C_{STR} at $t = T$. Values for x and y are thus calculated from the time-activity curves, generating a point for every frame, where T is the frame mid-time. The measured data are shown with a line of best fit that is made to a subset of the data. The slope of the line represents the total distribution volume ratio, a measure related to the binding potential.

of free tracer in tissue, is the parameter most widely used (Innis et al. 2007). Likewise, for tracers that undergo trapping (e.g., due to an irreversible enzymatic reaction) or irreversible binding on the timescale of the PET scan, the tracer uptake rate constant with respect to the tracer concentration in plasma is commonly evaluated. Increasing degrees of complexity of the models can be considered, which may allow the extraction of several parameters from the fit; however, the more complex the model, the more susceptible to noise the parameter determination becomes, and often a useful compromise between information detail and the robustness of the parameter estimates needs to be achieved.

According to the degree of complexity desired or possible, several approaches to fitting the data to a compartmental model exist, ranging from kinetic, to equilibrium, to graphical models. In the latter, the measured data are often transformed so that the slope of a linear fit to the transformed data yields the parameter of interest (Figure 8.12). For new radiotracers, a systematic approach to model development is crucial if the parameter outcomes are to retain biological relevance.

Generally, the graphical models can provide a good compromise between robustness and accuracy. Several graphical approaches are currently in use for tracers that bind reversibly and irreversibly (Patlak et al. 1983; Patlak and Blasberg 1985; Logan et al. 1996). Kinetic modeling can provide a parameter estimate for each rate constant, often using a nonlinear fitting method. The results from such methods can be imprecise unless some simplification in the number of parameters can be reasonably made and can depend on the initial guess. An assumption that is valid for many tracers is the kinetic similarity between the nonspecific binding compartment and the free tracer concentration in tissue, allowing them to be considered as a single compartment (Lammertsma and Hume 1996; Gunn et al. 1997) and reducing the number of rate constants to be estimated.

An important consideration in small animal imaging is the ability to use a tissue input function (from a reference region) as opposed to plasma input function. The latter requires arterial blood sampling and analysis, which can be technically challenging. Tissue-derived input functions can be appropriate when imaging ligands for which a region of tissue has no specific binding but for which the kinetics for uptake of the free radiotracer (and its nonspecific binding) is similar to that in the target region. This constraint is valid for several, but not all, tracers and needs to be verified for each tracer separately. Once an appropriate model is developed, it can be applied to time activity curves (TACs) determined on a region of interest basis or on a voxel-by-voxel basis. The latter approach, sometimes called parametric imaging, leads to so-called parametric maps, where each voxel value represents the value of a particular parameter obtained from the fit of a single-voxel TAC. Parametric maps are a convenient way to condense a large amount of temporal information into a single, quantitatively accurate biologically relevant image.

8.5 APPLICATION-SPECIFIC REQUIREMENTS

8.5.1 BRAIN IMAGING

Brain imaging is possibly the most challenging application due to the small size of the structures of interest and of the tight limits on the mass of the injected tracer required to avoid inducing saturation effects that would lead to biased estimates of the parameters of interest. As many studies aim to determine tracers' binding parameters or uptake rate constants, dynamic imaging is most often performed. The time frame length is chosen as a compromise between the need to accurately capture the radiotracer time course and still collect enough counts in each time frame to allow for a statistically robust image; the typical range is between 30 s and 10 min. The length of the entire scanning sequence is determined primarily by the half-life of the biological process under investigation—it is thus important that the radiotracer physical half-life is a good match to the biological half-life. Dynamic imaging can thus be quite challenging especially when ^{11}C is used due to its relatively short, 20.4 min half-life. Nevertheless, brain imaging is very widely used in many rodent models of several diseases, such as Parkinson's, Alzheimer's, and stroke (Virdee et al. 2012). Unique features are the ability to perform noninvasive longitudinal studies and to explore different neurochemical aspects by using a multitude of tracers.

It is normally a requirement that animals are kept under anesthesia during the PET scan to prevent animal motion. It is unfortunate that many anesthetics interfere with the aspects of brain function under study or interfere with the binding of the radiotracers. Developments to enable awake animal imaging are ongoing; the RatCAP camera, which consists of a ring of detectors fixed around the rat head is a unique approach (Schulz and Vaska 2011), while other groups are exploring methods to track and correct for head motion in "real time" (Kyme et al. 2011).

8.5.2 WHOLE BODY IMAGING

Whole body imaging is used primarily in cardiology and oncology applications. Here, heart and respiratory motion can be additional sources of image blurring, and implementation of gated data acquisition (synchronized to a particular phase of the heart position or breathing cycle) is often used to decrease such blurring. This comes at the cost of an increased complexity in data acquisition and image reconstruction, since data need to be acquired in separate histograms for each part of the cycle, and reconstructed separately.

8.5.3 RATIONALE FOR MULTIMODALITY IMAGING

Finally, it is important to place PET in the context of other available imaging modalities that can be used to study animal models of disease. PET can explore biochemical function with a very large number of probes—there are currently hundreds of PET tracers used in research that track different sites/processes and several have been approved for diagnostic purposes. PET excels for its chemical sensitivity as it can detect picomolar concentrations of a radiolabeled substance. On the other hand, the timing resolution achievable by PET is not commensurate with timescales at which neurons operate, nor is there generally sufficient anatomical information to identify the location of a tracer uptake site with respect to the animal's body. In addition, it is also important to remember that different imaging modalities are naturally best suited at providing fundamentally different information. In order to synergistically enhance the amount and quality of achievable information, multimodality imaging approaches that combine strengths from different imaging modalities are being developed. Although many combinations might prove useful, the combination of functional and structural information is often crucial. PET/CT scanners are currently the most common combined imaging modality and are available both for human and for preclinical use. There is, however, reason to believe that the recent advancement in light detector technology will enable rapid development of hybrid PET/MRI scanners, which would be especially beneficial for preclinical imaging as the soft tissue contrast achievable with MRI is much superior than CT and there is no additional radiation exposure (particularly relevant for longitudinal studies). Furthermore, MRI is capable of additional functional imaging, for instance, perfusion, BOLD effect, spectroscopy, diffusion, and others. A combined PET/MRI thus offers a very diverse and comprehensive imaging environment where several parameters can be imaged at once with, in principle, perfect image co-registration.

REFERENCES

Alessio, A. M., C. W. Stearns et al. (2010). Application and evaluation of a measured spatially variant system model for PET image reconstruction. *IEEE Trans. Med. Imaging* **29**(3): 938–949.

Ambrosini, V., M. Fani et al. (2011). Radiopeptide imaging and therapy in Europe. *J. Nucl. Med.* **52**(Suppl. 2): 42S–55S.

Bezrukov, I., F. Mantlik et al. (2013). MR-based PET attenuation correction for PET/MR imaging. *Semin. Nucl. Med.* **43**(1): 45–59.

Bing, B. and P. D. Esser. (2010). The effect of edge artifacts on quantification of Positron Emission Tomography. *Nucl. Sci. Symp. Conf. Rec.* **2010**: 2263–2266.

Blinder, S. A., K. Dinelle et al. (2012). Scanning rats on the high resolution research tomograph (HRRT): A comparison study with a dedicated micro-PET. *Med. Phys.* **39**(8): 5073–5083.

Boellaard, R., N. C. Krak et al. (2004). Effects of noise, image resolution, and ROI definition on the accuracy of standard uptake values: A simulation study. *J. Nucl. Med.* **45**(9): 1519–1527.

Carson, R. E., M. A. Channing et al. (1993). Comparison of bolus and infusion methods for receptor quantitation: Application to [18F]cyclofoxy and positron emission tomography. *J. Cereb. Blood Flow Metab.* **13**(1): 24–42.

Casey, M. E. and R. Nutt. (1986). A multicrystal two dimensional BGO detector system for positron emission tomography. *IEEE Trans. Nucl. Sci.* **33**(1): 460–463.

Chang, T., G. Chang et al. (2012). Reliability of predicting image signal-to-noise ratio using noise equivalent count rate in PET imaging. *Med. Phys.* **39**(10): 5891–5900.

Cherry, S. R. (2006). *PET: Physics, Instrumentation, and Scanners.* Springer, New York.

Cunningham, V. J., R. N. Gunn et al. (2004). Quantification in positron emission tomography for research in pharmacology and drug development. *Nucl. Med. Commun.* **25**(7): 643–646.

Daube-Witherspoon, M. E., J. S. Karp et al. (2002). PET performance measurements using the NEMA NU 2-2001 standard. *J. Nucl. Med.* **43**(10): 1398–1409.

deKemp, R. A. and C. Nahmias. (1994). Attenuation correction in PET using single photon transmission measurement. *Med. Phys.* **21**(6): 771–778.

Epstein, J., N., A. Benelfassi et al. (2004). A 82Rb infusion system for quantitative perfusion imaging with 3D PET. *Appl. Radiat. Isot.* **60**(6): 921–927.

Eric, V., C. Marie-Laure et al. (2007). Monte Carlo modelling of singles-mode transmission data for small animal PET scanners. *Phys. Med. Biol.* **52**(11): 3169.

Goertzen, A. L., Q. Bao et al. (2012). NEMA NU 4-2008 comparison of preclinical PET imaging systems. *J. Nucl. Med.* **53**(8): 1300–1309.

Green, M. V., H. G. Ostrow et al. (2010). Experimental evaluation of depth-of-interaction correction in a small-animal positron emission tomography scanner. *Mol. Imaging* **9**(6): 311–318.

Gunn, R. N., A. A. Lammertsma et al. (1997). Parametric imaging of ligand-receptor binding in PET using a simplified reference region model. *Neuroimage* **6**(4): 279–287.

Hong, I. K., S. T. Chung et al. (2007). Ultra fast symmetry and SIMD-based projection-backprojection (SSP) algorithm for 3-D PET image reconstruction. *IEEE Trans. Med. Imaging* **26**(6): 789–803.

Innis, R. B., V. J. Cunningham et al. (2007). Consensus nomenclature for in vivo imaging of reversibly binding radioligands. *J. Cereb. Blood Flow Metab.* **27**(9): 1533–1539.

Jødal, L., C. L. Loirec et al. (2012). Positron range in PET imaging: An alternative approach for assessing and correcting the blurring. *Phys. Med. Biol.* **57**(12): 3931.

Joshua, S., C. Michael et al. (2013). Clinical impact of time-of-flight and point response modeling in PET reconstructions: A lesion detection study. *Phys. Med. Biol.* **58**(5): 1465.

Kyme, A. Z., V. W. Zhou et al. (2011). Optimised motion tracking for positron emission tomography studies of brain function in awake rats. *PLoS One* **6**(7): e21727.

Lammertsma, A. A. and S. P. Hume. (1996). Simplified reference tissue model for PET receptor studies. *Neuroimage* **4**(3 Pt 1): 153–158.

Laruelle, M., M. Slifstein et al. (2002). Positron emission tomography: Imaging and quantification of neurotransporter availability. *Methods* **27**(3): 287–299.

Leahy, R. M. and J. Y. Qi. (2000). Statistical approaches in quantitative positron emission tomography. *Stat. Comput.* **10**(2): 147–165.

Lecomte, R., J. Cadorette et al. (1990). High resolution positron emission tomography with a prototype camera based on solid state scintillation detectors. *IEEE Trans. Nucl. Sci.* **37**(2): 805–811.

Lecomte, R., C. M. Pepin et al. (2001). Performance analysis of phoswich/APD detectors and low-noise CMOS preamplifiers for high-resolution PET systems. *IEEE Trans. Nucl. Sci.* **48**(3): 650–655.

Levin, C. S. and E. J. Hoffman. (1999). Calculation of positron range and its effect on the fundamental limit of positron emission tomography system spatial resolution. *Phys. Med. Biol.* **44**(3): 781–799.

Liu, X. and R. Laforest. (2009). Quantitative small animal PET imaging with nonconventional nuclides. *Nucl. Med. Biol.* **36**(5): 551–559.

Logan, J., J. S. Fowler et al. (1996). Distribution volume ratios without blood sampling from graphical analysis of PET data. *J. Cereb. Blood Flow Metab.* **16**(5): 834–840.

Melcher, C. L. (2000). Scintillation crystals for PET. *J. Nucl. Med.* **41**(6): 1051–1055.

Olivier, B., T. A. Carpenter et al. (2005). Monte Carlo simulation and scatter correction of the GE Advance PET scanner with SimSET and Geant4. *Phys. Med. Biol.* **50**(20): 4823.

Otte, A. N., J. Barral et al. (2005). A test of silicon photomultipliers as readout for PET. *Nucl. Instrum. Methods Phys. Res.* **545**: 705–715.

Patlak, C. S., R. G. Blasberg et al. (1983). Graphical evaluation of blood-to-brain transfer constants from multiple-time uptake data. *J. Cereb. Blood Flow Metab.* **3**(1): 1–7.

Patlak, C. S. and R. G. Blasberg. (1985). Graphical evaluation of blood-to-brain transfer constants from multiple-time uptake data: Generalizations. *J. Cereb. Blood Flow Metab.* **5**(4): 584–590.

Pepin, C. M., P. Berard et al. (2004). Properties of LYSO and recent LSO scintillators for phoswich PET detectors. *IEEE Trans. Nucl. Sci.* **51**(3): 789–795.

Pichler, B., G. Böning et al. (1998). Studies with a prototype high resolution PET scanner based on LSO-APD modules. *IEEE Trans. Nucl. Sci.* **45**(3): 1298–1302.

Pichler, B. J., H. F. Wehrl et al. (2008). Latest advances in molecular imaging instrumentation. *J. Nucl. Med.* **49**(Suppl. 2): 5S–23S.

Qi, J. and R. M. Leahy. (2006). Iterative reconstruction techniques in emission computed tomography. *Phys. Med. Biol.* **51**(15): R541–R578.

Rahmim A, J. Qi et al. (2013). Resolution modeling in PET imaging: Theory, practice, benefits, and pitfalls. *Med. Phys.* **40**: 064301.

Schulz, D. and P. Vaska. (2011). Integrating PET with behavioral neuroscience using RatCAP tomography. *Rev. Neurosci.* **22**(6): 647–655.

Soret, M., S. L. Bacharach et al. (2007). Partial-volume effect in PET tumor imaging. *J. Nucl. Med.* **48**(6): 932–945.

Strother, S. C., M. E. Casey et al. (1990). Measuring PET scanner sensitivity: Relating countrates to image signal-to-noise ratios using noise equivalents counts. *IEEE Trans. Nucl. Sci.* **37**(2): 783–788.

Virdee, K., P. Cumming et al. (2012). Applications of positron emission tomography in animal models of neurological and neuropsychiatric disorders. *Neurosci. Biobehav. Rev.* **36**(4): 1188–1216.

Wadas, T. J., E. H. Wong et al. (2010). Coordinating radiometals of copper, gallium, indium, yttrium, and zirconium for PET and SPECT imaging of disease. *Chem. Rev.* **110**(5): 2858–2902.

Watson, C. C. (2000). New, faster, image-based scatter correction for 3D PET. *IEEE Trans. Nucl. Sci.* **47**(4): 1587–1594.

Webb, P. P., R. J. McIntyre et al. (1974). Properties of avalanche photodiodes. *RCA Rev.* **35**: 234–278.

Zhang, Y., H. Hong et al. (2011). PET tracers based on Zirconium-89. *Curr. Radiopharm.* **4**(2): 131–139.

Section III
Small Animal Imaging
Nonionizing Radiation

9

MR Imaging

Dmitri Artemov

9.1   PHYSICAL PRINCIPLES OF MAGNETIC RESONANCE IMAGING AND MAGNETIC RESONANCE SPECTROSCOPY

9.1.1   MR Phenomenon, Observation of MR Signals

Magnetic resonance imaging (MRI) and magnetic resonance spectroscopy (MRS) are important technologies for studying small animals *in vivo*. They are based on the physical phenomenon of nuclear magnetic resonance (NMR), which was discovered in 1938 by I. Rabi. Nuclei with nonzero magnetic moment determined by the nuclear spin once placed in a static magnetic field B_0 have resonant absorbance of electromagnetic energy at the resonance frequency $\omega_0 = \gamma B_0$, where γ is the gyromagnetic ratio of the nucleus. For protons, the most abundant magnetic nuclei in biological systems, the gyromagnetic ratio is $\gamma = 42.6$ MHz/T. At typical magnetic field strengths of $B_0 = 4.7$–11.7 T, which are used in small animal MR imaging and spectroscopy, the resonance radio frequency (RF) or magnetic resonance frequency is in the range of $\omega_0 = 200$–500 MHz. Other magnetic nuclei frequently used for *in vivo* MR applications include ^{31}P, ^{23}Na, ^{19}F, and ^{13}C. Compared to protons, these nuclei have lower gyromagnetic ratios and are present at significantly lower concentrations that limit detection and often require sophisticated methods to improve the efficiency of detection.

Currently, practically all MR studies are performed using pulse excitation and Fourier transform-based detection of the NMR signal. Briefly, an excitation RF pulse applied at the resonance frequency ω_0 induces rotation of the sample macroscopic magnetization M_0 around the x-axis, parallel to the applied RF field B_1 in the reference frame, and rotating around the z-axis with a frequency ω_0, as shown in Figure 9.1.

Depending upon the duration and power of the RF pulse, the angle of rotation can be described as $\theta = \gamma B_1 t$, where B_1 is the amplitude of the RF field, and t is the duration of the RF pulse. Pulse angles most frequently used are a 90° pulse that rotates the M_0 magnetization from its initial equilibrium state parallel to the B_0 field to the orthogonal plane, and a 180° pulse that inverts magnetization to the state antiparallel to B_0. The NMR RF signal is generated only by the persistent transversal magnetization components $M_{x,y}$, which are orthogonal to the B_0 field. These components precess in the orthogonal (xy) plane at the resonance frequency ω_0 and produce an electric signal in the RF coil, which is amplified and detected by the receiver system of the instrument. From this description, it becomes clear that, for example, the 90° excitation pulse produces the largest transversal magnetization and the strongest NMR signal, whereas no signals are observed following a 180° pulse applied to the equilibrium magnetization M_0. While a single RF coil with an appropriate transmit/receive switch can be used both to transmit the excitation RF pulse and to receive NMR signal,

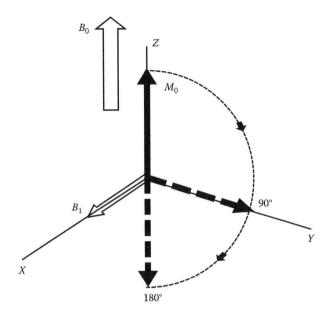

FIGURE 9.1   Effect of 90° and 180° RF pulses on the macroscopic magnetization M_0 in the rotating reference frame.

many current experimental setups include separate RF coils optimized to transmit and receive high-power RF pulses and very weak MR signals, respectively. After application of the RF pulse(s), the magnetization is disturbed from its equilibrium state parallel to B_0 and returns back to equilibrium with characteristic relaxation times, which will be discussed in following sections.

9.1.2   GENERAL PRINCIPLES OF MR SPECTROSCOPY

The resonance frequency of an individual spin is determined by the static magnetic field present at the position of the spin $B = B_0 + \Delta B$, where B_0 is the field generated by the magnetic system of the MR spectrometer and ΔB is the local field, which is composed of several components including the so-called chemical shift, which depends on the spin microenvironment and is sensitive to the chemical structure of neighboring magnetic nuclei as well as local magnetic fields or B_0 inhomogeneities. For high-resolution MR spectroscopy experiments, B_0 inhomogeneities are compensated by adjusting currents in multiple shimming coils, the so-called shimming process, so as to minimize the contribution of these fields to the resulting NMR signal. The NMR signal produced by a macroscopic sample after an RF pulse (90°, for instance) contains a combination of signals generated by multiple spins precessing at different unique frequencies $\omega_i = B_0 + \Delta B_i$. To reconstruct the frequency spectrum of the sample, the recorded time domain RF signal after the pulse, the so-called free-induction decay or FID, is amplified by the receiver with a phase detector to remove ω_0 high-frequency modulation, and a Fourier transform is applied to extract spectral components of this time-domain signal. A typical proton *in vivo* spectrum of a rat brain is shown in Figure 9.2, which

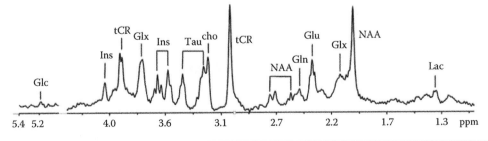

FIGURE 9.2   *In vivo* ^{1}H spectrum acquired from a representative rat brain at 9.4 T with a voxel size of $6 \times 6 \times 4$ mm and a 6 min signal averaging with 200 scans. (Adapted from Du, F. et al., *J. Cerebr. Blood Flow Metabol.*, 32(9), 1778, 2012. With permission.)

demonstrates well-resolved signals from multiple metabolites including among the others choline (Cho), NAA, lactate (Lac), creatine (tCr), and glutamine (Glu). The water signal at 4.7 ppm has been suppressed by the pulse sequence and is not shown in the spectrum (Du et al. 2012).

Water protons are present in biological system at high concentrations, approaching 100 M. Therefore, to detect signals of compounds that are present at much lower concentrations of ~1 mM, the water signal needs to be selectively attenuated or suppressed to improve the dynamic range of spectroscopy. The most popular approach is to irradiate protons at the water frequency with a selective RF field either applied continuously or as a train of 90° pulses followed by "crusher" gradients that completely destroy magnetization in the transversal plane before acquiring the spectrum (Haase et al. 1985). In more sophisticated methods of water suppression, computer-optimized excitation pulses with variable flip angles are used (VAPOR (Tkac et al. 1999)). Multidimensional MR spectroscopy takes advantage of an additional encoding of nuclei signals with different chemical shifts using various mechanisms, which may include chemical shift, scalar or spin–spin, and dipole–dipole interactions (van de Ven 1995). Multidimensional Fourier transform of these data produces the main diagonal peaks and cross peaks between the interacting spins.

9.1.3 GENERAL PRINCIPLES OF MR IMAGING

MRI can be considered as an extension of multidimensional MR spectroscopy and utilizes gradients of the static B_0 magnetic field to achieve spatial encoding of MR signals. MRI was discovered in 1973 by Dr. Lauterbur, who subsequently won the Nobel Prize with Dr. Mansfield for their "discoveries concerning magnetic resonance imaging." Briefly, the application of a linear "read" gradient G_x along the x-axis results in the spatial encoding of the resonance frequency ω_0^x as a function of the distance x according to the equation $\omega_0^x = \gamma B_0^x = (\gamma B_0 + xGx)$. Fourier transformation of the encoded time-domain signal renders the profile or spatial distribution of the spins along the x-axis. Subsequent encoding of the signal along y and optionally z orthogonal axes allow generation of two- or three-dimensional MR images. Typically, $G_{y,z}$ gradients are incremented across the range of values from $-G_{y,z}^{MAX}$ to $G_{y,z}^{MAX}$ and experiments are repeated for each combination of these so-called phase-encoding gradients. For 2D MRI, an additional slice selection is typically used to only detect signals from nuclei located in a thin slice parallel to the imaging plane, as shown in Figure 9.3.

Slice selection can be achieved by using a "shaped" excitation pulse, such as a sinc function modulated pulse envelope, which approximates a rectangular excitation profile in the frequency domain $\Delta\Omega$. When such a pulse is applied to the specimen while the gradient of magnetic field G_z is "on" in the direction perpendicular to the imaging plane (xy), only spins in a thin slice that resonate in the range of $\omega_0 \pm \Delta\Omega/2$ are rotated by the pulse and contribute to the measured MR signal.

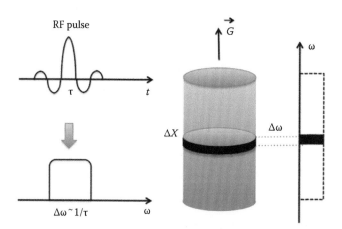

FIGURE 9.3 Principle of slice selection for 2D MRI using a frequency-selective shaped RF pulse and slice selection gradient. The slice thickness $\Delta x \sim \Delta\omega/\gamma G$ is controlled both by the strength of the slice selection gradient G and by the duration of the selection RF pulse τ.

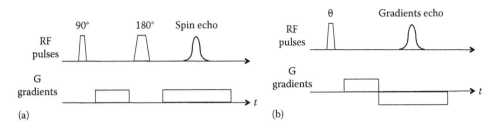

FIGURE 9.4   Spin echo (a) and gradient echo (b) pulse sequences used in T_2 and T_2^* MR imaging, respectively.

Multiple slices can be consequently excited by repeating the imaging sequence while changing the central frequency of the excitation pulse.

To improve the quality of MR images, the signal is often measured in the form of a spin or gradient echo rather than FID. A spin echo is generated by applying an inverting 180° pulse at time TE/2 following the excitation pulse. This pulse inverts the magnetization of spins precessing in the transversal plane and completely refocuses all effects of B_0 inhomogeneity at time TE/2 after the pulse. The spin-echo is therefore formed at time TE (echo time) after the initial excitation pulse (Figure 9.4a). The gradient echo is formed by switching the polarity of the read gradient so that the echo signal is formed at time TE after the pulse when the integral value of the inverted gradient becomes equal to that of the positive gradient (Figure 9.4b). A gradient echo does not refocus static B_0 inhomogeneities and therefore is very sensitive to magnetic susceptibility gradients present in the sample.

Acquisition parameters of MR imaging sequences such as repetition time, echo time, and excitation pulse flip angle can be adjusted so that the intensity of the resulting image is primarily determined by the spin density, MR relaxation times (discussed later), or chemical exchange to generate an appropriate image contrast used to reveal certain biological properties of the subject for diagnostics or other applications.

9.1.4   MR SPECTROSCOPIC IMAGING

Adding spectroscopy or the frequency chemical shift domain to traditional MRI (Buchthal et al. 1989) has enabled the visualization of important chemical information such as absolute concentrations of metabolites in brain that can be digitally superimposed with morphological images (Soher et al. 1996). MR spectroscopic imaging (MRSI) is performed in a manner similar to regular MRI except that the MR signal is acquired in the absence of gradients. To enable spatial encoding, incremented phase encoding gradients are applied in two or three dimensions to produce 2D or 3D spectroscopic MR images. These measurements have to be repeated through all combinations of incremented phase gradient values and therefore the total acquisition time is significantly longer than for traditional MRI. To complete the acquisition within an acceptable measurement time (usually within 60 min), a reduced number of increment steps and correspondingly decreased spatial resolution is used. Novel acquisition schemes based on fast MRI such as echo-planar spectroscopic imaging have been developed as an alternative approach to MRSI, which can result in significantly reduced measurement time while retaining relatively high spatial and chemical shift resolution (Posse et al. 1997).

9.1.5   INTRODUCTION TO MR RELAXATION

Three distinct relaxation processes with corresponding unique relaxation times characterize NMR experiments and can be specifically used as contrast mechanisms to control the intensity of MR images. T_1 relaxation characterizes the recovery of the longitudinal magnetization M_z to its equilibrium value M_0 with an orientation parallel to the static B_0 magnetic field. This is the longest relaxation time, and the intrinsic values for T_1 water relaxation in biological systems are in the range of 1–3 s. T_2 relaxation characterizes the decay of the transversal magnetization from its initial state to zero in spin-echo type of MR experiments. The T_2 time ($T_2 \leq T_1$) is not affected by magnetic field inhomogeneities, and typical values vary from seconds in pure liquids to several milliseconds for lipids and fat. T_2^* relaxation time ($T_2^* \leq T_2$) is defined as the decay constant of the transversal magnetization in gradient-echo acquisition. This relaxation time strongly depends on the local filed magnetic inhomogeneities, which generally are due to the variation of magnetic susceptibility at tissue interfaces. The most important practical significance of the relaxation imaging contrast is that the relaxation times can be modified *in vivo* by exogenous contrast agents that create dramatic changes in contrast and have found an increasingly large number of applications in preclinical and clinical MRI.

9.2 CONTRAST MECHANISMS IN MR IMAGING

9.2.1 T_1 AND T_2 RELAXATION MECHANISMS, DYNAMICS OF MAGNETIC MOMENTS

Nuclear magnetic relaxation involves dynamic changes in the orientation and amplitude of the macroscopic magnetic moment created by precessing magnetic spins of individual nuclei. From the quantum mechanical perspective, it corresponds to changes in occupancy and coherences between quantum levels and transitions of the spin system placed in a magnetic field. The only type of mechanism that can induce these transitions in the spin system, which on macroscopic level results in relaxation of the magnetization, is spin interaction with external fluctuating magnetic fields. Different relaxation mechanisms are sensitive to different frequency components of these external variable magnetic fields. Thus, T_1 relaxation is induced by frequencies close to the NMR frequency of the spins $\omega_0 = \gamma B_0$, whereas T_2 relaxation is also caused by low-frequency components. The various sources of these fluctuating magnetic fields correspond to different relaxation mechanisms. One of the most important relaxation mechanisms for biological systems is the dipole–dipole proton relaxation. In a water molecule, the energy of the magnetic dipole–dipole interaction between two neighboring protons I_j and I_k is described by the following equation:

$$H = -\frac{\mu_0 \gamma_j \gamma_k h^3}{4\pi r_{jk}^3}(3(I_j e_{jk})(I_k e_{jk}) - I_j I_k)$$

where

   γ_j and γ_k are gyromagnetic ratios of interacting spins
   r_{jk} is the distance between the spins
   μ_0 is magnetic permeability of free space
   e_{jk} is a unit vector connecting the two dipoles

In the assumption of high magnetic field B_0, all nonsecular terms in the dipole–dipole interaction can be ignored and the Hamiltonian can be defined as

$$H_{jk} = -\frac{\mu_0 \gamma_j \gamma_k h^3}{4\pi r_{jk}^3} \times \frac{1 - 3\cos^2\theta}{2} \times (3I_{jz}I_{kz} - I_j I_k)$$

where θ is the angle between the vector connecting the protons. Thermal tumbling of the water molecule results in rapid changes of θ, and therefore the B_{jk} field generated by spin j at the position of spin k is modulated with high frequency of approximately $1/\tau$, where $\tau \approx 10^{-12}$ s is the rotational correlation time of water at room temperature. Frequency components of B_{12} at 0, ω_0 and $2\omega_0$ contribute to T_2 and T_1 relaxation, respectively, as described by equations of Bloembergen–Purcell–Pound theory (BPP theory) (Ernst et al. 1987):

$$\frac{1}{T_1} = K\left[\frac{\tau_c}{1 + \omega_0^2 \tau_c^2} + \frac{4\tau_c}{1 + 4\omega_0^2 \tau_c^2}\right]$$

$$\frac{1}{T_2} = \frac{K}{2}\left[3\tau_c + \frac{5\tau_c}{1+\omega_0^2\tau_c^2} + \frac{2\tau_c}{1+4\omega_0^2\tau_c^2}\right]$$

where $K = (3\mu_0^2/160)(\hbar^2\gamma^4/\pi^2 r^6)$ is the constant for spin-½ nuclei such as protons.

9.2.2 T_1 AND T_2 CONTRAST: PARAMAGNETIC (GADOLINIUM-BASED) AND SUPERPARAMAGNETIC (IRON OXIDE NANOPARTICLE-BASED) CONTRAST AGENTS

MRI contrast agents are extensively used to modify the intrinsic contrast of MR images by enhancing relaxation of a specific pool of water molecules that provide contribution to the measured MRI signals. From the molecular perspective, relaxation rates are significantly increased as a result of additional local magnetic fields generated by the large fluctuating magnetic moment of the unpaired electron in paramagnetic metals (approximately 700-fold higher then magnetic moment of a proton). In metals that are efficient T_1 relaxation agents, the correlation time of the electron spin is close to $1/\omega_0$. Manganese and copper are examples of such metals, but gadolinium is the metal of choice for the development of clinical T_1 paramagnetic contrast agents. Gadolinium has several important features that make it invaluable for clinical applications. Its electron relaxation time is sufficiently long, so it provides efficient T_1 enhancement over the whole range of magnetic fields used in preclinical and clinical MRI (B_0 = 0.5–14 T). Gadolinium has nine electrons that can be coordinated by chelating compounds such as a linear chelator DTPA (diethylenetriamine pentaacetate) or a cyclic chelator DOTA (1,4,7,10-tetraazacyclododecane-1,4,7,10-tetraacetic acid) to form stable complexes (Runge et al. 2011). Also, the molecular radius of the gadolinium atom is ideal for the formation of stable complexes. Typically, eight electrons of gadolinium are coordinated by the chelating compound, whereas a single electron participates in contact interaction with water molecules. The specific relaxivity of a standard gadolinium-based contrast agent depends on the magnetic field and is in the range of 2–5 (s·mM[Gd])$^{-1}$. One strategy to improve the relaxation properties of Gd-based contrast agents is to use complexes with increased number of electrons available to interact with water molecules. However, these compounds with decreased number of coordinating sites typically have lower stability *in vivo*, which is a serious concern because of the potential biological toxicity of free gadolinium ions (Werner et al. 2008). Paramagnetic gadolinium-based contrast agents are typically used as T_1 contrast agents, and their accumulation in tissues accelerates T_1 relaxation resulting in increased signal and brighter images in T_1-weighted MRI acquisition. On the other hand, high local concentrations of a gadolinium agent, which, for instance, can be present in the blood after a rapid bolus administration significantly, reduce T_2^* relaxation time due to the large difference in magnetic susceptibility between the tissue and the gadolinium-filled blood capillaries.

Another class of relaxation agent for preclinical MRI is represented by superparamagnetic iron oxide nanoparticles (SPIO). These nanoparticles generate extremely strong local magnetic fields but, due to their relatively large size of ~50 nm and above, their correlation time is relatively long $1/\tau_c < \omega_0$ and T_1 effects of SPIO are typically observed for ultrasmall magnetic nanoparticles (USPIO) (Islam and Wolf 2009) at low magnetic fields that are not common for preclinical MRI studies. At a high magnetic field of $B_0 \geq 4.7$ T, the major contribution of SPIO is a dramatic shortening of T_2 and T_2^* relaxation times. T_2/T_2^*-weighted images acquired with spin-echo and gradient-echo pulse sequences, respectively, have reduced intensities in regions where SPIO are present even at low micromolar concentrations. Interestingly, water molecules in the close vicinity of SPIO nanoparticles experience a magnetic field B' that can be significantly different from B_0 due to the strong magnetic susceptibility of iron oxide. This effect can be used to produce MR images with positive contrast (increased intensities) from SPIO by detecting off-resonance water signals and suppressing the central water resonance at $\omega_0 = \gamma B_0$ frequency (Stuber et al. 2007).

9.2.3 Water Exchange, Magnetization Transfer, and CEST-Based Contrast in MR Imaging

Another contrast-generating mechanism that is increasingly used in preclinical MRI includes methods based on water exchange dynamics. Briefly, this approach uses standard MR images acquired while applying a continuous or pulsed saturation field at a certain frequency band typically shifted from the resonance frequency of bulk water ω_0 by $\Delta\omega$. The saturation rapidly dephases magnetic moments of the resonating spins and brings the total magnetization to zero. Magnetization recovery is relatively slow and takes several T_1s for complete recovery. If there are molecules that contain chemical groups with exchangeable protons such as –NH, –NH$_2$, –OH, and so on, that resonate at the saturation frequency $\omega_0' = \omega_0 + \Delta\omega$, the resulting bulk water signal intensity will be reduced because some water protons that are in continuous exchange with the chemical groups will become saturated while residing at the positions with resonance frequency (or chemical shift) of ω_0'. Originally, this technique was applied for magnetization transfer MRI using saturation of solid macromolecules such as proteins with exchangeable groups that have very broad spectra and can be saturated at large offsets $\Delta\omega$ of several kilohertz (Filippi and Rocca 2004). More recently, the method was adapted to detect water exchange with soluble molecules and was called "chemical exchange saturation transfer" or "CEST" (Ward et al. 2000; van Zijl and Yadav 2011). Interestingly, CEST provides a built-in sensitivity enhancement mechanism, and the method can sense the presence of low concentrations of exchangeable protons in the micromolar range with molar sensitivity via the bulk water signal used in MRI. Typically, CEST MRI involves acquisition of so-called z-spectra by recording a series of MR images and measuring water signal intensity as a function of the saturation frequency offset $\Delta\omega$. The presence of exchangeable groups is detected as an asymmetry in the z-spectrum, as shown in Figure 9.5. The method was extended to the detection of compounds with paramagnetic metals (PARACEST) (Woods et al. 2006; Hancu et al. 2010) and liposomes encapsulating chemical shift reagents that induce a significant shift or water resonance frequency in the liposomal compartment (LIPOCEST) (Terreno et al. 2008a,b). A novel preclinical application of CEST includes the detection of endogenous agents such as hydroxyl protons on glycogen (glycoCEST; van Zijl et al. 2007) and D-glucose (glucoCEST; Chan et al. 2012).

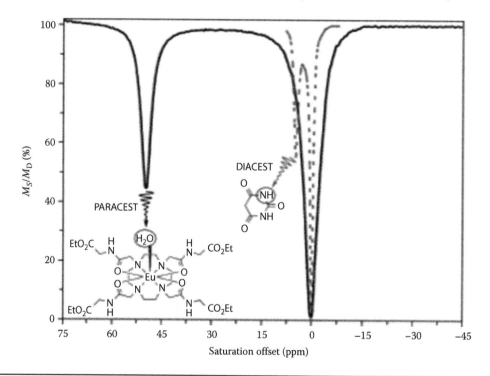

FIGURE 9.5 Comparison of CEST z-spectra for typical DIACEST and PARACEST agents. The dramatically larger $\Delta\omega$ displayed by the PARACEST agent makes activation of the agent by an applied RF pulse less ambiguous. (From Hancu, I. et al., *Acta Radiologica*, 51(8), 910, 2010. With permission.)

9.2.4 BEYOND PROTONS: HETERONUCLEAR MRS/MRI, HYPERPOLARIZATION

Use of heteronuclei or nuclei other than protons in preclinical experiments is quite challenging due to the significantly reduced sensitivity because of their lower gyromagnetic ratios and/or low concentrations of magnetic isotopes in biological systems. The relative sensitivity of MR for different spins is summarized in Table 9.1.

Endogenous ^{31}P and ^{23}Na at natural abundance are most often used for *in vivo* MRS and MRI animal studies. Phosphorus-31 (^{31}P) MRS provides important metabolic information as well as enables detailed studies of phosphoethers such as phosphocholine and phosphoethanolamine, which are important in cancer biology (Wijnen et al. 2010; Glunde et al. 2011). Low intrinsic sensitivity of ^{31}P results in a relatively coarse spatial resolution of about 250 mm^3 in typical high-field MRSI experiments at 7 T (in 't Zandt et al. 2004).

Sodium MRI is highly sensitive to ^{23}Na compartmentalization and can provide early markers of cell damage in neurological and treatment response studies (Schepkin et al. 2012). ^{23}Na MRI has relatively high sensitivity due to large γ and short T_1 of ^{23}Na spins. On the other hand, an extremely short ^{23}Na T_2 time in biological system requires application of ultrashort echo-time imaging sequences, and currently a novel strategy of spatial encoding such as twisted projection has been developed for ^{23}Na MRI (Yushmanov et al. 2009).

^{13}C and ^{19}F MRS and MRI have been used in preclinical studies to detect signals of exogenous compounds. Studies of glucose metabolism (Artemov et al. 1995a; Nabuurs et al. 2008), drug delivery (Artemov et al. 1995b) and conversion (Li et al. 2010), tissue oxygenation (Mason et al. 1996; Procissi et al. 2007), and detection of transplanted cells (Ahrens et al. 2005) have been performed using these nuclei. The high gyromagnetic ratio of ^{19}F, which is close to that of protons, and the absence of ^{19}F background signals in biological systems render this nucleus an ideal probe for noninvasive MR detection.

^{13}C has relatively low γ, and the natural abundance of about 1% results in low sensitivity of ^{13}C MRS/MRI. To enable *in vivo* detection, substrates enriched with >99% of the magnetic isotope are typically used in combination with signal amplification techniques that allow observation of ^{13}C spin via signals of spin-coupled protons using various schemes of polarization transfer (Kato et al. 2010).

Hyperpolarization of nuclear spins is a novel approach to ^{13}C MRI/MRS, which dramatically increases the population difference between quantum levels of nuclear spins in the magnetic field far above the Boltzmann distribution. Indeed, even at the highest magnetic field of 21.1 T available for *in vivo* MR experiments (National High Magnetic Field Laboratory, Tallahassee, FL), the equilibrium difference between the lower (N_{lower}) and upper (N_{upper}) spin-½ states given by the Boltzmann factor $N_{upper}/N_{lower} = \exp^{(-\delta E/KT)} = \exp^{(-\gamma hB_0/KT)}$ is less than ~7.1 × 10^{-5}. Hyperpolarization theoretically can increase this difference to >50% (Viale and Aime 2010). Two alternative technologies have been developed for nuclei hyperpolarization in liquids. Hydrogenation of unsaturated chemical bonds with para-hydrogen, followed by polarization transfer to ^{13}C, results in almost complete polarization of the product (Bowers and Weitekamp 1986). This PASADENA technique requires ^{13}C-labeled compounds with a specific arrangement of unsaturated chemical bonds, which may limit the range of its applications. Dynamic nuclear polarization (DNP) utilizes Overhauser enhancement of nuclear magnetization via electron spin resonance, and experiments are performed with frozen solid specimens at a relatively high magnetic field (~3 T) and very low temperature (≤1 K) (Ardenkjaer-Larsen et al. 2003). These techniques have been tested *in vivo* with exogenous precursor molecules enriched with the ^{13}C isotope (Day et al. 2007; Chekmenev et al. 2008). Hyperpolarization provides

TABLE 9.1					
MR Properties of Magnetic Nuclei Used in Preclinical *In Vivo* Experiments					
Nuclei	**^{1}H**	**^{19}F**	**^{23}Na**	**^{31}P**	**^{13}C**
Gyromagnetic ratio (MHz/T)	42.57	40.05	11.26	17.23	10.71
Natural abundance (%)	99.98	100	100	100	1.108
Relative sensitivity	1.00	0.83	9.25×10^{-2}	6.63×10^{-2}	1.46×10^{-4}

dramatically increased MR sensitivity, but the lifetime of the hyperpolarized state is limited by the nuclear T_1 time, which is in the range of 10–100 s for ^{13}C compounds. Therefore, the experimental procedure should be optimized to provide a rapid thawing, delivery, and administration of the hyperpolarized compound to the animal and fast MR acquisition. An important example of ^{13}C hyperpolarized MRI is noninvasive detection of the conversion of ^{13}C-pyruvate to ^{13}C-lactate to monitor tumor glycolysis, which can be used as a marker of malignancy and therapeutic response.

9.3 TECHNICAL ASPECTS AND INSTRUMENTATION FOR MR IMAGING IN SMALL ANIMALS

9.3.1 ESSENTIAL COMPONENTS OF A SMALL ANIMAL MR SYSTEM: MAGNET, GRADIENT SYSTEM, RF COILS, AND RECEIVER

High-field MR systems for preclinical small animal studies are uniquely positioned at the watershed between clinical MRI systems and research instruments for high-resolution NMR spectroscopy. While they provide a similar functionality to clinical MRI in terms of automatic adjustment of multiple imaging parameters, graphic prescription of scan geometry, synchronization of acquisition with respiration and the heart cycle, and online image processing, preclinical MRI instruments are typically based on a modified research hardware and software. Therefore, in comparison to clinical systems, it is significantly more convenient and less time consuming to develop and implement new pulse sequences and methods and to have precise control over multiple acquisition and processing parameters.

Preclinical MR systems are typically equipped with horizontal superconducting magnets with a bore size in the range of 30–50 cm and magnetic field strengths ranging from 4.7 to 11 T with corresponding proton resonance frequencies of 200–500 MHz, respectively. Horizontal magnets offer significantly more convenient handling of live animals because they provide easy access to the subject and straightforward placement of anesthesia and catheter lines, temperature control, circulating water blankets, and animal monitoring devices including temperature sensors, breathing pads with pressure transducers, and electrocardiography (ECG) electrodes. Also, the horizontal animal position is physiologically preferable for small animal imaging (Foley et al. 2005). Modern MRI magnets can run practically maintenance free, as they have liquid-nitrogen-free design and helium recuperation systems.

Gradient systems for preclinical small animal imaging are characterized by strong available maximum gradients of up to 1000 mT/m and fast gradient rise time. Inner diameters vary from 16 cm, which is suitable for imaging of small rodents, to ~30 cm, which is sufficient for experiments with small mammals such as rabbits and cats. While preclinical MR instruments typically have a complete set of high-order shim coils for precise correction of magnetic field inhomogeneities, first-order shimming is often done using X-, Y-, and Z-gradient coils. This arrangement allows for much higher shim currents and correspondingly increased shimming range, and the heat produced is dissipated by the gradient water cooling system.

There are multiple choices for RF coils for preclinical imaging offered either by the supplier of the MR system or by independent companies. Also, many research groups develop their own RF resonators specifically optimized for a particular type of MR experiments. Briefly, RF coils can be separated into two large groups, which include volume resonators that have cylindrical shape and generate uniform RF field suitable for whole body and head imaging, and surface coils that provide improved sensitivity for MR imaging and spectroscopy in small local regions. Volume resonators typically operate in the transmit–receive mode, meaning that the same coil is used to produce high-power excitation RF pulses and receive much weaker MR signals. Surface coils can operate both in transmit–receive and receive-only modes if used in combination with a transmit volume coil. In the latter case, a pin-diode RF switch needs to be used to detune coils to prevent RF coupling during the pulse and during the acquisition of MR signals. The system also includes RF transmitters, receivers, and preamplifiers with appropriate RF filters optimized for proton and/or heteronuclear experiments. Multiple preamplifiers and receivers are required for multi-coil MRI.

9.3.2 MULTICHANNEL ACQUISITION: IMPROVED SENSITIVITY AND REDUCED MEASUREMENT TIME: OPTIMAL CHOICE OF RF COILS AND RF CHANNEL CONFIGURATION

Several overlapping surface coils can be arranged in an array to either increase field of coverage while retaining intrinsically high signal-to-noise ratio SNR (phased-array mode) (Roemer et al. 1990), or to enable parallel MR imaging, for example, using SENSE technology (Schneider et al. 2011). In the latter case, the number of phase-encoding steps can be reduced in the directions where field of views of multiple RF coils are overlapping. Parallel imaging is especially important for time-sensitive applications such as cardiac MRI (CMRI) to quantify cardiac function with noninvasive MR imaging. One example is the use of an eight-element coil array in a 9.4 T system to image the mouse heart. A threefold acceleration of data acquisition was achieved without compromising quality of imaging data. An example of CMRI performed with this array is shown in Figure 9.6.

Virtually all research modern preclinical MR scanners are equipped or can be upgraded with a multi-receiver system and image acquisition reconstruction software that supports parallel imaging with coil arrays. Parallel imaging with multiple coils is warranted wherever fast acquisition is required to record functional information such as cardiac functions, brain activity, or contrast uptake kinetics. One potential problem with a multiple coil arrangement is that tuning and matching of each individual coil can be difficult, and fixed tuned arrays may not provide optimal performance for all samples.

9.3.3 CRYOGENIC RF COILS: THEIR ROLE IN MR MICROIMAGING

Cryogenically cooled RF coils have recently become available for high-field horizontal-bore small animal MR systems. Their main advantage is significantly reduced RF thermal noise arising from the coil conductor especially when coupled with a low-noise cryogenic preamplifier. However, there is a concern that in a typical MR experiment with biological samples such as live animals (or human beings), the conductive sample itself is a major source of thermal noise, and therefore experimental increase in the SNR of cryogenic coils in comparison to room-temperature RF coils may not be that dramatic. Cryogenic RF coils, on the other hand, offer a clear advantage for MR

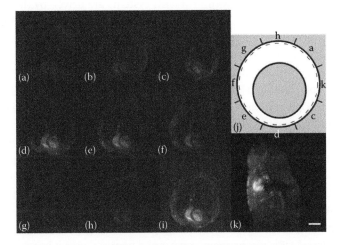

FIGURE 9.6   (a–h) Fully encoded axial *in vivo* gradient echo images of the individual coil elements. The sum-of-squares reconstruction of (i) axial and (k) sagittal coil images illustrate excellent sensitivity to cover the entire heart in each relevant direction. The schematic in panel (j) depicts the location of the individual coil elements relative to the mouse indicated by the gray circle. Scale bar: 5 mm. (From Schneider, J.E. et al., *Magn. Reson. Med.*, 65(1), 60, 2011. With permission.)

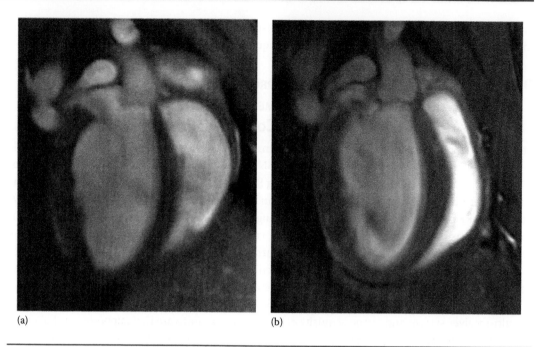

(a)   (b)

FIGURE 9.7 Comparison between high-resolution (69 x 115 x 800 μm³) images of a four-chamber view of mouse heart acquired with the birdcage resonator in conjunction with a four-channel receive-only cardiac surface coil array at room temperature (a) and the CryoProbe (b). (From Wagenhaus, B. et al., *PLoS One*, 7(8), e42383, 2012. With permission.)

microimaging of small samples where the coil loading by the sample and, hence, the RF noise generated by the specimen are minimal (Ratering et al. 2008). Experimental comparison of a commercial Bruker cryogenic coil operating at 30 K with a chilled preamplifier at 77 K with a standard four-channel receive-only mouse cardiac phased array was reported by Wagenhaus et al. (2012). Experiments performed at 9.4 T demonstrated a considerable gain in SNR for the cryogenic coil by a factor of 3–5, as shown in Figure 9.7.

9.3.4 STATE-OF-THE-ART MR SYSTEMS: ULTRAHIGH-FIELD AND DESKTOP LOW-FIELD SCANNERS

Recent trends in the development of preclinical MRI systems are driven by the demands of studies in animal models of human diseases. One trend is to produce systems with very high magnetic fields to deliver the highest sensitivity and resolution for applications in molecular imaging and heteronuclear spectroscopy, along with the development of state-of-the-art MR technology. The vertical-bore system with the highest magnetic field available for preclinical imaging of small animals is currently installed at the National High Magnetic Field Laboratory (NHMFL), Tallahassee, Florida, and includes a unique combination of the high magnetic field (21.1 T) and wide bore (105 mm). An example of sodium imaging of a normal mouse (C57BL/6J) head acquired with custom-designed RF probes at frequencies of 237/900 MHz for sodium and proton, respectively, is shown in Figure 9.8. Sodium MRI resolution of ~0.125 μL was achieved by using a 3D backprojection pulse sequence. In comparison to the more traditional B_0 field for small animal studies of 9.4 T, the ultrahigh-field MRI resulted in an SNR gain of ~3 for sodium and of ~2 for proton imaging (Schepkin et al. 2010).

Another interesting development in preclinical MRI is the availability of low-cost, low-field MRI systems that are especially suitable for high-throughput screening of animals for drug development studies. They are equipped with a permanent magnet and have a compact benchtop maintenance-free design that would ideally suit many non-MR imaging laboratories. A prototype 21 MHz (~ 0.5 T) benchtop MRI system from Oxford Instruments was tested for analysis of tumor progression in nude mouse xenograft models of human cancer (Caysa et al. 2011). Benchtop MRI provided an acceptable image quality and allowed reliable identification of the lesion on Gd contrast-enhanced scans (Figure 9.9).

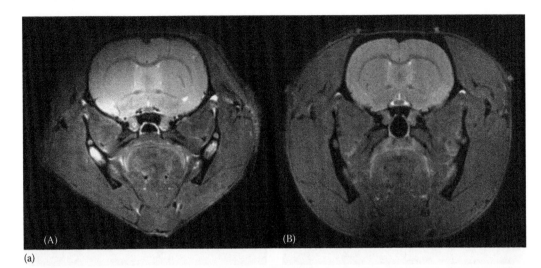

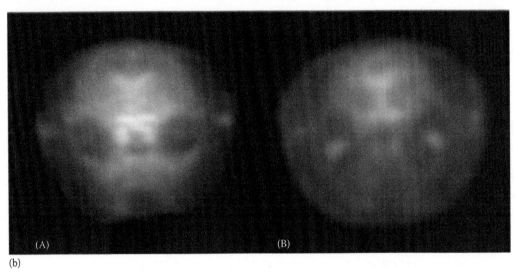

FIGURE 9.8 (a) *In vivo* proton MRI of a rat head at 21.1 T (A) and 9.4 T (B). Both MR images were acquired using spin-echo pulse sequence and the same imaging parameters. The resolution of images was 0.137 × 0.137 × 0.41 mm³. (b) *In vivo* sodium MRI of the rat head at 21.1 (A) and 9.4 T (B). Both MRI images were acquired using backprojection pulse sequence with the same imaging parameters. The isotropic resolution of images was ~1 mm³. (From Schepkin, V.D. et al., *Magn. Reson. Imaging*, 28(3), 400, 2010. With permission.)

9.4 EXAMPLES OF MRI/MRSI APPLICATIONS IN SMALL ANIMALS

The range of applications of preclinical MRI in model animal systems is truly vast and is continuously expanding into new areas such as imaging reporters, image-guided therapy, and combination of complementary imaging modalities such as MRI/PET. Here, to demonstrate tremendous potential of the technology, we review several representative examples of preclinical MRI applied to important biological and medical problems.

9.4.1 MRI/MRS in Animal Models of Human Cancer

Important application of preclinical MR in cancer research can be generally separated into four main areas. Morphological MRI enables noninvasive detection of cancer and determination of the lesion volume and size of the primary tumor and disseminated metastases (Schmidt et al. 2007). These experiments are essential for studies of tumorigenesis and therapeutic response. On the other

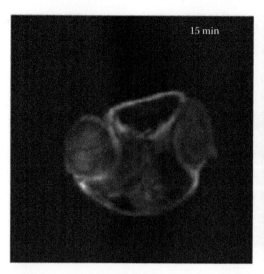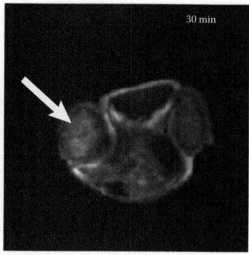

FIGURE 9.9 Transaxial NMR images of a mouse (face-down position) bearing two s.c. HT29 xenografts 15 and 30 min after i.v. application of Gd-BOPTA. One tumor showed strong contrast enhancement, and an interior structuring can be observed (white arrow). (Adapted from Caysa, H. et al., *J. Exp. Clin. Cancer Res.*, 30(1), 69, 2011. With permission.)

hand, novel highly efficient anticancer therapies have been developed to target the tumor microenvironment, and therefore, direct measurement of the tumor volume may not provide a reliable marker of the treatment response. For these treatment strategies, MRI and MRS offer sensitive detection of multiple tumor microenvironmental parameters. Tumor perfusion and vascular function can be measured by dynamic contrast-enhanced MRI (DCE-MRI), which is an important tool to monitor effects of antiangiogenic and antivascular therapies (Barrett et al. 2007). Tumor hypoxia and pH are very important prognostic factors that are critical for tumor progression and resistance to therapy and can be also used as potential targets for therapy (Stasinopoulos et al. 2011). Several noninvasive MRI and MRS methods have been developed to assess these parameters using ^{19}F-perfluorocarbon probes to measure oxygen tissue tension (Mason et al. 1996) and proton, ^{31}P, ^{19}F, and CEST-based probes to measure pH (Bhujwalla et al. 2002; Gillies et al. 2004; Wu et al. 2010; Liu et al. 2011). Preclinical MRS has been successfully applied to tumor metabolic studies including glycolysis by standard and hyperpolarized ^{13}C MR and choline and phospholipid metabolism by ^{1}H and ^{31}P MRS, which can provide key information regarding tumor aggressiveness and response (Glunde et al. 2010). Imaging of drug delivery to cancer and elucidating the role of state-of-the-art delivery systems (Kaida et al. 2010; Onuki et al. 2010) and prodrug activation approaches (Li et al. 2006, 2008) comprise another highly important area of preclinical applications of MRI/MRS. An important aspect of noninvasive MR of cancer in animals is a clear pathway for translation of methods and techniques developed in preclinical models of human cancer to the clinic.

9.4.2 NEUROIMAGING IN SMALL ANIMAL MODELS

MR neuroimaging is one of the most important areas of application of preclinical MRI, and many preclinical MR systems have dedicated animal holders and RF coils optimized for brain studies. These applications include studies of brain morphology and morphogenesis (Mori et al. 2006; Zhang et al. 2006), brain functional imaging (Ferris et al. 2011), cerebral perfusion and metabolism, and stem cell research using progenitor cells labeled with various contrast agents for noninvasive MRI tracking. Stem cell tracking technology can be applied to studies of brain development as well as for brain tissue repair and regeneration (Walczak and Bulte 2007). *In vivo* MR spectroscopy provides brain proton spectra of high quality and information content with multiple resolved metabolites in comparison to other tissues due to the relative homogeneity of the brain tissue and low lipid peak (see Figure 9.2). Three-dimensional rendering of vascular structures of a fixed mouse brain presented in Figure 9.10 demonstrates a unique capability of micro-MRI (μMRI) to reveal minute features of brain vasculature.

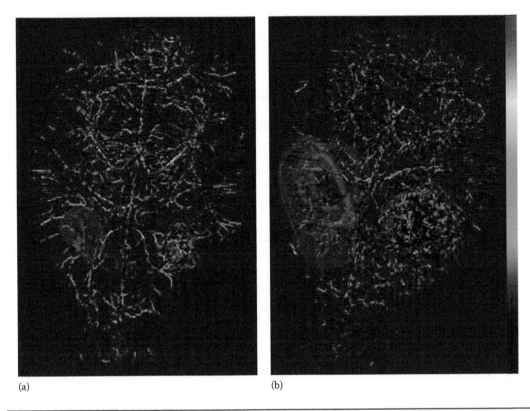

(a)   (b)

FIGURE 9.10   Vascular tree images extracted from μMRI of representative brains with implanted 9L tumors at day 12 (a) and day 17 (b) post inoculation. Three-dimensional vascular structures color-coded by vessel radius and volume rendered ROIs (red for the tumor and green for the contralateral brain) corresponding to the μMRI data. (Adapted from Kim, H. et al., *J. Cereb. Blood Flow Metab.*, 31(7), 1623, 2011. With permission.)

9.4.3   MOLECULAR MRI

MR imaging provides intrinsically high spatial resolution and high contrast, which can be modified in a substantial degree by various classes of contrast agents, some of which have been discussed above. On the other hand, the net sensitivity of MRI is quite limited in comparison with other imaging modalities such as nuclear and optical imaging. For instance, the sensitivity limit of *in vivo* proton MR spectroscopy is in the low millimolar range. Paramagnetic gadolinium contrast agents can be reliably detected at concentrations of ≤ 100 μM of Gd; and iron oxide and CEST contrast agents have significantly lower detection threshold in the low micromolar range (McMahon et al. 2008). However, to enable detection of molecular events at the cellular level, significant improvement in detection sensitivity is required. For example, the average concentration of cell surface receptors expressed at the level of 10^6 receptors per cell can be as low as ~0.5 μM (Artemov et al. 2003); the concentration of intracellular targets is typically even lower. Therefore, some variants of signal amplification mechanisms need to be applied to enable MR detection at molecular levels. In recent years, several efficient strategies have been developed for molecular MRI, including two-step labeling using the avidin–biotin approach (Artemov et al. 2003), polymerization and deposition of monomeric contrast agent molecules (Shazeeb et al. 2011), and magnetically labeled nanoparticles that provide a platform for development of vascular targeting agents and for cell labeling for MRI cell tracking studies (Frank et al. 2003; Geninatti Crich et al. 2006). A novel area of research is the development of activated MR agents that can theoretically provide low imaging background from the nonactive form and a significant increase in contrast once the agent is activated by the target biological process (Louie et al. 2000; Kato and Artemov 2009). Intensive research in the development of multimodality imaging such as preclinical and clinical MRI/PET may lead to a rethinking of our molecular imaging strategies. Imaging protocols and imaging agents can be optimized to explore the advantages of the individual imaging modalities such as high spatial resolution and functional and spectroscopy information provided by MR, combined with very high sensitivity and specificity of PET tracers.

9.5 DISCUSSION AND CONCLUSIONS

9.5.1 ROLE OF PRECLINICAL SMALL ANIMAL MRI STUDIES IN DRUG/AGENT DEVELOPMENT AND IN ADVANCING OF IMAGING TECHNOLOGY

Preclinical MR imaging and spectroscopy is a powerful technology with tremendous scope to address biomedical problems. Among these multiple applications, we can emphasize two important directions where preclinical MRI can provide a major impact both for basic research and for translation studies.

1. Development of targeted imaging agents and image-guided therapeutic platforms that can be used in novel theranostic (combined therapy and diagnostics) applications is a high-priority task. The rational design of these compounds, which must have high specificity of the target recognition combined with an efficient generation of imaging contrast and minimal toxicity, is a formidable pursuit. Preclinical models are the ideal testbed for the development and validation of these new agents, which need to be completed before the agent can be pushed to clinical trials. Preclinical MRI is, therefore, an indispensable technology that facilitates these studies and so far has been successfully applied to several important problems including the development of a nanoplatform for delivery and MRI/MRS monitoring of the combination prodrug therapy with cytosine-deaminase/5-fluorocytosine and siRNA (Li et al. 2008) and molecular imaging of cancer with targeted (Artemov et al. 2003) and activated MR agents (Olson et al. 2010; Onuki et al. 2010; Shazeeb et al. 2011).

2. Preclinical MRI is essential for the development of novel imaging technology. The main motivation that drives the extensive use of small animal systems is their higher availability, lower operation costs, fewer risks involved in the development of novel methods, and less restrictive and more compact codes for development of pulse programs and imaging methods. While the translation of imaging methods developed on high-field preclinical systems to clinical MRI is not straightforward, the development loop can typically be significantly shortened and simplified by the use of initial programming, debugging, optimization, and testing new methods and resolving possible issues at the preclinical stage.

9.5.2 CLINICAL MRI VERSUS SMALL ANIMAL MRI: WHICH PARAMETERS ARE IMPORTANT?

Significant difference in B_0 magnetic field used for preclinical versus clinical MR exams has traditionally been a major stumbling block for direct comparison of preclinical data or translation of preclinical experiments to clinical studies. While this trend is certainly continuing as modern ultrahigh-field systems become available for *in vivo* animal MR studies, generally the gap between a typical animal MR system operating at the field of 4.7–9.4 T and clinical scanners that currently operate at 3 T and several 7 T scanners available at many MR centers becomes less distinct. Clinical scanners are typically equipped with multiple RF transmit–receive channels and fast and strong gradients (45 mT/m maximum amplitude and a slew rate of 200 T/m/s for Magnetom Trio 3 T system and 80 mT/m for Philips Achieva 3 T); therefore, their experimental capabilities are similar to or exceed those of the preclinical MR instruments. Scaling experiments from small animals to humans usually require RF pulses that operate at high RF power levels and may present significant problems with the specific absorption rates (SARs) permissible at clinical systems. Currently, the SAR limit for the first level controlled mode under medical supervision is set at the level of 4 W/kg for whole-body RF irradiation and 3.2 W/kg for the head. These limits are almost never an issue for preclinical MRI; however, careful control and optimization (often provided by the scanner software) is required for the clinical translation of methods developed on a small animal MR system.

9.5.3   FUTURE DIRECTIONS OF SMALL ANIMAL MRI IN THE CONTEXT OF CLINICAL TRANSLATION

Small animal MR has had a longstanding role in preclinical studies, and some of its relevant applications have been reviewed in this and other publications. While the rapid progress in all imaging modalities that is currently taking place can cause dramatic changes in the scope of potential applications and problems that can be addressed, major directions to advance preclinical MR imaging and spectroscopy can be envisioned. From the instrumentation perspective, further development and integration of multichannel acquisition and dynamic and functional MRI and MRS imaging matching clinical system capability into a user-friendly interface can be expected. Also, there will be a constant push toward even higher magnetic fields of ≥ 11 T, which will result in significantly improved sensitivity and spectral separation. Multimodality imaging will also become available on preclinical platforms, and MRI will provide morphological and functional information matched to the molecular imaging capabilities of micro PET or SPECT and possibly optical imaging technologies. The range of applications will continue to expand, and we anticipate that high-resolution morphological imaging of elaborate animal models of human diseases combined with functional and molecular imaging will become an active area of research. Another growing area will be the development and validation of image-guided therapy and theranostics, where MRI will be used to monitor the delivery of therapy and optional activation of the therapeutic agents followed by monitoring of the treatment response longitudinally and noninvasively.

ACKNOWLEDGMENTS

The author thanks Dr. Z.M. Bhujwalla for helpful discussions and critically reading the chapter and M. Holt for help with the preparation of the manuscript. This work was supported in part by S.G. Komen grant KG 100594 and NIH/NCI RO1 R01 CA154738.

REFERENCES

Ahrens, E. T., R. Flores et al. (2005). In vivo imaging platform for tracking immunotherapeutic cells. *Nat. Biotechnol.* **23**(8): 983–987.

Ardenkjaer-Larsen, J. H., B. Fridlund et al. (2003). Increase in signal-to-noise ratio of > 10,000 times in liquid-state NMR. *Proc. Natl. Acad. Sci. USA* **100**(18): 10158–10163.

Artemov, D., Z. M. Bhujwalla et al. (1995a). In vivo selective measurement of (1–13C)-glucose metabolism in tumors by heteronuclear cross polarization. *Magn. Reson. Med.* **33**(2): 151–155.

Artemov, D., Z. M. Bhujwalla et al. (1995b). Pharmacokinetics of the 13C labeled anticancer agent temozolomide detected in vivo by selective cross-polarization transfer. *Magn. Reson. Med.* **34**(3): 338–342.

Artemov, D., N. Mori et al. (2003a). MR molecular imaging of the Her-2/neu receptor in breast cancer cells using targeted iron oxide nanoparticles. *Magn. Reson. Med.* **49**(3): 403–408.

Artemov, D., N. Mori et al. (2003b). Magnetic resonance molecular imaging of the HER-2/neu receptor. *Cancer Res.* **63**(11): 2723–2727.

Barrett, T., M. Brechbiel et al. (2007). MRI of tumor angiogenesis. *J. Magn. Reson. Imaging* **26**(2): 235–249.

Bhujwalla, Z. M., D. Artemov et al. (2002). Combined vascular and extracellular pH imaging of solid tumors. *NMR Biomed.* **15**(2): 114–119.

Bowers, C. R. and D. P. Weitekamp. (1986). Transformation of symmetrization order to nuclear-spin magnetization by chemical reaction and nuclear magnetic resonance. *Phys. Rev. Lett.* **57**(21): 2645–2648.

Buchthal, S. D., W. J. Thoma et al. (1989). In vivo T1 values of phosphorus metabolites in human liver and muscle determined at 1.5 T by chemical shift imaging. *NMR Biomed.* **2**(5–6): 298–304.

Caysa, H., H. Metz et al. (2011). Application of Benchtop-magnetic resonance imaging in a nude mouse tumor model. *J. Exp. Clin. Cancer Res.* **30**(1): 69.

Chan, K. W., M. T. McMahon et al. (2012). Natural D-glucose as a biodegradable MRI contrast agent for detecting cancer. *Magn. Reson. Med.* **68**(6): 1764–1773.

Chekmenev, E. Y., J. Hovener et al. (2008). PASADENA hyperpolarization of succinic acid for MRI and NMR spectroscopy. *J. Am. Chem. Soc.* **130**(13): 4212–4213.

Day, S. E., M. I. Kettunen et al. (2007). Detecting tumor response to treatment using hyperpolarized 13C magnetic resonance imaging and spectroscopy. *Nat. Med.* **13**(11): 1382–1387.

Du, F., Y. Zhang et al. (2012). Simultaneous measurement of glucose blood-brain transport constants and metabolic rate in rat brain using in-vivo (1)H MRS. *J. Cereb. Blood Flow Metab.* **32**(9): 1778–1787.

Ernst, R. R., G. Bodenhausen et al. (1987). *Principles of Nuclear Magnetic Resonance in One and Two Dimensions.* Clarendon Press, Oxford, U.K.

Ferris, C. F., B. Smerkers et al. (2011). Functional magnetic resonance imaging in awake animals. *Rev. Neurosci.* **22**(6): 665–674.

Filippi, M. and M. A. Rocca. (2004). Magnetization transfer magnetic resonance imaging in the assessment of neurological diseases. *J. Neuroimaging* **14**(4): 303–313.

Foley, L. M., T. K. Hitchens et al. (2005). Murine orthostatic response during prolonged vertical studies: Effect on cerebral blood flow measured by arterial spin-labeled MRI. *Magn. Reson. Med.* **54**(4): 798–806.

Frank, J. A., B. R. Miller et al. (2003). Clinically applicable labeling of mammalian and stem cells by combining superparamagnetic iron oxides and transfection agents. *Radiology* **228**(2): 480–487.

Geninatti Crich, S., B. Bussolati et al. (2006). Magnetic resonance visualization of tumor angiogenesis by targeting neural cell adhesion molecules with the highly sensitive gadolinium-loaded apoferritin probe. *Cancer Res.* **66**(18): 9196–9201.

Gillies, R. J., N. Raghunand et al. (2004). pH imaging. A review of pH measurement methods and applications in cancers. *IEEE Eng. Med. Biol.* **23**(5): 57–64.

Glunde, K., D. Artemov et al. (2010). Magnetic resonance spectroscopy in metabolic and molecular imaging and diagnosis of cancer. *Chem. Rev.* **110**(5): 3043–3059.

Glunde, K., Z. M. Bhujwalla et al. (2011). Choline metabolism in malignant transformation. *Nat. Rev. Cancer* **11**(12): 835–848.

Haase, A., J. Frahm et al. (1985). ^{1}H NMR chemical shift selective (CHESS) imaging. *Phys. Med. Biol.* **30**(4): 341–344.

Hancu, I., W. T. Dixon et al. (2010). CEST and PARACEST MR contrast agents. *Acta Radiologica* **51**(8): 910–923.

in 't Zandt, H. J., W. K. Renema et al. (2004). Cerebral creatine kinase deficiency influences metabolite levels and morphology in the mouse brain: A quantitative in vivo ^{1}H and 31P magnetic resonance study. *J. Neurochem.* **90**(6): 1321–1330.

Islam, T. and G. Wolf. (2009). The pharmacokinetics of the lymphotropic nanoparticle MRI contrast agent ferumoxtran-10. *Cancer Biomarkers* **5**(2): 69–73.

Kaida, S., H. Cabral et al. (2010). Visible drug delivery by supramolecular nanocarriers directing to single-platformed diagnosis and therapy of pancreatic tumor model. *Cancer Res.* **70**(18): 7031–7041.

Kato, Y. and D. Artemov. (2009). Monitoring of release of cargo from nanocarriers by MRI/MR spectroscopy (MRS): Significance of T2/T2* effect of iron particles. *Magn. Reson. Med.* **61**(5): 1059–1065.

Kato, Y., D. A. Holm et al. (2010). Noninvasive detection of temozolomide in brain tumor xenografts by magnetic resonance spectroscopy. *Neuro Oncol.* **12**(1): 71–79.

Kim, H., J. Zhang et al. (2011). Vascular phenotyping of brain tumors using magnetic resonance microscopy (µMRI). *J. Cereb. Blood Flow Metab.* **31**(7): 1623–1636.

Li, C., M. F. Penet al. (2008). Image-guided enzyme/prodrug cancer therapy. *Clin. Cancer Res.* **14**(2): 515–522.

Li, C., M. F. Penet et al. (2010). Nanoplex delivery of siRNA and prodrug enzyme for multimodality image-guided molecular pathway targeted cancer therapy. *ACS Nano* **4**(11): 6707–6716.

Li, C., P. T. Winnard, Jr. et al. (2006). Multimodal image-guided enzyme/prodrug cancer therapy. *J. Am. Chem. Soc.* **128**(47): 15072–15073.

Liu, G., Y. Li et al. (2011). Imaging in vivo extracellular pH with a single paramagnetic chemical exchange saturation transfer magnetic resonance imaging contrast agent. *Molecular Imaging* **11**(1): 47–57.

Louie, A. Y., M. M. Huber et al. (2000). In vivo visualization of gene expression using magnetic resonance imaging. *Nat. Biotechnol.* **18**(3): 321–325.

Mason, R. P., W. Rodbumrung et al. (1996). Hexafluorobenzene: A sensitive 19F NMR indicator of tumor oxygenation. *NMR Biomed.* **9**(3): 125–134.

McMahon, M. T., A. A. Gilad et al. (2008). New "multicolor" polypeptide diamagnetic chemical exchange saturation transfer (DIACEST) contrast agents for MRI. *Magn. Reson. Med.* **60**(4): 803–812.

Mori, S., J. Zhang et al. (2006). Magnetic resonance microscopy of mouse brain development. *Methods Mol. Med.* **124**: 129–147.

Nabuurs, C. I., D. W. Klomp et al. (2008). Localized sensitivity enhanced in vivo 13C MRS to detect glucose metabolism in the mouse brain. *Magn. Reson. Med.* **59**(3): 626–630.

Olson, E. S., T. Jiang et al. (2010). Activatable cell penetrating peptides linked to nanoparticles as dual probes for in vivo fluorescence and MR imaging of proteases. *Proc. Natl. Acad. Sci. USA* **107**(9): 4311–4316.

Onuki, Y., I. Jacobs et al. (2010). Noninvasive visualization of in vivo release and intratumoral distribution of surrogate MR contrast agent using the dual MR contrast technique. *Biomaterials* **31**(27): 7132–7138.

Posse, S., S. R. Dager et al. (1997). In vivo measurement of regional brain metabolic response to hyperventilation using magnetic resonance: Proton echo planar spectroscopic imaging (PEPSI). *Magn. Reson. Med.* **37**(6): 858–865.

Procissi, D., F. Claus et al. (2007). In vivo 19F magnetic resonance spectroscopy and chemical shift imaging of trifluoro-nitroimidazole as a potential hypoxia reporter in solid tumors. *Clin. Cancer Res.* **13**(12): 3738–3747.

Ratering, D., C. Baltes et al. (2008). Performance of a 200-MHz cryogenic RF probe designed for MRI and MRS of the murine brain. *Magn. Reson. Med.* **59**(6): 1440–1447.

Roemer, P. B., W. A. Edelstein et al. (1990). The NMR phased array. *Magn. Reson. Med.* **16**(2): 192–225.

Runge, V. M., T. Ai et al. (2011). The developmental history of the gadolinium chelates as intravenous contrast media for magnetic resonance. *Invest. Radiol.* **46**(12): 807–816.

Schepkin, V. D., F. C. Bejarano et al. (2012). In vivo magnetic resonance imaging of sodium and diffusion in rat glioma at 21.1 T. *Magn. Reson. Med.* **67**(4): 1159–1166.

Schepkin, V. D., W. W. Brey et al. (2010). Initial in vivo rodent sodium and proton MR imaging at 21.1 T. *Magn. Reson. Imaging* **28**(3): 400–407.

Schmidt, G. P., H. Kramer et al. (2007). Whole-body magnetic resonance imaging and positron emission tomography-computed tomography in oncology. *Top. Magn. Reson. Imaging* **18**(3): 193–202.

Schneider, J. E., T. Lanz et al. (2011). Accelerated cardiac magnetic resonance imaging in the mouse using an eight-channel array at 9.4 Tesla. *Magn. Reson. Med.* **65**(1): 60–70.

Shazeeb, M. S., C. H. Sotak et al. (2011). Targeted signal-amplifying enzymes enhance MRI of EGFR expression in an orthotopic model of human glioma. *Cancer Res.* **71**(6): 2230–2239.

Soher, B. J., P. C. van Zijl et al. (1996). Quantitative proton MR spectroscopic imaging of the human brain. *Magn. Reson. Med.* **35**(3): 356–363.

Stasinopoulos, I., M. F. Penet et al. (2011). Exploiting the tumor microenvironment for theranostic imaging. *NMR Biomed.* **24**(6): 636–647.

Stuber, M., W. D. Gilson et al. (2007). Positive contrast visualization of iron oxide-labeled stem cells using inversion-recovery with ON-resonant water suppression (IRON). *Magn. Reson. Med.* **58**(5): 1072–1077.

Terreno, E., A. Barge et al. (2008a). Highly shifted LIPOCEST agents based on the encapsulation of neutral polynuclear paramagnetic shift reagents. *Chem. Commun.* (5): 600–602.

Terreno, E., D. D. Castelli et al. (2008b). First ex-vivo MRI co-localization of two LIPOCEST agents. *Contrast Media Mol. Imaging* **3**(1): 38–43.

Tkac, I., Z. Starcuk et al. (1999). In vivo ^{1}H NMR spectroscopy of rat brain at 1 ms echo time. *Magn. Reson. Med.* **41**(4): 649–656.

van de Ven, F. J. M. (1995). *Multidimensional NMR in Liquids: Basic Principles and Experimental Methods.* VCH, New York, 1995.

van Zijl, P. C. and N. N. Yadav. (2011). Chemical exchange saturation transfer (CEST): What is in a name and what isn't? *Magn. Reson. Med.* **65**(4): 927–948.

van Zijl, P. C. M., C. K. Jones et al. (2007). MRI detection of glycogen in vivo by using chemical exchange saturation transfer imaging (glycoCEST). *Proc. Natl. Acad. Sci. USA* **104**(11): 4359–4364.

Viale, A. and S. Aime. (2010). Current concepts on hyperpolarized molecules in MRI. *Curr. Opin. Chem. Biol.* **14**(1): 90–96.

Wagenhaus, B., A. Pohlmann et al. (2012). Functional and morphological cardiac magnetic resonance imaging of mice using a cryogenic quadrature radiofrequency coil. *PLoS One* **7**(8): e42383.

Walczak, P. and J. W. Bulte. (2007). The role of noninvasive cellular imaging in developing cell-based therapies for neurodegenerative disorders. *Neurodegener. Dis.* **4**(4): 306–313.

Ward, K. M., A. H. Aletras et al. (2000). A new class of contrast agents for MRI based on proton chemical exchange dependent saturation transfer (CEST). *J. Magn. Reson.* **143**(1): 79–87.

Werner, E. J., A. Datta et al. (2008). High-relaxivity MRI contrast agents: Where coordination chemistry meets medical imaging. *Angew. Chem. Int. Ed. Engl.* **47**(45): 8568–8580.

Wijnen, J. P., T. W. Scheenen et al. (2010). 31P magnetic resonance spectroscopic imaging with polarisation transfer of phosphomono- and diesters at 3 T in the human brain: Relation with age and spatial differences. *NMR Biomed.* **23**(8): 968–976.

Woods, M., D. E. Woessner et al. (2006). Paramagnetic lanthanide complexes as PARACEST agents for medical imaging. *Chem. Soc. Rev.* **35**(6): 500–511.

Wu, Y., T. C. Soesbe et al. (2010). A responsive europium(III) chelate that provides a direct readout of pH by MRI. *J. Am. Chem. Soc.* **132**(40): 14002–14003.

Yushmanov, V. E., B. Yanovski et al. (2009). Sodium mapping in focal cerebral ischemia in the rat by quantitative (23)Na MRI. *J. Magn. Reson. Imaging* **29**(4): 962–966.

Zhang, J., L. J. Richards et al. (2006). Characterization of mouse brain and its development using diffusion tensor imaging and computational techniques. *Conf. Proc. IEEE Eng. Med. Biol. Soc.* **1**: 2252–2255.

10

Optical Imaging

Matthew A. Lewis

10.1 INTRODUCTION

Imaging with nonionizing photons in the visible and infrared regions of the electromagnetic spectrum is a firmly established method in preclinical imaging. While some imaging and tomographic techniques in the near-infrared have been translated from mouse to humans, preclinical optical imaging has also produced some extremely useful modalities with little or no potential for clinical translation. The advantages of optical imaging include excellent sensitivity, good contrast, lower cost, and the use of nonionizing radiation. In general, however, optical imaging suffers from poor resolution, even in comparison with other high-sensitivity modalities in nuclear medicine imaging.

10.2 CLASSIFICATIONS

Optical imaging can be classified into a variety of families. The most interesting classification of optical imaging mirrors those in clinical imaging: that is, the distinction between transmission and emission imaging. In transmission imaging, some form of energy is transmitted into the body, where it interacts and then leaves the body, to be detected. Emission imaging encompasses the passive modalities that capture photons generated inside the body via the conversion of energy from other sources to optical light. Before differentiating these differences, though, classifications based on differences in data acquisition are highlighted, and appropriate mathematical models for light propagation in small animal models are presented.

10.3 DATA ACQUISITION

10.3.1 CONTACT VERSUS NONCONTACT

The earliest systems for optical imaging in preclinical models were contact-based using fiber optics attached to the animal being imaged. There are several advantages of this approach. First, from a modeling perspective, the source term in any transmission-based optical imaging technique is fairly well understood; in principle, the input light can be modeled as an impulse function on the surface. Similarly, the response of a fiber-based sensor attached to the surface can be easily modeled. The use of point source and point detector geometries gives rise to the infamous "banana" plot, which is a map of the sensitivity of the geometry to heterogeneities in the media. Although contact-based systems are compatible with continuous wave (CW), frequency-domain, and time-domain modes of operation, the disadvantage is the low number of data channels. There is an associated cost with adding each additional fiber sensor.

More recently, the rapid commercial expansion of scientific-grade, charge-coupled device (CCD)-based cameras has given rise to noncontact, lens-coupled acquisition systems. The advantage of noncontact imaging is a larger degree of sampling and the *potential* for higher resolution imaging. One disadvantage is the relative inefficiency of lens coupling (Swindell 1991; Liu et al. 1994; Yu and Boone 1997). While most noncontact CCD camera–based systems operate as CW imaging systems, both frequency-domain and time-domain imaging are possible with CCD-based detectors through the integration of a time-dependent filter.

For noncontact imaging, the data from cameras must first be returned to the surface of the animal in order to perform the inverse problem for the mathematical model (Meyer et al. 2007). This step is one area where proper modeling of the index of refraction may be important. Along with the surface curvature relative to the camera lens, the index of refraction can modulate the apparent light fluence on the animal surface, but fortunately there are tools to account for these effects (Lasser et al. 2008; Chen et al. 2010; Sarasa-Renedo et al. 2010).

10.3.2 TIME DOMAIN, FREQUENCY DOMAIN, AND CW/DC

Considering the generation of photons and their subsequent detection after propagating through tissue, there are three classifications of interest. In CW or DC optical imaging, the sources of optical photons are either constant or slowly changing compared to the scale of the imaging experiment. CW is generally the least expensive form of optical imaging as it may utilize integrating detectors such as scientific-grade CCD cameras. The emission optical imaging modalities (bioluminescence imaging (BLI) and Cerenkov luminescence imaging (CLI)) are always considered as CW since the sources are nearly in steady state. Although dynamic studies in time are routinely utilized, the luminescence image acquired during a given exposure is taken as the average of the signal emitted during that time period. One disadvantage of the CW approach is the inability to separate the effects of scatter and absorption in the image reconstruction formulation.

Frequency-domain imaging is commonly utilized in transmission optical imaging modalities such as diffuse optical imaging and tomography (DOT) and fluorescence molecular tomography (FMT). Instead of a steady-state light source, the light that is injected into the animal is amplitude-modulated at frequencies above 10 MHz. This amplitude modulation has several advantages. First, with a lock-in amplifier, it can increase the sensitivity on the detector side of the imaging apparatus. Second, through appropriate analysis during image reconstruction the absorption and scattering maps can be independently reconstructed. Frequency-domain imaging can be accomplished with integrating detectors such as CCDs provided that a filter with fast temporal dynamics is coupled to the camera. Unlike CW imaging, frequency-domain imaging can involve wave-like phenomena that propagate through tissue. These diffuse-photon density waves (DPDWs) can interact with tissue in a manner that can be modeled as wave scattering and diffraction. Wave-based image reconstruction methods such as diffraction tomography have been explored for this mode of optical imaging (Li et al. 1997; Liu et al. 1999; Matson and Liu 1999a,b).

Time-domain optical imaging is both the most technically challenging and expensive form of transmission optical imaging. If a short pulse of light is injected into an animal, some of the detected photons will have experienced less scattering events during propagation compared to others. These *ballistic* or *snake* photons travel the short optical path length from source to detector, and therefore appear earliest in the pulse generated in the detector electronics. In general, time-domain optical imaging cannot be realized with inexpensive integrating detectors and requires specialized fast optical detectors such as photomultiplier tubes and Kerr filters.

These three forms of optical imaging are in fact not equivalent. For instance, they differ in sensitivity to inhomogeneities in absorption and scatter coefficients in the imaged object. Qualitatively, the previously described banana plot indicates the typical paths that photons take between source and detector. Quantitatively, the banana plot is related to the photon-measurement density function (PMDF), a perturbative sensitivity map for the system Green's function. Compared to CW imaging, where the system is most sensitive to inhomogeneities near the source or the detector, time-domain imaging provides a more uniform sensitivity in the space between detector and source. The sensitivity for frequency domain is somewhere in between, and many measurements made at different frequencies are related to time-domain optical imaging by a Fourier transform.

10.4 MATHEMATICAL MODELS

10.4.1 TRANSPORT EQUATION

The master equation for the propagation of visible and near-infrared photons through complex media such as tissues is the radiative transport equation (RTE). The RTE is a integrodifferential equation for the radiance $\varphi\left(\vec{r},\hat{n},E,t\right)$, which describes the number of photons at position $\vec{r}$ in space, propagating in direction $\hat{n}$, and with energy E at time t:

$$\left\{\hat{n}\cdot\nabla+\mu_a+\mu_s+\frac{\partial}{\partial t}\right\}\varphi\left(\vec{r},\hat{n},E,t\right)=S\left(\vec{r},\hat{n},E,t\right)+\mu_s\int\sigma\left(\hat{n},E,\hat{n}',E'\right)\varphi\left(\vec{r},\hat{n},E,t\right)d\Omega' \qquad (10.1)$$

In the case of monochromatic light or integrating detectors, the dependence on energy (or equivalently wavelength) can be dropped. In general, the RTE is too difficult to solve analytically, and Monte Carlo (MC) methods can be used to represent the physical processes for each term in the RTE (Wang et al. 1995). An MC package specific to modeling light transport in mice is publically available (Li et al. 2004). In some cases, however, the RTE can be simplified to another model. For example, the RTE is also valid for imaging with x-rays or γ-rays. Under a few assumptions, the Radon transform can be derived from the RTE.

The standard approach in the tissue optics literature to avoid the complexity of the RTE is to move to the diffusion approximation. Since this derivation exists in uncountable papers in the field, only a sketch of the derivation is appropriate: the radiance $\varphi(\vec{r},\hat{n},t)$ is expanded in terms of spherical harmonics, and the angular dependence of the scattering kernel is factored out. Under the assumption that the scattering coefficient is much larger than the absorption coefficient, one may truncate the series expansion of the radiance at the first order, called the P1 approximation. Finally, the resulting radiance is integrated over the unit sphere to remove the direction parameter, giving a diffusion approximation to the RTE in terms of the fluence or intensity $\phi(\vec{r},t)$.

10.4.2 DIFFUSION APPROXIMATION

The resulting model is a well-studied partial differential equation

$$D\nabla^2\phi\left(\vec{r},t\right)-\mu_a\phi\left(\vec{r},t\right)=\frac{1}{c}\frac{\partial\phi\left(\vec{r},t\right)}{\partial t}-S\left(\vec{r},t\right) \qquad (10.2)$$

where the diffusion coefficient D is taken as

$$D=\frac{1}{3\left(\mu_s'+\mu_a\right)} \qquad (10.3)$$

In CW and frequency-domain imaging systems, one operates in the steady state, and given that in general the reduced scattering coefficient μ_s' is at least one order of magnitude larger that the absorption coefficient μ_a (otherwise diffusion approximation would not be valid), the model can be written in a simpler form as

$$(\nabla^2-k^2)\phi\left(\vec{r}\right)=-S\left(\vec{r}\right) \qquad (10.4)$$

where $k=\sqrt{\mu_a/D}\approx\sqrt{3\mu_a\mu_s'}$ and is, in general, a function of position.

This inhomogeneous modified Helmholtz equation illustrates several characteristics of optical imaging. First, the terms to be reconstructed in the appropriate inverse problem are evident.

For transmission optical imaging, the experimentalist controls the source terms through the placement of fiber optics or noncontact illumination sources. The task in the transmission case is to reconstruct the information in $k(\vec{r})$, which represents the heterogeneities in the preclinical animal, based on the fluence of light at the surface of the animal. In the emission imaging case (bioluminescence and Cerenkov luminescence tomography; BLT and CLT), both $k(\vec{r})$ and the source $S(\vec{r})$ are unknown. In general, it is too difficult to reconstruct both with the given data, and hence $k(\vec{r})$ is either assumed to be constant, mapped using an independent optical modality (DOT), or modeled using a priori information from an anatomical modality (computed tomography (CT) or magnetic resonance imaging (MRI)).

Second, it is not possible to separately image $\mu_a(\vec{r})$ and $\mu'_s(\vec{r})$ in CW transmission optical imaging with a monochromatic source; only the product of these parameters can be reconstructed. In general, the reduced scattering coefficient does not vary as much as the absorption, so this reconstruction often exhibits the gross features of the major absorbing structures in the animal, but the image is qualitative.

For fluorescence imaging, an appropriate model is two coupled diffusion approximations for the excitation and emission wavelengths. The inhomogeneous source term in the emission wavelength equation is proportional to the product of the fluorophore concentration and the excitation light fluence.

The diffusion approximation is known to be inaccurate near boundaries and sources. Higher order approximations to the RTE are commonly found in the literature for all types of optical imaging (Zhong et al. 2011c).

10.4.3 INVERSE PROBLEM

To numerically implement the image reconstruction in optical imaging, most algorithms take a similar approach. Let x be the unknown source distribution or attenuation map (emission and transmission imaging, respectively) that is sought in some discrete basis set such as pixels or other mesh. Using the mathematical model of choice for light propagation in tissue, a forward solver F is implemented to calculate the photon fluence ϕ:

$$\phi = Fx \tag{10.5}$$

The data acquisition system D is modeled as a mapping from the surface photon density to the sensor measurements:

$$y = D\phi \tag{10.6}$$

In the simplest form, all optical imaging can be modeled as a mapping from the unknown parameter map to the measured data:

$$y = Ax = DFx \tag{10.7}$$

A is generally a nonlinear operator, but in some cases the problem can be linearized to produce a matrix equation. The vast majority of optical image reconstruction algorithms are then a variation on the theme of the following problem:

$$\hat{x} = arg\min_{x \geq 0} \frac{1}{2}\|y - Ax\|^2 + \frac{\alpha}{2}x'Rx \tag{10.8}$$

The first term can be viewed as a data consistency term, while the second term is a regularization that deals with the stability issues common to most optical imaging techniques. The choice of the regularization parameter is often the key to the success of the image reconstruction method (Feng et al. 2011).

10.4.4 TRANSMISSION OPTICAL IMAGING

10.4.4.1 Diffuse Optical Imaging and Tomography

DOT is the most mature form of optical imaging with a rich theoretical and experimental literature extending back three decades, including clinical exploration in the brain, breast, and neonate. The principal idea is to inject near-infrared photons into tissue and measure the intensity of photons that leave the tissue at other locations. Using these data, the image reconstruction task is to map the spatial distribution of the two major chromophores in mammalian tissue: oxy- and deoxy-hemoglobin. Minor chromophores such as cytochrome *c* are occasionally targeted, but in general oximetry is the principal application of DOT. If two or more wavelengths of light can be utilized, then absorption due to these two species of chromophores can be separated.

DOT is primarily of interest since it provides a foundation for the development and exploration of the other optical imaging modalities to follow. At present, there are no noncontact commercial preclinical imaging systems optimized for straight DOT. Fiber-based systems for CW and frequency-domain acquisition can be assembled quite readily from off-the-shelf components. As described in the section on hybrid imaging, the contrast mechanism in photoacoustic/optoacoustic imaging is related to DOT with the advantage of dramatically improved spatial resolution. In practice, many applications of DOT are now being covered using photoacoustics.

10.4.4.2 Fluorescence Molecular Tomography

FMT (also referred to as FDOT, i.e., fluorescence diffuse optical tomography) is a molecular imaging version of DOT, which relies on exogenous contrast agents rather than endogenous contrast from chromophores. Compared to DOT, FDOT is much more applicable to the investigator interested in preclinical imaging. Typically, a targeting moiety will be labeled with a dye, fluorophore, or, more recently, a quantum dot. Since fluorophores require excitation, many investigators have focused on reporter systems that are shifted toward the near-infrared, both in excitation and emission. In some sense, FMT is a hybrid of transmission and emission tomography. The transport of the excitation photons through the animal tissue is identical to DOT: the sources are limited to the animal surface. The transport of the photon emitted by the fluorophores is similar to the inverse source problem in emission optical imaging. One complication is that excitation and emission photons experience correlated but slightly different attenuation due to the Stokes shift between excitation and emission. In practice, though, FMT is somewhat easier than DOT and BLI due to the high contrast of the agents, the modulation control one has on the excitation source, and the a priori knowledge of the fluorescence spectrum.

Since the ideal fluorescence contrast agent would have high specificity and accumulate only in the targeted tissue, a significant degree of spatial sparsity may be inherent to preclinical fluorescence imaging. That said, one disadvantage of fluorescence imaging is the autofluorescence produced by endogenous fluorophores, which contributes to a diffuse background in the image (Soubret and Ntziachristos 2006). In that sense, a fluorophore with a large Stokes shift is advantageous. Nevertheless, sparsity-exploiting methods such as compressive sensing have been applied to the FMT imaging reconstruction problem (Jin et al. 2014).

An interesting solution to the difficult image reconstruction problem can be constructed by exploiting the time course of signals obtained during dynamic optical imaging. Since different organs have different pharmacokinetic uptakes, it was demonstrated that the optical fluence on the surface of the mouse is dependent upon the organ that dominates the contrast for that particular surface area. By applying principal component analysis (PCA) to an image series following indocyanine green (ICG) injection, the heart, small intestine, left and median lobes of liver, kidneys, spleen, large and small intestines, and the brain were all differentiated in the light emitted on the animal surface over 400 s (Hillman and Moore 2007).

Using similar mathematical tools but applied to spectral data rather than time series, unmixing of multiple fluorophores in the same animal is possible. This experiment is more complicated than the former since each fluorophore truly requires an independent image reconstruction due to differential absorption at different emission wavelengths. Recently, Radrich et al. described methods for simultaneous imaging with both IntegriSense 680 and AngioSense 750 in the KRAS mouse model (Radrich et al. 2014).

The majority of commercial preclinical optical imaging systems are capable of CW, multispectral, fluorescence imaging. Both epi- and trans-excitation illumination are common. Fully tomographic FMT systems are typically built using mirror-based (Guggenheim et al. 2013) or multi-camera approaches, as shown in Figures 10.1 and 10.2, respectively. Simon Cherry and colleagues described a system for simultaneous small animal positron emission tomography (PET) and fluorescence imaging using mirrors (Li et al. 2009, 2011). This approach is attractive for fluorescence reporter systems that are dependent upon other factors such as pH. If the fluorescence contrast agent is also radiolabeled, then the PET image can give a concentration map and the secondary parameter can be reconstructed instead. In recent work, the Nziachristos group described efforts to combine a preclinical FMT system with a phase-contrast micro-CT system (Mohajerani et al. 2014). The fused images, as shown in Figure 10.3, demonstrate a remarkable delineation of the mouse anatomy with specific uptake of IntegriSense 680 fluorescence contrast agent in the pancreatic tumor.

Mirroring the migration from DOT to photoacoustics, the recent explosion in preclinical multispectral optoacoustic imaging with fluorescence contrast agents will undoubtedly come at the expense of traditional FMT.

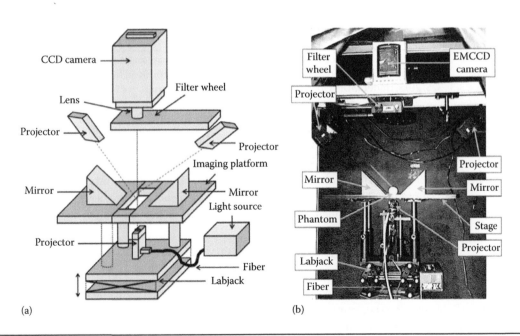

(a) (b)

FIGURE 10.1 Single-camera, multispectral, multimodality optical imaging system capable of noncontact, continuous wave fluorescence molecular imaging and bioluminescence imaging. Mirrors enable tomographic coverage of the animal surface. Trans-illumination is supplied by an optical fiber from the bottom. (a) Schematic and (b) implementation. (Reproduced from Guggenheim, J.A. et al., Multi-modal molecular diffuse optical tomography system for small animal imaging, *Meas. Sci. Technol.*, 24(10), 105405. Copyright 2013 by permission of IOP Publishing. All rights reserved.)

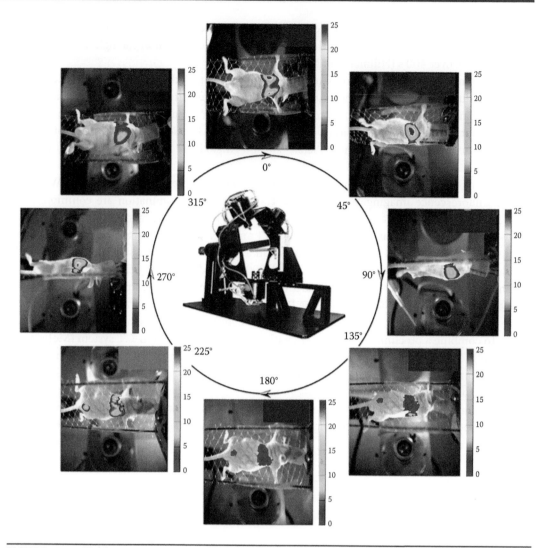

FIGURE 10.2 Multicamera, noncontact, light emission system for bioluminescence tomography in mice. Note the mesh bed for light propagation from the mammary fat pad. (Reproduced from Lewis, M. et al., *Diagnostics*, 3(3), 325, 2013. Under MDPI Open Access Policy. With permission.)

10.4.5 EMISSION OPTICAL IMAGING

10.4.5.1 Bioluminescence Imaging

BLI is a firmly established optical emission imaging modality that is routinely employed in the biomedical sciences laboratory. By far the most common bioluminescence reporter system is the North American firefly (*Photinus pyralis*) luciferase-catalyzed oxidation of the substrate D-luciferin.

$$\text{ATP} + \text{D-Luciferin} + \text{O}_2 + \text{Mg}^{2+} \xrightarrow{\text{luciferase}} \text{Oxyluciferin} + \text{AMP} + \text{CO}_2 + \gamma \qquad (10.9)$$

This multistep reaction is now well understood (Marques and Esteves da Silva 2009), and the modulation of the light emission spectrum via modification of the substrate is actively being explored. The reaction product oxyluciferin is in an excited state, which decays to the ground state through the spontaneous emission of a visible light photon with a mean wavelength of 570 nm. While luciferin-regenerating enzymes (LREs) have been identified that convert oxyluciferin back to the original substrate (Gomi and Kajiyama 2001; Gomi et al. 2002), expression of a regenerative system has been limited to *E. coli*, and it remains to be seen if brighter, longer lasting reporter system using this approach can be developed for preclinical models.

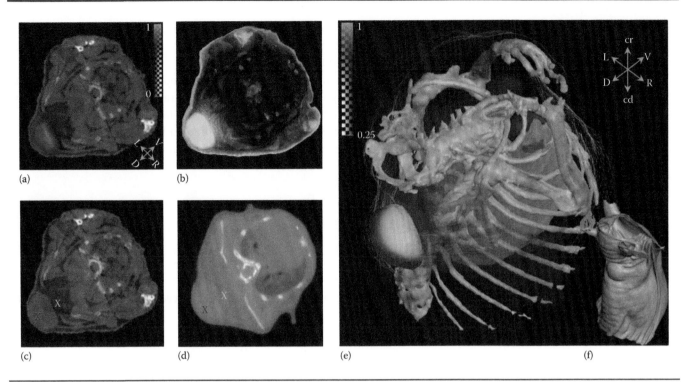

FIGURE 10.3 Example of high-resolution fluorescence molecular tomography fused with phase-contrast micro-CT in mouse model: (a) fluorescence reconstruction superimposed on CT slice, (b) validation with cryoslice imaging post mortem, (c) phase-contrast CT slice, (d) normal CT slice, (e) fused 3D visualization, and (f) surface rendering. (Reproduced with permission from Mohajerani, P. et al., FMT-PCCT: Hybrid fluorescence molecular tomography-x-ray phase-contrast CT imaging of mouse models, *IEEE Trans. Med. Imaging*, 33(7), 1434–1446. © 2014 IEEE.)

BLI is now a firmly established modality for preclinical imaging in oncology (Thorne and Contag 2005), infection, and other disease models. While an extremely weak autoluminescence due to inherent metabolism has been reported as detectable with extremely long integration times (Troy et al. 2004), there is essentially no background in BLI and consequently very weak signals can be detected. It is reported that less than 1000 cells can be detected either subcutaneously or after intraperitoneal implantation (Klerk et al. 2007). When compared to tumor volume as assessed by MRI or calipers, a strong correlation with the planar bioluminescence signal is observed (Szentirmai et al. 2006; Lewis et al. 2013), although extensive tumor vascularization can result in decreased light output to due absorption by hemoglobin. As the temperature is increased, the broad spectral emission for firefly bioluminescence shifts (~100 nm full width at half-maximum) to the red, producing more near-infrared photons that have less absorption and a greater probability of exiting the animal model (Zhao et al. 2005). While efforts to mutate the firefly reporter system and further red shift the emission are ongoing, ab initio quantum chemistry computer simulations are now bringing additional insight into the molecular mechanism of light production at the level of individual protein residues (Tagami et al. 2009).

Although BLI has been purported to be in the same sensitivity range as the nuclear medicine modality PET, there is some concern that the bioluminescence reporter system may not follow the tracer principle and may in fact perturb the biological system. High-level expression of the transfected luciferase enzyme has been shown to inhibit tumor growth (Brutkiewicz et al. 2007), presumably since the cellular machinery is so involved in the expression of the reporter that proliferation is downregulated. Furthermore, the substrate luciferin appears to be actively pumped out of cancer cells, leading to a reduction in the reporter system efficiency. While MDR1, MRP1, and MRP2 (common multidrug resistance pumps) do not recognize D-luciferin as a substrate, the ATP-binding cassette (ABC) family transporter AVCG2/BCRP has been shown to modulate the bioluminescence output (Zhang et al. 2007; Huang et al. 2011). In the opposing realm of thought, Contag and colleagues have attempted to engineer a releasable luciferin–transporter conjugate that allows cargo release and bioluminescence only after entry into the cell. This system "emulates drug-conjugate delivery into a cell, drug release, and drug turnover by an intracellular target" (Jones et al. 2006). Furthermore, continuous, steady-state bioluminescence has been reported in rat using implanted micro-osmostic pumps (Gross et al. 2007).

Initial efforts to reconstruct bioluminescent light sources at depth in tissue focused on point sources modeled using the diffusion approximation to the radiative transport equation (Comsa et al. 2006). An important consideration in this model is the appropriate boundary condition at the surface of the tissue. A popular boundary condition for BLI is the extrapolated boundary condition that accounts for the refractive index mismatch at the surface of the animal. In this model, the surface fluence can be solved exactly (Comsa et al. 2006):

$$\phi(r) = \frac{1}{4\pi D}\left[\frac{\exp(-\mu_{eff}r_1)}{r_1} - \frac{\exp(-\mu_{eff}r_2)}{r_2}\right] \tag{10.10}$$

where

r is the distance from the source to the surface

$D = 1/3(\mu_s' + \mu_a)$ is the diffusion coefficient (dominated by reduced scatter if diffusion approximation is valid)

$$\mu_{eff} = \sqrt{\mu_a/D}$$

$$r_1 = \sqrt{r^2 + d^2}$$

$$r_2 = \sqrt{r^2 + (d + 2z_b)^2}$$

$z_b = 2\dfrac{1 + R_{eff}}{1 - R_{eff}}D$ is the extrapolated boundary that is slightly off-set from the true boundary.

The effective internal reflection coefficient (R_{eff}) can be calculated from the refractive index mismatch at the surface. Provided that the tissue attenuation parameters are known and the reflection coefficient can be approximated with reasonable accuracy, the source depth d can be estimated from the data using a nonlinear fitting algorithm such as the Levenberg–Marquardt algorithm. In addition, this model is reasonably accurate for point sources near curved tissue surfaces, as can be found in a small animal. Analytical expressions for the forward problem for spherical surface and solid sources are also available (Cong et al. 2004).

Another popular boundary condition is the partial-current BC, which also has a closed form for parameter fitting. To improve the source estimation, multispectral data may also be considered. Since the absorption coefficient is strongly dependent upon wavelength (scattering coefficient is not as variable), information on the depth of the source is encoded in the spectrum of the light that escapes the surface.

10.4.5.2 Bioluminescence Tomography

A more ambitious application of BLI involves the complete mapping of the surface intensity of light and the attempt to reconstruct the source distribution using this tomographic approach. The original BLT system was developed (patented but not published) by Ge Wang during his time at the University of Iowa, who combined CW optical imaging with a micro-CT for anatomical coregistration. The combination with a second modality was soon proven useful for the extremely difficult imaging reconstruction problem in BLT.

In BLT, one is primarily interested in reconstructing the source term on the right-hand side of the diffusion equation 10.2 above. While it is possible to formulate an inverse problem for reconstructing the source, absorption map, and scatter map solely from the light fluence on the surface of the animal (Han et al. 2007), this problem (which is akin to the identification problem in single-photon emission computed tomography (SPECT)) is too much to ask from the limited data. Instead, almost all approaches to the BLT image reconstruction problem assume that the attenuation maps are known, perhaps from a model generated by a second modality such as CT (Lv et al. 2006a; Klose and Beattie 2009; Ma et al. 2013; Naser et al. 2014; Wu et al. 2014), MRI (Allard et al. 2007; Klose and Beattie 2009; Zhang et al. 2014), or DOT (Zhang et al. 2008; Naser and Patterson 2010; Naser et al. 2012).

Such model-based reconstruction has been a feature of BLT since its introduction (Gu et al. 2004). Models have been complemented by MC forward solvers to estimate the surface fluence for a given distribution (Côté et al. 2005; Kumar et al. 2007). The Mouse Optical Simulation Environment (MOSE) is one of several, publically available MC platforms that can be used to simulate emission tomography (Li et al. 2004, 2007). Since MC is typically slow, a more practical approach is to use the finite element method as the forward solver for iteration image reconstruction methods for BLT (Cong et al. 2005; Lv et al. 2006, 2007). Klose and collaborators have thoroughly investigated the use of higher order approximations to the radiative transport equation that can be implemented effectively and utilized as forward solvers (Klose and Larsen 2006), and similar reports have subsequently been published by others with higher order spherical harmonic approximations (Lu et al. 2009a), the full RTE (Lu et al. 2009b; Gao and Zhao 2010a,b), as well as other approximations to the RTE such as the phase approximation model (Cong and Wang 2010).

The full machinery of the applied mathematics of inverse problems has also been applied to BLT (Wang et al. 2004; Han et al. 2006; Han and Wang 2007, 2008). In particular, BLT without using multispectral data is underdetermined and therefore nonunique. The simplest example of this nonuniqueness is identical data produced by a point source or a concentric spherical source at the center of a spherical turbid body. While this limitation can be mitigated with multispectral BLT (msBLT) (Cong and Wang 2006), the ill-posedness due to stability that is inherent to all optical imaging modalities remains, and reconstructions will always be low-resolution, low-pass-filtered versions of the true source distribution

A variety of algorithms have demonstrated some progress with this difficult imaging problem. Cong and colleagues reported using a Born-type approximation to invert the diffusion equation (Cong et al. 2006). Recently, it has been shown that a Born series solution to problems in optical imaging has better convergence than in wave propagation problems (Markel and Schotland 2007; Moskow and Schotland 2009; Arridge et al. 2012). While Cong reported inversion of data from a phantom, the higher order Born series approach has not been investigated. Another approach is to limit the volume for valid reconstruction of the source term. Since the bioluminescent cells were typically implanted in an animal model by the investigator, a priori information on the possible location can be incorporated as a *permissible source region* (Cong and Wang 2006; Feng et al. 2008; Naser and Patterson 2011a,b; Qin et al. 2011), *trust region* (Zhang et al. 2010), or *blocking-off* (Klose et al. 2010). Spatial sparsity of the source term has also been exploited using l1 norm approaches adapted from compressive sensing (Lu et al. 2009c; Cong and Wang 2010; Gao and Zhao 2010a,b; Basevi et al. 2012; Feng et al. 2012; Jin and He 2012; Zhang et al. 2012a). Statistical iterative algorithms common in nuclear medicine emission tomography can also be adapted to the BLT problem (Slavine et al. 2006; Behrooz et al. 2013). Iterative algorithms for solving the unconstrained optimization BLT image reconstruction formulation have been studied in terms of solver type, including gradient projection, preconditioned conjugate gradient, coordinate descent, Landweber, modified Newton, graph cuts, and incremental gradient (related to ordered subsets methods in nuclear medicine) (Ahn et al. 2008; Shi and Mao 2013). When a complex model of the animal anatomy is available, a Bayesian approach to the image reconstruction can be made, as was done by Feng et al. to incorporate the a priori information from the model (Feng et al. 2009).

In some cases, the light fluence may not be available on the entire animal surface, but image reconstruction from partial measurement is still possible (Jiang et al. 2007). In one paper of note, Wang, Shen, and Cong considered the temperature dependence of bioluminescence and analyzed the possibility of improving BLT by looking for modulations due to localized heating at depth from therapeutic ultrasound (Wang et al. 2006b; Wang et al. 2008a). In a more recent development, ultrasound tagging of bioluminescent light emission (analogous to acoustooptic imaging) has been proposed (Huynh et al. 2013), and the image reconstruction problem has been analyzed (Bal and Schotland 2014). Since typical BLI intensities are five orders of magnitude or more lower than fluorescence (where ultrasound modulation is a more active area of research), it is doubtful that this approach will be adopted for widespread use. Another type of modulation that may improve the image reconstruction is the varying of boundary conditions, as proposed by Soloviev (2007).

For some time, a commercial BLT system was marketed by Xenogen, but the majority of BLI in the literature is planar imaging. Many research groups have developed and published BLT systems (Wang et al. 2006a; Yan et al. 2012), including multicamera (Lewis et al. 2013) and multispectral (Dehghani et al. 2006, 2008) modifications of planar system with mirrors (Chaudhari et al. 2005; Wang et al. 2008b), and even a system proposed to combine small animal PET with BLT (Alexandrakis et al. 2005, 2006).

10.4.5.3 Cerenkov Luminescence Imaging and Tomography

In 2009, Robertson et al. (2009) reported using instrumentation designed for BLI to detect PET tracers in vivo. Shortly afterward, others reported imaging of ^{18}F in a microfluidics platform (Cho et al. 2009). Both groups characterized the light emission as Cerenkov radiation, a classical phenomenon in which a charged particle traveling with a velocity greater than the speed of light in a medium produces optical photons. Mathematically, this condition is satisfied when the product of the relativistic velocity ratio ($\beta = v_p/c$) and the local index of refraction n is greater than 1.

$$\beta n > 1 \tag{10.11}$$

The particle velocity and hence β can be related to the relativistic kinetic energy of the charged particle:

$$KE = m_0 c^2 \left[\frac{1}{\sqrt{1-\beta^2}} - 1 \right] \tag{10.12}$$

So, the inequality above can also be written as

$$n \sqrt{1 - \left(\frac{m_0 c^2}{KE + m_0 c^2} \right)^2} > 1 \tag{10.13}$$

Since β-particles associated with PET and SPECT tracers are emitted with a spectrum of kinetic energies characterized by the end-point or maximum kinetic energy, only a subset of decays will produce charged particles that can produce Cerenkov radiation. The Cerenkov kinetic energy threshold can be found by solving the inequality* for the charged particle kinetic energy:

$$KE > m_0 c^2 \left[\left(1 - \frac{1}{n^2} \right)^{-\frac{1}{2}} - 1 \right] \tag{10.14}$$

Furthermore, since the charged particle will continuously lose kinetic energy in a nearly linear relationship with the propagation distance, it may quickly drop below the threshold for Cerenkov production.

---

* For β-particles, $m_0 c^2 = 0.511$ MeV. For $n = 4/3$ (as in water), the quantity in brackets is 0.511..., which is entirely coincidental.

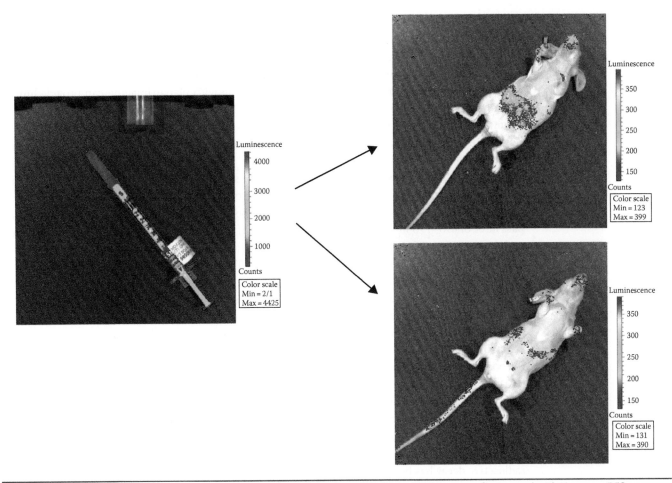

FIGURE 10.4   Cerenkov luminescence imaging pre- and post-injection of FDG into MCF7 xenografts. Imaged with Xenogen IVIS Spectrum by the author.

Although ^{18}F does produce Cerenkov luminescence, it is far from optimal because a large portion of positrons fall below the threshold for Cerenkov production. The production and efficiency of Cerenkov light production in both water and soft tissue from medical radionuclides have been studied in the context of the Frank–Tamm formula by several groups (Beattie et al. 2012). In terms of efficiency among the cyclotron-produced PET isotopes, ^{18}F is better than only [64]Cu (0.1338 vs. 0.0583; 550–570 nm photons per disintegration). In fact, radioisotopes that possess high positron end-point kinetic energy, which is typically associated with lower image quality due to expanded point-spread function, are better for CLI due to the increased sensitivity. As shown in Figure 10.4, the Cerenkov luminescence from 100 μCi of [18]FDG is detectable both in the shipping syringe and post injection in the preclinical tumor model.

Robertson et al. demonstrated FDG uptake in colon tumors implanted in the mouse flank and correlated the Cerenkov luminescence with whole-body micro-PET imaging of the same animals. The authors commented that, unlike preclinical PET, multiple mice could be imaged simultaneously in a commercial optical imaging system producing potential high-throughput screening with CLI. They estimated that CLI was 4–30 times faster than small animal PET imaging. By the end of 2009, Spinelli and colleagues in Italy reported dynamic CLI in a mouse using ^{18}F-FDG (Spinelli et al. 2010) and Cerenkov luminescence observed with [68]Ga. A comprehensive comparison of PET and CLI as high-throughput quantitative modalities in the evaluation of new cancer therapeutics was subsequently presented by the original Cerenkov authors (Robertson et al. 2011). While still a relatively new preclinical optical imaging modality, a least four reviews on Cerenkov imaging have already appeared in the literature (Xu et al. 2011; Qin et al. 2012; Thorek et al. 2012; Chin et al. 2013).

Because CLI has essentially the same acquisition geometry as planar BLI, there is a lack of information on the depth of the radioisotope source. As with bioluminescence, there are two approaches for reconstructing the source term $S(x)$ in the appropriate inverse problem model. First, multispectral imaging using filters exploits the fact that Cerenkov photons at different wavelengths will experience different absorption. Using multispectral imaging with planar imaging can provide limited depth information. Spinelli et al. have translated depth estimation methods for planar imaging in a semi-infinite medium from BLI (Kuo et al. 2007) to CLI (Spinelli et al. 2010, 2011b; Spinelli and Boschi 2012). A more detailed treatment of the relevant physics for Cerenkov production and detection can be found in the study of Das et al. (2014).

The second approach is to attack the full inverse source problem and attempt Cerenkov luminescence tomography (CLT). Li et al. used two mirrors placed adjacent to the mouse so that the entire surface light emission could be captured by the CCD camera (Li et al. 2010). Using a bandpass filter of 695–770 nm, only Cerenkov photons with potentially deep penetration were detected, and the source distribution for CLT was solved using a finite element, DC diffusion approximation model and a preconditioned conjugate gradient with Tikhonov regularization numerical inversion. While the CLT source reconstruction was of lower resolution than the micro-PET image in the same mouse, the CLT and PET activity reconstructions were spatially co-located. For single-photon gamma emitters, imaging validation with preclinical SPECT is appropriate (Hu et al. 2010). Many image reconstructions methods developed for BLT have subsequently been adapted to CLT (Zhong et al. 2011a–c; Ding et al. 2014). For β-emitters that also emit γ-rays, it may be possible to improve the poor resolution of the Cerenkov imaging by using data from a secondary gamma camera system (Hu et al. 2012).

CLI has been employed for a variety of applications. In the study of Liu et al. (2010a), skeletal uptake of $Na^{18}F$ in the mouse spine was demonstrated using CLI and compared with micro-PET. In addition, the authors reported in vivo imaging with ^{90}Y, a pure β-emitter that is difficult to image with nuclear medicine techniques.[*] For ^{90}Y, a total emission greater than 50 photons per radioactive decay has been estimated through simulation. Ruggiero and colleagues at the Memorial Sloan–Kettering Cancer Center reported *in vivo* preclinical imaging with ^{89}Zr-labeled monoclonal antibodies targeting prostate-specific membrane antigen (PSMA) (Ruggiero et al. 2010). Using the β⁻ emitter ^{32}P, Spinello and Boschi used PSA to classify the time–activity curves (TACs) in dynamic CLI (Spinelli and Boschi 2011a). Recently, ^{177}Lu was radioimmunoconjugated and used to image in a disseminated murine model for leukemia (Balkin et al. 2014), and ^{124}I-CLI was demonstrated with the human sodium iodide symporter.

Unlike fluorophores, which have a peak emission, the emission spectrum for Cerenkov luminescence is proportional to $1/\lambda^2$ and is therefore heavily weighted in the ultraviolet and blue, where the mean free path for photons is short. In order to modulate Cerenkov luminescence to more penetrating longer wavelengths, several groups have investigated the interaction of Cerenkov radioisotopes with quantum dots (Dothager et al. 2010; Liu et al. 2010b; Boschi and Spinelli 2012; Carpenter et al. 2012), fluorophores (Lewis et al. 2010; Axelsson et al. 2011), and photoactivatable probes for imaging and therapy (Ran et al. 2011; Kotagiri et al. 2013). While some investigators have reported noise in the CCD sensor due to direct hits by 511 keV γ-rays, the Spinelli group demonstrated little or no signal from a mouse covered with black paper (Boschi et al. 2011). Although Cerenkov luminescence endoscopy (CLE) is more invasive than CLI and CLT, a custom-made fiber-bundle-based endoscope has been described for use in rodents (Carpenter et al. 2014) with the advantage of increased sensitivity.

In addition to Cerenkov luminescence produced by radioisotopes, Cerenkov radiation is associated with external beam radiation therapy. In this specialized application, Cerenkov luminescence has been used to validate dosimetry and to estimate tissue oxygenation in phantoms

---

[*] Before CLI, pure β-emitting radioisotopes could only be imaged by detecting bremsstrahlung photons produced by the β-particles (Clarke et al. 1992; Shen et al. 1994).

(Axelsson et al. 2011, 2012; Glaser et al. 2012, 2013a–c; Zhang et al. 2012b). These methods are anticipated to be deployed in small animal imaging in the near future.

10.5 COMMERCIAL PRECLINICAL OPTICAL IMAGING SYSTEMS

As stated earlier, CW optical imaging is both the simplest and most economical approach for preclinical imaging, and it is not surprising that the current selection of commercial, preclinical optical imaging system is almost exclusively CW. By using scientific-grade thermoelectrically cooled CCD cameras, today's investigator has an excellent selection from which to choose, including the Bruker In-Vivo line, the LI-COR Pearl Impulse, the Trifoil InSyTe FLECT, Spectral Instruments Lago and Ami lines, and three product lines from Perkin Elmer (formerly Caliper Life Sciences and originally Xenogen): the IVIS Spectrum, Lumina, and FMT. All of these systems are capable of all modes of luminescence and epi-illumination fluorescence. Some systems also support trans-illumination fluorescence and are available with an option to coregister a planar x-ray image or CT. In terms of data and image processing, typical optional software for these systems might include spectral unmixing, analysis of dynamics, and/or tomographic reconstruction.

While time-domain optical imaging theoretically produces the best quality images, there are no commercial offerings for this type of system as of 2014. In the past, GE offered the eXplore Optix (Keren et al. 2008), a time-domain system originally designed by the Canadian firm ART. That said, the price of components for time-domain imaging have reduced somewhat in the past decade, and such systems developed in individual labs from off-the-shelf components are periodically reported in the literature, including for fluorescence diffuse tomography with CT priors (Patwardhan and Culver 2008; Tichauer et al. 2012).

10.5.1 APPLICATIONS

Here, various applications of preclinical optical imaging are highlighted, with an emphasis on the utilization of diverse modalities for the same task.

10.5.1.1 Tumor Burden

The most common application of BLI is the monitoring of tumor burden. In a recent multimodality study of tumor angiogenesis in a breast cancer lung metastasis model (Zhang et al. 2013b), the investigators aimed to develop a dual-modality PET and near-infrared fluorescence agent to target CD105 expression in metastatic breast cancer cells. Through serial imaging, the tumor development could be tracked in time with the accumulation of the contrast agent. While preclinical PET is normally accompanied by micro-CT, BLI imaging gives better sensitivity and contrast for soft-tissue tumors than x-rays. Tracking of lung tumor model was also reported in three dimensions using planar multispectral BLI with vendor-supplied reconstruction at depth (Iochmann et al. 2012).

10.5.1.2 Infection

Preclinical models of infection can be imaged using a variety of optical imaging modalities. Van Oosten et al. (2013) recently reported FMT with fluorescently labeled vancomycin in a mouse myositis model. Vancomycin labeled with a near-infrared fluorophore was shown to specifically target Gram-positive infection in the mouse. Likewise, the infectious bacterial strain can be transfected with a bioluminescence reporter for imaging (Niska et al. 2012). Since FDG-PET is used clinically to evaluate infection, FDG-CLI in the preclinical infection model will most likely prove fruitful.

10.5.1.3 Reporter Genes

The sodium iodide symporter (NIS) is a commonly used reporter gene that can be used with several free radioisotopes, avoiding the complexity of cell targeting. While NIS is most often targeted in preclinical SPECT and PET application with the appropriate β-emitting isotope, the reporter was recently employed with Cerenkov luminescence to visualize mesenchymal stem cells in a mouse xenograft (Wolfs et al. 2014). The CLI imaging of the NIS reporter can be validated in systems that coexpress a fluorescent protein such as mCherry (Kim et al. 2011b). The beauty of this experiment is that fluorescence and Cerenkov luminescence can often be performed in the same system.

10.5.1.4 Herceptin

For a given target such as the her2/neu receptor, numerous optical imaging approaches may apply. For example, Park et al. (2011) in Korea radiolabeled iodo-beads with ^{124}I, typically a PET radiotracer, and targeted the beads by labeling with herceptin. Using Cerenkov imaging in a Xenogen IVIS 200 system, they correlated the optical signal in tumors in the flank and shoulder with micro-PET imaging. Building upon their previous experience in prostate cancer, the Memorial Sloan–Kettering group also imaged her2/neu using Cerenkov luminescence, but with the agent ^{89}Zr-DFO-trastuzumab (Holland et al. 2011).

10.5.1.5 Monitoring Cancer Therapy

Since FDG and FLT uptake and retention often decrease in response to chemotherapeutics, CLI has the potential for high-throughput evaluation of novel compounds. As a proof of principle, both agents were evaluated using CLI as indicators of antitumor efficacy of bevacizumab (Avastin) (Xu et al. 2012a,b). BLT has been utilized in a dual-modality study of tumor response to the essential chemotherapeutic agent cyclophosphamide (Ma et al. 2011).

10.5.1.6 Metabolic Imaging

The *in vivo* assessment of the metabolic status of brown fat has recently been an important research topic in molecular imaging. While FDG-PET is the gold standard imaging modality for this task, CLI of FDG in intrascapular adipose tissue in the mouse has been reported in the literature (Zhang et al. 2013a). This development is notable since there are many more metabolism and nutrition labs with access to preclinical optical imaging than preclinical PET.

10.5.1.7 Brain Imaging

While oncological studies make up a large segment of preclinical optical imaging, brain imaging in rodent models is also an important application for all variations of optical imaging. For example, 2-[^{18}F]fluoro-CP-118,954 has been reported as a Cerenkov imaging agent in mouse models of Alzheimer's disease (Kim et al. 2011a). BLI has also been used to characterize both stroke pathology (no bioluminescence if luciferin substrate cannot be transported to tissue) and implanted stem cells (Aswendt et al. 2014).

10.5.1.8 The Awake Animal

For emission optical imaging, BLI in moving rodents has been described in the literature (Roncali et al. 2008). The experimental system used in this work required an additional optical amplifier to detect the weak bioluminescence with short integration times. Similarly, for transmission imaging, diffuse optical tomography has been used to investigate hemodynamics during acute seizures in an awake rat model (Zhang et al. 2014).

10.6 HYBRID MODALITIES

In order to overcome the poor resolution limitations associated with *in vivo* optical imaging, many groups have explored the development of the so-called hybrid modalities that combine a high-resolution/low-contrast imaging modality with a low-resolution/high-contrast technique to produce a high-resolution/high-contrast modality. By far, the most successful example of this approach is photoacoustics (also referred to as optoacoustics). The photoacoustic effect was first observed experimentally by Alexander Graham Bell in the late 1800s. In its modern preclinical imaging form, a nanosecond laser with energy greater than 10 mJ per pulse and operating at repetition rates on the order of 10–100 Hz is used to illuminate soft tissue. The endogenous optical contrast due to chromophores such as hemoglobin in the tissue will lead to local energy absorption, tissue heating, expansion, and the subsequent generation of an acoustic wave. Since the acoustic wave does not experience multiple scattering as a fluorescent photon would, the subsequent coherent hybrid image is of higher resolution than its strictly incoherent optical counterpart.

There are two common geometries used in photoacoustics: photoacoustic imaging (PAI) and photoacoustic tomography (PAT). In PAI, a traditional rastered-scanned focused transducer or an ultrasound array is used to detect the acoustic emission from the illuminated tissue. To form an image using the pressure waveforms acquired by an array transducer, traditional beamforming algorithms such as those used in B-mode imaging can be utilized. Because of the recent proliferation of ultrasound research interfaces (Verasonics, Cephasonics, Ultrasonix, Zonare, Samplify, etc.), many groups have developed home-build PAI systems using compact pulsed Nd:YAG lasers. In practice, the transmission profile for the ultrasound subsystem must be nulled, and then the laser fired instead, using an appropriate synchronization. With a more expensive, off-the-shelf, tunable laser, multispectral PAI is also possible and well suited for exploring exogenous contrast agents in PAI. On the commercial side, the Visualsonics LAZR system is based around a preclinical veterinary ultrasound system using a 20 MHz phased array. One advantage of this approach is that the ultrasound subsystem can also be used to generate coregistered B-mode or Doppler images.

By surrounding the animal with a ring of transducers, PAT can be realized (Rosenthal et al. 2013). In tomographic geometries, it is possible to include information of the variable speed of sound in the tissue during the image reconstruction. The a priori information on the speed of sound and possibly attenuation could arise either from ultrasound tomography (perhaps as a dual modality) or from a model segmented from another high-resolution modality such as CT or MRI. In either case, full-wave inversion of the acoustic sources has been demonstrated using a variety of mathematical and numerical methods (Chao et al. 2013; Rosenthal et al. 2013). As of fall 2014, iThera Medical is marketing the MSOT inSight and inVision preclinical imaging systems. Although these systems use a tomographic ultrasound detector array, the acoustic subsystem is a receive-only system: they lack the coregistered anatomical image that is common with PAI systems.

10.7 FURTHER INFORMATION

For additional, focused coverage of specific types of optical imaging, the interested reader is directed to the various excellent reviews covering diffuse optical imaging (Arridge and Hebden 1997; Arridge and Schweiger 1997; Hebden et al. 1997; Arridge 1999; Boas et al. 2001; Schweiger et al. 2003; Gibson et al. 2005; Arridge and Schotland 2009; Arridge 2011), fluorescence imaging (Weissleder and Ntziachristos 2003; Graves et al. 2004; Stuker et al. 2011; Darne et al. 2014), bioluminescence imaging (Wang et al. 2008a; Darne et al. 2014; Qin et al. 2014), Cerenkov imaging (Qin et al. 2012; Thorek et al. 2012; Das et al. 2014), and instrumentation (Ntziachristos et al. 2005; Zhang 2014).

REFERENCES

Ahn, S., A. J. Chaudhari et al. (2008). Fast iterative image reconstruction methods for fully 3D multispectral bioluminescence tomography. *Phys. Med. Biol.* **53**(14): 3921–3942.

Alexandrakis, G., F. R. Rannou et al. (2005). Tomographic bioluminescence imaging by use of a combined optical-PET (OPET) system: A computer simulation feasibility study. *Phys. Med. Biol.* **50**(17): 4225–4241.

Alexandrakis, G., F. R. Rannou et al. (2006). Effect of optical property estimation accuracy on tomographic bioluminescence imaging: Simulation of a combined optical-PET (OPET) system. *Phys. Med. Biol.* **51**(8): 2045–2053.

Allard, M., D. Cote et al. (2007). Combined magnetic resonance and bioluminescence imaging of live mice. *J. Biomed. Opt.* **12**(3): 034018.

Arridge, S., S. Moskow et al. (2012). Inverse Born series for the Calderon problem. *Inverse Probl.* **28**(3): 035003.

Arridge, S. R. (1999). Optical tomography in medical imaging. *Inverse Probl.* **15**(2): R41–R93.

Arridge, S. R. (2011). Methods in diffuse optical imaging. *Phil. Trans. Roy. Soc. A* **369**(1955): 4558–4576.

Arridge, S. R. and J. C. Hebden. (1997). Optical imaging in medicine. 2. Modelling and reconstruction. *Phys. Med. Biol.* **42**(5): 841–853.

Arridge, S. R. and J. C. Schotland. (2009). Optical tomography: Forward and inverse problems. *Inverse Probl.* **25**(12): 123010.

Arridge, S. R. and M. Schweiger. (1997). Image reconstruction in optical tomography. *Phil. Trans. Roy. Soc. B* **352**(1354): 717–726.

Aswendt, M., J. Adamczak et al. (2014). A review of novel optical imaging strategies of the stroke pathology and stem cell therapy in stroke. *Front. Cell. Neurosci.* **8**: 226.

Axelsson, J., S. C. Davis et al. (2011). Cerenkov emission induced by external beam radiation stimulates molecular fluorescence. *Med. Phys.* **38**(7): 4127–4132.

Axelsson, J., A. K. Glaser et al. (2012). Quantitative Cherenkov emission spectroscopy for tissue oxygenation assessment. *Opt. Express* **20**(5): 5133–5142.

Bal, G. and J. C. Schotland. (2014). Ultrasound-modulated bioluminescence tomography. *Phys. Rev. E Stat. Nonlin. Soft Matter Phys.* **89**(3): 031201.

Balkin, E. R., A. Kenoyer et al. (2014). In vivo localization of 90Y and 177Lu radioimmunoconjugates using Cerenkov luminescence imaging in a disseminated murine leukemia model. *Cancer Res.* **74**(20): 5846–5854.

Basevi, H. R. A., K. M. Tichauer et al. (2012). Compressive sensing based reconstruction in bioluminescence tomography improves image resolution and robustness to noise. *Biomed. Opt. Express* **3**(9): 2131–2141.

Beattie, B. J., D. L. J. Thorek et al. (2012). Quantitative modeling of Cerenkov light production efficiency from medical radionuclides. *PLoS One* **7**(2): e31402.

Behrooz, A., C. Kuo et al. (2013). Adaptive row-action inverse solver for fast noise-robust three-dimensional reconstructions in bioluminescence tomography: Theory and dual-modality optical/computed tomography in vivo studies. *J. Biomed. Opt.* **18**(7): 76010.

Boas, D. A., D. H. Brooks et al. (2001). Imaging the body with diffuse optical tomography. *IEEE Signal Process. Mag.* **18**(6): 57–75.

Boschi, F., L. Calderan et al. (2011). In vivo (1)F-FDG tumour uptake measurements in small animals using Cerenkov radiation. *Eur. J. Nucl. Med. Mol. Imaging* **38**(1): 120–127.

Boschi, F. and A. E. Spinelli. (2012). Quantum dots excitation using pure beta minus radioisotopes emitting Cerenkov radiation. *RSC Adv.* **2**(29): 11049–11052.

Brutkiewicz, S., M. Mendonca et al. (2007). The expression level of luciferase within tumour cells can alter tumour growth upon in vivo bioluminescence imaging. *Luminescence* **22**(3): 221–228.

Carpenter, C. M., X. Ma et al. (2014). Cerenkov luminescence endoscopy: Improved molecular sensitivity with beta-emitting radiotracers. *J. Nucl. Med.* **55**(11): 1905–1909.

Carpenter, C. M., C. Sun et al. (2012). Radioluminescent nanophosphors enable multiplexed small-animal imaging. *Opt. Express* **20**(11): 11598–11604.

Chao, H., W. Kun et al. (2013). Full-wave iterative image reconstruction in photoacoustic tomography with acoustically inhomogeneous media. *IEEE Trans. Med. Imaging* **32**(6): 1097–1110.

Chaudhari, A. J., F. Darvas et al. (2005). Hyperspectral and multispectral bioluminescence optical tomography for small animal imaging. *Phys. Med. Biol.* **50**(23): 5421–5441.

Chen, X. L., X. B. Gao et al. (2010). 3D reconstruction of light flux distribution on arbitrary surfaces from 2D multi-photographic images. *Opt. Express* **18**(19): 19876–19893.

Chin, P. T., M. M. Welling et al. (2013). Optical imaging as an expansion of nuclear medicine: Cerenkov-based luminescence vs fluorescence-based luminescence. *Eur. J. Nucl. Med. Mol. Imaging* **40**(8): 1283–1291.

Cho, J. S., R. Taschereau et al. (2009). Cerenkov radiation imaging as a method for quantitative measurements of beta particles in a microfluidic chip. *Phys. Med. Biol.* **54**(22): 6757–6771.

Clarke, L. P., S. J. Cullom et al. (1992). Bremsstrahlung imaging using the gamma-camera—Factors affecting attenuation. *J. Nucl. Med.* **33**(1): 161–166.

Comsa, D. C., T. J. Farrell et al. (2006). Quantification of bioluminescence images of point source objects using diffusion theory models. *Phys. Med. Biol.* **51**(15): 3733–3746.

Cong, A. X. and G. Wang. (2006). Multispectral bioluminescence tomography: Methodology and simulation. *Int. J. Biomed. Imaging* **2006**: 1–7.

Cong, W., K. Durairaj et al. (2006). A Born-type approximation method for bioluminescence tomography. *Med. Phys.* **33**(3): 679–686.

Cong, W. and G. Wang. (2006). Boundary integral method for bioluminescence tomography. *J. Biomed. Opt.* **11**(2): 020503.

Cong, W. and G. Wang. (2010). Bioluminescence tomography based on the phase approximation model. *J. Opt. Soc. Am. A Opt. Image Sci. Vis.* **27**(2): 174–179.

Cong, W., G. Wang et al. (2005). Practical reconstruction method for bioluminescence tomography. *Opt. Express* **13**(18): 6756 -6771.

Cong, W., L. V. Wang et al. (2004). Formulation of photon diffusion from spherical bioluminescent sources in an infinite homogeneous medium. *Biomed. Eng.* **3**(1): 12.

Côté, D., M. Allard et al. (2005). Three-dimensional light-tissue interaction models for bioluminescence tomography. In *Photonic Applications in Biosensing and Imaging*, Toronto, Ontario, Canada.

Darne, C., Y. J. Lu et al. (2014). Small animal fluorescence and bioluminescence tomography: A review of approaches, algorithms and technology update. *Phys. Med. Biol.* **59**(1): R1–R64.

Das, S., J. Grimm et al. (2014). Cerenkov imaging. *Adv. Cancer Res.* **124**: 213–234.

Dehghani, H., S. C. Davis et al. (2006). Spectrally resolved bioluminescence optical tomography. *Opt. Letters* **31**(3): 365–367.

Dehghani, H., S. C. Davis et al. (2008). Spectrally resolved bioluminescence tomography using the reciprocity approach. *Med. Phys.* **35**(11): 4863–4871.

Ding, X., K. Wang et al. (2014). Probability method for Cerenkov luminescence tomography based on conformance error minimization. *Biomed. Opt. Express* **5**(7): 2091–2112.

Dothager, R. S., R. J. Goiffon et al. (2010). Cerenkov radiation energy transfer (CRET) imaging: A novel method for optical imaging of PET isotopes in biological systems. *PLoS One* **5**(10): e13300.

Feng, J., K. Jia et al. (2009). Three-dimensional bioluminescence tomography based on Bayesian approach. *Opt. Express* **17**(19): 16834–16848.

Feng, J. C., K. B. Jia et al. (2008). An optimal permissible source region strategy for multispectral bioluminescence tomography. *Opt. Express* **16**(20): 15640–15654.

Feng, J. C., C. H. Qin et al. (2011). An adaptive regularization parameter choice strategy for multispectral bioluminescence tomography. *Med. Phys.* **38**(11): 5933–5944.

Feng, J. C., C. H. Qin et al. (2012). Total variation regularization for bioluminescence tomography with the split Bregman method. *Appl. Opt.* **51**(19): 4501–4512.

Gao, H. and H. Zhao. (2010a). Multilevel bioluminescence tomography based on radiative transfer equation. Part 1: l1 regularization. *Opt. Express* **18**(3): 1854–1871.

Gao, H. and H. Zhao. (2010b). Multilevel bioluminescence tomography based on radiative transfer equation. Part 2: Total variation and l1 data fidelity. *Opt. Express* **18**(3): 2894–2912.

Gibson, A. P., J. C. Hebden et al. (2005). Recent advances in diffuse optical imaging. *Phys. Med. Biol.* **50**(4): R1–R43.

Glaser, A. K., S. C. Davis et al. (2013a). Projection imaging of photon beams by the Cerenkov effect. *Med. Phys.* **40**(1): 012101.

Glaser, A. K., S. C. Davis et al. (2013b). Projection imaging of photon beams using Cerenkov-excited fluorescence. *Phys. Med. Biol.* **58**(3): 601–619.

Glaser, A. K., W. H. Voigt et al. (2013c). Three-dimensional Cerenkov tomography of energy deposition from ionizing radiation beams. *Opt. Lett.* **38**(5): 634–636.

Glaser, A. K., R. X. Zhang et al. (2012). Time-gated Cherenkov emission spectroscopy from linear accelerator irradiation of tissue phantoms. *Opt. Lett.* **37**(7): 1193–1195.

Gomi, K., K. Hirokawa et al. (2002). Molecular cloning and expression of the cDNAs encoding luciferin-regenerating enzyme from Luciola cruciata and Luciola lateralis. *Gene* **294**(1–2): 157–166.

Gomi, K. and N. Kajiyama. (2001). Oxyluciferin, a luminescence product of firefly luciferase, is enzymatically regenerated into luciferin. *J. Biol. Chem.* **276**(39): 36508–36513.

Graves, E. E., R. Weissleder et al. (2004). Fluorescence molecular imaging of small animal tumor models. *Curr. Mol. Med.* **4**(4): 419–430.

Gross, S., U. Abraham et al. (2007). Continuous delivery of D-luciferin by implanted microosmotic pumps enables true real-time bioluminescence imaging of luciferase activity in vivo. *Mol. Imaging* **6**(2): 121–130.

Gu, X., Q. Zhang et al. (2004). Three-dimensional bioluminescence tomography with model-based reconstruction. *Opt. Express* **12**(17): 3996–4000.

Guggenheim, J. A., H. R. Basevi et al. (2013). Multi-modal molecular diffuse optical tomography system for small animal imaging. *Meas. Sci. Technol.* **24**(10): 105405.

Han, W., W. Cong et al. (2006). Mathematical theory and numerical analysis of bioluminescence tomography. *Inverse Probl.* **22**: 1659–1675.

Han, W., K. Kazmi et al. (2007). Bioluminescence tomography with optimized optical parameters. *Inverse Probl.* **23**(3): 1215–1228.

Han, W. and G. Wang. (2007). Theoretical and numerical analysis on multispectral bioluminescence tomography. *IMA J. Appl. Math.* **72**: 67–85.

Han, W. M. and G. Wang. (2008). Bioluminescence tomography: Biomedical background, mathematical theory, and numerical approximation. *J. Comput. Math.* **26**(3): 324–335.

Hebden, J. C., S. R. Arridge et al. (1997). Optical imaging in medicine.1. Experimental techniques. *Phys. Med. Biol.* **42**(5): 825–840.

Hillman, E. M. and A. Moore. (2007). All-optical anatomical co-registration for molecular imaging of small animals using dynamic contrast. *Nat. Photon.* **1**(9): 526–530.

Holland, J. P., G. Normand et al. (2011). Intraoperative imaging of positron emission tomographic radiotracers using Cerenkov luminescence emissions. *Mol. Imaging* **10**(3): 177–186, 171–173.

Hu, Z., J. Liang et al. (2010). Experimental Cerenkov luminescence tomography of the mouse model with SPECT imaging validation. *Opt. Express* **18**(24): 24441–24450.

Hu, Z. H., X. L. Chen et al. (2012). Single photon emission computed tomography-guided Cerenkov luminescence tomography. *J. Appl. Phys.* **112**(2): 024703.

Huang, R. M., J. Vider et al. (2011). ATP-binding cassette transporters modulate both coelenterazine- and D-luciferin-based bioluminescence imaging. *Mol. Imaging* **10**(3): 215–226.

Huynh, N. T., B. R. Hayes-Gill et al. (2013). Ultrasound modulated imaging of luminescence generated within a scattering medium. *J. Biomed. Opt.* **18**(2): 20505.

Iochmann, S., S. Lerondel et al. (2012). Monitoring of tumour progression using bioluminescence imaging and computed tomography scanning in a nude mouse orthotopic model of human small cell lung cancer. *Lung Cancer* **77**(1): 70–76.

Jiang, M., T. Zhou et al. (2007). Image reconstruction for bioluminescence tomography from partial measurements. *Opt. Express* **15**(18): 11095–11116.

Jin, A., B. Yazici et al. (2014). Light illumination and detection patterns for fluorescence diffuse optical tomography based on compressive sensing. *IEEE Trans. Image Process.* **23**(6): 2609–2624.

Jin, W. M. and Y. H. He. (2012). Iterative reconstruction for bioluminescence tomography with total variation regularization. *Opt. Health Care Biomed. Opt. V* **8553**: 855333.

Jones, L. R., E. A. Goun et al. (2006). Releasable luciferin-transporter conjugates: Tools for the real-time analysis of cellular uptake and release. *J. Am. Chem. Soc.* **128**(20): 6526–6527.

Keren, S., O. Gheysens et al. (2008). A comparison between a time domain and continuous wave small animal optical imaging system. *IEEE Trans. Med. Imaging* **27**(1): 58–63.

Kim, D. H., Y. S. Choe et al. (2011a). Binding of 2-[18F]fluoro-CP-118,954 to mouse acetylcholinesterase: MicroPET and ex vivo Cerenkov luminescence imaging studies. *Nucl. Med. Biol.* **38**(4): 541–547.

Kim, K. I., J. J. Park et al. (2011b). Gamma camera and optical imaging with a fusion reporter gene using human sodium/iodide symporter and monomeric red fluorescent protein in mouse model. *Int. J. Radiat. Biol.* **87**(12): 1182–1188.

Klerk, C. P. W., R. M. Overmeer et al. (2007). Validity of bioluminescence measurements for noninvasive in vivo imaging of tumor load in small animals. *Biotechniques* **43**(1): 7–13.

Klose, A. D. and B. J. Beattie. (2009). Bioluminescence tomography with CT/MRI co-registration. *Conf. Proc. IEEE Eng. Med. Biol. Soc.* **2009**: 6327–6330.

Klose, A. D., B. J. Beattie et al. (2010). In vivo bioluminescence tomography with a blocking-off finite-difference SP3 method and MRI/CT coregistration. *Med. Phys.* **37**(1): 329–338.

Klose, A. D. and E. W. Larsen. (2006). Light transport in biological tissue based on the simplified spherical harmonics equations. *J. Comput. Phys.* **220**(1): 441–470.

Kotagiri, N., D. M. Niedzwiedzki et al. (2013). Activatable probes based on distance-dependent luminescence associated with Cerenkov radiation. *Angew. Chem. Int. Ed. Engl.* **52**(30): 7756–7760.

Kumar, D., W. X. Cong et al. (2007). Monte Carlo method for bioluminescence tomography. *Indian J. Exp. Biol.* **45**(1): 58–63.

Kuo, C., O. Coquoz et al. (2007). Three-dimensional reconstruction of in vivo bioluminescent sources based on multispectral imaging. *J. Biomed. Opt.* **12**(2): 024007.

Lasser, T., A. Soubret et al. (2008). Surface reconstruction for free-space 360 degrees fluorescence molecular tomography and the effects of animal motion. *IEEE Trans. Med. Imaging* **27**(2): 188–194.

Lewis, M., E. Richer et al. (2013). A multi-camera system for bioluminescence tomography in preclinical oncology research. *Diagnostics* **3**(3): 325–343.

Lewis, M. A., V. D. Kodibagkar et al. (2010). On the potential for molecular imaging with Cerenkov luminescence. *Opt. Lett.* **35**(23): 3889–3891.

Li, C., G. S. Mitchell et al. (2010). Cerenkov luminescence tomography for small-animal imaging. *Opt. Lett.* **35**(7): 1109–1111.

Li, C., G. Wang et al. (2009). Three-dimensional fluorescence optical tomography in small-animal imaging using simultaneous positron-emission-tomography priors. *Opt. Lett.* **34**(19): 2933–2935.

Li, C., Y. Yang et al. (2011). Simultaneous PET and multispectral 3-dimensional fluorescence optical tomography imaging system. *J. Nucl. Med.* **52**(8): 1268–1275.

Li, H., J. Tian et al. (2004). A mouse optical simulation environment (MOSE) to investigate bioluminescent phenomena in the living mouse with the Monte Carlo method. *Acad. Radiol.* **11**(9): 1029–1038.

Li, H., J. Tian et al. (2007). Design and implementation of an optical simulation environment for bioluminescent tomography studies. *Progr. Nat. Sci.* **17**(1): 87–94.

Li, X. D., T. Durduran et al. (1997). Diffraction tomography for biochemical imaging with diffuse-photon density waves. *Opt. Lett.* **22**(8): 573–575.

Liu, H., A. Karellas et al. (1994). Methods to calculate the lens efficiency in optically coupled CCD x-ray-imaging systems. *Med. Phys.* **21**(7): 1193–1195.

Liu, H., G. Ren et al. (2010a). Molecular optical imaging with radioactive probes. *PLoS One* **5**(3): e9470.

Liu, H., X. Zhang et al. (2010b). Radiation-luminescence-excited quantum dots for in vivo multiplexed optical imaging. *Small* **6**(10): 1087–1091.

Liu, H. L., C. L. Matson et al. (1999). Experimental validation of a backpropagation algorithm for three-dimensional breast tumor localization. *IEEE J. Sel. Top. Quant. Electron.* **5**(4): 1049–1057.

Lu, Y., A. Douraghy et al. (2009a). Spectrally resolved bioluminescence tomography with the third-order simplified spherical harmonics approximation. *Phys. Med. Biol.* **54**(21): 6477–6493.

Lu, Y., H. B. Machado et al. (2009b). Experimental bioluminescence tomography with fully parallel radiative-transfer-based reconstruction framework. *Opt. Express* **17**(19): 16681–16695.

Lu, Y. J., X. Q. Zhang et al. (2009c). Source reconstruction for spectrally-resolved bioluminescence tomography with sparse a priori information. *Opt. Express* **17**(10): 8062–8080.

Lv, Y., J. Tian et al. (2006a). MicroCT-guided bioluminescence tomography based on the adaptive finite element tomographic algorithm. *Conf. Proc. IEEE Eng. Med. Biol. Soc.* 1: 381–384.

Lv, Y., J. Tian et al. (2006b). A multilevel adaptive finite element algorithm for bioluminescence tomography. *Opt. Express* **14**(18): 8211–8223.

Lv, Y., J. Tian et al. (2007). Spectrally resolved bioluminescence tomography with adaptive finite element analysis: Methodology and simulation. *Phys. Med. Biol.* **52**(15): 4497–4512.

Ma, X. B., K. X. Deng et al. (2013). Novel registration for microcomputed tomography and bioluminescence imaging based on iterated optimal projection. *J. Biomed. Opt.* **18**(2): 26013.

Ma, X. B., Z. F. Liu et al. (2011). Dual-modality monitoring of tumor response to cyclophosphamide therapy in mice with bioluminescence imaging and small-animal positron emission tomography. *Mol. Imaging* **10**(4): 278–283.

Markel, V. A. and J. C. Schotland. (2007). On the convergence of the Born series in optical tomography with diffuse light. *Inverse Probl.* **23**(4): 1445–1465.

Marques, S. M. and J. C. Esteves da Silva. (2009). Firefly bioluminescence: A mechanistic approach of luciferase catalyzed reactions. *IUBMB Life* **61**(1): 6–17.

Matson, C. L. and H. L. Liu. (1999a). Analysis of the forward problem with diffuse photon density waves in turbid media by use of a diffraction tomography model. *J. Opt. Soc. Am. A* **16**(3): 455–466.

Matson, C. L. and H. L. Liu. (1999b). Backpropagation in turbid media. *J. Opt. Soc. Am. A* **16**(6): 1254–1265.

Meyer, H., A. Garofaiakis et al. (2007). Noncontact optical imaging in mice with full angular coverage and automatic surface extraction. *Appl. Opt.* **46**(17): 3617–3627.

Mohajerani, P., A. Hipp et al. (2014). FMT-PCCT: Hybrid fluorescence molecular tomography-x-ray phase-contrast CT imaging of mouse models. *IEEE Trans. Med. Imaging* **33**(7): 1434–1446.

Moskow, S. and J. C. Schotland. (2009). Numerical studies of the inverse Born series for diffuse waves. *Inverse Probl.* **25**(9): 095007.

Naser, M. A. and M. S. Patterson. (2010). Algorithms for bioluminescence tomography incorporating anatomical information and reconstruction of tissue optical properties. *Biomed. Opt. Express* **1**(2): 512–526.

Naser, M. A. and M. S. Patterson. (2011a). Bioluminescence tomography using eigenvectors expansion and iterative solution for the optimized permissible source region. *Biomed. Opt. Express* **2**(11): 3179–3192.

Naser, M. A. and M. S. Patterson. (2011b). Improved bioluminescence and fluorescence reconstruction algorithms using diffuse optical tomography, normalized data, and optimized selection of the permissible source region. *Biomed. Opt. Express* **2**(1): 169–184.

Naser, M. A., M. S. Patterson et al. (2012). Self-calibrated algorithms for diffuse optical tomography and bioluminescence tomography using relative transmission images. *Biomed. Opt. Express* **3**(11): 2794–2808.

Naser, M. A., M. S. Patterson et al. (2014). Algorithm for localized adaptive diffuse optical tomography and its application in bioluminescence tomography. *Phys. Med. Biol.* **59**(8): 2089–2109.

Niska, J. A., J. A. Meganck et al. (2012). Monitoring bacterial burden, inflammation and bone damage longitudinally using optical and μCT imaging in an orthopaedic implant infection in mice. *PLoS One* **7**(10): e47397.

Ntziachristos, V., J. Ripoll et al. (2005). Looking and listening to light: The evolution of whole-body photonic imaging. *Nat. Biotechnol.* **23**(3): 313–320.

Park, J. C., G. I. An et al. (2011). Luminescence imaging using radionuclides: A potential application in molecular imaging. *Nucl. Med. Biol.* **38**(3): 321–329.

Patwardhan, S. V. and J. P. Culver. (2008). Quantitative diffuse optical tomography for small animals using an ultrafast gated image intensifier. *J. Biomed. Opt.* **13**(1): 011009.

Qin, C., S. Zhu et al. (2011). Comparison of permissible source region and multispectral data using efficient bioluminescence tomography method. *J. Biophoton.* **4**(11–12): 824–839.

Qin, C. H., J. C. Feng et al. (2014). Recent advances in bioluminescence tomography: Methodology and system as well as application. *Laser Photon. Rev.* **8**(1): 94–114.

Qin, C. H., J. H. Zhong et al. (2012). Recent advances in Cerenkov luminescence and tomography imaging. *IEEE J. Sel. Top. Quant. Electron.* **18**(3): 1084–1093.

Radrich, K., P. Mohajerani et al. (2014). Limited-projection-angle hybrid fluorescence molecular tomography of multiple molecules. *J. Biomed. Opt.* **19**(4): 046016.

Ran, C., Z. Zhang et al. (2011). In vivo photoactivation without "light": Use of Cherenkov radiation to overcome the penetration limit of light. *Mol. Imaging Biol.* **14**: 156–162.

Robertson, R., M. S. Germanos et al. (2009). Optical imaging of Cerenkov light generation from positron-emitting radiotracers. *Phys. Med. Biol.* **54**(16): N355–N365.

Robertson, R., M. S. Germanos et al. (2011). Multimodal imaging with 18F-FDG PET and Cerenkov luminescence imaging after MLN4924 treatment in a human lymphoma xenograft model. *J. Nucl. Med.* **52**(11): 1764–1769.

Roncali, E., M. Savinaud et al. (2008). New device for real-time bioluminescence imaging in moving rodents. *J. Biomed. Opt.* **13**(5): 054035.

Rosenthal, A., V. Ntziachristos et al. (2013). Acoustic inversion in optoacoustic tomography: A review. *Curr. Med. Imaging Rev.* **9**(4): 318–336.

Ruggiero, A., J. P. Holland et al. (2010). Cerenkov luminescence imaging of medical isotopes. *J. Nucl. Med.* **51**(7): 1123–1130.

Sarasa-Renedo, A., R. Favicchio et al. (2010). Source intensity profile in noncontact optical tomography. *Opt. Lett.* **35**(1): 34–36.

Schweiger, M., A. Gibson et al. (2003). Computational aspects of diffuse optical tomography. *Comput. Sci. Eng.* **5**(6): 33–41.

Shen, S., G. L. Denardo et al. (1994). Planar gamma-camera imaging and quantitation of Y-90 bremsstrahlung. *J. Nucl. Med.* **35**(8): 1381–1389.

Shi, S. K. and H. Mao. (2013). A generalized hybrid algorithm for bioluminescence tomography. *Biomed. Opt. Express* **4**(5): 709–724.

Slavine, N. V., M. A. Lewis et al. (2006). Iterative reconstruction method for light emitting sources based on the diffusion equation. *Med. Phys.* **33**(1): 61–68.

Soloviev, V. Y. (2007). Tomographic bioluminescence imaging with varying boundary conditions. *Appl. Opt.* **46**(14): 2778–2784.

Soubret, A. and V. Ntziachristos. (2006). Fluorescence molecular tomography in the presence of background fluorescence. *Phys. Med. Biol.* **51**(16): 3983–4001.

Spinelli, A. E. and F. Boschi. (2011a). Unsupervised analysis of small animal dynamic Cerenkov luminescence imaging. *J. Biomed. Opt.* **16**(12): 120506.

Spinelli, A. E. and F. Boschi. (2012). Optimizing in vivo small animal Cerenkov luminescence imaging. *J. Biomed. Opt.* **17**(4): 040506.

Spinelli, A. E., D. D'Ambrosio et al. (2010). Cerenkov radiation allows in vivo optical imaging of positron emitting radiotracers. *Phys. Med. Biol.* **55**(2): 483–495.

Spinelli, A. E., C. Kuo et al. (2011b). Multispectral Cerenkov luminescence tomography for small animal optical imaging. *Opt. Express* **19**(13): 12605–12618.

Stuker, F., J. Ripoll et al. (2011). Fluorescence molecular tomography: Principles and potential for pharmaceutical research. *Pharmaceutics* **3**(2): 229–274.

Swindell, W. (1991). The lens coupling efficiency in megavoltage imaging. *Med. Phys.* **18**(6): 1152–1153.

Szentirmai, O., C. H. Baker et al. (2006). Noninvasive bioluminescence imaging of luciferase expressing intracranial U87 xenografts: Correlation with magnetic resonance imaging determined tumor volume and longitudinal use in assessing tumor growth and antiangiogenic treatment effect. *Neurosurgery* **58**(2): 365–372.

Tagami, A., N. Ishibashi et al. (2009). Ab initio quantum-chemical study on emission spectra of bioluminescent luciferases by fragment molecular orbital method. *Chem. Phys. Lett.* **472**(1–3): 118–123.

Thorek, D. L. J., R. Robertson et al. (2012). Cerenkov imaging - a new modality for molecular imaging. *Am. J. Nucl. Med. Mol. Imaging* **2**(2): 163–173.

Thorne, S. H. and C. H. Contag. (2005). Using in vivo bioluminescence imaging to shed light on cancer biology. *Proc. IEEE* **93**(4): 750–762.

Tichauer, K. M., R. W. Holt et al. (2012). Computed tomography-guided time-domain diffuse fluorescence tomography in small animals for localization of cancer biomarkers. *J. Vis. Exp.* **65**: e4050.

Troy, T., D. Jekic-McMullen et al. (2004). Quantitative comparison of the sensitivity of detection of fluorescent and bioluminescent reporters in animal models. *Mol. Imaging* **3**(1): 9–23.

van Oosten, M., T. Schafer et al. (2013). Real-time in vivo imaging of invasive- and biomaterial-associated bacterial infections using fluorescently labelled vancomycin. *Nat. Commun.* **4**: 2584.

Wang, G., W. Cong et al. (2006a). In vivo mouse studies with bioluminescence tomography. *Opt. Express* **14**(17): 7801–7809.

Wang, G., W. X. Cong et al. (2008a). Overview of bioluminescence tomography—A new molecular imaging modality. *Front. Biosci.* **13**: 1281–1293.

Wang, G., Y. Li et al. (2004). Uniqueness theorems in bioluminescence tomography. *Med. Phys.* **31**(8): 2289–2299.

Wang, G., H. Shen et al. (2006b). Temperature-modulated bioluminescence tomography. *Opt. Express* **14**(17): 7852–7871.

Wang, G., H. Shen et al. (2008b). Digital spectral separation methods and systems for bioluminescence imaging. *Opt. Express* **16**(3): 1719–1732.

Wang, L., S. L. Jacques et al. (1995). MCML—Monte Carlo modeling of light transport in multi-layered tissues. *Comput. Methods Programs Biomed.* **47**(2): 131–146.

Weissleder, R. and V. Ntziachristos. (2003). Shedding light onto live molecular targets. *Nat. Med.* **9**(1): 123–128.

Wolfs, E., B. Holvoet et al. (2014). Optimization of multimodal imaging of mesenchymal stem cells using the human sodium iodide symporter for PET and Cerenkov luminescence imaging. *PLoS One* **9**(4): e94833.

Wu, P., Y. F. Hu et al. (2014). Bioluminescence tomography by an iterative reweighted l(2)-norm optimization. *IEEE Trans. Biomed. Eng.* **61**(1): 189–196.

Xu, Y., H. Liu et al. (2011). Harnessing the power of radionuclides for optical imaging: Cerenkov luminescence imaging. *J. Nucl. Med.* **52**(12): 2009–2018.

Xu, Y., H. Liu et al. (2012a). Cerenkov Luminescence Imaging (CLI) for cancer therapy monitoring. *J. Vis. Exp.* **69**: e4341.

Xu, Y. D., E. Chang et al. (2012b). Proof-of-concept study of monitoring cancer drug therapy with Cerenkov luminescence imaging. *J. Nucl. Med.* **53**(2): 312–317.

Yan, H., Y. T. Lin et al. (2012). A gantry-based tri-modality system for bioluminescence tomography. *Rev. Sci. Instrum.* **83**(4): 043708.

Yu, T. and J. M. Boone. (1997). Lens coupling efficiency: Derivation and application under differing geometrical assumptions. *Med. Phys.* **24**(4): 565–570.

Zhang, B., X. Yang et al. (2010). A trust region method in adaptive finite element framework for bioluminescence tomography. *Opt. Express* **18**(7): 6477–6491.

Zhang, J., D. F. Chen et al. (2014). Incorporating MRI structural information into bioluminescence tomography: System, heterogeneous reconstruction and in vivo quantification. *Biomed. Opt. Express* **5**(6): 1861–1876.

Zhang, Q. T., X. L. Chen et al. (2012a). Comparative studies of l(p)-regularization-based reconstruction algorithms for bioluminescence tomography. *Biomed. Opt. Express* **3**(11): 2916–2936.

Zhang, Q. Z., L. Yin et al. (2008). Quantitative bioluminescence tomography guided by diffuse optical tomography. *Opt. Express* **16**(3): 1481–1486.

Zhang, R. X., A. Glaser et al. (2012b). Cerenkov radiation emission and excited luminescence (CREL) sensitivity during external beam radiation therapy: Monte Carlo and tissue oxygenation phantom studies. *Biomed. Opt. Express* **3**(10): 2381–2394.

Zhang, T., J. Zhou et al. (2014). Detecting epileptic seizures in awake rats with diffuse optical tomography. In *Biomedical Optics 2014*, Miami, FL.

Zhang, X. (2014). Instrumentation in diffuse optical imaging. *Photonics* **1**(1): 9–32.

Zhang, X. L., C. Kuo et al. (2013a). In vivo optical imaging of interscapular brown adipose tissue with F-18-FDG via Cerenkov luminescence imaging. *Plos One* **8**(4): e62007.

Zhang, Y., J. P. Bressler et al. (2007). ABCG2/BCRP expression modulates D-Luciferin based bioluminescence imaging. *Cancer Res.* **67**(19): 9389–9397.

Zhang, Y., H. Hong et al. (2013b). Imaging tumor angiogenesis in breast cancer experimental lung metastasis with positron emission tomography, near-infrared fluorescence, and bioluminescence. *Angiogenesis* **16**(3): 663–674.

Zhao, H., T. C. Doyle et al. (2005). Emission spectra of bioluminescent reporters and interaction with mammalian tissue determine the sensitivity of detection in vivo. *J. Biomed. Opt.* **10**(4): 41210.

Zhong, J., C. Qin et al. (2011a). Fast-specific tomography imaging via Cerenkov emission. *Mol. Imaging Biol.* **14**: 286–292.

Zhong, J., C. Qin et al. (2011b). Cerenkov luminescence tomography for in vivo radiopharmaceutical imaging. *Int. J. Biomed. Imaging* **2011**: 641618.

Zhong, J., J. Tian et al. (2011c). Whole-body Cerenkov luminescence tomography with the finite element SP(3) method. *Ann. Biomed. Eng.* **39**(6): 1728–1735.

Section IV
Hybrid Imaging

11

Optical-CT Imaging

Xueli Chen, Dongmei Chen, FengLin Liu,
Wenxiang Cong, Ge Wang, and Jimin Liang

11.1 INTRODUCTION

As a small-animal molecular imaging technology, optical imaging has been attracting increased attention and has been a rapidly developing biomedical imaging field because of its significant advantages in temporal resolution, imaging contrast and sensitivity, nonionizing radiation, and cost-effectiveness (Weissleder and Ntziachristos 2003; Ntziachristos et al. 2005; Tian et al. 2008). The goal of optical imaging is to depict noninvasive *in vivo* cellular and molecular processes sensitively and specifically, such as monitoring multiple molecular events, cell trafficking, and targeting (Bhaumik and Gambhir 2002; Contag and Bachmann 2002; Massoud and Gambhir 2003). However, optical imaging has been in a planar mode and largely a qualitative imaging tool, which limits its applications. To overcome the limitations, a tomographic counterpart such as optical tomography has been developed and has become a valuable tool in the biomedical imaging field, which integrates multiple optical acquisitions from optical imaging, geometrical structures from computed tomography (CT), and tissue's optical properties to directly reconstruct the optical probe distribution inside a small living animal. Optical tomography (OT), also called hybrid optical-CT imaging, is capable of three-dimensional (3D) recovery of the location and concentration of the optical probe inside a small living animal (Arridge 1999; Ntziachristos et al. 2002; Gibson et al. 2005). Among the modalities of OT, two main categories can be addressed according to whether the optical probe receives an external excitation source. One category is called spontaneous optical tomography (SOT), in which the optical probes emit luminescent light in the absence of an external excitation source, such as bioluminescence tomography (BLT) and Cerenkov luminescence tomography (CLT) (Wang et al. 2003, 2004; Hu et al. 2010; Li et al. 2010). The other is called passive optical tomography (POT), in which the optical probes can only emit fluorescent light in case of an external excitation source, such as fluorescence-mediated tomography (FMT) and x-ray luminescence computed tomography (XRLT). This chapter aims to review the four typical kinds of OT modalities stated earlier, including the mechanism, the imaging principle, the mathematical model and related reconstruction algorithm, and the prototype system and its biomedical applications.

11.1.1 SPONTANEOUS OPTICAL TOMOGRAPHY

SOT is an active imaging technique, in which the biological cells targeted with optical probes emit luminescent light in the absence of an external excitation source. Generally speaking, luminescence is the emission of light by a substance caused by a physical or chemical means; examples include chemiluminescence from a chemical reaction and bioluminescence from a biochemical enzyme-driven reaction in living organisms and animals (Wang et al. 2008). There are two typical categories of SOT: BLT and CLT (Wang et al. 2003, 2004; Hu et al. 2010; Li et al. 2010).

Bioluminescent light is produced from the expression of luciferase enzymes, such as firefly luciferase, Renilla luciferase, and jellyfish luciferase. Cells encoded with the luciferase enzymes can serve as bioluminescent probes, which allow bioluminescent light emission. The BLT technique uses the luciferase enzyme gene as a reporter (Contag and Bachmann 2002). Oxidation of luciferin, which is injected into the animal, leads to the emission of bioluminescent light and occurs only in cells in which the luciferase enzyme is expressed by the reporter gene. The major attraction of this approach is that although absolute light levels are low, signal is produced only where luciferase is present, leading to extremely low background signals (Troy et al. 2004). BLT was first proposed by Wang's group in 2003 (Wang et al. 2003). In mathematics, it is an inverse source problem. Based on an accurate light transport model, BLT aims to reconstruct the 3D spatial distribution and concentration of the bioluminescent probes inside a small living animal.

Cerenkov luminescent light is produced based on Cerenkov radiation, which brings about emerging *in vivo* optical imaging of small living animals, or Cerenkov luminescence imaging (CLI) (Spinelli et al. 2010; Holland et al. 2011; Mitchell et al. 2011; Robertson et al. 2011; Xu et al. 2012). Cerenkov radiation is a well-known phenomenon in which the visible and near-infrared (NIR) light can be emitted when the charged and high-energy particle travels faster than the speed of light in the dielectric medium (Jelley 1955). Because of the advantages of cost-effectiveness, high sensitivity, and high throughput compared with the existing nuclear instrumentation, CLI has become a promising and preclinical molecular imaging technique (Liu et al. 2010a), which has obtained wide applications in monitoring the uptake of ^{18}F-FDG and ^{131}I, detecting thyroid cancer, and evaluating the treatment of lymphoma (Boschi et al. 2011; Jeong et al. 2011; Robertson et al. 2011; Hu et al. 2012c). Based on the promising CLI technology, CLT was developed to resolve source depth measurement using the same detection system and reconstruction strategy as in the BLT technique (Li et al. 2010; Zhong et al. 2011, 2012; Hu et al. 2012b). Because only the signal generation mechanism is different between the CLT and BLT technologies, the same image acquisition system, mathematical model, and reconstruction algorithms can be used for both the techniques. As a result, advances of the BLT technique will be selected as the representative type of SOT and be described in detail in the following section.

11.1.2 PASSIVE OPTICAL TOMOGRAPHY

POT is a passive imaging technique in which biological cells labeled with optical probes emit fluorescent light in case of an external excitation source. Fluorescence described here is the emission of light from a material that has been excited by an external excitation source; examples include photoluminescence excited by a light beam such as fluorescence and cathodoluminescence excited by an electron beam such as in x-ray luminescence (XRL) in living organisms and animals. There are two typical categories of POT presented in this section: FMT and XRLT (Wu et al. 1997; Ntziachristos et al. 2004; Pratx et al. 2010b; Cong et al. 2011).

When fluorophores, fluorochromes, or fluorescent dyes are excited by an external excitation source, such as a laser, their molecules would absorb light and raise their energy levels to a brief excited state. As they decay from this excited state, they emit fluorescent light (Biosciences 2002). Compared with bioluminescence, fluorescence requires an external light source to stimulate the emission of light with larger signals from probe. However, these signals are usually accompanied by autofluorescence that brings down the quality of image (Zavattini et al. 2006). In particular, FMT is an imaging method aimed at resolving fluorescence distribution in tissues *in vivo*. The method scans focused light at multiple positions on an animal surface at one or more wavelengths and collects excited fluorescence photons propagating through the tissue for each focal spot.

XRL is the emission of characteristic "secondary" (or fluorescent) x-rays from a material that has been excited by bombarding it with high-energy x-rays or gamma rays. Nanoparticles (NPs) are a class of materials generally ranging in size from 1 to 100 nm that are emerging as potentially powerful probes for *in vivo* XRL imaging in medical and biological diagnostics. A technique using x-ray excitable nanophosphors has been suggested for simultaneous imaging of anatomical and molecular features (Pereira et al. 2007; Carpenter et al. 2010; Cheong et al. 2010). Several NP-based contrast agents have been used as probes in XRLT systems. XRLT aims to generate molecular contrast by detecting the characteristic x-rays from the interaction between the x-ray and the probe(s) containing a high atomic number element, such as gold NPs (Jones and Cho 2011; Sun et al. 2011; Carpenter et al. 2012). The 3D distribution of the high-Z contrast agent within the object is tomographically reconstructed based on the measured fluorescence sinogram data obtained using photon-counting detectors. In this modality, characteristic x-rays (i.e., x-ray fluorescence photons) are initially induced from a sample containing one or more elements in question by x-rays, and subsequently, the spatial distribution and concentration of each element are determined through a full 3D image reconstruction process (Pratx et al. 2010; Carpenter et al. 2011; Bazalova et al. 2012).

11.1.3 HYBRID OT–MCT

The complex geometry of small animals has a tremendous influence on the accuracy of OT reconstruction, and this complex structural information cannot be acquired by OT alone. Structural information from other modalities, such as microcomputed tomography (MCT), ultrasound, and magnetic resonance imaging (MRI), is introduced into the OT reconstruction (Lin et al. 2011;

Stuker et al. 2011; Ale et al. 2012). The information provides accurate boundary conditions for the forward calculations and an accurate constraint condition to reduce the ill-posedness and increase the convergence speed and accuracy in the inverse reconstruction. For ultrasound, acoustic attenuation may influence the quantification and image fidelity, and the MRI/optics system demands that the optical system uses fiber coupling or low-performance magnetic resonance–compatible cameras, which allows fewer sources and detectors. Combined with CT, an optical system can be engineered onto the rotating gantry, because of which a high-performance charge-coupled device (CCD) can be applied (Ntziachristos and Weissleder 2002; Quan et al. 2012). Therefore, dual modality OT/MCT can provide both structural and functional information for small animals in biological research.

11.2 HYBRID IMAGING SYSTEM DESIGN

The hybrid imaging system is designed with a concept of optical imaging–centric multimodality molecular imaging, which is also called a hybrid OT–MCT prototype imaging system. Resolving the system, the hybrid imaging system comprises a BLT/CLT prototype subsystem, an FMT prototype subsystem, an XRLT prototype subsystem, and a MCT prototype subsystem, as well as some other control components placed on an optical bench, as shown in Figure 11.1. These three prototype subsystems use a communal high-precision rotation stage with an integrated high-precision x–y translation stage to hold the small animal.

The BLT/CLT prototype subsystem employs a high-sensitivity CCD camera to acquire the luminescence signal. The high-sensitivity CCD camera should be cooled to below –70°C, which reduces the background and dark current noise to an extreme. Additionally, an electron multiplying CCD (EMCCD) is commonly adopted in the CLT prototype subsystem to strengthen the weak Cerenkov luminescence signals. In the FMT prototype subsystem, a laser source and a high-sensitivity CCD camera are provided with the illumination and detector. Considering the principle of resource sharing, the CW mode FMT prototype subsystem is equipped in the hybrid imaging system, in which a CW laser source and a high-sensitivity CCD or EMCCD are used. A set of bandpass filters are fixed before the CCD to allow light transmission at the emission wavelength. Substituting the CW laser source with an x-ray source, the FMT prototype subsystem can be adapted into an XRLT prototype subsystem. The MCT prototype subsystem involves a micro-focus x-ray source, a rotation stage with an x–y translation stage, and 2D x-ray detectors mounted exactly opposite to the source. To implement the hybrid imaging system, several prototype subsystems share some devices. For example, the BLT/CLT, FMT, and XRLT prototype subsystems can adopt one CCD camera; the XRLT and MCT prototype subsystem can utilize one

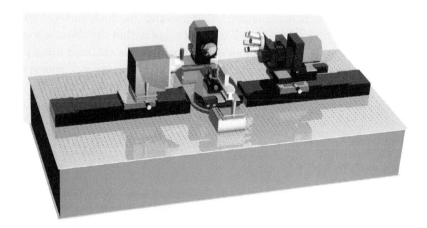

FIGURE 11.1 Design concept of the hybrid imaging system.

x-ray source. Thus, in the designed hybrid imaging system, we can only use an EMCCD camera, a CW laser source, a micro-focus x-ray source, and an x-ray panel detector to provide five kinds of imaging modalities, which are BLT, CLT, FMT, XRLT, and MCT.

11.3   BLT IMAGE RECONSTRUCTION

Implementation of BLT image reconstruction involves two aspects: the construction of an animal model and the development of the source reconstruction algorithm. The construction of an animal model is a key component for BLT image reconstruction. In this context, the animal model includes (1) a physical model of the photon migration, (2) a geometrical model of the animal anatomy, and (3) optical properties of different animal structures.

11.3.1   PHYSICAL MODELING AND METHODOLOGY

Cells encoded with the luciferase enzymes can serve as bioluminescent probes, which allow bioluminescent light emission. Because of the turbid nature of biological tissues, bioluminescent light is absorbed and scattered by the tissues when the light propagates in them. In a statistical sense, the propagation of the bioluminescent light can be accurately simulated with the Monte Carlo (MC) method (Wang et al. 1995; Li et al. 2004; Ren et al. 2010, 2013; Shen and Wang 2010). However, the MC method is time consuming. Recently, the massive parallel approach using general purpose graphic processing units has been adopted to speed up the MC simulation. The radiative transfer equation (RTE) is an analytic model to well describe photon propagation in biological tissues (Arridge 1999; Kim 2004; Klose and Larsen 2006; Cong et al. 2007; Cong and Wang 2010). Because the scattering of the photon transmission in the biological tissues predominates over the absorption, the propagation of bioluminescent photons can be described thoroughly by a diffusion approximation model (Cong et al. 2004a). The diffusion approximation model has been successfully applied for diffuse optical tomography (DOT), BLT, and FMT. Because the internal bioluminescent light is continuously on during the measurement, BLT operates only in the CW mode. In this case, the photon propagation in biological tissues can be described accurately by the steady-state diffusion equation (DE) and the Robin boundary condition (Cong et al. 2004b):

$$\begin{cases} -\nabla \cdot [D(\mathbf{r})\nabla\Phi(\mathbf{r})] + \mu_a(\mathbf{r})\Phi(\mathbf{r}) = S(\mathbf{r}) & \mathbf{r} \in \Omega \\ \Phi(\mathbf{r}) + 2A(\mathbf{r})D(\mathbf{r})(\upsilon(\mathbf{r}) \cdot \nabla\Phi(\mathbf{r})) = 0 & \mathbf{r} \in \partial\Omega \end{cases} \tag{11.1}$$

where

    $\Phi(\mathbf{r})$ is the photon fluence rate (W/mm^2) at location $\mathbf{r}$
    $S(\mathbf{r})$ is the density of the bioluminescent source distribution (W/mm^2)
    $D(\mathbf{r}) = (3(\mu_a(\mathbf{r}) + \mu_s'(\mathbf{r})))^{-1}$ is the diffusion coefficient (mm^{-1})
    $\mu_a(\mathbf{r})$ is the absorption coefficient (mm^{-1})
    $\mu_s'(\mathbf{r})$ is the reduced scattering coefficient (mm^{-1})
    Ω is the support for the animal body
    $\partial\Omega$ is the body surface of the animal
    $A(\mathbf{r})$ is the mismatch coefficient of different refractive indices across $\partial\Omega$

With an adequate exposure time, a significant amount of bioluminescent photons can reach the body surface of the animal and be detected with a highly sensitive CCD camera. The measured quantity is the photon current on the body surface of the animal:

$$\tilde{\Phi}(\mathbf{r}) = -D(\mathbf{r})(\upsilon \cdot \nabla\Phi(\mathbf{r})) \quad \mathbf{r} \in \partial\Omega \tag{11.2}$$

where υ denotes the unit outer normal to $\partial\Omega$. Based on Equations 11.1 and 11.2, a linear system that links the bioluminescent source distribution and the boundary measurement data can be obtained as

$$AS = \tilde{\Phi}$$

(11.3)

where

S is the bioluminescent source distribution

A is the weighing matrix consisting of N-dimensional column vectors about the source distribution

$\tilde{\Phi}$ is an N-dimensional vector of bioluminescent data measured on the body surface of the mouse

The BLT problem reconstructs S from data $\tilde{\Phi}$ (Gu et al. 2004; Alexandrakis et al. 2005, 2006; Chaudhari et al. 2005; Cong et al. 2005, 2006, 2010; Cong and Wang 2006; Li et al. 2006a,b; Lv et al. 2006, 2007; Wang et al. 2006a; Jiang et al. 2007; Kuo et al. 2007; Soloviev 2007; Ahn et al. 2008; Feng et al. 2008; Qin et al. 2008; Zhang et al. 2008, 2010; Han et al. 2009; Lu et al. 2009c; Qin et al. 2009; He et al. 2010; Huang et al. 2010; Liu et al. 2010d,e, 2011; Naser and Patterson 2010, 2011; Okawa and Yamada 2010; Yu et al. 2010). To reduce the modeling error in BLT reconstruction, researchers have also developed high-order approximations of the RTE-based photon migration model, such as the phase approximation and the simplified spherical harmonics approximation (SP_N) based reconstructions (Lu et al. 2009a,b; Cong and Wang 2010; Klose et al. 2010; Liu et al. 2010c).

11.3.2 GEOMETRIC MODELING

A small animal can be scanned by micro-CT and/or micro-MRI anatomically. The acquired images can be segmented into major regions (heart, lungs, liver, muscle, spleen, and so on). We are also evaluating the extent to which this process may be simplified using the small animal atlas deformable matching technique. Because small animal anatomy is rather complex, it is difficult to reflect all of the features in a geometrical model. Also, numerical computation will become impractical given an overcomplicated geometrical model. Hence, one approach would be to segment an image volume into the major organ regions at an approximately ~0.5 mm resolution. The commercial software Amira 4.1 (Mercury Computer Systems, Inc. Chelmsford, MA) is available for segmentation of images and construction of the small animal geometrical model. This will significantly simplify geometric modeling. We acknowledge that there are differences between the geometrical model and true small animal anatomy. However, in the small animal modeling process, optical parameters can be optimally determined to compensate for the geometric mismatches as shown in the following section.

11.3.3 ATTENUATION MAPS

Photon propagation in biological tissues depends on not only the geometrical model but also the optical properties of the small animal. Based on the geometrical model of the small animal, the optical parameters (absorption, scattering coefficient, and anisotropic coefficient) are considered as variables that are piecewise constants within each organ region. Then, the optical parameters will be reconstructed in the spectral bands of interest using a DOT technique. In this procedure, the multi-excitation and multi-detection strategies shall be employed to enhance numerical stability. We typically use the finite element method (FEM) for DOT (Elisee et al. 2010). From the finite element theory, the DE and the boundary condition can be formulated into a finite-element-based matrix equation. An objective function is defined to measure the total variation between the model-predicted photon density and measured photon density on the body surface of the small animal. The adjoint approach will be used as an effective and efficient way to calculate the gradient of the objective function (Wang et al. 2009b). The Quasi-Newton method and an active set strategy will be used to solve the minimization problem subject to simple constrains. Because the optical parameters are constrained to a piecewise constant related to different organ regions, the reconstruction of optical parameters will be numerically more robust.

11.3.4 Source Reconstruction

BLT is used to reconstruct a 3D distribution of the bioluminescent source inside a small living animal based on the bioluminescent signals on the body surface of a small animal. In general, the number of unknown variables of the underlying bioluminescent source distribution is much more than the number of independent measures at the surface nodes. Hence, BLT is a typical underdetermined problem, strong *a priori* knowledge must be incorporated into the reconstruction process to overcome the ill-posedness of the problem effectively (Wang et al. 2004). In this context, the permissible source region is used to reduce the number of unknown variables and thus enhance the stability of the BLT algorithm (Feng et al. 2008; Naser and Patterson 2011). High-value clusters on the body surface of a small animal clearly indicate the permissible source region. Then, a relatively large region was specified as the initial permissible region and iteratively refined as necessary. The reconstructed results were obtained in reference to the permissible source region. On the other hand, multispectral measurements were another kind of *a priori* knowledge to improve the ill-posedness of the reconstruction problem (Chaudhari et al. 2005; Wang et al. 2006b; Kuo et al. 2007; Ahn et al. 2008; Feng et al. 2008; Qin et al. 2011).

Mathematically, given the solving domain $\mathbf{\Omega}_S$, the BLT reconstruction is equivalent to the following optimization subject to regularization (Chaudhari et al. 2005; Cong et al. 2005, 2010; Cong and Wang 2006; Lv et al. 2006, 2007; Ahn et al. 2008; Feng et al. 2008; Han et al. 2009; Lu et al. 2009c; Qin et al. 2009; He et al. 2010; Huang et al. 2010; Liu et al. 2010d,e, 2011; Okawa and Yamada 2010; Yu et al. 2010; Zhang et al. 2010):

$$\min_{\substack{0 \le S \le U \\ S \in \Omega_S}} \left\| AS - \tilde{\Phi} \right\|_W^2 + \xi \eta(S) \tag{11.4}$$

where

η is the stabilizing function
ξ is the regularization parameter
W is the weighing matrix

$$\left\| V \right\|_W^2 = V^T W V$$

U denotes the upper bound for the source density to be physically meaningful

On the basis of the regularization-based objective function, researchers have proposed optimization methods to reconstruct a 3D source distribution, such as a modified Newton method with a specific Hessian matrix or active set strategy, a spectral projected gradient-based large-scale optimization algorithm, the level set method, the generalized graph cuts method, the preconditioned conjugate gradient algorithm, the trust region method, the truncated Newton interior-point method, the incomplete variables–truncated conjugate gradient method, the limited memory variable metric bound constrained quasi-Newton method, the stage-wise fast least absolute shrinkage and selection operator algorithm, and so on.

11.3.5 Typical Experimental Results

In the past few years, although the most efforts have been focused on the algorithmic development of BLT image reconstruction, the BLT technology has been successfully applied to biomedical experiments based on the developed prototype imaging systems and the related reconstruction algorithms. The successful applications mainly referred to the studies of cancer imaging. In 2006, Dr. Wang's group first applied the emerging BLT technology to biomedical experiments and conducted the monitoring of prostate cancer metastasis (Wang et al. 2006a) (typical results are shown in Figure 11.2). Dr. Tian's group studied the quantitative relationship between the number of cancer cells and BLT-reconstructed signals based on the fact that the reconstructed signals reflect the total number of cancer cells, and further illustrated the superiority of BLT over the planar bioluminescence imaging (Liu et al. 2010b) (typical results are shown in Figure 11.3). In the past 2 years, extension applications in the detection of solid and cavity cancer were presented, such as the detection of liver cancer (Ma et al. 2011) and gastric cancer (Chen et al. 2012; Hu et al. 2012a); the related results are shown in Figures 11.4 and 11.5.

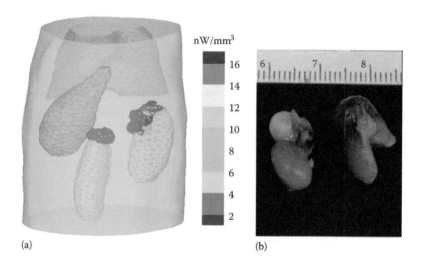

(a)   (b)

FIGURE 11.2   Biomedical applications in monitoring prostate cancer metastasis. (a) Two bioluminescent sources reconstructed on the two kidneys respectively; (b) two tumors at the same locations on the dissected kidneys. (From Wang, G. et al., *Opt. Express*, 14(17), 7801, 2006. With permission.)

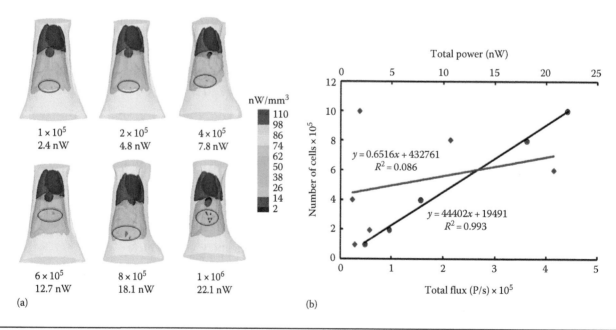

(a)   (b)

FIGURE 11.3   Quantification results of PC3-Luc prostate cancer. (a) Reconstructed total power of mice with different numbers of injected cells; the region in the red ellipse contains the reconstructed tumors and (b) relationship between the total power and tumor cell number in BLT (blue line) and that in planar bioluminescence imaging (red line). (From Liu, J. et al., *Opt. Express*, 18(12), 13102, 2010. With permission.)

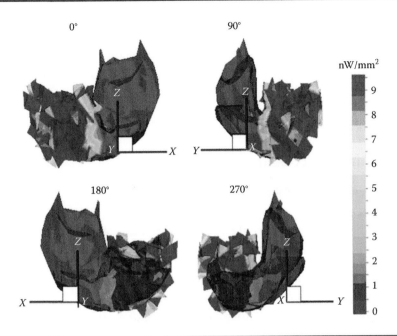

FIGURE 11.4 Reconstructed results based on the calibrated bioluminescent light intensity. The dark blue area is the liver (half-moon shaped). (From Ma, X. et al., *Appl. Opt.*, 50(10), 1389, 2011. With permission.)

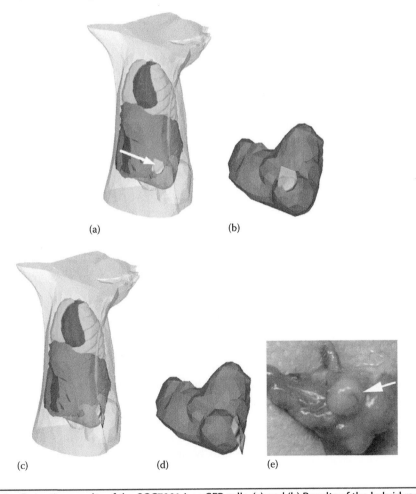

FIGURE 11.5 Detection results of the SGC7901-Luc-GFP cells. (a) and (b) Results of the hybrid radiosity-diffusion model based algorithm, (c) and (d) results of the diffusion equation model based algorithm, (a) and (c) 3D views of the results, (b) and (d) enlarged views of the stomach, and (e) necropsy observation of the affected stomach. (From Chen, X. et al., *J. Biomed. Opt.*, 17(6), 066015, 2012. With permission.)

11.4 POT IMAGE RECONSTRUCTION

As a passive imaging technique, POT needs an external excitation source to excite the biological cells labeled with optical probes. Thus, in this imaging modality, two photon propagation processes are involved, excitation and emission, which is the critical difference from BLT image reconstruction. Advances in the reconstruction algorithm of FMT are first described in this section and then those of XRLT are presented.

11.4.1 FMT IMAGE RECONSTRUCTION

11.4.1.1 Physical Models of Fluorescence Tomography

When fluorophores, fluorochromes, or fluorescent dyes are excited by an external excitation source such as a laser, their molecules absorb light and raise their energy levels to a brief excited state. As they decay from this excited state, they emit fluorescent light. Thus, the physical model of FMT involves two processes: light from the laser propagates in biological tissues to the fluorescent probe and the emitted fluorescent light transports to the body surface of the small animal. Considering the computational complexity of the RTE, the DE, the first-order approximation of the RTE, and its Robin boundary condition are commonly used in FMT to describe the transport characteristics for both the excitation and emission processes (Arridge 1999; Tan and Jiang 2008; Wang et al. 2009a):

$$
\begin{cases}
\nabla \cdot (D_x(r)\nabla\Phi_x(r)) - \mu_{ax}(r)\Phi_x(r) = -\Theta\delta(r - r_s) \\
\nabla \cdot (D_m(r)\nabla\Phi_m(r)) - \mu_{am}(r)\Phi_m(r) = -\Phi_x(r)\eta\mu_{af}(r)
\end{cases}
\quad (r \in \Omega)
\tag{11.5}
$$

where

subscript x and m denote excitation light and emission light, respectively

D is the diffusion coefficient

μ_a is the absorption coefficient

Φ denotes the photon density

$\eta\mu_{af}(r)$ is the fluorescent yield to be reconstructed, which is denoted as $X(r)$ in the following part of this article

The absorption coefficient due to the fluorophore μ_{af} is directly related to the fluorophore concentration by the formula $\mu_{af} = In(10)\varepsilon N$, where ε is the molar extinction coefficient and N is the concentration of the fluorophore (Naser and Patterson 2011). Here, the Robin-type boundary conditions are implemented on the boundary (Schweiger et al. 1995).

$$
\Phi_{x,m}(r) + 2BD_{x,m}(r)(n(r) \cdot \Phi_{x,m}(r)) = 0 \quad (r \in \partial\Omega)
\tag{11.6}
$$

where

n represents the outward normal vector to the surface

B defines the internal reflection of the light at the tissue boundary due to the index of refraction mismatch between the tissues and the air

Apart from DE approximation, discrete ordinates (S_N), spherical harmonics (P_N) equation, and simplified spherical harmonics (SP_N) have also been utilized to approximate RTE (Klose and Larsen 2006).

11.4.1.2 Image Reconstruction Algorithm

There are many methods for solving DE, such as the Green function, the MC method, the finite differential method, the FEM, and the boundary element method (Arridge et al. 1993; Ripoll et al. 2001;

Ripoll and Ntziachristos 2003; Hielscher et al. 2004). The FEM is widely used in optical tomography due to its ability of approximating arbitrary shape. When using the FEM to solve Equation 11.5, the coupled equations of FMT have the following matrix forms:

$$K_x \Phi_x = b_x$$
$$K_m \Phi_m = FX \tag{11.7}$$

where

$K_{x,m}$ is the system matrix

F is obtained by discretizing the unknown fluorescent yield distribution

For each excited point source at $r_s(l = 1,2,...,S)$, the corresponding Φ_x is obtained from solving Equation 11.7. For total S excitation point sources, we have the final weighted matrix:

$$\Phi_m = AX \tag{11.8}$$

This is a linear relationship between the measured photon flux density and the unknown fluorescent yield. Detailed descriptions are presented in Wang et al. (2009a).

Because of the strong scattering of light in tissues and the limited surface data, the inverse problem is usually an ill-posed problem. Regularization methods are commonly employed to obtain a stable solution, such as the Tikhonov regularization, sparsity regularization, total variation regularization, and hybrid regularization method (Joshi et al. 2004; Lee et al. 2007; Zacharopoulos et al. 2009; Han et al. 2010; Baritaux et al. 2011; Behrooz et al. 2012; Dutta et al. 2012). These methods usually have the following formula:

$$\min_X \left\{ \left\| AX - \Phi_m \right\|_2^2 + \lambda_p \left\| X \right\|_p \right\} \tag{11.9}$$

where

λ_p is the regularization parameter

$\left\| X \right\|_p$ is the p-norm of X

When $p = 2$, Equation 11.9 becomes the Tikhonov regularization that is also called an l_2-norm regularization problem; when $p = 1$, it is the l_1-norm regularization problem, which is a good sparsity measure of the image to be reconstructed. The l_2-norm regularization problem can be solved efficiently using standard minimization methods, such as the Newton method and conjugate gradient method. In the few past years, the l_1-norm regularization method has received considerable attention due to the development of the compressive sensing theory, and it can be solved by a primal-dual interior-point, orthogonal matching pursuit, and iterative shrinkage thresholding (Kojima et al. 1993; Davis et al. 1997; Daubechies et al. 2004). Apart from the regularization method, the iterative algorithm has also been used in FMT, such as the algebraic reconstruction technique and Landweber iterative regularization method (Intes et al. 2002; Song et al. 2007).

11.4.2 XRLT Image Reconstruction

XRLT, which is similar to x-ray fluorescence computed tomography (Boisseau 1986; Cesareo and Mascarenhas 1989; La Riviere and Vargas 2006), was first proposed by Guillem Pratx based on a selective excitation mechanism (Pratx et al. 2010). X-ray excitable nanophosphors can be fabricated to produce the visible or near infrared luminescent light when excited by x-rays and gamma rays (Liang et al. 2011; Sun et al. 2011). In the schematic diagram of the XRLT, the sample is irradiated by a sequence of narrow x-ray beams at certain locations or cone-beam x-rays (Pratx et al. 2010;

Carpenter et al. 2011; Chen et al. 2013). There are two different mathematical methods in the process of XRLT. The first kind of method is suitable for the case that the sample is excited by a narrow beam x-ray. When the sample is irradiated by a narrow beam x-ray, it is obvious that those photons are emitted from somewhere on the path of the x-ray beam. Considering the CCD camera as a "single-pixel detector," a projection sinogram of XRLT is formed by summing all of the pixels together, no matter where the photons are detected (Pratx et al. 2010a,b; Carpenter et al. 2011). Reconstructed images can be obtained by back-projecting the sinogram along the path of the x-ray beam as conventional CT does. In such a scheme, the reconstruction results are not affected by the scattering of photons, meanwhile, the spatial resolution is determined by the beam size and sampling (Pratx et al. 2010a,b; Carpenter et al. 2011).

The other approach considers the scattering characteristics of NIR light and adopts the reconstruction ideology that is similar to FMT, which can accurately describe the activities of the x-ray beam and the emitted NIR light. This kind of approach can be divided into the following three steps (Cong et al. 2011; Chen et al. 2013).

Step 1: X-rays are emitted from the x-ray source and travel through the sample following Beer's law. The x-ray intensity distribution can be expressed as follows:

$$X(\mathbf{r}) = X_0 \exp\left\{ -\int_{\mathbf{r}_0}^{\mathbf{r}} \mu_t(\tau) d\tau \right\} \tag{11.10}$$

where
  X_0 is the x-ray source intensity at the initial position $\mathbf{r}_0$
  μ_t is the x-ray attenuation coefficient at position τ

Step 2: Once the x-ray is transported to the sample, the sample emits NIR light:

$$S(\mathbf{r}) = \varepsilon X(\mathbf{r}) \rho(\mathbf{r}) \tag{11.11}$$

where
  $S(\mathbf{r})$ is the light source
  $X(\mathbf{r})$ is the x-ray intensity
  $\rho(\mathbf{r})$ is the concentration of nanophosphors at position $\mathbf{r}$
  ε is the light yield

Step 3: Similar to the FMT reconstruction, the diffusion equation and its Robin boundary condition are selected to describe the propagation of NIR light, as shown in Equation 11.1.

Based on the finite element theory, the following system matrix equation can be obtained by transforming Equations 11.1, 11.10, and 11.11 to the weak form and discretizing the domain with the shape function:

$$\mathbf{M}\Phi = \mathbf{F}\varepsilon X\rho \tag{11.12}$$

The reconstruction of XRLT is to recover the concentration ρ of nanophosphors from the measured photon flux density Φ on the body surface of a small animal.

Once the system matrix equation was formed, the reconstruction problem became easier because of thorough studies on the inverse problem of BLT/FMT. For example, in the maximum-likelihood expectation-maximization (MLEM) (Pratx et al. 2010), the l_1 term is based on the compressed sensing technique (Cong et al. 2011), the Levenberg–Marquardt algorithm (Carpenter et al. 2011), and the incomplete variables–truncated conjugate gradient method (Chen et al. 2013), which can all be implied in the reconstruction of XRLT.

11.5 OTHER OPTICAL TOMOGRAPHY MODALITIES

Apart from the four typical optical tomography modalities discussed, there are others, such as DOT (Gibson et al. 2005; Arridge 2011; Flexman et al. 2012), ultrasound-modulated optical tomography (Wang 2001; Sakadzic and Wang 2004), optical projection tomography (OPT) (Sharpe et al. 2002; Sharpe 2003, 2004; Lee et al. 2006; Alanentalo et al. 2007, 2010; Boot et al. 2008; Hajihosseini et al. 2009; Martínez-Estrada et al. 2009; Birk et al. 2010; Kumar et al. 2010; Thomas et al. 2010; Reiner et al. 2011; Cheddad et al. 2012), scanning laser optical tomography (SLOT) (Huisken et al. 2004; Dodt et al. 2007; Heidrich et al. 2011; Lorbeer et al. 2011; Kellner et al. 2012), pH-sensitive tomography (Mahoney et al. 2003; Andreev et al. 2007; Li et al. 2012), and photoacoustic tomography (Razansky et al. 2009; Xu et al. 2011; Wang and Hu 2012). In the following, we briefly review some of the modalities because of the limitation of chapter's length and details of the others can be obtained from the references provided.

11.5.1 OPTICAL PROJECTION TOMOGRAPHY

OPT is a valuable tool for the study of small specimens, typically smaller than 10 mm in size (Wang 2001; Gibson et al. 2005; Arridge 2011; Flexman et al. 2012). It can close the gap between diffraction-limited imaging, such as confocal laser scanning microscopy, and the various macro-scale imaging techniques used in medical imaging, such as DOT, FMT, MRI, CT, and so on. Because OPT provides good optical resolution at a reasonable imaging/penetration depth, it has seen steady growth in recent years in terms of both technical improvements and the development of applications (Sharpe et al. 2002; Sharpe 2003, 2004; Sakadzic and Wang 2004; Boot et al. 2008; Thomas et al. 2010).

The apparatus used for OPT data capture shares many similarities with a MCT scanner. A uniform back-lighting source is usually used in OPT imaging, which is best performed using a white-light diffuser, a stepper motor is used to rotate the specimen, which is usually embedded in matching liquid, and a 2D array detector (CCD camera) coupled with a microlens is used to capture the signals (Martínez-Estrada et al. 2009). OPT has the significant advantage of producing 3D images in both transmission and emission modes. In transmission mode, photons are projected through the specimen from a light source on one side and captured by the array detector on the other, very much like a MCT scanner, and the data captured essentially represent linear projections through the sample. In the emission mode, fluorescent dyes or proteins are excited by light and then emit photons with a longer wavelength, which will be captured by the CCD camera. The emission mode of OPT is similar to the SPECT scanner.

Applications of OPT are limited to samples that allow photons to traverse them along substantially straight paths. If too much photon scattering occurs within the sample, then predictable, deterministic paths are no longer possible and photon transport becomes probabilistic. In the past 3 years, successful applications with a number of whole organs taken from the adult mouse have been presented, such as the brain, pancreas, lungs, and kidneys (Alanentalo et al. 2010; Kumar et al. 2010; Reiner et al. 2011). The most important role of OPT imaging is the discovery, analysis, and mapping of gene expression patterns. Such information is a vital clue in our attempts to assign functions to the thousands of genes in the genome. The speed of OPT made it feasible to generate comprehensive maps for all of these genes in a reasonable time.

11.5.2 SCANNING LASER OPTICAL TOMOGRAPHY

In order to image specimens of sizes between 1 and 10 mm, several optical techniques have been developed in the last decade (Lee et al. 2006; Alanentalo et al. 2007; Birk et al. 2010; Arridge 2011). SLOT was introduced by scientists at Laser Zentrum Hannover (Hajihosseini et al. 2009). It simultaneously records transmissive, scattered, and fluorescent light. Samples can thus be imaged with a 3D resolution of at least 1/1000th of the object size in a short time. Compared with conventional OPT, SLOT increases the photon collection efficiency 100-fold, thus making it be used for objects larger than 2 mm as one of the most sensitive optical microscopy techniques. SLOT can be apart from homogeneous lighting with a 300 times higher photon exploitation and a high signal-to-noise ratio of 10–90 dB and avoid ring artifacts and speckles due to 1D detection.

In SLOT, the laser source is spatially cleaned by a pinhole within a telescope to obtain a flat and homogeneous wavefront that is directed onto a set of galvo-scanning mirrors. A diaphragm is used to adjust the aperture of the beam, which determines the numerical aperture of the optical system. The laser is scanned through the imaging lens into a glass cuvette filled with immersion liquid to acquire projection images of the specimen. A photodiode (PD) on the optical axis of the system behind the cuvette detects the transmitted laser light to acquire projection images of the extinction. Simultaneously, the autofluorescent signal is collected by a lens system below the cuvette and detected by a photomultiplier tube (PMT) to obtain projection images of the autofluorescence. Therefore, SLOT allows simultaneous acquisition of projection images by using transmitted light (via PD) and autofluoresence (via PMT). The optical filter blocks the excitation light of the laser source. By rotating the specimen 360° with a rotation stage, the full angle projection images can be obtained. Custom software is applied to a filtered back projection algorithm (known as OPT and CT) to reconstruct all projection images and then volumetric data are created to represent a 3D image of the specimen. The volumetric data can be used for producing sectional views and 3D segmentation. All projection datasets and volumetric data stacks can be clipped and visualized with the imaging software.

In comparison with other imaging modalities, SLOT is a very fast and convenient method that enables 3D imaging of recording both fluorescent and nonfluorescent signals from intact samples of a size of at least several millimeters. Now, SLOT has been used in serotonin immunofluorescence in the nervous systems of larger locusts and smaller *Drosophila* larvae, where the visualization of living biofilms is done in 3D, as well as the study of the internal structure of the mouse lung (Huisken et al. 2004; Cheddad et al. 2012). Thus, SLOT might be applied for monitoring single cell responses in whole organisms during pharmacological drug screening, where the structural and volumetric investigation of biofilm formation on implants with sizes up to several millimeters provides rapid screening of volumetric plasticity in insect brains.

11.5.3   pH-Sensitive Tomography

Monitoring cellular pH changes can provide diverse physiological and pathological processes, including studies related to cancer, cell proliferation, endocytotic, and other physiological processes (Dodt et al. 2007; Lorbeer et al. 2011). A common strategy is to monitor the pH of biochemical events by using pH-sensitive fluorescent molecular probes. The technique of imaging pH-sensitive fluorescent probes is highly sensitive and can be useful for imaging disease regions where pH-value is abnormal by using microscopy and whole-body animal imaging. Near-infrared diffuse fluorescence tomography (DFT) combines the model-based DOT and pH-sensitive NIR molecular probe and has been emerging as a powerful *in vivo* small animal imaging tool (Ntziachristos et al. 2002). Based on the DFT system, Gao et al. developed pH-sensitive tomography to accurately localize the target with a quantitative resolution to pH-sensitive variation of the fluorescent yield, which might provide a promising alternative method of pH-sensitive fluorescence imaging in addition to fluorescence-lifetime imaging (Lorbeer et al. 2011).

The commonly used DFT system modes are continuous-wave (CW) mode and time-domain (TD) mode. In the CW-DFT mode, a CT-analogous photon counting system is used for its well-established ultrahigh sensitive and ultralow-noise detection techniques. As the light source, a fiber-coupled laser diode emits the excitation light whose intensity is adjusted appropriately by a variable attenuator, collimated and coupled into an input fiber. Then, the collimated light beam impinges the boundary of the phantom through a collimation lens. The emission light is collected by eight detection fibers equally placed from 90° to 270° opposite to the incidence. The detection fibers are connected to an 8:1 fiber-optic switch whose output is collimated by the lens, routed on normal incidence to the aforementioned motorized filter wheel, and then directly sent into a PMT photon counting head coupled with a counting unit to quantify photon numbers, respectively, for the fluorescent excitation and emission signals. By rotating the phantom at an angular interval and lifting it at a vertical displacement, the system achieves 3D scanning process with high-density spatial sampling. In the TD-DFT mode, a multi-channel time-correlated single photon counting system is used and a pulsed diode laser is adopted as the excitation light source. Then, the 3D scanning process is similar to the CW mode.

The related phantom experiments demonstrate that the pH-sensitive DFT system achieves the high-speed, ultrahigh sensitivity and dense spatial sampling measurements. The pH-sensitive tomography might provide a promising alternative to the fluorescence-lifetime imaging method for probing pH-involved pathology and physiology of diseases in some small-animal-based investigations.

ACKNOWLEDGMENTS

X. Chen, D. Chen, and J. Liang acknowledge funding support from the Program of the National Basic Research and Development Program of China (973) under grant No. 2011CB707702, the National Natural Science Foundation of China under grant Nos. 81090272, 81227901, and 81101083, and the Fundamental Research Funds for the Central Universities.

REFERENCES

Ahn, S., A. J. Chaudhari et al. (2008). Fast iterative image reconstruction methods for fully 3D multispectral bioluminescence tomography. *Phys. Med. Biol.* **53**(14): 3921–3942.

Alanentalo, T., A. Asayesh et al. (2007). Tomographic molecular imaging and 3D quantification within adult mouse organs. *Nat. Methods* **4**(1): 31–33.

Alanentalo, T., A. Hornblad et al. (2010). Quantification and three-dimensional imaging of the insulitis-induced destruction of beta-cells in murine type 1 diabetes. *Diabetes* **59**(7): 1756–1764.

Ale, A., V. Ermolayev et al. (2012). FMT-XCT: In vivo animal studies with hybrid fluorescence molecular tomography-x-ray computed tomography. *Nat. Methods* **9**(6): 615–620.

Alexandrakis, G., F. R. Rannou et al. (2005). Tomographic bioluminescence imaging by use of a combined optical-PET (OPET) system: A computer simulation feasibility study. *Phys. Med. Biol.* **50**(17): 4225–4241.

Alexandrakis, G., F. R. Rannou et al. (2006). Effect of optical property estimation accuracy on tomographic bioluminescence imaging: Simulation of a combined optical-PET (OPET) system. *Phys. Med. Biol.* **51**(8): 2045–2053.

Andreev, O. A., A. D. Dupuy et al. (2007). Mechanism and uses of a membrane peptide that targets tumors and other acidic tissues in vivo. *Proc. Natl. Acad. Sci. USA* **104**(19): 7893–7898.

Arridge, S. R. (1999). Optical tomography in medical imaging. *Inverse Probl.* **15**(2): R41–R93.

Arridge, S. R. (2011). Methods in diffuse optical imaging. *Phil. Trans. Roy. Soc. A* **369**(1955): 4558–4576.

Arridge, S. R., M. Schweiger et al. (1993). A finite element approach for modeling photon transport in tissue. *Med. Phys.* **20**(2 Pt. 1): 299–309.

Baritaux, J.-C., K. Hassler et al. (2011). Sparsity-driven reconstruction for FDOT with anatomical priors. *IEEE Trans. Med. Imaging* **30**(5): 1143–1153.

Bazalova, M., Y. Kuang et al. (2012). Investigation of x-ray fluorescence computed tomography (XFCT) and K-edge imaging. *IEEE Trans. Med. Imaging* **31**(8): 1620–1627.

Behrooz, A., H.-M. Zhou et al. (2012). Total variation regularization for 3D reconstruction in fluorescence tomography: Experimental phantom studies. *Appl. Opt.* **51**(34): 8216–8227.

Bhaumik, S. and S. S. Gambhir. (2002). Optical imaging of Renilla luciferase reporter gene expression in living mice. *Proc. Natl. Acad. Sci. USA* **99**(1): 377–382.

Biosciences, A. (2002). *Fluorescence Imaging: Principles and Methods.* Amersham Biosciences, Piscataway, NJ.

Birk, U. J., M. Rieckher et al. (2010). Correction for specimen movement and rotation errors for in-vivo optical projection tomography. *Biomed. Opt. Express* **1**(1): 87–96.

Boisseau, P. (1986). Determination of three dimensional trace element distributions by the use of monochromatic x-ray microbeams. PhD dissertation, Massachusetts Institute of Technology, Department of Physics, Cambridge, MA.

Boot, M. J., C. H. Westerberg et al. (2008). In vitro whole-organ imaging: 4D quantification of growing mouse limb buds. *Nat. Methods* **5**(7): 609–612.

Boschi, F., L. Calderan et al. (2011). In vivo F-18-FDG tumour uptake measurements in small animals using Cerenkov radiation. *Eur. J. Nucl. Med. Mol. Imaging* **38**(1): 120–127.

Carpenter, C. M., G. Pratx et al. (2011). Limited-angle x-ray luminescence tomography: Methodology and feasibility study. *Phys. Med. Biol.* **56**(12): 3487–3502.

Carpenter, C. M., C. Sun et al. (2010). Hybrid x-ray/optical luminescence imaging: Characterization of experimental conditions. *Med. Phys.* **37**(8): 4011–4018.

Carpenter, C. M., C. Sun et al. (2012). Radioluminescent nanophosphors enable multiplexed small-animal imaging. *Opt. Express* **20**(11): 11598–11604.

Cesareo, R. and S. Mascarenhas. (1989). A new tomographic device based on the detection of fluorescent x-rays. *Nucl. Instrum. Methods Phys. Res. A* **277**(2): 669–672.

Chaudhari, A. J., F. Darvas et al. (2005). Hyperspectral and multispectral bioluminescence optical tomography for small animal imaging. *Phys. Med. Biol.* **50**(23): 5421–5441.

Cheddad, A., C. Svensson et al. (2012). Image processing assisted algorithms for optical projection tomography. *IEEE Trans. Med. Imaging* **31**(1): 1–15.

Chen, D., S. Zhu et al. (2013). Cone beam x-ray luminescence computed tomography: A feasibility study. *Med. Phys.* **40**(3): 031111–031111.

Chen, X., D. Yang et al. (2012). Comparisons of hybrid radiosity-diffusion model and diffusion equation for bioluminescence tomography in cavity cancer detection. *J. Biomed. Opt.* **17**(6): 066015.

Cheong, S.-K., B. L. Jones et al. (2010). X-ray fluorescence computed tomography (XFCT) imaging of gold nanoparticle-loaded objects using 110 kVp x-rays. *Phys. Med. Biol.* **55**(3): 647–662.

Cong, A., W. Cong et al. (2010). Differential evolution approach for regularized bioluminescence tomography. *IEEE Trans. Biomed. Eng.* **57**(9): 2229–2238.

Cong, W., H. Shen et al. (2007). Modeling photon propagation in biological tissues using a generalized Delta-Eddington phase function. *Phys. Rev. E* **76**(5): 051913.

Cong, W., H. Shen et al. (2011). Spectrally resolving and scattering-compensated x-ray luminescence/fluorescence computed tomography. *J. Biomed. Opt.* **16**(6): 066014.

Cong, W. and G. Wang. (2006). Boundary integral method for bioluminescence tomography. *J. Biomed. Opt.* **11**(2): 020503.

Cong, W. and G. Wang. (2010). Bioluminescence tomography based on the phase approximation model. *J. Opt. Soc. Am. A* **27**(2): 174–179.

Cong, W., L. V. Wang et al. (2004a). Formulation of photon diffusion from spherical bioluminescent sources in an infinite homogeneous medium. *Biomed. Eng.* **3**(1): 12.

Cong, W. X., K. Durairaj et al. (2006). A Born-type approximation method for bioluminescence tomography. *Med. Phys.* **33**(3): 679–686.

Cong, W. X., D. Kumar et al. (2004b). A practical method to determine the light source distribution in bioluminescent imaging. *Proc. SPIE* **5535**: 679–686.

Cong, W. X., G. Wang et al. (2005). Practical reconstruction method for bioluminescence tomography. *Opt. Express* **13**(18): 6756–6771.

Contag, C. H. and M. H. Bachmann. (2002). Advances in vivo bioluminescence imaging of gene expression. *Annu. Rev. Biomed. Eng.* **4**: 235–260.

Daubechies, I., M. Defrise et al. (2004). An iterative thresholding algorithm for linear inverse problems with a sparsity constraint. *Comm. Pure Appl. Math.* **57**(11): 1413–1457.

Davis, G., S. Mallat et al. (1997). Adaptive greedy approximations. *Constr. Approx.* **13**(1): 57–98.

Dodt, H.-U., U. Leischner et al. (2007). Ultramicroscopy: Three-dimensional visualization of neuronal networks in the whole mouse brain. *Nat. Methods* **4**(4): 331–336.

Dutta, J., S. Ahn et al. (2012). Joint L-1 and total variation regularization for fluorescence molecular tomography. *Phys. Med. Biol.* **57**(6): 1459–1476.

Elisee, J. P., A. Gibson et al. (2010). Combination of boundary element method and finite element method in diffuse optical tomography. *IEEE Trans. Biomed. Eng.* **57**(11): 2737–2745.

Feng, J., K. Jia et al. (2008). An optimal permissible source region strategy for multispectral bioluminescence tomography. *Opt. Express* **16**(20): 15640–15654.

Flexman, M. L., F. Vlachos et al. (2012). Monitoring early tumor response to drug therapy with diffuse optical tomography. *J. Biomed. Opt.* **17**(1): 016014.

Gibson, A. P., J. C. Hebden et al. (2005). Recent advances in diffuse optical imaging. *Phys. Med. Biol.* **50**(4): R1–R43.

Gu, X. J., Q. H. Zhang et al. (2004). Three-dimensional bioluminescence tomography with model-based reconstruction. *Opt. Express* **12**(17): 3996–4000.

Hajihosseini, M. K., R. Duarte et al. (2009). Evidence that fgf10 contributes to the skeletal and visceral defects of an apert syndrome mouse model. *Dev. Dynam.* **238**(2): 376–385.

Han, D., J. Tian et al. (2010). A fast reconstruction algorithm for fluorescence molecular tomography with sparsity regularization. *Opt. Express* **18**(8): 8630–8646.

Han, R., J. Liang et al. (2009). A source reconstruction algorithm based on adaptive hp-FEM for bioluminescence tomography. *Opt. Express* **17**(17): 14481–14494.

He, X., J. Liang et al. (2010). Sparse reconstruction for quantitative bioluminescence tomography based on the incomplete variables truncated conjugate gradient method. *Opt. Express* **18**(24): 24825–24841.

Heidrich, M., M. P. Kuehnel et al. (2011). 3D imaging of biofilms on implants by detection of scattered light with a scanning laser optical tomograph. *Biomed. Opt. Express* **2**(11): 2982–2994.

Hielscher, A. H., A. D. Klose et al. (2004). Sagittal laser optical tomography for imaging of rheumatoid finger joints. *Phys. Med. Biol.* **49**(7): 1147–1163.

Holland, J. P., G. Normand et al. (2011). Intraoperative imaging of positron emission tomographic radiotracers using cerenkov luminescence emissions. *Molecular Imaging* **10**(3): 177–186.

Hu, H., J. Liu et al. (2012a). Real-time bioluminescence and tomographic imaging of gastric cancer in a novel orthotopic mouse model. *Oncol. Rep.* **27**(6): 1937–1943.

Hu, Z., X. Chen et al. (2012b). Single photon emission computed tomography-guided Cerenkov luminescence tomography. *J. Appl. Phys.* **112**(2): 024703.

Hu, Z., J. Liang et al. (2010). Experimental Cerenkov luminescence tomography of the mouse model with SPECT imaging validation. *Opt. Express* **18**(24): 24441–24450.

Hu, Z., X. Ma et al. (2012c). Three-dimensional noninvasive monitoring iodine-131 uptake in the thyroid using a modified Cerenkov luminescence tomography approach. *PLoS One* **7**(5): e37623.

Huang, H., X. Qu et al. (2010). A multi-phase level set framework for source reconstruction in bioluminescence tomography. *J. Comput. Phys.* **229**(13): 5246–5256.

Huisken, J., J. Swoger et al. (2004). Optical sectioning deep inside live embryos by selective plane illumination microscopy. *Science* **305**(5686): 1007–1009.

Intes, X., V. Ntziachristos et al. (2002). Projection access order in algebraic reconstruction technique for diffuse optical tomography. *Phys. Med. Biol.* **47**(1): N1–N10.

Jelley, J. (1955). Cerenkov radiation and its applications. *Br. J. Appl. Phys.* **6**(7): 227.

Jeong, S. Y., M.-H. Hwang et al. (2011). Combined Cerenkov luminescence and nuclear imaging of radioiodine in the thyroid gland and thyroid cancer cells expressing sodium iodide symporter: Initial feasibility study. *Endocr. J.* **58**(7): 575–583.

Jiang, M., T. Zhou et al. (2007). Image reconstruction for bioluminescence tomography from partial measurement. *Opt. Express* **15**(18): 11095–11116.

Jones, B. L. and S. H. Cho. (2011). The feasibility of polychromatic cone-beam x-ray fluorescence computed tomography (XFCT) imaging of gold nanoparticle-loaded objects: A Monte Carlo study. *Phys. Med. Biol.* **56**(12): 3719–3730.

Joshi, A., W. Bangerth et al. (2004). Adaptive finite element based tomography for fluorescence optical imaging in tissue. *Opt. Express* **12**(22): 5402–5417.

Kellner, M., M. Heidrich et al. (2012). Imaging of the mouse lung with scanning laser optical tomography (SLOT). *J. Appl. Physiol.* **113**(6): 975–983.

Kim, A. D. (2004). Transport theory for light propagation in biological tissue. *J. Opt. Soc. Am. A* **21**(5): 820–827.

Klose, A. D., B. J. Beattie et al. (2010). In vivo bioluminescence tomography with a blocking-off finite-difference SP3 method and MRI/CT coregistration. *Med. Phys.* **37**(1): 329–338.

Klose, A. D. and E. W. Larsen. (2006). Light transport in biological tissue based on the simplified spherical harmonics equations. *J. Comput. Phys.* **220**(1): 441–470.

Kojima, M., N. Megiddo et al. (1993). Theoretical convergence of large-step primal–dual interior point algorithms for linear programming. *Math. Program.* **59**(1–3): 1–21.

Kumar, V., E. Scandella et al. (2010). Global lymphoid tissue remodeling during a viral infection is orchestrated by a B cell-lymphotoxin-dependent pathway. *Blood* **115**(23): 4725–4733.

Kuo, C., O. Coquoz et al. (2007). Three-dimensional reconstruction of in vivo bioluminescent sources based on multispectral imaging. *J. Biomed. Opt.* **12**(2): 024007.

La Riviere, P. J. and P. A. Vargas. (2006). Monotonic penalized-likelihood image reconstruction for X-ray fluorescence computed tomography. *IEEE Trans. Med. Imaging* **25**(9): 1117–1129.

Lee, J. H., A. Joshi et al. (2007). Fully adaptive finite element based tomography using tetrahedral dual-meshing for fluorescence enhanced optical imaging in tissue. *Opt. Express* **15**(11): 6955–6975.

Lee, K., J. Avondo et al. (2006). Visualizing plant development and gene expression in three dimensions using optical projection tomography. *Plant Cell* **18**(9): 2145–2156.

Li, C., G. S. Mitchell et al. (2010). Cerenkov luminescence tomography for small-animal imaging. *Opt. Lett.* **35**(7): 1109–1111.

Li, H., J. Tian et al. (2004). A mouse optical simulation environment (MOSE) to investigate bioluminescent phenomena in the living mouse with the Monte Carlo method. *Acad. Radiol.* **11**(9): 1029–1038.

Li, J., X. Wang et al. (2012). Towards pH-sensitive imaging of small animals with photon-counting difference diffuse fluorescence tomography. *J. Biomed. Opt.* **17**(9): 0960111.

Li, S., W. Driessen et al. (2006a). Bioluminescence tomography based on phantoms with different concentrations of bioluminescent cancer cells. *J. Opt. A Pure Appl. Opt.* **8**(9): 743–746.

Li, S., Q. Zhang et al. (2006b). Two-dimensional bioluminescence tomography: Numerical simulations and phantom experiments. *Appl. Opt.* **45**(14): 3390–3394.

Liang, H., H. Lin et al. (2011). Luminescence of Ce3+ and Pr3+ doped Sr2Mg(BO3)(2) under VUV-UV and X-ray excitation. *J. Lumin.* **131**(2): 194–198.

Lin, Y., M. T. Ghijsen et al. (2011). A photo-multiplier tube-based hybrid MRI and frequency domain fluorescence tomography system for small animal imaging. *Phys. Med. Biol.* **56**(15): 4731–4747.

Liu, H., G. Ren et al. (2010a). Molecular optical imaging with radioactive probes. *PLoS One* **5**(3): e9470.

Liu, J., Y. Wang et al. (2010b). In vivo quantitative bioluminescence tomography using heterogeneous and homogeneous mouse models. *Opt. Express* **18**(12): 13102–13113.

Liu, K., Y. Lu et al. (2010c). Evaluation of the simplified spherical harmonics approximation in bioluminescence tomography through heterogeneous mouse models. *Opt. Express* **18**(20): 20988–21002.

Liu, K., J. Tian et al. (2010d). A fast bioluminescent source localization method based on generalized graph cuts with mouse model validations. *Opt. Express* **18**(4): 3732–3745.

Liu, K., J. Tian et al. (2011). Tomographic bioluminescence imaging reconstruction via a dynamically sparse regularized global method in mouse models. *J. Biomed. Opt.* **16**(4): 046016.

Liu, K., X. Yang et al. (2010e). Spectrally resolved three-dimensional bioluminescence tomography with a level-set strategy. *J. Opt. Soc. Am. A* **27**(6): 1413–1423.

Lorbeer, R.-A., M. Heidrich et al. (2011). Highly efficient 3D fluorescence microscopy with a scanning laser optical tomograph. *Opt. Express* **19**(6): 5419–5430.

Lu, Y., A. Douraghy et al. (2009a). Spectrally resolved bioluminescence tomography with the third-order simplified spherical harmonics approximation. *Phys. Med. Biol.* **54**(21): 6477–6493.

Lu, Y., H. B. Machado et al. (2009b). Experimental bioluminescence tomography with fully parallel radiative-transfer-based reconstruction framework. *Opt. Express* **17**(19): 16681–16695.

Lu, Y., X. Zhang et al. (2009c). Source reconstruction for spectrally-resolved bioluminescence tomography with sparse a priori information. *Opt. Express* **17**(10): 8062–8080.

Lv, Y., J. Tian et al. (2006). A multilevel adaptive finite element algorithm for bioluminescence tomography. *Opt. Express* **14**(18): 8211–8223.

Lv, Y., J. Tian et al. (2007). Spectrally resolved bioluminescence tomography with adaptive finite element analysis: Methodology and simulation. *Phys. Med. Biol.* **52**(15): 4497–4512.

Ma, X., J. Tian et al. (2011). Early detection of liver cancer based on bioluminescence tomography. *Appl. Opt.* **50**(10): 1389–1395.

Mahoney, B. P., N. Raghunand et al. (2003). Tumor acidity, ion trapping and chemotherapeutics I. Acid pH affects the distribution of chemotherapeutic agents in vitro. *Biochem. Pharmacol.* **66**(7): 1207–1218.

Martínez-Estrada, O. M., L. A. Lettice et al. (2009). Wt1 is required for cardiovascular progenitor cell formation through transcriptional control of Snail and E-cadherin. *Nat. Genet.* **42**(1): 89–93.

Massoud, T. F. and S. S. Gambhir. (2003). Molecular imaging in living subjects: Seeing fundamental biological processes in a new light. *Gene Dev.* **17**(5): 545–580.

Mitchell, G. S., R. K. Gill et al. (2011). In vivo Cerenkov luminescence imaging: A new tool for molecular imaging. *Phil. Trans. Roy. Soc. A* **369**(1955): 4605–4619.

Naser, M. A. and M. S. Patterson. (2010). Algorithms for bioluminescence tomography incorporating anatomical information and reconstruction of tissue optical properties. *Biomed. Opt. Express* **1**(2): 512–526.

Naser, M. A. and M. S. Patterson. (2011). Improved bioluminescence and fluorescence reconstruction algorithms using diffuse optical tomography, normalized data, and optimized selection of the permissible source region. *Biomed. Opt. Express* **2**(1): 169–184.

Ntziachristos, V., J. Ripoll et al. (2005). Looking and listening to light: The evolution of whole-body photonic imaging. *Nat. Biotechnol.* **23**(3): 313–320.

Ntziachristos, V., E. A. Schellenberger et al. (2004). Visualization of antitumor treatment by means of fluorescence molecular tomography with an annexin V-Cy5.5 conjugate. *Proc. Natl. Acad. Sci. USA* **101**(33): 12294–12299.

Ntziachristos, V., C.-H. Tung et al. (2002). Fluorescence molecular tomography resolves protease activity in vivo. *Nat. Med.* **8**(7): 757–761.

Ntziachristos, V. and R. Weissleder. (2002). Charge-coupled-device based scanner for tomography of fluorescent near-infrared probes in turbid media. *Med. Phys.* **29**(5): 803–809.

Okawa, S. and Y. Yamada. (2010). Reconstruction of fluorescence/bioluminescence sources in biological medium with spatial filter. *Opt. Express* **18**(12): 13151–13172.

Pereira, G. R., M. J. Anjos et al. (2007). Computed tomography and x-ray fluorescence CT of biological samples. *Nucl. Instrum. Methods Phys. Res. A* **580**(2): 951–954.

Pratx, G., C. M. Carpenter et al. (2010a). Tomographic molecular imaging of x-ray-excitable nanoparticles. *Opt. Lett.* **35**(20): 3345–3347.

Pratx, G., C. M. Carpenter et al. (2010b). X-ray luminescence computed tomography via selective excitation: A feasibility study. *IEEE Trans. Med. Imaging* **29**(12): 1992–1999.

Qin, C., J. Tian et al. (2008). Galerkin-based meshless methods for photon transport in the biological tissue. *Opt. Express* **16**(25): 20317–20333.

Qin, C., X. Yang et al. (2009). Adaptive improved element free Galerkin method for quasi- or multi-spectral bioluminescence tomography. *Opt. Express* **17**(24): 21925–21934.

Qin, C., S. Zhu et al. (2011). Comparison of permissible source region and multispectral data using efficient bioluminescence tomography method. *J. Biophoton.* **4**(11–12): 824–839.

Quan, G., K. Wang et al. (2012). Micro-computed tomography-guided, non-equal voxel Monte Carlo method for reconstruction of fluorescence molecular tomography. *J. Biomed. Opt.* **17**(8): 086006.

Razansky, D., M. Distel et al. (2009). Multispectral opto-acoustic tomography of deep-seated fluorescent proteins in vivo. *Nat. Photon.* **3**(7): 412–417.

Reiner, T., G. Thurber et al. (2011). Accurate measurement of pancreatic islet beta-cell mass using a second-generation fluorescent exendin-4 analog. *Proc. Natl. Acad. Sci. USA* **108**(31): 12815–12820.

Ren, N., J. Liang et al. (2010). GPU-based Monte Carlo simulation for light propagation in complex heterogeneous tissues. *Opt. Express* **18**(7): 6811–6823.

Ren, S., X. Chen et al. (2013). Molecular Optical Simulation Environment (MOSE): A platform for the simulation of light propagation in turbid media. *PLoS One* **8**(4): e61304.

Ripoll, J. and V. Ntziachristos. (2003). Iterative boundary method for diffuse optical tomography. *J. Opt. Soc. Am. A* **20**(6): 1103–1110.

Ripoll, J., V. Ntziachristos et al. (2001). Kirchhoff approximation for diffusive waves. *Phys. Rev. E* **64**(5): 051917.

Robertson, R., M. S. Germanos et al. (2011). Multimodal imaging with F-18-FDG PET and Cerenkov luminescence imaging after MLN4924 treatment in a human lymphoma xenograft model. *J. Nucl. Med.* **52**(11): 1764–1769.

Sakadzic, S. and L. H. V. Wang. (2004). High-resolution ultrasound-modulated optical tomography in biological tissues. *Opt. Lett.* **29**(23): 2770–2772.

Schweiger, M., S. R. Arridge et al. (1995). The finite element method for the propagation of light in scattering media: Boundary and source conditions. *Med. Phys.* **22**(11 PART 1): 1779–1792.

Sharpe, J. (2003). Optical projection tomography as a new tool for studying embryo anatomy. *J. Anat.* **202**(2): 175–181.

Sharpe, J. (2004). Optical projection tomography. *Annu. Rev. Biomed. Eng.* **6**: 209–228.

Sharpe, J., U. Ahlgren et al. (2002). Optical projection tomography as a tool for 3D microscopy and gene expression studies. *Science* **296**(5567): 541–545.

Shen, H. and G. Wang. (2010). A tetrahedron-based inhomogeneous Monte Carlo optical simulator. *Phys. Med. Biol.* **55**(4): 947–962.

Soloviev, V. Y. (2007). Tomographic bioluminescence imaging with varying boundary conditions. *Appl. Opt.* **46**(14): 2778–2784.

Song, X., D. Wang et al. (2007). Reconstruction for free-space fluorescence tomography using a novel hybrid adaptive finite element algorithm. *Opt. Express* **15**(26): 18300–18317.

Spinelli, A. E., D. D'Ambrosio et al. (2010). Cerenkov radiation allows in vivo optical imaging of positron emitting radiotracers. *Phys. Med. Biol.* **55**(2): 483–495.

Stuker, F., C. Baltes et al. (2011). Hybrid small animal imaging system combining magnetic resonance imaging with fluorescence tomography using single photon avalanche diode detectors. *IEEE Trans. Med. Imaging* **30**(6): 1265–1273.

Sun, C., G. Pratx et al. (2011). Synthesis and radioluminescence of PEGylated Eu^{3+}-doped nanophosphors as bioimaging probes. *Adv. Mater.* **23**(24): H195–H199.

Tan, Y. and H. Jiang. (2008). DOT guided fluorescence molecular tomography of arbitrarily shaped objects. *Med. Phys.* **35**(12): 5703–5707.

Thomas, A., J. Bowsher et al. (2010). A comprehensive method for optical-emission computed tomography. *Phys. Med. Biol.* **55**(14): 3947–3957.

Tian, J., J. Bai et al. (2008). Multimodality molecular imaging - Improving image quality. *IEEE Eng. Med. Biol. Mag.* **27**(5): 48–57.

Troy, T., D. Jekic-McMullen et al. (2004). Quantitative comparison of the sensitivity of detection of fluorescent and bioluminescent reporters in animal models. *Mol. Imaging* **3**(1): 9–23.

Wang, D., X. Liu et al. (2009a). A novel finite-element-based algorithm for fluorescence molecular tomography of heterogeneous media. *IEEE Trans. Inf. Technol. Biomed.* **13**(5): 766–773.

Wang, G., W. Cong et al. (2006a). In vivo mouse studies with bioluminescence tomography. *Opt. Express* **14**(17): 7801–7809.

Wang, G., W. Cong et al. (2008). Overview of bioluminescence tomography-a new molecular imaging modality. *Front. Biosci.* **13**: 1281–1293.

Wang, G., E. Hoffman et al. (2003). Development of the first bioluminescent CT scanner. *Radiology* **229**: 566.

Wang, G., Y. Li et al. (2004). Uniqueness theorems in bioluminescence tomography. *Med. Phys.* **31**(8): 2289–2299.

Wang, G., H. Shen et al. (2006b). The first bioluminescence tomography system for simultaneous acquisition of multiview and multispectral data. *Int. J. Biomed. Imaging* **2006**: 58601.

Wang, L., S. L. Jacques et al. (1995). MCML-Monte Carlo modeling of light transport in multi-layered tissues. *Comput. Meth. Programs Biomed.* **47**(2): 131–146.

Wang, L. H. V. (2001). Mechanisms of ultrasonic modulation of multiply scattered coherent light: An analytic model. *Phys. Rev. Lett.* **87**(4): 043903.

Wang, L. V. and S. Hu. (2012). Photoacoustic tomography: In vivo imaging from organelles to organs. *Science* **335**(6075): 1458–1462.

Wang, Z. G., L. Z. Sun et al. (2009b). Modeling and reconstruction of diffuse optical tomography using adjoint method. *Comm. Numer. Meth. Eng.* **25**(6): 657–665.

Weissleder, R. and V. Ntziachristos. (2003). Shedding light onto live molecular targets. *Nat. Med.* **9**(1): 123–128.

Wu, J., L. Perelman et al. (1997). Fluorescence tomographic imaging in turbid media using early-arriving photons and Laplace transforms. *Proc. Natl. Acad. Sci. USA* **94**(16): 8783–8788.

Xu, X., H. Liu et al. (2011). Time-reversed ultrasonically encoded optical focusing into scattering media. *Nat. Photon.* **5**(3): 154–157.

Xu, Y., E. Chang et al. (2012). Proof-of-concept study of monitoring cancer drug therapy with Cerenkov luminescence imaging. *J. Nucl. Med.* **53**(2): 312–317.

Yu, J., F. Liu et al. (2010). Fast source reconstruction for bioluminescence tomography based on sparse regularization. *IEEE Trans. Biomed. Eng.* **57**(10): 2583–2586.

Zacharopoulos, A. D., P. Svenmarker et al. (2009). A matrix-free algorithm for multiple wavelength fluorescence tomography. *Opt. Express* **17**(5): 3025–3035.

Zavattini, G., S. Vecchi et al. (2006). A hyperspectral fluorescence system for 3D in vivo optical imaging. *Phys. Med. Biol.* **51**(8): 2029–2043.

Zhang, B., X. Yang et al. (2010). A trust region method in adaptive finite element framework for bioluminescence tomography. *Opt. Express* **18**(7): 6477–6491.

Zhang, Q., L. Yin et al. (2008). Quantitative bioluminescence tomography guided by diffuse optical tomography. *Opt. Express* **16**(3): 1481–1486.

Zhong, J., C. Qin et al. (2012). Fast-specific tomography imaging Cerenkov emission. *Mol. Imaging Biol.* **14**(3): 286–292.

Zhong, J., J. Tian et al. (2011). Whole-body Cerenkov luminescence tomography with the finite element SP3 method. *Ann. Biomed. Eng.* **39**(6): 1728–1735.

12

PET/CT

Mohammad Reza Ay and Nafiseh Ghazanfari

12.1 INTRODUCTION

The advent of dual-modality PET/CT imaging has revolutionized the practice of clinical and preclinical studies by improving lesion localization and the possibility of accurate quantitative analysis. In addition, the use of CT images for CT-based attenuation correction (CTAC) of PET data allows to decrease the overall scanning time and to create a noise-free attenuation map. The near simultaneous data acquisition in a fixed combination of a PET and a CT scanner in a hybrid PET/CT imaging system with a common table minimizes spatial and temporal mismatches between the modalities by eliminating the need to move the object in between exams. The purpose of this chapter is to introduce the principles of PET/CT imaging systems and describe the sources of error and artifact in CTAC algorithm. This chapter also focuses on currently available PET/CT systems and their areas of application. It should be noted that due to limited space the references contained herein are for illustrative purposes and are not inclusive, and not to imply that those chosen are better than others that are not mentioned.

12.2 ANIMAL MOLECULAR IMAGING

In the last decades, advanced demand in biomedical research and interest in knowing more about the base function of the human body, like functional operation of the cells' system or their organism, has increased the interest in understanding pathophysiologic and physiological processes that advent a new field of research as molecular imaging. Utilizing different imaging modalities accelerate the advancement of investigation for new imaging markers and procedure of probing disease. Different molecular imaging systems and techniques such as positron emission tomography (PET), single photon emission computed tomography (SPECT), computed tomography (CT), magnetic resonance imaging (MRI) in its various forms such as magnetic resonance spectroscopy (MRS), diffusion tensor imaging (DTI), functional MRI (fMRI), and ultrasound (US), optical bioluminescence/fluorescence have been developed in the past few years for animal studies purposes (Figure 12.1). Applications of the mentioned imaging techniques vary accordingly to their ability in detecting the structure of considered tissue (Sossi 2011; Levin 2012).

Among the different imaging approaches, some utilize in vitro and in vivo application, which are invasive and need sacrificing animals or analytical dissections in inanimate animals, and some are appropriate for noninvasive studies, which have more applicability in human investigations; this categorization makes a virtual edge between preclinical and clinical imaging realm (Figure 12.2).

The need for small animal imaging is being driven to a large extent by the unique role that small animals have come to play in various aspects of modern biomedical research. As anatomical imaging systems can provide structural information, functional imaging systems can bridge the gap between preclinical "pharmaceutical" studies in animals and phase-one trials in humans, by allowing pharmacokinetic and pharmacodynamics studies and the so-called administration,

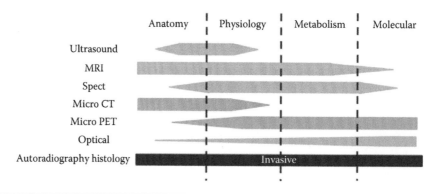

FIGURE 12.1　The information acquired by different modalities in the area of small animal imaging.

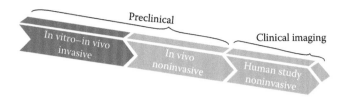

FIGURE 12.2 The steps toward clinical evaluation of a drug or technique.

distribution, metabolism, and excretion studies to be performed in a much easier way. Mainly, molecular imaging is aiming at developing and testing novel tools, reagents, and methods to image-specific molecular pathways in vivo, particularly those that are key targets in disease processes. It is worthwhile to report the descriptions provided by American College of Radiology (ACR) and Society of Nuclear Medicine and Molecular Imaging (SNMMI) for molecular imaging as follows:

ACR description: "Molecular Imaging is a growing research discipline aimed at developing and testing novel tools, regents and methods to image specific molecular pathways in vivo."

SNMMI description: "Molecular Imaging is a technique which directly or indirectly monitors and records the spatiotemporal distribution of molecular and cellular process for biochemical, biologic, diagnosis or therapeutic applications."

12.3 PREDILECTION OF NUCLEAR MEDICINE IMAGING

It must be considered that each imaging modality according to its nature and detection of penetrated signal through tissue has several technical constraints, so the information can be gained from an individual system in essence different from the others. Among all the proposed imaging systems in Figure 12.1, a combination taking the advantages from nuclear medicine modalities (PET or SPECT) with other modalities such as x-ray CT, MRI, and US or even OI, are more popular. It should be noted that radionuclide imaging and optical imaging systems have the ability to cover wide realm of imaging from functional information to genetic visualization. Moreover, PET (from the deviation of radionuclide imaging systems), in contrast to optical imaging, has the ability to receiving signal from subtle molecular process within the considered subject of interest and also need low concentration of probe (radio tracer) in comparison to optical imaging. Despite the fact that the radionuclide imaging needs low concentration of imaging probe, it could gain more signal from the imaged structure by providing more accurate spatial and temporal resolution. SPECT, similar to PET, can be categorized in radionuclide imaging; both rely on external detection of molecular tracer but PET imaging is based on the coincidence detection of simultaneous emission of back-to-back gamma rays emitted as a result of positron annihilation. In such manner, data acquisition in PET would not need any physical collimations such as the one available in the SPECT. This causes the substantial advantage of PET (up to 10%) sensitivity over SPECT (~0.01%). Other parameters that make PET and SPECT distinct from each other are isotopes used for labeling molecular probes, for example, the size of the probes and their diffusions, and half-life, as SPECT radioisotopes have a longer life than PET. Therefore, each one of these imaging systems relies on the features of their radioisotope, which is used for difference applications. However, PET scanners, despite their exquisite sensitivity, have limited spatial and temporal resolution as well as poor anatomical information. Other imaging techniques, such as CT and MRI, have higher spatial and temporal resolution but are less sensitive and convey less specific molecular information. Combination of these anatomical imaging systems (CT and MRI) with PET could provide much more complete information from subject under study (Meikle et al. 2006; Loudos et al. 2011).

12.4 PET/CT PHYSICS AND INSTRUMENTATION

12.4.1 THE PHYSICS OF PET

PET imaging relies on the nature of positron decay; when a nucleus undergoes positron decay, the result is a new nuclide with one less proton and one more neutron as well as the emission of a positron and a neutrino. As positrons pass through matter, they experience the same interactions as electrons, including the loss of energy through ionization and excitation of nearby atoms and molecules. After losing enough energy and traveling a given distance in matter (depending on the initial energy of the positron), the positron will annihilate with a nearby electron and two photons are emitted in opposite directions with energy of 511 keV each. These photons are the basis of coincidence detection and coincidence imaging. PET imaging systems detect annihilation events by means of several rings of photon detectors that surround a patient or an object under study. When two matching photons originating from the same annihilation event are recorded within nanoseconds of each other, two opposite detectors in two opposite side of the bed register a coincidence event along the line between both detectors (line of response, or LOR). The PET system, then, registers all LOR between each detector pair registering a coincidence event during the scan. At the end of the acquisition, there will be areas of overlapping lines, which indicate more highly concentrated areas of radioactivity, according to the tracer distribution within the object. Then, the raw data can be reconstructed to create cross-sectional images representing radioactivity distribution within the tissues (Turkington 2001; Larobina et al. 2006).

12.4.2 THE PHYSICS OF CT

X-ray computed tomography is an imaging modality that produces cross-sectional images representing the x-ray attenuation properties of the body. Unlike conventional tomography, CT does not suffer from interference from structures in the object outside the slice being imaged. This is achieved by irradiating only thin slices of the object. Compared to planar radiography, CT images have superior contrast resolution, that is, they are capable of distinguishing very small differences between tissues' attenuation properties. Two steps are necessary to derive a CT image: firstly, physical measurements of the attenuation of x-rays traversing the object under study in different directions and, secondly, mathematical calculations of the linear attenuation coefficients (μ) all over the slice. The object remains on the examination table while the x-ray tube rotates in a circular or spiral orbit around it, in a plane perpendicular to the length axis. The data acquisition system in CT is an array of several small separate detectors placed on the opposite side of the object under study and x-ray tube. The arrangements of the x-ray tube and detectors have changed over the years for achieving the better resolution, which is necessary for preclinical imaging. When readings from detectors have been stored in the computer of the scanner, the x-ray tube is rotated to another angle and a new projection profile is measured. After a complete rotation, it is possible to calculate the average linear attenuation coefficient (μ) for each pixel. This procedure is called image reconstruction from measured projections. In modern CT scanners, pixels in images represent the CT number, which is expressed in Hounsfield units (HU). The CT number is defined as

$$\text{CT number} = \frac{\mu - \mu_{H_2O}}{\mu_{H_2O}} \times 1000 \tag{12.1}$$

where μ is the average linear attenuation coefficient for the material in a given pixel. With this definition, air and water have a CT number of −1000 and 0 HU, respectively.

12.4.3 DUAL MODALITY PET/CT IMAGING

The advantages of the integrated dual-modality imaging systems for clinical and preclinical applications and biological research cause riveting in the use (Levin and Zaidi 2007). The PET/CT scanner combines premier technology from two modalities, functional imaging using PET and anatomical imaging using CT, making it possible to reveal detailed anatomy and biological processes and metabolic information at the molecular level of internal organs and tissues from one single noninvasive procedure. The combined PET/CT scanner is the most powerful imaging tool available for localizing, evaluating, and therapeutic monitoring. Separately, PET and CT do not provide images with the necessary combination of clear structural definition and metabolic activity, which is achieved with PET/CT (Townsend and Cherry 2001). The first development of combined PET and CT scanners in the same gantry followed the same trend in the later part of the 1990s by investigators from the University of Pittsburgh and has revolutionized the practice of clinical PET (Beyer et al. 2000). Historically, the first dual-modality system was a combination of SPECT and CT in the pioneering works of Hasegawa et al. (1991) and Lang et al. (1992). They combined anatomical and functional images by using a single detector for both modalities. In the few last years, there have been significant advances in both CT and PET technology for animal imaging and, consequently, these advances were incorporated into current preclinical generation of PET/CT scanners. In current preclinical PET/CT designs, the two scanners (PET and CT) are physically kept separate with the CT positioned anterior to the PET, similar to the clinical ones, in the same cover. The advantage of this minimal hardware integration is that each system can use the latest technology independently. In some systems, a single housing is placed over the two modules, whereas in others, the modules have separate covers but are positioned very close to one another. To overcome the challenges of aligning independently acquired PET and CT image sets, several ad hoc concepts of integrating PET and CT imaging in a single device have been proposed (Townsend et al. 2004). Regardless of the widespread interest and advantages of PET/CT dual modality, only a few prototype platforms of PET/CT have been developed. A group at the University of California, Davis, (UC Davis) developed a small animal imaging system for anatomic and molecular imaging of the mouse as a micro-CT/micro-PET dual modality (Goertzen et al. 2002). In designing the micro-PET, detectors took advantages from lutetium yttrium orthosilicate (LSO) scintillator coupled through a fiber optic taper to a position sensitive photomultiplier tube. The included micro-CT used a micro-focus x-ray tube and an amorphous selenium detector coupled to a flat-panel readout array of thin-film resistors (Andre et al. 1998). The same group also has developed a typical micro-CT scanner based on photodiode detectors that have a flexible C-arm gantry design with flexible detector positioning, integrated within the micro-PET II scanner (Chatziioannou 2002; Bérard et al. 2007). The group at the University of Sherbrooke developed the LabPET scanner, which was brought to market by Gamma Medica-Ideas Inc. It had the advanced CT capability that allows anatomic images to be attained using the same detector channels and electronics. PET and CT scanning can be accomplished simultaneously. Distinctive x-ray photons can be discriminated and counted in CT mode by sampling the analog signal using high-speed analog-to-digital converters and digital processing in field-programmable gate arrays (Fontaine et al. 2005). The parallel architecture and fast digital processing make high count rates possible for both PET and CT modes, whereas the modularity of the system design allows the number of channels to be extended up to 10^4 or more. Even though most commercial dual-modality systems have been constructed as SPECT/CT or PET/CT scanners, several prototype dual-modality systems that combine various imaging technologies such as SPECT and PET (Marshall 2001), PET and OI (Rannou et al. 2004; Alexandrakis et al. 2005; Chaudhari et al. 2005; Gulsen et al. 2006; Zavattini et al. 2006), and PET and MR imaging have been proposed (Cho et al. 2007; Pichler et al. 2008a,b). Moreover, efforts have been made to develop tri-modality preclinical systems integrated in a single gantry including SPECT/PET/CT (Parnham et al. 2006) and SPECT/PET/OI (Zaidi 2009) (Figure 12.3).

12.4.4 CURRENTLY AVAILABLE PRECLINICAL PET/CT SCANNERS

At present, several dedicated small animal PET/CT are available on the market. These systems show differences in the parameters that characterize their performances. This section attempts to name some of those systems, which are commercially available. U-PET+ is developed by MILABS (Utrecht, the Netherlands) with spatial resolution of 0.75 mm coupled to VECT and has the ability of coupling to any other CT system to have U-PET+/CT. It has the potential of upgrading to the multimodality

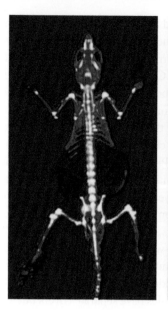
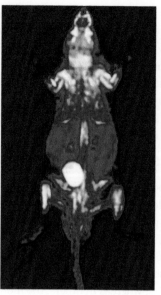
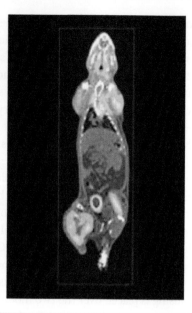

FIGURE 12.3 Micro-PET/CT dual-modality small animal imaging. (From Tubingen, U., ed. *Handbook of Sixth Animal Imaging Workshop*, Eberhard Karls University, Tubingen, Germany, 2011.)

system by adding SPECT module as well. The other dual-modality system, called nanoScan PET/CT, is manufactured by Mediso (Budapest, Hungary). Mediso nanoScan PET/CT with cerium-doped lutetium yttrium orthosilicate (LYSO) crystal size $1.12 \times 1.12 \times 13$ mm^3 and sub-half mm^3 PET volumetric resolution is a full ring geometry, with 12 cm PET transaxial field of view (FOV), 9% absolute sensitivity made to have one of the largest FOV among currently available PET/CT imagers that minimizes parallax error. The focal spot size of CT scanner in this system is 7 μm. The PET scanner in Super Argus PET/CT from SEDECAL (Madrid, Spain) with dynamic FOV bigger than 45 cm and phoswich depth of interaction (DOI) has a resolution less than 1 mm. The CT scanner in this system is dual energy and dual exposure, while its focal spot varies from 15 to 80 micron. Inveon detector technology is a larger 20×20 array detection with LSO crystal pixel spacing 1.6 mm × 1.6 mm developed by Siemens Company (Erlangen, Germany) for Inveon PET/CT and Inveon docked PET/CT. The Inveon delivers the spatial resolution of 1.4 mm FWHM at the center of FOV. In a docked mode, the two scanners can operate independently or as a single multimodality system under the control of a single workstation and have the ability to efficiently configure the multimodal system. The CT module in this hybrid scanner has a focal spot with the size less than 50 μm. The eXplore Vista small animal PET/CT developed by General Electric Company (Milwaukee, USA) is one of the multimodality small animal imaging systems that can provide high-sensitivity functional imaging along with anatomical images within a single instrument. It has a 7 cm bore, which is suitable for mice and rats that weigh up to 400 g. Currently, 18F-FDG, Cu-64, and 124I are approved and available for use, but studies requiring other longer-lived positron emitters may be discussed. EXplore Vista introduced by General Electric Company (Milwaukee, USA) has a phoswich detection array with an inner dimension of 13×13 and a dual-layer phoswich (front layer: LYSO; back layer: gadolinium oxyorthosilicate (GSO)) consists of element dimensions $1.45 \times 1.45 \times 7$ mm^3 for LYSO and $1.45 \times 1.45 \times 8$ mm^3 for GSO. The PET scanner in this system is a dual ring, 18 detectors per ring, with axial resolution of about 1.2 mm in the center of FOV and an axial FOV of 4.6 cm with the CT x-ray that has source energy of 4–50 kVp and tube current between 0 and 1 mA. Figure 12.4 shows the various commercially available PET/CT scanners.

All imaging modalities according to constrain and restraint in detection operating mode have their own strengths and weaknesses that make them succeed in specific realm of imaging. The strength and intrinsic aptitude of each modality rely on different factors such as system detection performance (e.g., spatial resolution, temporal resolution, level of signal-to-noise ratio, contrast resolution, and size of probe) and properties of imaged tissue (e.g., the time scale of the procedure and diffusion or penetration of probe in tissue). Hence, all mentioned factors could change the preference

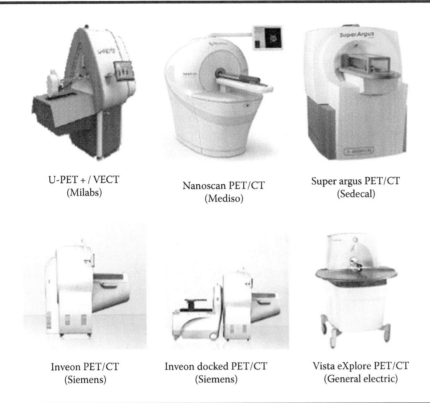

U-PET + / VECT
(Milabs)

Nanoscan PET/CT
(Mediso)

Super argus PET/CT
(Sedecal)

Inveon PET/CT
(Siemens)

Inveon docked PET/CT
(Siemens)

Vista eXplore PET/CT
(General electric)

FIGURE 12.4   Some of the commercially available preclinical PET/CT scanners.

of the choice of imaging modality for a specific application. On the other hand, cooperating different modalities increases not only the gain of each modality system but also the reliability in the interpretation of the generated images in comparison to each enhanced single modality. Although this integration of functional and such exquisite anatomical details is more beneficial in molecular imaging, this simultaneous imaging poses many challenges to the quality of images from different modalities.

12.5   ATTENUATION CORRECTION IN PRECLINICAL PET/CT IMAGING

One of the confounding factors that affects image quality and quantitative accuracy in PET imaging is the attenuation of photons in tissues. Attenuation correction (AC) in preclinical imaging is potentially important due to presenting accurate estimation of tracer uptake in high-resolution images of a small animal (Prasad et al. 2011). Reliable attenuation correction methods for PET require the determination of an accurate attenuation map, which represents the spatial distribution of linear attenuation coefficients at 511 keV for the region under study. When attenuation map is generated, it can be incorporated into image reconstruction algorithms in order to correct the emission data for errors contributed by photon attenuation. The methods for generating the attenuation maps can be categorized into two main classes: transmission-less methods and transmission-based methods (Zaidi and Hasegawa 2003). The transmission-less methods are based on the calculation of boundary, distribution of attenuation coefficients by means of approximate mathematical methods and statistical modeling for simultaneous estimation of attenuation and emission distribution and consistency conditions criteria. It is generally difficult to generate accurate attenuation map using transmission-less methods especially in imaging with more complex positions of media with different attenuation properties and irregular contours in a small animal. Therefore, the use of transmission-less techniques have considerable limitations in preclinical or clinical studies, where specific attenuation map is mandatory for accurate quantification of PET data. Historically, in standalone PET systems, the attenuation map was provided by the application of an external source for transmission scan to generate attenuation map. The transmission map provided by this method was quite noisy and not sufficient for small animal studies due to the fact that attenuation lengths of 511 keV photons are larger than the thickness of small animals, so the majority of photons pass through the object without any interaction. This behavior makes the attenuation map noisy and not useful for implementation of accurate attenuation correction of PET data. Although the introduction of different transmitter source (e.g., ^{57}Co) with lower gamma emitter energy and the consequent smaller attenuation length slightly reduced this strain, challenges in small animal studies remained unresolved. The advent of elaborating CT images offers the advantage of higher photon fluence rates and faster transmission scans, in addition to true anatomic imaging and localization capability that cannot be obtained using traditional radionuclide transmission scans (Zaidi and Hasegawa 2003). In practice, the attenuation coefficient distribution in an object under study is not known a priori, and for areas of inhomogeneous attenuation such as the chest, more adequate methods (transmission-based methods) must be used to generate the attenuation map. This includes transmission scanning using x-ray CT scanning and segmented MRI data. In the latter method, the T1-weighted MR images are realigned to preliminary reconstructed PET data using an automatic algorithm and then segmented to classify the tissues in different categories depending on their density and composition. Then, the theoretical tissue-dependent attenuation coefficients are assigned to the related voxels in order to generate an appropriate attenuation map (Zaidi et al. 2003).

12.5.1 CT-Based Attenuation Correction of PET Data

The concept of generating CT images with an x-ray tube that transmits radiation through the object, with transmitted intensity recorded by an array of detector elements, is identical to the attenuation maps generated for AC, which have traditionally been obtained using external radionuclide sources. The transmission data can, then, be reconstructed using a tomographic algorithm that inherently calculates the attenuation coefficients at each point in the reconstructed slice. The reconstructed CT image contains pixel values that are related to the linear attenuation coefficient at that point in the object, calculated from the effective CT energy at operational tube voltage of scanner. However, the attenuation map at 511 keV can be generated from the CT images to correct the PET emission data for photon attenuation. CT-based AC offers four significant advantages (Kinahan et al. 2003): firstly, the CT data offers much lower statistical noise; secondly, the CT scan can be acquired much more quickly than a radionuclide transmission scan; thirdly, the ability to collect uncontaminated postinjection transmission scan; and fourthly, using the x-ray transmission scan eliminates the need for PET transmission hardware and periodic replacing positron sources. The other potential benefit is exploring the direct incorporation of anatomical information derived from the CT data into the PET image reconstruction process and correction for partial volume effect.

As noted here, CT inherently provides an object-specific measurement of the linear attenuation coefficient at each point in the image. However, the linear attenuation coefficient measured with CT is calculated at the x-ray energy rather than at the 511 keV. It is, therefore, necessary to convert the linear attenuation coefficients obtained from the CT scan to those corresponding to the 511 keV (Zaidi and Prasad 2009). The CT scan, with proper consideration for differences between the spectrum of x-ray energies generated in CT and the monoenergetic 511 keV photons detected in PET and many other physical factors, can be used to correct for photon attenuation in PET (Zaidi and Hasegawa 2003; Zaidi and Prasad 2009). In CTAC, the attenuation map is calculated by converting CT numbers derived from low-energy polyenergetic x-ray spectra to linear attenuation coefficients at 511 keV. Through knowledge of the attenuation properties of the object being scanned, the measurement along each line of response can be corrected for photon attenuation. Several strategies have been reported in the literature for converting CT images to the linear attenuation coefficients at 511 keV, which are called energy mapping techniques such as segmentation (Kinahan et al. 2003), scaling (Beyer et al. 1994), hybrid (segmentation/scaling) (Kinahan et al. 1998), bilinear or piece-wise scaling (Bai et al. 2003), and dual-energy decomposition methods (Bai et al. 2003). In the following, a short description of the nominated methods is presented.

Scaling: The scaling approach estimates the attenuation image at 511 keV by multiplying the CT image by the ratio of attenuation coefficients of water at CT and PET energies. A single effective energy is chosen to represent the CT spectrum.

Segmentation: This method forms the attenuation image at 511 keV by segmenting the reconstructed CT image into different tissue types. The CT image value for each tissue type is, then, replaced with appropriate attenuation coefficients at 511 keV. Typical choices for tissue types are soft tissue, bone, and lung.

Hybrid: This method appears to be the most promising and is based on a combination of the scaling and segmentation methods using the fact that for most materials except bone, the ratio of the linear attenuation coefficient at any two photon energies is essentially constant.

Piece-wise linear: In this method, a series of CT scans from a known material (e.g., K_2HPO_4 solution) with different concentrations is performed. A calibration curve is, then, generated in which the measured CT number is plotted against the known attenuation coefficients at 511 keV. The resulting calibration curve is piece-wise linear and covers the range of linear attenuation coefficients commonly encountered in the body. It should be noted that most commercially available PET/CT scanners use the bilinear and recently the three linear calibration curve method.

Dual-energy decomposition: A technically challenging approach is to acquire the CT image at two different photon energies (e.g., 40 and 80 keV) and use these data to extract the individual photoelectric and Compton contributions to linear attenuation coefficients. The different contributions can then be scaled separately in energy in order to accurately generate attenuation map in 511 keV.

In addition to the energy conversion, there are other issues that must be considered when using CT to generate attenuation maps for the correction of emission data. CT fundamentally has a higher spatial resolution and is reconstructed in a finer image matrix than PET. Typically, 512 × 512, CT images can be downsampled to the same image size (e.g., 64 × 64, 128 × 128, 256 × 256) as that is used for the reconstruction of PET emission data. CT images must also be smoothed with a Gaussian filter using an appropriate kernel to match the spatial resolution of emission data (Guy et al. 1998). Figure 12.5 shows different steps in PET/CT imaging including the process for CTAC of PET data.

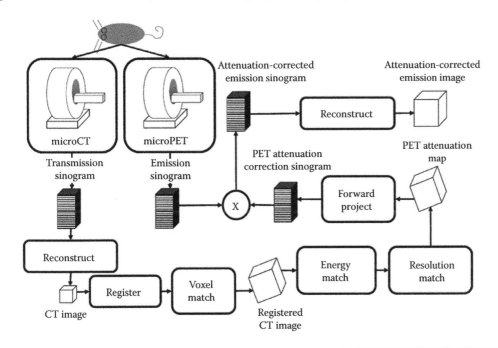

FIGURE 12.5 Principle of CT-based attenuation method on commercial PET/CT scanners. (Reprinted with permission from Chow, P.L. et al., *Phys. Med. Biol.*, 50, 1837, 2005.)

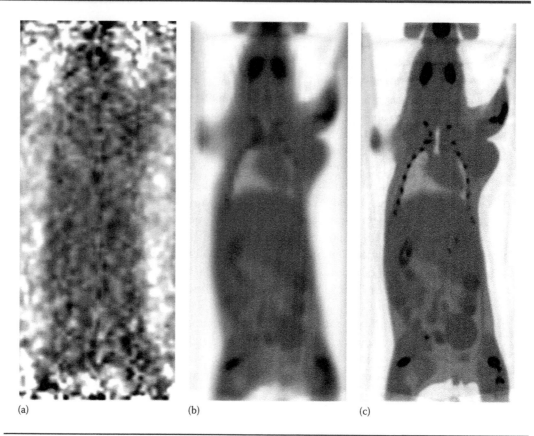

(a)   (b)   (c)

FIGURE 12.6   Attenuation map generated using (a) transmission scan and (b) CT image; (c) original CT image. (Reprinted with permission from Chow, P.L. et al., *Phys. Med. Biol.*, 50, 1837, 2005.)

Figure 12.6 compares the quality of the attenuation maps generated using transmission scan with radionuclide and the one generated from CT image. The difference in the level of noise in the generated attenuation maps is obvious. The generated attenuation map from CT images has a lower noise level and more detailed information from the anatomy of object under the study, which definitely increases the accuracy of AC procedure.

12.6   SOURCES OF ERROR AND ARTIFACT IN CTAC

Despite the fact that the use of CT images in cooperation of PET data could provide more benefit, it is well known that CTAC method could be a source of some errors due to misalignment between PET and CT images object motion, respiratory motion, truncation, beam hardening, contrast medium, and metallic implants. Some source of error such as beam hardening is not dominant in small animal imaging due to small animal size and lower absorption of x-ray photons when passing through the animal body.

12.6.1   RESPIRATORY MOTION IN SMALL ANIMAL IMAGING

In collaboration of multimodality imaging (PET with CT or MRI), accurate calibrations of the data acquisition sequences and prevention of subject movement during data acquisition process have decisive roles. In spite of all preventions, refrain from natural physiological motion of lung and heart due to respiratory and heart beating is not possible and poses some difficulties. These existing challenges originate from the fact that data acquisition in micro-CT is fast, images from multiple slices can be acquired in a fraction of a second, whereas data acquisition of PET for the same slices as CT needs several minutes per projection. The distorted images after data acquisition from PET and CT scanners lead to an underestimate/overestimate amount of uptake in parts of heart wall in cardiovascular investigations or mis-registration the accurate localization of lungs' lesion in thoracic deliberation as image quantification is influenced by the amplitude and pattern of motion. On the other hand, the capability in gating (respiratory gating, heart beating gating, or dual gating) promoted the accuracy of registering different datasets by detracting the influence of cardiac or respiratory phase motion on data acquisition (Patton 2010).

As discussed before, respiratory motion can reduce contrast and quantitative accuracy in terms of recovered contrast activity concentration and functional volumes. Several methods suggested for reducing the efficacy of respiratory motion have been based on the development of respiratory-gated data acquisitions in PET scanners. Principally, 4D datasets from PET have low statistic counts and high image noise in contrast to static 3D datasets. Due to the fact that the same number of coincidence events must be divided into many different respiratory phase bins or time frames in 3D PET, the number of coincidence events per image bin is decreased significantly. Since there is a trade-off between noise and temporal resolution, this leads to a higher image noise and a lower signal-to-noise ratio in 4D PET images. Prolonging the duration of each frame causes low noise but blurred as a result of loss of temporal resolution (Branco et al. 2011).

On the other hand, data acquisition in CT imaging takes much longer time than the term of cardiac or respiratory phase, so the image quality and time resolution be confined by the motion of heart and lung. This can be intensified in integration and registration of different dataset from diverse modalities. It should be noted that the heart rate in rats and mice is 250–600 and 300–600 min^{-1}, respectively, and also it has been reported that the respiratory rate is around 60–100 and 80–230 min^{-1} for rats and mice (Bartling et al. 2008, 2010). By exploiting cardiac and respiratory gating, it is expected to achieve better image quality and more accurate quantitative assessments such as tumor size and uptake in small animal PET/CT imaging. The difference between each gating approach relies on the type of scanner, the animal under investigation, and diagnostic problem, which is covered by the exam. The type of scanners are noticeable from the side that system at which specific angular positions perform data acquisition, so every geometry design (e.g., continues over a certain distance and step and shoot) has its own advantages and disadvantages in small animal gating. Gating could take into account various concepts as described in the following:

Prospective/retrospective gating: In prospective gating, data acquisition is influenced by the motion-gating signal, and retrospective gating is performed independently of any motion gating and all motion correction can be applied during post-processing.

Amplitude-based/phase-based gating: Amplitude-based gating, resorting of the projection, is modulated with respect to the amplitude of the motion. Gated signal acquisition must be in correlation with the aspect of the organ motion. This is more advantageous if reproducibility of the organ motion is close to the resolution of the scanner. Amplitude-based gating is used for respiratory gating and phase-based gating is commonly used in cardiac gating rebining in respect to gating signals.

Intrinsic/extrinsic gating: Extrinsic gating needs hardware system to derive signal acquisition, while intrinsic do not need any additional hardware for gathering data and need numerous post-processing algorithms adopted to the scanner features and characteristics (Martinez-Möller et al. 2007; Bartling et al. 2008, 2010) (Figure 12.7).

12.6.2 CONTRAST MEDIUM

Proper photon flux is required for having a good signal-to-noise ratio and image quality in small animal CT images, but limited x-ray focal spot imposes restraints on them. In order to overcome these challenges, it is necessary to prolong the data acquisition time or using contrast agent media. The use of a contrast agent substance (e.g., iodine) in small animal PET/CT imaging could improve anatomic referencing and tumor delineation but may introduce inaccuracies in the AC of the PET images (Patton 2010). Naturally, the presence of positive contrast agents in dual-modality PET/CT systems significantly overestimates the attenuation map in some cases and may generate artifacts

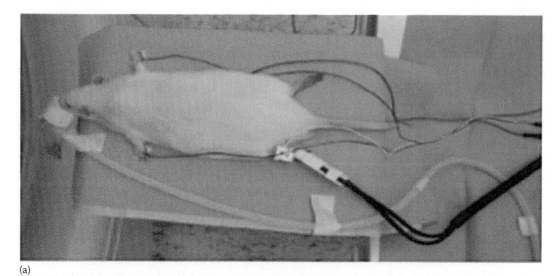

(a)

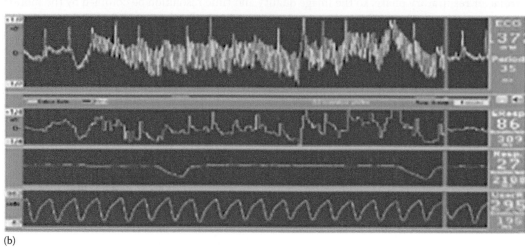

(b)

FIGURE 12.7 The effect of an extrinsic gating signal. A setup to derive a motion-gating signal extrinsi-cally is shown (a). The rat is equipped with electrodes to derive an ECG signal (red, black, and white wire), a pneumatic cushion (of which the blue tubing is visible), and an infrared-based pulse-meter (large white clip). (b) Output of the proprietary gating monitor system. The first row shows an electrical ECG signal, which is severely disturbed by electrical interferences from the scanner system. The third row shows the respiratory motion as detected by the pneumatic cushion. Here gasping respiratory motion is visible with long phases of virtually no motion. The last row shows the pulse wave as detected by the infrared pulse meter that is not dis-turbed by electrical interference. (Reprinted with permission from Bartling, S.H. et al., *Methods*, 50(1), 42, 2010.)

during CTAC (Antoch et al. 2004). This is due to high attenuation coefficient of these materials at low effective energy of the corresponding x-ray spectra that results in high CT numbers in the region of contrast agent accumulation through misclassification with high-density cortical bone (Nehmeh et al. 2003). Currently available algorithms for conversion from CT numbers to linear attenuation coefficients at 511 keV are based on the assumption that information in the CT image is contributed by mixtures of air, soft tissue, and bone. The presence of contrast medium complicates this process since two regions that have the same image contrast may indeed have different compositions, for example, contributed by bone and soft tissue in one case and iodine/barium contrast and soft-tissue in another case. These artifacts are most severe in cases where the contrast media is concentrated. In this case, the higher densities contributed by the contrast media can lead to an overestimation of PET activity. Lanon et al. (2013) evaluated the diagnostic performance and accuracy of quantitative values in contrast-enhanced small-animal PET/CT and concluded that the use of iodinated con-trast media for small animal PET imaging significantly improved tumor delineation and diagnostic

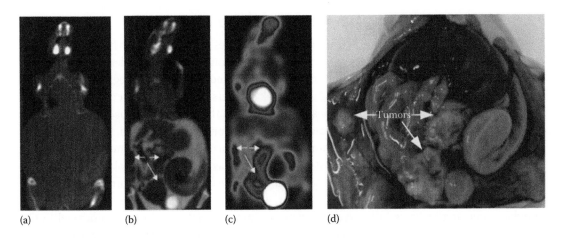

(a)   (b)   (c)   (d)

FIGURE 12.8   Impact of media enhancement on tumor localization. Representative coronal slices for unenhanced CT (a), contrast-enhanced SA-PET/CT (b and c), and necropsy photograph of a mouse (d). Tumors are well defined on the contrast-enhanced CT slice (yellow arrows), including a necrotic lesion located near the bladder harboring a low 18F-FDG uptake and a central photopenic area on an SA-PET slice. (Reprinted with permission from Lasnon, C. et al., *EJNMMI Res.*, 3(1), 1, 2013.)

performance, without any significant influence on the quantification of PET images corrected for attenuation using CTAC (Lasnon et al. 2013). However, the issue of whether the use of contrast medium in dual-modality PET/CT scanning produces significant artifacts is still controversial with some studies corroborating (Dizendorf et al. 2003) and others contradicting (Ay and Zaidi 2006) the fact that the presence of contrast medium can be a source of errors and artifacts when the CT data are used for AC on PET images (Figure 12.8).

12.6.3   METALLIC IMPLANTS

In some specific research in small animal PET/CT imaging, there are artificial metallic implants inside the body of the animal under study. In standard PET transmission scanning with radionuclides, metal implants cause a little or no artifact while these artifacts can be significant in CT energies due to the significantly higher x-ray absorption of high-Z materials (e.g., metals) compared to the low-Z materials (e.g., tissues). The presence of streak artifacts caused by metallic implants in CT images may mislead the quantification of PET/CT images, particularly when lesions are present in the vicinity of metallic implants (Bockisch et al. 2004). Several authors addressed the impact of using CTAC on quantitative analysis of PET/CT images in the presence of metallic implants (Kamel et al. 2003). Although the quantification error induced by metallic artifacts is still not fully addressed in small animal studies, there is an active area of research in this field.

12.6.4   TRUNCATION ARTIFACTS

During CT imaging of obese animals, part of the anatomy may extend beyond the boundaries of the CT field of view and is not reconstructed in CT. This truncation artifact will propagate errors to the CT-based AC, which is based on fully reconstructed CT images including all anatomies appearing in PET images. In the presence of truncation errors in CT images, the reconstructed emission images appear to be masked by the truncated CT. The tracer distribution is, then, only partially recovered outside the CT field of view as some bias of the reconstructed activity distribution inside the FOV is observed (Sureshbabu and Mawlawi 2005; Mawlawi et al. 2006). There are two approaches for truncation artifact correction. In the software approach, several algorithms have been suggested to extend the truncated CT projections to recover truncated parts of the attenuation map. In the hardware approach, most manufacturers offer PET/CT scanners with increased CT bore size in order to avoid truncation of CT images for most of the object even when imaging obese rats.

12.6.5 BEAM HARDENING AND X-RAY SCATTERED RADIATION

The polyenergetic x-ray spectra used during CT imaging makes it subject to beam hardening artifact caused by the absorption of low-energy x-rays as they pass through the object under study. The direct consequence is that the linear attenuation coefficient calculated for thick body regions is lower than thin regions. This effect generates cupping and streak artifacts in the reconstructed CT image and makes it unacceptable for diagnostic purposes. Furthermore, the resulting erroneous CT-based AC subsequently propagates the error to the calculated activity concentration in PET images. Although beam hardening effect correction algorithms (Joseph and Ruth 1997) are implemented as a part of CT reconstruction software, this effect is still visible when having dense material on the FOV. This effect is not significant in small animal imaging but should be considered when imaging obese rats.

The contamination of CT data with scattered radiation reduces reconstructed CT numbers and introduces cupping artifacts in the reconstructed images. This effect is more pronounced in CT scanners with large area flat-panel detectors that seem to be candidates as CT modules in the next generation of PET/CT scanners with panel-based PET module (Townsend et al. 2004). Ay and Zaidi (Ay and Zaidi 2006; Zaidi and Ay 2006) quantified the contribution of x-ray scatter during the CTAC procedure for commercially available multi-slice CT and prototype large-area, flat-panel, detector-based cone-beam CT scanners using MC simulation for human PET/CT scanners; the study for the contribution of scattered radiation in small animal PET imaging is on the way. They reported that the magnitude of scatter in CT images for the cone-beam geometry is significant in human imaging and might create cupping artifacts in reconstructed PET images during CTAC. However, its effect is small for current-generation small animal PET/CT due to the smaller size of both object and detector area.

12.6.6 MISALIGNMENT BETWEEN PET AND CT IMAGES

In order to achieve the utmost benefit from visualization and quantification of acquired datasets in synergistic collaboration of different modalities, robust approaches are demanded to overlay images. Recently, progress in simultaneous data acquisitioning at the same physical position for different modalities reduced the amount of displacement. Registration is one of these approaches for synchronizing data from different modalities. It is a concept based on the determination of special transformation matrix. However, most image-registration techniques were originally developed for human studies (Hasegawa et al. 1991; Hawkes 2005), but in animal studies some dedicated effort is needed. The registration process can be categorized into two main approaches: software-based and hardware-based approaches (Figure 12.9).

Software-based approaches: Corresponding to the type of performed study and demanded information, algorithms of co-registration can vary from inter-subject approach accordance to acquired images from different subject with a single modality to intra-subject approach correspondence to acquired images from the same subject with multimodality. The latter approach is the one generally requested in the area of multimodality imaging. In selecting a method of co-registration, the type of body motion for the under-investigation tissue, rigid (transformation imaged almost stable during data acquisition) and non-rigid (transformation imaged structure will change during data acquisition) must be taken into account. So, co-registration methods can be categorized into different groups: landmark measures, surface and edge measures, and voxel intensity measures (Levin and Zaidi 2007; Sossi 2011).

Hardware-based approaches: Despite the widespread interest in software-based co-registration methods in animal studies, some constraints of these algorithms provide a tendency to the development of hardware-based approaches. The advantage in this method is the independency of the molecular imaging probe. These methods of registration rely on particular design for mounting on imaging modalities. The application of the hardware registration is characterized by Jan et al. (2005) in integration with small animal PET scanner with a combined SPECT/CT scanner. A common mouse holder is used manually to align the position of

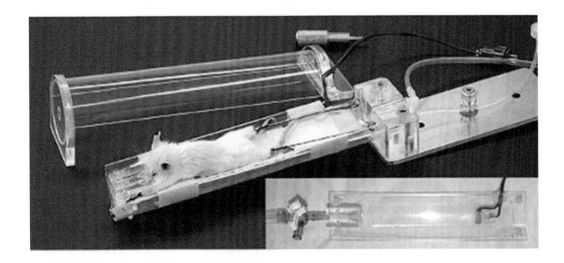

FIGURE 12.9 Image of the animal holder used in hardware-based registration. (Reprinted with permission from Chow, P.L. et al., *Phys. Med. Biol.*, 51(2), 379, 2006.)

markers and a bubble level. The other methods were applied using a custom-made imaging chamber positioned separately on animal CT and PET scanners and also a 3D-grid phantom with 1288 lines designed to work in combination with chambers to derive the spatial transformation matrix from software registration by using a 15-parameter perspective model (Chow et al. 2005, 2006).

12.7 CONCLUSION

Hybrid imaging and particularly PET/CT technology has not yet reached its final performance and still needs substantial improvements in order to cover current pitfalls in CTAC and also to increase the accuracy of quantitative analysis on PET/CT images. This chapter has focused on describing the physics of PET/CT and introducing challenges and pitfalls in this imaging modality.

It is well known that AC is one of the most pertinent corrections on PET data. The ability to accurately carry out AC with validated hardware and software solutions, in parallel with dedicated quality control protocols, enhances the interpretive confidence and accuracy of molecular PET/CT imaging. Despite the remarkable progress in CTAC achieved during the last couple of years, there is still scope for further research to address the mentioned challenges in CTAC. There are many active research areas as the field is very open to novel ideas in order to move this field forward more rapidly in the near future.

REFERENCES

Alexandrakis, G., F. R. Rannou et al. (2005). Tomographic bioluminescence imaging by use of a combined optical-PET (OPET) system: A computer simulation feasibility study. *Phys. Med. Biol.* **50**(17): 4225.

Andre, M. P., B. Spivey et al. (1998). An integrated CMOS-selenium x-ray detector for digital mammography. *Proc. SPIE* **3336**: 204–220.

Antoch, G., L. S. Freudenberg et al. (2004). To enhance or not to enhance? 18F-FDG and CT contrast agents in dual-modality 18F-FDG PET/CT. *J. Nucl. Med.* **45**(Suppl. 1): 56S–65S.

Ay, M. R. and H. Zaidi. (2006). Assessment of errors caused by X-ray scatter and use of contrast medium when using CT-based attenuation correction in PET. *Eur. J. Nucl. Med. Mol. Imaging* **33**(11): 1301–1313.

Bai, C., L. Shao et al. (2003). A generalized model for the conversion from CT numbers to linear attenuation coefficients. *IEEE Trans. Nucl. Sci.* **50**(5): 1510–1515.

Bartling, S. H., J. Dinkel et al. (2008). Intrinsic respiratory gating in small-animal CT. *Eur. Radiol.* **18**(7): 1375–1384.

Bartling, S. H., J. Kuntz et al. (2010). Gating in small-animal cardio-thoracic CT. *Methods* **50**(1): 42–49.

Bérard, P., J. Riendeau et al. (2007). Investigation of the LabPET™ detector and electronics for photon-counting CT imaging. *Nucl. Instrum. Methods Phys. Res. A* **571**(1): 114–117.

Beyer, T., P. Kinahan et al. (1994). The use of X-ray CT for attenuation correction of PET data. *Nucl. Sci. Symp. Med. Imaging Conf. Rec.* **4**: 1573.

Beyer, T., D. W. Townsend et al. (2000). A combined PET/CT scanner for clinical oncology. *J. Nucl. Med.* **41**(8): 1369–1379.

Bockisch, A., T. Beyer et al. (2004). Positron emission tomography/computed tomography—Imaging protocols, artifacts, and pitfalls. *Mol. Imaging Biol.* **6**(4): 188–199.

Branco, S., P. Almeida et al. (2011). Evaluation of the respiratory motion effect in small animal PET images with GATE Monte Carlo simulations. In: *Applications of Monte Carlo Methods in Biology, Medicine and Other Fields of Science.* InTech, Croatia. http://www.intechopen.com/books/applications-of-monte-carlo-methods-in-biology-medicine-and-other-fields-of-science.

Chatziioannou, A. F. (2002). Molecular imaging of small animals with dedicated PET tomographs. *Eur. J. Nucl. Med.* **29**(1): 98–114.

Chaudhari, A. J., F. Darvas et al. (2005). Hyperspectral and multispectral bioluminescence optical tomography for small animal imaging. *Phys. Med. Biol.* **50**(23): 5421.

Cho, Z. H., Y. D. Son et al. (2007). A hybrid PET-MRI: An integrated molecular-genetic imaging system with HRRT-PET and 7.0-T MRI. *Int. J. Imaging Syst. Technol.* **17**(4): 252–265.

Chow, P. L., F. R. Rannou et al. (2005). Attenuation correction for small animal PET tomographs. *Phys. Med. Biol.* **50**: 1837.

Chow, P. L., D. B. Stout et al. (2006). A method of image registration for small animal, multi-modality imaging. *Phys. Med. Biol.* **51**(2): 379.

Dizendorf, E., T. F. Hany et al. (2003). Cause and magnitude of the error induced by oral CT contrast agent in CT-based attenuation correction of PET emission studies. *J. Nucl. Med.* **44**(5): 732–738.

Fontaine, R., F. Belanger et al. (2005). Architecture of a dual-modality, high-resolution, fully digital positron emission tomography/computed tomography (PET/CT) scanner for small animal imaging. *IEEE Trans. Nucl. Sci.* **52**(3): 691–696.

Goertzen, A. L., A. K. Meadors et al. (2002). Simultaneous molecular and anatomical imaging of the mouse in vivo. *Phys. Med. Biol.* **47**(24): 4315.

Gulsen, G., O. Birgul et al. (2006). Combined diffuse optical tomography (DOT) and MRI system for cancer imaging in small animals. *Technol. Cancer Res. Treat.* **5**(4): 351–363.

Guy, M., I. Castellano-Smith et al. (1998). DETECT-dual energy transmission estimation CT-for improved attenuation correction in SPECT and PET. *IEEE Trans. Nucl. Sci.* **45**(3): 1261–1267.

Hasegawa, B. H., B. Stebler et al. (1991). A prototype high-purity germanium detector system with fast photon-counting circuitry for medical imaging. *Med. Phys.* **18**: 900.

Hawkes, D. J. (2005). Coregistration of structural and functional images. In Hawkes, D. J., Hill, D. L. G., Hallpike, L., and Bailey, D. L. (eds.), *Positron Emission Tomography*, Springer, New York, pp. 161–177.

Jan, M.-L., K.-S. Chuang et al. (2005). A three-dimensional registration method for automated fusion of micro PET-CT-SPECT whole-body images. *IEEE Trans. Med. Imaging* **24**(7): 886–893.

Joseph, P. M. and C. Ruth. (1997). A method for simultaneous correction of spectrum hardening artifacts in CT images containing both bone and iodine. *Med. Phys.* **24**: 1629.

Kamel, E. M., C. Burger et al. (2003). Impact of metallic dental implants on CT-based attenuation correction in a combined PET/CT scanner. *Eur. Radiol.* **13**(4): 724–728.

Kinahan, P., D. Townsend et al. (1998). Attenuation correction for a combined 3D PET/CT scanner. *Med. Phys.* **25**: 2046.

Kinahan, P. E., B. H. Hasegawa et al. (2003). X-ray-based attenuation correction for positron emission tomography/computed tomography scanners. *Semin. Nucl. Med.* **33**(3): 166–179.

Lang, T. F., B. H. Hasegawa et al. (1992). Description of a prototype emission transmission computed tomography imaging. *J. Nucl. Med.* **33**(10): 1881–1887

Larobina, M., A. Brunetti et al. (2006). Small animal PET: A review of commercially available imaging systems. *Curr. Med. Imaging Rev.* **2**(2): 187.

Lasnon, C., E. Quak et al. (2013). Contrast-enhanced small-animal PET/CT in cancer research: Strong improvement of diagnostic accuracy without significant alteration of quantitative accuracy and NEMA NU 4–2008 image quality parameters. *EJNMMI Res.* **3**(1): 1–11.

Levin, C. S. (2012). Molecular imaging instrumentation. *Mol. Imaging Probes Cancer Res.* 29.

Levin, C. S. and H. Zaidi. (2007). Current trends in preclinical PET system design. *PET Clin.* **2**(2): 125–160.

Loudos, G., G. C. Kagadis et al. (2011). Current status and future perspectives of in vivo small animal imaging using radiolabeled nanoparticles. *Eur. J. Radiol.* **78**(2): 287–295.

Marshall, E. (2001). Celera assembles mouse genome; public labs plan new strategy. *Science* **292**(5518): 822–823.

Martinez-Möller, A., D. Zikic et al. (2007). Dual cardiac–respiratory gated PET: Implementation and results from a feasibility study. *Eur. J. Nucl. Med. Mol. Imaging* **34**(9): 1447–1454.

Mawlawi, O., J. J. Erasmus et al. (2006). Truncation artifact on PET/CT: Impact on measurements of activity concentration and assessment of a correction algorithm. *Am. J. Roentgenol.* **186**(5): 1458–1467.

Meikle, S. R., F. J. Beekman et al. (2006). Complementary molecular imaging technologies: High resolution SPECT, PET and MRI. *Drug Discov. Today Tech.* **3**(2): 187–194.

Nehmeh, S. A., Y. E. Erdi et al. (2003). Correction for oral contrast artifacts in CT attenuation-corrected PET images obtained by combined PET/CT. *J. Nucl. Med.* **44**(12): 1940–1944.

Parnham, K., S. Chowdhury et al. (2006). Second-generation, tri-modality pre-clinical imaging system. In *Nuclear Science Symposium Conference Record, 2006 IEEE*, San Diego, CA.

Patton, J. A. (2010). History and principles of hybrid imaging. In D. Delbeke and O. Israel (eds.), *Hybrid PET/CT and SPECT/CT Imaging*. Springer, New York, pp. 3–33.

Pichler, B. J., M. S. Judenhofer et al. (2008a). Multimodal imaging approaches: PET/CT and PET/MRI. *Mol. Imaging* **I**: 109–132.

Pichler, B. J., M. S. Judenhofer et al. (2008b). PET/MRI hybrid imaging: Devices and initial results. *Eur. Radiol.* **18**(6): 1077–1086.

Prasad, R., M. R. Ay et al. (2011). CT-based attenuation correction on the FLEX triumph preclinical PET/CT scanner. *IEEE Trans. Nucl. Sci.* **58**(1): 66–75.

Rannou, F. R., V. Kohli et al. (2004). Investigation of OPET performance using GATE, a Geant4-based simulation software. *IEEE Trans. Nucl. Sci.* **51**(5): 2713–2717.

Reza Ay, M. and H. Zaidi. (2006). Simulation-based assessment of the impact of contrast medium on CT-based attenuation correction in PET. In *Nuclear Science Symposium Conference Record, 2006 IEEE*, San Diego, CA.

Sossi, V. (2011). Multi-modal imaging and image fusion. In Kiessling, F. and Pichler, B. J. (eds.), *Small Animal Imaging: Basics and Practical Guide*. Springer Science & Business Media, 2011, pp. 293–314.

Sureshbabu, W. and O. Mawlawi. (2005). PET/CT imaging artifacts. *J. Nucl. Med. Technol.* **33**(3): 156–161.

Townsend, D. W., J. P. Carney et al. (2004). PET/CT today and tomorrow. *J. Nucl. Med.* **45**(Suppl. 1): 4S–14S.

Townsend, D. W. and S. R. Cherry. (2001). Combining anatomy and function: The path to true image fusion. *Eur. Radiol.* **11**(10): 1968–1974.

Tubingen, U. (ed.). (2011). *Handbook of Sixth Animal Imaging Workshop*. Eberhard Karls University, Tubingen, Germany.

Turkington, T. G. (2001). Introduction to PET instrumentation. *J. Nucl. Med. Technol.* **29**(1): 4–11.

Zaidi, H. (2009). Navigating beyond the 6th dimension: A challenge in the era of multi-parametric molecular imaging. *Eur. J. Nucl. Med. Mol. Imaging* **36**(7): 1025–1028.

Zaidi, H. and M. R. Ay. (2006). Impact of x-ray scatter when using CT-based attenuation correction in PET: A Monte Carlo investigation. In *Proceedings of IEEE Nuclear Science Symposium & Medical Imaging Conference*, San Diego, CA.

Zaidi, H. and B. Hasegawa. (2003). Determination of the attenuation map in emission tomography. *J. Nucl. Med.* **44**(2): 291–315.

Zaidi, H., M.-L. Montandon et al. (2003). Magnetic resonance imaging-guided attenuation and scatter corrections in three-dimensional brain positron emission tomography. *Med. Phys.* **30**: 937.

Zaidi, H. and R. Prasad. (2009). Advances in multimodality molecular imaging. *J. Med. Phys.* **34**(3): 122.

Zavattini, G., S. Vecchi et al. (2006). A hyperspectral fluorescence system for 3D in vivo optical imaging. *Phys. Med. Biol.* **51**(8): 2029.

13

Introduction to Combining MRI with PET

Volkmar Schulz, Jakob Wehner, and Yannick Berker

The combination of magnetic resonance imaging (MRI) with positron emission tomography (PET) into an integrated MRI–PET device is considered to be enriching the field of molecular imaging. The main advantages of this new combination are the outstanding soft-tissue contrast of the anatomical and functional information from MRI combined with the high sensitivity of PET, which allows the observation of molecular (e.g., metabolic) processes. A good illustration of the gain in soft-tissue contrast is given in Figure 13.1, in which single-modality images of the same object (mouse) from PET, MRI, and CT are reproduced (Wehrl et al. 2009). In this figure, the high soft-tissue contrast of the MR image clearly shows details of the kidney anatomy, such as its cortex, which matches well with the ^{64}Cu-labeled monoclonal antibody uptake in the PET images. Compared to the MRI, the CT image in Figure 13.1 only shows very minor soft-tissue contrast and is thus lacking anatomical details of the kidneys. In the fused MRI–PET image, the high uptake of ^{64}Cu could be clearly assigned to the cortex of the kidney.

The contribution of preclinical MRI is, however, beyond anatomical imaging. Functional MRI (fMRI), MR spectroscopy (MRS), and chemical shift imaging (CSI) have evolved to become promising tools for detecting changes in blood flow, tissue oxygenation, or concentrations of endogenous molecules such as lactate, choline, *N*-acetylaspartate (NAA), or deoxyglucose uptake and metabolism (glucoCEST) (Rivlin et al. 2013).

In recent years, the spatial resolution of PET has been continuously improved. Image resolution of less than 1 mm has already been reported even for combined MRI–PET devices (Weissler et al. 2015). Elastic image-registration techniques, for example, in combination with fiducial markers, have improved significantly over the last years. However, these techniques have their limitations especially if object motion comes along with long-term deformation as observable in the gastrointestinal tract and in the bladder. Therefore, the estimation and correction of cardiac, breathing, bulk motion, and deformation, supported by MRI, will become a very powerful tool for simultaneous

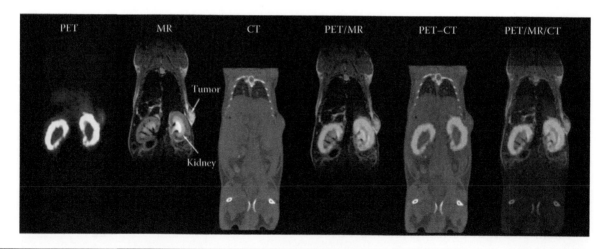

FIGURE 13.1 Coronal planes of PET, MR, CT, and fused MRI–PET and PET–CT images (left to right). The MR image shows an excellent soft tissue contrast, and the cortex of the kidneys can be clearly identified. Unlike the CT image, the MR image also shows a clear delineation between the tumor and the connective tissues. The fused images show the advantage of MRI–PET over PET–CT. Highly sensitive functional information from the PET is matched with high soft-tissue contrast and high-resolution morphological MR images. (With kind permission from Springer Science+Business Media: *Eur. J. Nucl. Med. Mol. Imaging*, Pre-clinical PET/MR: Technological advances and new perspectives in biomedical research, 36(Suppl. 1), 2009, S56–S68, Wehrl, H.F., Judenhofer, M.S. et al.)

MRI and PET. The enhanced quantification of PET due to significantly improved spatial and temporal registration of both modalities are to be expected in the coming years (Catana et al. 2011; Chun et al. 2012). Furthermore, motion estimation and compensation support studies in which the animal is only under weak anesthesia. Finally, time-resolved functional information from MRI can be correlated with dynamic PET data, for example, for studies of pharmacokinetics.

Stimulated by the pioneering work by Shao et al. (1997) and Pichler et al. (2006), research groups all over the world started focusing on combining PET and MRI systems for preclinical as well as clinical applications. Since that time, several system designs have been proposed using different detector technologies and concepts showcasing different levels of system integration (Shao et al. 1997; Bieniosek et al. 2013; Nagy et al. 2013; Weissler et al. 2015). However, until now, not many commercial preclinical systems are available (Nagy et al. 2013). This might be due to several aspects, such as the complexity of system integration of these two imaging modalities, the required sensor technology allowing for hybridization, and the high cost for prototyping and production. Furthermore, a deep understanding of various technical aspects of both modalities is required to cope with the wide range of interference phenomena of these two imaging modalities. The interference depends on a plurality of design choices on the sensor and readout technology, the system concepts, and the targeted protocols, and is as such not straightforward. In general, influences could be expected on all kinds of aspects of PET (such as degradation of sensitivity, spatial resolution, quantification) and MRI (degradation of signal-to-noise ratio (SNR) and image artefacts such as distortion, ghosting, and spurious signals). Unfortunately, these influences depend on the used MR sequences and the PET radiation dose. It is therefore crucial to acquire a sufficient level of understanding of all these potential influences before any study is executed on these devices. This chapter will throw some light on the various mentioned aspects to support successful operation of this exciting new hybrid modality.

13.1 CHALLENGES AND BENEFITS OF MRI–PET

Three major challenges have to be overcome for a successful integration of PET and MRI. The first challenge is about understanding and reducing the electronic interference between these modalities. The earliest published attempt at solving the interference problem made in 1997 by Shao et al. used long optical fibers connecting the scintillator crystals with the photomultiplier tube (PMT) readout system (Shao et al. 1997). Thereby, first simultaneous images could be acquired. Nevertheless, the presented system concept showed limited PET performance due to its limited axial field of view (FOV) and the transfer losses of the scintillation photons via the optical fibers. These losses resulted in poor energy, spatial, and timing resolution of the PET system. Since then, several research groups have tried to prevent this degradation by proposing alternative system concepts for PET using solid-state sensors (SSSs) instead of PMTs. These devices could operate in high static magnetic fields, and the sensors could be located in proximity of the scintillator crystals. Nine years after Shao's first experiments, the first simultaneous MRI–PET system for preclinical applications using avalanche photodiodes (APDs) was demonstrated by Pichler et al. (2006). However, the integration of the PET detector electronics into the MRI bore led to a different kind of degradation of PET and MRI signals. Several research groups have worked on solving the electromagnetic interference by choosing different system concepts and detector technologies. These concepts range from using light guides of several tens of centimeters between the SSS and the scintillator, over optical readout of the SSS by modulation of lasers, to integration of all electronics locally to the sensor (Catana et al. 2006; Schulz et al. 2011; Grant and Levin 2012). Details of this challenge will be given in Section 13.2.

The second challenge is related to system integration, in particular, spatial constraints. The proposed system concepts range from abutted PET and MRI systems to simultaneous systems in which the outer diameter of the radio frequency (RF) coil is reduced to generate a gap between the RF coil and the gradient coil to locate the PET detector. In such a way, this integration step already led to a degradation of the SNR of MRI due to the reduction in sensitivity of the RF coil (Oceguera and Rodríguez 2006). This problem especially exists in small-bore MR gantries, in which the distance between the resonator structure and the RF screen is already limited. In order to keep this effect as small as possible, the PET readout electronics needs to be as thin as possible to leave the required space for the crystals and the sensors. Very often, compromises on both sides (i.e., MRI and PET) are made to balance the degradation in image quality. Details of these challenges will be discussed in Section 13.3.

With the existence of new systems following the aforementioned design approaches, a final challenge appears at the horizon of MRI–PET: the need for MR-based attenuation and scatter correction. While the correction maps for attenuation and scatter in PET–CT could be extracted with sufficient accuracy using piecewise linear scaling from the CT image, the MR acquisition does not (so far) offer a direct measurement of the correction map. This is reflected by the fact that the MR image contains information of the spin relaxation properties and the density of nuclei (e.g., ^{1}H) but no information about the electron density. Therefore, new technologies need to be invented to derive the correction maps for preclinical and clinical MRI–PET. A logical step is to use the actual anatomical information that is provided by the MRI and try to complete this information with prior knowledge or information provided by PET. In addition, MRI enables measuring the motion of the object, which might enable the recovery of quantitative tracer uptake values. This, however, requires motion-aware attenuation correction. We will focus on the challenge of attenuation correction in Section 13.4.

In summary, developments in the new field of MRI–PET are oriented along several directions that are all related to the three major technical challenges:

1. Understanding and reducing the electronic interference
2. Finding suitable approaches to system integration
3. Implementing accurate and reliable attenuation and motion-correction techniques

The effort required to address the challenges is enormous, but will be worthwhile, as the combination of PET and MRI results in the following benefits:

1. *Soft-tissue contrast for no additional radiation dose*: Compared to CT, MRI offers a large variety of soft-tissue contrasts, which could be acquired at no additional radiation dose over PET. By contrast, the dose of the CT could be an issue, especially if multiple investigations in a longitudinal study are necessary (Kalender et al. 2011).
2. *MR-Spectroscopic Imaging (MRSI)*: Additional information that is offered by the use of MRSI is a powerful complement to conventional MRI contrasts. It offers spatially resolved, specific chemical information about the biochemistry of low-molecular-mass metabolites. MRSI can be used in combination with different nuclei such as phosphorus (^{31}P), hydrogen (^{1}H), carbon (^{13}C), nitrogen (^{15}N), fluorine (^{19}F), and sodium (^{23}Na).
3. *Functional MRI (fMRI)*: Besides the known soft-tissue contrast given by T1, T2, T2*, and proton density, MRI offers the measurement of several functional parameters such as blood-oxygen-level dependent (BOLD) perfusion imaging, arterial spin labeling (ASL), and diffusion MRI.
4. *Similar scan time to PET–CT*: Simultaneous acquisition of PET and MRI can keep the overall scan time similar to PET–CT. As the time for anatomical conventional T1- or T2-weighted scans is usually shorter than the overall whole-body PET scan, additional information (fMRI, MRS, etc.) could be measured, which would advance MRI–PET over PET–CT.
5. *Improved co-registration*: Simultaneous MRI and PET would offer significantly improved spatial co-registration of both imaging modalities compared to PET–CT. This feature will make an MRI–PET investigation more robust and stable compared to sequential acquisition modalities.
6. *MR-guided motion-corrected PET*: With the possibility of motion detection, the PET reconstruction and thus the overall quantification of the PET tracer could be significantly improved. This will be of particular importance if the resolution of PET is to be further increased.
7. *Transversal spatial resolution improvement*: PET acquisitions in a simultaneous MRI–PET system would highly benefit from the high magnetic field due to the reduction of the

transversal positron range. This is a particular advantage for tracers with high kinetic energy, such as ^{15}O or ^{82}Rb, and within tissue classes of low density, such as lung tissue. Section 13.5 will give some insight into this specific topic.

While the first three items are independent of the kind of system integration, the last four items are dedicated to simultaneous acquisitions. Unfortunately, there is currently no commercial preclinical MRI–PET system available that offers simultaneous acquisition. Experience with simultaneous systems in the preclinical field has been acquired only with research or prototype systems or with commercial clinical systems, which do not offer the desired high spatial resolution. It is, therefore, essential to give the users of these prototypes and future commercial MRI–PET systems a critical view on the available technologies and correction algorithms. Thus, the following sections will address the three major challenges in more detail.

13.2 FIRST CHALLENGE: ELECTROMAGNETIC INTERFERENCE

13.2.1 FUNDAMENTALS OF MRI–PET INTERFERENCE

Initially, the first challenge received the highest priority as it appears that mastering it enables simultaneous acquisition. We will, therefore, give a brief introduction to why the first challenge is faced and why it is still challenging to solve it.

In order to do so, let us briefly recall the main components of an MRI system, with focus on the specific topic of interference. As described in Chapter 10, an MRI systems needs three major components, as depicted in Figure 13.2.

The *main magnet*, which produces a homogeneous static magnetic field, the so-called B_0 field (~1–10 T) within the FOV to induce a net magnetization.

The *gradient coils*, which are used for spatial encoding of the MRI signal. These coils generate linear field changes of the longitudinal field component (usually called B_z) along the three (x, y, z) axes. The gradient strength is given in mT/m. Values in the range of 300–700 mT/m are quite common for preclinical MRI. The gradients are rapidly switched on and off with a certain slew rate, given in T/m/s. Values in the preclinical area in the order of 3000 T/m/s are common, while clinical systems offer slew rates of just about 100-200 T/m/s.

An *RF coil*, with a preferably homogeneous high-frequency field with a field strength of the order of 10–30 μT (peak-to-peak value). The frequency is determined by the gyromagnetic ratio, which is different for different nuclei: for example, for single protons (^{1}H nuclei) this ratio is about

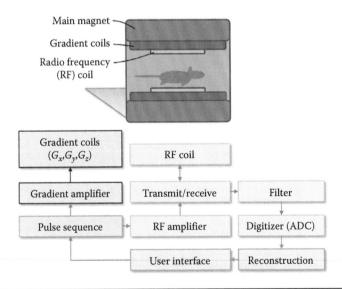

FIGURE 13.2 Main components of an MRI system with the corresponding block diagram.

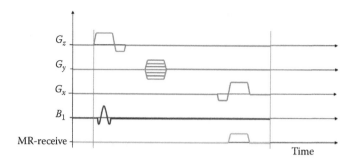

FIGURE 13.3   Schematic pulse diagram describing the procedure of an MRI acquisition over time.

42.567 MHz/T. Thus, a system with a main magnet of 9.4 T has an RF center frequency of about 400 MHz and a bandwidth of a few hundred kilohertz.

Figure 13.3 describes a schematic pulse diagram of an MRI acquisition. All three gradients and the RF excitation are switched by the pulse sequence unit in a specific way. For the understanding of interferences, the specific meaning of the involved gradient and RF pulses is not important, and the reader can refer to Chapter 10 for details. However, for the interference it is important to note that none of the components—gradient fields, RF excitation (B_1), or receiver—is continuously active. We will see later that this is important with respect to simultaneous PET and MRI acquisition. The fraction of the time period in which these components are active is called their duty cycle ($\tau/TR < 1$), where τ is the time in which the component (gradient or, RF excitation, or RF receive) is active, and TR is the duration of the acquisition cycle. The duty cycles have a strong dependence on the sequence being used. However, they are always known before start of the acquisition.

Now let us consider the main components of a PET scanner (shown in Figure 13.4). In order to detect high-energy gamma photons, highly dense scintillator materials are used that convert the high-energy gamma photon into a number (10,000–20,000) of optical scintillation photons. The attached sensor will measure as many scintillation photons as possible (currently 1500–2000 detected optical photons per gamma photon) in order to estimate the position, energy, and time of arrival of the photon shower. The operation of the sensors involves a well-controlled power supply. A synchronization signal has to be provided to the detector unit in order to allow the precise measurement of the time of arrival. After identification of the location, time of arrival, and energy of a single gamma event (we will call this a single), coincidence filtering is applied. (Refer to Chapter 16 for details.) In Figure 13.4, the orange components are usually located near the sensor elements.

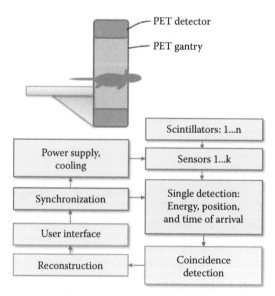

FIGURE 13.4   Block diagram of a PET system.

In contrast to an MRI scanner, a PET scanner measures continuously, meaning that the detector elements are active all the time to enable the detection of as many gamma events as possible (Gebhardt et al. 2015).

After having recalled the main system components, we will now give a phenomenological explanation on why interference occurs. We will divide our discussion into two main parts: quasi-static and electromagnetic interference. For more details and theoretical background, the reader may refer to the book by Jackson (1998).

13.2.1.1   Quasi-Static Interference

As known from Maxwell's equations, every conductor carrying a current produces an electromagnetic field, which consists of an electric field $\vec{E}$ and a magnetic field $\vec{H}$. As we consider quasi-static interference in this section, we can neglect the electric field, as the main cause of coupling between PET and MRI is magnetic in nature. The magnetic flux density $\vec{B}$ caused by a current distribution $\vec{J}(\vec{x})$ at any point $\vec{x}$ is given by the Biot–Savart law.

$$\vec{B}(\vec{x}) = \iiint_V \vec{J}(\vec{x}') \times \frac{\vec{x} - \vec{x}'}{|\vec{x} - \vec{x}'|^3} dv' \tag{13.1}$$

Here, we used the generalized formulation for arbitrary current distributions, which leads to a volume integral. However, this equation is true only if no magnetic material or any other sources generating additional magnetics fields are present. Equation 13.1 is used in the design of an MRI system to find a current distribution $\vec{J}$ that leads to the targeted main field $\vec{B}$, which in MRI terms is called B_0. In particular, in MRI the B_0 field needs to be as homogeneous as possible within the FOV to synchronize the net magnetizations along B_0. However, if we bring ferromagnetic materials such as iron or ferrites, both often used in electronics, close to the FOV, we will disturb the homogeneity of the main field. If this disturbance is substantial, MRI acquisition around this added material will be degraded or even impossible.

13.2.1.2   Electrodynamic Interference

In a static field arrangement without any relative movement between the field-generating coils and the disturbing material or coil, no energy will be transferred between these two systems. However, energy can be transferred if the field changes over time, as caused by the gradient and RF coils. This phenomenon is described by Faraday's law of induction:

$$u(t) = \oint_{\partial A} \vec{E}(\vec{x}',t) \cdot d\vec{s}' = -\iint_A \frac{\partial}{\partial t} \vec{B}(\vec{x}',t) \cdot d\vec{a}' \tag{13.2}$$

According to this law, a voltage u could be measured along the path ∂A, which is similar to the change over time of the magnetic flux density $\vec{B}$ through the corresponding area A. For instance, the path ∂A could be represented by a conductive wire, which will then pick up the voltage. To gain an understanding of the relevance of this effect, we consider a simple example of a rectangular loop of $10 \times 10 \text{ mm}^2$ located in a homogeneous field that increases linearly with 1000 T/s, which is about the order of magnitude we can expect for an induced voltage u_G from the gradient system of an MRI system far from the isocenter.

$$|u_G| = \left(10^{-2} \text{ m}\right)^2 1000 \frac{\text{V s}}{\text{m}^2 \text{ s}} = 0.1 \text{ V} \tag{13.3}$$

For RF pulses, the magnitude of the induced voltage around the same area, assuming a system with 3 T and an RF pulse of 10 µT, equals

$$\left| u_{RF} \right| = \left(10^{-2} \text{ m} \right)^2 10^{-5} \frac{\text{V s}}{\text{m}^2} 2 \cdot \pi \cdot 3\text{T} \cdot 42.6 \text{ MHz/T} = 0.8 \text{ V} \tag{13.4}$$

Besides these two effects, the other important physical effect to be mentioned here is the Lorentz force.

$$\vec{F}_q = q \left(\vec{E} + \vec{v}_q \times \vec{B}_0 \right) \tag{13.5}$$

The equation reflects that every charged particle q can be accelerated by an electric field $\vec{E}$. However, if a particle moves within a magnetic field with speed $\vec{v}_q$, the force will lead to a spiral trajectory around this magnet field $\vec{B}_{ex}$. As a result of this equation, every moving electron (e.g., within a conductive wire) that is placed inside a magnetic field will experience a force. This essentially leads to effects such as the Hall effect or magnetoresistance. Furthermore, induction in a highly conductive surface will create high currents, which will lead to Lorentz forces causing vibrations. In addition, these local currents will generate local heat. Unfortunately, both effects will degrade the operation and lifetime of the PET detector electronics.

13.2.1.3   Particular Interference Problems with a PET System

Both PET and MR imaging involve active and passive electronic circuits, which usually operate at a very low noise level. As described in Chapter 9 and the previous section, the acquisition principles of MRI involve a strong, homogeneous but static magnetic field, gradient fields for spatial encoding, and an RF field for excitation. The measurement of the MRI signal occurs in a narrow band around the Larmor frequency. Influences on characteristics of these fields will also influence the MRI acquisition, which will lead to degradation of the received signal (lower SNR), ghosting, image distortion, and so on. Following the discussion in the previous section, these influences could be caused by the presence of ferromagnetic material, by static stray fields due to the power lines on PCBs or cables, by high conductive shielding material that causes high eddy currents with long life times, and by switching magnetic fields with high amplitude. The RF field of the MRI system is used for nuclei excitation, and its amplitude is related to image intensity in many MRI techniques. Influences on this signal can cause significant signal drop.

In addition, the PET electronics is sensitive to electromagnetic fields across the entire frequency range. Depending on the sensor design and technology and readout electronics, PET could emit electromagnetic (EM) radiation over a large frequency domain. Especially, EM radiation from quartz oscillators, which are usually used in analog-to-digital converters (ADCs) or even on the sensor (e.g., in case of digital silicon photomultipliers [SiPMs]) could cause significant image disturbance or even artifacts. Figure 13.5 summarizes the frequency domains that we discussed so far.

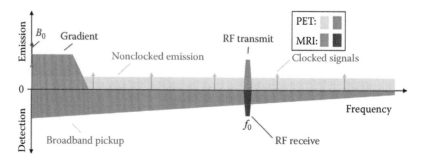

FIGURE 13.5   Interference frequency diagram summarizing the electromagnetic emission of PET and MRI (top half) and the areas where PET and MRI pick up signals (lower half).

13.2.2 POSSIBLE **MRI–PET** INTERFERENCE PROBLEMS

As described in the previous section, introducing new detector systems (e.g., a preclinical PET insert) is a challenging task, since it has the potential to interfere with all these subsystems of the MRI system (a strong, static magnetic field, a gradient system, and an RF system), and vice versa. While the previous section described the principles of electromagnetic interference on a more theoretical level, this section will focus on the particular problems that might occur as a result of the combination of a PET detector with an MRI system.

13.2.2.1 B_0 Field

A solenoid, typically represented by a superconducting coil in the case of MRI, produces the B_0 field (vectorial: $\vec{B}_0 \parallel \vec{e}_z$) with a field homogeneity of a few ppm (parts per million) over the entire FOV. This high homogeneity is essential since the precession frequency of the nuclear spins is proportional to B_0 and thus inhomogeneities in the field will lead to image distortions and fast decaying transversal magnetization (signal loss). Additional materials inside the MRI such as the PET detector or even phantoms alter the field distribution. The local field $\vec{B}(\vec{x})$ can be calculated by the following dipole approximation (Koch et al. 2006):

$$\vec{B}(\vec{x}) \propto \int \frac{1}{|\vec{x} - \vec{x}'|^3} \left(3 \frac{\vec{M}(\vec{x}') \cdot (\vec{x} - \vec{x}')}{|\vec{x} - \vec{x}'|^2} (\vec{x} - \vec{x}') - \vec{M}(x') \right) dv' \tag{13.6}$$

The magnetization $\vec{M}$, contributing to the local $\vec{B}$ field, results from the same B_0 field and depends on the susceptibility of the material. To maximize the homogeneity, shimming (active or passive) can be used to flatten the field distribution, but these techniques are applicable only in cases in which the inhomogeneities are not too localized and thus contain only low order contributions of spherical harmonics. To control and optimize the homogeneity, the susceptibility distribution has to be modified. This gives rise to design rules for the component choice and positioning of the PET detector's electronics. All components should have a very small (e.g., nonferromagnetic) susceptibility and be spatially distributed so as to minimize their impact on the homogeneity of B_0; for example, positioning utilizing symmetries of the scanner could lead to a significant improvement of the B_0 distortion.

On the PET side, the strong magnetic field limits the repertoire of usable sensors. Photomultiplier tubes, for example, are highly sensitive to external magnetic fields and thus are not suitable as PET detectors inside an MRI system. Fortunately, solid-state detectors such as APDs or SiPMs show almost no performance dependence on the applied magnetic field strength, which makes them a suitable choice.

13.2.2.2 Gradient System

The gradient system with gradient strengths up to 300–700 mT/m and slew rates of a few thousand T/m/s is responsible for the spatial alteration of the local magnetic field and thus the spatial encoding of the acquired MR signal. During the switching of the gradients, the local field varies as a function of time and induces eddy currents in any conducting material. The magnetic field of these currents, according to Lenz's law, is opposed to the eddy currents' origin, namely, the change of gradient strength. As a consequence, the local field does not reach the intended value immediately, and the compensational effect of the eddy currents leads to different effectively applied gradient fields. Figure 13.6 illustrates the problem: the desired temporal evolution of the gradient fields (Figure 13.6a, orange curve with sharp edges) is smoothed as a result of the presence of eddy currents.

Since the gradients are intended for spatial encoding of the MR image, image distortions and ghost artifacts (Figure 13.6b) might occur in MR sequences heavily relying on gradient switching (e.g., EPI, spiral) (Le Bihan et al. 2006). The reason for this sensitivity originates from the data acquisition in k-space, which is the Fourier transform of the MR image space. Therefore, an accurate assignment of the measured MR signal to the intended k-space coordinate is of great importance.

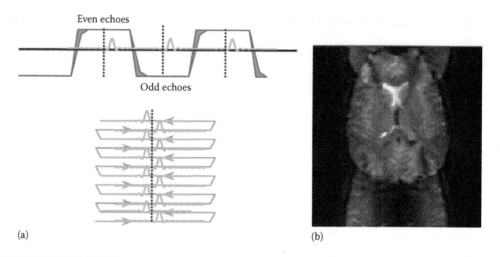

(a)   (b)

FIGURE 13.6 Example for the effects of eddy currents. (a) The shape of the gradient profile is smoothened as a result of the eddy currents, which leads consequently to shifts in the k-space sampling. (b) The resulting reconstructed image shows a so-called ghost artifact (N/2 ghost, shifted by half the FOV). (From Le Bihan, D., Poupon, C. et al.: Artifacts and pitfalls in diffusion MRI. *J. Magn. Reson. Imaging.* 2006. 24(3). 478–488. Copyright Wiley-VCH Verlag GmbH & Co. KGaA. Reproduced with permission.)

The k-space coordinates are determined by the time evolution of the gradients $\vec{G} = [G_x, G_y, G_z]$ (see Chapter 9) and are given by

$$\vec{k}(t) = \frac{\gamma}{2\pi} \int_{t_0}^{t} \vec{G}(\tau) d\tau \tag{13.7}$$

Here, γ is called the gyromagnetic ratio. Thus, any deviation from the ideal gradient switching, namely, a variation of $\vec{G}(t)$ due to eddy currents, leads to misplacements in the k-space (Figure 13.6a, bottom) and thus might produce image distortions/artifacts (Figure 13.6b). Since the eddy currents are induced in the conducting parts of the MRI system, this problem is ubiquitous, but modern gradient systems compensate this problem by adapting their switching characteristics; or, with known distortions, the k-space allocation can be corrected. Problems might occur if new conducting detector materials are inserted into the MRI system, producing additional eddy currents.

On the PET side, the strong and fast switching gradients cause additional requirements for the system design. Induced eddy currents are dissipated in detector components (e.g., RF shielding, heat sinks for electronics) and could cause substantial heating, which tightens the requirements for the cooling system. Since the performance characteristics of solid-state detectors depend strongly on the temperature, the cooling requirements are of fundamental importance. Besides the heating problem, the induced eddy currents interact with the B_0 field, leading to mechanical forces working on the PET components. Theses mechanical forces manifest themselves as vibrations and cause additional stress to the PET modules. Therefore, components such as plug connections should be chosen cautiously.

The gradient switching can also induce currents in the conducting paths of the printed circuit boards (PCBs) or wire connections. If so, this could negatively impact the stability of the PET modules, namely, the power supply of the detector elements and data communication. Induced voltage peaking on data transmission lines can cause misinterpretation in the data processing and thereby lead to faulty data. Also, variations in the supply voltage can temporarily shift the operating point of the particle detectors and might lead to a degradation of the PET detector performance.

13.2.2.3 RF System

The RF system, namely, the RF coil, produces the B_1 field, which is used to flip the longitudinal magnetization ($\vec{M} \parallel \vec{e}_z$) into the transversal plane ($\vec{M}_{xy}$) and is responsible for the MR signal acquisition. As described in Chapter 10, the flip angle at a given location is proportional to B_1 (more precisely to the time integral of the emitted B_1 pulse). Since the produced B_1 field is, depending on the RF coil, not homogenously distributed over the FOV, the applied flip angle varies spatially. This is of great interest since the alteration of the flip angle modifies the contrast and quality of an image. New components inside the MRI, especially conducting materials, can influence the B_1 distribution of the used RF coil, leading to additional distortions of the B_1 map.

Conducting materials can also cause a degradation of the SNR in the MR images. The SNR describes how well a desired signal can be distinguished from background noise and is proportional to the power sensitivity P_s, which describes how much electric power P is needed to produce a certain B_1 field (Ocegueda and Rodríguez 2006):

$$SNR \propto P_s \propto \frac{B_1}{\sqrt{P}} \tag{13.8}$$

Since the B_1 field induces electric currents in any conducting materials, energy is dissipated in these materials, leading to increased power consumption to produce a certain field. As a consequence, the SNR decreases.

As mentioned earlier, the RF system is responsible for the MR signal acquisition. Since the measurable MR signal is very small, the RF system has to be very sensitive and, consequently, is extremely sensitive to external sources emitting spurious electromagnetic fields. Therefore, a PET detector with its electronics has to be designed very carefully: the electronics should emit as little EM radiation as possible and the emitted radiation should not interact with the RF resonator, meaning almost no energy is transferred from the PET electronics to the RF system. To reduce the emitted EM radiation, RF screens are utilized, but one has to take special care of the shielding concept since large conducting areas, which are suitable as RF screens, possibly interfere with other subsystems of the MRI system (e.g., the gradient system). Advanced shielding designs use slits to suppress eddy currents; or, the choice of suitable shielding materials (less conductive materials) might help produce an effective RF screen which is mostly gradient-transparent (Duppenbecker et al. 2012). Another very useful approach to reduce interference between both imaging modalities is to optimize the frequency spectrum of the EM emission in a way that the emission spectrum overlaps with frequency ranges where the MR resonator has no resonance and is thus less sensitive for external EM sources. Also, modifying the emission pattern (spatial distribution of the emitted field) might help in suppressing coupling to the resonator (Gebhardt et al. 2013).

As the gradient system, the RF system produces time-varying magnetic fields inducing eddy currents, which can lead to the problems described earlier. The generated B_1 field is about three orders of magnitude smaller than the gradient fields, but in the frequency range of a few hundred megahertz (in comparison to kilohertz for the gradient switching), the difference in amplitude is compensated by the very short rise times of the field, leading to approximately the same order of magnitude for dB/dt (see examples in Section 13.2). Due to the smaller duty cycle and lower penetration depth (skin effect) in comparison to the gradient fields, the eddy current production through conducting areas is much lower than for gradient fields. Consequently, heating and vibration problems are unlikely to occur. However, RF excitation pulses might induce data transmission problems due to induced voltage peaks. Since the skin depth δ is much smaller for RF frequencies than for the gradient switching frequencies ($\delta \propto f^{-1/2}$), the shielding is much more effective for RF frequencies and thus the PET electronics can be shielded much easier.

13.2.3 CHARACTERIZATION TECHNIQUES TO QUANTIFY THE INTERFERENCE

Since the combination of PET and MRI is a relatively new hybrid modality, there exist no standard quality and characterization protocols to quantify the interference of both systems. As described previously, a PET detector can potentially interfere with all three subsystems of the MRI system, and vice versa. Thus, an interference characterization on all three subsystems is necessary to obtain a complete picture of MRI compatibility of a PET detector and might help to improve the system design.

13.2.3.1   Characterization Techniques to Quantify the Influence on the MRI System

The B_0 distortion map is measurable within the MRI using the phase images of two gradient echo (GRE) sequences with two different echo times $T_{E,1}$ and $T_{E,2}$ (TE extension: $\Delta T_E = T_{E,2} - T_{E,1}$). A large phantom filling the FOV under investigation is used. In a given pixel, the phase advance between both measurements is given by $\Delta\phi = 2\pi\Delta f\Delta T_E$, whereby Δf is proportional to ΔB_0, the local B_0 modification. Thus, via subtraction of both phase images and rescaling, the B_0 distortion map can be reconstructed. However, the B_0 map contains information only about the field homogeneity and not about the absolute B_0 field at a given point. The absolute scale of B_0 could be shifted and has to be measured separately by determining the resonance frequency using, for example, spectroscopic methods. A well-defined phantom material with a clear resonance spectrum (e.g., no peak splitting due to chemical shifts) is preferred to quantitatively measure any shifts. Tetramethylsilane (TMS), for example, is widely accepted in NMR spectroscopy as reference material since it provides single-peak spectra for ^{1}H, ^{13}C, and ^{29}Si nuclei due to its highly symmetric structure and thus might be a suitable choice (Mohrig et al. 2006).

Various methods exist to investigate eddy-current-induced gradient distortions. These methods either use eddy-current-sensitive phase-mapping techniques or are based on measuring the temporal frequency response of an MR signal in dedicated small signal sources (phantom) (Liu et al. 1994; Terpstra et al. 1998; Spees et al. 2011). The basic idea of the latter is to apply a long gradient pulse (of duration several times longer than the eddy-current decay time) and, after switching off the gradient, a 90° degree RF excitation pulse (Figure 13.7a). The resulting free induction decay (FID) of the test sample (e.g., a small phantom sphere featuring a single peak spectrum) is modulated since the test sample experiences a decaying residual gradient caused by the decaying eddy current components in conducting materials. The caused frequency offset is measured as function of time (starting from the gradient's switch-off), and a mathematical model containing the eddy currents amplitudes and characteristic decay times is fitted to the data (Figure 13.7b) (Liu et al. 1994). As a result, the contributing eddy current components are measured at a given location. Moving the test sample in the FOV results in an eddy-current-induced gradient distortion map. This basic approach is very time consuming because the frequency shift has to be measured for various time differences and the test sample has to be moved over the FOV. To overcome this limitation, an array of microcoils (example of a single microcoil is shown in Figure 13.8) with point-like test samples can be utilized to measure the position dependence (Hammer 1996; De Zanche et al. 2008).

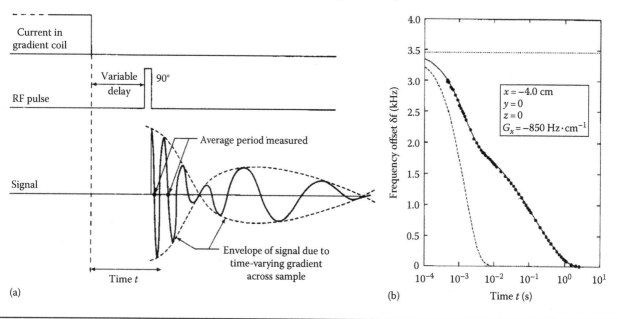

(a)   (b)

FIGURES 13.7   (a) Measurement principle: a 90° excitation is executed after a variable delay t after switching off the gradient. The FID is sampled and the frequency offset is extracted. (b) Resulting measurement: the frequency offset is shown as function of delay time t. A mathematical model containing delay constants and amplitudes is fitted to the data. (Reprinted with permission from Liu, Q., Hughes, D.G. et al., Quantitative characterization of the eddy current fields in a 40-cm bore superconducting magnet, *Magn. Reson. Med.*, 31(1), 73–76. Copyright 1994 by Williams & Wilkins.)

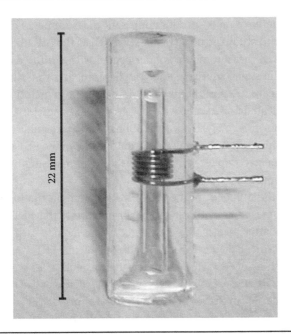

FIGURE 13.8 Example of a microcoil. The probehead is based on a cydohexane droplet with doped D_2O plugs and epoxy casing. (From De Zanche, N., Barmet, C. et al.: NMR probes for measuring magnetic fields and field dynamics in MR systems. *Magn. Reson. Med.* 2008. 60(1). 176–186. Copyright Wiley-VCH Verlag GmbH & Co. KGaA. Reproduced with permission.)

The basic idea to measure B_1 maps is to encode the local B_1 field in measurable quantities, namely, the signal intensity or signal phase. B_1 mapping methods can thus be classified into two main categories: signal-phase-based and signal-magnitude-based methods, with the majority of methods falling in the latter category (Sacolick et al. 2010). Methods to reconstruct the B_1 map cover a whole variety of techniques, such as signal nulling at certain flip angles (Dowell and Tofts 2007), data fitting for increasing flip angles (Hornak et al. 1988), and comparisons of signal ratios for different flip angles (Cunningham et al. 2006). One simple method is the double-angle method (Stollberger and Wach 1996) in which two spin echo images (magnitude image) are recorded with different flip angles α_1 and α_2. The signal intensity $I(\vec{x})$ in a certain pixel is given by

$$I(\vec{x}) \propto \rho(\vec{x}) S(\vec{x}) \sin\big(\alpha(\vec{x})\big) \sin^2\left(\frac{\beta(\vec{x})}{2}\right) R_1\big(\alpha(\vec{x}),\beta(\vec{x}),T_R,T_1\big) e^{-\frac{T_E}{T_2(\vec{x})}} \tag{13.9}$$

Here

$\rho(\vec{x})$ is the spin density
$S(\vec{x})$ is the coil's sensitivity
$\beta(\vec{x})$ is the flip angle of the refocusing pulse
R_1 is a function describing the longitudinal relaxation

Calculating the ratio of the signal intensities $I_1(\vec{x})/I_2(\vec{x})$ yields the following equation:

$$\frac{I_1(\vec{x})}{I_2(\vec{x})} = \frac{\sin(\alpha_1(\vec{x}))}{\sin(\alpha_2(\vec{x}))} \frac{R_1(\alpha_1(\vec{x}),\ldots)}{R_1(\alpha_2(\vec{x}),\ldots)} \tag{13.10}$$

For long repetition times (e.g., $T_R > 5T_1$), the longitudinal relaxation terms approaches 1.0, and thus the ratio becomes almost independent of relaxation processes. Choosing $\alpha_2 = 2\alpha_1$ finally allows the reconstruction of the B_1 map (via the spatial flip angle distribution) using

$$\alpha_1(\vec{x}) = \arccos\left(\frac{I_2(\vec{x})}{2I_1(\vec{x})}\right) \tag{13.11}$$

Concerning the RF interference (noise production), SNR measurements can be performed with or without the PET detector to study the noise increase as well as the MR image quality loss due to the PET electronics. Since SNR measurements are performed with a limited receiver bandwidth, spatial encoding, and a strong signal source, namely, a phantom, and thus always around the Larmor frequency, the resulting scans are useful only to a limited extent to quantitatively study the induced noise-floor increase. Dedicated noise scans with larger receiver bandwidth, multiple center frequencies (away from the Larmor frequency), and without phantom as a signal source are helpful to cover a larger frequency bandwidth to scan for noise contributions from the PET side and to find the origin of a noise source. Also, spectroscopy as a quantitative measurement method might be a useful tool to quantify the PET system's noise contribution.

In order to evaluate the influence of the active MRI system on the PET system and its performance, characterization measurements can be performed with the PET system inside the MRI scanner and outside. The latter case can be considered as a reference case. The comparison of both scenarios allows an assessment of the MRI compatibility of the PET component. Especially, performance parameters such as energy, timing, and spatial resolution, as well as gain and count rate parameters are of interest. Influence of the B_0 field on the PET detector can be investigated directly by comparing these two scenarios. In that case, the PET modules' orientation with respect to the B_0 field might have an influence (e.g., due to the Hall effect) and should be examined (Chaudhari et al. 2009).

To check whether a gradient- or RF-dominated MR environment is harming the PET performance, sequences providing the desired MR environment can be prepared and run during simultaneous acquisition while monitoring the PET system's performance parameters. Thereby, a correlation between the current MRI environment (e.g., gradient switching, RF excitation pulses) and the corresponding performance parameters on the PET side has to be established to allow conclusions about potential sensitivities of the PET electronics to the MRI environment. Dedicated sensors for external field variations on the PET modules, as well as a signal input from the MRI over which information about the MRI system's actions can be transmitted, might help in studying the robustness of the electronics in more detail.

13.3   SECOND CHALLENGE: MRI–PET SYSTEM INTEGRATION

As stated earlier, the fundamental physical quantities used for image information of PET and MRI are different and to some extent independent. Therefore, truly simultaneous operation of both modalities with a shared FOV is in principle possible. Nevertheless, as interference could exist at the level of readout electronics and the required volume of both modalities, different levels of system integration have been proposed to alleviate these problems. For example, interference could be reduced by separating the image acquisitions of both modalities in space (same vs. different FOV) and/or in time (concurrent vs. consecutive). This leads to essentially four different combinations of hybrid schemes, three of which are currently used in research prototypes and commercial systems. Figure 13.9 presents three system variants which we call inline system (a), abutted system (b), and integrated system (c). Table 13.1 shows the MRI–PET system variants with appropriate sensor technology. The use of the combination of concurrent acquisition at different FOVs is not discussed here, as currently no research prototype takes this combination into account.

The inline systems, as shown in Figure 13.9a, essentially benefit from a completely independent operation of the PET and MRI device. Image registration is ideally supported by a proper animal handling system, which allows a quick and easy exchange between the two modalities, as appropriate for the study protocol, and image fusion software to register one modality to the other during postprocessing. Sometimes, dual-modality markers are integrated into the animal handling systems to simplify the acquisition and the registration process.

The abutted system benefits from the need for only minor modification of the PET detector from PET–CT, depending on static stray field at the PET detector location. In some cases, additional μ-metal (high permeability of about $\mu_r \sim 50,000–150,000$) or RF shielding is required, allowing an entire reuse of the PET detector from PET–CT. Images are co-registered by simple geometric

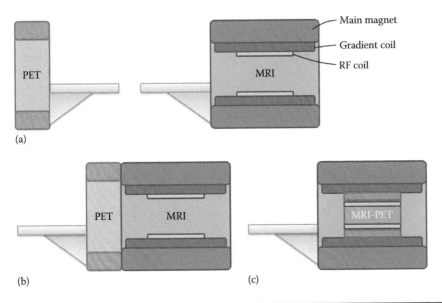

FIGURE 13.9 Different levels of MRI and PET system integration. (a) Shuttle system with an object table that could be moved into both devices. (b) Abutted system with PET at one of the flanges of MRI. (c) Integrated system that allows simultaneous acquisition.

TABLE 13.1
Different MRI–PET System Variants and Required Detector Technologies

MRI–PET System Variants	Different FOV (a, b)	Same FOV (c)
Consecutive acquisition	Sequential MRI–PET	Semi-simultaneous MRI–PET (c-1)
	Inline MRI–PET (PMT)	Field cycle MRI–PET(PMT)
	Abutted MRI–PET (PMT/SSS)	Alternating of PET with MRI within the MRI pulse sequence (SSS, see Figure 13.10)
Concurrent acquisition	—	Simultaneous MRI–PET (c-2)
		Optical fiber approach (PMT)
		Fully integrated PET detector (SSS)

translation along the longitudinal axis. As the integration effort is still moderate, both modalities may show similar image performance as in a separate system configuration. The main advantage is that animal handling is simplified and co-registration is improved. In comparison with simultaneous PET-MR, downsides of the abutted system implementation are potentially slightly larger dimensions and the extended scan times due to consecutive acquisition of both images. In this configuration, we assume no concurrent operation of both modalities, which would prevent interference of the two systems. However, additional magnetic shielding for the sensors and RF shielding of the PET electronics might be required, which leads to additional material between the detectors and the object under investigation (Nagy et al. 2013). This material will unfortunately lead to additional attenuation and scattering of the gamma photons, which could degrade the PET performance.

In general, two different levels of implementations of the integrated system are known that allow imaging of the same FOV: first, consecutively (c-1), and second, simultaneously (c-2), which is more challenging. We will call the option c-1 semi-simultaneous MRI–PET and the option c-2 simultaneous MRI–PET. In a semi-simultaneous system configuration, some interference aspects can be avoided in order to reduce the system complexity. There are various levels of semi-simultaneous systems possible, and in some cases it may be hard to make a distinction between a semi-simultaneous and a fully simultaneous system. An obvious way for a semi-simultaneous system is a field-cycle MRI system as proposed by Gilbert et al. (2009), in which the main B_0 field is switched off during a PET image acquisition cycle. This technique is limited to low field strengths. The main advantage of this approach at this point in time is that conventional PMTs could be used as sensor elements. In the mentioned paper, an MRI acquisition cycle takes about 1.2 s. The main drawback of this configuration

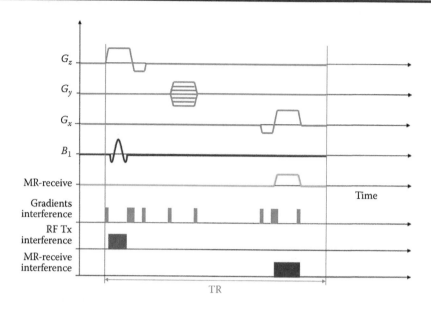

FIGURE 13.10 Interference avoidance in a semi-simultaneous system in which PET is prone to gradient or RF interference, or MR is prone to PET operation.

is the low SNR due to the low field strength of the MRI and the interleaved acquisitions, leading to an overall acquisition time that is comparable to a sequential MRI–PET. Furthermore, motion-compensated PET and MRI acquisitions are possible but limited to slower processes.

A more interleaved version of a semi-simultaneous system could be realized if solid-state sensors are used. These sensors could operate at high static and quasi-static magnetic fields and thus do not require switching off of the main magnetic field. Depending on the type and severity of the remaining interference, for example, during RF emission, MRI signal reception, or gradient switching, the useful acquisition time of PET per MR measurement time could be reduced, which would lead to an overall extended PET acquisition time (see gradient or RF Tx interference in Figure 13.10). More precisely, if during RF or gradient switching the PET signal is degraded, PET data might get affected or lost. It is important to know that this influence is highly dependent on the MRI protocol, and precise knowledge about the MRI is required; otherwise PET quantification will be affected. In case the MRI is affected by the PET operation, it might be desirable to switch off PET during MRI acquisition (see MR-receive interference in Figure 13.10).

If the difference between PET during and without MRI is not noticeable, and vice versa, truly simultaneous operation can be achieved. Image co-registration of both modalities is intrinsically given in all four dimensions (space and time). New features such as MR-based motion compensation can be applied, which could highly improve the resolution and quantification of PET. Nevertheless, type-C systems have the main drawback that the MRI has to give up some volume for the PET detector, which especially for small-bore MRI–PET systems results in a noticeable reduction of the openness of RF coil or in a reduction of the MR sensitivity due to the reduction of the distance between the RF resonator and the screen (which leads to a reduction of the power sensitivity of any RF coil).

13.3.1 PET Sensor Technologies for MRI–PET

The system variants in the last section differ significantly in the way they address the electronic interference between PET and MRI. According to this section, MRI–PET devices can be classified in the three categories: sequential (inline and abutted), semi-simultaneous, and simultaneous MRI–PET systems. As mentioned, there are system concepts that allow the reuse of PMT-based PET detectors, while other solutions require SSSs, which allow operation in the high static and quasi-static magnetic fields. The mutual influence between the two imaging modalities strongly depends on the system concept, the used sensor technology, the readout architecture, and the operation mode (e.g., MRI sequences).

Very often, the term "MR compatible" is used in combination with sensor technology to indicate the possibility that this detector technology still works in a high static field, while other effects are neglected. A good example of potential interference is given in Chaudhari et al. (2009). Here, the influence of the static magnetic field on the event (detection of a 511 keV gamma photon) positioning was reported, which indicates a risk of losing spatial resolution if not compensated for. Other examples in Yamamoto et al. (2012) were bright spots in the MR image, indicating potential electromagnetic coupling from the PET readout electronics into the RF receive chain.

13.3.1.1 Photomultiplier-Tube-Based PET Detectors

As a truly simultaneous MRI–PET system is highly complex and its real benefit is still to be proven at the application level, a logical step for combining MRI and PET is the concept of the abutted system allowing the reuse of the state-of-the-art detector technology of preclinical PET–CT scanners while accepting other compromises in the workflow. As the level of system integration is lower compared to a simultaneous MRI–PET system, less interference and therefore higher PET and MRI image quality are more likely.

In order to assign a single event to one of the small crystals (1–2 mm pitch), position-sensitive photomultiplier tubes (PSPMTs) are a very powerful sensor technology for preclinical PET. The main advantage of the PMT technology is its high gain, its low noise, and its fast impulse response. The first single-stage PMT was demonstrated in 1935 by Harley Iams and Bernard Salzberg (1935), and the concept is still unchanged. The fundamental function of a PSPMT is identical to that of a PMT: generation of a photoelectron at the cathode and amplification via electron multiplication using acceleration of the electrons along a well-defined path that is defined via the electric field, as given by Lorentz force:

$$\vec{F}_q = q\vec{E} = m_q \cdot \dot{\vec{v}}_q \tag{13.12}$$

The electric field $\vec{E}$ causes a force $\vec{F}_q$, which accelerates the electron with the charge q and rest mass m_q (Lorentz force in absence of a magnetic field). As can be seen in Figure 13.11, in a PMT the amplification is realized in several cascades of dynodes. Along this structure, the electric potential difference between the cathode, dynodes, and the anode will lead to an electric field that will accelerate the electrons along this defined path. In such a way, an initial single photoelectron will lead to usually 10^6–10^8 electrons, read out via the anode. By means of structuring of the dynodes and anode into an array of dynodes and anodes, the concept of the PMT can be modified to a position-sensitive PMT (PSPMT) offering multiple individual channels. Typical configurations are 4×4 or 8×8 channels. PMT as well as PSPMTs are very powerful photosensors, but they have a drawback for MRI–PET, which we would like to look into in more detail. The force $\vec{F}_q$ in the presence of an external (static or dynamic) magnetic field $\vec{B}_{ex}$ also depends on the magnetic field:

$$\vec{F}_q = q\left(\vec{E} + \vec{v}_q \times \vec{B}_{ex}\right) = m_q \cdot \dot{\vec{v}}_q \tag{13.13}$$

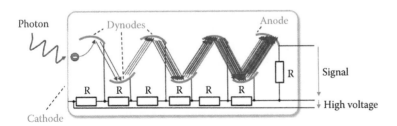

FIGURE 13.11 Amplification principle of a photomultiplier tube (PMT) with different dynode stages. The electron multiplication is very similar to a position-sensitive PMT.

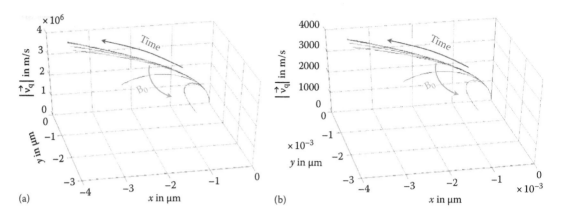

FIGURE 13.12 Acceleration trajectory (red curve) of an electron within a magnetic field $B_z = 0, 0.01,$ 0.05, 0.1, 0.5, 1.0, 5, 10 T for two electric fields of $E_x = 10^7$ (a) and $E_x = 10^4$ V/m (b). Initial velocity of the electron is 0 m/s.

This equation is a vector differential equation in $\vec{v}_q$. We will present some solutions for given constant electric and homogeneous magnetic fields. If the electric field is parallel to the magnetic field and the particle velocity is parallel to the electric field, the vector product of the particle velocity and the external magnetic field vanishes. The amplification process would be undisturbed in this case. Practically, the electric field is not homogeneous in a PMT, and thus the cross product cannot be neglected. Solutions of this vector differential equation for given electric fields orthogonal to magnetic flux densities are given in Figure 13.12. Here, we used electric fields of different strengths (10^4 and 10^7 V/m) and magnetic flux densities ranging from 0 up to 10 T. The two graphs are qualitatively similar. Only the spatial extent depends on the electric field strength. Both graphs show that a high magnetic flux density leads to a strong bending of the trajectory of the electron path, which affects the amplification process, potentially even the spatial resolution of a PSPMT. As indicated in this graph, a high electric field keeps the charged particle (here the photoelectron) closer to the trajectory calculated at 0 T.

Therefore, in a system approach (a), even though simultaneous operation of PET and MRI was excluded, the amplification of the PMT could be significantly degraded by the static fringe field of the main magnet. Nowadays, low-field MRI magnets for preclinical MRI with several teslas are not specially designed for very low ($<10^{-4}$ T) fringe fields. Thus for PSPMTs additional magnetic shielding might be required. In general, magnetic shielding consists of a ferromagnetic material (μ-metal) with a high permeability (μ_r ~50,000–150,000). The static magnetic field is essentially guided inside the shield around the PMT, which results in a magnetic-field-free volume at the location of amplification inside the PMT. Depending on the magnetic field strength, substantial amounts of material are needed for sufficient shielding. Therefore, magnetic shielding for higher field strength requires thicker shielding material, which may be a limiting factor for the use of PMTs close to main magnets with high field strength or high stray fields.

13.3.1.2 Solid-State-Sensor-Based PET Detectors

In the last decade, SSSs have been further developed and are an attractive alternative to PMT-based detectors. In particular, SSSs are of interest for MRI–PET, especially for simultaneous MRI–PET, as they are able to operate at very high magnetic fields. This is due to the fact that the amplification region of an SSS in which the amplification process takes place has small extensions of ~1 μm. The effects from B_0 are thus not significant. As we are detecting only a few (~1000–2000) optical photons from the scintillation process, we are interested in detectors with intrinsic high gain. Conventional photodiodes are unfortunately not suitable to detect this low amount. Therefore, SSSs are operated in higher gain mode (avalanche mode) or even in the so-called Geiger mode. To understand the difference of these two modes, Figure 13.13 depicts the schematic current–voltage curve of a diode cell in reverse mode. This curve shows two main characteristics: the first mode is the so-called

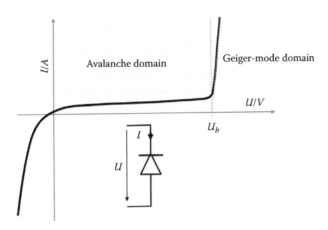

FIGURE 13.13 Reverse-bias current *I* as a function of reverse bias voltages *U*. A photodiode can be operated in the linear amplification mode (avalanche domain), or in the Geiger mode.

avalanche mode, in which the diode amplifies in a linear manner. A typical linear amplification is about 10^2–10^3. Devices that are operated in this domain are called APDs (avalanche photo diode). APDs typically consist of a single sensitive area, for example, 3×3 mm^2.

If we increase the bias voltages further, we pass the breakdown voltage. Operation beyond the breakdown voltage always leads, as indicated via this high current, to a spark of electron–hole pairs. As mentioned, this mode is called the Geiger mode. However, the operation of a diode in this mode requires two things: geometric structuring into several parallel connected diodes, and a quenching mechanism per cell to reset the process. This is normally achieved via series resistors (passive quenching). When the high current occurs, the bias voltage across the diode drops below the breakdown voltage because of the voltage drop across the series resistor. In order to achieve a response that is approximately proportional to the number of detected scintillation photons, the device is structured into an array of parallel-connected diode resistor cells (called SPAD). The overall device is called an SiPM (see Figure 13.14a). Although this is an analog device, its operation is essentially digital with one current spark per detected photon and cell. This has triggered

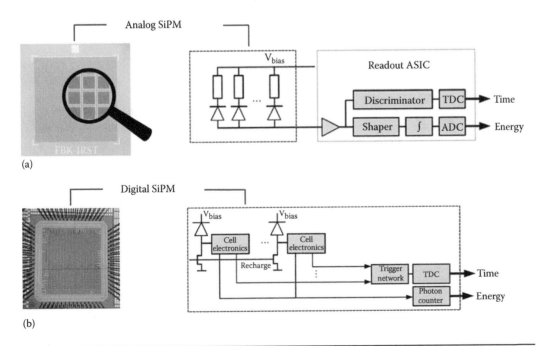

FIGURE 13.14 SiPM devices (left) with functional block diagrams (right). (a) Analog silicon photomultiplier (SiPM) with readout ASIC. (b) Digital SiPM (dSiPM) with integrated readout.

researchers to invent a digital version of the SiPM, which directly counts and actively quenches the number of cells that fire in a given time window (see Figure 13.14b).

For the APDs as well as SiPMs, position-sensitive models have been invented by various research groups (Ferri et al. 2015; McClish et al. 2010). These devices aim at simplifying the identification of the individual crystals without the need for additional light-spreading elements between the crystals and the sensor.

13.3.2 READOUT CONCEPTS FOR MRI–PET

Figure 13.15 shows four different implementations of the signal readout. Figure 13.15a uses no electronic component in the electromagnetically harsh inside of an MRI, and reflects the readout concept of the first published simultaneous MRI–PET system of Shao et al. in 1997. Optical fibers between the scintillator crystals and detector were used to transport the scintillation photons to the detector. The light losses in this concept, resulting in degradation of the energy, timing, and spatial resolution, called for local detection of optical photons.

Luckily, SSSs have been rapidly improved over the last decade by various research groups and have emerged as an alternative detector technology. Because these sensors are insensitive even to high static magnetic fields, they can be placed close to the crystals and thus inside the B_0 field. As these sensors require electric power and potentially local electronics, the challenge of electromagnetic interference between the two modalities was created. Current sensor concepts in the preclinical domain differ at the level of location of the readout electronics and signal transmission.

The concept of Figure 13.15b aims at an optoelectronic-to-optical signal conversion with the use of local laser diodes. Here, the signal of the SSS is used to modulate the current of a laser diode. The main advantages is that this chain of signal conversions is highly effective. Even time-of-flight measurements (Chapter 8) have been reported with this concept (Bieniosek et al. 2013). The concept of Figure 13.15c is based on the idea of proper shielding of the cables. Thus cable routing, shielding, grounding, and so on, need to be arranged very carefully to avoid interference between PET and

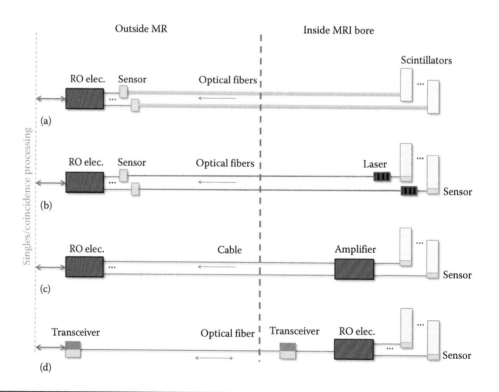

FIGURE 13.15 Four different readout topologies for a simultaneous MRI–PET system. The readout electronic (RO elec.) is kept outside the MRI in the first three concepts (a–c), while the last concept (d) shows the extreme of local detection, digitization, and processing.

MRI (Yamamoto et al. 2012). Local digitization offers many advantages over remote digitization, but has the risk of creating substantial disturbance of the MRI system.

All mentioned concepts have their advantages and disadvantages, and it is recommended to take a critical view on the interferences expected from the individual system concept.

13.4   THIRD CHALLENGE: PET ATTENUATION CORRECTION IN MRI–PET

The situation of attenuation correction (AC) in MRI–PET is closely linked to that in standalone PET (Chapter 9). While AC is straightforward in PET–CT (Chapter 12), because the transmission of x-rays (CT) and γ-rays (PET) is both related to electron density, it is much more challenging without an x-ray source. In standalone PET, this may be the result of design decisions considering space or resource constraints, and can in principle be resolved by employing PET–CT. In MRI–PET, however, MR compatibility of x-ray tubes is an additional concern, which needs to be addressed. A further aspect is the challenge to design MR-compatible motors required for a rotating movement of any transmission source, be it an X-ray tube or a radionuclide source. Thus, transmission measurements using moving sources do not appear to be a valid option in MRI–PET to date.

Static (nonrotating) radionuclide sources have been successfully tested in clinical MRI–PET recently (Mollet et al. 2012). In this approach, an annular transmission source can be used as a stationary source, and time-of-flight information can be considered to separate radionuclide transmission and PET emission in simultaneous acquisitions. However, in order to achieve the same separation in preclinical PET, a time-of-flight resolution in the low-double-digit picosecond regime (equivalent to a few centimeters of photon path) would be required, which has not been achieved so far. Summarizing, transmission measurements are especially challenging in preclinical MRI–PET and may require dedicated MRI–PET–CT trimodality imaging. More sophisticated approaches are thus needed to correct for photon attenuation by both the animal and the equipment within the FOV.

As to the necessity of correcting for photon attenuation in preclinical PET, we note that attenuation has a much more pronounced effect in clinical than in preclinical PET since the fraction of absorbed photons scales with the dimensions of the object studied. Many techniques have therefore been proposed, evaluated, and refined in the context of clinical imaging but rarely translated to preclinical imaging. Nevertheless, for quantitative imaging, AC may be required depending on the quantification accuracy required in a particular application. Furthermore, the major challenge of PET AC is the generation of an attenuation map. Since that attenuation map may also be used for other purposes, such as for scatter or positron range correction (see Section 13.5), a correction map may still be of use even if AC is not performed.

13.4.1   ANIMAL ATTENUATION

Many alternative techniques to incorporate the attenuation of the animal are thus similar to those employed in standalone PET, including uniform scaling of the reconstructed emission image (Chow et al. 2005). In this approach, the animal is approximated by a water-filled cylinder, and from the diameter of this cylinder, a global AC factor is computed. The reconstructed PET image is then rescaled by this factor. This technique can be compared with the transmission-based AC technique in Figure 13.16. The scaling method corrects the mean error, but does not take into account the different photon paths at different depths within the object—hence cupping artifacts are visible in the uniform profile.

As an alternative to this analytical approach, attenuation maps—with varying degrees of detail—can be used during image reconstruction. These can also be created without transmission measurements, for example, by filling the animal's outline with the linear attenuation coefficient of water (Yao et al. 2005).

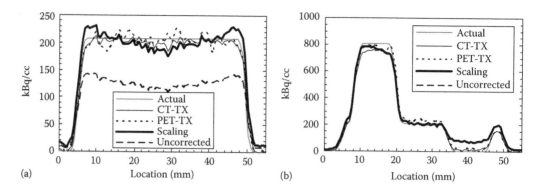

FIGURE 13.16   Transverse microPET emission profiles through (a) uniform and (b) nonuniform regions of a rat-sized phantom. Different profiles represent different attenuation correction techniques: actual (ground truth), CT-TX (x-ray transmission measurement), PET-TX (radionuclide transmission measurement), scaling (global scaling method), uncorrected (uniform profile only). (Reproduced by permission of IOP Publishing from Chow, P.L., Rannou, F.R. et al., Attenuation correction for small animal PET tomographs, *Phys. Med. Biol.*, 50(8), 1837–1850, 2005. All rights reserved. © Institute of Physics and Engineering in Medicine.)

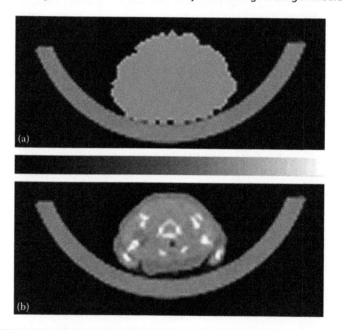

FIGURE 13.17   Transaxial slices through attenuation maps of a mouse obtained (a) from the outline the emission image and (b) from x-ray CT measurements. (Reprinted with permission from Yao, R.T., Seidel, J. et al., Attenuation correction for the NIH ATLAS small animal PET scanner, *IEEE Trans. Nucl. Sci.*, 52, 664–668. © 2005 IEEE.)

When this approach is used in standalone PET, the animal's outline needs to be determined from the non-attenuation-corrected emission image, as in Figure 13.17. In this case, the error compared to perfect AC may already be sufficiently small for many applications in very small animals, especially in tissues the attenuation properties of which resemble those of water (Figure 13.18).

In addition, this method can be significantly expanded using MR information, which is generally of higher spatial resolution and less noisy. Such expansion is driven by the development in clinical MRI–PET scanners in which AC is of much greater importance due to longer photon paths through the patients' bodies. Two approaches have received dedicated attention in recent years: template-based and segmentation-based methods.

Template-based methods employ a database of pairs of CT and MR images, each pair acquired in advance from a single animal. One or more of these MR images can then be mapped on the MR image of the current animal using an elastic image registration step. After applying the same

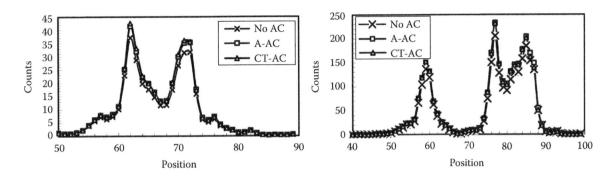

FIGURE 13.18   Transverse emission profiles through (a) the head and (b) heart of the mouse depicted in Figure 13.17. Different profiles represent different attenuation correction techniques: No AC (uncorrected), A-AC (analytical, water-filled outline), CT-AC (X-ray transmission measurement). (Reprinted with permission from Yao, R.T., Seidel, J. et al., Attenuation correction for the NIH ATLAS small animal PET scanner, *IEEE Trans. Nucl. Sci.*, 52, 664–668. © 2005 IEEE.)

transformation to the corresponding CT image, one obtains a CT image that is similar to the CT image that could have been acquired from the animal and that can be used to determine an attenuation map. The accuracy of this approach relies fundamentally on the similarity of the current animal with at least one from the database. Variations of this approach employ machine learning to infer a mapping from MR image intensities to CT image intensities in order to predict attenuation map values from small patches of MR images (Hofmann et al. 2008; Johansson et al. 2011; Navalpakkam et al. 2013). While this may be of use in clinical imaging, the required similarity may not be guaranteed in studies of most animal models unless a dedicated database is established first. Note that this database needs to consider the particularities of a studied animal model, for example, anatomical abnormalities in tumor-bearing models.

A second and more versatile group of techniques is based on image segmentation. There, MR images are segmented into a number of tissue classes (Schulz et al. 2011). Each tissue class (e.g., air, lungs, fat, nonfat soft tissue, bone) is then assigned a predefined attenuation coefficient. The information from multiple MR images can be combined in this image segmentation step. In general, the more MR images are available and the more differentiated the image contrasts are, the more tissue classes can be distinguished, with up to four or five classes being distinguished in existing literature (air, lungs, fat tissue, nonfat soft tissue, bone). For example, Dixon-like sequences allow the discrimination of the lungs as well as fat and nonfat soft tissue (Martinez-Möller et al. 2009). Ultrashort echo-time (UTE) sequences enable the detection of bone signal, which is not detectable in conventional MR sequences due to T2 relaxation times (see Section 9.2.1) being in the millisecond range (Catana et al. 2010; Keereman et al. 2010). In some cases, multiple MR image contrasts can be combined in a multi-echo MR sequence, for example, in a sequence featuring both Dixon-like and UTE properties (Berker et al. 2012).

One advantage of preclinical imaging over clinical imaging is that MR sequences may be applicable that are prohibitive in clinical settings. This comprises, for example, whole-body UTE sequences with long acquisition times, or extended ranges of allowed specific absorption rate (SAR) or gradient rise times (peripheral nerve stimulation, PNS).

13.4.2   EQUIPMENT ATTENUATION

The term "equipment" comprises all devices that are necessary to put and maintain the animal in a state and position appropriate for (extensive) imaging and to carry out the imaging operations. This includes the animal bed, anesthesia equipment, injection instruments, and MR coils. Depending on the equipment employed, especially their material composition and relation to the size of the animal, devices may or may not contribute the major part of photon attenuation.

In PET–CT, equipment may already be designed to interfere to the least degree possible with image acquisition. Still, PET and CT attenuation cannot be completely eliminated. However, all equipment is visible in the CT images according to their attenuation properties and is therefore easily included in CT-based PET attenuation correction. Thus, objects invisible in CT do not usually interfere with PET acquisitions. In MRI–PET, however, MR invisibility does not imply the absence of interference with PET. This can be appreciated by considering polymethylmethacrylate (PMMA), a material commonly used for MRI phantoms. Because of its solid structure, relaxation times are ~50 μs (Sinnott 1960); hence, PMMA is virtually invisible for standard MRI sequences. By contrast, with a mass attenuation coefficient of 0.0889 cm²/g and a mass density of 1.18 g/cm³ (Kucuk et al. 2013), PMMA has a linear attenuation coefficient (at 511 keV photon energy) of 0.105 cm⁻¹, which is higher than that of water (0.096 cm⁻¹). Each piece of equipment must therefore be separately evaluated for attenuation effects and, if this is found necessary to achieve a certain level of accuracy, taken care of in attenuation correction.

The shape and the attenuation distribution of some equipment must therefore be known beforehand. It can be determined, for example, from a separate CT scan or from information about device dimensions and material composition. The exact localization, however, has to be determined on a scan-by-scan basis. Two methods have been proposed for co-registration of a clinical MR coil attenuation template and the PET FOV (Paulus et al. 2012), which should also be applicable to devices in preclinical imaging. First, UTE-MR imaging may reveal some of the equipment's plastic housing and can be used for registration of, in that latter work, a previous CT scan. Second, and applied with more success, cod liver oil capsules that are highly MR visible can be attached to the surface to localize the device. In the case of devices composed by several connected rigid elements, each element needs to be localized separately; flexible equipment may need to be taken care of by deformable co-registration of the attenuation template or parameterization of the template, depending on the number of degrees of freedom.

13.5 A POTENTIAL BENEFIT: POSITRON RANGE EFFECTS IN MRI–PET

A physical effect unique to MRI–PET is the positron behavior in strong magnetic fields. Here, we first treat the case of positron emission in vacuum; the findings will then be combined with the effect of positron range limited by the presence of tissue.

Because of its (positive) electric charge, each positron emitted from a nucleus in vacuum is forced on a helical trajectory around a magnetic field line by the Lorentz force, which is $\vec{F} = e^+ \cdot \vec{v} \times \vec{B}$. In this equation, e^+ represents the positron's electric charge, $\vec{v}$ is its velocity vector, and $\vec{B}$ is the magnetic field vector. The radius of this helix, the *gyroradius*, is therefore determined not only by the magnetic field strength: it also depends on the positron's kinetic energy and its angle of emission relative to $\vec{B}$. Note that a statement similar to this holds true in a medium in non-MRI–PET, where the mean positron range is closely linked to the positron's mean kinetic energy. However, the underlying mechanism is fundamentally different: in non-MRI–PET, a positron's range is determined by the properties of the surrounding tissue in a stochastic process, and is not strictly bounded in any direction of space. In vacuum MRI–PET, however, the positron's range perpendicular to $\vec{B}$ is strictly limited to twice the gyroradius, which is given by $r = (m \cdot v_\perp)/(e \cdot B)$. In this equation, m and e represent the positron's mass and electric charge, respectively, $B = |\vec{B}|$ is the magnetic field strength, and $v_\perp$ is the positron's velocity component perpendicular to the magnetic field $\vec{B}$.

In the nonrelativistic case, the kinetic energy E_k is linked to the velocity v by the formula $E_k = 1/2\,mv^2$. Generally, $v_\perp \leq v$, expect in the case of positron emission perfectly perpendicular to $\vec{B}$, when $v_\perp = v$. In that special case, one finds the numerical value equation

$$\frac{r_g}{\text{mm}} = 3.37 \cdot \sqrt{\left(\frac{E_k}{\text{MeV}}\right)} \bigg/ \left(\frac{B}{T}\right) \tag{13.14}$$

for the gyroradius as a function of kinetic energy, and generally, $r \leq r_g$. This provides a first impression of the dimensions and asymptotic behavior of the positron range effect. Note that for accurate calculations, relativistic effects need to be considered in the MeV range, and instead of Equation 13.14, we must use (Raylman et al. 1996)

$$\frac{r_g}{mm} \leq 3.37 \cdot \sqrt{\left(\frac{E_k}{MeV}\right) + \frac{\left(\frac{E_k}{MeV}\right)^2}{1.022}} \bigg/ \left(\frac{B}{T}\right) \tag{13.15}$$

Inserting the positron's maximum kinetic energy (usually several MeV), which is a function of the chosen isotope, this equation will lead to the maximum possible gyroradius for this isotope; examples are given in Table 13.2. In both nonrelativistic and relativistic cases, the maximum r_g increases with maximum kinetic energy and decreases with increasing magnetic field strengths. It is the latter property of positron confinement by the magnetic field that is especially interesting in MRI–PET.

Since the decaying nucleus is on the helical positron trajectory, the *maximum* 2-D positron range (perpendicular to $\vec{B}$) is $2 \cdot r_g$, as noted previously, while the *mean* 2-D range in absence of annihilation is about $1.27 \cdot r_g$. Hence, combining this latter mean value with the mean value of the velocity component perpendicular to $\vec{B}$, which is $\bar{v}_\perp = 0.79 \cdot v$, we note that the value of r_g calculated using the maximum positron energy E_k represents the mean 2-D range (perpendicular to the magnetic field) in vacuum of isotropically emitted positrons of maximum kinetic energy.

In tissue, the positron range in MRI–PET is thus limited by a combination of tissue and magnetic field effects, both of which influence the achievable image resolution. If the gyroradius calculated earlier is significantly smaller than typical positron ranges in non-MRI–PET, then MRI–PET can be expected to yield improvements in terms of combined positron range in two of three spatial directions perpendicular to $\vec{B}$. If the positron range, in turn, is also on the order of (or larger than) the PET system's spatial resolution, decreasing the positron range can in fact yield improved spatial resolution (Hammer and Christensen 1995).

In contrast to the effects of photon attenuation, which are more prominent in clinical imaging, the positron range effect is more pronounced in preclinical than in clinical imaging for two reasons. First, magnetic field strengths in preclinical imaging may be higher than in clinical imaging, leading to increased confinement. Second, as stated previously, the positron range must be a dominant term in the system's spatial resolution for improvements to carry over. This is not the case in clinical imaging, while it may be so in preclinical imagers. In addition, the range of isotopes for both clinical and preclinical applications with higher kinetic energies than ^{18}F is slowly but steadily being

TABLE 13.2
Isotopes Commonly Used in PET with Some of Their Positron Properties

Isotope	Maximum Positron Kinetic Energy (in MeV) (Hammer et al. 1994)	Mean 1-D Positron Range (in mm) in Water/ Lung Tissue at 0 T (Soultanidis et al. 2011)	Maximum Relativistic Gyroradius r_g: Mean 2-D Range of Maximum-Kinetic-Energy Positrons in Vacuum (in mm) at 9.5 T (Equation 13.15)	Mean 1-D Positron Range (in mm) in Water/Lung Tissue at 9.5 T (Soultanidis et al. 2011)
^{18}F	0.64	0.27/0.73	0.36	0.16/0.19
^{11}C	0.97	0.44/1.21	0.49	0.21/0.25
^{15}O	1.74	0.90/2.66	0.77	0.32/0.34
^{68}Ga	1.90	0.93/2.74	0.83	0.33/0.35
^{82}Rb	3.15	1.87/6.07	1.27	0.50/0.48

Notes: Care must be taken in comparing different values reported in the literature, which may vary in terms of reporting 1-D vs. 2-D ranges, mean vs. maximum ranges, and for positrons of mean vs. maximum kinetic energies. Positron ranges reported here are indicative of the effect of a 9.5 T magnetic field on the level of radiation emission. Whether the positron confinement results in enhanced spatial resolution in the reconstructed images is a function of whether positron range is a major determinant for PET image resolution outside the MRI.

extended, for example, with [68]Ga (Al-Nahhas et al. 2007). For a list of commonly used PET isotopes and their positron kinetic energies, see Table 13.2.

The effect of magnetic field strengths on positron ranges in tissues had been quantified long before MRI–PET was technically possible (Iida et al. 1986). Magnetic fields of 5 or 10 T were found to successfully confine positrons with maximum kinetic energies of 2 or 3 MeV. Later measurements employing one or two detectors, especially using ^{68}Ga point sources, have confirmed these findings by significantly reduced widths of annihilation spread functions in high magnetic fields (Hammer et al. 1994; Hammer and Christensen 1995; Wirrwar et al. 1997). For example, a reduction from 3.5 to 3.0 mm has been measured at 4.5 T (Wirrwar et al. 1997).

For isotopes other than ^{68}Ga, the situation can be much different even when using the same measurement setup: for example, a 3.7 mm resolution of [82]Rb at 0 T can be drastically reduced to 2.7 mm in a 7 T simulation. For ^{18}F, by contrast, the 0 T resolution cannot be further reduced either by 7 T simulation or by measurements at 4.5 T. Similar conclusions can be drawn both from whole-system simulations (Raylman et al. 1996) and from a detailed particle simulation for a variety of isotopes and field strengths (Soultanidis et al. 2011). Some of these last results have also been put into context in Table 13.2.

Since the gyroradius is comparable to the positron range in absence of a magnetic field, the tissue or material surrounding the decaying nuclei can have a major impact on whether the positron confinement by the magnetic field results in measurable changes (Figure 13.19). As a borderline case, consider [15]O: while the positron range reduction in soft tissue (from 0.90 to 0.32 mm) will be hardly detectable using most equipment, the reduction in lung tissue (from 2.66 to 0.34 mm) should well be measurable.

In summary, magnetic fields up to 10 T are not likely to have a significant effect on imaging of isotopes with maximum positron kinetic energy less than 1 MeV, such as ^{18}F or [11]C. By contrast, for high-energy emitters, such as [82]Rb, the magnetic field might be an important enabler to high-resolution imaging, especially in lung tissue. In any case, care must be taken to consider the nonisotropic positron range in image reconstruction: in a first approximation, treating the body as composed of water, this effect can be included in the scanner's system matrix. If tissue-specific effects are to be considered, however, the effect is clearly object-dependent and therefore requires the use of an attenuation map (see Section 13.4).

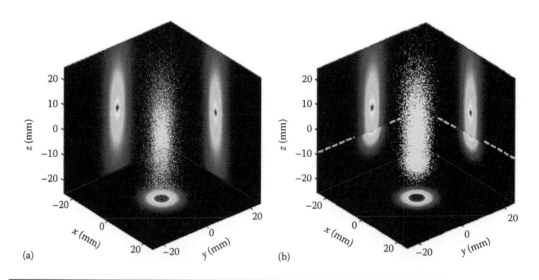

FIGURE 13.19 (a) Simulated annihilation endpoint coordinates for a [82]Rb point source in homogeneous lung tissue at 3 T along the z-direction. (b) Same configuration with a soft tissue region below the dashed line. (Reprinted with permission from Kraus, R., Delso, G. et al., Simulation study of tissue-specific positron range correction for the new biograph mMR whole-body PET/MR system, *IEEE Trans. Nucl. Sci.*, 59, 1900–1909. © 2012 IEEE.)

REFERENCES

Al-Nahhas, A., Z. Win et al. (2007). Gallium-68 PET: A new frontier in receptor cancer imaging. *Anticancer Res.* **27**(6B): 4087–4094.

Berker, Y., J. Franke et al. (2012). MRI-based attenuation correction for hybrid PET/MRI systems: A 4-class tissue segmentation technique using a combined ultrashort-echo-time/Dixon MRI sequence. *J. Nucl. Med.* **53**(5): 796–804.

Bieniosek, M. F., P. D. Olcott et al. (2013). Readout strategy of an electro-optical coupled PET detector for time-of-flight PET/MRI. *Phys. Med. Biol.* **58**(20): 7227–7238.

Catana, C., T. Benner et al. (2011). MRI-assisted PET motion correction for neurologic studies in an integrated MR-PET scanner. *J. Nucl. Med.* **52**(1): 154–161.

Catana, C., A. van der Kouwe et al. (2010). Toward implementing an MRI-based PET attenuation-correction method for neurologic studies on the MR-PET brain prototype. *J. Nucl. Med.* **51**(9): 1431–1438.

Catana, C., Y. Wu et al. (2006). Simultaneous acquisition of multislice PET and MR images: Initial results with a MR-compatible PET scanner. *J. Nucl. Med.* **47**(12): 1968–1976.

Chaudhari, A. J., A. A. Joshi et al. (2009). Spatial distortion correction and crystal identification for MRI-compatible position-sensitive avalanche photodiode-based PET scanners. *IEEE Trans. Nucl. Sci.* **56**(3): 549–556.

Chow, P. L., F. R. Rannou et al. (2005). Attenuation correction for small animal PET tomographs. *Phys. Med. Biol.* **50**(8): 1837–1850.

Chun, S. Y., T. G. Reese et al. (2012). MRI-based nonrigid motion correction in simultaneous PET/MRI. *J. Nucl. Med.* **53**(8): 1284–1291.

Cunningham, C. H., J. M. Pauly et al. (2006). Saturated double-angle method for rapid B_1+ mapping. *Magn. Reson. Med.* **55**(6): 1326–1333.

De Zanche, N., C. Barmet et al. (2008). NMR probes for measuring magnetic fields and field dynamics in MR systems. *Magn. Reson. Med.* **60**(1): 176–186.

Dowell, N. G. and P. S. Tofts. (2007). Fast, accurate, and precise mapping of the RF field in vivo using the 180 degrees signal null. *Magn. Reson. Med.* **58**(3): 622–630.

Duppenbecker, P. M., J. Wehner et al. (2012). Gradient transparent RF housing for simultaneous PET/MRI using carbon fiber composites. *IEEE Nucl. Sci. Symp. Med. Imaging Conf. Rec.* **2012**: 3478–3480.

Ferri, A., F. Acerbi et al. (2015). Characterization of linearly graded position-sensitive silicon photomultipliers. *IEEE Trans. Nucl. Sci.* **62**(3): 688–693.

Gebhardt, P., J. Wehner et al. (2013). RF interference reduction for simultaneous digital PET/MR using an FPGA-based, optimized spatial and temporal clocking distribution. *IEEE Nucl. Sci. Symp. Med. Imaging Conf. Rec.* **2013**.

Gebhardt, P., J. Wehner et al. (2015). RESCUE - Reduction of MRI SNR degradation by using an MR-synchronous low-interference PET acquisition technique. *IEEE Trans. Nucl. Sci.* **62**(3): 634–643.

Gilbert, K. M., T. J. Scholl et al. (2009). Evaluation of a positron emission tomography (PET)-compatible field-cycled MRI (FCMRI) scanner. *Magn. Reson. Med.* **62**(4): 1017–1025.

Grant, A. M. and C. S. Levin. (2012). Optical encoding and multiplexing of PET coincidence events. *IEEE Nucl. Sci. Symp. Med. Imaging Conf. Rec.* **2012**: 3112–3114.

Hammer, B. E. (1996). Magnetic field mapping with an array of nuclear magnetic resonance probes. *Rev. Sci. Instrum.* **67**(6): 2378–2380.

Hammer, B. E., N. L. Christensen et al. (1994). Use of a magnetic field to increase the spatial resolution of positron emission tomography. *Med. Phys.* **21**(12): 1917–1920.

Hammer, B. E. and N. L. Christensen. (1995). Measurement of positron range in matter in strong magnetic fields. *IEEE Trans. Nucl. Sci.* **42**(4): 1371–1376.

Hofmann, M., F. Steinke et al. (2008). MRI-based attenuation correction for PET/MRI: A novel approach combining pattern recognition and atlas registration. *J. Nucl. Med.* **49**(11): 1875–1883.

Hornak, J. P., J. Szumowski et al. (1988). Magnetic field mapping. *Magn. Reson. Med.* **6**(2): 158–163.

Iams, H. and B. Salzberg. (1935). The secondary emission phototube. *Proc. IRE* **23**(1): 55.

Iida, H., I. Kanno et al. (1986). A simulation study of a method to reduce positron annihilation spread distributions using a strong magnetic field in positron emission tomography. *IEEE Trans. Nucl. Sci.* **33**(1): 597–600.

Jackson, J. D. (1998). *Classical Electrodynamics*, 3rd edition. John Wiley & Sons, New York.

Johansson, A., M. Karlsson et al. (2011). CT substitute derived from MRI sequences with ultrashort echo time. *Med. Phys.* **38**(5): 2708–2714.

Kalender, W. A., P. Deak et al. X-ray and x-ray-CT. In: Kiessling, F. and B. J. Pichler. (2011). *Small Animal Imaging: Basics and Practical Guide*. Springer, Berlin, Germany, pp. 125–139.

Keereman, V., Y. Fierens et al. (2010). MRI-based attenuation correction for PET/MRI using ultrashort echo time sequences. *J. Nucl. Med.* **51**(5): 812–818.

Koch, K. M., X. Papademetris et al. (2006). Rapid calculations of susceptibility-induced magnetostatic field perturbations for in vivo magnetic resonance. *Phys. Med. Biol.* **51**(24): 6381–6402.

Kraus, R., G. Delso et al. (2012). Simulation study of tissue-specific positron range correction for the new biograph mMR whole-body PET/MR system. *IEEE Trans. Nucl. Sci.* **59**(5): 1900–1909.

Kucuk, N., M. Cakir et al. (2013). Mass attenuation coefficients, effective atomic numbers and effective electron densities for some polymers. *Radiat. Protect. Dosim.* **153**(1): 127–134.

Le Bihan, D., C. Poupon et al. (2006). Artifacts and pitfalls in diffusion MRI. *J. Magn. Reson. Imaging* **24**(3): 478–488.

Liu, Q., D. G. Hughes et al. (1994). Quantitative characterization of the eddy current fields in a 40-cm bore superconducting magnet. *Magn. Reson. Med.* **31**(1): 73–76.

Martinez-Möller, A., M. Souvatzoglou et al. (2009). Tissue classification as a potential approach for attenuation correction in whole-body PET/MRI: Evaluation with PET/CT data. *J. Nucl. Med.* **50**(4): 520–526.

McClish, M., P. Dokhale et al. (2010). Performance measurements of CMOS position sensitive solid-state photomultipliers. *IEEE Trans. Nucl. Sci.* **57**: 2280–2286.

Mohrig, J. R., C. N. Hammond et al. (2006). *Techniques in Organic Chemistry*, 2nd edition. W.H. Freeman, New York.

Mollet, P., V. Keereman et al. (2012). Simultaneous MR-compatible emission and transmission imaging for PET using time-of-flight information. *IEEE Trans. Med. Imaging* **31**(9): 1734–1742.

Nagy, K., M. Tóth et al. (2013). Performance evaluation of the small-animal nanoScan PET/MRI system. *J. Nucl. Med.* **54**(10): 1825–1832.

Navalpakkam, B. K., H. Braun et al. (2013). Magnetic resonance-based attenuation correction for PET/MR hybrid imaging using continuous valued attenuation maps. *Invest. Radiol.* **48**(5): 323–332.

Ocegueda, K. and A. O. Rodríguez. (2006). A simple method to calculate the signal-to-noise ratio of a circular-shaped coil for MRI. *Concepts Magn. Reson. A* **28A**(6): 422–429.

Paulus, D. H., H. Braun et al. (2012). Simultaneous PET/MR imaging: MR-based attenuation correction of local radiofrequency surface coils. *Med. Phys.* **39**(7): 4306–4315.

Pichler, B. J., M. S. Judenhofer et al. (2006). Performance test of an LSO-APD detector in a 7-T MRI scanner for simultaneous PET/MRI. *J. Nucl. Med.* **47**(4): 639–647.

Raylman, R. R., B. E. Hammer et al. (1996). Combined MRI-PET scanner: A Monte Carlo evaluation of the improvements in PET resolution due to the effects of a static homogeneous magnetic field. *IEEE Trans. Nucl. Sci.* **43**(4): 2406–2412.

Rivlin, M., J. Horev et al. (2013). Molecular imaging of tumors and metastases using chemical exchange saturation transfer (CEST) MRI. *Sci. Rep.* **3**: 3045.

Sacolick, L. I., F. Wiesinger et al. (2010). B$_1$ mapping by Bloch-Siegert shift. *Magn. Reson. Med.* **63**(5): 1315–1322.

Schulz, V., Y. Berker et al. (2013). Sensitivity encoded silicon photomultiplier—A new sensor for high-resolution PET-MRI. *Phys. Med. Biol.* **58**(14): 4733–4748.

Schulz, V., I. Torres-Espallardo et al. (2011). Automatic, three-segment, MR-based attenuation correction for whole-body PET/MR data. *Eur. J. Nucl. Med. Mol. Imaging* **38**(1): 138–152.

Schulz, V., B. Weissler et al. (2011). SiPM based preclinical PET/MR insert for a human 3T MR: First imaging experiments. *IEEE Nucl. Sci. Symp. Med. Imaging Conf. Rec.* **2011**: 4467–4469.

Shao, Y., S. R. Cherry et al. (1997). Simultaneous PET and MR imaging. *Phys. Med. Biol.* **42**(10): 1965–1970.

Sinnott, K. M. (1960). Nuclear magnetic resonance and molecular motion in polymethyl acrylate, polymethyl methacrylate, and polyethyl methacrylate. *J. Polym. Sci.* **42**(139): 3–13.

Soultanidis, G., N. Karakatsanis et al. (2011). Study of the effect of magnetic field in positron range using GATE simulation toolkit. *J. Phys. Conf. Ser.* **317**: 012021.

Spees, W. M., N. Buhl et al. (2011). Quantification and compensation of eddy-current-induced magnetic-field gradients. *J. Magn. Reson.* **212**(1): 116–123.

Stollberger, R. and P. Wach. (1996). Imaging of the active B$_1$ field in vivo. *Magn. Reson. Med.* **35**(2): 246–251.

Terpstra, M., P. M. Andersen et al. (1998). Localized eddy current compensation using quantitative field mapping. *J. Magn. Reson.* **131**(1): 139–143.

Wehrl, H. F., M. S. Judenhofer et al. (2009). Pre-clinical PET/MR: Technological advances and new perspectives in biomedical research. *Eur. J. Nucl. Med. Mol. Imaging* **36**(Suppl. 1): S56–S68.

Weissler, B., P. Gebhardt et al. (2015). A digital preclinical PET/MRI insert and initial results. *IEEE Trans. Med. Imaging* doi: 10.1109/TMI.2015.2427993.

Wirrwar, A., H. Vosberg et al. (1997). 4.5 Tesla magnetic field reduces range of high-energy positrons - potential implications for positron emission tomography. *IEEE Trans. Nucl. Sci.* **44**(2): 184–189.

Yamamoto, S., T. Watabe et al. (2012). Simultaneous imaging using Si-PM-based PET and MRI for development of an integrated PET/MRI system. *Phys. Med. Biol.* **57**(2): N1–N13.

Yao, R. T., J. Seidel et al. (2005). Attenuation correction for the NIH ATLAS small animal PET scanner. *IEEE Trans. Nucl. Sci.* **52**(3): 664–668.

14

Exotic Imaging Approaches

Maria Koutsoupidou and Irene S. Karanasiou

14.1 IMAGING PRINCIPLES

14.1.1 THE NEED FOR NEW MEDICAL IMAGING AND THE ROLE OF TERAHERTZ IMAGING

Medical diagnosis and treatment have been greatly promoted by medical imaging. Specially, medical research has been completely regenerated by the new levels of analysis that medical imaging has introduced, from the study of individual biomolecules microscopically to the actual visualization of living tissues. Additionally, the combination of data by different imaging techniques reveals important information, such as in neuroscience research, the main demand is to bridge biomolecular and cellular functions through anatomic and functional biomedical imaging. Generally, structural imaging providing only static images of tissue and organ states has been enhanced and evolved with functional imaging that provides almost real-time insights of the changes in metabolism, blood flow, and regional chemical composition. In this context, researchers try new methods to discover novel markers to potentially monitor the chemical functions of the cells within the organs and to image the real-time activity of genes and proteins within the cells. Next-generation imaging technologies, as in vivo subsurface imaging of tissue for early detection of disease (Pickwell-MacPherson and Wallace 2009), will combine macroscopic and microscopic imaging advances while providing also functional information.

In this milieu, terahertz (THz) imaging, which is able to provide biomolecular spectral fingerprints (Siegel 2004), is a promising choice for contributing to this effort by enabling the coupling of macroscopic and microscopic imaging while actually leading the ways toward nanoscopy; in parallel, THz imaging enables the acquisition of both structural and functional information from intermolecular interactions (Tonouchi 2007; Qasymeh 2012).

More specifically, biomolecules naturally vibrate at THz frequencies. In other words, each of these weak vibrations that is directly associated to the molecular structure has a unique THz fingerprint (MacPherson 2013). Additionally, hydrogen bonds present significant absorption of THz waves (Cherkasova et al. 2011). As a result, THz radiation is strongly attenuated when it travels through mediums with high concentration in water. On the other hand, from a macroscopic point of view, it can sense small changes in the water content of tissues, such as increased blood flow due to tumor occurrence (MacPherson 2013). Still on a macroscopic level, the structural changes and the disruption that cancer creates at the cellular mesh of a tissue can be detected by THz imaging (Ashworth et al. 2009). On a microscopic level, since these changes on the tissue structure are potentially caused by cross-links among specific proteins (MacPherson 2013), THz spectroscopy can detect the responsible proteins, providing also a biochemical profile based on their functions and interactions.

14.1.2 PROPERTIES OF TERAHERTZ LIGHT AND APPLICATIONS

The THz electromagnetic spectrum lies between the microwave and optical frequencies, typically ranging from 0.1 to 10 THz. In the past decade, there has been a raised interest in THz radiation as its unique properties render THz technology very appealing for numerous applications in various fields, spanning from homeland security and telecommunications to material science and biomedicine (Siegel 2004). THz radiation and its characteristics were well known since the 1890s when a bolometer had been used to measure black body radiation (Nichols 1897; Rubens and Nichols 1897). However, the lack of effective THz sources and detectors and the difficulties to approach this electromagnetic regime with electronic or optical methods left these frequencies unexplored for a long time.

In the field of biomedicine and diagnostic medicine, THz technology is a very promising option for many applications (Figure 14.1): the acquisition of spectral signatures of proteins, DNA molecules, and other biomolecules (He et al. 2006; Parrott et al. 2011); the observation of neurotransmitters associated to various brain functions (Bakopoulos et al. 2009); the identification of changes, abnormalities, and blood volume in biological tissues by in vivo and in vitro imaging (Ferguson et al. 2002; Sy et al. 2010) with significantly enhanced contrast; the early detection of diseases such

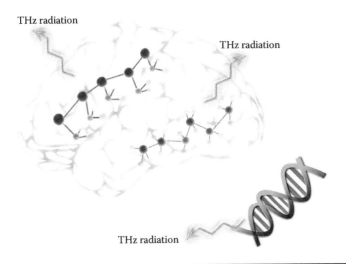

THz radiation

THz radiation

THz radiation

FIGURE 14.1   Large proteins, DNA molecules, brain neurotransmitters, and other biomolecules naturally vibrate at terahertz frequencies, and each has a distinct terahertz "fingerprint."

as dental carries (Pickwell and Wallace 2006); the examination of skin tissue to assess skin burn's magnitude and depth, wound healing and scarring, to determine hydration levels, and to detect subdermal carcinomas and to specify their extent (Cole et al. 2001; Pickwell-MacPherson and Wallace 2009; Yu et al. 2012).

THz radiation has been characterized as "nonionizing," as a THz photon carries approximately 0.4–41 meV, which is very low energy to ionize large biomolecules (Woolard et al. 2005). Moreover, the thermal effects caused by THz radiation are negligible considering that the output power of current THz sources are less than 1 μW (Woolard et al. 2005; MacPherson 2013).

Many materials and chemical substances exhibit characteristic spectral responses at THz frequencies, especially at the 0.5–3.0 THz region (Kawase et al. 2003; Ning et al. 2005). More importantly, as mentioned earlier, intermolecular vibrations and molecular rotations and vibrations of large molecules and biomolecules lie at THz regime providing unique THz "fingerprints" (Parrott et al. 2011). Generally, metals are opaque to THz waves; polar liquids (i.e., water) highly absorb them while most materials (i.e., ceramics, paper, wood, clothing) are almost transparent to THz radiation or it presents relatively little attenuation through them (Masson et al. 2006b; Lee 2009).

14.2   METHODS AND TECHNOLOGY

As THz regime lies between microwave and optical frequencies, both electronic and photonic technologies have contributed for generating and detecting THz radiation: solid state electronics, such as amplifiers and multipliers, with a chain of diodes (i.e., Gunn diode) have been used to reach frequencies from 200 GHz to 1 THz (Gallerano et al. 2004); on the other hand, photonics offered quantum cascade lasers that generate high frequencies over 2 THz (Faist et al. 1994; Williams et al. 2003; Gallerano et al. 2004; Valavanisa et al. 2013).

The two most common laser-based techniques for the generation and detection of THz waves rely on the nonlinear interaction of an optical signal with an appropriate material (Rogalski and Sizov 2011). The first is the photoconductive generation and detection (i.e., photoconductive antennas), where the gap of two parallel metals on a semiconductor, as silicon-on-sapphire or GaAs, biased by a DC voltage, is excited with a femtosecond laser pulse. The created transient current radiates at the THz range (Shan and Heinz 2004; Rogalski and Sizov 2011). The output frequency depends on the carrier recombination time of the semiconductor and the bandwidth of the metallic stripline (Gallerano et al. 2004). On the other hand, the generation of THz pulses through the electro-optic effect does not require photon absorption (Jiang and Zhang 1999). A transient polarization is produced when an ultrashort optical pulse comes through a medium, such as in the case of an electro-optic crystal like ZnTe, with a large second-order nonlinear susceptibility (Jiang and Zhang 1999; Dexheimer 2007). Detecting THz radiation with the same method (electro-optic sampling) is possible with the presence of optically induced carriers in the material (Dexheimer 2007).

A different way to detect a THz signals is to downconvert it to intermediate frequency and then measure it (Rogalski and Sizov 2011). Both signal phase and amplitude are preserved from the conversion and lately THz heterodyne detection is commonly used for high-resolution spectroscopy and imaging (Siegel and Dengler 2006). Some of the devices used for heterodyne detections systems are Schottky barrier diodes, tunnel junctions mixers, hot electron bolometers, and superlattices (Rogalski and Sizov 2011). On the contrary, direct or incoherent detection only measure the amplitude of the THz signal and includes Golay cells, pyroelectric detectors, and other thermal detectors, that is, bolometers (Dobroiu et al. 2004; Karpowicz et al. 2005a; Hammar et al. 2011; Rogalski and Sizov 2011).

14.2.1   THz System Types

THz systems can be divided into two main categories based on the THz source used: THz pulsed and continuous wave (CW) systems. CW techniques offer high spectral resolution at the order of a few MHz. On the other hand, pulsed systems are used when the application demands broad spectral information (Karpowicz et al. 2005b). Pulsed systems can measure THz waves in the time domain, providing amplitude and phase information. This important characteristic renders pulsed THz systems useful for numerous applications.

With a few exception (Nahata et al. 2002; Loffler et al. 2007), coherent detection is not possible for CW systems, as the THz source is not phase-locked to a laser source that is used for sampling (Chan et al. 2007). As a result, only the intensity of the radiation is measured. Thus, the system structure is compact and inexpensive, as it can be solely constructed by laser diodes and the postdetection process is quite simple (Sun et al. 2011a). Constant wave THz systems can be tuned at a narrow spectrum and the CW measurements present low signal-to-noise ratio (Smith and Arnold 2011).

14.2.2   Imaging and Spectroscopy

The advancements in the development of THz systems have allowed their implementation in practical applications and occasionally in commercial use (TeraView). Relatively to the type of the system used, THz imaging can be also divided into CW imaging and pulsed imaging or THz time domain spectroscopy. The combination of these techniques in transmission and reflection geometries provides different kind of information for the probed specimen.

However, for measuring biological samples with THz radiation, more aspects should be considered except the THz imaging technique that will be used. One of the most challenging problems that has arisen, is the biosample preparation and maintenance of a concrete experimental protocol (MacPherson 2013) that will allow an efficient comparison among the results and a safe extraction of conclusions.

14.2.2.1   Terahertz Time-Domain Spectroscopy

In the THz time-domain spectroscopy (THz-TDS), the examined sample is probed with short THz pulses and the detection scheme measures both the amplitude and the phase of the received radiation. Thus, more information is acquired regarding the spectral properties and thickness of the sample (McClatchey et al. 2002). The electro-optic effect and the photoconductive generation and detection are the most commonly employed methods for THz-TDS (Chan et al. 2007). THz 2D pulsed imaging is performed as an extension of THz-TDS, when the generated beam spatially scans the sample. Additionally, three-dimensional (3D) images can be obtained if the radiated beam focuses in different layers of the sample. Also, the development of efficient THz systems has led THz time domain techniques combined with advanced signal processing to be used in real-time imaging (Mittleman et al. 1996).

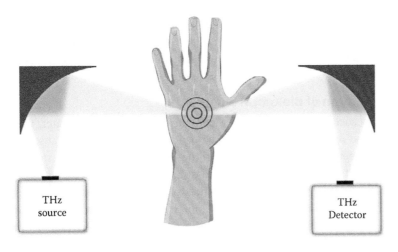

FIGURE 14.2   Block diagram of a THz imaging system with a reflection geometry for detection of skin burns and carcinomas.

A THz-TDS spectrometer functions efficiently at room temperature, as these methods do not detect thermal radiation. Additionally, THz-TDS exhibits a broad bandwidth of operation, larger than an order of magnitude in wavelength, which is a very important quality of spectrometers (Sun et al. 2011a). Also, when a wide bandwidth is used for spectroscopy, the created image is of high quality, as the scattered wavelets from the rough surface are strongly suppressed (Karpowicz et al. 2005b).

14.2.2.2   Time of Flight Technique

Generally, as most materials are transparent to THz radiation, the transmission geometry is usually applied on THz systems. However, in many cases, where the examined sample is opaque to THz waves, that is, metals and living tissue because of the high water content, or it is composed of different layers of materials, a reflection geometry is preferred (Chan et al. 2007).

Time of flight technique for imaging requires a reflection geometry (Figure 14.2) and it is combined with pulsed THz systems, which can provide wide bandwidth: the incident THz pulse is reflected on each layer of the targeted sample; each reflected signal reaches the detector with a different time delay. This method allows the imaging of each layer of the sample and it is used for 3D imaging and THz tomography (Chan et al. 2007; Takayanagi et al. 2009).

14.2.2.3   CW Terahertz Imaging

For THz imaging performed by CW systems, an incoherent detector is usually used. This can be a Golay cell, a bolometer, or an array of detectors such as an array of microbolometers and pyroelectric cameras (Miller et al. 2004). In case that the source is tunable over a broad spectrum, the imaging results can be combined to spectroscopic information of the sample (Chan et al. 2007). Moreover, a single frequency can be selected in order to improve the quality of the imaging: for example, for imaging a biosample in vivo, a wavelength that is not significantly absorbed by water is chosen.

14.2.2.4   Terahertz Near-Field Imaging

The aforementioned THz imaging techniques suffer from limited spatial resolution, which is comparable to a wavelength—objects smaller than a wavelength are not visible with this techniques. In order for the resolution to drop below the diffraction limit, propagating and evanescent waves must be collected by the detector. However, the probe must be located close to the sample, within a wavelength distance, in order to measure the evanescent waves and, as a result the detection takes place in the near field. This technique is combined with a sub-wavelength aperture for the detector's probe and even a dielectric protrusion to improve the aperture's transmission of the evanescent waves. At THz near-field imaging, the resolution depends on the probe's aperture that can be much smaller than the wavelength and even a spatial resolution of $\lambda/200$ has been achieved

(Mitrofanov et al. 2001). Ion flow in heart tissue of a frog has been studied with near-field imaging with small probe aperture (Masson et al. 2006a). However, this method is limited by the power of the radiation transmitted through the aperture that decreases quickly with the aperture's diameter (Hunsche et al. 1998).

14.2.2.5 Preparation of Biological Samples

Biological samples, such as tissues and cells, contain concentrations of water, while biomolecule samples are obtained in liquid form. As already mentioned, water strongly absorbs THz radiation and THz waves can detect even small changes in the water content of the biosamples (MacPherson 2013). In case of in vitro measurements, where the sample has been excised, it is dehydrated if not preserved correctly. In order to prevent dehydration and resemble in vivo conditions, freeze drying (lyophilization) should be applied on fresh thin samples. Another method for preparing biomolecules in solution is the membrane method: the sample is inserted and dried on a polymer membrane filter. Moreover, for comparing the measurements of different biosamples, they must all have been prepared and stored identically and exposed at the same environmental conditions, such as temperature and humidity (MacPherson 2013).

14.3 CURRENT APPLICATIONS

Biological and biomedical applications of THz technology have significantly progressed over the last decade attracting considerable attention (Siegel 2004; MacPherson 2013). The nonionizing and safe, for research even at cellular level, characteristics of THz radiation renders this part of the electromagnetic spectrum an advantageous solution for imaging. Given also the high sensitivity to absorption of water molecules at THz frequencies, THz waves have long been advocated to have a valuable potential in biomedical imaging (Siegel 2004; Pickwell and Wallace 2006; Arbab et al. 2011). In order to guide THz biomedical applications toward actual clinical viable solutions of THz imaging, significant research effort is mainly focused along two main pathways:

1. An issue that still remains is the definition of the optical properties of biomedical samples and tissues at THz frequencies; although differences between tissues have already been demonstrated, there is still only limited information available about the human tissue properties in the THz regime. For example, Fitzgerald et al. showed that distinct human tissues differ in their optical properties at THz frequencies, including, for example, striated muscle, adipose tissue, and skin (Fitzgerald et al. 2003; Siegel 2004; MacPherson 2013). Also, Pickwell and colleagues found that a double Debye model, based principally on the effective water content of different skin layers, may be used to estimate the THz reflection response of healthy human skin (Pickwell et al. 2004a,b; Arbab et al. 2011).

2. In parallel, a number of THz imaging systems have already been developed and used in a variety of imaging applications spanning from skin wound and cancer imaging, dental carries imaging, breast to colon and liver imaging, and biomolecule and pharmaceutical imaging, based mainly on differential measurements (Pickwell-MacPherson and Wallace 2009; Pickwell-MacPherson 2010).

In this section, the main clinical applications of THz imaging focusing also on small animal THz imaging will be discussed.

14.3.1 Skin Burn Imaging

THz imaging has been strongly proposed as a potential clinical technique for skin imaging and especially for burn imaging being among the first biomedical applications of THz waves (Mittleman et al. 1999; Pickwell-MacPherson and Wallace 2009). Skin is easily accessible and reflection geometry in vivo THz measurements can be made to determine information about the skin such as its moisture content (Huang et al. 2009b) or thickness (Huang et al. 2009a). The tissue with severe burned wounds that was first chosen for THz imaging was chicken muscle tissue (Mittleman et al. 1999; Arbab et al. 2011). Recently, large differences in THz reflectivity between normal and second-degree burns on excised porcine skin were detected (Taylorh et al. 2008). The strong sensitivity of water absorption in this frequency regime was once more proven valuable; water removal from the burned parts of the excised skin samples resulted in lower THz reflectivity compared to those of healthy tissue (Taylor et al. 2008).

Human skin in vivo studies have also been already reported (Pickwell and Wallace 2006). The first in vivo skin measurements were carried out by Cole et al. (2001), who used a THz imaging setup in reflection mode to measure changes in skin hydration caused by occlusion. Also in another recent in vivo study (Pickwell et al. 2004a), THz signals reflected from the palm of the hand were measured to quantify the thickness of the outermost skin layer (Pickwell and Wallace 2006).

14.3.2   TEETH IMAGING

THz imaging has been recently proven as a viable solution in dentistry to cover the void of the necessity of a nonionizing imaging modality to detect early tooth decay (Figure 14.3). The first efforts toward this direction were performed by Crawley et al. who demonstrated the ability of THz light to detect regions of teeth caries in excised teeth samples (Crawley et al. 2003a,b; MacPherson 2013). Follow-up studies correlated the THz refractive index profile with mineral content lost as a function of depth into the enamel (Pickwell et al. 2006). THz imaging has been applied not only for decay monitoring but also for monitoring the curing process of dental composites, where with the application of THz, time-domain spectroscopy may detect significant changes of the THz dielectric parameters during stepwise therapy (Schwerdtfeger et al. 2012).

14.3.3   CANCER DETECTION

14.3.3.1   Skin Cancer

Imaging systems for skin condition diagnosis with the exception of infrared imaging for melanoma detection are not very common in clinical practice. The main form of diagnosis is visual inspection and biopsy, which is time consuming in general. THz waves, due to the wavelength at these frequencies, can penetrate the outermost layer of the skin and, therefore, any lesion may be observed beneath the skin surface. Differences in the fundamental THz properties of skin may produce valuable

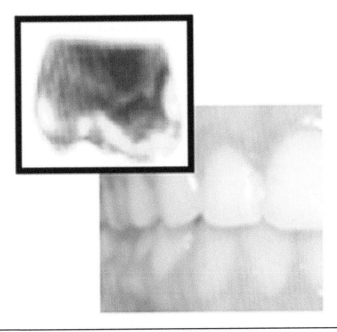

FIGURE 14.3   THz image of a tooth revealing cavities and lesions.

contrast in THz images of skin cancer providing real-time quantitative diagnosis of various skin conditions (Pickwell-MacPherson and Wallace 2009; MacPherson 2013). The main research efforts so far include comparison of THz images of skin cancer (basal cell carcinoma, BCC) and healthy tissue (Wallace et al. 2004). Differences in refractive index and absorption coefficient of BCC have been detected and larger differences were found compared to healthy skin from the same patient (Wallace et al. 2006). The underlying mechanisms causing the observed contrast are in parallel under investigation using numerical analysis; it has been verified though that once again changes in water content and tissue structure attributed to tumor characteristics (blood supply by increased vascularity) cause changes in THz imaging coefficients (Pickwell et al. 2004b, 2005). Research is ongoing in order to reveal the feasibility of THz imaging as a technique to classify various types of skin cancer into cancerous and noncancerous lesions.

14.3.3.2 Breast Cancer

Mammography and ultrasound are well established and successfully used in clinically screening breast cancer (Figure 14.4a). Even microwave techniques have been used as nonionizing methods complementary to these techniques. Nevertheless, noncalcified tumors are nonpalpable and consequently often missed intraoperatively while they are also not well detected by x-rays. THz imaging could be a useful technique for clinical practice during surgery in such cases (Pickwell-MacPherson and Wallace 2009).

This is particularly significant because such tumors are often missed during breast surgery as they are nonpalpable and do not show up on x-ray images. Because of this finding, researchers are keen to develop the technique such that THz imaging can be used by clinicians. To determine the feasibility of such a technique, the main THz properties, namely the absorption coefficient and refractive index of breast tissue, have been measured in transmission spectroscopy (Figure 14.4b) (Ashworth et al. 2009), while ex vivo studies of breast cancer show the ability of THz imaging to detect tumors of the aforementioned noncalcified form (Fitzgerald et al. 2006). The results are promising showing both good spatial resolution and detection sensitivity (Pickwell-MacPherson and Wallace 2009; MacPherson 2013; Peter et al. 2013).

14.3.4 PHARMACEUTICALS, CHEMICALS, AND BIOMOLECULE IMAGING

As mentioned earlier, THz is sensitive to interactions between molecules and to small changes in molecular structure and concentrations of proteins. Recent advances in THz technology improved the sensitivity of THz spectroscopy leading the way toward THz chemical imaging. It allows molecular recognition based on the spectroscopy of molecular networks providing in parallel dynamic functional information. THz chemical imaging can reveal hydrogen bond distributions and has been used in a variety of applications in this area showing significant potential (Ajito and Ueno 2011; Sun et al. 2011b). Several studies have been reported using THz spectroscopy for identifying intermolecular hydrogen bonds in biological samples, such as organic acids (Ueno and Ajito 2007),

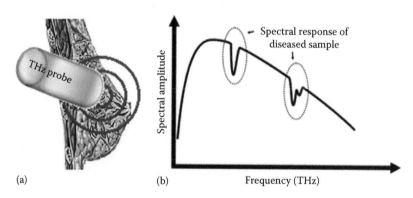

(a) (b)

FIGURE 14.4 (a) Breast cancer imaging and (b) spectral response of a diseased sample detected with THz spectroscopy.

amino acids (Korter et al. 2006; Rungsawang et al. 2006a), sugars (Rungsawang et al. 2006b), pharmaceuticals (Ajito et al. 2011), polypeptides (Yamamoto et al. 2005; Plusquellic et al. 2007), DNA (Markelz et al. 2000; Luo et al. 2006), proteins (Ebbinghaus et al. 2008), and cancer cells (Ashworth et al. 2009; Oh et al. 2009). Amino acids and pharmaceuticals have hydrogen bonds, which have specific fingerprints in THz spectra that enable also quantitative analyses (Ueno et al. 2006; Nguyen et al. 2007; Ajito and Ueno 2011; Sun et al. 2011b).

14.3.5 POTENTIAL OF THz IMAGING IN BRAIN IMAGING APPLICATIONS

In medical research and especially in neuroscience research, current and future trends aim at connecting anatomic and function biomedical imaging in order to combine biomolecular information and neural function. Bridging these fields is possible by the rapidly advancing research of neuroimaging, molecular medicine, and genetics. In this milieu, research effort aims to discover markers related to various psychiatric and neurological conditions and simultaneously using cutting-edge imaging technologies to real-time monitor inter- and intracellular chemical functions of genes and proteins.

In this context, THz technology may add significant knowledge to the understanding of brain function in health and disease by providing biochemical profiling of various neurotransmitters in various conditions. For example, recent findings suggest that a distress in the equilibrium of different excitatory and inhibitory neurotransmitters may be central to the mechanisms of bipolar disorder (Karanasiou and Uzunoglu 2004). Recent findings support the aforementioned claims; in a novel study, healthy and diseased snap-frozen tissue samples obtained from three regions of the human brain were distinguished using THz spectroscopy (Karanasiou and Uzunoglu 2004). Real-time THz spectroscopy was used to detect biomolecule processes associated with neurodegenerative phenomena (Karanasiou et al. 2008; Karathanasis et al. 2010). Also, recently it has been shown that THz spectroscopy can be used to differentiate soft protein microstructures that highly interest medical researchers since they form from naturally occurring proteins suggested to be involved in several human diseases, such as Alzheimer's disease (Karanasiou et al. 2008).

14.3.6 SMALL ANIMAL IN VIVO STUDIES

Several studies have been reported regarding in vivo measurements using THz waves in animal models. In several studies, the novel methodology was compared with well-standardized or more conventional ones such as mammography, near-infrared imaging photoacoustic, laser Doppler imaging, and thermal imaging. THz molecular imaging by using nanoparticle probes was compared with respective results from conventional near-infrared (NIR) absorption imaging (Oh et al. 2011) using a mouse model. The mouse was initially injected with nanoprobes that were exclusively delivered to an epidermal growth factor receptor carcinoma cell. In vivo images were acquired with both THz and near-infrared imaging (Figure 14.5). Due to its low sensitivity, the latter was unable to exploit the enhancement by nanoprobes for detection of the small tumor due to its low sensitivity, whereas THz imaging enabled classification of targeted nanoprobes quantitatively (Oh et al. 2011).

In a subcutaneous xenograft mouse study, T-ray breast imaging was introduced for the in vivo early detection of breast cancer (Chen et al. 2011). THz imaging exhibited the potential of detecting the early cancer before it could be sensed or imaged by other modalities due to the high THz absorption contrast between fatty and cancerous tissues.

As also mentioned in the previous sections, one of the most promising clinical applications of THz imaging is skin imaging either for cancer diagnosis or burning and wound characterization (Arbab et al. 2011; Tewari et al. 2012). THz measurements were performed on intensely burned skin of rat models. THz reflectivity at a frequency range 0.5–0.7 THz of second-degree burns was statistically significant higher than that of normal skin (Arbab et al. 2011). Due to its high sensitivity to water concentration and relative robustness to surface scatter, THz imaging may be able to delineate between different wound zones and help differentiate between viable and unviable tissues (Tewari et al. 2012).

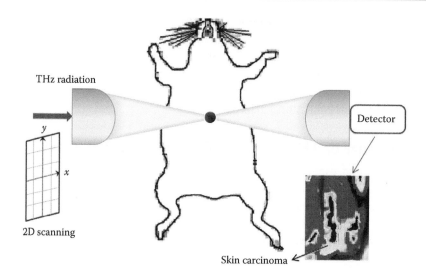

FIGURE 14.5 2D THz imaging in reflection geometry for skin cancer detection in mouse.

In vivo mouse models were also used during THz spectroscopy measurements in order to visualize and quantify topical transdermal drug delivery (Kim et al. 2012). Higher contrast was observed in drug-applied sites and unapplied skin in THz reflection images compared to that of visual inspection and images.

14.4 CONCLUSIONS AND FUTURE PROSPECTS

THz imaging and spectroscopy are employed in a wide range of scientific fields: biomedicine and diagnostic medicine, biology, telecommunication, material science, chemistry, astronomy. However, faster and more affordable THz systems are demanded so that these techniques be practically implemented in commercial use. New approaches are required in every step of generation and detection of THz radiation, such as the exclusion of the femtosecond laser from the THz source (Jansen et al. 2010).

More specifically, for establishing THz imaging and spectroscopy as reliable biomedical imaging modality, further developments should be performed and ambiguous points should be clarified. The contrast of THz biomedical images in many cases, such as in skin and breast cancer, is due to the changes in the tissue's optical properties. However, it is not certain whether the optical changes, because of increased water, are related to a disease (Pickwell-MacPherson and Wallace 2009). Additionally, although until now water content has been considered the main factor for the contrast mechanism in THz imaging, current research indicates that structural and biomolecular changes within the tissues may have an effect on the contrast of the resulting THz image (Parrott et al. 2011).

Additionally, presently, in vivo THz imaging is possible only for skin tissue (Pickwell-MacPherson and Wallace 2009), as the penetration depth of THz radiation is very limited. Further improvements should be performed for extending this method to different tissue types. Also, endoscopic probing with THz imaging is a promising potential. The aqueous environment of biomolecules for in vivo measurements is also a great challenge for THz imaging and spectroscopy because of the high absorption of the THz waves by water molecules (Parrott et al. 2011). As a result, THz sources with greater power output and more sensitive detectors are demanded.

The main reason for the difficulties to understand the interaction of THz radiation with biological samples is that the vibrational modes of the biomolecules cannot be easily distinct, because of the overlapping between their energy levels (Parrott et al. 2011). A significant contribution of the research of biomolecular imaging with THz radiation is the thorough study of their energy states and how these change while they undergo structural and conformational changes (Parrott et al. 2011). Computation modeling would give a first insight in this field before the experimental validation (Parrott et al. 2011).

An important area of research of biomedicine and biotechnology is the study of the clustering networks of biological molecules in high water content environments on which medication molecules are attached (Ajito and Ueno 2011). The molecular networks have been created because of the hydrogen bonding among biomolecules, such as single molecules in water, organic crystals, high-order protein structures, and DNA double strands. The study of these molecular networks is very important for the better understanding of the protein structure and the effectiveness of the medication. These bonds between the biomolecules and the medication molecules of the cluster vibrate at THz frequencies, and THz imaging and spectroscopy may aid to reveal the characteristics and functions of the complicated molecular structures (Ajito and Ueno 2011).

The aforementioned aspects are some important key points to be considered for establishing THz imaging and spectroscopy as a vital and efficient biomedical imaging modality. However, if these difficulties are surpassed, THz imaging will revolutionize biomedicine and will be an invaluable tool for diagnostic medicine.

REFERENCES

Ajito, K. and Y. Ueno. (2011). THz chemical imaging for biological applications. *IEEE Trans. THz Sci. Technol.* **1**(1): 293–300.

Ajito, K., Y. Ueno et al. (2011). Terahertz spectroscopic imaging of polymorphic forms in pharmaceutical crystals. *Mol. Cryst. Liq. Cryst.* **538**(1): 33–38.

Arbab, M. H., T. C. Dickey et al. (2011). Terahertz reflectometry of burn wounds in a rat model. *Biomed. Opt. Express* **2**(8): 2339–2347.

Ashworth, P. C., E. Pickwell-MacPherson et al. (2009). Terahertz pulsed spectroscopy of freshly excised human breast cancer. *Opt. Express* **17**(15): 12444–12454.

Bakopoulos, P., I. Karanasiou et al. (2009). A tunable continuous wave (CW) and short-pulse optical source for THz brain imaging applications. *Meas. Sci. Technol.* **20**(10).

Chan, W. L., J. Deibel et al. (2007). Imaging with terahertz radiation. *Rep. Prog.Phys.* **70**(8): 1325.

Chen, H., T.-H. Chen et al. (2011). High-sensitivity in vivo THz transmission imaging of early human breast cancer in a subcutaneous xenograft mouse model. *Opt. Express* **19**(22): 21552–21562.

Cherkasova, O. P., A. V. Kargovsky et al. (2011). The effect of the nature of hydrogen bonding on THz and Raman spectra of cyclopentaphenanthrene derivatives. In *2011 International Conference on Infrared, Millimeter, and Terahertz Waves*.

Cole, B. E., R. M. Woodward et al. (2001). Terahertz imaging and spectroscopy of human skin in vivo. *Int. Soc. Opt. Photon.* **4276**: 1–10.

Crawley, D., C. Longbottom et al. (2003a). Three-dimensional terahertz pulse imaging of dental tissue. *J. Biomed. Opt.* **8**(2): 303–307.

Crawley, D. A., C. Longbottom et al. (2003b). Terahertz pulse imaging: A pilot study of potential applications in dentistry. *Caries Res.* **37**(5): 352–359.

Dexheimer, S. L. (2007). *Terahertz Spectroscopy: Principles and Applications.* CRC Press, Boca Raton, FL.

Dobroiu, A., M. Yamashita et al. (2004). Terahertz imaging system based on a backward-wave oscillator. *Appl. Opt.* **43**(30): 5637–5646.

Ebbinghaus, S., S. J. Kim et al. (2008). Protein sequence-and pH-dependent hydration probed by terahertz spectroscopy. *J. Am. Chem. Soc.* **130**(8): 2374–2375.

Faist, J., F. Capasso et al. (1994). Quantum cascade laser. *Science* **264**(5158): 553–556.

Ferguson, B., S. Wang et al. (2002). Identification of biological tissue using chirped probe THz imaging. *Microelectron. J.* **33**(12): 1043–1051.

Fitzgerald, A. J., E. Berry et al. (2003). Catalogue of human tissue optical properties at terahertz frequencies. *J. Biol. Phys.* **29**(2–3): 123–128.

Fitzgerald, A. J., V. P. Wallace et al. (2006). Terahertz pulsed imaging of human breast tumors. *Radiol. Soc. North Am.* **239**(2): 533–540.

Gallerano, G. P., S. Biedron et al. (2004). Overview of terahertz radiation sources. In *Proceedings of the 2004 FEL Conference*, Trieste, Italy, pp. 216–221

Hammar, A., S. Cherednichenko et al. (2011). Terahertz direct detection in microbolometers. *IEEE Trans. THz Sci. Technol.* **1**(2): 390–394.

He, M., A. K. Azad et al. (2006). Far-infrared signature of animal tissues characterized by terahertz time-domain spectroscopy. *Opt. Commun.* **259**(1): 389–392.

Huang, S., P. C. Ashworth et al. (2009a). Improved sample characterization in terahertz reflection imaging and spectroscopy. *Opt. Express* **17**(5): 3848–3854.

Huang, S. Y., Y. X. J. Wang et al. (2009b). Tissue characterization using terahertz pulsed imaging in reflection geometry. *Phys. Med. Biol.* **54**(1): 149–160.

Hunsche, S., M. Koch et al. (1998). THz near-field imaging. *Opt. Commun.* **150**(1): 22–26.

Jansen, C., S. Wietzke et al. (2010). Terahertz imaging: Applications and perspectives. *Appl. Opt.* **49**(19): E48–E57.

Jiang, Z. and X.-C. Zhang. (1999). Terahertz imaging via electrooptic effect. *IEEE Trans. Microw. Theor. Tech.* **47**(12): 2644–2650.

Karanasiou, I. S., K. T. Karathanasis et al. (2008). Development and laboratory testing of a noninvasive intracranial focused hyperthermia system. *IEEE Trans. Microw. Theor. Tech.* **56**(9): 2160–2171.

Karanasiou, I. S. and N. K. Uzunoglu. (2004). Experimental study of 3D contactless conductivity detection using microwave radiometry: A possible method for investigation of brain conductivity fluctuations. *Conf. Proc. IEEE Eng. Med. Biol. Soc.* **3**: 2303–2306.

Karathanasis, K. T., I. A. Gouzouasis et al. (2010). Noninvasive focused monitoring and irradiation of head tissue phantoms at microwave frequencies. *IEEE Trans. Inf. Technol. Biomed.* **14**(3): 657–663.

Karpowicz, N., H. Zhong et al. (2005a). Non-destructive sub-THz CW imaging. *Int. Soc. Opt. Photon.* **5727**: 132–142.

Karpowicz, N., H. Zhong et al. (2005b). Comparison between pulsed terahertz time-domain imaging and continuous wave terahertz imaging. *Semicond. Sci. Technol.* **20**(7): S293–S299.

Kawase, K., Y. Ogawa et al. (2003). Non-destructive terahertz imaging of illicit drugs using spectral fingerprints. *Opt. Express* **11**(20): 2549–2554.

Kim, K. W., H. Kim et al. (2012). Terahertz tomographic imaging of transdermal drug delivery. *IEEE Trans. THz Sci. Technol.* **2**(1): 99–106.

Korter, T. M., R. Balu et al. (2006). Terahertz spectroscopy of solid serine and cysteine. *Chem. Phys. Lett.* **418**(1): 65–70.

Lee, Y.-S. (2009). *Principles of Terahertz Science and Technology: Proceedings of the International Conference, Held in Mainz, Germany, June 5–9, 1979.* Springer, New York.

Loffler, T., T. May et al. (2007). Continuous-wave terahertz imaging with a hybrid system. *Appl. Phys. Lett.* **90**(9): 091111–091111.

Luo, Y., B. Gelmont et al. (2006). Bio-molecular devices for terahertz frequency sensing. *Mol. Nano Electron.* **2**: 55–81.

MacPherson, E. (2013). Biomedical imaging. In K.-E. Peiponen, A. Zeitler, and M. Kuwata-Gonokami (eds.), *Terahertz Spectroscopy and Imaging.* Springer, Berlin, Germany, pp. 415–431.

Markelz, A. G., A. Roitberg et al. (2000). Pulsed terahertz spectroscopy of DNA, bovine serum albumin and collagen between 0.1 and 2.0 THz. *Chem. Phys. Lett.* **320**(1): 42–48.

Masson, J.-B., M.-P. Sauviat et al. (2006a). Ionic contrast terahertz time resolved imaging of frog auricular heart muscle electrical activity. *Appl. Phys. Lett.* **89**(15): 153904.

Masson, J.-B., M.-P. Sauviat et al. (2006b). Ionic contrast terahertz near-field imaging of axonal water fluxes. *Proc. Natl. Acad. Sci. USA* **103**(13): 4808–4812.

McClatchey, K., M. T. Reiten et al. (2002). Time resolved synthetic aperture terahertz impulse imaging. *Appl. Phys. Lett.* **79**(27): 4485–4487.

Miller, A. J., A. Luukanen et al. (2004). Micromachined antenna-coupled uncooled microbolometers for terahertz imaging arrays. *Int. Soc. Opt. Photon.* **5411**: 18–24.

Mitrofanov, O., M. Lee et al. (2001). Collection-mode near-field imaging with 0.5-THz pulses. *IEEE J. Sel. Top. Quant. Electron.* **7**(4): 600–607.

Mittleman, D. M., M. Gupta et al. (1999). Recent advances in terahertz imaging. *Appl. Phys. B* **68**(6): 1085–1094.

Mittleman, D. M., R. H. Jacobsen et al. (1996). T-ray imaging. *IEEE J. Sel. Top. Quant. Electron.* **2**(3): 679–692.

Nahata, A., J. T. Yardley et al. (2002). Two-dimensional imaging of continuous-wave terahertz radiation using electro-optic detection. *Appl. Phys. Lett.* **81**(6): 963–965.

Nguyen, K. L., T. Friščić et al. (2007). Terahertz time-domain spectroscopy and the quantitative monitoring of mechanochemical cocrystal formation. *Nat. Mater.* **6**(3): 206–209.

Nichols, E. F. (1897). A method for energy measurements in the infra-red spectrum and the properties of the ordinary ray in quartz for waves of great wave length. *Phys. Rev. (Series I)* **4**: 297.

Ning, L., J. Shen et al. (2005). Study on the THz spectrum of methamphetamine. *Opt. Express* **13**(18): 6750–6755.

Oh, S. J., J. Choi et al. (2011). Molecular imaging with terahertz waves. *Opt. Express* **19**(5): 4009–4016.

Oh, S. J., J. Kang et al. (2009). Nanoparticle-enabled terahertz imaging for cancer diagnosis. *Opt. Express* **17**(5): 3469–3475.

Parrott, E. P. J., Y. Sun et al. (2011). Terahertz spectroscopy: Its future role in medical diagnoses. *J. Mol. Struct.* **1006**(1): 66–76.

Peter, St. B., S. Yngvesson et al. (2013). Development and testing of a single frequency terahertz imaging system for breast cancer detection. *IEEE J. Biomed. Health Inform.* **17**(4): 785–797.

Pickwell, E., B. E. Cole et al. (2004a). In vivo study of human skin using pulsed terahertz radiation. *Phys. Med. Biol.* **49**(9): 1595–1607.

Pickwell, E., B. E. Cole et al. (2004b). Simulation of terahertz pulse propagation in biological systems. *Appl. Phys. Lett.* **84**(12): 2190–2192.

Pickwell, E., A. J. Fitzgerald et al. (2005). Simulating the response of terahertz radiation to basal cell carcinoma using ex vivo spectroscopy measurements. *J. Biomed. Opt.* **10**(6): 064021–064021.

Pickwell, E. and V. P. Wallace. (2006). Biomedical applications of terahertz technology. *J. Phys. D Appl. Phys.* **39**(17): R301–R310.

Pickwell, E., V. P. Wallace et al. (2006). A comparison of terahertz pulsed imaging with transmission micro-radiography for depth measurement of enamel demineralisation in vitro. *Caries Res.* **41**(1): 49–55.

Pickwell-MacPherson, E. (2010). Practical considerations for in vivo THz imaging. *THz Sci. Technol.* **3**: 163–171.

Pickwell-MacPherson, E. and V. P. Wallace. (2009). Terahertz pulsed imaging—A potential medical imaging modality? *Photodiagn. Photodyn. Therap.* **6**(2): 128–134.

Plusquellic, D. F., K. Siegrist et al. (2007). Applications of terahertz spectroscopy in biosystems. *ChemPhysChem* **8**(17): 2412–2431.

Qasymeh, M. (2012). Terahertz generation in an electrically biased optical fiber: A theoretical investigation. *Int. J. Opt.* **2012**: 1–6.

Rogalski, A. and F. Sizov. (2011). Terahertz detectors and focal plane arrays. *Opto-Electron. Rev.* **19**(3): 346–404.

Rubens, H. and E. F. Nichols. (1897). Heat rays of great wave length. *Phys. Rev. (Series I)* **4**(4): 314–323.

Rungsawang, R., Y. Ueno et al. (2006a). Angle-dependent terahertz time-domain spectroscopy of amino acid single crystals. *J. Phys. Chem. B* **110**(42): 21259–21263.

Rungsawang, R., Y. Ueno et al. (2006b). Terahertz notch filter using intermolecular hydrogen bonds in a sucrose crystal. *Opt. Express* **14**(12): 5765–5772.

Schwerdtfeger, M., S. Lippert et al. (2012). Terahertz time-domain spectroscopy for monitoring the curing of dental composites. *Biomed. Opt. Express* **3**(11): 2842–2850.

Shan, J. and T. F. Heinz. (2004). Terahertz radiation from semiconductors. In K. T. Tsen (ed.), *Ultrafast Dynamical Processes in Semiconductors.* Springer, Berlin, Germany, pp. 1–56.

Siegel, P. H. (2004). Terahertz technology in biology and medicine. *Int. Microw. Symp.* **3**: 1575–1578.

Siegel, P. H. and R. J. Dengler. (2006). Terahertz heterodyne imaging. Part I: Introduction and techniques. *Int. J. Infrared Millimeter Waves* **27**(4): 465–480.

Smith, R. M. and M. A. Arnold. (2011). Terahertz time-domain spectroscopy of solid samples: Principles, applications, and challenges. *Appl. Spectros. Rev.* **46**(8): 636–679.

Sun, Y., M. Y. Sy et al. (2011a). A promising diagnostic method: Terahertz pulsed imaging and spectroscopy. *World J. Radiol.* **3**(3): 55–65.

Sun, Y., Y. Zhang et al. (2011b). Investigating antibody interactions with a polar liquid using terahertz pulsed spectroscopy. *Biophys. J.* **100**(1): 225–231.

Sy, S., S. Huang et al. (2010). Terahertz spectroscopy of liver cirrhosis: Investigating the origin of contrast. *Phys. Med. Biol.* **55**(24): 7587–7596.

Takayanagi, J., H. Jinno et al. (2009). High-resolution time-of-flight terahertz tomography using a femtosecond fiber laser. *Opt. Express* **17**(9): 7533–7539.

Taylor, Z. D., R. S. Singh et al. (2008). Reflective terahertz imaging of porcine skin burns. *Opt. Lett.* **33**(11): 1258–1260.

Tewari, P., C. P. Kealey et al. (2012). In vivo terahertz imaging of rat skin burns. *J. Miomed. Opt.* **17**(4): 0405031–0405033.

Tonouchi, M. (2007). Cutting-edge terahertz technology. *Nat. Photon.* **1**(2): 97–105.

Ueno, Y. and K. Ajito. (2007). Terahertz time-domain spectra of aromatic carboxylic acids incorporated in nano-sized pores of mesoporous silicate. *Anal. Sci.* **23**(7): 803–807.

Ueno, Y., R. Rungsawang et al. (2006). Quantitative measurements of amino acids by terahertz time-domain transmission spectroscopy. *Anal. Chem.* **78**(15): 5424–5428.

Valavanisa, A., P. Deana et al. (2013). Transient analysis of substrate heating effects in a terahertz quantum cascade laser using an ultrafast NbN superconducting detector. *International Conference on Infrared, Millimeter, and Terahertz Waves, IRMMW-THz,* Rhine, Germany.

Wallace, V. P., A. J. Fitzgerald et al. (2004). Terahertz pulsed imaging of basal cell carcinoma ex vivo and in vivo. *Br. J. Dermatol.* **151**(2): 424–432.

Wallace, V. P., A. J. Fitzgerald et al. (2006). Terahertz pulsed spectroscopy of human basal cell carcinoma. *Appl. Spectros.* **60**(10): 1127–1133.

Williams, B. S., S. Kumar et al. (2003). Terahertz quantum-cascade laser operating up to 137 K. *Appl. Phys. Lett.* **83**(25): 5142–5144.

Woolard, D. L., E. R. Brown et al. (2005). Terahertz frequency sensing and imaging: A time of reckoning future applications? *Proc. IEEE* **93**(10): 1722–1743.

Yamamoto, K., K. Tominaga et al. (2005). Terahertz time-domain spectroscopy of amino acids and polypeptides. *Biophys. J.* **89**(3): L22–L24.

Yu, C., S. Fan et al. (2012). The potential of terahertz imaging for cancer diagnosis: A review of investigations to date. *Quant. Imaging Med. Surg.* **2**(1): 33–45.

Section V
Imaging Agents

15

X-Ray, MRI, and Ultrasound Agents
Basic Principles

Michael F. Tweedle, Krishan Kumar, and Michael V. Knopp

This chapter describes the basic chemistry and biology of the contrast agents (CAs) used as imaging pharmaceuticals in the X-ray, MRI, and ultrasound (US) imaging modalities. As other chapters deal with the instrumentation used to create images, we will move directly to the pharmaceuticals themselves. Sections are divided by modality and include brief sections before and after "Basic Principles" to describe, respectively, the history and future directions. The primary emphasis will be the required and desired chemical and biological characteristics of the CA molecules as exemplified by the approved human variants. We include references throughout, but will not repeat the excellent published, often exhaustive, reviews of technical literature, preferring rather to teach the principles and cite one or more references for further details and nuances. Controversial issues are few, but, when encountered, we will label opinions clearly.

To be useful to patients (human or veterinary), pharmaceuticals must be developed to the satisfaction of regulatory authorities in accordance with published guidelines, and then commercialized. For this reason, we will focus on existing and imminently available commercial pharmaceuticals. Many noncommercial examples of CAs exist, and can generally be located in the reviews cited. It is presumed that most animal clinical veterinary use agents will be the human commercial ones.

All imaging modalities make use of exogenously administered CAs, but the actual purpose varies widely according to the particular strengths and weaknesses of the modality, particularly its spatial and natural soft tissue contrast resolution, the time it takes to collect and display images, and the limit of detection (LOD) for its contrast agents. To aid in the description of these parameters for each class of agent, we show in Table 15.1 a comparison. Of the approved clinical CAs, there are no examples targeted to receptors, as are typical of the numerous radioactive pharmaceutical tracers available to nuclear medicine. This is clearly the result of the higher LOD, millimolar to micromolar, for X-ray and MR CAs. For the US microbubbles, a single bubble can be imaged in vivo when it is in motion in a capillary, and about 10–20 can be visualized in ideal conditions when stationary (Klibanov et al. 2004). There are biochemically targeted US agents in human clinical trials, but no commercial agents as of this writing.

15.1   X-RAY CONTRAST AGENTS (XRCAs)

15.1.1   History

In X-ray imaging, a high-energy beam of X-rays is absorbed to varying degrees by tissues depending on their average density and atomic number. Soft tissues and fluids are primarily carbon (C), hydrogen (H), nitrogen (N), and oxygen (O) and absorb less X-radiation than bone, which contains

TABLE 15.1
Comparison of Modalities and Their Approved CA

Modality	Typical Speed	Ionizing Radiation	Typical CA	Approx. Size (nm)	Dose (mmol/kg)	[CA][a]	LD_{50}/Dose[b]	LOD (μM)
X ray	Seconds	Yes	Iodine	<1	1–2	1	>50	1000
MRI	Minutes	No	Gd	<1	0.1–0.3	1	>30	10
US	Real time	No	Gas bubble	2000	10^{6c}	$10^{6\,c}$	$>10^2$	1–10 bubble
Nuclear	Minutes	Yes	^{18}F	<1	10^{-3}	10^{-3}	$>10^3$	10^{-6}

a mol/L in maximum administered dose.
b Mice intravenously/human dose.
c No. of ~2 μm bubbles.

large amounts of calcium (Ca) and phosphorus (P). XRCAs are, therefore, composed of elements heavier than those in soft tissues, primarily barium (Ba) and iodine (I). The discovery of X-rays by Wilhelm Roentgen in 1895 led almost immediately to X-ray images of human body parts, famously his wife's hand, and from there by 1896 to injection of heavy metals such as Pb into cadaver veins. Rontgen himself concluded after testing various materials, that X-ray absorption increased more with higher atomic number than with density, and was not dependent on their state (liquid or solid). Nonmetals like iodine-containing antiseptics were noticed by 1896, followed by oral use of encapsulated Bi, Ag, and other metals and nonmetals, in the next decades, mostly for oral, rectal, and other nonsystemic uses. Ultimately, these studies led to the widespread dominance of $BaSO_4$ for nonsystemic X-ray imaging, and this substance still dominates that usage today, with only very rare serious adverse reactions due to inadvertent systemic escape due to unknown perforations. $BaSO_4$ is today expertly formulated to coat bowel and other lumen surfaces to diagnose abnormalities mainly through highlighting the normal lumen, most commonly the gastrointestinal tract.

By the 1920s, the desire for systemic agents that could demonstrate renal physiology led to experimentation with intravenous iodide and other salts, and ultimately to the first widely compelling intravascular CA, Uroselectan®, a monoiodinated pyridone solubilized via a carboxylic acid, launched in 1930 by Schering-Kahlbaum AG (Table 15.2). It represented a true breakthrough in tolerance, but this molecule still caused serious reactions on use. The high doses and concentrations necessary for XRCA, combined with large potential markets, led to continuous research with the goal of improving tolerance for iodine. The evolution of the iodinated X-ray agents used today proceeded from Uroselectan to Iodopyracet, which is a similar pyridone with two iodine atoms per molecule, through acetrizoate, a benzenoid with two iodine atoms, to the widely successful and long-lived diatrizoate, a carboxylate solublized tri-iodobenzenoid launched in 1954, and still in some selected uses today. The final and highly significant improvement over diatrizoate was made in the 1970s by Torsten Almen and Nyegaard A/S, called metrizamide, by replacing the carboxylic acid with hydroxylated hydrocarbons to impart the necessary high solubility. This reduced osmolality

TABLE 15.2
Prototypical Iodinated XRCA

Uroselectan 1930   Acetrizoate 1952   Diatrizoate 1954   Metrizamide 1973

Note: The first three are shown as anions that were counterbalanced by sodium ions. Metrizamide was the first nonionic agent.

and toxicity dramatically. These "nonionic" compounds were so well tolerated that they could even be used intrathecally among sensitive neural tissues. Modern variants of this early agent have nearly now replaced the ionically solublized XRCA, as well as metrizamide. The history of intravenous XRCA is much more richly described by Grainger (1982).

15.1.2   BASIC PRINCIPLES

Iodinated XRCAs absorb X-rays, creating a shadow image. Most of their uses require ≥ 1 mM concentrations to be seen clearly above surrounding tissues. Most uses involve intravenous injections by either bolus or drip infusion, although many of the nonsystemic $BaSO_4$ indications are moving to iodinated agents, and intrathecal and other cavities also benefit from opacification. One of the widespread and demanding uses is in angiography, which requires a rapid bolus injection into arteries at the highest concentrations. Administration of 50–100 g of the drug at 1 M concentrations imposes severe constraints on the chemistry. The agents must (1) have extremely high water solubility, (2) have high chemical stability, (3) have extremely low systemic acute toxicity, (4) not be metabolized to any significant degree, and (5) be fully and rapidly excreted. The modern XRCAs shown in Table 15.3 are now amazingly close to these ideals thanks to the more than 50 years of dedicated research that produced them (Sovak 1984).

15.1.3   CHEMISTRY OF XRCAS

We refer to Hoey et al. for a thorough and highly detailed review of the chemistry and primary screening of candidate molecules that led to the commercially preferred ones (Hoey 1984). All modern X-ray CAs are 2,4,6-tri-iodinated benzenoids. Table 15.3 lists the commercially available molecules, along with their generic names, copyrighted trade names, and physical chemical properties that are important in formulation and use. The ring positions not occupied by iodine are used to ensure the water solubility of the molecule and to inhibit protein binding that is assumed to negatively affect tolerance at the very high doses required. The C(I)–CH–C(I) sequence (e.g., in acetrizoate) tends to encourage protein binding and, in many compounds, hepatobiliary excretion. Such molecules were marketed to evaluate the liver, but were eventually removed from use due mainly to higher toxicity. The mechanism for protecting the highly polarizable iodine from protein interactions in vivo is assumed to be mainly the steric bulk of the hydroxylated substituents. Hoey et al. tabulate and reference hundreds of molecules that were screened mainly from the 1950s through 1970s. The required solubility of ~1 M is irregularly associated with various hydroxylated substituents, and not predictable in any given molecule in a series.

During manufacture, the easiest means to sterilize solutions is through autoclaving of the sealed vials at temperatures >100°C. Therefore high chemical stability is highly desirable. This latter requirement was a primary flaw of metrizamide (Table 15.2). Despite its dramatically improved tolerance relative to its predecessors, imparted by its nonionized nature, a lack of shelf stability in solution meant that metrizamide had to be lyophilized sterile and the user was required to dissolve it in bicarbonate solution. It was therefore an easy target to be replaced by molecules (e.g., iopamidol and iohexol in Table 15.3) that were more stable. Interestingly, XRCAs can achieve their 1 M solubility requirement through innate solubility or through supersaturation, a state of nonequilibrium that is stable for lack of a path to precipitation such as a seed crystal. Terminal heat sterilization also aids in ensuring that microscopic seed crystals are not present in the vial through accidental local evaporation. So, in addition to extreme water solubility, extreme chemical/thermal stability is required to survive the sterilization procedure, plus withstand any hydrolysis over a 2-year expected shelf-life in aqueous solution. De-iodination is inhibited by the >50 kcal aromatic C–I bonds. Note that the water for injection used in these formulations must be strictly free of metals such as copper (Cu), which can catalyze de-iodination. With three I per aromatic ring and relatively compact substituents, I atoms now represent ~50% of the molecular weight of the CA molecules. The leading molecules are now bulk-synthesized in extremely automated manufacturing plants at the 1000 ton/year scale.

TABLE 15.3
Tri-Iodinated XRCA, Including the Osmolality and Viscosity When Formulated at or near 300 mg I/mL and 37°C

Generic Name	Trade Name	Osmolality (mOsmol/kg-Water)	Viscosity (mPa s)	Maximum (mg I/mL)
Diatrizoic acid	Hypaque, Renografin	1550	2.4	370
Iothalamic acid	Conray	1700	3.0	370
Ioxitalamic acid	Telebrix	1500	5.2	350
Ioxaglic acid	Hexabrix	600 (320 mg I/mL)	7.5	320
Ioversol	Optiray	651	5.5	350
Iopromide	Ultravist	607	4.9	370
Iopentol	Imagopaque	383	6.5	300

(Continued)

TABLE 15.3 (*Continued*)
Tri-Iodinated XRCA, Including the Osmolality and Viscosity When Formulated at or near 300 mg I/mL and 37°C

	Generic Name	Trade Name	Osmolality (mOsmol/kg-Water)	Viscosity (mPa s)	Maximum (mg I/mL)
	Iopamidol	IsoVue, Iopamiro	616	4.7	370
	Iohexol	Omnipaque	709	6.8	350
	Iomeprol	Iomeron	521	4.5	400
	Ioxilan	Oxilan	291	8.5	320
	Iotrolan	Isovist, Osmovist	291	8.5	320

(Continued)

TABLE 15.3 (Continued)
Tri-Iodinated XRCA, Including the Osmolality and Viscosity When Formulated at or near 300 mg I/mL and 37°C

Generic Name	Trade Name	Osmolality (mOsmol/kg-Water)	Viscosity (mPa s)	Maximum (mg I/mL)
Iodixanol	Visipaque	290 (270 mg I/mL)	8.5	320
N-methyl-glucaminium	NMGH⁺	Counter ion		

Notes: Osmolality and viscosity vary slightly for the ionic CAs (first four) depending upon their formulation counterions, either or both of sodium and N-methylglucamine (last table entry) (Fischer 1986; Krause et al. 1994). Very little, if any, protein binding exists. The trade names are protected by copyright, and other trade names may exist.

15.1.4 Biology of XRCAs

In vivo, XRCA should be noninteractive, demonstrating no physiological effects other than distribution and excretion. Biologically, these CAs are now nearly devoid of protein binding. On injection into mammals, they distribute in a few minutes through the full extracellular space, without crossing cell membranes, and are thereafter almost 100% excreted unmetabolized by the renal route into the urine. Left for days with live cells, they will penetrate into cells, probably by the cells' natural means of sampling their environment, but in vivo the excretion is generally complete with roughly 2 h elimination half-times in humans. This structure type evolved as the most tolerated one. The acute intravenous tolerances are among the highest known for exogenous substances. Measured by LD_{50} (the administered intravenous dose that is fatal in 50% of small rodents), acute intravenous LD_{50} value, for the best agents (nonionic monomers and dimmers) is >50 mmol/kg (~50 times the human dose). Biological interaction of the nonionic variants is extraordinarily low with almost no protein binding, and chemo- and neurotolerance are high enough for intra-arterial and intrathecal administration, although only selected agents are approved for the latter use.

15.1.4.1 Chemotoxicity

CAs have among the lowest chemotoxicity of any exogenously administered agents, but some chemo/immunotoxicity must exist to explain the rare and unpredictable side effects, many of which resemble anaphylaxis (Shehadi 1969; Shehadi et al. 1975). It is reasonable that weak and transient protein binding arises from nonspecific interactions, in part determined by (1) a small quantity of protonated agent existing at equilibrium (only for ionic agents), (2) the lipophilic character of portions of the molecule, and/or (3) unpredictable molecular characteristics, such as hydrogen-bonding interactions between proteins and hydroxylated or other side-chain moieties, possibly encouraged or regulated by self-association of the CA (Lasser 2011). Chemotoxicity can damage the endothelium, bind to and alter blood cells, and inhibit enzymes. LD_{50} values are not well correlated with such events, but for what they represent, crude and general indicators, the values of the current nonionic agents that have much reduced risk of adverse events are about twice as high as those of the former generation of ionic predecessor agents (Shehadi 1969; Golman 1984).

15.1.4.2 Osmolality

To achieve the necessary concentrations in vivo in a bolus, XRCAs are formulated at concentrations up to 1 M (~350–400 mg I/mL). At these concentrations, colligative properties, those that depend on the number of particles in solution rather than their chemical properties, can cause acute toxicity. The osmolality of CA would ideally match the physiological osmolality of ~300 mOsmol/kg-water, but with few exceptions the XRCAs are hyperosmolar compared to blood (Table 15.3) ranging up to >2000 mOsmol/kg (ionic monomers at 370 mg I/mL). CAs are initially concentrated in the blood by intravenous or intra-arterial administration. This causes local bulk water to rapidly diffuse into the vessel from intracellular and extravascular spaces, creating elevated systemic blood volume and peripheral blood flow, depressed systemic resistance and blood pressure, and crenation of blood cells and platelets that can release complement activating entities. Nonionic molecules (e.g., iopamidol) have one particle per three I atoms, while the older ionic molecules (e.g., diatrizoate) have two particles per three I atoms because they are solubilized by carboxylic acids that require positive counterions like Na^+ or $NMGH^+$, which also contribute to osmolality. For the ionic molecules, osmolality of the 1 M solutions is ~2000 mOsmol/kg-water, while the nonionic monomer molecules have osmolalities of 1 M solutions ~700–900 mOsmol/kg-water, which is still higher than physiological but far safer and less apt to cause patient discomfort (Golman 1984). There are also nonionic "dimers" that have physiological osmolality at the mid concentrations (~300 mg I/mL), but these are too viscous to formulate at higher iodine concentrations. XRCAs have osmolality below the expected theoretical numbers based on concentrations of dissociated ions or molecules. This suggests that the molecules self-associate at high formulation concentrations, producing fewer particles in solution. The nonionic dimeric agents actually are formulated with added salts to bring the osmolality up to physiological values at the most popular concentration (~300 mg I/mL).

Osmolality was the primary driver for commercializing the dimers, but they are more expensive to produce than monomers, are still relatively new, and have viscosities high enough to prevent use at the highest useful concentrations. A final word on their usefulness awaits the years of routine experience that will generate sufficient numbers of adverse events for comparison among the CA types. Certainly, nothing as dramatically obvious as the change from ionic to nonionic agents is obvious at this juncture.

15.1.4.3 Viscosity

For hand injection, a high viscosity requires more pressure to deliver the CA through a syringe needle. While this is still a concern in a minor number of applications, power injection is generally used in computed tomography, the most routine application in humans, but much less often in rodents and other smaller animals, suggesting that attention be paid to the rate of hand injection in these species. Chemically, and crudely stated, a more spherical molecule tends to have a lower viscosity. But the hydroxylated solubilizing/protecting groups now used in XRCAs are good hydrogen binders and attract water molecules to associate with the CAs, making estimates of sphericity tenuous. Nevertheless, the smaller ionic monomer agents are less viscous than the nonionic monomers, and the nonionic dimers are the most viscous, as would be expected of an elongated molecule with more hydroxyl groups on it. Viscosity is also a strong and nonlinear function of the solution concentration. The new dimers are too viscous for use above the 320 mg I/mL concentrations, even when warmed (higher temperature reduces viscosity, and warming to 37°C is standard procedure). There is controversial speculation that both osmolality and viscosity can contribute to contrast-induced nephropathy (CIN) secondary to natural concentration in the collecting system during excretion (Stacul et al. 2011). However, the rates of CIN in the safest molecules appear to be the same and not differentiated by colligative properties or structure type (monomer or dimer) in humans (Aspelin et al. 2003; Solomon et al. 2007).

15.1.5   Future Directions

XRCAs have progressed to an incredibly high state of development over 100+ years, involving hundreds of researchers and hundreds of millions of dollars spent on their development. Most of the research that produced the currently used nonionic molecules was done within companies, but at this point, very little laboratory research is being done toward new molecular types in industry or academia. The most significant problem in the field is the very rare, deadly, and largely unpredictable anaphylactoid reactions. Animal models do not well mimic the human reactions, although hyperimmune rats have been used (Lasser et al. 1995). Any attempt to create a new XRCA of the iodinated benzenoid type would require a highly significant advance to justify the probable $200–$300 million to develop a commercial example (Nunn 2006). The reasons for the high cost are fourfold: (1) the existing CA are extremely well refined; (2) manufacturing plants must be extremely large capacity to serve 1000-ton annual markets; (3) each clinical indication must be separately proven to be effectively dealt with by a new contrast agent; and (4) the anaphylactoid reactions have no really well-validated animal model and are so rare that many thousands of patients would need to be screened to prove an advantage, which seems unlikely for any but a genuinely new structure. A new structure, for example, a heavy metal chelate, would require considerable work to refine to even the level of acute tolerance of the XRCA (see in the following) and would then require all of the aforementioned hurdles be overcome before a successful candidate could be imagined. There is an ancillary possibility for improving outcomes if an in vitro skin, blood, genetic, or other rapid test can be developed that predicts the serious adverse events, but so far none has appeared, other than older pilot studies (Eaton et al. 1988).

15.2   MRI CONTRAST AGENTS

15.2.1   History

MRI is the in vivo version of nuclear magnetic resonance spectroscopy, for the discovery of which Felix Bloch and Edward Purcell won the 1952 Nobel Prize in Physics. Bloch famously detected a proton signal from his finger on his spectrometer. It required many years of effort before Paul Lauterbur and Peter Mansfield won the Nobel Prize in Physiology or Medicine in 2003 for the work that made MRI possible in the 1970s. All imaging modalities so far had benefited from exogenous CAs, and MRI was no exception. Paramagnetic metals were well known to powerfully affect water proton signals in solutions, and these would prove to dominate the field of MRCA as well. This fact was obvious to Lauterbur in 1978 when he published the first paramagnetic MRCA images of canine hearts (Lauterbur et al. 1978; Brady et al. 1982). Paramagnetism is generated by unpaired electrons, and the highest useful paramagnetism was found to be with Gd(III) with seven unpaired electrons per ion. Gd is a heavy metal that precipitates at in vivo pH values and also binds proteins, with a very acute toxic effect at imaging doses. Therefore, it was, analogous to the case of XRCA, solublized and isolated from the endogenous environment. In the case of a metal, it was known that chelating agents affected this kind of protection, in part from studies on metal chelates as X-ray CAs. Gadopentetate was the first acceptable MRI contrast agent and was launched in 1989 (Table 15.4) by Schering AG, followed in 1992 by the nonionic agents gadoteridol and gadodiamide (Weinmann et al. 1984; Greco et al. 1990; Dischino et al. 1991; Runge et al. 1991).

It is of interest that progress from the ionic gadopentetate to the two nonionic monomers was rapid, due to existing experience in the XRCA field where nonionics were already successful as safer agents. The expected success of this change, however, was not realized because MRCAs were delivered intravenously only at 0.5 M in only 14 mL, compared to 1 M in 50+ mL for XRCA, both intravenously and intra-arterially. The large *systemic* effects and pain that resulted from the

TABLE 15.4
MRCA: In Solution, Each Structure Also Maintains One Gd-Coordinated Water Molecule

Abbreviation	Gd(DTPA)²⁻	Gd(DTPA-BMA)	Gd(HP-DO3A)	Gd(DOTA)⁻	Gd(BT-DO3A)	Gd(DTPA-BMEA)	Gd(BOPTA)²⁻	Gd(DTPA-EOB)²⁻	MS-325
Generic name	Gadopentetate dimeglumine	Gadodiamide	Gadoteridol	Gadoterate	Gadobutrol	Gadoversetamide	Gadobenate dimeglumine	Gadoexetate disodium	Gadofosveset
Trade name	Magnevist	Omniscan	ProHance	Dotarem, Magnescope	Gadovist, Gadavist	OptiMARK	MultiHance	Eovist, Primovist	Ablavar, Vasovist
Structure	Ionic linear	Nonionic linear	Nonionic macrocyclic	Ionic macrocyclic	Nonionic macrocyclic	Nonionic linear	Ionic linear	Ionic linear	Ionic linear
Protein bound	None	None	None	None	None	None	Slight, ~5%	Moderate, ~10%	Significant, ~90%
Excretion	Renal	Renal	Renal	Renal	Renal	Renal	Renal, 4% hepatic	Renal, 50% hepatic	Renal, 5% hepatic
[Gd], M	0.5	0.5	0.5	0.5	1.0	0.5	0.5	0.25	0.25
Osmolality, 37°C mOsmo/kg-water	1960	789	630	1350	1603	1110	1970	688	825
Viscosity, mPa s	2.9	1.4	1.3	2.0	5.0	2.0	5.3	1.2	2.1
Log K_{eq}; log K_{eq}' M⁻¹	22.5; 18.4	16.9; 14.9	23.8; 17.1	25.6; 19.3	21.8; 15.5	16.6; 15.0	22.6; 18.4	23.5; 18.7	22.1; 18.9
Human Serum stability at 15 days, %dissociated	1.2–2.0	22–26	0	0	0	26–32	1.3–2.1	0.76–12	1.4–1.9
r_{12} plasma 1.5 T	3.9–4.1; 4.6–5.3	4.3; 5.2	4.1; 5.0	3.6; 4.3	4.7–5.2; 6.1–7.5	4.7; 5.2	6.3–7.9; 8.7–18.9	6.9; 8.7	19; 34
r_{12} plasma 3.0 T mM⁻¹ s⁻¹	3.7–3.9; 5.2	4.0; 5.6	3.7; 5.7	3.5; 4.9	4.5–5.0; 6.3–7.1	4.5; 5.9	5.5–5.9; 11.0–17.5	6.2; 11.0	9.9; 60

Notes: Physical data are at 37°C but stability constants are at 25°C. Protein binding is highly dependent on conditions of concentration; data are approximate. References: Formulation data from Package Inserts; Log *K* data from (Uggeri et al. 1995; Caravan et al. 1999; Schmitt-Willich et al. 1999; Port et al. 2008); Human serum stability data from (Frenzel et al. 2008); relaxivity data from (Rohrer et al. 2005; Pintaske et al. 2006). The trade names are protected by copyright; alternative trade names may exist (Hao 2012).

hyperosmolality of ionic XRCA agents were not manifest in the much smaller volumes of gado-pentetate delivered. Since 1992, certain clinical indications have favored nonionics (high dose, rapid power injection), but other uses favor the ionics (intra-articular). A second innovation was a chelating agent structure known as "macrocyclic" that offered more resistance to the release of Gd ions from their chelates under stress. Toxicologically, the MRCAs are less well tolerated in acute systemic studies than the best XRCAs, showing LD_{50} values generally less than or equal to that of the older ionic XRCAs, but the dose is so much lower in MRI relative to the molecule's chemotoxicity that adverse events occur at lower rates with MRCAs. This fact contributed to the eventual substitution of MRI scans for X-ray scans in patients with severe kidney disease at risk of CIN. MRCA use in severely renally impaired patients led to very rare adverse events now known as nephrogenic systemic fibrosis (NSF) that were connected with MRCA scans in 2006 (Marckmann et al. 2006). In this context, the more inert macrocyclic agents are now preferred in these patients.

While the Gd-based intravenous agents dominate, there is some rare use of various solutions, such as paramagnetic green tea, as oral agents whose primary function, as in XRCAs with barium sulfate, is to mark the location of the bowel. Powerful T_2 relation agents in the form of iron oxide nanoparticles reached clinical use and commercialization briefly for oral use and in liver imaging, but low usage rates and greater toxicity, primarily minor naturally resolved reactions such as transient low back pain, led to the withdrawal of the intravenously administered agents. Lumirem®, an iron oxide particle for oral use, remains available but has limited distribution. The same fate befell mangafodapir (Teslscan®), an interesting Mn(II)-based agent that deliberately released free Mn(II) ions that highlighted myocardium and pancreas through natural sequestration in those organs; it caused facial flushing from vasodilation. Overall, in a short period relative to XRCA, the MRCA evolved to a similar level of sophistication, and coalesced to a single dominant element, Gd, and molecular structure type, the chelating agent. Unlike the XRCA, and probably due to the lower doses of MRCA, some molecules are in clinical use that are safely excreted through the hepatobiliary route, and also some bound to proteins to various degrees. Otherwise, there has been no commercially significant evolution of the structure of the Gd agents since the 1980s, and each Gd agent that has been commercialized so far is still in use. Use of MRCAs has steadily increased in MRI and is currently estimated at ~25% of MRI scans.

15.2.2 Basic Principles

MRCAs have many of the same requirements as XRCAs. They must be effective at their contrast function, be highly stable for heat sterilization, have shelf-life in solution of 2 years, have extreme water solubility (usually 0.5 M solutions), be nontoxic and inert physiologically, and be rapidly distributed and excreted unchanged in vivo. To a remarkable degree, the existing MRCAs accomplish these goals.

15.2.2.1 Mechanism and Relaxivity

While XRCAs operate by a simple absorbance principle, the MRCAs' operative mechanism is far more complex. The theory that best describes this mechanism has been discussed in great detail (Lauffer 1987; Merbach and Tóth 2001). We will summarize them here sufficiently to make the structure–functional relationships clear. MRI images are composites of volume elements, or voxels, each of which has signal intensity represented digitally by a number from 1 to 256 displayed visually from black to white. The number depends on complex equations with user-adjustable parameters, but those most important in the context of MRCA are the T_1 and T_2 relaxation times of the body's water protons. Normally, protons excited by radio frequency bursts in the MRI relax to their ground state in 0.1 to >1 s, but any water proton bound chemically to a paramagnetic Gd ion relaxes about a million-fold faster. The site for attachment of the water molecule to the Gd ion is labile, rapidly exchangeable with the bulk water around it in the voxel, at the level of a million exchanges per

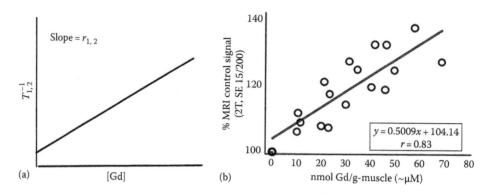

FIGURE 15.1 (a) Relaxivity is obtained by plotting reciprocal relaxation times versus [Gd] and taking the slope. (b) Plot of MRI signal versus [Gd], taken to determine the LOD for a Gd-based MRCA in mice. ^{153}Gd-Gd(HP-DO3A) was titrated intravenously while measuring MRI signal in skeletal muscle of nephrectomized mice, followed by sacrifice and counting the radioactivity. Limit of detection was [Gd] = 30 μM. (From Wedeking, P.R. et al., *Magn. Reson. Imaging*, 17(4), 569, 1999.)

second. Hence, the Gd ion becomes an effective water proton relaxation catalyst, and the volume element in which the Gd exists has high signal on MRI, proportional to the extent of the Gd concentration (within limits). The catalytic effectiveness (called relaxivity) depends on some molecular subtleties, which make one Gd agent potentially different from another, although in practice, with minor exceptions, the chemistry has been mostly optimized to a relative sameness. Endogenous paramagnets such as Fe(III) also affect the brightness in organs like liver and in diseases where paramagnetic ions are present in excess (Fe in hemochromatosis or Cu in Wilson's disease), but Gd is the most effective ion on a molar basis. Relaxivity is a second-order rate constant measured simply by plotting the reciprocal of the measured relaxation time (s^{-1}) against [Gd] in mM and taking the slope (in s^{-1} mM^{-1}) (Figure 15.1, Equation 15.1). The greater the relaxivity, the more effective the MRCA. Water protons relax in a dual mechanistic way, however, and the two relaxation times, T_1 and T_2, enter the signal intensity equations in opposite ways, with catalyzed (shortened) T_1 increasing signal on typical "T_1-weighted" (i.e., by the operator) MRI scans, and catalyzed T_2 decreasing the signal. With very few exceptions, MRI scanners are set up choosing TE and TR in Equation 15.1 to heavily emphasize T_1 when Gd-enhanced scans are recorded. T_2-weighting is not completely removed however, and at high enough [Gd], for example, >1 mM, in the renal collection system, Gd images can turn black instead of white (Tweedle 1989).

$$\text{MRI signal intensity} \sim e^{-\frac{TE}{T_2}} \times \left(1 - e^{-\frac{TR}{T_1}} \right) \tag{15.1}$$

TE and TR are instrumental variables, and can be adjusted by the operator

$$T^{-1}_{1,2\,\text{Gd}} - T^{-1}_{1,2\,\text{solv}} = r_{1,2}\,[\text{Gd}] \tag{15.2}$$

where the first term is the reciprocal relaxation time in Gd solution, the second is the reciprocal relaxation time in solvent, $r_{1,2}$ is the relaxivity, and [Gd] is the Gd concentration, traditionally in mM.

15.2.2.2   Why Gd?

The best MRCA needs a large unpaired electron count plus a spherical distribution of the unpaired electrons around the nucleus. Organic radicals have only one unpaired electron per molecule, but metals can have up to seven. Gd ion, Gd(III) or Gd^{3+}, has the largest number of unpaired electrons, seven electrons in 7f orbitals, which maximizes its catalytic effectiveness (Weinmann et al. 1984; Lauffer 1987). Organic radicals called nitroxides were tried, but were weak relaxers due to having only one unpaired electron. A MRCA must have very high "lability," that is, water molecules need to cycle in and out of bonding distance with the metal ion at least a million times a second. This reduced the consideration to Gd(III), Fe(III), and Mn(II). Mn(II) and Fe(III) have five unpaired electrons compared to Gd(III) with seven, and also the Mn(II) and Fe(III) have multiple oxidation states available in vivo, namely Mn(III) and Fe(II). Moving electrons in vivo at the 0.1 mM concentrations needed for MRI contrast could potentially cause intolerance, even though these two metals are naturally present in vivo in low concentrations. While Gd(III) has no known physiological role, it has only one possible oxidation state.

15.2.2.3 Chelation Chemistry

Ionic Gd^{3+} is not soluble above pH 6, and is therefore toxic. Gd^{3+} that becomes freed from its chelation complex is not only acutely toxic but, because it is similar in size to Ca^{2+} and has a higher charge, it also tends to bind irreversibly to the many natural Ca^{2+} binding sites in vivo. The rate of excretion of unchelated Gd^{3+} in vivo is extremely slow, with deposits in rodents being ultimately highest in bones (Wedeking and Tweedle 1988). Even now, little is really known about the long-term Gd^{3+} tolerance. The logical solution (to a chemist) is to encapsulate Gd^{3+} with a carrier or chelating ligand (from the Greek, *chelè*, meaning claw) that wraps tightly around a metal ion. A chelating ligand in this context is an organic chemical that binds multiple times, strongly and simultaneously, to the Gd^{3+}. The requirements for a pharmaceutical in this class are the most powerful chelating agents, whose function it is to protect the body from the freed Gd^{3+} ion until the Gd–chelate complex is excreted. It follows that the strongest chelating agents and rapid excretion are important requirements for MRCA. In addition, at least one very labile water molecule binding site on the Gd^{3+} is needed for it to function well as a water proton relaxation catalyst. Diethylenetriaminepentaacetic acid (DTPA) was the first successful chelated Gd agent (see Table 15.4). It existed in the literature before MRI was created as $^{99m}Tc(DTPA)$ and $^{xxx}Ln(DTPA)$ (Ln represents the periodic table neighbor lanthanides of Gd) nuclear medicine agents that were known to be rapidly excreted renally and very stable, and all Gd^{3+} MRCAs today are of the same amino-carboxylate general structure.

Table 15.4 lists the commercial MRCAs, showing their chemical structures, generic and copyrighted trade names, and their most important properties. Each of the molecules is made up of an amino-carboxylate multidentate ligand with N and O donor atoms bridged through hydrocarbon links and bound to the Gd^{3+} ion. Two prototype chemical backbones exist, one with three N and five O donor atoms in a linear array, and one with four N and four O donor atoms in a cyclic (macrocyclic, or macrocycle) array. All examples are highly water soluble and formulated from 0.25 to 1.0 M, aligned with their purpose and dose. Acute LD_{50} values, intravenous in rodents, are >50 times the human base dosage of 0.1 mmol/kg. Heat sterilization and several years of shelf-life are standard. Colligative properties, namely osmolality and viscosity, are determined by the charge type and formulation excipients. In all of the MRCAs, Gd^{3+} is chelated by ligands that have eight donor atoms bonded to the Gd ion. Nonionic molecules have three negatively charged O^- donor atoms (deprotonated carboxylic acids) in the backbone exactly balancing the three positive charges inherent in Gd^{3+}. Hydroxylic or amide O and aliphatic N makes up the remaining eight donor atoms. Ionic molecules have a surplus of negative charge having either four or five negatively charged O^- donors in the backbone, so the resulting $Gd(Ligand)^{n-}$ has an overall net negative charge that must be balanced with one or two $HNMG^+$ or Na^+ ions in the formulation. A ninth Gd^{3+} coordination position is always a labile water molecule's charge-neutral O (Caravan et al. 1999).

15.2.2.4 Colligative Properties and Tolerance

The fact that some MRCA molecules are ionic and some are nonionic produces about the same magnitude of range in osmolality as for XRCAs, about 600–2000 mOsmol/kg-water. Compare gadopentetate (2000 mOsmol/kg-water) and gadoteridol (616 mOsmol/kg-water) to diatrizoate (2100 mOsmol/kg-water) and iopamidol (796 mOsmol/kg-water). However, the impact on the tolerance, dramatic in the XRCAs, has been minimal in the MRCAs, because (1) the volumes of the concentrated fluids administered are 3- to 10-fold lower in MRI than in X-ray imaging, (2) the administration route in MRI is exclusively intravenous, while in X-ray more sensitive arterial and intrathecal routes are also used, and (3) the Gd^{3+} chelates as a class produce lower rates of idiosyncratic adverse reactions (not zero, however). There is one area of impact when rapid bolus injections are made via power injectors which are now almost universally used. In the rare event of misplacement of the needle such that the MRCA was extravagated, serious tissue damage resulted (even loss of limb). High osmolality is known to cause tissue necrosis, but the nonionic variants have not had this problem. The higher viscosity of some of the MRCAs makes them theoretically more difficult to hand-push through small catheters and syringes, but most MRCA injections are performed now with power injectors well capable of injecting all agents.

15.2.2.5 Distribution and Pharmacokinetics

Aside from the protein binders and liver agents discussed in the following, the Gd^{3+} chelates have almost identical tissue distribution rates and excretion, whether in animals or humans. In humans, distribution is very rapid (half-time ~12 min) after administration to the entire extracellular space, but these highly hydrophilic agents (log $P < -3$) remain extracellular. Excretion is nearly 100% through the renal route into urine with a half-time of ~1.6 h in patients with healthy kidneys and a normal glomerular filtration rate. In severely renally impaired patients, the elimination times can increase to as high as 50 h, requiring dialysis.

15.2.2.6 Stability and Chronic Tolerance

Stability of the CA is a far more important feature in MRCAs than it is in XRCAs. Shelf-life specifications prevent instability of the CAs in vials from being a significant concern, and unused agents are replaced by manufacturers. But in vivo stability (e.g., metabolism) and its effects are much harder to control and predict. The most toxic expected breakdown products from tri-iodinated benzenoids, aromatic amines, are further metabolizable and would be expected to be rapidly excreted. Iodide itself is relatively nontoxic in this context, and is indeed a required biological ion avidly sequestered in the thyroid gland. Any metabolic reaction of a Gd^{3+} chelate involving its backbone structure would result in the liberated Gd^{3+} ion and a metabolizable amino acid. The organic chelating agent from an MRCA can deplete Zn^{2+} ion and is known to do that in highly elevated dosing in rodents, and minor Zn^{2+} chelation and excretion were measured in humans (Puttagunta et al. 1996), but these are unlikely to play a significant role in chronic intolerance. Liberated Gd^{3+}, however, is not natural in vivo, is extremely long-lived in vivo, is a powerful calcium antagonist (Molgo et al. 1991), and has unknown long-term toxicity.

15.2.2.7 Stability of Gd Chelates

There are two types of chemical structures: linear and macrocyclic. Stability, to a chemist, is a technical abbreviation of the more precise term "thermodynamic stability," which in the case of a metal–chelate complex in our context refers to the energy involved in the equilibrium reaction shown in Equation 15.3:

$$Gd^{3+} + Ligand^{n-} \rightleftharpoons Gd(Ligand)^{3-n}$$

$$Keq = \frac{[Gd(Ligand)^{3-n}]}{[Gd^{3+}] \times [Ligand^{n-}]} \tag{15.3}$$

All Gd^{3+} chelate MRCAs exist in equilibrium; the binding reaction is reversible. The equilibrium stability constant, Keq, is defined in Equation 15.3. Keq values define the theoretical strength by which the pure chelating ligand holds onto the Gd^{3+} ion, favoring the chelate complex over the free Gd^{3+} ion. In aqueous solution, however, the chelating ligands' Gd^{3+} binding donor atoms are acids and bases, and so at physiological pH some of the ligand is protonated, and therefore, in whole or

TABLE 15.5
Comparison of Equilibrium Stability Constants, LD_{50}, and In Vivo Gd Retention

	Keq (M^{-1})	Keq' (M^{-1})	LD_{50} (mmol/kg)	%Gd in Mice at 7 Days
$Gd(H_2O)_9$	—	—	0.1	~90
$Gd(EDTA)(H_2O)_3^-$	$10^{17.7}$	$10^{14.9}$	0.3	~50
$Gd(DO3A)(H_2O)$	$10^{21.0}$	$10^{14.5}$	6	≤1
$Gd(DTPA-BMA)$	$10^{16.8}$	$10^{14.9}$	15	≤1

Sources: Weinmann, H.J. et al., *AJR Am. J. Roentgenol.*, 142(3), 619, 1984; Wedeking, P. et al., *Magn. Reson. Imaging*, 10(4), 641, 1992.

part, unable to bind to the Gd^{3+} ion. In other words, protons (H^+) compete with Gd^{3+} for binding to the chelating ligands. So in aqueous media, the practical *conditional equilibrium constant* Keq' is pH dependent and must be defined at physiological pH to account for the different ligands' differing affinity for H^+. Both of these constants are listed for the MRCAs in Table 15.4.

The structures of the agents in Table 15.4 are fully consistent with the Keq and Keq' numbers. The N atom donors in all of the ligands have similar thermodynamic contributions to the overall Keq values, while for the O donor atoms the negatively charged $-COO^-$ has a slightly stronger attractive interaction for positively charged Gd^{3+} than the uncharged O donors in $-CONH$ and COH. This can be seen quantitatively by comparing gadopentetate (five $-COO^-$) with gadodiamide (three $-COO^-$ and two $-CONH$). This kind of retro-analysis, explaining the measured numbers on the basis of the structure, helps chemists to be comfortable that they understand the molecular forces at work and also helps them to design and test new molecules in a simpler way than testing each new molecule for stability in vivo.

Unfortunately, Keq and Keq' can be quite misleading with respect to prediction of in vivo stability. They are measured in water, and not ex vivo in biological fluids or in vivo. Table 15.5 shows data on Gd^{3+} ion and three Gd^{3+} chelates—all powerful multidentate amino carboxylates—Keq and Keq', their acute LD_{50} values, and the amount of Gd^{3+} that remains unexcreted in mice measured 7 days after intravenous injection. Gd(DTPA-BMA) is an approved clinical MRCA in Table 15.4; Gd(DO3A) has a high LD_{50} and low Gd dechelation, but failed subacute tolerance testing due to kidney damage to rodents; and $Gd(EDTA)^-$ is virtually useless as a protective agent, dissociating half of its Gd^{3+} rapidly on injection (Wedeking et al. 1992).

Water is a very simple environment compared to that found in vivo, where a multitude of competing ions are present. A few of these are Cu^{2+}, Zn^{2+}, Ca^{2+}, and Fe^{3+}, which compete with Gd^{3+} to bind the ligands, and OH^-, PO_4^{3-} and CO_3^{2-} that compete with the ligand to bind to Gd^{3+}—in these cases to form insoluble salts such as $GdPO_4$. In addition, proteins can interact with small quantities of the agents in subtle and unpredictable ways (e.g., causing the rare idiosyncratic adverse reactions). Another complication is that thermodynamic parameters such as Keq and Keq' define only the beginning and ending states, but are silent on the time it takes to arrive at that end state. In MRCAs, the beginning state is a fully formed MRCA. In a biological milieu, given sufficient time, it is virtually certain that the exogenous ions competing for the ligand and for the Gd, plus the metabolizing enzymes in the body working to break down the organic ligand, would eventually succeed, so the rate at which the Gd dissociates from the Gd(Ligand) is clearly important, as is the rate at which the Gd(Ligand) is excreted from the body. The rate at which Gd leaves the ligand is referred to as dissociation kinetics (dissociation of Gd from its chelating ligand). Most MRCAs are excreted renally into urine, with a blood clearance half-life of about 90 min. So any dissociation reaction that is slow on that time scale will be truncated on excretion. Certain Gd(ligands), namely the macrocycles, are so slow to dissociate that even under highly stressful conditions, such as in the presence of competing ions, heat, lower pH, or in vivo for long periods, they powerfully resist dissociation. The technical reason for this is that the cyclic structure does not allow the four Gd–N bonds to break all at the same time. Linear structures can peel away and begin binding to, for example, Cu^{2+} ion, breaking O and a single Gd–N bond initially on the path to complete dissociation (Margerum et al. 1978). For this reason, despite similar Keq' values, the dissociation of the linear molecules, especially the linear nonionic ones, is significantly greater over long times than dissociation of the macrocyclic molecules (Tweedle et al. 1995; Frenzel et al. 2008). In certain patient populations with severely limited renal capacity, the faster dissociating molecules are, in fact, contraindicated.

15.2.2.8 Formulations

Formulations of MRCAs can contain excipients in the form of buffers and excess chelating ligands. Confusion often arises over the excess chelating ligands because there are different purposes associated with different amounts of excess chelating ligands present in formulations. The general concept is that, for an equilibrium, as in the reaction shown by Equation 15.1, excess chelating ligand can assist in keeping the equilibrium shifted to the right if the product of the reaction, the Gd(ligand), comes under stress, which might cause the equilibrium to move left: that is, toward dissociation. Excess chelating ligand is added for three different reasons in different MRCA: (1) in any of the agent classes, in very small amounts (~0.1 mol%) to guard against competing metals that can leach from the glass vials during sterilization or over shelf storage times (Schmidtt-Willich 2007); (2) in some linear agents, also in small amounts, to reduce post-administration elevated serum iron and bilirubin to levels within the normal human range (Niendorf et al. 1990); and (3) at much larger concentrations, 5–10 mol%, to prevent the toxicity caused by the freed Gd in vivo (Cacheris et al. 1990). There is no published evidence that item (1) is necessary. Item (2) was a deliberate formulation change made to fix an unexpected problem encountered with the first MRCA, Magnevist (and later Omniscan (Vanwagoner et al. 1991)). Patients experienced asymptomatic transient serum iron and bilirubin elevation. While officially of unknown origin or significance, this author's opinion is that the fact that excess chelating ligand reversed the behavior suggests a mechanism involving small amounts of free Gd either injected or formed in vivo immediately post administration, possibly related to erythrocyte interference. Item (3) is encountered only in the nonionic linear chelates, where large amounts are required to insure against the release of Gd in vivo (Cacheris et al. 1990). The effects are noteworthy, since many animal imaging experiments use the human formulations, and, of course, the response of small animals might be contrary to that of humans.

15.2.2.9 Nephrogenic Systemic Fibrosis (NSF)

CIN was and remains a considerable concern in radiology. MRI scans were, from the beginning, capable of renal evaluation, and as the understanding of the CIN problem grew, it led in the late 1990s to the transfer of renally impaired patients (GFR < 30 mL/min × 1.73 m^2) to MRI with MRCA evaluation to take advantage of the lower doses of renally excreted CA needed in MRI to reduce CIN risk. Especially, magnetic resonance angiography (MRA) was used at 0.2–0.3 mmol/kg, which is several times the usual dose used in the CNS examinations. In 2006, a rare disease of uncontrolled fibrosis, now known as NSF, was connected to the use of several of the linear Gd-chelates. The number of real cases will probably remain unknown, but 500 cases is a rough order estimate (Cowper 2001–2009). (Reference is made to the Cowper website for up-to-date information on NSF.) The disease occurs only in severely renal-impaired patients, and only following an MRI examination with certain MRCA. The disease is quite severe eventually, unless it resolves, leading to severely disabled joints, cachexia, and even death. The starting symptom is usually a small skin plaque or plaques on a lower extremity discovered anywhere from hours to years after exposure to the MRCA. Circumstantial evidence has accumulated that, in most opinions, now points to the release of free Gd, probably through transmetallation, as a necessary, but insufficient element of the disease's etiology: (1) by a wide margin, gadodiamide has more unconfounded cases (i.e., only one agent used in the patient that develops NSF) relative to their market shares, compared to gadopentetate, a stronger and less labile chelate; (2) macrocyclic agents, the least labile, are very nearly devoid of cases relative to their market share; (3) studies in animals and humans demonstrate that linear chelating agents are more labile, and release more Gd, than macrocyclic ones; and (4) no new cases have developed since regulatory authorities contraindicated the aforementioned linear agents in susceptible patients, while the macrocyclic agents are still used in this patient population. Not all severely renal-impaired patients develop the disease after these linear MRCAs are administered, so some unknown element is also involved. One commonly discussed hypothesis is that freed Gd and some other unknown agent, event, or condition interferes with circulating fibrocytes, which in turn accelerates fibrosis (Vakil et al. 2009).

Transmetallation of MRCAs was first discussed in 1988, in the context of its ability to encourage more release of free Gd in linear versus macrocyclic agents, even to the point of reference to renal-impaired patients (Tweedle et al. 1988b). Transmetallation was thought to be important because the long-term tolerance of Gd was (and is) unknown, and (1) the molecules being developed at the time were almost entirely renally excreted; (2) radiolabeled chelate studies in mice demonstrated a difference in the retention of ^{153}Gd-linear versus ^{153}Gd-macrocyclic agents; (3) macrocycles did not react with endogenously available ions while linear molecules did (Tweedle et al. 1991); and (4) renal impairment could obviously exacerbate the problem by at least 10-fold, giving the body more time to metabolize the drugs. Preclinical studies demonstrated the correlation between the amount of Gd release and the *kinetics* of Gd release from linear and macrocyclic agents (Tweedle et al. 1988a), that free ligand in the formulation reduced but did not eliminate the release (Tweedle et al. 1995), that Gd definitely dissociated from its DTPA ligand in rats (Kasokat and Urich 1992), that endogenous ions could stimulate release only in linear chelates (Tweedle et al. 1991), and that the differences observed in animals were evident in humans having been administered a linear nonionic versus a macrocyclic agent (White et al. 2006). The message in the early studies was simply that kinetics is a factor to be considered along with thermodynamics (i.e., Keq). The association of NSF with MRCA occurred more than a decade later, but a recent and definitive study of kinetics of Gd dissociation from Gd-chelated MRCA in human serum, in the context of NSF, is important and clearly demonstrates the ordering of macrocycles >> linear > nonionic linear structures with respect to the temporal release of Gd (Frenzel et al. 2008).

Among the reasons why no particular concern over freed Gd occurred earlier was that no important acute tolerance differences were observed among the different MRCAs, so arguments could only be theoretical until a related disease was actually discovered. No diseases related to Gd were known, and the stability constants of even the weakest of the MRCAs are enormous by any objective criterion. In the clinic, most patients with severe renal impairment are extremely sick to begin with, and this can mask subtle problems such as a small skin rash (not all NSF is severe). The afflicted population occurred 18 years after the agents were first marketed, allowing a general and well-deserved high degree of confidence in their use to grow, even in very ill patients, through multiple academic generations of users. Radiologists found Gd agents generally much less toxic in all apparent ways compared to iodinated XRCAs. Finally, the general concern over freed Gd was made by companies or academics, mainly a few chemists, who preferred or studied the macrocyclic molecules. The U.S. FDA and other regulatory agencies now classify MRCAs into low, intermediate, and high risk, as shown in Table 15.4. New cases of NSF have apparently all but disappeared as of this writing. The only known animal to have some similar skin lesions is the rat (Sieber et al. 2008).

15.2.2.10 Protein Binding MRCA

Chelating agents with an aromatic (or probably other lipophilic entities) present in the structure are more prone to binding proteins in vivo, and also to be partially excreted by the hepatobiliary route. In particular, human serum albumin (HSA) is the binding target. HSA has the function of binding exogenous small molecules and is present at about 600 μM in plasma. Table 15.4 shows the three existing MRCAs with this property. Gd gains a factor of ~5 in r_1 when bound to albumin (vide infra) or other proteins. The protein binders are each based upon the linear ionic structure of DTPA. Their uses, however, are quite different and dependent on the relative amounts of protein binding. Each of them binds HSA only weakly, in the 10^3–10^4 range for Keq. The exact amount of binding in the context of a MRCA injection is variable with time and concentration of the agents because nonequilibrium conditions exist in vivo as the agents are excreted rapidly. Initially after intravenous administration, the MRCAs are in higher concentration than the HSA. For comparison, under the same concentration conditions of human use, gadobenate binds HSA only slightly (~5%), gadoexetate binds ~10%, and gadofosveset is ~90% bound. The agents are excreted via the hepatobiliary route to 4%, 50%, and 5%, respectively, indicating no obvious quantitative correlation between the protein binding and excretion pattern. It is also important to note that animal and human biodistribution and protein binding can be considerably different for the protein binders, while it is not for the rest of the MRCAs and XRCAs (Planchamp et al. 2004).

Gadoexetate and gadobenate add a simple aromatic group to the chelate backbone or one arm, respectively, with qualitatively similar but quantitatively different results. Gadoexetate, with 50% human hepatobiliary excretion, is used as an hepatobiliary agent for imaging the liver and bile ducts (Weinmann et al. 1996; Gschwend et al. 2011). Gadobenate, however, is a full indication agent in human use that competes with the agents that do not bind any protein, as an extracellularly distributed agent for CNS tumor detection (by far the largest indication for MRCAs (Maravilla 2006)), along with other body indications, including MR angiography (Woodard et al. 2012). Gadofosveset (Lauffer et al. 1998; Goyen 2008) is derivatized with a multiply charged amphiphilic arm that encourages much higher protein binding. This retards distribution and excretion, increasing signal from blood vessels relative to background tissue based on albumin distribution. In the growing indication of MRA (angiography), all MRCAs are capable, but the signal generated per Gd is highest for gadofosveset > gadobenate > non-protein binders because of differing protein bound fraction that has a far greater r_1 than the unbound MRCA. Gadobenate also produces somewhat greater signal in brain lesions in humans due to slight protein binding in the lesion (Maravilla et al. 2006). Gadofosveset is not used in the CNS in humans, probably because the extensive plasma albumin binding retards tumor uptake that occurs by passive diffusion through tumor blood vessels that lack a blood–brain barrier. Clearly, because of species differences in protein binding, all animal studies using the protein-binding class of MRCAs must be interpreted with knowledge of the protein-binding characteristics of the MRCAs in use in that species.

15.2.3 FUTURE DIRECTIONS

The future of MRCAs is less certain than for the much more mature field of XRCA. MRI itself is more complex and leaves room in its development for new classes of MRCAs based on different atoms such as ^{19}F (Wolf et al. 2000), ^{13}C (hyperpolarized) (Brindle et al. 2011; Gallagher et al. 2011), and other Ln as PARACEST agents (Woods et al. 2006; Hancu et al. 2010). Different or more tightly bound Gd chelates remain an active research area (Sethi et al. 2012).

The mechanism by which MRCAs generate a signal is complex, and currently unoptimized for protein bound agents, leaving open the tantalizing possibility of biological targeting analogous to what is achievable in radiopharmaceuticals. Gadolinium chelates are currently detectable at about 10 µM, but practically most useful at 5- to 10-fold higher concentration. The practical structures imageable at these concentrations are not receptors that exist in the nanomolar to picomolar range, but rather biological polymers such as fibrin and extracellular matrix. These targets have so far not been deemed worthy of imaging with MRCA despite advanced examples of Gd-based MRCA having been created (Botnar et al. 2004; Overoye-Chan et al. 2008). Theoretical r_1 relaxivities rise to about a factor of 10 over existing unbound MRCA, on a per Gd-OH$_2$ basis in current best examples. Research interest exists to try to achieve what is theoretically possible, $r_1 > 100 \, mM^{-1} \, s^{-1}$, depending on the magnetic field of the imager. This occurs according to theory for dipole–dipole interaction of the paramagnetic Gd^{3+} with the water proton bound to it (Figure 15.2).

The theory and chemical approaches to optimization have been thoroughly, but approachably, discussed in detail (Lauffer 1987; Caravan et al. 1999). Optimizing MRCA relaxation effectiveness devolves to optimizing the rate constants for certain molecular motions. A useful way to understand this is through the simplified equation

$$r_1 \alpha Q \frac{\tau_c}{a^6} \quad \text{and} \quad \tau_c^{-1} = \tau_r^{-1} + \tau_s^{-1} + \tau_m^{-1} \tag{15.4}$$

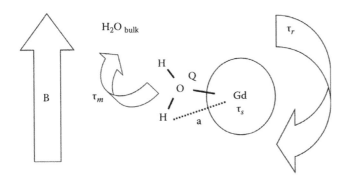

FIGURE 15.2 Schematic showing the molecular parameters that govern the relaxivity of each Gd MRCA. B is the main magnetic field, Q is the number of bound water molecules exchanging with bulk water, a is the distance between the center of the proton and center of the Gd atom, and τ_r, τ_s, and τ_m are the correlation times.

where Q is the number of water molecules bound to the Gd^{3+} ion. Obviously structures can be devised that increase this from the current number, 1, but reduction in stability is to be expected in all but the best cases; so far none has proven worthy of development (Datta and Raymond 2009). Here, a is the distance between the Gd ion and the relaxed proton on the water bound to the Gd atom. The shorter it is, the higher the relaxivity, and the effect runs as a power of 6. The reciprocal τ terms are first-order rate constants that represent certain processes: τ_r is molecular rotation, τ_s is electronic, and is τ_m is the exchange rate between coordinated water and bulk water. The τ_c influence operates like a series electrical circuit, so that one fast process (a small, short τ term) tends to dominate until it is slowed down, exposing a new dominant τ term. Current agents tumble too rapidly through solution, making the vector that connects the center of the Gd atom to the center of the coordinated water proton move rapidly relative to the primary external magnetic field. Optimization will involve eliminating overall molecular motions that move the vector. Immobilizing the Gd chelate rigidly at a large target protein would do this as long as subtle twisting or spinning motions are eliminated (Polasek and Caravan 2013). The target optimal rate constant for τ_c^{-1} is the Larmor frequency at which the imager operates. Most routine imaging magnets currently operate at 60 or 120 MHz, but clinical experimental magnets are up to 280 MHz, so the goal is unfortunately a moving target. There are also many lower field magnets operating. The rate of water exchange between the water coordinated to Gd and the bulk water propagates the Gd atom relaxation of the bulk water and must be rapid relative to the relaxation time of each water proton while the water is coordinated. Current agents do this, but as r_1 increases in an optimized r_1 agent, r_1 grows by, for example, 10-fold, and the exchange rates of current Gd chelates will become suboptimally slow. Changing the water exchange rate mechanism by adding more than one water and lengthening the Gd–O bond are current strategies to increase the rate of water exchange, the former having the additional benefit of doubling the overall r_1 on a per Gd basis. Once τ_m is optimized, the electronic τ_s will be exposed. τ_s is related to the T_1 of the unpaired electrons in the Gd nucleus, and depends on Gd-donor atom symmetry and ligand structural rigidity that would affect the Gd-donor atom vibrations. Maximum symmetry (e.g., a perfect cube with identical donor atoms) and rigidity is optimal, but difficult to achieve when both O and N donor atoms are necessary to maintain stability.

Unfortunately, even an optimized Gd chelate ($\sim 150 \ mM^{-1} s^{-1}(H_2O)^{-1}$) would fall 10- to 100-fold short of the required sensitivity to image most biological receptors (Nunn et al. 1997). Research is therefore directed in part at amplification of the signal through targeting many Gd per receptor, ultimately many r_1 optimized Gd chelates. Polymeric Gd agents will, however, inevitably suffer from target delivery restrictions as their size increases past ~ 5000 Da, relative to the spaces between capillary endothelial cells, between which they must pass to arrive at most of the useful targets.

15.3 ULTRASOUND CONTRAST AGENTS

An oscillating sound pressure wave with a frequency greater than the upper limit of the human hearing range, approximately 20 kHz (20,000 Hz) in healthy, young adults, is defined as ultrasound (US). Therefore, ultrasound devices operate with frequencies from 20 kHz up to several gigahertz. The potential for ultrasonic imaging with high frequency (3 GHz), with low resolution and contrast, was recognized as early as 1939, but saw little use in medicine. The technique has improved significantly over the decades, and ultrasound imaging is in routine use today to visualize muscles, tendons, and many internal organs, to capture their size and structure and to follow moving blood. The technique has several advantages compared to CT, MRI, and nuclear imaging: (1) It has become relatively inexpensive and portable, so equipment can be brought to the patient's bedside, emergency rooms, surgery suites, ambulances, field hospitals, and private practices. (2) Examinations are performed in real time and can visualize motion. (3) The long-term effects due to ultrasound exposure are still partially unknown, but ultrasound imaging has been used for over 20 years and has an excellent safety record, even for pediatric and neonatal imaging. The first two are the most important drivers of use. US imaging uses nonionizing radiation, so it is perceived as lower risk than X-ray and nuclear imaging. The limitations of US imaging include the following: (1) US imaging is not a whole-body imaging technique in large animals and humans due to the small field of view (20–30 cm). But obtaining a high-resolution full body scan in a mouse is not problematic with a small animal imager. (2) An experienced sonographer is needed for interpreting images. (3) The technique is mostly two-dimensional unless a rare and expensive quasi 3-D system is used. (4) US image quality is generally grainier and less "anatomic" than that of MRI or CT, although edge detection is more precise. (4) US images are prone to attenuation by bone, and do not image the lungs well due to the inability of air to uniformly transmit the sound waves. In the gastrointestinal tract, objects like air bubbles create large artifacts.

15.3.1 HISTORY

The history of the field has been discussed in detail by Klibanov (2002). Ultrasound CAs (USCAs) were proposed nearly four decades ago as a result of the detection of air bubbles in the blood stream after injection of agitated aqueous solutions. Since then, significant progress has been made in the discovery, development, and manufacture of novel USCAs and in the ability of ultrasound equipment to detect these agents with high sensitivity. The earliest convincing USCAs were simply air bubbles created by agitating solutions, among the best being those with high concentrations of XRCAs. These air bubbles included a wide dispersion of sizes and tended to disappear on the first pass through the lungs. The size of the effective air bubbles, >2 μm, keeps them inside the blood vessels. A frequent target organ for US to evaluate anatomy and function via wall motion was the heart, where agitated solutions were used to image right to left cardiac shunts. An important goal became the visualization of also the passage of the USCA through the right heart, then lungs, then left heart chambers, and finally through the myocardial capillaries to evaluate both of the chambers and the myocardial perfusion (only nuclear medicine was able to evaluate perfusion at that time). This goal stimulated research into coatings to protect the bubbles against dispersion and destruction in the lungs. Schering AG developed an initial commercial agent from finely pulverized galactose and dextran powders, which produced microbubbles when rapidly dissolved in water. Echovist® and then Levovist® (von Bibra et al. 1995) were marketed, with the latter containing a surfactant to protect the bubble. Relatively high doses of these were needed, however. A better strategy was developed by Feinstein et al. (Feinstein et al. 1990), who solved the primary problem using US itself in a preparative mode to partially denature human serum albumin at the air–water interface, which led to 2–3 μm air bubbles protected by a protein shell that survived the passage through the lung and left heart. This innovation made the first commercially convincing USCA, Albunex® (initially by Molecular Biosystems, Currently by GEHC). Eventually, the relatively water-soluble O_2/N_2 gases were replaced with water-insoluble perfluorocarbon gasses, which extended the blood lifetimes of the microbubbles from seconds to a few minutes. A second type of USCA was created using a

TABLE 15.6
Widely Available Ultrasound Contrast Agents

Trade Name	Manufacturer	Human Use	Gas	Shell	Size (µm)	Indications
Optison	GE Healthcare	Yes	C_3F_8	HSA	3.9–4.5; 95% < 10	Left ventricle
Lumason, SonoVue	Bracco SpA	Yes	SF_6	Phospholipids	1.5–2.5; 99% < 10	Left ventricle, Doppler of large vessels
Definity	Lantheus Medical Imaging	Yes	C_3F_8	Phospholipids	1.1–3.3; 98% < 10	Left ventricle
Targestar, Visistar, Targesphere	Targeson	No	C_4F_{10}	Lipid-s-avidin	1.0–2.0 2.0 3.0	Biotinylated ligands, VEGFR2, $\alpha_v\beta_3$, VCAM-1, P-selectin
Vevo MicroMarker	Visualsonics/ Bracco SpA	No	C_4F_{10}/N_2	Lipid-s-avidin	~2.5 ave	Biotinylated ligands

Note: Trade names are protected by copyright. Other trade names may exist.

formulation of nontoxic lipids. In one formulation (now called Definity®, Lantheus) Fritz et al. (1997) used a solution of fluorocarbon gas and lipids agitated on a portable shaker just prior to use; the energy of the shaking allows the formation of a protective lipid monolayer around the microbubbles. The gas used is perfluoropropane, to avoid the dissolution of the gas in blood water. In a second formulation, created by Schneider for Bracco and known as SonoVue® (Morel et al. 2000), small amounts of powdered excipients and lipids are stored in a lyophilized state along with SF_6. Adding normal saline dissolves the solids, and the energy of dissolution is used to form the lipid-coated microbubble for injection. These three cleverly evolved agents (Table 15.6), one albumin-based and two lipid-based, now represent the state of the current art for clinically successful and widely available USCAs. Commercially, the USCAs as a class are less often used than the XRCAs and MRCAs. Of the more than 400,000,000 annual diagnostic imaging procedures worldwide, >40% are US procedures, but only ~1 million (Faez et al. 2013) of these are enhanced with USCAs. About one-third of >200,000,000 XRCA and MRCA procedures are enhanced with XR or MRCA.

15.3.2 BASIC PRINCIPLES

USCAs are sold with about 10^8–10^{10} microbubbles per milliliter of injectable solution. The chemical compositions of USCAs are complex. For example, Definity contains perfluorpropane and three complex lipids, DPPA, DPPC, and MPEG5000 DPPE (Figure 15.3), along with perfluoropropane gas (Definity Package Insert 2011). The complex formulations for the lipid-based shells resulted from years of testing many small variations, which tended to alter the final products when scaled up, making the ultimate manufacturing of USCAs a specialized science similar to that of liposomes (lipid bilayer shells filled with liquids).

The ideal USCA should have following characteristics: (1) The agent should have acceptable stability (on the shelf and in vivo), sufficient circulation time to perform the procedure, and an excellent safety profile, that is, the ratio of the toxic dose of the agent to the dose administered for the procedure should exceed 100. (2) The agent must be of a size sufficient (1–4 nm) to scatter the US effectively but not so large as to affect tissue perfusion, for example, by lodging in capillaries. (3) It is important for safety that the size distribution be fairly tight as demonstrated by a reproducible synthesis/production process. (4) The components of the agents' shells and the gas inside must either be excreted from the body intact or be easily metabolized, and their tolerance, metabolism, and excretion characteristics should be known. (5) There should not be any long-term deposition of the agent or its components. If new agents are created for experimental use in animals, the parameters will be slightly more relaxed than those for human use, as needed for preclinical development in an animal protocol. Finally, as in the other discussed agents, it is important to have a consistent and well-understood relationship between the contrast agent's concentration in vivo and the signal generated, especially for quantitative imaging.

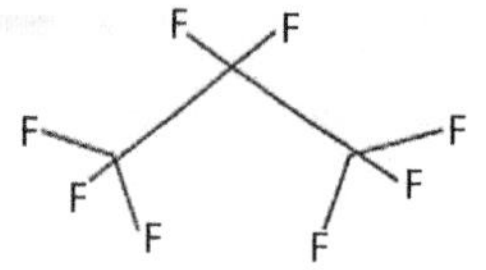

1,1,1,2,2,3,3,3-octafluoropropane (PFP)

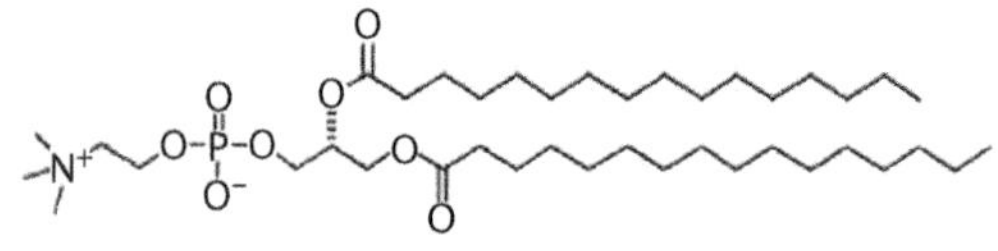

(R)-4-hydroxy-*N,N,N*-trimethly10-oxo-7-[(1-oxohexadecyl)oxy]-3,4,9-trioxa-4-phosphapentacosan-1-aminium, 4-oxide, inner salt (DPPC)

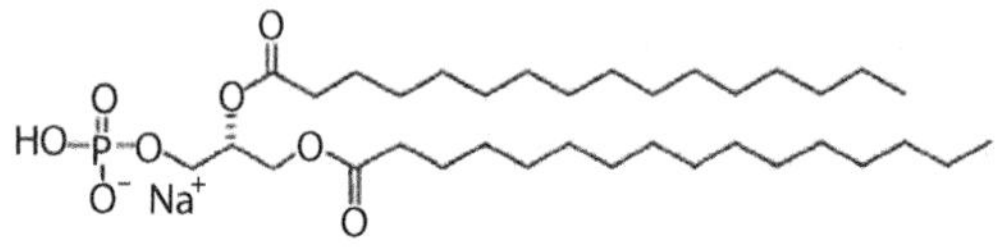

(R)-hexadecanoic acid, 1-[(phosphonoxy)methyl]-1,2ethanediyl ester, monosodium salt (DPPA)

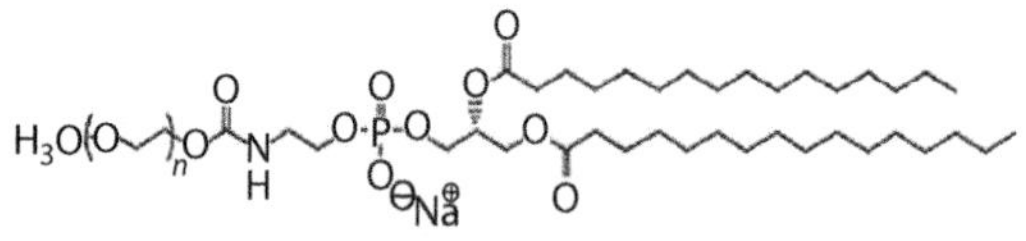

(R)- ∞[6-hydroxy-6-oxido-9-[(1-oxohexadecyl)oxy]5,7,11-trioxa-2-aza-6-phosphahexacos-1-yl]-ω-methoxypoly(ox-1,2-ethanediyl), monosodium salt (MPEG5000 DPPE)

Vial contents: 6.52 mg/mL PFP (headspace). Each mL of unshaken clear liquid: 0.045 mg DPPA, 0.401 mg DPPC, 0.304 mg MPEG5000 DPPE, 103.5 mg propylene glycol, 126.2 mg glycerin, 2.34 mg sodium phosphate monobasic monohydrate, 2.16 mg sodium phosphate dibasic heptahydrate, 4.87 mg sodium chloride, water for injection. pH = 6.2–6.8

FIGURE 15.3 Ingredients used in the clinical phospholipid type USCA, Definity (Lantheus) (http://www.definityimaging.com/pdf/DEFINITY%20Prescribing%20Information%20515987-0413.pdf).

15.3.2.1 Advantages and Disadvantages of Ultrasound Contrast Agents

Unlike XRCAs and MRCAs, USCAs are made from nontoxic natural or synthetic biodegradable materials and a small amount of an inert low-water-soluble gas. Current commercial USCAs that use perfluorocarbon and SF_6 gases shelled with human serum albumin or lipid monolayers must have highly reproducible microbubble size distributions. They are not monodispersed but possess tight dispersions for both safety and efficacy reasons, typically (using data from the Definity Package Insert) with a mean diameter of 1.1–3.3 μm with 98% less than 10 μm. USCAs are therefore small enough to freely circulate in the arterial and venous blood stream and through capillaries of similarly sized red blood cells. But unlike the much smaller XRCAs and MRCAs, they are too large to penetrate the capillary endothelium to extravascular spaces, in general. The advantage of this is that the agent distributes only into a small distribution volume, reducing the dose needed to achieve a detectable concentration, and that the agents are true blood pool agents, defined practically as agents that remain inside the vasculature to >90% as they pass through a tissue. MRCAs and XRCAs diffuse out of blood through capillary fenestrae to ~30%–40% per pass through (i.e., extraction fraction), which complicates attempts to quantitate blood perfusion of tissues. True blood pool agents like USCAs allow, with an accurate detection of the bubbles, an image that is more reliably and directly related to perfusion, rather than the detected signal having a complex relationship between perfusion, blood volume, and other parameters, such as those found in dynamic contrast-enhanced (DCE) MRI. Contrast-enhanced ultrasound can therefore be used to image blood perfusion in organs, and to measure blood flow rate in the heart and other organs.

Circulation of current microbubbles is fairly short-lived (i.e., a few minutes), which makes their application less convenient for lengthy multi-plane exams, requiring multiple injections. They have low circulation residence times because they either are sequestered by immune system cells or by the liver or spleen, even when they are experimentally coated with poly(ethylene glycols) (PEGs) known to resist these mechanisms. Additionally, microbubbles burst at low ultrasound frequencies and at high mechanical indices (MI), which is a measure of the acoustic power output of the ultrasound imaging system. Increasing MI tends to increase the image quality, but results in some microbubble destruction, whose energy can cause local microvasculature ruptures and hemolysis. Interestingly, microbubbles can also be deliberately destroyed by high-MI application, a technique that is used experimentally to increase target to background ratio for USCAs targeted to biological receptors. It is accomplished by allowing the time for the targeted bubbles to reach their endpoints, and then destroying the remaining bubbles with high-MI application remote from the targets in the blood.

From the regulatory and manufacturing perspectives, the USCAs are somewhat challenging. The gas-filled microbubbles, while delivered at tiny doses, are currently approached by physicians and regulatory agencies with caution, as are all insoluble microstructures. Establishment of manufacturing facilities is at a much smaller scale than for MRI and XRCA, the latter being in ton quantities, but is made expensive by the requirement that the microbubbles actually form at their site of use, making the formulation development and controls extremely technical and specialized.

15.3.2.2 Echogenicity

The ability of microbubbles to reflect ultrasound waves from objects in vivo known as echogenicity, analogous to the "relaxivity" or MRCA. Under the influence of the US pressure wave, the microbubbles change their radius, expanding and contracting in a pulsatile manner. Fortunately, the resonance frequency of the microbubbles in the range of diameters that can pass through capillaries lies within the clinical US frequency range of 1 to ~10–15 MHz. What is seen in the images is the scattered signal. The equations that describe this phenomenon are complex (Goldberg et al. 1994) (de Jong 1997) and indicate a detected signal dependence of microbubble radius to the fourth to sixth power. Modern US imagers are also now capable of imaging multiple harmonics of the microbubbles signal, making the micobubble physics of echogenicity in use complex to as great or greater degree than the relaxivity of the MRCA.

Like the MRCA, each commercial USCA is probably slightly different in this ability, but in practice, also like MRCA and XRCA, the commercial variants are similar. The echogenicity derives from the difference in the speed of sound between the gas in the microbubbles and the soft tissue surroundings of the body, which is immense, and it is responsible for producing contrast in the images. A very small mass in a fraction of a milliliter dose (sub-milligram quantities of actual shell material and microliters of gas), intravenously administered, is all that is generally needed for imaging. The shell thickness of the bubble is also important because it must be thin enough to allow the bubble to flex as it passes through capillaries. US creates a pressure wave that causes the bubble to expand and contract in a pulsatile manner, and therefore thicker shells have poorer echogenicity and can change the frequency-dependent response of the bubble.

15.3.2.3 Tolerance

USCAs are safe provided one bears in mind that any intravenous injection carries some risks. The quantities of material injected are much smaller than those of XRCAs and MRCAs and are on the order of what is administered in the field of nuclear medicine. Serious adverse events are therefore quite rare. The US FDA did once add a "black box" warning for the risk of serious cardiopulmonary reactions in some patients (Jakobsen et al. 2005). However, the risk was later recognized to be the same as or less than for other intravenously administered CA, with mortality unaffected (Piscaglia et al. 2006; Kusnetzky et al. 2008; Main et al. 2008).

15.3.3 Future Directions

In fact, only a very small amount of contrast agent in molar quantity of gas and lipids is needed for detection by US. That makes them attractive for targeted contrast-enhanced ultrasound imaging. Microbubble limit of detection (LOD) is so remarkably low that biological receptors at nanomolar concentrations can be readily detected (Figure 15.4) (Klibanov et al. 2004). Even a single bubble can be detected in vitro (Figure 15.5), and even in vivo if it is in motion, although it appears to be larger than it really is due to detection of the scatter rather than the bubble itself.

However, the only biological receptors that the microbubble can access are on the endothelium of the blood vessels or blood cells, and this severely limits the range of targets available to those

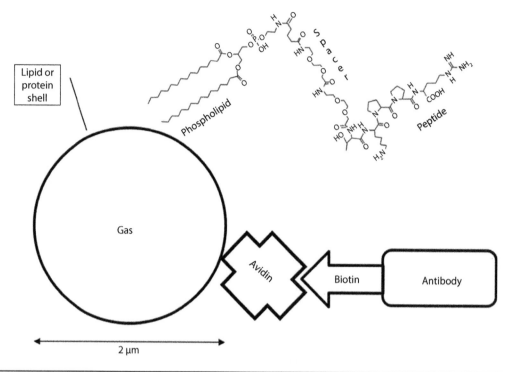

FIGURE 15.4 Schematic of biological receptor targeted USCA.

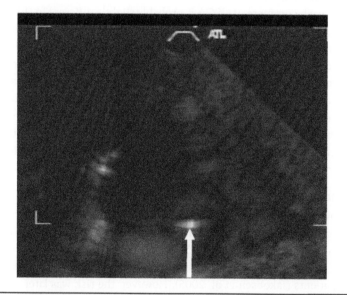

FIGURE 15.5 Ultrasound imaging of an individual targeted microbubble immobilized on a Petri dish, using a Philips HDI 5000 system, equipped with a P4-2 transducer. White arrow points to microbubble. (Reprinted with permission from Klibanov, A.L. et al., *Invest. Radiol.*, 39(3), 187, 2004.)

expressed on endothelial cells, mainly. A few dozen 2 μm diameter bubbles could fit on one side of an endothelial cell, so receptor targeted imaging is nevertheless a recognized and vibrant field of research in USCA. For attachment of the targeting ligands (e.g., antibodies, peptides) to the microbubble surface, well-known techniques, such as carbodiimide, maleimide, and biotin–streptavidin coupling are used (see Figure 15.5). Biotin–streptavidin is a popular coupling strategy because biotin's binding affinity for streptavidin is very high (>10^{15} M^{-1}). Several commercial research companies supply these microbubbles ready to label with biotinylated targeting vectors (see Table 15.6).

Potential applications of targeted USCAs are those for inflammation, cardiac events, cancer, gene, and drug delivery (Unnikrishnan and Klibanov 2012). For inflammation imaging, for example, contrast agents may be designed to bind to certain proteins that become expressed in inflammatory diseases such as Crohn's disease, atherosclerosis, and even acute coronary artery events. The inflamed blood vessels express certain receptors, functioning as cell adhesion molecules such as VCAM-1, ICAM-1, and E-selectin. If microbubbles are targeted with ligands that bind these molecules, they can be used in contrast echocardiography to detect the onset of inflammation immediately following, for example, a coronary thrombosis, in vitro blood detection of which would take hours. Cancer cells can stimulate the expression of specific receptors on nearby endothelial cells, for example, receptors that encourage angiogenesis. If microbubbles are targeted with ligands that bind receptors such as VEGF-R2, they can potentially noninvasively and specifically identify areas of active cancer, angiogenesis being one of the hallmarks of aggressive malignancies. Theoretically, this could make such targeted USCAs more specific for the angiogenesis process than nontargeted USCAs or MRCAs that can image only the existing tumor vasculature, that is, the angiogenesis that occurred in days and weeks past. Such an agent could theoretically identify cancer within the prostate gland, and be used to guide prostate biopsies, currently a notoriously random procedure subject to significant sampling error. The first targeted USCA to be clinically tested is such an angiogenesis targeted agent, binding to the VEGF R2 receptor, and is in human testing at this writing (BR55, Bracco Group).

REFERENCES

Aspelin, P., P. Aubry et al. (2003). Nephrotoxic effects in high-risk patients undergoing angiography. *New Engl. J. Med.* **348**(6): 491–499.

Botnar, R. M., A. Buecker et al. (2004). In vivo magnetic resonance imaging of coronary thrombosis using a fibrin-binding molecular magnetic resonance contrast agent. *Circulation* **110**(11): 1463–1466.

Brady, T. J., M. R. Goldman et al. (1982). Proton nuclear magnetic-resonance imaging of regionally ischemic canine hearts—Effect of paramagnetic proton signal enhancement. *Radiology* **144**(2): 343–347.

Brindle, K. M., S. E. Bohndiek et al. (2011). Tumor imaging using hyperpolarized 13C magnetic resonance spectroscopy. *Magn. Reson. Med.* **66**(2): 505–519.

Cacheris, W. P., S. C. Quay et al. (1990). The relationship between thermodynamics and the toxicity of gadolinium complexes. *Magn. Reson. Imaging* **8**(4): 467–481.

Caravan, P., J. J. Ellison et al. (1999). Gadolinium(III) chelates as MRI contrast agents: Structure, dynamics, and applications. *Chem. Rev.* **99**(9): 2293–2352.

Cowper, S. E. (2001–2009). Nephrogenic systemic fibrosis [ICNSFR Website]. Accessed May 19, 2013.

Datta, A. and K. N. Raymond. (2009). Gd-hydroxypyridinone (HOPO)-based high-relaxivity magnetic resonance imaging (MRI) contrast agents. *Acc. Chem. Res.* **42**(7): 938–947.

de Jong, N. (1997). *Physics of Microbubble Scattering.* Kluwer, Dordrecht, the Netherlands.

Dischino, D. D., E. J. Delaney et al. (1991). Synthesis of nonionic gadolinium chelates useful as contrast agents for magnetic resonance imaging: 1,4,7-tris(carboxymethyl)-10-substituted-1,4,7,10-tetraazacyclododecanes and their corresponding gadolinium chelates. *Inorg. Chem.* **30**(6): 1265–1269.

Eaton, S. M., J. J. Hagan et al. (1988). A predictive test for adverse reactions to contrast media: Preliminary results. *Invest. Radiol.* **23**(Suppl. 1): S206–S208.

Faez, T., M. Emmer et al. (2013). 20 Years of ultrasound contrast agent modeling. *IEEE Trans. Ultrason. Ferroelectr. Freq. Control* **60**(1): 7–20.

Feinstein, S. B., J. Cheirif et al. (1990). Safety and efficacy of a new transpulmonary ultrasound contrast agent: Initial multicenter clinical results. *J. Am. Coll. Cardiol.* **16**(2): 316–324.

Fischer, H. W. (1986). Catalog of intravascular contrast media. *Radiology* **159**(2): 561–563.

Frenzel, T., P. Lengsfeld et al. (2008). Stability of gadolinium-based magnetic resonance imaging contrast agents in human serum at 37°C. *Invest. Radiol.* **43**(12): 817–828.

Fritz, T. A., E. C. Unger et al. (1997). Phase I clinical trials of MRX-115. A new ultrasound contrast agent. *Invest. Radiol.* **32**(12): 735–740.

Gallagher, F. A., S. E. Bohndiek et al. (2011). Hyperpolarized 13C MRI and PET: In vivo tumor biochemistry. *J. Nucl. Med.* **52**(9): 1333–1336.

Goldberg, B. B., J. B. Liu et al. (1994). Ultrasound contrast agents—A review. *Ultrasound Med. Biol.* **20**(4): 319–333.

Golman, K. A., T. (1984). Urographic contrast media and methods of investigative uroradiology. In M. Sovak (ed.), *Radiocontrast Agents.* Springer Verlag, New York, Vol. 73, pp. 127–191.

Goyen, M. (2008). Gadofosveset-enhanced magnetic resonance angiography. *Vasc. Health Risk Manag.* **4**(1): 1–9.

Grainger, R. G. (1982). Intravascular contrast-media—The past, the present and the future. *Br. J. Radiol.* **55**(649): 1–18.

Greco, A., M. T. McNamara et al. (1990). Gadodiamide injection: Nonionic gadolinium chelate for MR imaging of the brain and spine—Phase II-III clinical trial. *Radiology* **176**(2): 451–456.

Gschwend, S., W. Ebert et al. (2011). Pharmacokinetics and imaging properties of Gd-EOB-DTPA in patients with hepatic and renal impairment. *Invest. Radiol.* **46**(9): 556–566.

Hancu, I., W. T. Dixon et al. (2010). CEST and PARACEST MR contrast agents. *Acta Radiologica* **51**(8): 910–923.

Hao, D. A. T., Goerner, F., Hu, X., Runge, V. M., and Tweedle, M. F. (2012). MRI contrast agents: Basic chemistry and safety. *J. Magn. Reson. Imag.* **35**: 1060–1071.

Hoey, G. B. S. K. R., S. El-Antably, and G. P. Murphy. (1984). Chemistry of x ray contrast media. In M. Sovak (ed.), *Radiocontrast Agents*. Springer-Verlag, New York, Vol. 73, pp. 23–114.

Insert, P. (2011). Definity. Retrieved April 2, 2014, from http://dailymed.nlm.nih.gov/dailymed/lookup.cfm?setid=8ab9c79c-1b5c-4e86-899c-cc74686f070a.

Jakobsen, J. A., R. Oyen et al. (2005). Safety of ultrasound contrast agents. *Eur. Radiol.* **15**(5): 941–945.

Kasokat, T. and K. Urich. (1992). Quantification of dechelation of gadopentetate dimeglumine in rats. *Arzneimittel-Forschung/Drug Res.* **42-1**(6): 869–876.

Klibanov, A. L. (2002). *Ultrasound Contrast Agents: Development of the Field and Current Status*. In W. Krause (ed.), *Control Agents II*. Springer, Berlin, Germany, pp. 73–106.

Klibanov, A. L., P. T. Rasche et al. (2004). Detection of individual microbubbles of ultrasound contrast agents—Imaging of free-floating and targeted bubbles. *Invest. Radiol.* **39**(3): 187–195.

Krause, W., H. Miklautz et al. (1994). Physicochemical parameters of x-ray contrast media. *Invest. Radiol.* **29**(1): 72–80.

Kusnetzky, L. L., A. Khalid et al. (2008). Acute mortality in hospitalized patients undergoing echocardiography with and without an ultrasound contrast agent—Results in 18,671 consecutive studies. *J. Am. Coll. Cardiol.* **51**(17): 1704–1706.

Lasser, E. C. (2011). X-ray contrast media mechanisms in the release of mast cell contents: Understanding these leads to a treatment for allergies. *J. Allergy* **2011**: 1–5.

Lasser, E. C., G. E. Lamkin et al. (1995). A role for nitric-oxide in x-ray contrast material toxicity. *Acad Radiol.* **2**(7): 559–564.

Lauffer, R. B. (1987). Paramagnetic metal-complexes as water proton relaxation agents for nmr imaging—Theory and design. *Chem. Rev.* **87**(5): 901–927.

Lauffer, R. B., D. J. Parmelee et al. (1998). MS-325: Albumin-targeted contrast agent for MR angiography. *Radiology* **207**(2): 529–538.

Lauterbur, P.C., M. Mendonca-Dias et al. (1978). Augmentation of tissue water proton spin-lattice relaxation rates by in vivo addition of paramagnetic ions. *Front. Biol. Energ.* **1**: 752–759.

Magnevist. (2013). Package Insert, MAGNEVIST® (brand of gadopentetate dimeglumine) injection. http://labeling.bayerhealthcare.com/html/products/pi/Magnevist_PI.pdf.

Main, M. L., A. C. Ryan et al. (2008). Acute mortality in hospitalized patients undergoing echocardiography with and without an ultrasound contrast agent (multicenter registry results in 4,300,966 consecutive patients). *Am. J. Cardiol.* **102**(12): 1742–1746.

Maravilla, K. R. (2006). Gadobenate dimeglumine-enhanced MR imaging of patients with CNS diseases. *Eur. Radiol.* **16**(Suppl. 7): M8–M15.

Maravilla, K. R., J. A. Maldjian et al. (2006). Contrast enhancement of central nervous system lesions: Multicenter intraindividual crossover comparative study of two MR contrast agents. *Radiology* **240**(2): 389–400.

Marckmann, P., L. Skov et al. (2006). Nephrogenic systemic fibrosis: Suspected causative role of gadodiamide used for contrast-enhanced magnetic resonance imaging. *J. Am. Soc. Nephrol.* **17**(9): 2359–2362.

Margerum, D. W., G. R. Calyley et al. (1978). KInetics and Mechanisms of Complex Formation and Ligand Exchange. In A. E. Martell (ed.), *Coordination Chemistry*. American Chemical Society, Washington, DC, Vol. 2, pp. 165.

Merbach, A. E. and E. Tóth. (2001). *The Chemistry of Contrast Agents in Medical Magnetic Resonance Imaging*. John Wiley & Sons, New York.

Molgo, J., E. del Pozo et al. (1991). Changes of quantal transmitter release caused by gadolinium ions at the frog neuromuscular junction. *Br. J. Pharmacol.* **104**(1): 133–138.

Morel, D. R., I. Schwieger et al. (2000). Human pharmacokinetics and safety evaluation of SonoVue, a new contrast agent for ultrasound imaging. *Invest. Radiol.* **35**(1): 80–85.

Niendorf, H. P., J. C. Dinger et al. (1990). Tolerance of Gd-DTPA: Clinical experience. In G. M. Bydder, R. Felix et al. (eds.), *Contrast Media in MRI*. Medicom Europe, Bussum, the Netherlands, pp. 31–40.

Nunn, A. D. (2006). The cost of developing imaging agents for routine clinical use. *Invest. Radiol.* **41**(3): 206–212.

Nunn, A. D., K. E. Linder et al. (1997). Can receptors be imaged with MRI agents? *Q. J. Nucl. Med.* **41**(2): 155–162.

Overoye-Chan, K., S. Koerner et al. (2008). EP-2104R: A fibrin-specific gadolinium-based MRI contrast agent for detection of thrombus. *J. Am. Chem. Soc.* **130**(18): 6025–6039.

Pintaske, J., P. Martirosian et al. (2006). Relaxivity of Gadopentetate Dimeglumine (Magnevist), Gadobutrol (Gadovist), and Gadobenate Dimeglumine (MultiHance) in human blood plasma at 0.2, 1.5, and 3 Tesla. *Invest. Radiol.* **41**(3): 213–221.

Piscaglia, F., L. Bolondi et al. (2006). The safety of Sonovue (R) in abdominal applications: Retrospective analysis of 23188 investigations. *Ultrasound Med. Biol.* **32**(9): 1369–1375.

Planchamp, C., M. Gex-Fabry et al. (2004). Gd-BOPTA transport into rat hepatocytes: Pharmacokinetic analysis of dynamic magnetic resonance images using a hollow-fiber bioreactor. *Invest. Radiol.* **39**(8): 506–515.

Polasek, M. and P. Caravan. (2013). Is macrocycle a synonym for kinetic inertness in Gd(III) complexes? Effect of coordinating and noncoordinating substituents on inertness and relaxivity of Gd(III) chelates with DO3A-like ligands. *Inorg. Chem.* **52**(7): 4084–4096.

Port, M., J. M. Idee et al. (2008). Efficiency, thermodynamic and kinetic stability of marketed gadolinium chelates and their possible clinical consequences: A critical review. *Biometals* **21**(4): 469–490.

Puttagunta, N. R., W. A. Gibby et al. (1996). Human in vivo comparative study of zinc and copper transmetallation after administration of magnetic resonance imaging contrast agents. *Invest. Radiol.* **31**(12): 739–742.

Rohrer, M., H. Bauer et al. (2005). Comparison of magnetic properties of MRI contrast media solutions at different magnetic field strengths. *Invest. Radiol.* **40**(11): 715–724.

Runge, V. M., W. G. Bradley et al. (1991). Clinical safety and efficacy of gadoteridol: A study in 411 patients with suspected intracranial and spinal disease. *Radiology* **181**(3): 701–709.

Schmidtt-Willich, H. (2007). Stability of linear and macrocyclic gadolinium based contrast agents. *Br. J. Radiol.* **80**: 581–582.

Schmitt-Willich, H., M. Brehm et al. (1999). Synthesis and physicochemical characterization of a new gadolinium chelate: The liver-specific magnetic resonance imaging contrast agent Gd-EOB-DTPA. *Inorg. Chem.* **38**(6): 1134–1144.

Sethi, R., Y. Mackeyev et al. (2012). The Gadonanotubes revisited: A new frontier in MRI contrast agent design. *Inorg. Chim. Acta* **393**: 165–172.

Shehadi, W. H. (1969). On adverse reactions to contrast media, and their incidence. *Am. J. Roentgenol. Radium Ther. Nucl. Med.* **107**(1): 207-&.

Shehadi, W. H., R. C. Pfister et al. (1975). Adverse reactions to intravascularly administered contrast-media—Comprehensive study based on a prospective survey. *Am. J. Roentgenol.* **125**(4): 1001–1001.

Sieber, M. A., H. Pietsch et al. (2008). A preclinical study to investigate the development of nephrogenic systemic fibrosis: A possible role for gadolinium-based contrast media. *Invest. Radiol.* **43**: 65–75.

Solomon, R. J., M. K. Natarajan et al. (2007). Cardiac angiography in renally impaired patients (CARE) study—A randomized double-blind trial of contrast-induced nephropathy in patients with chronic kidney disease. *Circulation* **115**(25): 3189–3196.

Sovak, M. (1984). Introduction: State of the art and design principles of contrast media. In M. Sovak (ed.), *Radiocontrast Agents.* Springer-Verlag, New York, Vol. 73, pp. 1–22.

Stacul, F., A. J. van der Molen et al. (2011). Contrast induced nephropathy: Updated ESUR Contrast Media Safety Committee guidelines. *Eur. Radiol.* **21**(12): 2527–2541.

Tweedle, M. F. (1989). Relaxation agents in NMR imaging. In J. C. G. Bünzli and G. R. Choppin (eds.), *Lanthanide Probes in Life, Chemical and Earth Sciences, Theory and Practice.* Elsevier, Amsterdam, the Netherlands, pp. 127–179.

Tweedle, M. F., S. M. Eaton et al. (1988a). Comparative chemical structure and pharmacokinetics of MRI contrast agents. *Invest. Radiol.* **23**(Suppl. 1): S236–S239.

Tweedle, M. F., G. T. Gaughan et al. (1988b). Considerations involving paramagnetic coordination-compounds as useful NMR contrast agents. *Nucl. Med. Biol.* **15**(1): 31–36.

Tweedle, M. F., J. J. Hagan et al. (1991). Reaction of gadolinium chelates with endogenously available ions. *Magn. Reson. Imaging* **9**(3): 409–415.

Tweedle, M. F., P. Wedeking et al. (1995). Biodistribution of radiolabeled, formulated gadopentetate, gadoteridol, gadoterate, and gadodiamide in mice and rats. *Invest. Radiol.* **30**(6): 372–380.

Uggeri, F., S. Aime et al. (1995). Novel contrast agents for magnetic resonance imaging. Synthesis and characterization of the ligand BOPTA and its Ln(III) complexes (Ln = Gd, La, Lu). X-ray structure of disodium (TPS-9-145337286-C-S)-[4-Carboxy-5,8,11-tris(carboxymethyl)-1-phenyl-2-oxa-5,8,11-triazatridecan-13-oato(5-)]gadolinate(2-) in a mixture with its enantiomer. *Inorg. Chem.* **34**: 633–642.

Unnikrishnan, S. and A. L. Klibanov. (2012). Microbubbles as ultrasound contrast agents for molecular imaging: preparation and application. *Am. J. Roentgenol.* **199**(2): 292–299.

Vakil, V., J. J. Sung et al. (2009). Gadolinium-containing magnetic resonance image contrast agent promotes fibrocyte differentiation. *J. Magn. Reson. Imaging* **30**(6): 1284–1288.

Vanwagoner, M., M. Otoole et al. (1991). A phase-I clinical-trial with Gadodiamide injection, a nonionic magnetic-resonance-imaging enhancement agent. *Invest. Radiol.* **26**(11): 980–986.

von Bibra, H., G. Sutherland et al. (1995). Clinical evaluation of left heart Doppler contrast enhancement by a saccharide-based transpulmonary contrast agent. The Levovist Cardiac Working Group. *J. Am. Coll. Cardiol.* **25**(2): 500–508.

Wedeking, P., K. Kumar et al. (1992). Dissociation of gadolinium chelates in mice—Relationship to chemical characteristics. *Magn. Reson. Imaging* **10**(4): 641–648.

Wedeking, P., R. Shukla et al. (1999). Utilization of the nephrectomized mouse for determining threshold effects of MRI contrast agents. *Magn. Reson. Imaging* **17**(4): 569–575.

Wedeking, P. and M. Tweedle. (1988). Comparison of the biodistribution of Gd-153-labeled Gd(Dtpa)2-, Gd(Dota)-, and Gd(Acetate)N in mice. *Nucl. Med. Biol.* **15**(4): 395–402.

Weinmann, H. J., H. Bauer et al. (1996). Mechanism of hepatic uptake of gadoxetate disodium. *Acad. Radiol.* **3**(Suppl. 2): S232–S234.

Weinmann, H. J., R. C. Brasch et al. (1984). Characteristics of gadolinium-DTPA complex: A potential NMR contrast agent. *AJR Am. J. Roentgenol.* **142**(3): 619–624.

White, G. W., W. A. Gibby et al. (2006). Comparison of Gd(DTPA-BMA) (Omniscan) versus Gd(HP-DO3A) (ProHance) relative to gadolinium retention in human bone tissue by inductively coupled plasma mass spectroscopy. *Invest. Radiol.* **41**(3): 272–278.

Wolf, W., C. A. Presant et al. (2000). 19F-MRS studies of fluorinated drugs in humans. *Adv. Drug Deliv. Rev.* **41**(1): 55–74.

Woodard, P. K., T. L. Chenevert et al. (2012). Signal quality of single dose gadobenate dimeglumine pulmonary MRA examinations exceeds quality of MRA performed with double dose gadopentetate dimeglumine. *Int. J. Cardiovasc. Imaging* **28**(2): 295–301.

Woods, M., E. W. C. Donald et al. (2006). Paramagnetic lanthanide complexes as PARACEST agents for medical imaging. *Chem. Soc. Rev.* **35**(6): 500–511.

Radiochemistry for Preclinical Imaging Studies

Sven Macholl and Matthias Glaser

16.1   RADIOTRACER DESIGN

16.1.1   INTRODUCTION

Initially, imaging agents used in single photon emission computed tomography (SPECT) and positron emission tomography (PET) were radioisotopes or basic modifications thereof (ions, very basic small molecules, or aggregates like colloids). Since then, modern synthetic radiochemistry has created a large and increasing choice of more complex radiotracer molecules. Radiochemistry nowadays can be considered one of the main motors of PET and SPECT research. To facilitate and promote a quicker, more widespread use of developed radiotracers in different imaging labs and clinics, radiosynthesis automation is another key advancement for nonmetal radiotracers.

Small animal radionuclear imaging can be performed with established radiotracers, but more often it is a means to study and validate novel radiolabeled compounds generated by teams of medicinal chemists and radiochemists with input by biologists and applied imaging scientists. Experimental results obtained from these imaging and other biological experiments allow chemists to further optimize the radiotracer molecule and/or its formulation. Because of the intertwined, multidisciplinary nature of this area of research, imaging scientists and biologists will benefit from a good understanding of the radiochemistry concepts. This chapter first covers the production of radioisotopes and then the physicochemical properties, synthesis, formulation, and quality control of radiotracers. Separate sections are dedicated to the most popular radioisotopes that are used in SPECT and PET.

16.1.2   NOMENCLATURE

Radiochemists synthesize radiotracers and base their names on the molecular structure. However, these names may be lengthy and cumbersome. Therefore, radiochemists and the application scientists like PET/SPECT imaging scientists instead use acronyms, common names, pharmaceutical compound codes, trade names, International Nonproprietary Names (INN), etc. Still, the isotopical modification needs to be conveyed unequivocally, which is achieved by conventional rules based on the IUPAC recommended nomenclature of isotopically modified compounds (IUPAC Commission on the Nomenclature of Organic Chemistry 1978; Ferneliums et al. 1981). We will use the following rules:

1. Molecular formulae: the ordinary chemical symbol of the isotopically modified element is replaced by the isotope symbol, in its original position. This isotope symbol is enclosed in square brackets to indicate a specifically labeled compound as per IUPAC recommendations. Examples: $Na[^{123}I]$, $Na[^{99m}Tc]O_4$, $[^{111}In]^{3+}$.
2. Chemical names: the isotope symbol is added to the name of the isotopically unmodified compound. The isotope symbol is written as in rule (1) enclosed in square brackets, and it directly precedes the modified entity. Examples: sodium $[^{123}I]$iodide, sodium $[^{99m}Tc]$pertechnetate, $[^{111}In]$indium(III).

3. Common/trivial names, codes and acronyms of compounds containing radio-nonmetals and radio-metalloids: similar to rule (2), the isotope symbol in square brackets is added to the name of the isotopically unmodified compound, preceding the name of the modified entity. Examples: [^{123}I]ioflupane, [^{75}Se]tauroselcholic acid, [^{11}C]PK11195.

4. Names of radiometal complexes: these complexes consist of the radiometal and an isotopically unmodified organic molecule, which is usually given a common/trivial name, a code, or an acronym. The isotope symbol is written without parentheses or brackets, followed by a hyphen and the organic molecule name. Examples: ^{99m}Tc-bicisate, ^{99m}Tc-TRODAT-1. This is very common in the scientific community (Ballinger 2005), although contrary to the IUPAC rules that would result in names like [^{99m}Tc]technetium bicisate.

The IUPAC recommendations define a specifically labeled compound as the isotopically unmodified compound (i.e., mixture of isotopologues based on the naturally occurring composition of isotopes) labeled with the same compound exhibiting a single isotopical modification. Note that here the compound batch rather than a single molecule is considered to be labeled. For simplicity, this concept is used in the rules discussed even for compounds that are produced carrier-free, that is, which are made up of only a single isotopologue (with regard to the isotopically modified atom[s]) and which, therefore, in principle could be written without square brackets as per IUPAC rules, for example, Na^{123}I.

16.2 SPECT RADIOTRACERS FOR PRECLINICAL IMAGING

16.2.1 INTRODUCTION

A range of SPECT radiotracers is currently in clinical use in a large number of clinics. About 80% of all routine nuclear medical diagnostic imaging procedures are based on ^{99m}Tc-labeled radiotracers making this the workhorse of radiopharmacy (Alberto and Abram 2011). Selected current and more historic SPECT radiotracers that are listed more frequently in review papers and monographs (e.g., Schwochau 2000; Keidar et al. 2003; Agdeppa and Spilker 2009; Pimlott and Sutherland 2011; Adak et al. 2012) are summarized in Table 16.1 grouped by application.

In addition to the established SPECT radiotracers as exemplified in Table 16.1, new SPECT imaging agents are continuously devised in research labs with preclinical imaging being conducted as part of mechanistic, developmental, and translational studies. In particular, ^{99m}Tc labeling of investigational compounds for preclinical biological and imaging studies can be more straightforward than the generally more resource-intensive ^{18}F or ^{11}C PET radioisotope labeling. The generation of the radioisotope ^{99m}Tc, the ^{99m}Tc labeling of the precursor and subsequent formulation often can be done in a convenient way and with rather basic radiochemistry training. However, it may be challenging for some small molecules to attach a ^{99m}Tc chelate and still retain adequate affinity for the molecular target.

The remainder of this section on SPECT radiotracers is organized as follows: first, radiochemistry and imaging-related characteristics of the most common SPECT radioisotopes are discussed, and then concepts and examples for the production of SPECT radiotracers are given by radioisotope. SPECT radiotracers that involve little radiochemistry (e.g., pertechnetate, and Ga^{3+} and Tl$^+$ aquo complexes) and radiotherapeutic agents (e.g., ^{131}I labeled antibodies, and theranostic companions like $^{186/188}$Re paired with ^{99m}Tc, or ^{90}Y paired with ^{111}In) are mentioned in less detail.

16.2.2 PROPERTIES AND PRODUCTION OF COMMON SPECT RADIOISOTOPES

16.2.2.1 Physical Half-Life and Specific Activity

Fundamental physical parameters for radionuclear imaging are the physical half-life and the energy ranges and intensities of the main gamma-ray emission peaks. The most common SPECT isotopes exhibit physical half-lives between 6 h and a few days, which makes them suitable for a certain

TABLE 16.1

Examples of Established (Current and More Historic) SPECT Imaging Agents and Their Fields of Application as per Anatomical Therapeutic Chemical (ATC) Classification System

ATC First-Level Code, Anatomical Main Group	Exemplary SPECT Imaging Agents
A, central nervous system	^{99m}Tc-bicisate (ECD, Neurolite)
	^{99m}Tc-exametazime (HMPAO, Ceretec)
	^{99m}Tc-TRODAT-1
	[^{123}I]ioflupane (DaTSCAN)
	^{111}In-pentetate
B, skeleton	^{99m}Tc-oxidronate (TechneScan HDP)
	^{99m}Tc-medronate (MDP)
C, renal system	^{99m}Tc-pentetate
	^{99m}Tc-mertiatide (TechneScan MAG3)
D, hepatic and reticuloendothelial system	^{99m}Tc-mebrofenin
	[^{75}Se]tauroselcholic acid (SeHCAT)
E, respiratory system	^{99m}Tc-pentetate
F, thyroid	Sodium [^{99m}Tc]pertechnetate
	Sodium [^{123}I]iodide
G, cardiovascular system	^{99m}Tc-sestamibi (MIBI, Cardiolite)
	^{99m}Tc-tetrofosmin (Myoview)
	[^{201}Tl]thallium chloride
H, inflammation and infection detection	^{99m}Tc-exametazime labeled cells
	^{111}In-oxinate labeled cells
	[^{67}Ga]gallium citrate
I, tumor detection	[^{123}I]iobenguane (AdreView)
	^{111}In-pentetreotide (OctreoScan)
X, other	[^{75}Se]norcholesterol

Source: WHO, WHOCC—ATC/DDD index, retrieved March 14, 2014, from http://www.whocc.no/atc_ddd_index/?code=V09, 2013.
Notes: International Nonproprietary Names (INN) are used, followed by common alternative names and trade names in parentheses.

range of clinical applications; see Table 16.2. Another selection criterion for SPECT radioisotopes is a gamma-ray emission peak at energy within the sensitive range of the detector. Commercial SPECT detectors are usually based on sodium iodide crystals doped with thallium, NaI(Tl), and more recently also on cadmium zinc telluride (CdZnTe, CZT) crystals (Seo et al. 2008).

There are a number of direct and indirect consequences of these two parameters on radioisotope selection and radiochemistry that will be explored in the following. If a much shorter half-life is desired for the clinical application, a range of PET radionuclides may be better suited. Thus the half-life may even influence the decision on the modality SPECT or PET for a radiotracer development program.

The radiopharmaceutical preparation is directly affected by the physical half-life. This includes the delivery of the radioisotope to the chemistry lab, radiosynthesis, purification, quality control, and the delivery of the radiolabeled product to the imaging lab. These processes need to be optimized particularly for the mostly rather short-lived PET radiotracers. But nuclear decay may also be significant in some ^{99m}Tc radiotracer preparations if, for example, the purification process and delivery to the imaging lab take hours. The radiochemist may be challenged to speed up the purification process or to devise a new radiosynthesis route to obtain an acceptable purity profile directly after synthesis. Finally, significant nuclear decay may reduce the radiochemical activity concentration (RAC) of the radiotracer preparation to a level below the requirements of an imaging study. In turn, the RAC at the end of synthesis may have to be increased leading to accelerated autoradiolysis. Again radiochemistry can provide a solution by adding a radiostabilizer to the formulation.

TABLE 16.2
Radionuclear and Radiochemical Characteristics of SPECT Radiotracers

Radioisotope	Physical Half-Life (h)	Max. Molar Specific Activity (TBq/µmol)[a]	Main Nuclear Decay Route, Probability	Main Gamma Emission Peak(s) (keV), Relative Intensity	Reference[b]	Production	Product	Radiochemistry[c]
^{99m}Tc	6.0	19.3	Internal conversion of this metastable (m) nuclear isomer, 99%	140.5, 89%	Vol. 1	Generator (parent ^{99}Mo from nuclear reactor)	[^{99m}Tc]pertechnetate	Radiometal-coordination complex
^{123}I	13.2	8.8	Electron capture, 97%	159.0, 83.3%	Vol. 1	Cyclotron	[^{123}I]iodide	Halogen chemistry
^{111}In	67.3	1.7	Electron capture, 100%	171.3, 90.6% 245.4, 94.1%	Vol. 3	Cyclotron	[^{111}In]indium(III)	Radiometal-coordination complex
^{201}Tl	73.0	1.6	Electron capture, 100%	167.5, 10%	Vol. 2	Cyclotron	[^{201}Tl]thallium(I)	Directly as Tl^+ ion
^{67}Ga	78.3	1.5	Electron capture, 100%	93.3, 38.1% 184.6, 21.0% 300.2, 16.6%	Vol. 7	Cyclotron	[^{67}Ga]gallium(III)	Radiometal-coordination complex (see ^{68}Ga)
^{75}Se	2875	0.04	Electron capture, 100%	121.1, 16.9% 136.0, 57.7% 264.6, 58.8% 279.5, 24.9%	Vol. 5	Nuclear reactor, cyclotron	[^{75}Se]selenious acid or elemental ^{75}Se	Organoselenium chemistry

[a] Assuming a single radiolabel per radiotracer molecule and being carrier-free; T = tera = 10^{12}.

[b] Radioactive decay data from http://www.nucleide.org/DDEP_WG/DDEPdata.htm (Laboratoire National Henri Becquerel, C.E.A. Saclay—91191 Gif-sur-Yvette Cedex, France), Monographie BIPM-5—Table of Radionuclides; volume given in table.

[c] Ions like TcO_4^-, Tl^+, and Ga^{3+} for SPECT without synthetic chemistry are out of scope for the rest of this review.

Specific activity A_{spec} is the radioactivity at a certain time point in relation to the mass (or amount) of the compound material. Simple examples in the literature refer to pure, carrier-free radionuclides. Then, the maximal molar specific activity is calculated via the half-life of the radionuclide as $A_{\text{spec,mol,max}} = N_A \ln(2)/T_{1/2}$. Consequently, for this theoretical maximum of the specific activity, the same arguments apply as for the half-life discussed earlier. See Table 16.2 for the values of $A_{\text{spec, mol, max}}$ for pure radionuclides or for hypothetical carrier-free radiotracers with a single radiolabel, ignoring decay products and any other chemicals like solvents and impurities. Multiple (n) radiolabels of the same kind in a radiotracer molecule simply multiply the specific activity, $A_{\text{spec,mol,max,multiple}} = nA_{\text{spec,mol,max}}$. From all chemical species that might be present in the formulation, some can be ignored in the calculation of specific activity while others need to be included. The amount of radiotracer decay products (e.g., ^{99}Tc-labeled radiotracer from $^{99\text{m}}$Tc-labeled radiotracer) is negligible if carrier is present. In a no-carrier-added (n.c.a) formulation, such a decay product can be considered a carrier itself, here being continuously and naturally produced rather than added on purpose by the radiochemist; see de Goeij and Bonardi (2005). Carriers are discussed further in the following sections. For other radionuclides in n.c.a. formulations, the decay product may be considered an impurity. For example, ^{123}I decay to stable ^{123}Te leads to disintegration of the original radiotracer molecule into smaller, inactive molecules. These and other chemical species may be present in the formulation in considerable quantities compared to the radiotracer, but they are irrelevant for the specific activity calculations. These species include any unlabeled impurities (quantified via a chemical purity profile instead), radiolabeled impurities (accounted for in the calculation of radiochemical purity [RCP]), and solvents (quantified in the RAC).

Those chemical species that need to be included in the specific activity calculation are isotopic carriers, nonisotopic carriers, and chemically nearly identical precursors. They all behave chemically at least very similar, if not identical, to the radiotracer, but they are present in much larger amounts and thus dominate the formulated compound material. For a detailed review on slightly different definitions of the carrier, see the review by de Goeij and Bonardi (2005) on the IUPAC recommendations. The relative amount of carrier (and the radiotracer if present in a nonnegligible amount) decreases the specific activity by a scaling factor: $A_{\text{spec,mol}} = A_{\text{spec,mol,max}}(N_{\text{spec}}/(N_{\text{tracer}} + N_{\text{carrier}}))$. Normally (but not always, see ref. de Goeij and Bonardi (2005)) $N_{\text{carrier}} \gg N_{\text{tracer}}$, leading to the simplification $A_{\text{spec,mol}} = A_{\text{spec,mol,max}}(N_{\text{tracer}}/N_{\text{carrier}})$. In case of a small molecule, any precursor will likely be chemically quite different. But in case of a macromolecule, the presence of $^{99\text{m}}$Tc ion in the complete radiotracer molecule or its absence (macromolecule-chelator precursor) may have very little influence on the pharmacokinetics and the target affinity (i.e., efficacy in blocking the target). Still, to distinguish this case from that with the usual carrier-added formulation, the specific activity is called pseudo-specific activity $A_{\text{pseudo-spec,mol}} = A_{\text{spec,mol,max}}(N_{\text{tracer}}/N_{\text{precursor}})$. Further concepts are the apparent and the effective specific activities, which are discussed in Eckelman et al. (2008). In summary, the amount of carrier (or, in some cases, the precursor) is an additional factor on the experimental specific activity that can be tuned by the radiochemist. High specific activity is often desired to provide radiotracer properties, meaning that (1) the physiological equilibrium of endogenous metabolites should not be perturbed noticeably by the administered radiopharmaceutical and (2) there should be minimal blocking of the target by the administered formulation. Only in some instances like radioligand binding assays with long-lived isotopes a reduced specific activity may be advisable, for example, to lessen autoradiolysis.

Other aspects related to the physical half-life are more concerned with the imaging process and application, for example, imaging sensitivity, clinical application, and radiation dose (both to researcher/radiographer and animal/patient) requirements. They indirectly impact on the radiochemistry involved and are discussed in the next section.

16.2.2.2 Biological Time Scales Compared to Physical Half-Life

To follow a biological process by imaging with a radiotracer, the physical half-life needs to be at least of the same order of magnitude as that of the biological process. For imaging with targeted radiotracers, the primary biological processes are uptake into and retention at the target. Uptake is very fast and clearance somewhat slower for an optimized radiotracer allowing in principle the usage of a wide range of radioisotopes with respect to their physical half-lives. However, certain secondary biological processes are also important, namely, nonspecific binding of the radiotracer and of undesired radiometabolites in the tissue of interest, in surrounding tissues and in the blood pool. Here, the uptake and metabolism may again be very fast and clearance slow, which therefore causes significant background signal over a relatively long period of time. Therefore, the "biological process" underlying SPECT and PET imaging is actually a combination of many processes with their individual rates. A simple approach to combine them is the ratio of (specific) signal in the tissue of interest to the sum of background signals. This may be optimal at a late time point postinjection (p.i.) for a given radiotracer, for example, several hours or a day p.i. In such cases, ^{99m}Tc and ^{123}I are well-suited radioisotopes, also considering additional factors like patient comfort, safety, and logistics. Note also that for some radiotracers especially the clearance may be slowed down significantly when comparing rodents with humans, often modeled by allometric scaling. Thus, a radioisotope with slower physical decay may be advantageous when translating a radiotracer from preclinical experiments to the clinic. This may, however, come at a cost of increased radiation dose, and the patient radiation dose needs to be kept within an acceptable, safe limit.

16.2.2.3 Effects of Energy Spectrum on Radiochemistry

Detection sensitivity is not only affected by the amount of radioisotope administered and its physical half-life but also by (1) photopeak position(s) in the detector efficiency spectrum, (2) radiochemical factors like RCP, (3) biological factors like the radiotracer concentration in the tissue of interest (determined by receptor density, pharmacokinetics, metabolism), and (4) physical parameters like the scanner hardware and configuration (collimator design). Factors (1) and (2) are discussed further here.

With regard to the detectors, mostly NaI(Tl) crystals are in use currently, which show an approximately linearly decreasing detection efficiency in the energy range relevant for the isotopes listed in Table 16.2, for example, from ca. 90% at 150 keV to ca. 65% at 300 keV for the detector setup studied in Dewan et al. (2005). Inspecting the photopeak energies and relative intensities in Table 16.2 and ignoring differences in photopeak widths, it can be concluded that ^{99m}Tc is the best choice for a SPECT radioisotope from a detection sensitivity point of view, closely followed by ^{123}I and ^{111}In.

The radiochemistry-related factors of the injectate that directly affect the detection sensitivity include the purity of the radiotracer (i.e., RCP), the (pseudo-)specific activity, the radioactivity concentration (RAC) and radiolysis. Typically a robustly high RCP can be achieved by an optimized radiosynthesis method especially for ^{99m}Tc-labeled radiotracers. The (pseudo-)specific activity and RAC may vary more depending on the radioisotope production and, for example, the need for product purification. For small animal imaging, these parameters are important because the volume of injectate is limited for physiological reasons, that is, to reduce perturbation of physiological equilibria and to avoid discomfort to the animal. The limit may be set at 5% of the blood volume in case of an intravenous bolus injection, for example, ca. 0.1 mL for an adult mouse. With regards to reducing radiolysis of high RAC preparations, radiostabilizer may be added as mentioned earlier.

16.2.2.4 Physicochemical Properties and Complex Stability

A number of physicochemical properties of radiotracers affect the interaction with biological matter and thus the pharmacokinetics, for example, molecular shape and size, charge, lipophilicity, hydrogen bond donors and acceptors, acidity, and moieties susceptible to metabolism. These also define the pharmacophore and with it the binding affinity to a receptor, the probability to cross biological barriers including cell membranes, etc. Especially for small molecules, these physicochemical parameters may be strongly influenced by the radioisotope selection. For example, replacing a hydrogen atom by an iodine atom increases molecule lipophilicity and size. Attaching a technetium chelate complex to a small molecular scaffold increases the size considerably and often introduces charge. However, note that even such complexes can be developed for brain imaging; see ^{99m}Tc-TRODAT-1. Another field where physicochemical parameters play a role is the pretargeting approach, but this is out of scope here.

With regard to radiometal complexes, radiochemists often study and optimize complex stability. Thermodynamic complex stability is defined through the equilibrium reaction

$$C \rightleftharpoons Z + L; \quad K_D = \frac{[Z][L]}{[C]}; \quad K_{st} = \frac{[C]}{[Z][L]}$$

between the complex C on one hand, and the central atom Z and ligand L (for simplicity, a single chelator in this example) on the other hand. This reaction equation leads to the definition of the dissociation constant K_D. Its inverse is the stability constant K_{st}, usually reported as the decadic logarithm thereof, $\log_{10}(K_{st})$. For a radiotracer preparation at equilibrium, $[Z] = [C]/(K_{st}[L])$, and with a *per definitionem* small concentration [C] of a very stable complex, the concentration of radiometal [Z] is very small. It becomes even smaller if more ligand L is added or potentially formed by physical decay of a radiometal with subsequent complex dissociation (except for ^{99m}Tc, which decays to long-lived, chemically practically identical ^{99}Tc).

Of more concern is the situation after intravenous *in vivo* administration. Two effects may be considered: (1) a systemic distribution of the free radiometal different to that of the intact radiotracer and (2) the dilution of the injectate within the blood pool. An example for effect (1) is free technetium metal being readily oxidized to pertechnetate, which then accumulates in glands like the thyroid. Mechanistically, the thyroid can be considered a sink for technetium meaning that further continuous, minute release of free radiometal from the intact radiotracer in blood is promoted as per the dynamic equilibrium. There might be also delayed metal release from radiotracer clearing from background tissues such as muscle. Finally, this leads to a change of the safety profile (toxicity and radiation dose) and of the SPECT image compared to the theoretical situation of a fully intact radiotracer.

Effect (2) causes a shift toward dissociation because of the dilution of the injectate within the blood pool. This can be estimated as follows. Consider a formulation with negligible free ligand and free metal, thus $[Z] \approx [L]$ and $K_D = [Z]^2/[C]$, and a volume of 0.1 mL to be injected into a typical mouse or rat with 2.0 and 15 mL total blood volume, respectively. Introducing the dilution factor f for the complex leads to the new initial concentration *in vivo* $[C'] = f[C]$. Considering the definition of the constant K_D, the new free metal and ligand concentrations are $[L'] \approx [Z'] = \sqrt{f}[Z]$. Overall, the complex dissociation equilibrium shifts toward an increased fraction of free metal as per $([C']/[C])/([Z']/[Z]) = \sqrt{f}$ which is ca. 5 for a mouse and ca. 12 for a rat. In other words, the intravenous injection (and thus dilution of complex) leads to a loss of chelate complex by about one order of magnitude. This is significant, but not of concern with a typical, small K_D, often on the order of 10^{-20} mol/L. In conclusion, effect (1) is the main reason why radiometal complexes with high thermodynamic stability are desired.

16.2.2.5 Auxiliary Compounds

Classes of auxiliary compounds that may be components of the radiotracer preparation include (1) blocking agents to prevent harm by the radioisotope in sensitive tissues that are otherwise irrelevant for the diagnostic test, (2) radiostabilizers in case autoradiolysis is of concern as discussed earlier, and (3) antioxidants, solubilizers, radiometal stabilizing ligands, etc. (see Section 16.2.3.1.1). An example for a blocking agent is nonradioactive ("cold") iodide. Radioiodine can be cleaved from certain iodine radiotracers *in vivo* by enzymes, and the blocking agent can minimize harmful radioiodine uptake in the thyroid.

16.2.2.6 Production

^{99m}Tc (as [^{99m}Tc]pertechnetate, [^{99m}Tc]O$_4^-$) can be produced from ^{99}Mo through its β^- decay route (half-life $T_{1/2}(^{99}$Mo$) = 66.0$ h). ^{99}Mo is currently generated in research nuclear reactors using highly enriched uranium. ^{235}U is irradiated with neutrons yielding ^{99}Mo and other fission products. After several days of irradiation, ^{99}Mo is chemically extracted from the uranium target in a hot cell. A risk to a sustainable ^{99m}Tc production capacity for medical diagnostics purposes is the small number and advanced age of research reactors that are available for ^{99}Mo production. Apart from the solution to build new reactors to mitigate this risk, alternative production modes have been proposed or are being developed. These include the use of ^{98}Mo and low-enriched uranium targets, use of accelerators instead of reactors, and irradiation of a liquid solution of an uranium salt in a reactor (Marck et al. 2010; Green 2012).

In convenient, commercially available ^{99m}Tc generators, ammonium [^{99}Mo]molybdate (NH$_4$[^{99}Mo]O$_4$) is adsorbed onto alumina columns. Its nuclear decay leads to [^{99m}Tc]pertechnetate. This product can be extracted by the radiochemist from the alumina column by elution, for example, with normal saline, and this solution is used for subsequent radiolabeling. The solubility of molybdate in normal saline is very low and thus ^{99}Mo remains available on the column for later ^{99m}Tc elutions. Pertechnetate is stable in aqueous solution with respect to its oxidation state due to its rather low redox potential (standard electrode potential) of 0.78 V when forming TcO$_2$ in acidic solution, compared to the same reaction for more reactive permanganate with a redox potential of 1.68 V (Lide 2008).

The two consecutive processes in the ^{99m}Tc generator are (subatomic particles and photons omitted): ^{99}Mo $\xrightarrow{k_1 (\beta^- \text{ decay})}$ ^{99m}Tc $\xrightarrow{k_2 (\text{internal conversion})}$ ^{99}Tc. The rate constants k_1 and k_2 can be expressed via the half-life $T_{1/2}$ of the corresponding decay; see Table 16.1 and preceding text for the values, as $k_1 = \ln(2)/T_{1/2}(^{99}\text{Mo})$ and $k_2 = \ln(2)/T_{1/2}(^{99m}\text{Tc})$. Solving the corresponding system of three rate equations for $\partial c(^{99}\text{Mo},t)/\partial t$, $\partial c(^{99m}\text{Tc},t)/\partial t$, and $\partial c(^{99}\text{Tc},t)/\partial t$ yields the isotope concentrations c as function of time after the last elution. This step is controlled by the radiochemist and, therefore, the calculation of the concentrations warrants further discussion here. For $c(^{99}\text{Mo},t)$, the result is the expected monoexponential decay function with the ^{99}Mo half-life given earlier in the expression for k_1, $c(^{99}\text{Mo},t) = e^{-k_1 t} \cdot c(^{99}\text{Mo},t_0)$. For the other two isotopes, each of the equations contains both nuclear decay rate constants k_1 and k_2:

$$c\left(^{99m}\text{Tc},t\right) = \frac{k_1}{k_2 - k_1}\left(e^{-k_1 t} - e^{-k_2 t}\right) \cdot c\left(^{99}\text{Mo},t_0\right)$$

and

$$c\left(^{99}\text{Tc},t\right) = \left(1 + \frac{k_1 e^{-k_2 t} - k_2 e^{-k_1 t}}{k_2 - k_1}\right) \cdot c\left(^{99}\text{Mo},t_0\right).$$

Here, t_0 denotes the point of time of the last elution and t is the delay after that event. Figure 16.1 shows a plot of these two concentrations as function of the time interval after the last elution,

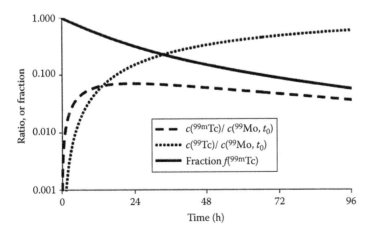

FIGURE 16.1 Decay characteristics of a ^{99}Mo-based ^{99m}Tc generator. Semi-log plot of the concentrations of ^{99m}Tc (dashed line) and ^{99}Tc (dotted line) normalized to the initial concentration of ^{99}Mo, as function of time after elution at $t = 0$. The straight line depicts the fraction of ^{99m}Tc of the sum of all Tc species present.

normalized to the initial ^{99}Mo concentration $c(^{99}$Mo$, t_0)$. These two equations also allow to predict the specific activity of ^{99m}Tc as function of the elution interval. This is the theoretical maximum specific activity (e.g., expressed per mole as in Table 16.1: $S(^{99m}$Tc, max$)=19.3$ TBq/µmol), scaled by the ratio of radioactivity from ^{99m}Tc to that of ^{99m}Tc and the isotopic carrier ^{99}Tc combined. Here, it is assumed that technetium has been perfectly separated from molybdenum. We can call this ratio the relative specific activity or the fraction of technetium species of interest f. This again is a function of the elution interval t

$$S\left(^{99m}\text{Tc},t\right) = S\left(^{99m}\text{Tc,max}\right) \cdot f\left(^{99m}\text{Tc},t\right)$$

and

$$f\left(^{99m}\text{Tc},t\right) = \frac{c\left(^{99m}\text{Tc},t\right)}{c\left(^{99m}\text{Tc},t\right) + c\left(^{99}\text{Tc},t\right)}.$$

Figure 16.1 shows the fraction f decreasing from nominally 1 at t very close to 0 (or practically >99.9% in the first minute after elution) to 90% at 1.85 h, 80% at 4 h, 50% at 13.4 h, 32% at 24 h, 15% at 48 h, and 5.7% at 96 h. This information can be taken into consideration for ^{99m}Tc-labeled preparations optimal for those imaging studies where high specific activity is required.

A sufficiently high radiochemical concentration is required for the injection of an adequate activity dose into small animals where the injection volume is limited. The RAC of the ^{99m}Tc eluate can be several hundred MBq/mL or even >1 GBq/mL (Zolle 2007) depending also on the specific activity. A decrease in the RAC during formulation with a freeze-dried kit is expected to be minor, provided no further purification is necessary. The RCP of the ^{99m}Tc eluate must be high for further use in imaging studies, that is, only very little ^{99}Mo in the eluate is accepted (Zolle 2007).

All other radioisotopes listed in Table 16.2 are produced via a cyclotron, and in case of ^{75}Se, another commercially employed route is via ^{74}Se or natural Se in a nuclear reactor. For ^{123}I, often high-enriched ^{124}Xe is used as target in a cyclotron generating aqueous [^{123}I]iodide with high specific activity.

16.2.3 PREPARATION OF **SPECT** RADIOTRACERS

The methods of the preparation of SPECT radiotracers can be divided into several classes:

1. Directly employing the radioisotope as inorganic ion. Examples are [^{67}Ga]gallium(III), [^{99m}Tc]O$_4^-$, and [^{201}Tl]thallium(I).
2. Mixing an organic molecule (precursor) with the radioisotope, which leads to coordination complex formation. In case of ^{99m}Tc labeling, the help of a reducing agent is required. Examples include ^{99m}Tc and ^{111}In chelate complexes.
3. Performing organic syntheses, which leads to covalent bond formation. Examples are ^{75}Se and ^{123}I.

Class 1 methods do not involve synthetic radiochemistry as such and, therefore, they are not discussed further here. For class 2, ^{99m}Tc is used as example and is discussed in the next section, followed by Section 16.2.3.2, which deals with ^{123}I as example for class 3.

The coordination chemistry for ^{111}In is dominated by utilizing 1,4,7,10-tetraazacyclododecane-1,4,7,10-tetraacetic acid (DOTA) as chelator. A list of examples of ^{111}In-labeled antibodies can be found in Khalil et al. (2011).

Established ^{75}Se-labeled radiotracers include selenocholesterol (Scintidren), selenomethionine, and tauroselcholic acid (SeHCAT).

16.2.3.1 ^{99m}Tc Radiochemistry

16.2.3.1.1 *^{99mTc} Radiolabeling*

Ca. 400 entries of ^{99m}Tc-labeled imaging agents were registered in the MICAD database by January 2013 (National Center for Biotechnology Information 2004–2013). Historically, some peptide-based imaging agents have been obtained simply by labeling directly with ^{99m}Tc making use of cysteine or other residues (e.g., see ubiquicidin (Welling et al. 2000)) that can directly coordinate the radiometal ion. Another class of compounds are ^{99m}Tc-labeled colloids. But labeling of biologically active molecules can be done in a much more versatile manner by conjugating a chelator to the biologically active domain with the radiometal being inserted into that chelator. ^{99m}Tc radiolabeling can be done as one-pot synthesis involving two processes: reduction of $[^{99m}Tc]O_4^-$ to, for example, a Tc^V species as in linear TcO_2^+, followed by complexation thereof with the ligand/chelator to form a coordination complex. With some excess ligand, the last step often achieves almost "quantitative yields" (i.e., a yield close to 100%).

For this ^{99m}Tc radiolabeling procedure, a minimum of three components are required: the precursor, $[^{99m}Tc]$pertechnetate, and a reducing agent (e.g., Sn^{II}). The reducing agent is usually included in a lyophilized, sterile kit containing additional excipients such as a radiostabilizer (e.g., sodium para-aminobenzoate, sodium ascorbate), antioxidants (e.g., sodium ascorbate, gentisic acid (Handeland and Sundrehagen 1987)), or Sn^{II} solubilizer (Edwards et al. 2008). This is kept in a vial sealed under nitrogen gas. Furthermore, the pH needs to be adjusted via a buffer (e.g., carbonate, acetate, or phosphate buffer) or with HCl or NaOH prior to lyophilization. Finally, for ligand-exchange preparations, another ligand like gluconate may be added to the kit. Such kits are commercially available to allow for a simple one-step preparation of the radiotracer injectate by adding the generator eluate under aseptic conditions—this preparation and administration are referred to as "shake and shoot" in lab jargon.

16.2.3.1.2 *Design of ^{99mTc} Radiotracers*

^{99m}Tc complexes can be classified in many different ways. For example, some authors define up to three generations of ^{99m}Tc radiotracers. Historically, labeling was rather random in terms of number and positions of ^{99m}Tc atoms per macromolecule, but this is the exception nowadays. Examples for this type of direct labeling are ^{99m}Tc-macroaggregated albumin (^{99m}Tc-MAA) and ^{99m}Tc-labeled antibodies (Carroll et al. 2012). In this context, the pretinning method (Rhodes et al. 1986) and the Schwarz–Steinstrasser method (Mather and Ellison 1990) had been developed and employed. Often it is preferred that the complete molecular structure of the radiotracer is well characterized and that the labeling is site specific, in particular for small molecules, but also for peptides. Those second and third generations of ^{99m}Tc radiotracers can be classified based on the complex type: (1) with more than one ligand (in the chemical sense) to form a coordination complex as in ^{99m}Tc-MDP (see next section) and (2) a single chelator like bis(aminoethanethiol) (Kung et al. 1985) (i.e., multidentate ligand) covalently conjugated to a targeting domain and forming a chelate with the metal ion (including TcO^{3+}, TcO_2^+, etc.). In other words, either there are multiple ligands that in themselves incorporate certain biological activity, or a chelator is used to carry the signal generating label, while the biological activity of the radiotracer is dominated by a separate entity of that molecule. In the latter case, the molecular link between the chelate and the targeting vector may be considered an additional, third molecular moiety. This linker may exhibit negligible biological activity or it may be deliberately employed as additional biomodifier, which alters the pharmacokinetics, for example, of through-membrane transport and background (nontarget) tissue clearance. The linker may also be employed to separate the chelate spatially from the targeting vector in order to minimize expected, undesired interference in the docking process to the target, that is, to prevent the reduction of the binding affinity.

Also, the chelator may be equipped with a reactive site that allows the coupling of the ^{99m}Tc chelate complex to the targeting vector after radiolabeling the chelator. This is an indirect labeling approach and such a chelating agent is termed bifunctional. Effectively the order of the two chemical synthesis steps (coupling of the chelator to the targeting domain and complexation with radiometal) is reversed compared to the previously discussed case where the complete precursor (targeting vector, linker, chelator) is first assembled and then radiolabeled (direct labeling). The so-called pretargeting involves the administration of a targeting vector to the subject first, followed by the delayed administration of the radiotracer to couple to the vector *in vivo* with an appropriate functionality. This can be viewed as a special case of indirect labeling.

In summary, the design of metal chelate radiotracers can be categorized with respect to the following criteria:

◆ Specific labeling or random labeling
◆ Chelate integrated radiotracer or chelate coupled to biologically active (targeting) domain
◆ Optionally, pharmacokinetics modifying groups present
◆ Direct labeling or indirect labeling

Some of these criteria apply to other radioisotopes as well. A further, more rare approach is to design a prodrug-like agent that contains a metabolic cleavage point. Then the desired radiotracer is formed *in vivo* in the presence of relevant enzymes.

16.2.3.1.3 Coordination Complex Types and Chelators for ^{99m}Tc

A fairly large number of ligands and chelators have been developed and explored, drawing from the field of coordination chemistry. This variety is possible because technetium can adopt a large range of oxidation states (Tc^V, Tc^{III}, and Tc^I being particularly popular in ^{99m}Tc complexes (Carroll et al. 2012)), and different coordination numbers and geometries can be realized by using different ligands.

The main criteria for an optimal chelator/ligand from a radiochemist's point of view are ease and low cost of precursor synthesis, precursor stability, rapid complexation with ^{99m}Tc at ambient temperature, and high complex stability. High complex stability in the formulation solution can be achieved by matching the ligand field strength with the hardness of the metal ion. For example, a soft metal ion and a weak field ligand match well. The reader is referred to the concepts of crystal field theory, ligand field theory, and kinetic stability for further information. One way to increase thermodynamic stability of complexes is to increase the denticity of the ligands, for example, chelate complexes are more stable than monodentate ligands. The higher the number of ligands, the higher the gain in entropy upon complex degradation. With entropy differences dominating the Gibbs free energy, this is synonymous to higher thermodynamic instability with monodentate ligands compared to multidentate chelators.

Further parameters that modulate the pharmacokinetics or stability *in vivo* (i.e., resistance against metabolism) are, for example, charge, size (molecular weight), and lipophilicity of the chelate complex, in particular for smaller compounds. These factors again vary with the technetium oxidation state and the chelator/ligand type.

Differences in the complex stability can also be exploited by first preparing a more labile precursor or intermediate complex. Subsequently, the ligands are exchanged to form a more stable complex. In case of chelator/chelator exchange, this is also called transchelation. See Fritzberg et al. (1986) for an example of gluconate being replaced by MAG_3.

With regard to nomenclature, coordination complexes and organometallic compounds are distinguished. ^{99m}Tc-sestamibi is an organometallic compound because at least one ligand forms a carbon–metal bond (for ^{99m}Tc-sestamibi, it is actually all six ligands). Many other studied ^{99m}Tc complexes exhibit a mix of bonds to the metal, for example, at least one covalent bond to oxygen and coordination bonds to further ligands. An example is ^{99m}Tc-tetrofosmin with TcO_2^+ core and bonds to four phosphine phosphorus atoms. A pure coordination complex is ^{99m}Tc-mebrofenin (Ghibellini et al. 2008) (Choletec) composed of a Tc^{III} center with bonds to two tertiary amino nitrogen and four carboxylate oxygen atoms (N_2O_4).

Another aspect in the description of the ligands is to distinguish between (1) oxo and nitrodo groups that can be viewed as forming a core unit (e.g., TcO_2 unit), (2) the chelator, and (3) coligands that are required if the aforementioned ligand types leave any coordination sites unsaturated (e.g. chloride in the HYNIC complex in Fig. 16.2e).

More recently developed complex types include Tc-nitrido, Tc-HYNIC, and Tc-tricarbonyl complexes; see Figure 16.2e through g. The nitrido core ($Tc \equiv N$) has similarities to the oxo core but provides higher stability. It can be obtained from a hydrazino derivative in the presence of Sn^{II} and pertechnetate (International Atomic Energy 2008). The other two complex types are discussed here in more detail.

HYNIC (2-hydrazino nicotinamide (Meszaros et al. 2010)) is a bifunctional chelator. First, it can be conjugated to nucleophilic groups in a biomolecule, such as the amine groups of lysine residues,

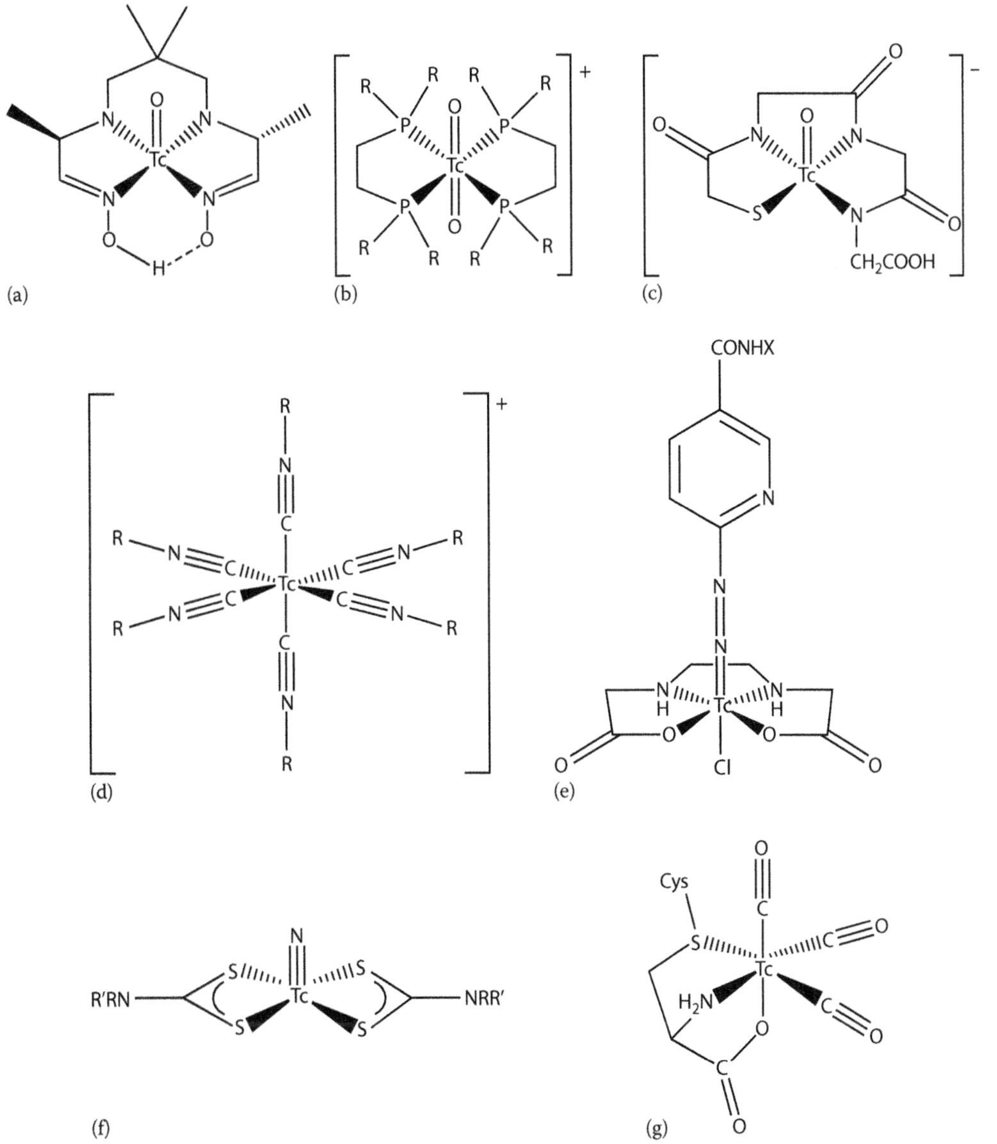

FIGURE 16.2   Molecular structures of technetium complexes. (a) Diamino dioxime chelator, example exametazime, Tc^V; (b) Tc-tetrofosmin [R = EtOEt], Tc^V; (c) Tc-MAG$_3$, Tc^V; (d) Tc-sestamibi [R = CH$_2$CMe$_2$(OMe)], Tc^I; (e) Tc-HYNIC-EDDA-Cl [X = peptide, for example], Tc^V (Liu 2004), but note the bidentate chelate structure proposed more recently (Meszaros, Dose et al. 2011); (f) TcN-NOET, Tc^V (Liu 2004); (g) Tc(CO)$_3$-type complex with tridentate chelator (*S*-functionalized cysteine, Tc^I). (From Alberto, R. et al., *Biopolymers*, 76(4), 324, 2004. With permission.)

via its carboxylic acid group. The reactivity of the carboxylic acid can be increased, for example, by preparing the ester with *N*-succinimidyl as leaving group. Furthermore, when incorporating HYNIC during the peptide synthesis, the hydrazine functional group may need to be protected because it itself is reactive toward activated carbonyl functions, forming a hydrazone. This may include bond formation between the unprotected hydrazine of one HYNIC molecule and the ester function of another HYNIC unit, leading to oligomerization (Meszaros et al. 2010). A popular protecting group is *N-tert*-butoxycarbonyl. In the subsequent ^{99m}Tc-labeling step, the hydrazido ligand binds to the radiometal and the pyridyl nitrogen usually coordinates with the radiometal as well forming a bidentate chelate complex, at least in the crystalline state (Meszaros et al. 2010). To complete the coordination complex, additional coligands are required. A range of mono- and oligodentate coligands have been studied for this, and among them are aminocarboxylates such as ethylenediamine-*N,N'*-diacetic acid (EDDA, see Figure 16.2e) and tricine, phosphines, *N*-heteroaromatics, polyalcohols, and combinations thereof yielding ternary complexes (Meszaros et al. 2010). The HYNIC approach has been developed for larger proteins (e.g., antibodies and serum albumin) and smaller proteins (e.g., annexin V). For a protein, ^{99m}Tc labels may be distributed between several potential coupling sites, but site-specific labeling can be achieved in certain cases as in the case of annexin V (Meszaros et al. 2010).

^{99m}Tc-sestamibi with its isonitrile ligands belongs to the class of organometallic compounds. More recently carbonyl ligands have been employed as well. For this, pertechnetate is reduced by sodium borohydride (NaBH$_4$) in the presence of CO (Alberto et al. 1998) or boranocarbonate Na$_2$H$_3$BCO$_2$ is used as a combination of reducing agent and *in situ* source of CO (Alberto et al. 2001). This gives the intermediate complex *fac*-[^{99m}Tc(CO)$_3$(H$_2$O)$_3$]$^+$ (*fac* for facial isomer). The three carbonyl ligands coordinate strongly with the Tc^I center, which provides higher kinetic stability than technetium in other oxidation states. The three weakly bound water molecules easily undergo ligand exchange, for example, with a tridentate chelator such as histidine or cysteine (Alberto et al. 2004b). Each of these amino acids can also act as bidentate chelator if they are the terminal residue of a peptide. The first of many examples for this type of compounds involves a bidentate chelator and a single chloride ligand (Alberto et al. 1999) but tridentate chelators provide generally more stable complexes (Kluba and Mindt 2013). Also, click-chemistry-based synthetic routes have been explored more recently, for example, coupling chelator and targeting domain equipped with alkyne and azido functions into a triazole (Carroll et al. 2012).

Moreover, analytical and biological studies of the imaging agent may require a nonradioactive ("cold") variant of the test item. Rhenium belongs to the same group in the periodic table as technetium, and due to the lanthanide contraction, many properties are similar. Therefore, nonradioactive Re and the therapeutic radioisotopes ^{186}Re and ^{188}Re form generally good surrogate coordination complexes of the corresponding ^{99m}Tc-labeled complexes and can be used for investigative studies. However, not all chelators form complexes with both ^{99m}Tc and Re such as bis(amine oxime) chelators. A small but significant difference between rhenium and technetium is observed in the redox potential (standard electrode potential) which is even lower for ReO$_4^-$/ReO$_2$ in acidic solution at 0.51 V compared to 0.78 V for the corresponding TcO$_4^-$/TcO$_2$ (Lide 2008). In other words, perrhenate is more difficult to reduce than pertechnetate to form a complex, and equally rhenium in a lower oxidation state in a complex is more easily oxidized compared to technetium. Ligands that can act in themselves as reducing agents, such as thiolates and phosphines, seem to stabilize Re^V well (Carroll et al. 2012).

16.2.3.1.4   Examples of ^{99m}Tc-Labeled Small Molecules

Table 16.3 lists the details of selected small molecules in order to give an overview of the typical range of the technetium oxidation state, ligand types, and the overall charge of established ^{99m}Tc imaging agents. Many more ^{99m}Tc radiotracers have been reviewed elsewhere (e.g., Dilworth and Parrott 1998; Schwochau 2000; International Atomic Energy 2008). Figure 16.2 illustrates some technetium coordinate complexes mentioned in Table 16.3, extended by further examples to cover the chelate types discussed in the preceding section.

TABLE 16.3
^{99m}Tc-Labeled Small Molecules

Name	Alias	Tc Oxidation State	Complex Core, Coordinating Ligand Atoms	Overall Charge
^{99m}Tc-sestamibi	Cardiolite	+1	Tc, C_6	+
^{99m}Tc-pentetate	^{99m}Tc-diethylenetriaminepentaacetic (DTPA), Techneplex	+4; or +5[a]	Tc, N_3O_3; or TcO, N_3O_2[a]	0[a]
^{99m}Tc-TRODAT-1	^{99m}Tc-Tropane as Dopamine Transporter-imaging agent, IUPAC name in Meegalla et al. (1997)	+5	TcO, N_2S_2	0
^{99m}Tc-bicisate	^{99m}Tc-ethylcysteinate dimer (ECD), Neurolite	+5	TcO, N_2S_2	0
^{99m}Tc-mertiatide	^{99m}Tc- mercaptoacetyltriglycine (MAG$_3$), Technescan MAG3	+5	TcO, N_3S	−1
^{99m}Tc-exametazime	^{99m}Tc-hexamethylpropyleneamineoxime (HMPAO), Ceretec	+5	TcO, N_4	0
^{99m}Tc-tetrofosmin	Myoview	+5	TcO_2, P_4	+
^{99m}Tc-DMSA	^{99m}Tc-dimercaptosuccinic acid (DMSA) (Blower et al. 1991)	+5	TcO, S_4	—

Notes: For the overall charge calculations, side chain functions (e.g., COOH in MAG$_3$) are considered neutral. Trademarks are not explicitly mentioned.

[a] Tentative assignments for ^{99m}Tc-pentetate (e.g., Handeland and Sundrehagen (1987) suggests oxidation state +4).

16.2.3.1.5 Examples of ^{99m}Tc-Labeled Macromolecules

Here we describe some selected macromolecules with their chelators, linkers, and any additional bio-modifiers (i.e., pharmacokinetics-modifying groups). The first example is ^{99m}Tc-demobesin 1 (Nock et al. 2003). A tetraamine chelator is employed, which is neutral so that the complex with TcO_2^+ gives a +1 charge. The linker, attached to the N terminus of the eight amino acid peptide, contains glutaramide and *para*-phenylene. The C terminus is capped as a small ethylamide (Figure 16.3).

^{99m}Tc-maraciclatide (previously known as ^{99m}Tc-NC100692) is an eight amino acid peptide based on the RGD motif (Edwards et al. 2008). The RGD unit binds to $\alpha_v\beta_3$ and $\alpha_v\beta_5$ integrins, which are biomarkers of angiogenesis. This core unit is flanked by a pair of cysteine residues, which together form an intramolecular disulfide bond. This leads to a cyclic, more rigid peptide exposing the RGD unit, which is favorable for binding to the integrins. A second bridge is formed by another pair of outer amino acids forming a thioether; see Figure 16.4. The chelator is a diamine dioxime with similarities to that in ^{99m}Tc-exametazime; see Figure 16.2a. The linker is glutaramide extending the sidechain of the N terminal lysine residue. The C terminus is decorated with a chain of two biomodifiers: a polyethyleneglycol (PEG) oligomer with terminal amino groups followed by diglycolamide. PEGylation generally increases molecular hydrophilicity and the systemic clearance rate *in vivo*.

Octreotide is a cyclic peptide of eight amino acids (Pohl, et al. 1995). The bifunctional HYNIC has been used for ^{99m}Tc labeling and a number of different coligands have been explored to optimize labeling yield, stability and lipophilicity, and to study isomerism (Decristoforo and Mather 1999a,b;

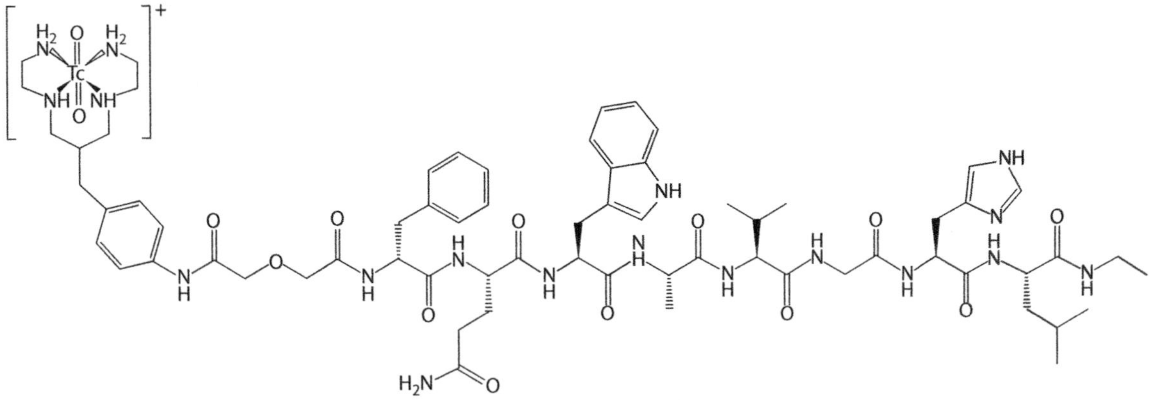

FIGURE 16.3   Molecular structure of ^{99m}Tc-demobesin 1. (From Nock, B. et al., *Eur. J. Nucl. Med. Mol. Imaging*, 30(2), 247, 2003. With permission.)

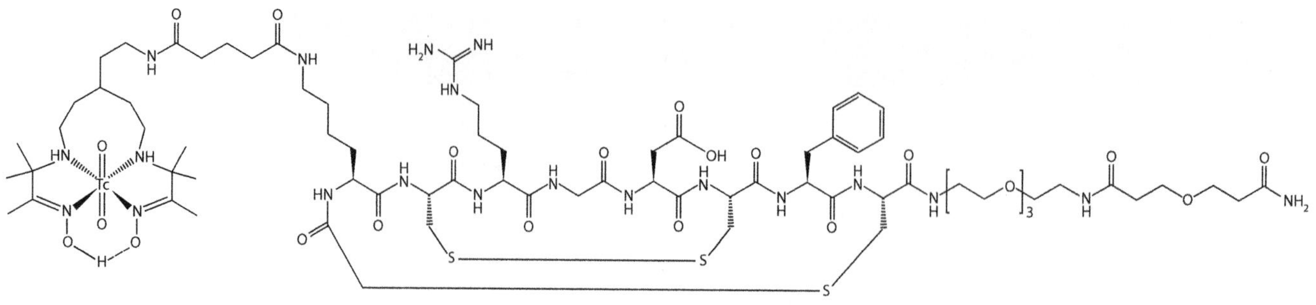

FIGURE 16.4 Molecular structure of ^{99m}Tc-maraciclatide.

Decristoforo et al. 2000). These compounds were also compared to octreotide ^{99m}Tc labeled via the benzoyl MAG$_3$—this had inferior properties as an imaging agent in this case (Decristoforo and Mather 1999; Decristoforo et al. 2000).

An example for a ^{99m}Tc-labeled protein is annexin V, which is used to image apoptosis. The 36 kDa protein can be produced by recombinant techniques in *Escherichia coli*. An established labeling technique is via the bifunctional HYNIC described in more detail earlier (Abrams et al. 1990; Blankenberg et al. 2006).

FDA and/or EMEA approved, targeted macromolecular ^{99m}Tc imaging agents include ^{99m}Tc-apcitide, ^{99m}Tc-arcitumomab, ^{99m}Tc-depreotide, and ^{99m}Tc-sulesomab (Bolzati et al. 2012).

16.2.3.2 ^{123}I Radiochemistry

16.2.3.2.1 *^{123}I Radiolabeling*

Radioiodination has been reviewed multiple times over several decades, see, for example, Coenen et al. (2006) and references therein. [^{123}I]iodide in aqueous solution is readily available as starting material that can be inserted into organic compounds to form a covalent carbon–iodine bond (Kabalka and Varma 1989; Vallabhajosula and Nikolopoulou 2011; Adak et al. 2012). Among the common halogens, fluorine forms the strongest and iodine the weakest bond with carbons because of the steady decrease in electronegativity within the halogen group. Accordingly, de-iodination *in vivo* occurs with many organoiodine compounds, which is to be avoided with imaging agents. Carbon–iodine bonds are observed to be more stable for aromatic than for aliphatic carbons and, therefore, aromatic rings within the molecular scaffold of imaging agents are the more attractive sites for radioiodination (Kabalka and Varma 1989).

Incorporation of iodide into an organic compound is done by substitution, either directly with iodide (nucleophilic substitution) or with the monoiodo cation (I$^+$) or possibly iodo monochloride (ICl) obtained by *in situ* oxidation of iodide (electrophilic substitution).

16.2.3.2.2 *^{123}I Radiolabeling by Nucleophilic Substitution*

For nucleophilic substitution, the precursor contains a suitable leaving group such as chloride, bromide, or tosylate. An alternative is the preparation of the precursor with the stable iodine ^{127}I followed by isotopic substitution with the radioactive isotope ^{123}I. The corresponding two organic compounds are then called isotopologues. Halogen–halogen exchange reactions can be catalyzed by metals, an example is the aromatic Finkelstein reaction that is described in the following paragraph. However, this does not affect the equilibrium concentration ratio of the two organohalide species. In particular, for isotopologues with nearly identical chemical properties, this ratio is rather affected (1) by the small, relative mass difference of the two isotopologues and (2) by the absolute concentrations and their ratio of the two reactants employed in the reaction mixture (i.e., excess level of [^{123}I]iodide compared with the organic precursor). The concentrations can be increased by working with melts instead of solvents (Coenen et al. 2006). Also, the crude product is practically impossible to purify from residual precursor isotopologue because of the chemical similarity and very small molecular weight difference between the isotopologues. Practically, only a low to moderate specific activity can be achieved via the isotopic exchange route (Eersels et al. 2005).

In contrast, when employing two different halogens in the substitution reaction, the equilibrium can be shifted toward completion as in the well-known Finkelstein reaction (Finkelstein 1910) on alkyl halides, employing Le Chatelier's principle. In the original Finkelstein reaction, a solvent is employed that dissolves well both alkyl halides (reactant and product) and the sodium iodide (reactant), but the halide salt produced precipitates from the reaction solution. This physical removal of the side product drives the reaction toward completion. While the original Finkelstein reaction is applied to alkyl halides, an aromatic Finkelstein reaction with copper(I) or nickel(II) as catalyst has been developed and applied to ^{123}I-labeled compounds more recently (Klapars and Buchwald 2002; Cant et al. 2012). Additionally, the substitution site of an aromatic ring may have to be activated by other substituents (Coenen et al. 2006). High specific activity can be achieved because excess chlorine or bromine precursor is easier to separate from the radioiodinated product than an isotopologue discussed earlier in the isotopic substitution approach (Eersels et al. 2005; Coenen et al. 2006).

16.2.3.2.3   *^{123}I Radiolabeling by Electrophilic Substitution*

Even more common for ^{123}I labeling are electrophilic substitution reactions with iodine cation (I^+), which may be present transiently or be provided indirectly through a compound like iodine monochloride (ICl). The iodine cation can be produced *in situ* via the oxidation $I^- \rightarrow I^+ + 2e^-$ with many different agents including chloramine-T (*N*-chloro tosylamide) and peracids. Consequently, some authors refer to this method as oxidative radioiodination (Eersels et al. 2005).

Chloramine-T slowly releases hypochlorite, which oxidizes iodide to form different sources of I^+ depending on the reaction conditions (e.g., pH) (Coenen et al. 2006). To avoid excess chloramine-T in the solution, this agent has been immobilized on polymer beads (Pierce Iodination Beads, previously IODO-BEADS) or used as coating on the reaction vessel (Pierce Iodination Reagent, previously IODO-GEN) (Glaser et al. 2003; Adak et al. 2012). However, these *in situ* oxidation routes can still lead to side products with iodine inserted at different and/or multiple sites, with chlorine being introduced inadvertently into the substrate, and with inadvertent oxidation of susceptible parts of the substrate like methionine residues in a peptide (Coenen et al. 2006). The oxidation related problems are reduced with peracids at a low concentration, which serve as a milder oxidant. Also there is no chlorine present (Coenen et al. 2006). These peracids can be formed *in situ* from hydrogen peroxide and an organic acid, which transforms iodide probably into hypoiodate, IO^-, providing formally I^+. This is the preferred synthesis protocol. The oxidation reaction can be catalyzed by a peroxidase such as lactoperoxidase. This concept means that the oxidant concentration is low throughout the synthesis, which greatly reduces the probability of forming side products.

Several electrophilic substitution reaction schemes have been developed to introduce iodine into different types of molecule positions. Hydrogens in aromatic ring positions in arenes need to be activated by substituents exerting inductive and mesomeric effects; see Table 16.4. An example is the phenol functional group, which also naturally occurs in tyrosine: a hydrogen atom ortho to the phenol hydroxy group can be substituted by I^+ to give 3-iodotyrosine. This substitution pattern is called iododeprotonation.

Otherwise, tin (trialkylstannyl), silicon (trialkylsilyl), and boron (boronic acid) substituents have been incorporated in the precursor as leaving groups for radioiodination (Coenen et al. 2006; Adak et al. 2012). Accordingly, these reactions are called iodo-destannylation, iodo-desilylation, and iodo-deboronation. Still, to introduce these groups in the first place, the arene substrate usually needs to be activated. The subsequent iodination step, however, can be done at milder conditions and with high regiospecificity.

All these methods differ in factors like typical radiochemical yield and purity. Purity here relates to avoiding the production of more than one radioiodinated species (relevant in particular for small molecules) and avoiding an effect of the oxidant on other parts of the molecule (e.g., oxidizing sensitive parts of a peptide). Oxidative radioiodination can often be done at room temperature, which is more compatible with substrates than an environment of usually >100°C required for nucleophilic

TABLE 16.4
Selection of ^{123}I Labeled Small Molecules and Building Blocks

Imaging Agent	Application	Iodine Labeling site	Synthesis Method[a]
[^{123}I]IMPY	Amyloid plaque		Iodo-destannylation with peroxide (Kung et al. 2002)
[^{123}I]Iobenguane (meta-iodobenzylguanidine, *m*IBG, AdreView™)	Tumour imaging		Isotopic exchange (Eersels et al. 2005)
[^{123}I]Iodobenzamide (IBZM)	Dopamine D$_2$/D$_3$ receptor		Isotopic exchange; iodo-deprotonation with chloramine-T (Kung et al. 1988)
[^{123}I]Iododopa	Dopamine D$_2$/D$_3$ receptor		Nucleophilic substitution of bromine (Adam et al. 1990)
[^{123}I]Iofetamine (IMP, SPECTamine™)	Cerebral blood perfusion		Isotopic exchange; nucleophilic substitution of bromine (Najafi 1987)
[^{123}I]Ioflupane (DaTSCAN™)	Dopamine reuptake transporter		Iodo-destannylation with peracetic acid (Neumeyer et al. 1994)
[^{123}I]PK11195	PB/TSPO receptor		Iodo-destannylation with peracetic acid and chloramine-T (Pimlott et al. 2008)
[^{123}I]Tyrosine, also Bolton-Hunter reagent	Tyrosine: direct labeling of peptides; Bolton-Hunter: indirect labeling		Iodo-deprotonation with chloramine-T (Rudinger and Ruegg 1973)

Note: Rests R (alkyl) and Ar (aryl), respectively, differ between entries.
[a] Examples, other synthetic routes may be published in addition to these noted here.

substitutions (Eersels et al. 2005). Importantly, the electrophilic substitution can be performed in an n.c.a. manner, that is, achieving high specific activity.

For macromolecule (usually peptide) labeling, mild synthesis conditions are particularly important. High specific activity and all other typical aims such as speed of synthesis, yield, purity, and stability of the product apply as with any other radiolabeling method. If direct labeling requires too harsh conditions, indirect labeling can be employed; that is, ^{123}I labeling of a smaller fragment is performed at conditions that may be too harsh for the macromolecule, for example, a peptide that may include moieties that are sensitive to strong oxidizing agents. This preparation of the radiolabeled prosthetic group is followed by a mild coupling reaction to the macromolecule. A prominent example of a prosthetic group for indirect radioiodination is the Bolton–Hunter reagent (Bolton and Hunter 1973). A variant thereof is described in Glaser et al. (2001) where the cold precursor carries these two functions: (1) a trimethyltin group to be substituted by radioiodine and (2) an ester functional group with *N*-succinimidyl leaving group for subsequent coupling to primary amines (lysine side chains and *N*-terminal amino acid of a peptide) (Adak et al. 2012). Further methods are reviewed in Coenen et al. (2006).

16.2.3.2.4 Examples of ^{123}I-Labeled Small Molecules

A selection of well-known ^{123}I-labeled small molecules is compiled in Table 16.4. This list serves the purpose of illustrating typical labeling sites, a range of inductive or mesomeric effect exerting substituents present in arenes and the diverse synthesis methods studied.

16.2.3.2.5 Examples of ^{123}I-Labeled Macromolecules

The radioiodination protocols for ^{123}I and other iodine radionuclides matured and continuously improved for more than three decades (Dewanjee 1992). Those principles of small molecule radioiodination chemistry also apply for macromolecules. However, the macromolecules of interest tend to be more sensitive substrates requiring milder reaction conditions. The majority of peptides and proteins have been directly radioiodinated with ^{123}I at the electron-rich aromatic rings of tyrosines. The iodogen method has been the most favored approach so far; see Table 16.5 (Salacinski et al. 1981). Removal of the reaction solution terminates the radioiodination process and avoids potential oxidative damage to the substrate. Lactoperoxidase offers an even milder method for more delicate substrates. However, the enzyme can become radiolabeled itself, which can pose a purification problem. The best approach for mild radioiodination is to use the aforementioned pre-iodinated prosthetic groups such as the Bolton–Hunter reagent, the related *N*-succinimidyl ^{123}I]iodobenzoate (^{123}I]SIB), or TFP-^{123}I]I-PEA; see Figure 16.5 and examples in Table 16.5.

If there are multiple reactive sidechain groups in a peptide or protein substrate, then it will reduce the chemoselectivity of both the direct or indirect radioiodination methods. This leads to

TABLE 16.5
Examples for ^{123}I-Labeled Macromolecules and Their Applications

Macromolecule	Application	Synthesis Method	Reference
c(RGDfE)K(DOTA)PLGVRY	Integrin $\alpha_v\beta_3$ receptor/MMP2, M21 and M21 L cell assay	Iodogen	Mebrahtu et al. (2013)
(RGD)-HAS-TIMP2	Integrin $\alpha_v\beta_3$ receptor/MMP2, human glioblastoma cancer U87MG xenografts	Iodogen	Choi et al. (2011)
c(YRKRLDRN)	Atherosclerosis	Iodobeads	Lee et al. (2008)
fNleLFNleYK	Model peptide	Iodobenzoate (^{123}I]SIB) or Iodovinylester (TFP-^{123}I]I-PEA) labeling reagents	Rossouw (2008)
GRKKRRQRRRPPQGYGC-anti-[p21[WAF-1/Cip-1]]	Human breast cancer MDA-MB-468 xenograft	Iodogen	Hu et al. (2007)
Tyr3-octreotide	Somatostatin receptor expressing tumors	Iodogen Chloramine-T	Schottelius et al. (2005) Bakker et al. (1991)
Antisauvagine-30 analogs	Corticotropin-releasing factor type 2; irritable bowel syndrome, gastric cancer	Chloramine-T	Rühmann et al. (2002)
Vasoactive intestinal peptide	Pancreatic cancer	Iodogen	Raderer et al. (1998)
Interleukin-1, fMLFK	Infection, inflammation	Bolton–Hunter reagent	Van Der Laken et al. (1998)
Carcinoembryonic antigen binding "minibody"	LS174T human colon carcinoma xenograft	Iodogen	Hu et al. (1996)
Atrial natriuretic peptide	Diabetic nephropathy	Lactoperoxidase	Lambert et al. (1994)
c(RGDyK-sugar amino acid)	Integrin $\alpha_v\beta_3$ receptor, angiogenesis	Iodogen	Johnson et al. (2008)
Annexin-V	Phosphatidyl serine, apoptosis	Iodogen, iodobeads	Lahorte (2001)

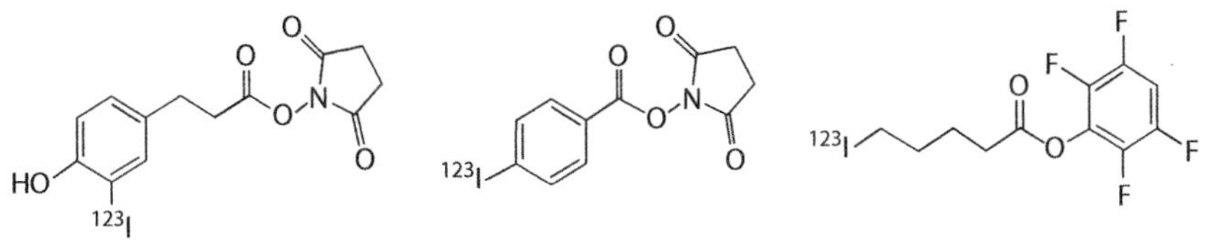

[123]I]Bolton–Hunter reagent ^{123}I]SIB TFP-^{123}I]I-PEA

FIGURE 16.5 Structures of indirect radioiodination reagents for targeting lysine amino groups of peptides or proteins.

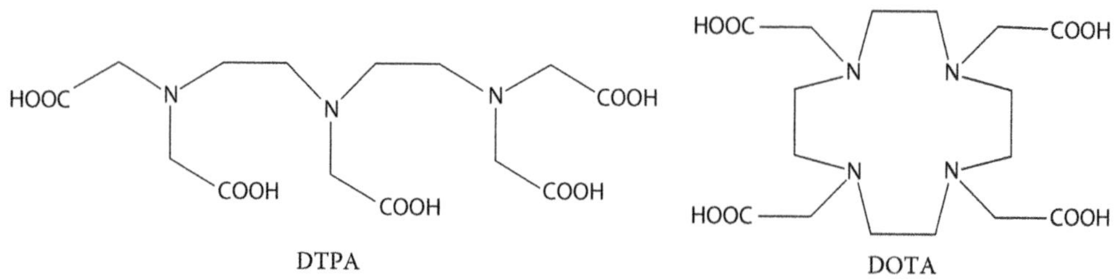

FIGURE 16.6 Chelators DTPA (linear) and DOTA (macrocyclic) to be coupled to substrates via a carboxylic acid group.

species each bearing, for example, one ^{123}I label albeit in different positions. In case these species exhibit significantly different biological activities, separation may be desired, but this is likely to be difficult for such macromolecules. In any case, the radioiodine labeling procedure itself will most certainly affect the biological activity compared to the unlabeled biomacromolecule. Therefore, the active fraction of the labeled compound (or compound mixture) needs to be measured before use in preclinical applications. The choice of such an assay will be determined largely by the size of the macromolecule. For instance, a smaller labeled peptide may be tested in a competitive radioligand binding experiment with nonlabeled peptide. Otherwise, the bioactivity of larger proteins can be verified by an enzyme-linked immunosorbent assay (Collingridge et al. 2002).

16.2.3.3 Examples of SPECT Imaging Agents with ^{67}Ga and [111]In

[67]Ga radiochemistry follows that of the more popular ^{68}Ga PET radioisotope. The principles are close to those for ^{99m}Tc labeling. DOTA , see Figure 16.6, is a widely used chelator for Ga labeling among others such as 1,4,7-triazacyclononane-1,4,7-triacetic acid (NOTA), 1,4,7,10-tetraazacylotetradecane-1,4,8,11-tetraacetato (4-) (TETA), and *cis,cis*-1,3,5-triaminocyclohexane (TACH) (Bandoli et al. 2009; Wadas et al. 2010; Fontes et al. 2011). NOTA is advantageous over DOTA because of higher thermodynamic stability of Ga complexes and radiolabeling at ambient versus elevated temperature (Velikyan et al. 2008).

Also, ^{111}In labeling is done via chelators such as the linear DTPA (diethylenetriaminepentaacetic acid) and the macrocyclic DOTA; see Figure 16.6. DTPA radiolabeling can be achieved at room temperature in contrast to DOTA, which requires higher temperature and an acidic environment.

Examples include ^{111}In-pentetreotide (^{111}In-DTPA-octreotide, OctreoScan) which is based on the octapeptide octreotide with DTPA chelator conjugated to the N terminal D-Phe residue. Variants include DOTA−octreotate (DOTATATE) and edotreotide ((DOTA[0]-Phe[1]-Tyr[3])octreotide, or DOTATOC). These can be considered as theranostic companions, for example, to ^{90}Y-DOTA-TOC and ^{177}Lu-DOTA-TATE therapeutics.

[111]In-oxine is a special case consisting of three oxine (8-hydroxyquinoline) ligands that are bidentate. Leukocytes are sampled from a patient and then labeled with a N_3O_3-type complex with [111]In. Presumably, ligand exchange takes place in the cells and a thermodynamically more stable, unknown complex forms while the oxine ligands are released (Roca et al. 2010). The labeled white blood cells are then readministered to the subject for imaging.

16.3 PET RADIOTRACERS FOR PRECLINICAL IMAGING

16.3.1 INTRODUCTION

The PET method offers accurate tracer quantification with a sensitivity that is two orders of magnitudes higher compared to SPECT (Jones 1996). Arguably, translating small molecule PET radiotracer research into clinical diagnostics appears to be more straightforward compared to SPECT, in particular for small molecules and considering the radioisotopes fluorine-18 or carbon-11 compared to technetium-99m. In an ideal scenario for a targeted small molecule radiotracer, a potent drug molecule can be translated into its radioactively labeled counterpart that features an identical chemical

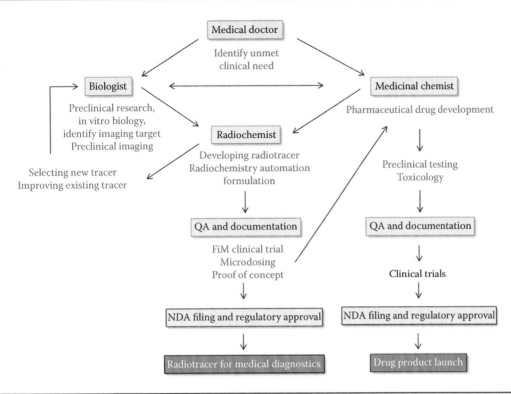

FIGURE 16.7   The central role of PET radiochemistry in translating preclinical research into clinical applications. *Note*: FiM, first-in-man study; QA, quality assurance; NDA, new drug application.

structure. This is an expanding area especially in neurology covering brain receptors, transporters, enzymes, and other targets such as amyloid plaques or neurofibrillar tangles (Heiss and Herholz 2006; Zimmer and Luxen 2012). Figure 16.7 shows a simplified, exemplary scheme that illustrates the relationship between preclinical radiotracer research and drug development. Pharmacophores identified by pharma industry and academia can serve as basis for PET radiolabeling development programs. After a preclinical phase, the new drug candidate is evaluated in a small first-in-man (FiM) clinical trial using PET. Here, the microdosing concept can be applied to streamline extensive toxicological evaluations (Bergström and Långström 2005). In addition, a well-characterized PET tracer can serve as gold standard in human studies to evaluate the efficacy of new drugs. Both approaches can therefore help to reduce time and cost in drug development.

There are certain demands to be fulfilled by preclinical PET radiotracer research. The short half-life of the common radionuclides requires a sufficiently high starting radioactivity. The first consequence of this is that the chosen radiochemical approach has to provide both a very rapid and efficient radiolabeling protocol. The second implication is the need to translate the radiolabeling work into a robust automated process when approaching clinical trials. The automation process usually includes preparing the radiolabeled molecule directly or via an intermediate, purifying the product, and its formulation into a suitable form ready for injection into patients. There are very active research areas such as microfluidics and there is an ever increasing number of potential new tracer applications. Impressive leaps of technology development can be observed as will be highlighted in the following. In the past decade, the preclinical PET radiotracer research has seen remarkable developments in labeling protocols and in new synthesizer platforms. We will discuss some basic PET radiochemistry first and then draw the reader's attention to the latest advances in the labeling technology.

16.3.2   PROPERTIES AND PRODUCTION OF COMMON PET RADIOISOTOPES

Table 16.6 gives a brief summary of popular PET radioisotopes including their properties and production. Fluorine-18 is probably the most-favored radionuclide for PET imaging. This is due to its half-life of about 2 h, which is optimal for a range of applications. In addition, fluorine-18 has

TABLE 16.6
Properties and Production of Common PET Radioisotopes

Radioisotope	Half-Life	Positron Energy $(E_{\beta+ \text{max}})$ (MeV)	Positron Abundance (%)	Production	Radiochemistry
^{18}F	110 min	0.64	97	Cyclotron: $^{18}O(p,n)^{18}F$, or ^{20}Ne (d,α) ^{18}F	Indirect or direct methods based on $[^{18}F]F^-$ or $[^{18}F]F_2$
^{11}C	20 min	0.97	99	Cyclotron: $^{14}N(p,\alpha)^{11}C$	Indirect or direct methods based on $[^{11}C]CO_2$ or $[^{11}C]CH_4$
^{68}Ga	68 min	1.90	89	$^{68}Ge/^{68}Ga$ generator	Directly as $[^{68}Ga]Ga^{3+}$
^{64}Cu	12.7 h	0.65	18	Cyclotron: $^{64}Ni(p,n)^{64}Cu$	Directly as $[^{64}Cu]Cu^{2+}$
^{124}I	4.2 days	2.14	25	Cyclotron: $^{124}Te(p,n)^{124}I$	Indirect or direct methods based on $[^{124}I]I^-$
^{89}Zr	3.27 days	0.897	23	Cyclotron: $^{89}Y(p,n)^{89}Zr$, or $^{89}Y(d, 2n)^{89}Zr$	Directly as $[^{89}Zr]Zr^{4+}$

Sources: Pagani, M. et al., *Eur. J. Nucl. Med.*, 24(10), 1301, 1997; Signore, A. et al., *Chem. Rev.*, 110, 3112, 2010.

optimal radionuclear properties with emitted positrons of relatively low energy and high abundance. Fluorine-18 is usually obtained by cyclotron irradiation of an oxygen-18 enriched target. The common production method provides n.c.a. $[^{18}F]$fluoride in water. Alternatively, although less widespread, $[^{18}F]F_2$ gas can be generated. The achievable concentration of ^{18}F does not allow the formation of $[^{18}F]F_2$ molecules consisting of two ^{18}F atoms. Therefore, this process depends on the addition of stable fluorine-19 carrier, reducing the specific radioactivity. Many PET applications demand a high specific radioactivity of the tracer to allow binding to molecular targets of low concentration such as brain receptors. Thus, the n.c.a.-based radionuclide production will be usually the method of choice.

The radionuclear properties of carbon-11 are favorable for PET with the main challenge of a relatively short half-life that requires the labeled compound to be prepared and applied in proximity of the cyclotron. Carbon-11 is usually provided as $[^{11}C]CO_2$ or $[^{11}C]CH_4$ gas. Low specific radioactivity is more of a concern compared to fluorine-18. This is because of a higher probability of carbon-12 contamination compared to the addition of fluorine-19. In general, specific radioactivities of radiotracers based on $[^{11}C]CH_4$ are superior to $[^{11}C]CO_2$. For a more in-depth discussion of specific radioactivity in PET radioisotopes, the reader is referred to a review by Lapi and Welch (2012).

Gallium-68 is available from commercial ^{68}Ge generator systems (Fani et al. 2008). Due to the long half-life of ^{68}Ge ($T_{1/2}$ = 270.8 days), such a generator can last up to 2 years. Gallium-68 has attracted renewed interest as a generator-based potential alternative to ^{99m}Tc, which is obtained via ^{99}Mo produced in nuclear reactors (Bartholomä et al. 2010). However, it should be noted that the combination of a relatively short half-life and high-energy positrons might be less favorable for optimal PET imaging in comparison with ^{18}F (Sanchez-Crespo 2013).

Copper-64 has been utilized in a number of preclinical PET studies (Wadas et al. 2010; Cutler et al. 2013). Limiting factors are high-energy positrons and their low abundance. Radioisotope production requires the handling of a solid target of enriched ^{64}Ni and a demanding wet chemistry extraction process. This procedure has recently been simplified by automation (Wadas et al. 2010).

Iodine-124 is the only positron-emitting iodine radioisotope that has found significant utility in preclinical PET. This radionuclide features a sufficiently long half-life for studying the biodistribution of larger labeled bioconjugates such as antibodies or fragments of antibodies. Whilst a long half-life can be seen as a useful property for preclinical research, the high positron energy combined with both a low abundance and a high ratio of gamma radiation seems to prevent this radioiodine nuclide from being widely accepted in the PET research community. This resulted in a poor acceptance as a diagnostic radionuclide in the clinic. Iodine-124 is routinely isolated by dry distillation of an irradiated solid target of $[^{124}Te]TeO_2$ (Knust et al. 2000; Glaser et al. 2004; Reischl et al. 2004). More recently, ^{89}Zr has started to attract researcher's attention as an alternative to ^{124}I, offering the benefits of a chelate-based labeling chemistry (Zeglis et al. 2013).

16.3.3 Labeling of Small Molecules for PET

16.3.3.1 Labeling Small Molecules Using Carbon-11

There are numerous carbon-11-labeled building blocks and labeling reagents that are accessible by using fast and efficient organic chemistry as well as sophisticated automation approaches. An overview of this rapidly growing field is shown in Figure 16.8 (Welch and Redvanly 2003; Allard et al. 2008; Miller et al. 2008). [^{11}C]CO$_2$ usually requires "wet" reactions with organometallic agents. This comes with a significant risk of reduced specific radioactivity by introducing carrier carbon dioxide with a contaminated reagent. [^{11}C]CH$_4$ may often be converted into a secondary labeling agent in a gas phase reaction (e.g., halogenation) without major possibility of introducing carrier. The recent trend is to use [^{11}C]CH$_4$ as the primary irradiation product because of its potential to deliver higher specific radioactivity and its scope for simple gas phase production of secondary labeling agents. The main focus for preclinical PET radiochemistry would be on reactive intermediates that quickly transfer ^{11}C from the gaseous starting materials into molecules of biological interest. Here, [^{11}C]iodomethane can be regarded as the most important building block. This versatile molecule is readily accessible to PET centers by a number of commercial platforms (Miller et al. 2008). [^{11}C] Iodomethane can be introduced into a wide variety of substrates by using alkylation reactions with acidic nitrogen, oxygen, or sulfur functional groups.

Figure 16.9 shows some selected examples for radiotracers that can be prepared using *N*-alkylation reactions. The compounds [^{11}C]*m*-hydroxyephedrine, [^{11}C]L-deprenyl, and [^{11}C]SCH23390 have found application as tracers for imaging noradrenaline uptake in heart (Rosenspire et al. 1990), brain monoamine oxidase (Fowler et al. 1988), and brain dopamine D1 receptors (Ram et al. 1989), respectively.

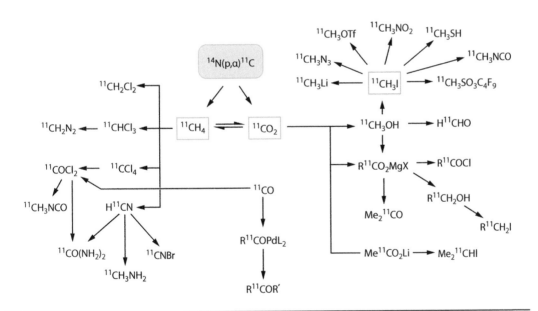

FIGURE 16.8   Preparation of carbon-11 based building blocks and labeling reagents.

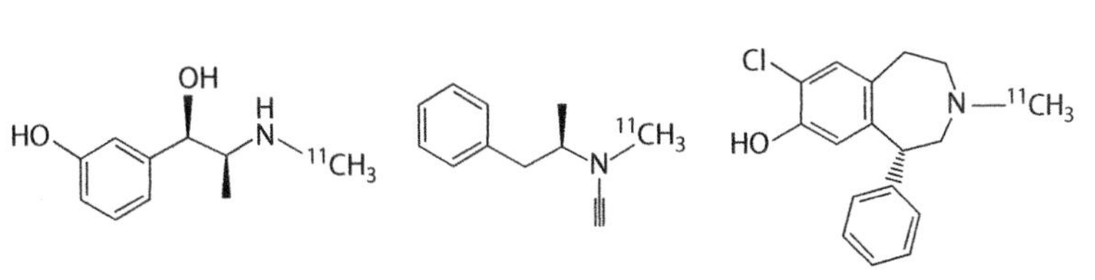

FIGURE 16.9   Examples of [^{11}C]methylated radiotracers obtained by *N*- and *O*-alkylation.

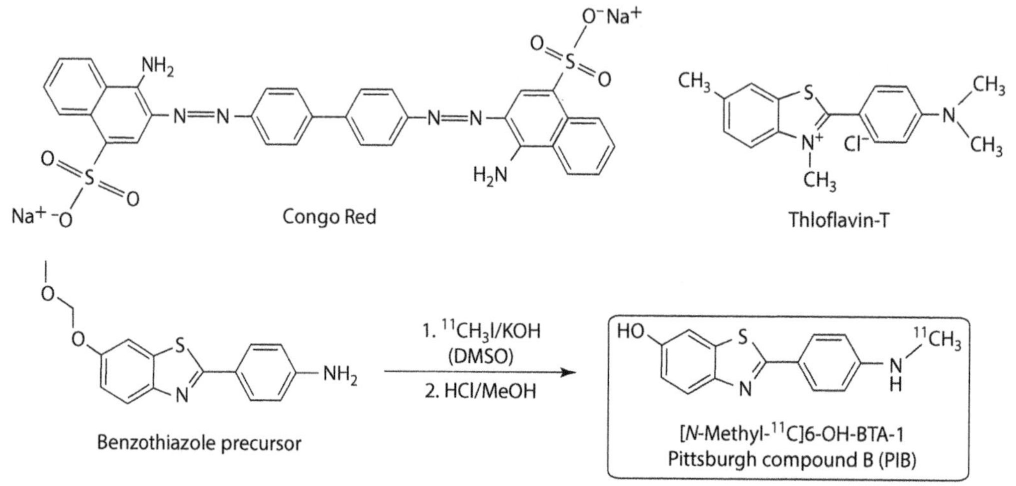

FIGURE 16.10 Chemical structures of fluorescent amyloid binders Congo red and thioflavin-T, and *N*-methylation of Pittsburgh compound B [¹¹C]PiB using [¹¹C]H₃I as a labeling reagent.

Another prominent representative example for *N*-methylated PET tracer is [¹¹C]PiB (Figure 16.10) (Klunk et al. 2004). This established diagnostic agent for imaging Aβ amyloid plaques in Alzheimer's disease has been developed starting from fluorescent markers. However, these original compounds cannot penetrate the blood–brain barrier due to their molecular weight (Congo red) and/or salt character (Congo red, thioflavin-T). Preclinical screening of a number of radiolabeled thioflavin derivatives eventually identified [¹¹C]PiB as a lead candidate, which has now well progressed into clinical trials and also inspired research in alternative amyloid binders (Klunk and Mathis 2008; Cohen et al. 2012; Svedberg et al. 2012). The radiochemical synthesis of [¹¹C]PiB is based on the ¹¹C-methylation of the aniline group of the hydroxy-protected benzothiazole precursor. This is followed by an acidic hydrolysis of the phenol-protecting group (Figure 16.10).

Likewise, a very common labeling approach is the O-alkylation using [¹¹C]iodomethane as a reagent. Figure 16.11 shows two typical examples: the opioid receptor marker [¹¹C]diprenorphine (Luthra et al. 1994) and the dopamine D2 receptor binding [¹¹C]raclopride (Ehrin et al. 1985).

A more recent example for the preparative use of [¹¹C]iodomethane is a *C*-alkylation reaction as demonstrated for the radiosynthesis of D2 receptor agonist [¹¹C]-(+)-PHNO (Figure 16.12) (Garcia-Arguello et al. 2013). The acetyl methyl group of a protected acetylamide precursor can be deprotonated by a strong base lithium bis(trimethylsilyl)amide (LHMDS). This allows the coupling with [¹¹C]iodomethane under mild conditions. The subsequent step with LiAlH₄ reduces the amide carbonyl functional group. Incidentally, the reducing agent here removes concomitantly the phenol protecting triisopropylsilyl moiety. This demanding labeling strategy for [¹¹C]-(+)-PHNO has been applied for a preclinical study (Egerton et al. 2010). A fully automated protocol was developed as well (Garcia-Arguello et al. 2013).

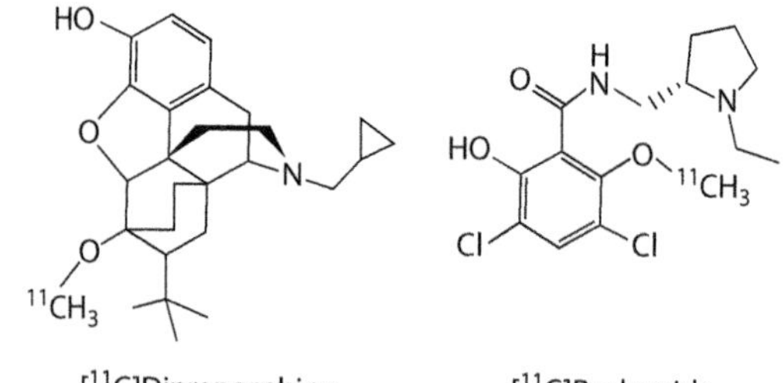

FIGURE 16.11 Chemical structures of [¹¹C]diprenorphine and [¹¹C]raclopride as examples for ¹¹C-labeled compounds obtained by *O*-methylation.

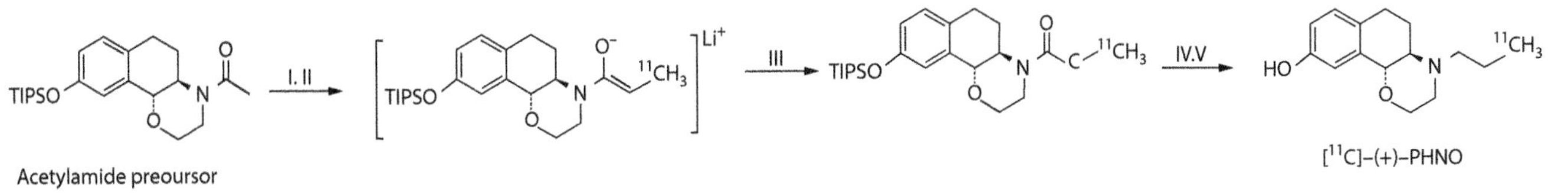

FIGURE 16.12 Radiosynthesis of [¹¹C]PHNO by ¹¹C-methylation of an acetyl amide precursor. Reagents and conditions: (I) −78 °C, THF, LHMDS, (II) warm up to room temperature for 3–7 min, [¹¹C]CH₃I in THF for 5 min at room temperature, (III) MeOH, (iv) LiAlH₄, 7 min at 60 °C, (v) HPLC purification (TIPS, trisisopropylsilyl protecting group).

Unfortunately, it will be not possible to give here a full list of all important examples for radiolabeling small molecules with carbon-11. Instead, we refer the interested reader to additional articles that report on the use of [¹¹C]iodomethane as starting material for [¹¹C]azidomethane in rapid click chemistry (Schirrmacher et al. 2008) and for Stille cross-coupling chemistry (Samuelsson and Långström 2003). There are also a number of noteworthy publications on using [¹¹C]carbon monoxide in catalyzed carbonylation reactions (Kihlberg 2002; Kealey et al. 2009; Kealey et al. 2011). There has been much interest in microfluidic devices as tools in the synthesis of carbon-11 labeled tracers (Wang et al. 2010; Kealey et al. 2011). Further comprehensive examples for preclinical ¹¹C radiochemistry can be gathered from the review by Allard et al. (2008).

16.3.3.2 Labeling Small Molecules Using Fluorine-18

Fluorine is an isostere for the hydroxy group. The replacement of aromatic hydrogen with fluorine renders the molecule more lipophilic, whereas the addition of fluorine to aliphatic moieties increases the polar character of the molecule (Purser et al. 2008). Fluorine-18 can be conjugated to molecules of interest in many ways. As mentioned earlier, the cyclotron provides fluorine-18 either as [¹⁸F]F₂ gas or as [¹⁸F]F⁻ ion. The former has the drawback of being carrier-added, that is, the reaction product will be diluted with nonradioactive fluorine-19. In addition, [¹⁸F]F₂ gas will be very reactive and thus show only little substrate selectivity. On the other hand, [¹⁸F]F⁻ shows only poor reactivity as a nucleophile and will have to be activated. The high reactivity of [¹⁸F]F₂ is usually reduced by producing a hypofluorite intermediate through reaction with acetic acid (Figure 16.13).

This method can be applied in the electrophilic aromatic fluorination with stannyl precursors. The radiosynthesis of the dopamine D2 tracer [¹⁸F]FDOPA is a well-known example. As shown in Figure 16.14, a precursor **1** is fluoro-destannylated using [¹⁸F]F₂. The latter can be obtained in high specific radioactivity using post-target produced [¹⁸F]F₂ following the method of Solin and coworkers (Forsback et al. 2008). The intermediate **2** is subsequently deprotected by HBr to form [¹⁸F]FDOPA.

However, for a number of reasons, the electrophilic fluorination approach can be regarded as a less-favored method in PET chemistry laboratories. This is mainly because of the demanding production of [¹⁸F]F₂, but also because of the limited utility of carrier-added radiotracers with their low-specific

$$[^{19}F]F_2 \xrightarrow{2\,AcO^-} AcO^{18}F/AcOF + {}^{18}F^-/F^-$$

FIGURE 16.13 Preparation of [¹⁸F]acetylhypofluorite.

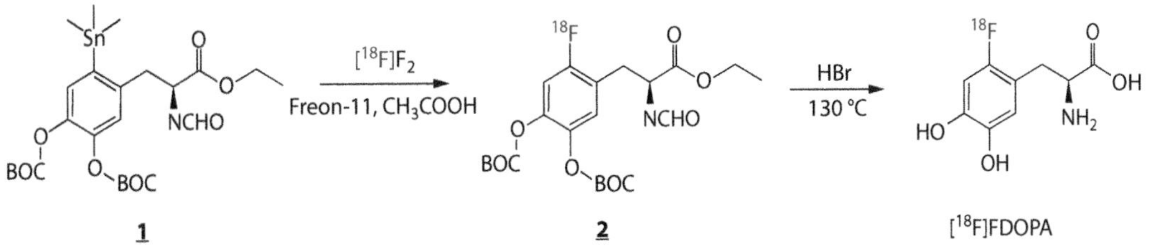

FIGURE 16.14 Electrophilic aromatic substitution reaction to prepare [¹⁸F]FDOPA.

radioactivity. Most applications of molecular imaging require a maximum of specific radioactivity that cannot be achieved by the carrier-added protocols. Finally, handling reactive $[^{18}F]F_2$ comes with certain constraints in terms of substrate selectivity and reaction parameters that narrow the field of applications. There is currently some active research in the area of electrophilic fluorination using palladium catalysis (Lee et al. 2011) and nucleophilic aromatic labeling with oxidants (Gao et al. 2012). However, at the time of writing this book chapter, these new approaches have not been utilized yet for preclinical PET.

The nucleophilicity of $[^{18}F]F^-$ can, however, be increased. For instance, potassium carbonate is added with potassium acting as a counter-cation to fluoride. The counter-cation can be complexed by a phase transfer catalyst such as Kryptofix. This creates the so-called naked fluoride and improves both its reactivity and solubility in the organic reaction solvent. Other phase-transfer systems can be used to modulate the strength of the base in case of sensitive labeling substrates. These $[^{18}F]F^-$ methods may rely on potassium bicarbonate, cesium carbonate, oxalate, tetrabutylammonium bicarbonate, 18-crown-6 ether, and others. Fluoride will preferably react with the protons of water and form $[^{18}F]HF$ with no nucleophilic activity. Therefore, in the classical protocol, any macroscopic amount of water will have to be carefully removed by concentration with solid-phase extraction on an ion exchange resin. This is followed by an azeotropic distillation using acetonitrile. This process serves also as the foundation for the preparation of 2-deoxy-2-$[^{18}F]$fluoro-D-glucose (Figure 16.15).

The anhydrous complex of $[^{18}F]KF$-Kryptofix/K_2CO_3 is reacted with the acetyl-protected mannose triflate **3** (Hamacher et al. 1986). The nucleophilic $[^{18}F]F^-$ ion replaces the trifluoromethanesulfonate ("triflate" or Tf)-leaving group to form the acetylated intermediate **4**. The acetyl-protecting groups are removed in the final hydrolysis step to give $[^{18}F]FDG$. The tracer molecule $[^{18}F]FDG$ can be regarded as the work-horse of PET. It has been estimated that more than 90% of all clinical PET scans rely on the manufacture of this compound (Coenen et al. 2010). This is due to the versatility of $[^{18}F]FDG$, which—as a marker of the local glucose metabolism rate—is being widely used for PET imaging in neurology, oncology, and cardiology (Anderson and Price 2000; Bar-Shalom et al. 2000; Shields 2006; Brooks 2010; Perumal et al. 2012; Saby et al. 2013). The radiochemistry of $[^{18}F]FDG$ has been evolved and matured from various protocols (Ribeiro Morais et al. 2013). The tracer synthesis is now well established on a number of commercial automated platforms.

Here, we have mentioned $[^{18}F]FDG$ as a representative example for radiofluorinating small molecules. These protocols frequently comprise steps to obtain anhydrous $[^{18}F]F^-$, to displace a suitable leaving group, and to remove a protecting group, and finally to apply a formulation method to provide the tracer in a solution ready for injection. Advanced protocols such as the $[^{18}F]FDG$ radiosynthesis do not require purification by HPLC. The vast majority of PET tracers is currently prepared by nucleophilic labeling reactions using n.c.a. $[^{18}F]F^-$ with specific radioactivities of >5 mCi/μmol (Cai et al. 2008). The reader interested in a more detailed discussion on the aspects of radiolabeling small molecules with $[^{18}F]F^-$ is referred to some excellent review articles (Ametamey et al. 2008; Cai et al. 2008; Miller et al. 2008; Ermert and Coenen 2010; Roeda and Dollé 2010; Littich and Scott 2012).

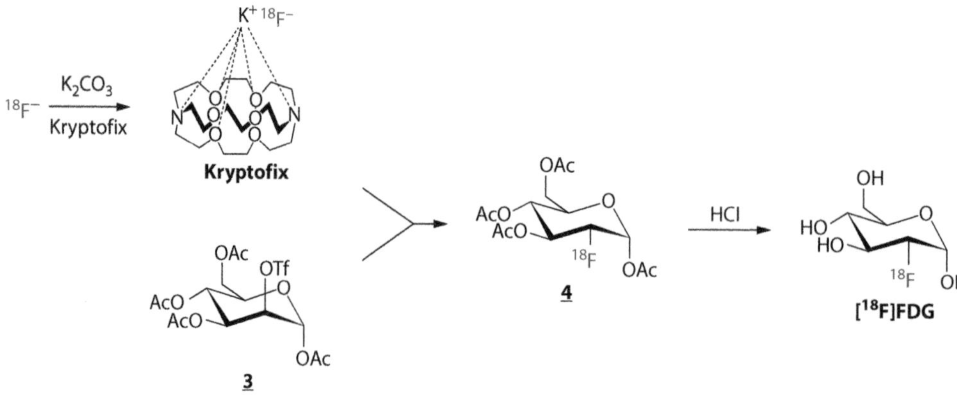

FIGURE 16.15 Nucleophilic aliphatic substitution reaction to prepare $[^{18}F]FDG$.

16.3.3.3   Labeling of Large Biomolecules

Larger biomolecules such as peptides or proteins are of particular interest to the PET radiochemist. After radiolabeling, these systems can be used to target receptors and enzymes with high selectivity and specificity. A preferred strategy is to prelabel a vector unit using bifunctional linker reagents (Figure 16.16). Occasionally, the step of conjugating the radionuclide with the linker reagent needs harsh reaction conditions. The radionuclide is, therefore, attached to the linker in a first reaction step. The resulting labeling reagent is then reacted with the vector molecule under milder conditions.

This protocol typically applies to radiolabeling peptides with fluorine-18. [^{18}F]*N*-succinimidyl fluoro-4-benzoate ([^{18}F]SFB) is a well-known prosthetic group with an activated ester function to primarily target lysine side chains (Vaidyanathan and Zalutsky 1992). The selectivity of the prelabeled reagent has been much improved with the use of [^{18}F]fluoro-4-benzaldehyde ([^{18}F]FBA) (Poethko et al. 2004). However, this approach required the coupling of an aminooxy reactive group to the vector molecule. In consequence, this labeling method allows for high chemoselectivity through the orthogonal reactant pair of an aldehyde and an aminooxy group. Although the aromatic ring introduces some lipophilicity into the vector, the aromatic fluorine-18 C-F bond is sufficiently stable as demonstrated in numerous preclinical studies. Figure 16.17 exemplifies the radiochemistry for the preparation of [^{18}F]fluciclatide, an RGD (arginylglycylaspartic acid) receptor–binding oncology marker that has already reached clinical trials (Pettitt et al. 2010). The labeling reagent [^{18}F]FBA can be obtained from the trifluoromethanesulfonate (triflate) precursor **5** after purification by a solid phase extraction cartridge (SPE). Following the coupling step with the aminooxy peptide precursor, [^{18}F]fluciclatide is purified by another SPE step. The full radiochemical preparation has been implemented on the FASTlab module and provides the tracer with a non-decay-corrected radiochemical yield of 20% (Pettitt et al. 2010).

Another more recent addition to the "toolbox" of bioconjugation methods for short-lived PET tracers is the so-called click reactions. The term "click chemistry" has been originally promoted by Barry Sharpless to identify some classes of very fast and efficient reactions (Kolb et al. 2001). The benefits of the copper(I) catalyzed 1,3-dipolar cycloaddition of an azide with an alkyne have been quickly recognized. This transformation opened a wide field of new applications spanning material sciences to biotechnology and medicine. The main reason for this phenomenon is an exceptional, high-yielding coupling reaction combined with a good chemoselectivity. The value of this useful reaction has been well appreciated by PET radiochemists. An impression on that area can be gleaned from various review papers (Glaser and Robins 2009; Mamat et al. 2009; Ross 2010; Wängler et al. 2010). Figure 16.18 shows an example based on the small labeling reagent [^{18}F]2-fluoroethylazide **8**. This compound is conveniently prepared from the corresponding tosylate precursor **7** with subsequent distillation into a reaction vessel that contains both a Cu(I)-bathophenanthroline disulfonate (BPDS) complex and an alkyne octreotate substrate. The addition of the copper(I) stabilizer BPDS improved labeling efficiencies of peptides significantly compared to the standard method that relied on sodium ascorbate only (Devaraj et al. 2009; Gill et al. 2009). The labeled peptide [^{18}F]FET-βAG-TOCA has been evaluated in preclinical tumor models (Iddon et al. 2011a,b; Leyton et al. 2011). The complete labeling protocol for the peptide has been also transferred to the FASTlab module (Iddon et al. 2011a).

In recent years, the click labeling research community has also explored a number of more reactive reagents that do not require metal catalysts (Zlatopolskiy et al. 2012a,b; Hausner et al. 2013).

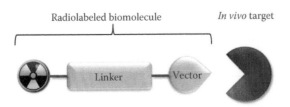

FIGURE 16.16   General principle of radioactive labeling reagents using biomolecule vectors to bind *in vivo* targets.

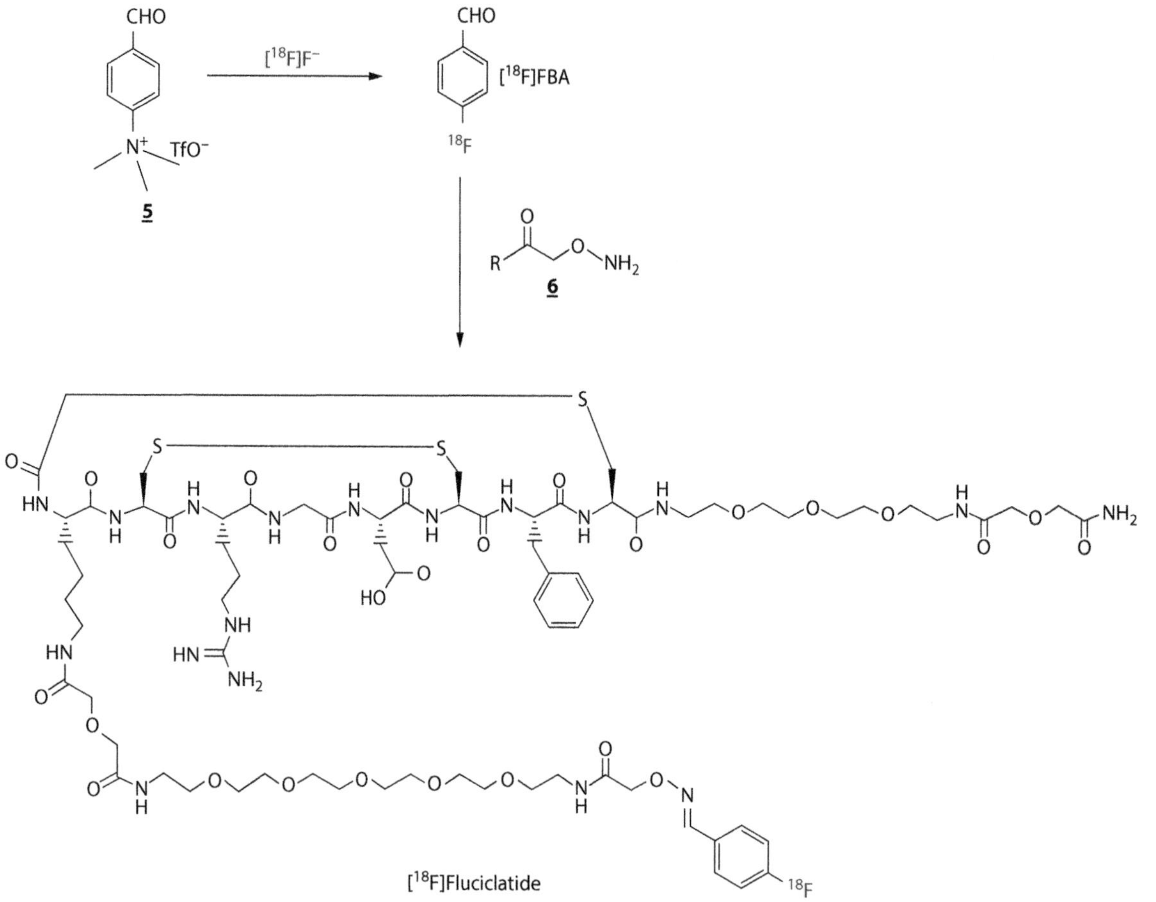

FIGURE 16.17   Radiosynthesis of [18F]fluciclatide.

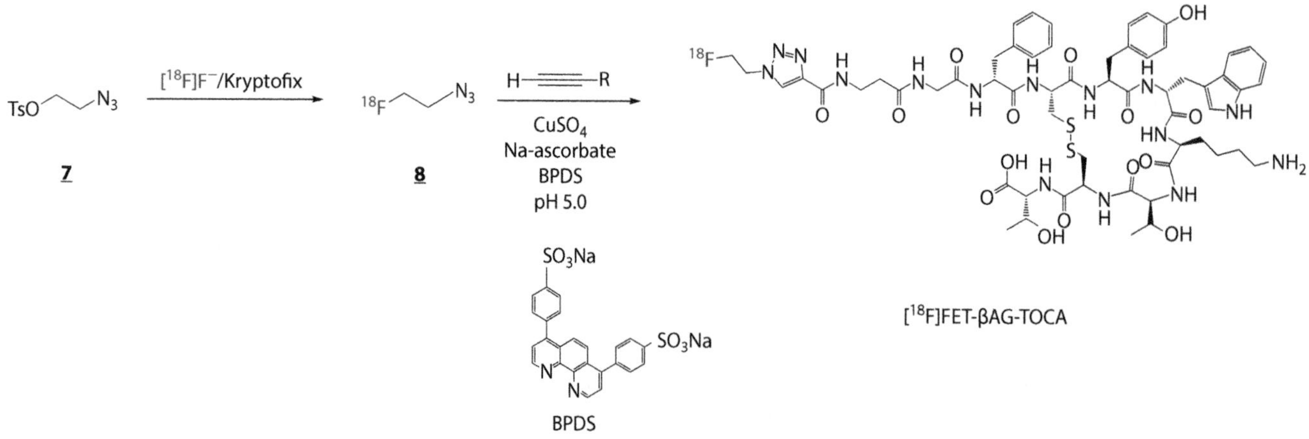

FIGURE 16.18   Radiosynthesis of [18F]FET-βAG-TOCA based on the labeling reagent 2-[18F]fluoroethylazide.

The rationale is that any trace of heavy metal ions can potentially have toxic side effects. Unfortunately, the price of that approach is to introduce large lipophilic units into the labeled biomolecule increasing nonspecific binding and slowing clearance of the radiotracer. Furthermore, until now, there is no means yet to control the formation of isomers upon conjugation.

As alluded here, incorporating [18F]fluoride into a molecule requires reaction conditions that might not be compatible with sensitive biomacromolecules, and hence the use of prosthetic groups or labeling reagents. Since this approach can be cumbersome and low-yielding, there have been also efforts to directly introduce the radiolabel into peptides. Four approaches deserve to be mentioned in this regard:

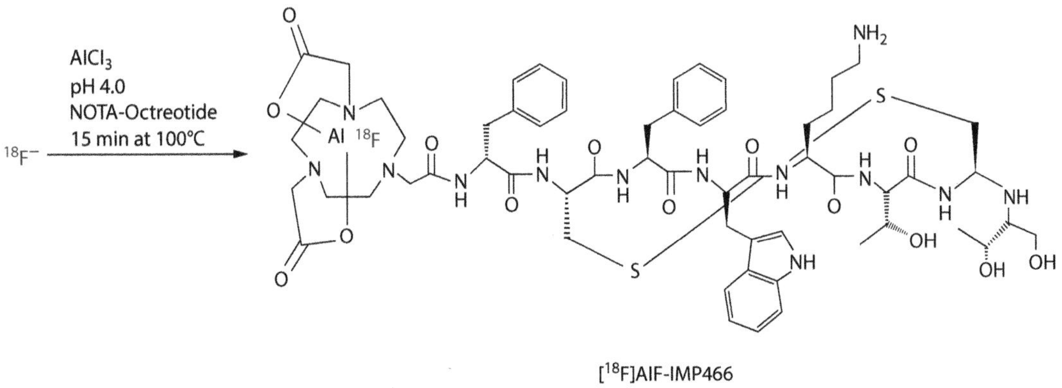

FIGURE 16.19 Radiosynthesis of [^{18}F]AlF-NOTA-octreotide IMP466.

McBride et al. developed a method that is based on binding [^{18}F]AlF^{2+} with NOTA chelator–modified peptides (Figure 16.19) (Laverman et al. 2012). The labeled octreotide was prepared using aqueous [^{18}F]fluoride in a total synthesis time of 45 min with 50% radiochemical yield. The tracer demonstrated plasma stability and high uptake in somatostatin receptor–expressing tumors.

Another direct peptide labeling protocol has been disclosed by the Schirrmacher group (Schirrmacher et al. 2006; Wängler et al. 2010). The idea was to use the isotopic $^{19/18}$F exchange reaction on silicon, also termed "SiFA" for silicone-fluoride acceptor systems. Technically, this chemistry is even more straightforward to carry out as there are only two components to be reacted (Figure 16.20). The reaction has been applied for labeling of octreotate in an imaging study of tumor-bearing mice (Wängler et al. 2010). However, a limiting factor of the SiFA chemistry is the added lipophilic character of the resulting labeled peptide. On the other hand, any size reduction of the bulky aliphatic silicone substituents increases the risk of hydrolytic sensitivity (Höhne et al. 2009; Balentova et al. 2011). Of concern is also the attainable specific radioactivity of the products. Another study reported the nucleophilic substitution of a silicon bound proton of bombesin-linked silane with [^{18}F]fluoride (Höhne et al. 2009). This tracer, which was prepared in anhydrous dimethyl sulfoxide (DMSO), was found to be stable in mouse and human plasma for 120 min. There was no *in vivo* degradation in the mouse for the first 30 min.

A third method to directly introduce [^{18}F]fluoride into a peptide is shown in Figure 16.21. Becaud et al. demonstrated that using 2-cyano-4-(methoxycarbonyl)-*N,N,N*-trimethylbenzenaminium trifluoromethanesulfonate–modified bombesin **9**, a direct fluorination of a peptide can be achieved. The product **10** was prepared with good radiochemical yield and specific radioactivity. This bombesin derivative showed a mouse plasma stability of 90% at 2 h post injection (Becaud et al. 2009).

A fourth protocol for the direct introduction of [^{18}F]fluoride into peptides uses the isotopic exchange reaction of fluoroborates (Auf dem Keller et al. 2010; Li et al. 2013; Liu et al. 2013). Again, disadvantages of this strategy are the need of carrier fluorine-19 and the impact of large lipophilic stabilizing ring systems on biodistribution.

All these direct fluorination methods mentioned clearly simplify the labeling procedures and can, therefore, easily be automated. The scope of the direct labeling strategies will be probed in the future. The available preliminary preclinical data seem to favor the [^{18}F]AlF-NOTA protocol, although the high temperature of the labeling procedure could limit applications.

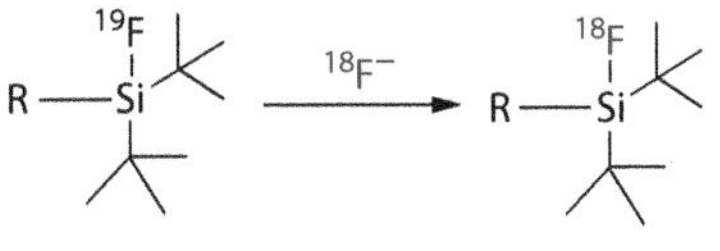

FIGURE 16.20 Radiolabeling of a biomolecule using isotopic exchange on SiFA functionality.

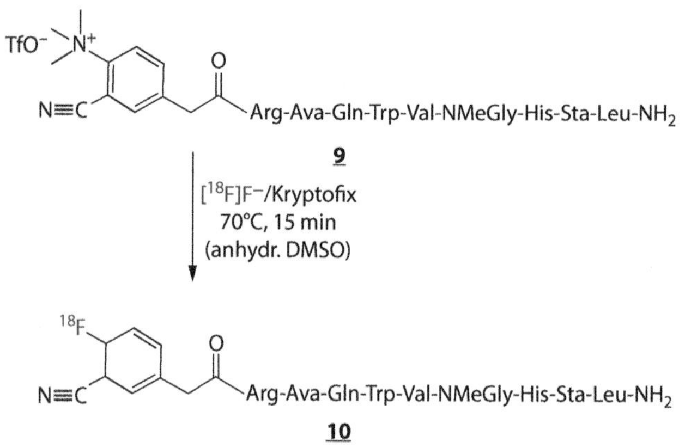

FIGURE 16.21 Radiolabeling of bombesin using nucleophilic aromatic substitution reaction.

16.3.3.4 Labeling of Biomacromolecules Using Gallium-68

Due to the shorter half-life of ^{68}Ga, the use of this positron emitter is limited to smaller biomolecules. The reason for this is that the half-life of the radionuclide dictates the timespan for the tracer to reach the target and clear the nontarget tissue. For example, a ^{68}Ga-radiolabeled antibody molecule of 140 kDa molecular weight would most likely not distribute quickly enough and exhibit a sufficiently fast systemic clearance rate to enable imaging studies.

Figure 16.22 gives a representative selection of the most popular chelating ligand systems for ^{68}Ga applications. DTPA is an established chelator for radiometals. This system has been recently

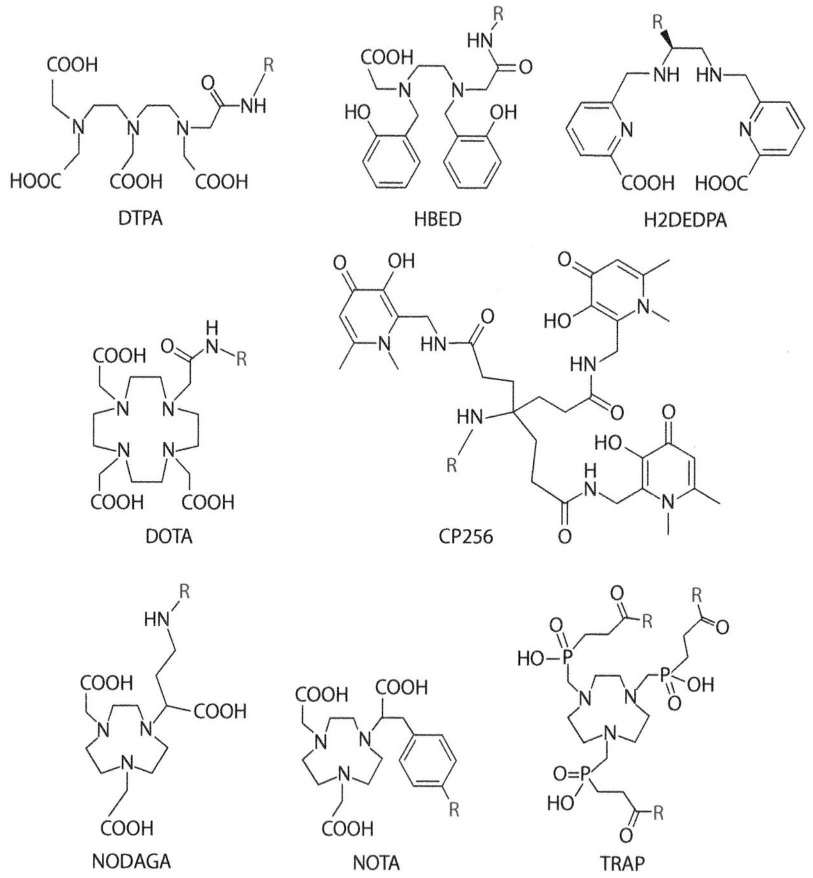

FIGURE 16.22 Chelator systems used for labeling biomacromolecules with ^{68}Ga (R = peptide).

applied for labeling of pamoic acid with ^{68}Ga as a necrosis marker (Prinsen et al. 2010). The macrocyclic DOTA served as a preferred binder for ^{68}Ga in numerous peptide labeling protocols (Stasiuk and Long 2013). The heterobifunctional compound *N,N'*-bis[2-hydroxy-5-(carboxyethyl) benzyl]ethylenediamine-*N,N'*-diacetic acid (HBED-CC) was found to be superior to DOTA as the ^{68}Ga complexing step occurred more efficient at room temperature (Eder et al. 2008). A related pyridine-based chelator suitable for chelation reaction at room temperature, H2DEDPA, has been reported by the group of Orvig et al. (Boros et al. 2012). The tripodal ligand CP256 is another promising new compound for rapid protein labeling under mild conditions (Berry et al. 2011).

More recently, the chelating NOTA has been successfully used in many instances of ^{68}Ga peptide labeling. For example, the angiogenesis specific RGD peptide could be conjugated with ^{68}Ga for tumor imaging (Jae et al. 2008). The more advanced ^{68}Ga chelator NODAGA has been developed as a potential replacement candidate for [^{18}F]Galacto-c(RGDfK) (Pohle et al. 2012).

The most recent addition to the "toolbox" of chelators has been TRAP (1,4,7-triazacyclononane-1,4,7-tris[(2-carboxyethyl)methylenephosphinic acid]), a phosphinic acid–modified NOTA macrocycle. TRAP has been applied for trimeric RGD peptide labeling. So far, this appears to be the most advanced chelator enabling the visualization of integrin receptors at low expression level (Notni et al. 2013).

16.3.3.5 Labeling of Biomacromolecules Using Copper-64

Although subject to much research, the PET radiochemistry of copper-64 is still dominating the preclinical field rather than the clinical field. There seems to be a combination of reasons for this. Presumably, it is because of a nontrivial radionuclide-manufacturing process, the limiting physical properties of ^{64}Cu, and a less favorable biodistribution of the tracers. The latter can be related to *in vivo* demetallation issues. Figure 16.23 compiles a short list of macrocycles suitable for complexing ^{64}Cu for conjugation to biomacromolecules. As in all radiometal complex compounds, *in vivo* stability is of particular concern. Here, the best results can be achieved if the macrocyclic ligand neutralizes the charge of Cu(II) to form a complex without net charge. For example, NOTA would be a better choice than DOTA or TETA. This has been shown for ^{64}Cu labeling of bombesin and in the corresponding *in vivo* studies (Prasanphanich et al. 2007). The best stability has been observed for ligands that allow for a higher level of metal ion shielding against hydrolysis, for instance, in the cross-linked CB-TE2A or the cryptand DIAMSAR. Both systems have been employed to prepare preclinical RGD peptide tracers (Wei et al. 2009). More recently, TRAP conjugates have been also recommended for labeling peptides with [64]Cu (Šimeček et al. 2012).

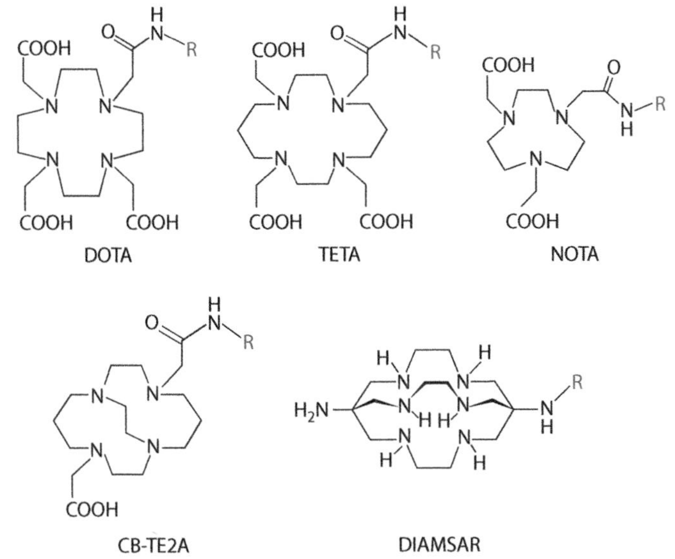

FIGURE 16.23 Chelator systems used for labeling biomacromolecules with [64]Cu (R = peptide/protein).

16.4   QUALITY CONTROL OF RADIOTRACERS

Quality control of preclinical PET radiotracers is an essential step in every *in vivo* experiment. The goal is to demonstrate authenticity of the product as well as to provide information on RCP and specific radioactivity. For short-lived compounds, the analytical methods are required to be efficient and fast to save valuable time. Standard analytical techniques are thin layer chromatography (TLC) and high-performance liquid chromatography. These systems are calibrated using nonradioactive reference compounds. The manufactured radiotracer is confirmed by co-elution with such a reference substance. Comparison of retention times of radioactivity and UV channels serves as proof for the identity of the product. However, not all species might be detected by UV absorption. Here, alternative methods are available: electrochemical, mass spectrometric, or refractometric detection. In the case of [^{18}F]FDG, there are established TLC protocols to visualize Kryptofix using a simple but sensitive staining method (Scott and Kilbourn 2007).

16.5   NEW RADIOCHEMISTRY APPROACHES AND TECHNIQUES

16.5.1   NANOTECHNOLOGY AND MULTIMODALITY

In the past decade, nanotechnology advances have been employed in preclinical PET radiochemistry research. If a nanometer-sized particle has been functionalized with multiple reactive groups, this will enable coupling with a high number of radioactive isotopes. Such a conjugation can be ideally achieved by using the aforementioned Click labeling approach. The resulting multilabeled construct will then increase the effective concentration of the radionuclide and thus help to amplify the detectable *in vivo* signals. These methods will, however, always create macromolecular radiotracers with a distribution of variants of labeled species. It is conceivable that this fact could later become a potential issue with regulatory bodies—if the compound is intended to be progressed into the clinical phase. One possible approach to address such regulatory issues would be to utilize highly biocompatible nanoparticles. For example, hydroxyapatide can be labeled with [^{18}F]fluoride. Also, aluminum hydroxide gel nanoparticles are very suitable as they are already used as adjuvants in vaccines (Jauregui-Osoro et al. 2011).

A number of laboratories are exploring supraparamagnetic iron oxide (SPIO) particles as imaging agents for magnetic resonance imaging (MRI). These particles can be coated using functionalized dextrans, proteins, or PEG (Glaus et al. 2010).

Sometimes, the aim is to create probes for multimodality imaging. For instance, copper-64 was conjugated via DOTA to a SPIO nanoparticle, which was found to be stable *in vivo*. The suitability of that dual PET/MR imaging probe could be established in a phantom study (Glaus et al. 2010).

In general, PET with its capability to quantify radiotracers on the picomolar level and MRI for imaging of morphology should be an ideal complementary pair of modalities. This still stands true for applying a combination of a PET tracer and an MRI contrast agent. But due to the very different sensitivity of these imaging methods, it will be hardly realistic to use a single tracer molecule both for PET and MRI (Jennings and Long 2009). Micromolar amounts that are mandatory for MRI would not be feasible for a PET tracer because here the molecular targets would become oversaturated.

Furthermore, PET tracers can be combined with chromophores to allow for optical imaging techniques. This hybrid approach has been widely used in basic preclinical PET research (Signore et al. 2010; Stasiuk and Long 2013). Although optical imaging has a better sensitivity than MRI, the intrinsic problems of quantification and tissue attenuation of the optical signal remain.

16.5.2   MICROFLUIDICS

Traditional radiosyntheses for PET tracers are performed in a batch reactor mode. The reaction vessels applied normally may contain volume approximately in the range 0.5–5 mL. Researchers have now demonstrated the feasibility to run automated radiolabeling chemistry also on a microliter level. The use of such miniaturized reactors offers a number of distinct advantages: reduced consumption of valuable precursors and reagents, better control of reaction parameters, shorter reaction times, option for supercritical reaction conditions, smaller footprint of the automated platforms, and less lead shielding required. An often-used reactor design in PET microfluidics is based on a flow reactor imbedded in a chip. In its simplest variant, this would mean to mix two reagents in a capillary system and after passage of the capillary to pump the crude mixture into a collecting vial (Wheeler et al. 2010; Zeng et al. 2013). Similar flow reactor systems are also commercially available and have been applied for the preparation of novel radiotracers (Pascali et al. 2010). These platforms are also very useful to rapidly explore different reaction conditions. Unfortunately, the achievable batch yields are not always sufficient enough for imaging purposes (Telu et al. 2011). This is partly caused by increased retention of radioactively labeled compounds in the capillary system. Developers address that issue by using miniaturized batch reactors. Such an advanced platform has been recently described by Lebedev and coworkers. The authors applied a disposable PEEK (polyether ether ketone) microreactor with a capacity of 50 μL for the radiosynthesis of clinical tracer doses using a one-step protocol (Lebedev et al. 2013). The purification and formulation of the product was, however, still performed by traditional "macroscopic" fluid systems. More advanced microfluidic chips can integrate several stages of a radiosynthesis. For example, [^{18}F]FDG has been obtained from a highly integrated single chip that allowed the concentration of fluoride, solvent exchange, fluorination, and removal of protecting groups (Lee et al. 2005).

An important concern in using microfluidic platforms is radiolysis because the higher level of RAC might also increase the risk of radiolytic effects. Nevertheless, by employing a rational reactor design, radiolysis by positrons can be minimized (Rensch et al. 2012).

In conclusion, PET radiochemistry by microfluidic approaches on the preclinical level remains a fast developing field that might shape future automated platforms for clinical use.

ACKNOWLEDGMENT

The authors thank Peter Iveson, GE Healthcare, for discussions leading to an improved manuscript.

DISCLOSURE STATEMENT

Both authors had been full-time employees of GE Healthcare (Research and Development, Amersham, UK) at the time of writing this book chapter.

REFERENCES

Abrams, M. J., M. Juweid et al. (1990). Technetium-99m-human polyclonal IgG radiolabeled via the hydrazino nicotinamide derivative for imaging focal sites of infection in rats. *J. Nucl. Med.* **31**(12): 2022–2028.

Adak, S., R. Bhalla et al. (2012). Radiotracers for SPECT imaging: Current scenario and future prospects. *Radiochim. Acta* **100**(2): 95–107.

Agdeppa, E. D. and M. E. Spilker. (2009). A review of imaging agent development. *AAPS J.* **11**(2): 286–299.

Alberto, R. and U. Abram. (2011). ^{99m}Tc: Labeling chemistry and labeled compounds. In A. Vértes, S. Nagy, Z. Klencsár, R. G. Lovas, and F. Rösch (eds.), *Handbook of Nuclear Chemistry*. Springer, Boston, MA, pp. 2073–2120.

Alberto, R., J. Kyong Pak et al. (2004a). Mono-, bi-, or tridentate ligands? The labeling of peptides with ^{99m}Tc-carbonyls. *Biopolymers* **76**(4): 324–333.

Alberto, R., K. Ortner et al. (2001). Synthesis and properties of boranocarbonate: A convenient in situ CO source for the aqueous preparation of [(99m)Tc(OH(2))3(CO)3]+. *J. Am. Chem. Soc.* **123**(13): 3135–3136.

Alberto, R., J. K. Pak et al. (2004b). Mono-, bi-, or tridentate ligands? The labeling of peptides with ^{99m}Tc-carbonyls. *Biopolymers* **76**(4): 324–333.

Alberto, R., R. Schibli et al. (1998). A novel organometallic aqua complex of technetium for the labeling of biomolecules: Synthesis of [99mTc(OH2)3(CO)3]+ from [99mTcO4]– in aqueous solution and its reaction with a bifunctional ligand. *J. Am. Chem. Soc.* **120**(31): 7987–7988.

Alberto, R., R. Schibli et al. (1999). First application of fac-[99mTc(OH2)3(CO)3]+ in bioorganometallic chemistry: Design, structure, and in vitro affinity of a 5-HT1A receptor ligand labeled with 99mTc. *J. Am. Chem. Soc.* **121**(25): 607606077.

Allard, M., E. Fouquet et al. (2008). State of art in C-11 labeled radiotracers synthesis. *Curr. Med. Chem.* **15**: 235–277.

Ametamey, S. M., M. Honer et al. (2008). Molecular imaging with PET. *Chem. Rev.* **108**: 1501–1516.

Anderson, H. and P. Price. (2000). What does positron emission tomography offer oncology? *Eur. J. Cancer* **36**: 2028–2035.

Auf Dem Keller, U., C. L. Bellac et al. (2010). Novel matrix metalloproteinase inhibitor [18F]marimastat- aryl-trifluoroborate as a probe for in vivo positron emission tomography imaging in cancer. *Cancer Res.* **70**: 7562–7569.

Balentova, E., C. Collet et al. (2011). Synthesis and hydrolytic stability of novel 3-[18F] fluoroethoxybis (1-methylethyl)silyl]propanamine-based prosthetic groups. *J. Fluorine Chem.* **132**: 250–257.

Ballinger, J. R. (2005). Nomenclature of 99mTc-Technetium-Labeled Radiopharmaceuticals. *J. Nucl. Med.* **46**(12).

Bandoli, G., A. Dolmella et al. (2009). Mononuclear six-coordinated Ga(III) complexes: A comprehensive survey. *Coord. Chem. Rev.* **253**(1–2): 56–77.

Bar-Shalom, R., A. Y. Valdivia et al. (2000). PET imaging in oncology. *Semin. Nucl. Med.* **30**: 150–185.

Bartholomä, M. D., A. S. Louie et al. (2010). Technetium and gallium derived radiopharmaceuticals: Comparing and contrasting the chemistry of two important radiometals for the molecular imaging era. *Chem. Rev.* **110**: 2903–2920.

Becaud, J., L. Mu et al. (2009). Direct one-step 18F-labeling of peptides via nucleophilic aromatic substitution. *Bioconjug. Chem.* **20**(12): 2254–2261.

Bergström, M. and B. Långström. (2005). Pharmacokinetic studies with PET. *Prog. Drug Res.* **62**: 281–317.

Berry, D. J., Y. Ma et al. (2011). Efficient bifunctional gallium-68 chelators for positron emission tomography: Tris(hydroxypyridinone) ligands. *Chem. Commun. (Camb)* **47**(25): 7068–7070.

Blankenberg, F. G., J.-L. Vanderheyden et al. (2006). Radiolabeling of HYNIC–annexin V with technetium-99m for in vivo imaging of apoptosis. *Nat. Protocol* **1**(1): 108–110.

Bolzati, C., D. Carta et al. (2012). Chelating systems for 99mTc/188Re in the development of radiolabeled peptide pharmaceuticals. *Anti Canc. Agents Med. Chem.* **12**(5): 428–461.

Boros, E., C. L. Ferreira et al. (2012). RGD conjugates of the H2dedpa scaffold: Synthesis, labeling and imaging with 68Ga. *Nucl. Med. Biol.* **39**(6): 785–794.

Brooks, D. J. (2010). Imaging approaches to Parkinson disease. *J. Nucl. Med.* **51**: 596–609.

Cai, L. S., S. Y. Lu et al. (2008). Chemistry with [F-18]fluoride ion. *Eur. J. Org. Chem.* 17: 2853–2873.

Cant, A. A., R. Bhalla et al. (2012). Nickel-catalysed aromatic Finkelstein reaction of aryl and heteroaryl bromides. *Chem. Commun.* **48**(33): 3993–3995.

Carroll, V., D. W. Demoin et al. (2012). Inorganic chemistry in nuclear imaging and radiotherapy: Current and future directions. *Radiochim. Acta* **100**(8–9): 653–667.

Coenen, H. H., P. H. Elsinga et al. (2010). Fluorine-18 radiopharmaceuticals beyond [18F]FDG for use in oncology and neurosciences. *Nucl. Med. Biol.* **37**: 727–740.

Coenen, H. H., J. Mertens et al. (2006). Methods of radioiodination. In H. H. Coenen, J. Mertens, and B. Mazière (eds.), *Radioionidation Reactions for Radio Pharmaceuticals*. Springer, Dordrecht, the Netherlands, pp. 29–72.

Cohen, A. D., G. D. Rabinovici et al. (2012). Using Pittsburgh Compound B for In Vivo PET imaging of fibrillar amyloid-beta. *Adv. Pharmacol. (San Diego, Calif.)* **64**: 27–81.

Collingridge, D. R., V. A. Carroll et al. (2002). The development of [(124)I]iodinated-VG76e: A novel tracer for imaging vascular endothelial growth factor in vivo using positron emission tomography. *Cancer Res.* **62**(20): 5912–5919.

Cutler, C. S., H. M. Hennkens et al. (2013). Radiometals for combined imaging and therapy. *Chem. Rev.* **113**: 858–883.

de Goeij, J. J. M. and M. L. Bonardi. (2005). How do we define the concepts specific activity, radioactive concentration, carrier, carrier-free and no-carrier-added? *J. Radioanal. Nucl. Chem.* **263**(1): 13–18.

Decristoforo, C. and S. J. Mather. (1999a). Preparation, 99mTc-labeling, and in vitro characterization of HYNIC and N3S modified RC-160 and [Tyr3]octreotide. *Bioconjug. Chem.* **10**(3): 431–438.

Decristoforo, C. and S. J. Mather. (1999b). Technetium-99m somatostatin analogues: Effect of labeling methods and peptide sequence. *Eur. J. Nucl. Med.* **26**(8): 869–876.

Decristoforo, C., L. Melendez-Alafort et al. (2000). 99mTc-HYNIC-[Tyr3]-octreotide for imaging somatostatin-receptor-positive tumors: Preclinical evaluation and comparison with 111In-octreotide. *J. Nucl. Med.* **41**(6): 1114–1119.

Devaraj, N. K., E. Keliher et al. (2009). 18F Labeled nanoparticles for in vivo PET-CT imaging. *Bioconjug. Chem.* **20**: 397–401.

Dewan, R., M. B. Saddi et al. (2005). Collision, scattering and absorption differential cross-sections in double-photon Compton scattering. *Ann. Nucl. Energ.* **32**(9): 1008–1022.

Dewanjee, M. K. (1992). *Radioiodination: Theory, Practice, and Biomedical Applications*. Springer, Boston, MA.

Dilworth, J. R. and S. J. Parrott. (1998). The biomedical chemistry of technetium and rhenium. *Chem. Soc. Rev.* **27**(1): 43.

Eckelman, W. C., M. Bonardi et al. (2008). True radiotracers: Are we approaching theoretical specific activity with Tc-99m and I-123? *Nucl. Med. Biol.* **35**(5): 523–527.

Eder, M., B. Wängler et al. (2008). Tetrafluorophenolate of HBED-CC: A versatile conjugation agent for Ga-68-labeled small recombinant antibodies. *Eur. J. Nucl. Med. Mol. Imaging* **35**: 1878–1886.

Edwards, D., P. Jones et al. (2008). ^{99m}Tc-NC100692—A tracer for imaging vitronectin receptors associated with angiogenesis: A preclinical investigation. *Nucl. Med. Biol.* **35**(3): 365–375.

Eersels, J. L. H., M. J. Travis et al. (2005). Manufacturing I-123-labeled radiopharmaceuticals. Pitfalls and solutions. *J. Labelled Comp. Rad.* **48**(4): 241–257.

Egerton, A., E. Hirani et al. (2010). Further evaluation of the carbon-11-labeled D2/3 agonist PET radiotracer PHNO: Reproducibility in tracer characteristics and characterization of extrastriatal binding. *Synapse* **64**: 301–312.

Ehrin, E., L. Farde et al. (1985). Preparation of ^{11}C-labeled raclopride, a new potent dopamine receptor antagonist: Preliminary PET studies of cerebral dopamine receptors in the monkey. *Int. J. Appl. Radiat. Isot.* **36**: 269–273.

Ermert, J. and H. H. Coenen. (2010). Nucleophilic ^{18}F-fluorination of complex molecules in activated carbocyclic aromatic position. *Curr. Rad.* **3**: 109–126.

Fani, M., J. P. André et al. (2008). ^{68}Ga-PET: A powerful generator-based alternative to cyclotron-based PET radiopharmaceuticals. *Contrast Media Mol. Imaging* **3**: 67–77.

Ferneliums, W. C., T. D. Coyle et al. (1981). Nomenclature of inorganic chemistry: II. 1—Isotopically modified compounds. *Pure Appl. Chem.* **53**: 1887–1900.

Finkelstein, H. (1910). Darstellung organischer Jodide aus den entsprechenden Bromiden und Chloriden. *Berichte der deutschen chemischen Gesellschaft* **43**(2): 1528–1532.

Fontes, A., M. I. M. Prata et al. (2011). Ga(III) chelates of amphiphilic DOTA-based ligands: Synthetic route and in vitro and in vivo studies. *Nucl. Med. Biol.* **38**(3): 363–370.

Forsback, S., O. Eskola et al. (2008). Electrophilic synthesis of 6-[F-18]fluoro-L-DOPA using post-target produced [F-18]F-2. *Radiochim. Acta* **96**: 845–848.

Fowler, J. S., A. P. Wolf et al. (1988). Mechanistic positron emission tomography studies: Demonstration of a deuterium isotope effect in the monoamine oxidase-catalyzed binding of [^{11}C]L-deprenyl in living baboon brain. *J. Neurochem.* **51**: 1524–1534.

Fritzberg, A. R., S. Kasina et al. (1986). Synthesis and biological evaluation of technetium-99m MAG3 as a hippuran replacement. *J. Nucl. Med.* **27**(1): 111–116.

Gao, Z., Y. H. Lim et al. (2012). Metal-free oxidative fluorination of phenols with [^{18}F]fluoride. *Angew. Chem. Int. Ed.* **51**: 6733–6737.

Garcia-Arguello, F. S., R. Fortt et al. (2013). Radiosynthesis of the D2/3 agonist [3–^{11}C]-(+)-PHNO using [^{11}C] iodomethane. *Appl. Rad. Isot.* **73**: 79–83.

Ghibellini, G., E. M. Leslie et al. (2008). Use of Tc-99m mebrofenin as a clinical probe to assess altered hepatobiliary transport: Integration of in vitro, pharmacokinetic modeling, and simulation studies. *Pharmaceut. Res.* **25**(8): 1851–1860.

Gill, H. S., J. N. Tinianow et al. (2009). A modular platform for the rapid site-specific radiolabeling of proteins with F-18 exemplified by quantitative positron emission tomography of human epidermal growth factor receptor 2. *J. Med. Chem.* **52**: 5816–5825.

Glaser, M., D. R. Collingridge et al. (2001). Preparation of [^{124}I]IBA-annexin-V as a potential pet probe for apoptosis. *J. Labelled Comp. Rad.* **44**(S1): S336–S338.

Glaser, M., D. R. Collingridge et al. (2003). Iodine-124 labeled annexin-V as a potential radiotracer to study apoptosis using positron emission tomography. *Appl. Rad. Isot.* **58**(1): 55–62.

Glaser, M., D. B. Mackay et al. (2004). Improved targetry and production of iodine-124 for PET studies. *Radiochim. Acta* **92**: 951–956.

Glaser, M. and E. G. Robins. (2009). 'Click labeling' in PET radiochemistry. *J. Labelled Comp. Rad.* **52**: 407–414.

Glaus, C., R. Rossin et al. (2010). In vivo evaluation of ^{64}Cu-labeled magnetic nanoparticles as a dual-modality PET/MR imaging agent. *Bioconjug. Chem.* **21**: 715–722.

Green, C. (2012). Technetium-99m production issues in the United Kingdom. *J. Med. Phys.* **37**(2): 66–71.

Hamacher, K., H. H. Coenen et al. (1986). Efficient stereospecific synthesis of NCA 2-[^{18}F]fluoro-2-deoxy-2-fluoro-D-glucose using aminopolyether supported nucleophilic substitution. *J. Nucl. Med.* **27**: 235–238.

Handeland, A. M. and E. Sundrehagen. (1987). Determination of the oxidation state of ^{99m}Tc in complexes with MDP and DTPA. *Int. J. Rad. Appl. Instrum. A* **38**(6): 479–484.

Hausner, S. H., R. D. Carpenter et al. (2013). Evaluation of an integrin $\alpha_v\beta_6$-specific peptide labeled with [^{18}F] fluorine by copper-free, strain-promoted click chemistry. *Nucl. Med. Biol.* **40**: 233–239.

Heiss, W. D. and K. Herholz. (2006). Brain receptor imaging. *J. Nucl. Med.* **47**: 302–312.

Höhne, A., L. Yu et al. (2009). Organofluorosilanes as model compounds for ^{18}F-labeled silicon-based PET tracers and their hydrolytic stability: Experimental data and theoretical calculations (PET = Positron Emission Tomography). *Eur. J. Chem.* **15**: 3736–3743.

Iddon, L., J. Leyton et al. (September 7–10, 2011a). Novel fluorine-18 click labeled octreotate analogues, including automation onto the FASTlab platform towards a clinical PET imaging agent. In *World Molecular Imaging Congress*, San Diego, CA.

Iddon, L., J. Leyton et al. (2011b). Synthesis and in vitro evaluation of [^{18}F]fluoroethyl triazole labeled [Tyr^3] octreotate analogues using click chemistry. *Bioorg. Med. Chem. Lett.* **21**: 3122–3127.

International Atomic Energy, A. (2008). *Technetium-99m Radiopharmaceuticals: Manufacture of Kits.* International Atomic Energy Agency, Vienna, Austria.

IUPAC Commission on the Nomenclature of Organic Chemistry (CNOC). (1978). Nomenclature of Organic Chemistry. Section H: Isotopically Modified Compounds. Recommendations 1977. *Eur. J. Biochem.* **86**(1): 9–25.

Jae, M. J., K. H. Mee et al. (2008). Preparation of a promising angiogenesis PET imaging agent: ^{68}Ga-labeled c(RGDyK)-isothiocyanatobenzyl-1,4,7-triazacyclononane-1, 4,7-triacetic acid and feasibility studies in mice. *J. Nucl. Med.* **49**: 830–836.

Jauregui-Osoro, M., P. A. Williamson et al. (2011). Biocompatible inorganic nanoparticles for [^{18}F]-fluoride binding with applications in PET imaging. *Dalton Trans.* **40**: 6226–6237.

Jennings, L. E. and N. J. Long. (2009). 'Two is better than one'-probes for dual-modality molecular imaging. *Chem. Commun.* 24: 3511–3524.

Jones, T. (1996). The role of positron emission tomography within the spectrum of medical imaging. *Eur. J. Nucl. Med.* **23**: 207–211.

Kabalka, G. W. and R. S. Varma. (1989). The synthesis of radiolabeled compounds via organometallic intermediates. *Tetrahedron* **45**(21): 6601–6621.

Kealey, S., P. W. Miller et al. (2009). Copper(I) scorpionate complexes and their application in palladium-mediated [C-11]carbonylation reactions. *Chem. Commun.* 25: 3696–3698.

Kealey, S., C. Plisson et al. (2011). Microfluidic reactions using [^{11}C]carbon monoxide solutions for the synthesis of a positron emission tomography radiotracer. *Org. Biomol. Chem.* **9**: 3313–3319.

Keidar, Z., O. Israel et al. (2003). SPECT/CT in tumor imaging: Technical aspects and clinical applications. *Semin. Nucl. Med.* **33**(3): 205–218.

Khalil, M. M., J. L. Tremoleda et al. (2011). Molecular SPECT imaging: An overview. *Int. J. Mol. Imaging* **2011**: 1–15.

Kihlberg T, K. F. (2002). [^{11}C]Carbon monoxide in selenium-mediated synthesis of ^{11}C-carbamoyl compounds. *J. Org. Chem.* **67**: 3687–3692.

Klapars, A. and S. L. Buchwald. (2002). Copper-catalyzed halogen exchange in aryl halides: An aromatic Finkelstein reaction. *J. Am. Chem. Soc.* **124**(50): 14844–14845.

Kluba, C. and T. Mindt. (2013). Click-to-chelate: Development of technetium and rhenium-tricarbonyl labeled radiopharmaceuticals. *Molecules* **18**(3): 3206–3226.

Klunk, W. E., H. Engler et al. (2004). Imaging brain amyloid in Alzheimer's disease with Pittsburgh Compound-B. *Ann. Neurol.* **55**: 306–319.

Klunk, W. E. and C. A. Mathis. (2008). The future of amyloid-beta imaging: A tale of radionuclides and tracer proliferation. *Curr. Opin. Neurol.* **21**: 683–687.

Knust, E. J., K. Dutschka et al. (2000). Preparation of ^{124}I solutions after thermodistillation of irradiated ^{124}TeO$_2$ targets. *Appl. Radiat. Isot.* **52**: 181–184.

Kolb, H. C., M. G. Finn et al. (2001). Click chemistry: Diverse chemical function from a few good reactions. *Angew. Chem. Int. Ed.* **40**: 2004–2021.

Kung, H. F., C. C. Yu et al. (1985). Synthesis of new bis(aminoethanethiol) (BAT) derivatives: Possible ligands for 99mTc brain imaging agents. *J. Med. Chem.* **28**(9): 1280–1284.

Lapi, S. E. and M. J. Welch. (2012). A historical perspective on the specific activity of radiopharmaceuticals: What have we learned in the 35 years of the ISRC? *Nucl. Med. Biol.* **39**: 601–608.

Laverman, P., C. A. D'Souza et al. (2012). Optimized labeling of NOTA-conjugated octreotide with F-18. *Tumour Biol.* **33**: 427–434.

Lebedev, A., R. Miraghaie et al. (2013). Batch-reactor microfluidic device: First human use of a microfluidically produced PET radiotracer. *Lab Chip* **13**: 136–145.

Lee, C. C., G. Sui et al. (2005). Multistep synthesis of a radiolabeled imaging probe using integrated microfluidics. *Science* **310**: 1793–1796.

Lee, E., A. S. Kamlet et al. (2011). A fluoride-derived electrophilic late-stage fluorination reagent for PET imaging. *Science* **334**: 639–642.

Leyton, J., L. Iddon et al. (2011). Targeting somatostatin receptors: Preclinical evaluation of novel ^{18}F-fluoroethyltriazole-Tyr^3-octreotate analogs for PET. *J. Nucl. Med.* **52**: 1441–1448.

Li, Z., F. P. Gabbai et al. (2013). Boron-based dual imaging probes, compositions and methods for rapid aqueous F-18 labeling, and imaging methods using the same. WO 2013/012754 A1. Patent US 20130189185.

Lide, D. R. (2008). *CRC Handbook of Chemistry and Physics: A Ready-Reference Book of Chemical and Physical Data: 2007–2008.* CRC Press, Boca Raton, FL.

Littich, R. and P. J. H. Scott. (2012). Novel strategies for fluorine-18 radiochemistry. *Angew. Chem. Int. Ed.* **51**: 1106–1109.

Liu, S. (2004). The role of coordination chemistry in the development of target-specific radiopharmaceuticals. *Chem. Soc. Rev.* **33**(7).

Liu, Z., Y. Li et al. (2013). Stoichiometric leverage: Rapid ^{18}F-aryltrifluoroborate radiosynthesis at high specific activity for click conjugation. *Angew. Chem. Int. Ed.* **52**: 2303–2307.

Luthra, S. K., F. Brady et al. (1994). Automated radiosyntheses of [6-O-methyl-^{11}C]diprenorphine and [6-O-methyl-^{11}C]buprenorphine from 3-O-trityl protected precursors. *Appl. Rad. Isot.* **45**: 857–873.

Mamat, C., T. Ramenda et al. (2009). Recent applications of click chemistry for the synthesis of radiotracers for molecular imaging. *Mini Rev. Org. Chem.* **6**: 21–34.

Marck, S. C., A. J. Koning et al. (2010). The options for the future production of the medical isotope ^{99}Mo. *Eur. J. Nucl. Med. Mol. Imaging* **37**(10): 1817–1820.

Mather, S. J. and D. Ellison. (1990). Reduction-mediated technetium-99m labeling of monoclonal antibodies. *J. Nucl. Med.* **31**(5): 692–697.

Meszaros, L. K., A. Dose et al. (2010). Hydrazinonicotinic acid (HYNIC)—Coordination chemistry and applications in radiopharmaceutical chemistry. *Inorg. Chim. Acta* **363**(6): 1059–1069.

Meszaros, L. K., A. Dose et al. (2011). Synthesis and evaluation of analogues of HYNIC as bifunctional chelators for technetium. *Dalton Trans. (Camb., Engl. 2003)* **40**(23): 6260–6267.

Miller, P. W., N. J. Long et al. (2008). Synthesis of C-11, F-18, O-15, and N-13 radiolabels for positron emission tomography. *Angew. Chem. Int. Ed.* **47**: 8998–9033.

National Center for Biotechnology Information (US). (2004–2013). Molecular imaging and contrast agent database (MICAD). National Center for Biotechnology Information, Bethesda, MD.

Nock, B., A. Nikolopoulou et al. (2003). [^{99m}Tc]Demobesin 1, a novel potent bombesin analogue for GRP receptor-targeted tumour imaging. *Eur. J. Nucl. Med. Mol. Imaging* **30**(2): 247–258.

Notni, J., K. Pohle et al. (2013). Be spoilt for choice with radiolabeled RGD peptides: Preclinical evaluation of ^{68}Ga-TRAP(RGD)3. *Nucl. Med. Biol.* **40**: 33–41.

Pagani, M., S. Stone-Elander et al. (1997). Alternative positron emission tomography with non-conventional positron emitters: Effects of their physical properties on image quality and potential clinical applications. *Eur. J. Nucl. Med.* **24**(10): 1301–1327.

Pascali, G., G. Mazzone et al. (2010). Microfluidic approach for fast labeling optimization and dose-on-demand implementation. *Nucl. Med. Biol.* **37**: 547–555.

Perumal, M., E. A. Stronach et al. (2012). Evaluation of 2-deoxy-2-[^{18}F]fluoro-d-glucoseand 3′-deoxy-3′-[^{18}F] fluorothymidine-positron emission tomography as biomarkers of therapy response in platinum-resistant ovarian cancer. *Mol. Imaging Biol.* **14**: 753–761.

Pettitt, R., J. Grigg et al. (2010). The development of an automated synthesis of [^{18}F]fluciclatide of the FASTlab synthesiser using chemometric design. *Q. J. Nucl. Med.* **54**: 26–27.

Pimlott, S. L. and A. Sutherland. (2011). Molecular tracers for the PET and SPECT imaging of disease. *Chem. Soc. Rev.* **40**(1): 149–162.

Poethko, T., M. Schottelius et al. (2004). Two-step methodology for high-yield routine radiohalogenation of peptides: ^{18}F-labeled RGD and Octreotide analogs. *J. Nucl. Med.* **45**: 892–902.

Pohl, E., A. Heine et al. (1995). Structure of octreotide, a somatostatin analogue. *Acta Cryst. D* **51**(Pt. 1): 48–59.

Pohle, K., J. Notni et al. (2012). ^{68}Ga-NODAGA-RGD is a suitable substitute for (18)F-Galacto-RGD and can be produced with high specific activity in a cGMP/GRP compliant automated process. *Nucl. Med. Biol.* **39**(6): 777–784.

Prasanphanich, A. F., P. K. Nanda et al. (2007). [^{64}Cu-NOTA-8-Aoc-BBN(7–14)NH$_2$] targeting vector for positron-emission tomography imaging of gastrin-releasing peptide receptor-expressing tissues. *Proc. Natl. Acad. Sci. USA* **104**: 12462–12467.

Prinsen, K., J. Li et al. (2010). Development and evaluation of a ^{68}Ga labeled pamoic acid derivative for in vivo visualization of necrosis using positron emission tomography. *Bioorg. Med. Chem.* **18**: 5274–5281.

Purser, S., P. R. Moore et al. (2008). Fluorine in medicinal chemistry. *Chem. Soc. Rev.* **37**: 320–330.

Ram, S., R. E. Ehrenkaufer et al. (1989). Synthesis of the labeled D1 receptor antagonist SCH 23390 using [^{11}C] carbon dioxide. *Appl. Rad. Isot.* **40**: 425–427.

Reischl, G., K. Henkova et al. (2004). Optimized, reliable production of iodine-124 and iodine-123 from the enriched tellurium dioxides. *J. Nucl. Med.* **45**: 471.

Rensch, C., B. Waengler et al. (2012). Microfluidic reactor geometries for radiolysis reduction in radiopharmaceuticals. *Appl. Rad. Isot.* **70**: 1691–1697.

Rhodes, B. A., P. O. Zamora et al. (1986). Technetium-99m labeling of murine monoclonal antibody fragments. *J. Nucl. Med.* **27**(5): 685–693.

Ribeiro Morais, G., R. A. Falconer et al. (2013). Carbohydrate-based molecules for molecular imaging in nuclear medicine. *Eur. J. Org. Chem.* **8**: 1401–1414.

Roca, M., E. F. J. Vries et al. (2010). Guidelines for the labeling of leucocytes with ^{111}In-oxine. *Eur. J. Nucl. Med. Mol. Imaging* **37**(4): 835–841.

Roeda, D. and F. Dollé. (2010). Aliphatic nucleophilic radiofluorination. *Curr. Rad.* **3**: 81–108.

Rosenspire, K. C., M. S. Haka et al. (1990). Synthesis and preliminary evaluation of carbon-11-meta-hydroxyephedrine: A false transmitter agent for heart neuronal imaging. *J. Nucl. Med.* **31**: 1328–1334.

Ross, T. L. (2010). The click chemistry approach applied to fluorine-18. *Cur. Rad.* **3**: 202–223.

Saby, L., O. Laas et al. (2013). Positron emission tomography/computed tomography for diagnosis of prosthetic valve endocarditis: Increased valvular ^{18}F-fluorodeoxyglucose uptake as a novel major criterion. *J. Am. Coll. Cardiol.* **61**: 2374–2382.

Salacinski, P. R., C. McLean et al. (1981). Iodination of proteins, glycoproteins, and peptides using a solid-phase oxidizing agent, 1,3,4,6-tetrachloro-3 alpha,6 alpha-diphenyl glycoluril (Iodogen). *Anal. Biochem.* **117**(1): 136–146.

Samuelsson, L. and B. Långström. (2003). Synthesis of 1-(2′-deoxy-2′-fluoro-b-D-arabinofuranosyl)-[Methyl-^{11}C] via a Stille cross-coupling reaction with [11C]methyl iodide. *J. Labelled Comp. Rad.* **46**: 263–272.

Sanchez-Crespo, A. (2013). Comparison of Gallium-68 and Fluorine-18 imaging characteristics in positron emission tomography. *Appl. Rad. Isot.* **76**: 55–62.

Schirrmacher, R., G. Bradtmöller et al. (2006). ^{18}F-Labeling of peptides by means of an organosilicon-based fluoride acceptor. *Angew. Chem. Int. Ed. Engl.* **45**: 6047–6050.

Schirrmacher, R., Y. Lakhrissi et al. (2008). Rapid in situ synthesis of [^{11}C]methyl azide and its application in C-11 click-chemistry. *Tetrahedron Lett.* **49**: 4824–4827.

Schwochau, K. (2000). *Technetium: Chemistry and Radiopharmaceutical Applications.* Wiley-VCH, Weinheim, Germany.

Scott, P. J. H. and M. R. Kilbourn. (2007). Determination of residual Kryptofix 2.2.2 levels in [F-18]-labeled radiopharmaceuticals for human use. *Appl. Rad. Isot.* **65**: 1359–1362.

Seo, Y., C. Mari et al. (2008). Technological development and advances in single-photon emission computed tomography/computed tomography. *Semin. Nucl. Med.* **38**(3): 177–198.

Shields, A. F. (2006). Positron emission tomography measurement of tumor metabolism and growth: Its expanding role in oncology. *Mol. Imaging Biol.* **8**: 141–150.

Signore, A., S. J. Mather et al. (2010). Molecular imaging of inflammation/infection: Nuclear medicine and optical imaging agents and methods. *Chem. Rev.* **110**: 3112–3145.

Šimeček, J., H. J. Wester et al. (2012). Copper-64 labeling of triazacyclononane-triphosphinate chelators. *Dalton Trans.* **41**: 13803–13806.

Stasiuk, G. J. and N. J. Long. (2013). The ubiquitous DOTA and its derivatives: The impact of 1,4,7,10-tetraazacyclododecane-1,4,7,10-tetraacetic acid on biomedical imaging. *Chem. Commun.* **49**: 2732–2746.

Svedberg, M. M., O. Rahman et al. (2012). Preclinical studies of potential amyloid binding PET/SPECT ligands in Alzheimer's disease. *Nucl. Med. Biol.* **39**: 484–501.

Telu, S., J. H. Chun et al. (2011). Syntheses of mGluR5 PET radioligands through the radiofluorination of diaryliodonium tosylates. *Org. Biomol. Chem.* **9**: 6629–6638.

Vaidyanathan, G. and M. R. Zalutsky. (1992). Labeling proteins with fluorine-18 using N-succinimidyl 4-[^{18}F] fluorobenzoate. *Nucl. Med. Biol.* **19**: 275–281.

Vallabhajosula, S. and A. Nikolopoulou. (2011). Radioiodinated metaiodobenzylguanidine (MIBG): Radiochemistry, biology, and pharmacology. *Semin. Nucl. Med.* **41**(5): 324–333.

Velikyan, I., H. Maecke et al. (2008). Convenient preparation of ^{68}Ga-based PET-radiopharmaceuticals at room temperature. *Bioconjug. Chem.* **19**(2): 569–573.

Wadas, T. J., E. H. Wong et al. (2010). Coordinating radiometals of copper, gallium, indium, yttrium, and zirconium for PET and SPECT imaging of disease. *Chem. Rev.* **110**(5): 2858–2902.

Wang, M. W., W. Y. Lin et al. (2010). Microfluidics for positron emission tomography probe development. *Mol. Imaging* **9**: 175–191.

Wängler, C., R. Schirrmacher et al. (2010). Click-chemistry reactions in radiopharmaceutical chemistry: Fast & easy introduction of radiolabels into biomolecules for in vivo imaging. *Curr. Med. Chem.* **17**: 1092–1116.

Wängler, C., B. Waser et al. (2010). One-step ^{18}F-labeling of carbohydrate-conjugated octreotate-derivatives containing a silicon-fluoride-acceptor (SiFA): In vitro and in vivo evaluation as tumor imaging agents for positron emission tomography (PET). *Bioconjug. Chem.* **21**: 2289–2296.

Wei, L., Y. Ye et al. (2009). ^{64}Cu-Labeled CB-TE2A and diamsar-conjugated RGD peptide analogs for targeting angiogenesis: Comparison of their biological activity. *Nucl. Med. Biol.* **36**: 277–285.

Welch, M. J. and C. S. Redvanly. (2003). *Handbook of Radiopharmaceuticals, Radiochemistry and Applications.* John Wiley & Sons Ltd, Chichester, U.K.

Welling, M. M., A. Paulusma-Annema et al. (2000). Technetium-99m labeled antimicrobial peptides discriminate between bacterial infections and sterile inflammations. *Eur. J. Nucl. Med.* **27**(3): 292–301.

Wheeler, T. D., D. Zeng et al. (2010). Microfluidic labeling of biomolecules with radiometals for use in nuclear medicine. *Lab on a Chip* **10**: 3387–3396.

WHO. (2013). WHOCC—ATC/DDD index. Retrieved March 14, 2014, from http://www.whocc.no/atc_ddd_index/?code=V09.

Zeglis, B. M., K. K. Sevak et al. (2013). A pretargeted PET imaging strategy based on bioorthogonal Diels-Alder click chemistry. *J. Nucl. Med.* **54**(8): 1389–1396.

Zeng, D., A. V. Desai et al. (2013). Microfluidic radiolabeling of biomolecules with PET radiometals. *Nucl. Med. Biol.* **40**: 42–51.

Zimmer, L. and A. Luxen. (2012). PET radiotracers for molecular imaging in the brain: Past, present and future. *NeuroImage* **61**: 363–370.

Zlatopolskiy, B. D., R. Kandler et al. (2012a). Beyond azide-alkyne click reaction: Easy access to ^{18}F-labeled compounds via nitrile oxide cycloadditions. *Chem. Commun.* **48**: 7134–7136.

Zlatopolskiy, B. D., R. Kandler et al. (2012b). C-(4-[^{18}F]fluorophenyl)-N-phenyl nitrone: A novel ^{18}F-labeled building block for metal free [3+2]cycloaddition. *Appl. Rad. Isot.* **70**: 184–192.

Zolle, I. (2007). *Technetium-99m Pharmaceuticals Preparation and Quality Control in Nuclear Medicine.* Springer, Berlin, Germany.

17

Molecular Targets and Optical Probes

Eleni K. Efthimiadou and George Kordas

17.1 NANOPARTICLES AS IMAGING AGENTS

This chapter focuses on the various strategies in the preparation, structure, and properties of naked and surface-functionalized inorganic nanoparticles (NPs) and their toxicity profile. NP synthesis with desired size/shape has enormous importance, especially in the emerging field of nanotechnology (Jatzkewitz 1955). Nowadays, nanotechnology plays a key role in different types of theragnostic applications exploiting several NP-based properties such as the extremely small size and the functional surface area, which enable the transfer of several compounds (i.e., drugs, probes, and targeting moieties). Furthermore, NPs display, by themselves, unique features due to their optical, electronic, and magnetic properties (Smith et al. 2008). In general, NPs used for biomedical research

315

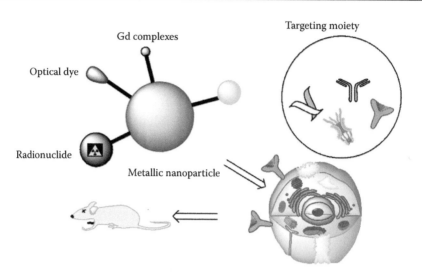

FIGURE 17.1 A specific multimodal imaging probe, for targeting, diagnosis, and treatment.

can be roughly categorized into three groups: (1) inorganic NPs [quantum dots (QDs), iron oxide NPs (IONPs), gold NPs], (2) polymeric NPs (dendrimers and amphiphilic NPs), and (3) lipid NPs [liposomes and solid lipid NPs (SLNs)] (Xue et al. 2011).

Many scientific groups are working on the domain of core/shell nanoparticle formation and functionalization either by coating and/or by modification with different active linkers, which strongly supports that this type of NPs will significantly contribute to new therapeutic approaches (Figure 17.1).

Noninvasive imaging techniques have shown great promise for monitoring pathological processes. As outlined by the members of the Society of Nuclear Medicine, molecular imaging provides the ability to visualize, characterize, and measure biological processes at the cellular and molecular levels in humans and other living organisms (Mankoff 2007). To develop effective NPs for *in vivo* imaging of molecular targets, one should (1) minimize uptake at the reticuloendothelial system (RES), (2) avoid *in vivo* aggregation due to their large diameters (several micrometers) that can be irreversibly trapped in the capillaries of the lungs, and finally, (3) prolong the plasma half-life times. Surface modification is required to camouflage the particles and prevent unwanted interactions (e.g., oxidation, protein adsorption, etc.) as well as to provide a location to anchor surface moieties.

Several imaging modalities have been applied either by incorporating imaging agents into nanomaterials or by taking advantage of the latters' properties to enable anatomical or functional imaging through computed tomography (CT), ultrasound (US), magnetic resonance imaging (MRI), molecular imaging via MRI, US, and optical imaging (OI), single-photon emission computed tomography (SPECT), and positron emission tomography (PET). MRI is capable of producing 3D images of soft tissues containing water with a high degree of spatial resolution but suffers from a relatively low sensitivity. In contrast, SPECT imaging techniques display higher sensitivity than MRI but reduced sensitivity relative to PET. SPECT remains the most widely used nuclear tracer method, because SPECT radionuclides are generally cheaper, are readily available, and have longer half-lives than PET radionuclides. Optical imaging is a very sensitive molecular imaging modality, although it is hampered by the very limited tissue penetration and low spatial resolution. Therefore, it is necessary to improve the visualization of diseases and elucidate NPs' biological behavior by combining the above-mentioned diagnostic imaging modalities (Lee et al. 2012).

In Table 17.1, some of the potential biomedical inorganic nanoparticles are provided in relation to the nature of nanomaterials and their physicochemical properties.

TABLE 17.1
Typical Inorganic NPs and Their Applications

Category	Examples	Intrinsic Properties	Applications
Metallic NPs	AuNPs	Surface plasmon resonance (SPR), surface reactivity, catalysis	Optical imaging, Raman probe, photothermal cancer therapy, drug delivery
Semiconductor NPs	QDs	Fluorescence, luminescence	High-resolution cellular imaging, long-term cell trafficking, diagnostics, and sensing based on energy transfer techniques
Magnetic NPs	IONPs	Supermagnetism	Magnetic separation, targeted delivery, MRI contrast agents, thermotherapy

17.2 INORGANIC NANOPARTICLES

17.2.1 QUANTUM DOTS (QDs)

Semiconductor particles in spherical shape, called QDs, have diameters between 2 and 10 nm and contain roughly 200–10,000 atoms. Due to their semiconducting nature and their size, QDs possess attractive properties for use in optoelectronic devices and as multi-sensing probes in biological systems (Alivisatos 1996). Their fluorescence properties are related to the size of synthesized nanocrystals. Small nanocrystals of semiconductors are characterized by a bandgap energy that is dependent on the particle size, allowing the optical characteristics of a QD to be tuned by adjusting its size. QDs are about 10–100 times brighter than other fluorescent dyes incorporated either into small biomolecules such as peptides or even into nanoparticles. This phenomenon can be attributed mainly to their large absorption cross sections, which are 100–1000 times more stable than the fluorescent molecules used to date, avoiding photobleaching, and show narrower and also more symmetric emission spectra. Moreover, a single light source can be used to excite QDs with different emission wavelengths, which can be tuned from the ultraviolet (Zhong et al. 2003) throughout the visible and near-infrared (Bailey and Nie 2003) and even into the mid-infrared (Hines and Scholes 2003) spectra. Bulk semiconductors are characterized by a composition-dependent bandgap energy, which is the minimum energy required to excite an electron to an energy level above its ground state, commonly through the absorption of a photon of energy greater than the bandgap energy. Relaxation of the excited electron back to its ground state may be accompanied by the fluorescent emission of a photon. QDs have been demonstrated to be excellent probes for optical imaging because of their bright and stable photoluminescence. More specifically, the tunable emission spectrum allows QD emission at near-infrared light (700–900 nm), which could effectively penetrate into deep tissue.

17.2.2 SYNTHESIS OF QDs

The synthesis of QDs was described in 1982 by Efros and Ekimov (Efros and Efros 1982; Ekimov and Onushchenko 1982). Nanocrystals and microcrystals of semiconductors were introduced into glass matrices and their optical properties were studied. Based on this work, a wide variety of synthetic methods have been developed for the synthesis of QDs in different solutions, including aqueous, high-temperature organic solvents, and solid materials (Bailey and Nie 2003; Zhong et al. 2003). It is well known that the synthesized QDs, after introducing semiconductor precursors, form crystals under thermodynamically favorable conditions. Semiconductor-binding agents kinetically control crystal growth, aiming at maintaining their size within the quantum size. A significant input to QDs' synthesis was introduced in 1993 by Murray et al. (1993), developing monodispersed QDs made from cadmium sulfide (CdS), cadmium selenide (CdSe), or cadmium telluride (CdTe). Based on this report, the synthetic chemistry of CdSe QDs generated brightly fluorescent QDs that could span the visible spectrum. As a result, CdSe has become the most common chemical composition for QD synthesis, especially for biological applications.

According to the literature, there are different coating agents for QDs that improve their colloidal stability, immune response, and targeting ability (Alivisatos 1996; Chan and Nie 1998).

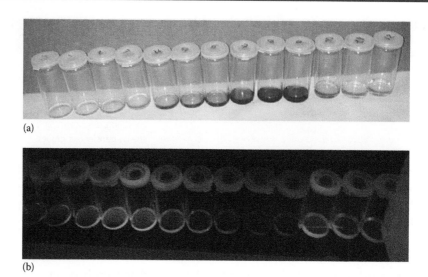

(a)

(b)

FIGURE 17.2 CdSe quantum dots (a) before and (b) after UV lighting (unpublished results).

More specifically, the synthesis of the CdSe core can initially take place in a nonpolar solvent, and subsequently a shell of zinc sulfide (ZnS) is added to the surface. The QDs are then transferred to an aqueous solution through encapsulation with an amphiphilic polymer, which can then be cross-linked to biomolecules to yield targeted molecular imaging agents. While synthesizing QDs as imaging probes, the selection of a QD core composition is determined by the desired wavelength of emission. For example, CdSe QDs may be size-tuned to emit in the 450–650 nm range, whereas CdTe QDs can emit in the 500–750 nm range.

A typical synthetic procedure of CdSe involves the use of selenium precursors (commonly trioctylphosphine selenide or tributylphosphine selenide) at room temperature. The solution of the selenium precursor is swiftly added to a hot solution (300°C) containing both a cadmium precursor (dimethylcadmium or cadmium oleate) and a coordinating ligand (trioctylphosphine oxide or hexadecylamine) under inert atmosphere (nitrogen or argon gas). It is expected that cadmium and selenium precursors react quickly at such a high temperature, forming CdSe nanocrystal nuclei. After introducing the coordinating ligands, the metal atoms on the nanocrystals' surface bind with them, leading to stabilized colloidal behavior in solution and controlling the rate of growth. The addition of a cool solution quickly reduces the temperature of the reaction mixture, causing nucleation to cease (Figure 17.2). When QDs have reached the desired size and emission wavelength, the reaction mixture is quenched by cooling to room temperature. The resulting QDs are coated using aliphatic coordinating ligands and highly hydrophobic polymers, allowing them to be purified through liquid–liquid extractions or via precipitation from a polar solvent (Chan and Nie 1998; Smith et al. 2008).

17.2.3 IRON OXIDE NANOPARTICLES

Magnetic nanoparticles, in particular IONPs, possess unique properties such as superparamagnetism, high coercivity, low Curie temperature, and high magnetic susceptibility along with inherent biocompatibility and inexpensiveness (Jun et al. 2007; Patel et al. 2008). IONPs serve as contrast agents in MRI, as sensors for metabolites and other biomolecules, or are broadly used in therapy (tissue regeneration, controlled drug release, targeted drug delivery, hyperthermia) and in magnetic separation (Pankhurst et al. 2003; McCarhty and Weissleder 2008). In relation to their hydrodynamic size, IONPs can be classified into ultrasmall paramagnetic iron oxide (USPIO) (5–40 nm), superparamagnetic iron oxide (SPIO) (60–150 nm), and micrometer-sized particles of iron oxide (MPIO) (0.3–3.5 μm). The stabilization of IONPs is crucial to obtain magnetic colloidal stability, biocompatibility, and reduced uptake by the RES (Laurent et al. 2008).

17.2.4   SYNTHESIS OF IONPs

Many scientific groups are working in the field of IONP synthesis for the purpose of medical imaging applications (Hines and Guyot-Sionnest 1996; Lee et al. 2014). For IONP manufacturing, the synthesis methodologies that have been established are co-precipitation (Figure 17.3) (Binh et al. 1998; Wu et al. 2007), thermal decomposition (Rockenberger et al. 1999; Sun and Zeng 2002; Woo et al. 2004), microemulsion (Solans et al. 2005; Vidal-Vidal et al. 2006), hydrothermal (Hu et al. 2007; Li 2007), and sonochemical synthesis (Vijayakumar et al. 2000; Bang and Suslick 2007), which can all be directed to synthesize high-quality IONPs. Moreover, these NPs can also be prepared by other methods such as electrochemical and aerosol synthesis (Pascal et al. 1999), laser pyrolysis techniques (Bomatí-Miguel et al. 2008), microorganism or bacterial synthesis, sol–gel reactions, polyol processes, and flow injection synthesis.

Surface coatings of IONPs not only provide colloidal suspendability and subsequent protection against the formation of aggregates either into a magnetic field or into the blood stream but also contribute to the binding of various ligands (fluorescent dyes for optical imaging, chelators for MRI/nuclear imaging, targeting moieties, or even drugs) to the NP surface (Titirici et al. 2006; Wang et al. 2007). The applied surface functionalization strategies include either inorganic or organic materials (Wu et al. 2008). In the case of inorganic coating, IONPs are modified at their outer shell by a layer such as silica, gold, gadolinium(III), or carbon. For instance, a surface enriched in silica enhances the presence of silanol groups, which can easily react with various coupling agents to covalently attach specific ligands to these magnetic particles. More specifically, the agents 3-aminopropyltriethoxysilane (APTES), *p*-aminophenyl trimethoxysilane (APTS), and mercaptopropyltriethoxysilane (MPTES) are mainly employed for introducing the amino and sulfhydryl groups (Shen et al. 2004). The mechanism of silane coating is depicted in Figure 17.4.

IONP stabilization by an organic layer can be performed either during nanoparticle synthesis (*in situ* coating) or following their synthesis (post-synthesis coating). The most commonly employed polymer-based stabilizers are either derived from natural materials such as dextran, starch, gelatin and chitosan, or from synthetic ones such as poly(ethylene glycol) (PEG), polyvinyl alcohol (PVA), poly(acrylic acid) (PAA), polymethylmethacrylate (PMMA), poly(lactide acid) (PLA), and alginate. Other polymers that have also been used are poly(ethyleneoxide)-*b*-poly(methacrylic acid),

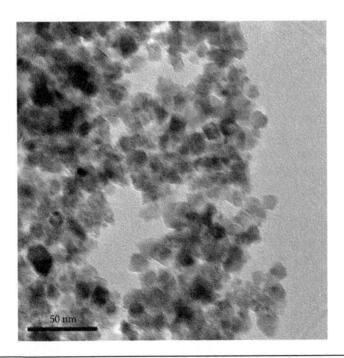

50 nm

FIGURE 17.3   TEM images of synthesized iron NPs, which were synthesized through the co-precipitation method (unpublished results).

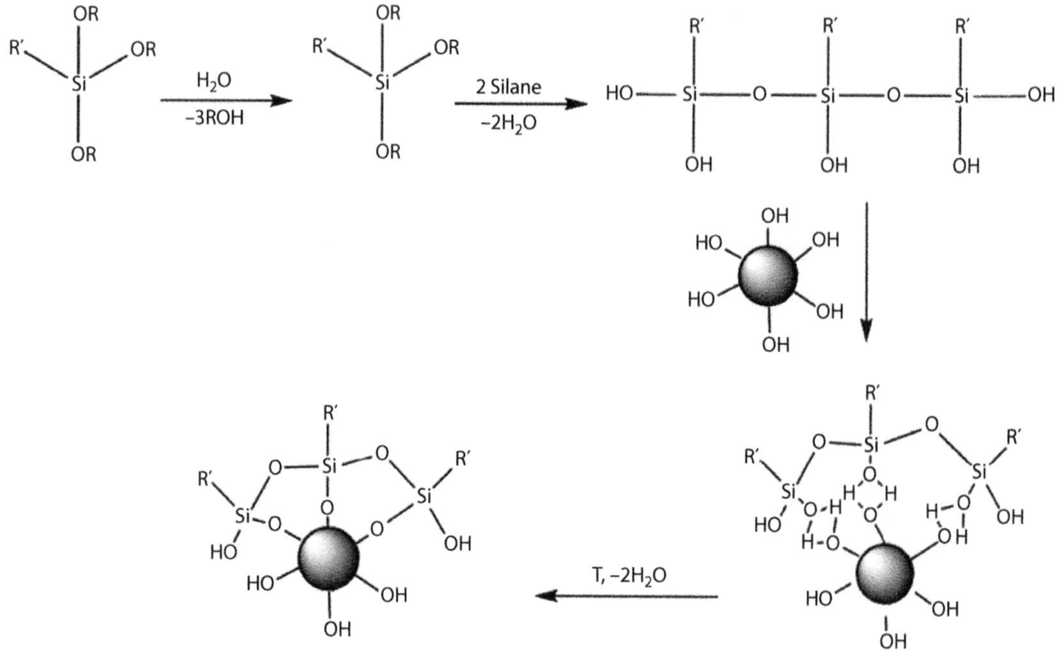

FIGURE 17.4 Mechanism of silane coating on the IONP surface.

polyvinylpyrrolidone (PVP), polyethylenimine (PEI), ethylcellulose, poly(ε-caprolactone), sulfonated styrene-divinylbenzene, and arabinogalactan (Laurent et al. 2008; Wu et al. 2008).

17.2.5 GOLD NANOPARTICLES

Gold nanoparticles (AuNPs) exhibit unique optical and electronic properties and have been investigated in various imaging approaches, such as CT and photoacoustic and surface-enhanced Raman spectroscopy (SERS) (Xie et al. 2010b). AuNPs can be relatively easily synthesized and are highly stable (Sandhu et al. 2002; Shiraishi et al. 2009; Xue et al. 2011; Lee et al. 2014). They also possess a high surface area to volume ratio, with excellent biocompatibility using appropriate coating (i.e., polymer stabilizers), functionalizing (i.e., thiols, mercaptans, phosphines, and amines), or targeting agents (i.e., oligonucleotides, proteins, and antibodies). These properties can be readily tuned by varying their shape (spheres, cubes, rods, cages, and wires), size, and composition. For example, spherical AuNPs can be fabricated with a high degree of monodispersity and in small sizes in contrast to other nanostructures. AuNPs have optical properties mainly in the visible region, and this characteristic limits their use in the treatment of deep tumors. AuNPs are also ideal scavengers for drug loading and release in specific areas due to their size. McIntosh et al. (2001) suggested a photodynamic controlled release of a loaded drug, 5-fluoracil, by using AuNPs. Finally, AuNPs offer a suitable platform for multifunctionalization with different biomolecules for the specific binding and detection of small molecules and biological targets (Saha et al. 2012).

17.2.6 SYNTHESIS OF AuNPs

The most common synthetic method is based on the citrate reduction of hydrogen tetrachloroaurate (HAuCl$_4$) in water (Turkevich et al. 1951). In this approach, citric acid not only acts as a reducing agent but also as a stabilizing agent, as it can be easily replaced by other capping agents such as thiols that bear an appropriate functional group for several ligands. In this typical citrate reduction, the reaction mechanism of AuNP formation is dependent on the concentration of the citrate ions (C$_6$H$_5$O$_7^{3-}$) and gold chloride (HAuCl$_4$), as well as the temperature during synthesis (Rodríguez-González et al. 2007). The elaborate chemical reaction can be presented as follows:

$$Au^{3+} + C_6H_5O_7^{3-} + 15OH^- \rightarrow Au + CO_2 + 10H_2O$$

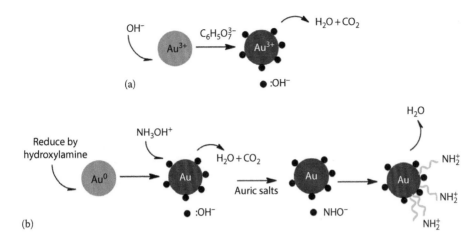

FIGURE 17.5 Schematic of AuNP formation during synthesis using (a) the citrate reduction and (b) the seeding growth methods.

The overall reaction mechanism of citrate reduction is illustrated in Figure 17.5a. However, citrate reduction results in the production of AuNPs with a broad size range (in the 10–120 nm range) and with particles exceeding 30 nm diameter, showing poor monodispersity.

In this context, Frens et al. (1973) initially tuned the AuNPs size by varying the gold salt to citrate ratio. Subsequently, an alternative synthesis was introduced Brown and Naten (1998) based on a hydroxylamine seed-mediated growth approach, producing monodisperse colloidal gold particles by "growing" the smaller particles (Figure 17.5b). In this method, hydroxylamine (NH_2OH) both, reduce and prevents the formation of new nuclei by accelerating the reduction of Au^{3+} on the surface of the seeds, so that the growth process can occur (Jana et al. 2001; Bauer et al. 2003).

After all seeds are covered by the NH_3O^- groups from hydroxylamine, the gold salts are added as a source of the Au^{3+} ions. These Au^{3+} ions then diffuse onto the surfaces of the seeds via reaction with NH_3O^-, which causes the formation of larger sized AuNPs covered with NH_2^-. The NH_2^- ions have an electrostatic charge and provide specific binding (amine bond), so that the AuNPs remain stable in the suspension (Maus et al. 2010).

Several other methods for the improvement of the size and shape through the careful selection of various experimental conditions (i.e., reducing agent, reaction time, temperature, and capping agent) have been reported (Baptista et al. 2008).

The AuNP dispersions can be structurally characterized using UV–vis spectroscopy (Figure 17.6), high-resolution transmission electron microcopy (HRTEM) (Figure 17.7), and dynamic light scattering (DLS).

17.3 BIO-FUNCTIONALIZATION OF NPs

The desired surface functionalization should satisfy criteria such as good water stability and chemical functionality to maintain the NPs' physical and chemical properties, as well as biocompatibility. These surface modification techniques can be roughly divided into two categories: ligand exchange and ligand addition (Xie et al. 2010a). As the first mentioned strategy may involve high affinity, hydrophilic ligands are used to replace the original hydrophobic coating. The second strategy includes the addition of a layer of inorganic material without the removal of any initial ligands (Figure 17.8).

The basic strategy for surface functionalization by biomolecules includes two steps: the synthesis of functional moieties (small molecule or polymer) and of NPs, and then the coupling of the biomolecules with NPs by chemical bond, physical adsorption, or biological interaction. Various biomolecules, such as proteins (Mikhaylova et al. 2004), peptides (Mikhaylova et al. 2004), antibodies (Tiefenauer et al. 1993), carbohydrates, nucleic acids, biotin (Weizmann et al. 2004), and other small

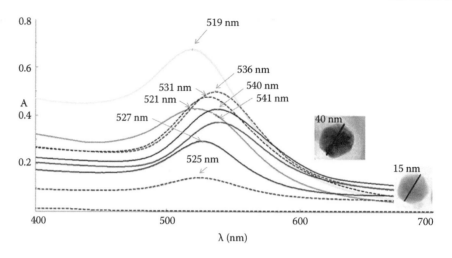

FIGURE 17.6 Normalized UV–vis absorption spectra for several Au/ligand molar ratios used for the growth of AuNPs with different sizes.

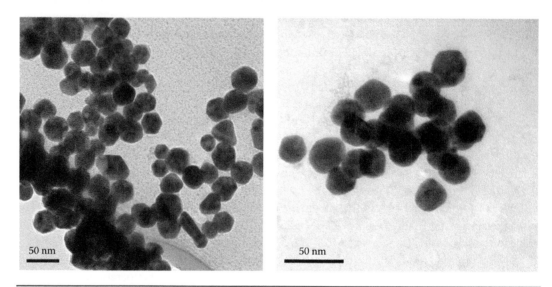

FIGURE 17.7 TEM images of spheroid Au NPs.

molecules such as folic acid, may also be bound to the surface of NPs directly or indirectly via some functional end groups (e.g., carboxylic acid, amines, thiols, etc.) (Figure 17.9), thereby improving their uptake by the cells. The main coupling strategies that are applied for the functionalization of NPs with targeting ligands, or even with chelating and therapeutic agents, are the following: (1) carbodiimide coupling for the covalent conjugation of carboxylic acids with amines via the coupling agent 1-ethyl-3-(3-dimethylaminopropyl) carbodiimide hydrochloride (EDC or EDAC) (Tsiapa et al. 2014); (2) maleimide coupling for the binding of primary amines with thiols (Mourtas et al. 2014); (3) click chemistry, which involves the covalent coupling of an alkyne to an azide providing a 1,2,3-triazole ring under catalyst of Cu (I); (4) disulfide bond formation between the NP and the relevant moiety facilitated by glutathione disulfide oxidation; (5) electrostatic interactions between oppositely charged NPs and functional moieties; (6) pre-conjugation of biomolecules to NPs for the formation of avidin–biotin bonds between biotinylated NPs and avidin or avidin-modified compounds and vice versa or of antibody–antigen biological interactions; and (7) encapsulation of drugs or amphiphilic polymers at hydrophobically overcoated NPs via hydrophobic interactions (Thanh and Green 2010).

Recently, Lee et al. (2006) developed a route for conjugating γ-Fe$_2$O$_3$ NPs with single-stranded oligonucleotides. Water-soluble magnetic NPs with carboxyl groups on their surfaces were prepared at first, and then by using EDC as a linker reagent, and a protein, streptavidin, was

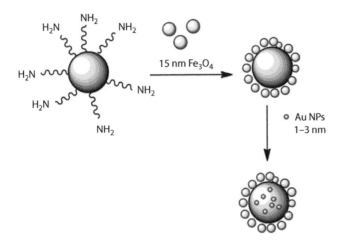

FIGURE 17.8   SiO$_2$ particles covered with silica-primed Fe$_3$O$_4$ NPs and heavily loaded with Au nanoparticle seeds (SiO$_2$–Fe$_3$O$_4$–Au seeds).

FIGURE 17.9   Cellular uptake by using confocal microscopy of MCF-7 cells treated with gold nanoparticles.

modified successfully on the surface of γ-Fe$_2$O$_3$ NPs. Streptavidin-functionalized Fe$_2$O$_3$ could catch biotin-labeled single-stranded oligonucleotides through the strong affinity between streptavidin and biotin.

17.4   TOXICITY OF NPs

Following the introduction of NPs into a living organism (Torchilin 2002), aspects such as biodistribution, circulation in the bloodstream, pharmacokinetics, and clearance by the organism, as well as possible cyto- and genotoxicity, are under investigation. It should be noted that there are only a few published data on NPs' biodistribution, toxicity, and elimination during the past 3–4 years (Maeda et al. 2000). A significant number of scientific works present a vast variation in the experimental methods (targeting ligand, contrast agent and the relative imaging modality, animal model, dose, particle components, surface functionalization, shape and size, number of studied time points, and competitive assay), leading to serious discrepancies in relation to toxicity estimates.

According to the literature, the organs of the RES serve as the main target for the accumulation of 10–100 nm NPs. The rapid reduction in particle concentration in the blood and their prolonged retention in the organism are associated with the hepatobiliary system. There are reports that it takes 3–4 months for the accumulated particles to be excreted from the liver and spleen, raising doubts about NPs' applications (Dykman and Khlebtsov 2011). In addition, NPs' penetration via the hemato–encephalic barrier is strongly dependent on their size, supporting that NPs of 5–20 nm afford the desired dimensions for bioimaging and therapeutic uses. Additionally, AuNPs in the size range 1–2 nm can be more toxic in contrast to bigger ones, due to their ability to conjugate on the biomacromolecules. More detailed in vitro experiments performed with bigger particles (3–100 nm) showed a threshold dose for in vitro toxicity of about 1012 particles/mL. In contrast to the relative in vitro aforementioned results, in vivo use of AuNPs seems inconsistent (Dykman and Khlebtsov 2011).

REFERENCES

Alivisatos, A. P. (1996). Semiconductor clusters, nanocrystals, and quantum dots. *Science* **271**(5251): 933–937.

Bailey, R. E. and S. Nie. (2003). Alloyed semiconductor quantum dots: Tuning the optical properties without changing the particle size. *J. Am. Chem. Soc.* **125**(23): 7100–7106.

Bang, J. H. and K. S. Suslick. (2007). Sonochemical synthesis of nanosized hollow hematite. *J. Am. Chem. Soc.* **129**(8): 2242–2243.

Baptista, P., E. Pereira, P. Eaton, G. Doria, A. Miranda, I. Gomes, P. Quaresma, and R. Francos. (2008). Gold nanoparticles for the development of clinical diagnosis method. *Anal. Bioanal. Chem.* **391**: 943–945.

Bauer, G., J. Hassman, H. Walter, J. Haglmüller, C. Mayer, and T. Schalkhammer. (2003). Resonant nano-cluster technology-from optical coding and high quality security features to biochips. *Nanotechnology* **14**: 1289–1311.

Binh, V. T., S. T. Purcell et al. (1998). Nanotips and nanomagnetism. *Appl. Surf. Sci.* **130**: 803–814.

Bomatí-Miguel, O., L. Mazeina et al. (2008). Calorimetric study of maghemite nanoparticles synthesized by laser-induced pyrolysis. *Chem. Mater.* **20**(2): 591–598.

Brown, K. R. and M. J. Natan. (1998). Hydroxylamine seeding of colloidal Au nanoparticles in solution and on surfaces. *Langmuir* **14**(4): 726–728.

Chan, W. C. W. and S. Nie. (1998). Quantum dot bioconjugates for ultrasensitive nonisotopic detection. *Science* **281**: 2016–2018.

Dykman, L. A. and N. G. Khlebtsov. (2011). Gold nanoparticles in biology and medicine: Recent advances and prospect. *Acta Nat.* **3**(2): 34–55.

Efros, A. L. and A. L. Efros. (1982). Interband absorption of light in a semiconductor sphere. *Sov. Phys. Semiconduct.* **16**(7): 772–775.

Ekimov, A. I. and A. A. Onushchenko. (1982). Quantum size effect in the optical-spectra of semiconductor micro-crystals. *Sov. Phys. Semiconduct.* **16**(7): 775–778.

Frens, G. (1973). Controlled nucleation for the regulation of the particle size in monodisperse gold suspensions. *Nat. Phys. Sci.* **241**: 20–22.

Hines, M. A. and P. Guyot-Sionnest. (1996). Synthesis and characterization of strongly luminescing ZnS-Capped CdSe nanocrystals. *J. Phys. Chem.* **100**(2): 468–471.

Hines, M. A. and G. D. Scholes. (2003). Colloidal PbS nanocrystals with size-tunable near-infrared emission: Observation of post-synthesis self-narrowing of the particle size distribution. *Adv. Mater.* **15**(21): 1844–1849.

Hu, X., J. C. Yu et al. (2007). Fast production of self-assembled hierarchical α-Fe$_2$O$_3$ nanoarchitectures. *J. Phys. Chem. C* **111**(30): 11180–11185.

Jana, N. R., L. Gearheart, and C. Murphy. (2001). Evidence for seed-mediated nucleation in the chemical reduction of gold salts to gold nanoparticles. *J Chem. Mater.* **13**: 2313.

Jatzkewitz, H. (1955). Peptamin (glycyl-L-leucyl-mescaline) bound to blood plasma expander (polyvinylpyrrolidone) as a new depot form of a biologically active primary amine (mescaline). *Z. Naturforsch.* **10**: 27–31.

Jun, Y. W., J. S. Choi et al. (2007). Heterostructured magnetic nanoparticles: Their versatility and high performance capabilities. *Chem. Commun. (Camb.)* **12**: 1203–1214.

Laurent, S., D. Forge et al. (2008). Magnetic iron oxide nanoparticles: Synthesis, stabilization, vectorization, physicochemical characterizations, and biological applications. *Chem. Rev.* **108**(6): 2064–2110.

Lee, C. W., K. T. Huang et al. (2006). Conjugation of γ-Fe$_2$O$_3$ nanoparticles with single strand oligonucleotides. *J. Magn. Magn. Mater.* **304**(1): e412–e414.

Lee, D.-E., H. Koo, I.-C. Sun, J. H. Ryu, K. Kim, and I. C. Kwon. (2012). Multifunctional nanoparticles for multimodal imaging and theragnosis. *Chem. Soc. Rev.* **41**(7): 2656–2672.

Lee, S. Y., S. I. Jeon et al. (2014). Targeted multimodal imaging modalities. *Adv. Drug Deliv. Rev.* **76**: 60–78.

Li, K. (2007). Average-case performance analysis of online non-clairvoyant scheduling of parallel tasks with precedence constraints. *Comput. J.* **51**(2): 216–226.

Maeda, H., J. Wu et al. (2000). Tumor vascular permeability and the EPR effect in macromolecular therapeutics: A review. *J. Control. Rel.* **65**(1–2): 271–284.

Mankoff, D. A. (2007). A definition of molecular imaging. *J. Nucl. Med.* **48**(6): 18N, 21N.

Maus, L., O. Dick et al. (2010). Conjugation of peptides to the passivation shell of gold nanoparticles for targeting of cell-surface receptors. *ACS Nano* **4**(11): 6617–6628.

McCarhty, J. R. and R. Weissleder. (2008). Multifunctional magnetic nanoparticles for targeted imaging and therapy. *Adv. Drug Deliv. Rev.* **60**(11): 1241–1251

McIntosh, C. M., E. A. Esposito et al. (2001). Inhibition of DNA transcription using cationic mixed monolayer protected gold clusters. *J. Am. Chem. Soc.* **123**(31): 7626–7629.

Mikhaylova, M., D. K. Kim et al. (2004). BSA immobilization on amine-functionalized superparamagnetic iron oxide nanoparticles. *Chem. Mater.* **16**(12): 2344–2354.

Mourtas, S., A. N. Lazar et al. (2014). Multifunctional nanoliposomes with curcumin-lipid derivative and brain targeting functionality with potential applications for Alzheimer disease. *Eur. J. Med. Chem.* **80**: 175–183.

Murray, C. B., D. J. Norris, and M. G. Bawendi. (1993). Synthesis and characterization of nearly monodisperse CdE (E = S, Se, Te) semiconductor nanocrystallites. *J. Am. Chem. Soc.* **115**: 8706–8715.

Pankhurst, Q. A., J. Connolly, S. K. Jones, and J. Dobson. (2003). Applications of magnetic nanoparticles in biomedicine. *J. Phys. D Appl. Phys.* **36**: R167–R181.

Pascal, C., J. L. Pascal et al. (1999). Electrochemical synthesis for the control of γ-Fe_2O_3 nanoparticle size: Morphology, microstructure, and magnetic behavior. *Chem. Mater.* **11**(1): 141–147.

Patel, D., J. Y. Moon et al. (2008). Poly(d,l-lactide-co-glycolide) coated superparamagnetic iron oxide nanoparticles: Synthesis, characterization and in vivo study as MRI contrast agent. *Colloid Surf. A* **313–314**: 91–94.

Rockenberger, J., E. C. Scher et al. (1999). A new nonhydrolytic single-precursor approach to surfactant-capped nanocrystals of transition metal oxides. *J. Am. Chem. Soc.* **121**(49): 11595–11596.

Rodríguez-González, B., P. Mulvaney et al. (2007). An electrochemical model for gold colloid formation via citrate reduction. *Zeitschrift für Physikalische Chemie* **221**(3): 415–426.

Saha, K., S. S. Agasti et al. (2012). Gold nanoparticles in chemical and biological sensing. *Chem. Rev.* **112**(5): 2739–2779.

Sandhu, K. K., C. M. McIntosh et al. (2002). Gold nanoparticle-mediated transfection of mammalian cells. *Bioconjug. Chem.* **13**(1): 3–6.

Shen, X.-C., X.-Z. Fang et al. (2004). Synthesis and characterization of 3-aminopropyltriethoxysilane-modified superparamagnetic magnetite nanoparticles. *Chem. Lett.* **33**(11): 1468–1469.

Shiraishi, K., K. Kawano et al. (2009). Preparation and in vivo imaging of PEG-poly(L-lysine)-based polymeric micelle MRI contrast agents. *J. Control. Rel.* **136**(1): 14–20.

Smith, A. M., H. Duan et al. (2008). Bioconjugated quantum dots for in vivo molecular and cellular imaging. *Adv. Drug Deliv. Rev.* **60**(11): 1226–1240.

Solans, C., P. Izquierdo et al. (2005). Nano-emulsions. *Curr. Opin. Colloid Interface Sci.* **10**(3–4): 102–110.

Sun, S. and H. Zeng. (2002). Size-controlled synthesis of magnetite nanoparticles. *J. Am. Chem. Soc.* **124**(28): 8204–8205.

Thanh, N. T. K. and L. A. W. Green. (2010). Functionalization of nanoparticles for biomedical applications. *Nano Today* **5**: 213–230.

Tiefenauer, L. X., G. Kuhne et al. (1993). Antibody-magnetite nanoparticles: In vitro characterization of a potential tumor-specific contrast agent for magnetic resonance imaging. *Bioconjug. Chem.* **4**(5): 347–352.

Titirici, M.-M., M. Antonietti et al. (2006). A generalized synthesis of metal oxide hollow spheres using a hydrothermal approach. *Chem. Mater.* **18**(16): 3808–3812.

Torchilin, V. P. (2002). PEG-based micelles as carriers of contrast agents for different imaging modalities. *Adv. Drug Deliv. Rev.* **54**(2): 235–252.

Tsiapa, I., E. K. Efthimiadou et al. (2014). Tc-labeled aminosilane-coated iron oxide nanoparticles for molecular imaging of alphabeta-mediated tumor expression and feasibility for hyperthermia treatment. *J. Colloid Interface Sci.* **433C**: 163–175.

Turkevich, J., P. C. Stevenson et al. (1951). A study of the nucleation and growth processes in the synthesis of colloidal gold. *Discuss. Faraday Soc.* **11**: 55.

Vidal-Vidal, J., J. Rivas et al. (2006). Synthesis of monodisperse maghemite nanoparticles by the microemulsion method. *Colloid Surf. A* **288**(1–3): 44–51.

Vijayakumar, R., Y. Koltypin et al. (2000). Sonochemical synthesis and characterization of pure nanometer-sized Fe_3O_4 particles. *Mat. Sci. Eng. A* **286**(1): 101–105.

Wang, S.-B., Y.-L. Min et al. (2007). Synthesis and magnetic properties of uniform hematite nanocubes. *J. Phys. Chem. C* **111**(9): 3551–3554.

Weizmann, Y., F. Patolsky et al. (2004). Magneto-mechanical detection of nucleic acids and telomerase activity in cancer cells. *J. Am. Chem. Soc.* **126**(4): 1073–1080.

Woo, K., J. Hong et al. (2004). Easy synthesis and magnetic properties of iron oxide nanoparticles. *Chem. Mater.* **16**(14): 2814–2818.

Wu, W., Q. He et al. (2008). Magnetic iron oxide nanoparticles: Synthesis and surface functionalization strategies. *Nanoscale Res. Lett.* **3**(11): 397–415.

Wu, W., H. L. Liu et al. (2007). Modeling dynamic cerebral blood volume changes during brain activation on the basis of the blood-nulled functional MRI signal. *NMR Biomed.* **20**(7): 643–651.

Xie, J., S. Lee et al. (2010a). Nanoparticle-based theranostic agents. *Adv. Drug Deliv. Rev.* **72**: 1054–1079.

Xie, J., S. Lee et al. (2010b). Nanoparticle-based theranostic agents. *Adv. Drug Deliv. Rev.* **62**(11): 1064–1079.

Xue, X. J., F. Wang et al. (2011). Emerging functional nanomaterials for therapeutics. *J. Mater. Chem.* **21**(35): 13107–13127.

Zhong, X., Y. Feng et al. (2003). Alloyed Zn(x)Cd(1-x)S nanocrystals with highly narrow luminescence spectral width. *J. Am. Chem. Soc.* **125**(44): 13559–13563.

Section VI
Therapeutic Research Platforms

18

Developing Technologies for Small Animal Radiotherapy

Frank Verhaegen, James Stewart, and David Jaffray

18.1 NEED FOR SMALL ANIMAL PRECISION RADIOTHERAPY

All over the world, the incidence of cancer is constantly increasing. Despite improvements in treatment methods and strategies, thousands suffer from this disease and a great number succumb to it. It is, therefore, clear that our understanding of the mechanisms of cancer and potential cures is far from complete. The arsenal of therapies is large, with the three most commonly applied being chemotherapy with cytotoxic drugs, surgical removal of the primary lesion, and radiotherapy which uses radiation in many forms to destroy tumor cells. Besides these, there is hormonal therapy, immunotherapy, gene therapy, and several others. Of these options, radiotherapy and surgery are local therapies, which can be effective only when the disease has not metastasized to regions beyond the primary tumor site. Radiation has been harnessed to treat cancer ever since x-rays were discovered in 1895. Radiation exists in many forms, and nearly all of them have been attempted as a cure for cancer. The most commonly employed forms of radiation these days are high-energy photon beams that kill tumors by putting them in the crossfire of beams from multiple directions, and brachytherapy where radioactive sources are implanted straight inside the tumor. But, also more exotic forms of radiation such as proton beams or heavier ion beams are under investigation.

Despite radiation's long history as a cancer-fighting tool, its action on cells, be it cancer cells or normal cells, is not completely understood owing to its complexity. From a vast body of radiobiology literature, the picture emerges that radiation acts on tumor cells by disrupting the DNA molecules in an irreparable way, leaving the cancer cell with no choice but death. Radiation acts in the same way on healthy cells, but these are thought to have a more efficient repair mechanism compared to cancer cells. So, radiotherapy is always a delicate balance between hitting the tumor cells hard and sparing the healthy cells to avoid debilitating side effects. Radiotherapy has always been a medical specialty that is highly technology driven. Many of the forms of imaging that are discussed in this volume are used to diagnose cancer and also to monitor its progress during the treatment. Historically, many efficient ways to deliver radiation to a tumor have been employed, and this has led to the current state of the art where radiation beams of complex shapes are tailored to the tumor shape, with very small margins. Today, radiation beams may change shape dynamically while arcing around a patient. It is even possible to adapt the beams to moving tumors, for example, in breathing lungs.

Many of the technological improvements were introduced without any actual proof that they would work better. Often, what can be described as the "sharp scalpel hypothesis" is used. It is obvious that a sharper scalpel cuts better than a blunt scalpel, therefore, whenever a sharper scalpel was invented (read, a more effective way to deliver a radiation dose to the tumor) it was deemed unethical to run a trial of the sharp versus the blunt scalpel on humans before introducing the new technology. For example, we introduce sharper, highly collimated radiation beams to deliver dose to the primary tumor, and keep the radiation away from the healthy tissue as far as possible. It was obvious that this would be a benefit over irradiating with larger rectangular fields exposing many healthy cells, which was the standard a few decades ago. However, with modern techniques, it is not unthinkable that microscopic tumor islands at the edge of the clearly visible primary site (e.g., on a CT image) are missed and that these will give rise to a new malignant growth if not irradiated. Therefore, the sharp scalpel may cut better but it may also have unforeseen side effects which may offset its sharpness. In radiotherapy, we have the additional problem that the interaction of radiation with cancer cells is not fully understood.

In an attempt to understand the mechanism of radiation action, cell culture and animal experiments have been performed by thousands of researchers. The former may serve to partially elucidate the interaction of radiation with cells, but it cannot provide much knowledge about radiotherapy in humans. The radiation equipment used to irradiate animals (most commonly mice or rats) often does not have the complexity of that used on humans. This means that the currently used radiation equipment has not been validated in animal models. There is a growing awareness that much may be learned and gained from experiments on animals with precision radiotherapy, whereby the radiation beams would be downscaled to animal size, thereby establishing conditions that would mimic human radiotherapy reasonably realistically. Just as human radiotherapy is often performed under image guidance (employing a variety of imaging techniques), small animal radiotherapy should then also be performed under high-resolution image guidance. Equipment that fulfills the need of combining precision radiotherapy with sophisticated imaging has been under development during the past decade and is now finally becoming mature. This will be the topic of this chapter.

Many questions remain in radiotherapy. What is the exact role of the oxygenation status of a tumor and how to best overcome the known radioresistance of poorly oxygenated tumors? Which radiation dose is optimal for a certain tumor, and which dose rate and fractionation scheme should one employ? Why do some peculiar radiotherapy schemes such as grid therapy (Penagaricano et al. 2010) work, whereas the common belief is that a tumor should get a large, uniform dose to all the cancer cells? What exactly is the optimal dose distribution? From positron emission tomography (PET) imaging we know that tumor metabolism often exhibits marked heterogeneity (Lambin et al. 2010), and should we then deposit more dose in the hyperactive regions? And most fundamental of all, what are the molecular mechanisms in a complex organism such as a mouse that lead to tumor death or regrowth? Of course, animal radiation studies can lead to valuable conclusions only if the findings can be successfully translated to human radiotherapy. The development of realistic animal models of tumor and healthy tissues is therefore paramount.

The growing understanding of tumor microenvironment factors and the development of transgenic animal models has led to a major effort to develop small animal imaging technologies (PET, single-photon emission computed tomography [SPECT], magnetic resonance imaging [MRI], ultrasound, computed tomography [CT], optical), which are mainly being investigated with respect to systemic cancer treatments and biomarkers to, for example, diagnose tumors. For a recent review, we refer the reader to Kagadis et al. (2010). These imaging techniques have also greatly aided our ability to longitudinally track the response of new pharmaceuticals in small animal disease models. In contrast, very little effort has been invested in animal radiation research despite the fact that vast numbers of cancer patients are treated with radiotherapy and that this modality is a highly imaging-driven technique. In fact, the majority of animal data acquired from radiation research was derived from large-field, single-beam irradiations, where the dose was usually crudely estimated. This bears almost no resemblance to modern fractionated clinical radiotherapy with multiple, conformal beams with a complex spatiotemporal dose pattern, where the dose is planned with sophisticated dose calculation algorithms and verified with an arsenal of modern techniques. Tumor response studies in animals were hampered by high dose to healthy tissue. It is, therefore, questionable to what extent the existing animal studies still have relevance for modern radiotherapy practice.

In recent literature, some effects of temporal and spatial dose variations were reported, such as the out-of-field radiation quality changes (Kirkby et al. 2007; Syme et al. 2009) and the influence of the prolonged treatment regime of intensity-modulated radiotherapy (Moiseenko et al. 2007). The significance of these findings for human radiotherapy is at present not clearly understood. The development of novel radiotherapeutic technologies has allowed the field to move into new, unexplored areas of radio-oncological practice; however, these advances have occurred without the corresponding research from the radiobiological perspective.

This discussion demonstrates that there is an acute need for small animal radiotherapy research platforms where human radiotherapy conditions can be faithfully reproduced. This means that radiation beams need to be downscaled in geometry, but also in energy (see Section 18.3.1.2). Such a platform could be used in conjunction with the myriad of small animal tumor models that were developed in recent years. Because of the small size of, for example, mice, the demands on technical precision of such a platform exceed those for human radiotherapy. The platform should also have imaging capabilities, similar to human radiotherapy, which means that the spatial resolution should be sufficiently high to resolve small organs and structures within. This would enable image-guided radiotherapy (IGRT) for small animals. Currently, there are several ongoing efforts to develop versatile small animal radiation research platforms that allow unprecedented animal radiotherapy studies. This is the topic of this chapter. Such platforms will enable a wide range of studies in fields such as radiotherapy, radiobiology, stem cell studies, immunology, and pharmacology, to name just a few.

In this chapter, we will first briefly discuss some animal radiation experimental work done before the advent of the sophisticated precision irradiator/imager systems for animals that are now becoming available. This is mostly fairly crude work, but there have also been some interesting efforts in precision irradiation. A section on the requirements for the irradiation and imaging capabilities of an animal radiotherapy platform follows. Then, an overview is given of the pioneering efforts that led to this new field of radiotherapy research. Finally, some thoughts on the use of the developing technologies for research are shared. This overview will not cover microbeam irradiators for cell radiobiology (Crosbie et al. 2010) as, for example, employed in the study of the bystander effect (Maeda et al. 2010).

18.2 EXPERIMENTAL ANIMAL RADIATION WORK BEFORE THE ADVENT OF PRECISION IMAGE GUIDANCE

It is probably fair to say that for many decades animal radiation studies were mostly performed using relatively crude experimental setups, for lack of precision image-guided technology. These studies used radiation fields that usually did not conform to only the target and therefore gave high doses to healthy tissue. The radiation sources employed were kilovolt (kV) x-ray units, ^{137}Cs or ^{60}Co gamma ray emitting isotopes, or linear accelerators producing megavolt (MV) x-rays, often specifically designed for the treatment of humans. These devices typically are only precise at the level of a few millimeters, which is sufficiently precise and accurate to deliver a dose in larger human structures at a certain depth under the skin, under conditions of sufficient dose buildup. But these devices cause much larger dose uncertainties in centimeter- or millimeter-sized structures in small animals. Many of these experiments were accompanied by little or no dose measurement. Again, very little or no consideration was given to the dose build-up region at beam entrance regions, which can extend up to several centimeters in MV photon beams. Similarly, little attention was commonly paid to the beam exit region, where missing backscatter can significantly influence the dose. A treatment planning system that was adequate for dose calculation in small animals was usually not available. Similar systems as used for human patients are unsuitable because of the use of MV beams in radiotherapy and also the much smaller beams and voxel sizes in animal systems. We will show later in this chapter that kV beams are needed to irradiate animals. Image guidance was mostly absent. Dose verification during beam delivery was missing. At best, only limited imaging information just before or during radiation delivery was available. In view of all this, it hardly needs stating that the animal radiation work in the past did not bear much resemblance to the sophisticated patient treatments that are now being used in radiotherapy or can be envisioned for the near future.

Much of the work in the past was limited by the lack of dedicated precision radiation equipment for small animals. The aim here is by no means to give a complete overview of older animal radiation experiments, but to show a few examples of the knowledge gained from experiments performed before the era of image-guided precision radiation research. Examples include studies of lung irradiation by exposing the partial/whole thorax cavity to kV x-rays (Raabe et al. 2001; van Eerde et al. 2001; Wiegman et al. 2003). Some of these studies involved a camera to image the specimen's position with respect to the beams. In other studies, combined chemo/radiation therapy was studied in head and neck cancers in mice by irradiating five mice simultaneously (this is not an uncommon procedure) in a ^{60}Co beam with some shielding provided (Hamstra et al. 1999). Many more similar studies can be found in the literature, but they all suffer from one or more of the shortcomings mentioned above. In the past, often tumors were implanted subcutaneously. These can be relatively easily targeted with a crudely shaped photon beam, at least when a homogeneous dose distribution is desired. Nowadays, the interest in targeting orthotopic tumors (i.e., "in the right place," e.g., a lung tumor in lung) is increasing, which demands higher targeting precision. Potential dose heterogeneity studies require advanced beam modulation technology.

A step toward more accurate radiation administration can be found in the recent literature. By making modifications to existing clinical radiation devices, often reasonably adequate dose distributions can be achieved. DesRosiers et al. (2003) built a dedicated positioning system for precision irradiation of a rat's eye lens using a GammaKnife unit with 201 ^{60}Co gamma ray sources. They combined the small precision collimators with Monte Carlo dose calculations and succeeded in selectively irradiating the small eye lens. While an accurate irradiation system for this application, each new application would require a dedicated setup. No image guidance was performed to visualize the targets.

Another example (Rodriguez and Jeraj 2008) was the attempt to implement some form of image guidance to irradiate small structures in anesthetized motionless zebra fish of a few millimeters in length. X-rays of 50 kV were applied through pinhole collimators, and a camera was used for monitoring the fish. The system could also be used to study very small organisms and cell cultures, but not for the study of radiation response in small rodents, which was therefore of limited use in radiotherapy research.

In an interesting series of experiments at the Universities of Groningen and Nijmegen, The Netherlands, the influence of radiation on the rat spine was investigated (Bijl et al. 2002, 2003, 2006; van Luijk et al. 2005). As an example, the difference in response (expressed as paralysis) after different irradiation schemes of the rat spine was compared: single versus split radiation fields, or bath and shower geometries. Figure 18.1 shows some of the irradiation schemes employed in one of their studies (Bijl et al. 2003), where the rat spine was targeted with a proton beam. They obtained significantly different ED$_{50}$ values (dose to achieve 50% of the maximal effect) for different radiation configurations. From this we can conclude that, in all likelihood, not only the total dose in radiotherapy is important but also the dose distribution plays an equally important role, at least in this example of healthy tissue toxicity. No sophisticated image guidance appears to have been used in this study. A study employing a similar animal model derived the relative biological effectiveness (RBE) for carbon beams compared to photon beams (Debus et al. 2003).

Interesting though some of the devices discussed in this section may be, none are fully flexible radiotherapy research systems with high precision and equipped for image guidance. There clearly is a need for such devices to encourage modern animal radiotherapy research.

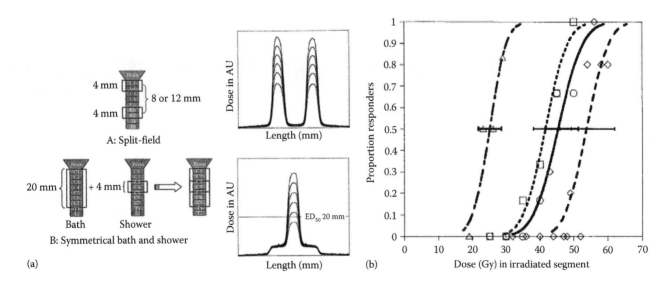

FIGURE 18.1   (a) Several irradiation schemes and dose distributions in the rat spinal cord with proton beams. (b) Dose–response curves for paralysis after split field irradiation (○: 12 mm spacing; □: 8 mm spacing) compared to uniform single field irradiations of 4 mm (◇) and 8 mm (△). (From Bijl, H.P. et al., *Int. J. Radiat. Oncol. Biol. Phys.*, 57(1), 274, 2003. With permission.)

18.3 REQUIREMENTS FOR PRECISION IMAGE-GUIDED SMALL ANIMAL RADIATION RESEARCH PLATFORMS

From the previous sections it is clear that to enable precision image-guided irradiation, a research platform must fulfill certain requirements. These are summarized in Table 18.1 and discussed in this section.

18.3.1 BEAM DELIVERY REQUIREMENTS

18.3.1.1 Targeting

Tumors in humans usually are on the order of a few centimeters when they receive radiation therapy. Radiation fields employed in radiotherapy for cancer treatment are rarely smaller than 1 cm. Most beam delivery systems have a precision of a few millimeters. To perform similar treatments in small lab animals, one must realize that, for example, a mouse lung measures only 1 or 2 cm at most in its largest dimension and that tumors within the lung are even smaller. Therefore, a precision (combined positioning, stability, and reproducibility) of a few millimeters is not sufficient for mouse radiotherapy studies. Instead, submillimeter precision is required, ideally on the order of 0.1 mm. For targeting subareas in tumors for boost studies, better precision may be needed. The possible size of the radiation beam should be 1 mm or smaller if substructure targeting is desired.

It should be possible to aim radiation beams from different directions at target structures, similar to clinical radiotherapy in humans. This is the preferred option over a stationary beam combined with a rotating animal target. Even an anesthetized animal will exhibit some stage motion-induced organ motion. In radiotherapy, motion of the radiation source is usually limited to an arc, sometimes combined with couch motion to enable noncoplanar irradiation. This should also be a minimum requirement in an animal system, which, in combination with 3D couch motion, offers many ballistic degrees of freedom. In addition, in radiotherapy, often the photon beam intensity and the beam shape are modulated during the arc motion. While the latter should also be possible for an animal platform, including even energy modulation when an x-ray tube is used, the former will be much harder to do because of the small beam sizes; for example, the equivalent of a human therapy multileaf collimator is difficult to envisage for much smaller beams. A fast shutter to block the beam temporarily should, on the other hand, be possible.

Arc motion of very small beams that have to be aimed precisely may pose strict mechanical demands on the system. If one considers that moving a radiation source such as a heavy x-ray tube around an animal easily introduces mechanical wobble of several millimeters, it becomes clear that

TABLE 18.1
Ideal Requirements for an Image-Guided Small Animal Irradiator Device

Ideal Requirements for a Small Animal Image-Guided Irradiator Device				Comments
Beam energy	100–300 kVp			X-ray tube potential
Min. dose rate	0.2 Gy/min			Based on 10 min for 2 Gy
	Rabbit	Rat	Mouse	
Image resolution (μm)	330–1650	150–750	65–330	Desired voxel size relative to 1–5 mm for patients
Beam diameter (mm)	1.6–3.3	0.76–1.53	0.33–0.66	Beam size diameter relative to smallest clinical beam diameter of 5–10 mm for patients
Targeting accuracy (mm)	±0.3	±0.2	±0.1	The ideal targeting accuracy has no mechanical positional error and is limited only by image resolution
Image temporal resolution	120 ms or 8 fps	75 ms or 13 fps	40 ms or 25 fps	Based on 1/10 of respiratory period of free breathing animals

Source: Verhaegen, F. et al., *Phys. Med. Biol.,* 56(12), R55, 2011. With permission.

extraordinary measures have to be taken to enable precision irradiation. Ideally, dynamic field-shape modulation should be possible. This places high demands on the mechanical precision and accuracy of the system, and will necessitate the adoption of a strict quality assurance program.

18.3.1.2 Beam Energy and Dose Distributions

MV photon beams (typically 4–25 MV) as used in external beam radiotherapy have several characteristics that are unsuitable for irradiating small targets in small animals. At the air–tissue interface in the entrance region of the beam, an MV photon beam exhibits dose buildup due to electronic disequilibrium (Attix 2004). The same phenomenon occurs behind, for example, the lung–tissue interface, which is often termed re-buildup (Shahine et al. 1999). At the downstream tissue side of the beam exit at a tissue–air interface, missing backscatter from the upstream region would cause a significant dose depression. The extent of the (re-)buildup regions corresponds roughly to the range of secondary electrons in tissue, which amounts to 1–2 cm for MV photon beams (Figure 18.2). Therefore, exposing small animals to such beams would cause dose build-up/down gradients of the order of the animal size itself. This would make it very challenging to, for example, deliver a uniform dose to a tumor. At the external air–tissue interface, build-up materials ("bolus") might be used to counteract these nonequilibrium effects but at internal interfaces this would be impossible. Adding bolus would also make the irradiation process much more cumbersome; for example, the bolus must be in exactly the same position during imaging and irradiation, which may be separated in time. Figure 18.2 also shows that the dose gradient past the buildup is steeper in kV than MV beams. This will put a restriction on the tube voltage range of about 100–300 kV. Also irradiations with, for example, ^{192}Ir, ^{137}Cs, or ^{60}Co sources with mean photon energies of around 380, 662, and 1250 keV, respectively, are suboptimal due to the extensive build-up regions.

Similar considerations apply to the lateral beam penumbra. These are caused by a combination of the finite source size, collimator photon transmission and scatter, secondary electrons from collimators, and the range of secondary electrons in tissue (Khan 1994). For MV photon beams, these may extend several millimeters beyond the geometric field, which may lead to unacceptable spillover of the dose distributions in structures outside the target, possibly leading to overdosed healthy tissue. This would be quite different from human radiotherapy, where the millimeter penumbras are small compared to the target. Therefore, again kV x-rays are needed to scale down the penumbras, resulting in sharp beam edges.

In summary, to avoid extensive disequilibrium dose regions and wide penumbras, the use of kV photons generated by an x-ray device is required. For accelerating potentials below 250 kV, the

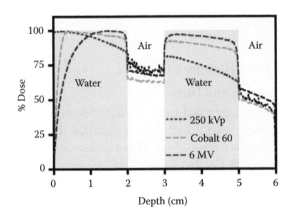

FIGURE 18.2 Monte Carlo calculations of depth dose profiles for a water/air/water/air slab phantom, modeling a low-density cavity embedded in tissue. Similar considerations would apply to a tissue/lung/tissue geometry. The beam was a 1 mm-diameter pencil beam, and dose scoring was performed over the whole width of the beam. For the 6 MV beam and the ^{60}Co photons, the buildup can be clearly seen behind both air/water interfaces. The kV beam exhibits only a minimal buildup effect. Missing backscatter is a more subtle effect but is discernible at both water/air interfaces for the ^{60}Co and 6 MV beam, but is again minimal for the kV beam. (From Verhaegen, F. et al., *Phys. Med. Biol.*, 56(12), R55, 2011. With permission.)

build-up regions in tissues are less than a few hundred micrometers. Penumbras are very sharp for kV x-rays, well below 1 mm. Missing backscatter at the exit side does occur, but its magnitude is limited and the effect persists for no more than a fraction of a millimeter.

A problem that arises when downscaling from the MV to the kV photon energy range is the increasing importance of the photoelectric effect. Tissues exhibit significant differences in their energy absorption at these low photon energies (see Chapter 20). Water equivalence of tissues, which is often assumed in high-energy photons, does not apply. Therefore, to avoid excessive problems with heterogeneous dose distributions due to tissue absorption, photon energies above 100 keV should be preferably used. Several irradiation devices that are now commercially available have opted for 225 kV as the x-ray tube potential (see further), but other devices have been built that operate at lower kVs. It should be emphasized that large dose increases or decreases in, for example, bone or adipose tissue in kV beams do not correspond to treatments with MV beams, which may invalidate preclinical studies. Conversely, kV irradiators may be used for studies of differential tissue absorption.

It is, therefore, important that dose calculation for small animal irradiators in the kV range be implemented carefully. Dose measurements also need to be performed with knowledge of the behavior of radiation detectors in this energy range.

Dose enhancement effects may also be noted when animals are irradiated in the presence of a high-atomic-number x-ray contrast medium, for example, an iodine-based contrast medium or gold nanoparticles (Lechtman et al. 2011, 2013). This is an uncommon situation for humans during radiotherapy (which is not administered as kV x-rays, anyway), but may be possible in combined imaging/irradiation studies in animals. On the other hand, this creates the possibility for studying contrast-enhanced radiotherapy (Verhaegen et al. 2005).

18.3.1.3 Photon Spectrum and Relative Biological Effectiveness

In addition to the differences in tissue energy absorption, one should also consider the fact that, when very low energy x-rays are used (say, below 50 kV) for irradiation of small animals, the RBE of the radiation in tissue exceeds unity. In other words, the dose required to cause a certain radiation effect (e.g., chromosome damage, cell death, paralysis, etc.) for low-energy kV x-rays is substantially below the dose required to lead to the same effect in, for example, MV x-rays (Nikjoo and Lindborg 2010). This diverging behavior of low-energy kV and high-energy MV photons may cause extrapolation difficulties from animal experiments to human radiotherapy. On the other hand, it opens up possibilities to perform RBE studies. Should this capability be desired, the beam delivery system needs to cover a range of photon spectral energies and distributions. This can be achieved by combining different kV values with different filtrations, for example, a K-edge filter to sharply attenuate photons below a certain energy.

18.3.2 IMAGING SYSTEM REQUIREMENTS

18.3.2.1 Spatial Resolution and Dose

CT imaging is the preferred imaging modality for the localization and treatment planning of localized malignancies for treatment with external beam radiation therapy. Though CT imaging has inherently poor soft tissue contrast, it is still a suitable tool for visualizing anatomical spatial orientation and is well suited to provide electron density data for treatment planning dose optimization. MRI imaging may be better at tumor segmentation, but the information contained in the voxels cannot be used for dose calculation. About a decade ago, kV cone-beam CT was introduced in radiotherapy (Jaffray et al. 2002); it features an imaging panel and x-ray tube mounted in-line on the gantry of a linear accelerator, perpendicular to the photon beam direction. This provides a tool for image-guided radiation therapy. Similarly, it would be desirable for a small animal irradiator to contain both the imaging x-ray beam and treatment beam in the same space and coordinate system. From the previous section, we saw that the treatment beam energy for small animals should be between 100 and 300 kV; these energies are also suitable for imaging, so both the imaging and irradiation can be done by a single x-ray tube.

Typical CT scans used for clinical treatment planning in radiotherapy contain a reconstructed image voxel (3D pixel) spacing of 1–5 mm in the xy transverse plane and 2–5 mm along the longitudinal direction (Supporting Information). For small animal imaging, the voxel sizes need to be scaled according to the animal's size (Table 18.1). Fortunately, nature provides us with convenient allometric ratios based on the mass (M) alone, which allows us to easily calculate equivalent physiological parameters for animals of differing weights. For instance, the body length scales as $M^{1/3}$ (Fahrig et al. 1997). Therefore, for equivalent voxel sizes 1–5 mm of a 70 kg patient, 330–1650 μm^3 for a 2.5 kg rabbit, 150–750 μm^3 for a 250 g rat, and 65–330 μm^3 for a 20 g mouse are required.

Modern day micro-CT scanners can routinely acquire image volumes with voxel sizes below 100 μm (Paulus et al. 2001; Ritman 2002). However, there is a nonlinear dose penalty when reducing voxel size for the same level of image noise. When the voxel size is reduced from 200 to 100 μm, a 16-fold increase in photon fluence, and therefore imaging dose, is required to maintain the same level of image noise (Ford et al. 2003; Kalender 2005). Consequently, micro-CT imaging of the smallest animals will have a proportionally higher imaging dose than the equivalent patient CT imaging; it is not uncommon for a single *in vivo* micro-CT scan of a mouse to deliver 0.5 Gy (Obenaus and Smith 2004; Willekens et al. 2010). Recent guidelines for the use of animals in cancer research caution that whole-body imaging doses exceeding 1 Gy can affect tumor growth (Workman et al. 2010). A whole-body radiation dose of 10 cGy is considered safe for repeated micro-CT imaging in longitudinal studies.

To achieve spatial resolutions of 100–200 μm, an additional important limiting consideration is the x-ray tube's focal spot size. Most commercial micro-CT scanners have focal spot sizes ranging from 5 to 50 μm (Wang and Vannier 2001) but these micro-focus x-ray tubes do not provide sufficient x-ray intensity to be used for therapy without causing undue target heating problems. Therefore, industrial x-ray tubes with larger focal spot sizes, of the order of 1 mm, are used for animal irradiators.

18.3.2.2 Dose–Response, Timing, and Artifacts

A micro-CT imaging system should be able to accommodate a wide range of tube potentials to achieve the optimal contrast-to-noise ratio (Huda et al. 2002). This would also be useful for future developments such as dual-energy imaging (Bazalova et al. 2008a,b). Also desirable is a detector with a relatively flat detector quantum efficiency and dose–response at treatment energies for dose reconstruction techniques. The x-ray detector should have a high temporal resolution of about 25 frames/s (Table 18.1) to resolve the peak inspiration for respiratory gating in murine models (Bazalova et al. 2009).

In addition to imaging requirements, accurate dose delivery and planning depends on reliable CT image data free from artifacts. CT image artifacts may influence the accuracy of dose calculations (Bazalova et al. 2007). The accuracy of the x-ray detector is an important component to achieve quantitative CT data. It may suffer from geometric distortion (Fahrig et al. 1997), image lag and ghosting (Siewerdsen and Jaffray 1999), and drift. Therefore, routine quality assurance should be implemented to ensure the accuracy, stability, and reliability of CT numbers (Bazalova et al. 2008).

18.4 RECENT DEVELOPMENTS IN PRECISION SMALL ANIMAL RADIOTHERAPY

The growing need for broad-purpose precision irradiation platforms for preclinical studies has led to the recent development of several devices, some of which are now commercially available. Most of the development took place in North America, but some recent work has emerged from European research centers as well. The development process is expected to continue for several more years, eventually leading to powerful research platforms equipped to address a wide range of radiation research questions potentially leading to progress in clinical radiotherapy. We will now address efforts by several groups to develop precision irradiation, often combined with high-resolution imaging. Microbeam irradiators based on photon beams generated by synchrotron technology are excluded from the overview (Slatkin et al. 1992; Laissue et al. 2007; Martinez-Rovira et al. 2012).

18.4.1   WASHINGTON UNIVERSITY SYSTEM

At the Washington University School of Medicine, a small animal irradiator was developed, based on an ^{192}Ir source (Stojadinovic et al. 2006, 2007; Kiehl et al. 2008). As explained in an earlier section, the photon energies of this isotope (ranging up to 800 keV, mean energy 380 keV) are slightly above the optimum to obtain sharp penumbras and avoid buildup over too large regions. Dose rates of 90 cGy/min are possible, presumably for a fresh ^{192}Ir source. Figure 18.3 shows their research platform, which has no onboard imaging system. The ^{192}Ir source was positioned by a standard brachytherapy remote afterloader at various short distances (1–8 cm) from the target. The device has to be placed in a shielded room. Several tungsten collimators were provided with field sizes of 5–15 mm. Fixed beams can be aimed from four different directions at a small animal, at 90° intervals. Positioning of the animal is done with a computer-controlled stage. Without an onboard imager, the device had to rely on the use of fiducial markers placed on the stage to register the CT treatment planning scans (acquired on a human CT scanner (Kiehl et al. 2008)). An irradiation accuracy of 0.3 mm was reported (Kiehl et al. 2008). Provided a high-dose rate brachytherapy afterloader is available, the original solution provided by the Washington University group allows small animal radiotherapy studies in a cost-effective manner. It does not fall in the category of image-guided small animal systems, though. No commercial system has been—as far as we are aware—developed out of these efforts.

18.4.2   STANFORD UNIVERSITY SYSTEM

Another pioneering effort to build a small animal research platform originated from Stanford University. They rebuilt a commercial micro-CT scanner (eXplore RS120 microCT scanner, GE Medical Systems, London, ON, Canada) to include a unique collimator with a variable aperture (Graves et al. 2007; Rodriguez et al. 2009; Zhou et al. 2010). Pseudo-circular field sizes (0.1–6 cm) could be produced with the aid of a sophisticated brass iris (Figure 18.4). The iris consists of 12 pentagonal lead/brass blocks arranged in 2 planes above each other, forming 2 stages of collimation. Each stage consists of sliding brass blocks that form a hexagonal aperture. The two hexagonal apertures are arranged coaxially, offset by 30°, each driven by a linear stepper motor. This results

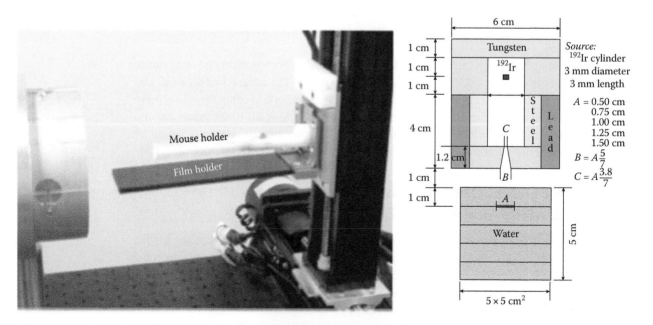

FIGURE 18.3   Small animal irradiator from Washington University, microRT. The left panel shows (from left to right) the collimator assembly which can accommodate the ^{192}Ir source and the computer-controlled mouse stage. The right panel shows a sketch of the tungsten collimator and the source. (From Stojadinovic, S. et al., *Med. Phys.*, 33(10), 3834, 2006; Stojadinovic, S. et al., *Med. Phys.*, 34(12), 4706, 2007. With permission.)

FIGURE 18.4 Stanford University small animal irradiator based on a micro-CT device showing, at the left panel, the 120 kV x-ray tube (bottom) the collimator (middle, in red), and the animal support system (top). The right panel shows details of the bilayer iris variable-aperture collimator. (Courtesy of Edward Graves, Stanford University, Stanford, CA.)

in a dodecagonal field shape, approaching a circular collimator. Beam "diameters" vary between 0 and 10.2 cm at the CT isocenter. Position sensors attached to each stage measure the aperture dimensions. The total shielding of the collimator exceeded 95%. This elegant device may allow field modulation during beam delivery, which in the case of small beams may be hard to achieve with multileaf-collimator technology, which is probably difficult to downscale. The x-ray tube operates between 70 and 120 kV. The radiation output of the device at 120 kV dropped sharply by about 30% going from a field diameter of 50–2 mm. This is a common feature of all x-ray-based small animal irradiators.

The micro-CT scanner on which the platform is based has a good spatial imaging resolution (0.1 mm). Small radiation beams can administer therapy level doses in a coplanar manner with an accuracy within 0.1 mm (Zhou et al. 2010). The system operates at 120 kV (could go down to 70 kV) with a maximum current of 50 mA and can irradiate animals in a CT-like arc or from discrete directions. To deliver therapy-level dose rates (~2 Gy/min), more work is needed to alleviate heating problems of the generator and the x-ray tube. These problems led to a delivery time of about 40 min for an 8 Gy dose (Rodriguez et al. 2009). The radiobiology of protracted pulsed irradiation should be investigated. The low accelerating voltage (120 kV) of the tube compared to the commercially available devices (225 kV) results in a steep dose falloff as soon as the beam enters an animal. The low x-ray tube voltage also results in a photon spectrum where photoelectric interactions will be important, which makes energy absorption very dependent on the tissue type (Bazalova and Graves 2011) (see Chapter 20).

Accurate positioning of the radiation beams was possible through careful calibration and alignment procedures involving a 3D stage. Animals could be anesthetized with a gas system. Submillimeter-sized beam penumbras were reported for small fields (0.5–2 cm) at 1 mm depth (Rodriguez et al. 2009); at greater depth, they increased by a small amount. To circumvent heating problems, the system may operate with a second larger focal spot, but this will result in larger beam penumbras.

In departments where a micro-CT scanner is already available, it may be possible to retrofit it as an irradiator in a cost-effective way, similarly to the Stanford University solution. No commercial product has resulted yet from this work, and, as far as we know, the development of the device has been terminated.

18.4.3 University of Texas Southwestern System

Recently, another homemade small animal research platform (Figure 18.5) was introduced that was designed primarily for stereotactic irradiation of, for example, brain and lung as well as for normal tissue studies (Song et al. 2010; Pidikiti et al. 2011). A fixed (nonrotating) 320 kV industrial x-ray

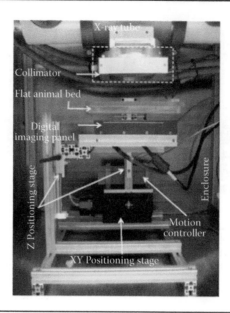

FIGURE 18.5   Small animal radiation research platform from University of Texas Southwestern. All components are fixed, but on the animal bed a rotating cylinder with an immobilized animal inside may be positioned, allowing coplanar irradiation. (From Song, K. et al., *Phys. Med. Biol.*, 55, 7345, 2010. With permission.)

tube, operating at 250 kVp (added filtration 1.65 mm Al) was capable of a high-dose rate of up to 20.8 Gy/min at an SSD of 20 cm. This was combined with small cylindrical collimators with a fixed aperture (1–10 mm diameter) to shape the stereotactic fields. For a large field of 50 mm diameter, the depth dose distribution in water resulted in a 50% dose decrease compared to the surface at a depth of about 27 mm (Pidikiti et al. 2011). For the smallest collimators, the 50% dose point was situated a few millimeter smaller.

Image guidance is done with a fixed high-resolution imaging panel operating at 20–30 kV, which would appear to be a very low energy for which the imaging dose may be an important issue. The animal is mounted on a 3D precision stage, which may also hold a cylinder in which the animal can be immobilized with a vacuum bag and made to rotate around its long axis, itself perpendicular to the vertical beam axis (although the x-ray tube and imaging panel are immobile in this device, the setup of the rotating animal turns it into a "gantry"). This setup allows administering stereotactic arc therapy in discrete steps. It also allows for repeated imaging at small, discrete increments, resulting in a series of projections in the relatively large imaging panel (13 × 13 cm; 1024 × 1024 pixels), from which a cone beam CT image could be reconstructed via standard backprojection techniques. The software for accurate animal positioning to locate the target at every discrete treatment angle was also developed. The operating mode of this device raises questions about organ motion during rotation of the animal due to gravity, which may occur even when the animal has been tightly immobilized.

18.4.4   Dresden System

A consortium of research institutes in Dresden (Germany) built an in-house animal research platform (Tillner et al. 2012), which resembles somewhat a compact version of the commercial X-RAD 225Cx system (discussed later in this chapter). It consists of a shielding cabinet, a 360° rotating arm with an x-ray tube, and a cone beam CT panel mounted on it. The animal stage has three degrees of freedom and is computer-controlled. The tube operates at 10–225 kV and has two foci of 1.0 and 5.5 mm. A range of fixed collimators with circular, oval, or customized apertures can be attached. Instead of nozzles, flat collimators are used. The imager is a CsI panel with 100 μm pixels in a 123 × 112 mm frame, with a frame rate of up to 88 Hz. The distance between the isocenter and the imaging panel is variable in this device.

18.4.5 Johns Hopkins University System

In this and the next section, we describe the only two platforms that led to commercial products to date. Figure 18.6 depicts the Small Animal Radiation Research Platform (SARRP) developed at the Johns Hopkins University, Baltimore, by a team of radiation physicists and engineers. The system is now commercially available from Xstrahl Ltd. (Surrey, United Kingdom) and continues to develop in partnership with the original Johns Hopkins research team. The SARRP consists of an x-ray source mounted on a motorized rotating arm, a dual imaging system for CT and planar x-ray imaging for target localization, and a set of robotically controlled stages for positioning the animal (Wong et al. 2008). The SARRP x-ray source is an industrial 225 kVp x-ray tube (dual focal spot sizes of ~0.4 and 3 mm according to the IEC 336 standard; maximum current of ~13 mA at 225 kV) and is used for both imaging (60–80 kVp, small focus, typically with 1 mm of Al filtration) and therapy (220 kVp, large focus, typically 0.15 mm of Cu filtration). A fixed 20×20 cm^2 (1024×1024 pixel) amorphous Si flat-panel detector is at 52.5 cm from the x-ray source when the latter is oriented horizontally. Cone-beam CT (CBCT) is performed while rotating the animal about a vertical axis, with the x-ray source and imaging panel stationary. The CBCT resolution is ~250–500 μm.

A second digital fluoroscopic imager located directly underneath the animal collects planar cine images when the source is pointed toward the floor. The robotic animal positioning system has four degrees of freedom: X and Y (cross-table), Z (vertical-stage), and θ (rotating table), where the latter is used for CBCT acquisition and noncoplanar arc irradiations. The radiation source arm and the table can rotate over 360° in two orthogonal planes.

The SARRP has two collimation systems, which attach to the primary imaging collimator (approximately 17×17 cm^2 at isocenter). A nozzle-shaped collimator with interchangeable inserts provides a range of field sizes (circular: 0.5 and 1 mm in diameter and rectangular: 3×3, 5×5 and 3×9 mm^2). These are used for high-precision conformal irradiations. Larger tray collimators

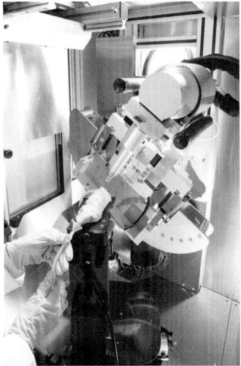

FIGURE 18.6 SARRP system provided by Xstrahl Ltd. and developed at Johns Hopkins University. The picture on the left shows the irradiation cabinet from the outside, with a window to allow observing the animals. The picture on the right shows the inside of the cabinet with the x-ray tube with a narrow collimator (top, pointing down at an angle), the animal with its nose in the anesthetics cone, and the CBCT imaging panel (left). (Image courtesy of Xstrahl Ltd., Surrey, United Kingdom.)

$(3 \times 3, 5 \times 5, 10 \times 10 \text{ cm}^2)$ are used for higher throughput, less sophisticated experiments (e.g., cell culture or whole/hemi-body irradiations). The source-to-aperture distance for the nozzle and tray collimator is about ~30 and 10 cm, respectively. Narrow radiation penumbras of about 250 μm are achieved for all apertures at shallow depths around the isocenter. This reduces to <100 μm when the small focal target spot is used with the 0.5 mm aperture, enabling precise microbeam applications. Radiation dose rates for shallow depths at 35 cm from the source (large focal spot) vary from ~2–3 Gy/min for the set of nozzle apertures to ~4 Gy/min with the tray collimators.

A robotic calibration procedure allows for high targeting accuracy and a tight radiation "sphere-of-delivery" during rotation of the x-ray source or the animal stage. The full procedure is described in Matinfar et al. (2009). For continuous arc deliveries (x-ray source or stage rotation), this methodology has been extended such that the virtual corrections are also continuous. The targeting accuracy of the SARRP is 200 μm and a stereotactic-arc delivery with the 0.5 mm aperture resulted in 1.07 mm-wide high-dose region (full width at half-maximum, FWHM) (Matinfar et al. 2009). The system also offers MRI-compatible immobilization devices.

18.4.6 PRINCESS MARGARET HOSPITAL SYSTEM

The X-RAD 225Cx small animal micro-irradiation platform, developed by researchers at the Princess Margaret Cancer Centre, ON, Canada, and commercialized through Precision X-Ray Inc. (North Branford, CT), is a self-shielded unit consisting of a dual-focal-spot industrial x-ray tube (COMET MXR 225/22; COMET Technologies, Stamford, CT), operating at potentials from 5 to 225 kV, and mounted opposite a flat panel detector (PerkinElmer XRD 0820 AN3-ES, PerkinElmer Optoelectronics, Fremont, CA) on a C-arm gantry. The 1024 × 1024 pixel (0.2 mm pixel pitch) imaging panel facilitates CBCT and fluoroscopic imaging for animal visualization purposes and image-guided positioning setup via a three-axis computer-controlled animal couch. As of 2015, this system also offers a rotating couch. As shown in Figure 18.7, the geometry of the system provides a cylindrical imaging field of view with 9.7 cm diameter and 9.7 cm length at the highest spatial resolution of 0.1 mm. A typical experimental workflow consists of image-based positioning of the target within the animal, followed by collimated dose delivery in a step-and-shoot or dynamic arc (from 0° to 360°) manner at dose rates up to 4 Gy/min. Dose delivery collimation is realized through insertion of an interchangeable collimator with dimensions ranging from a circular 1 mm diameter through a rectangular 40 × 40 mm field.

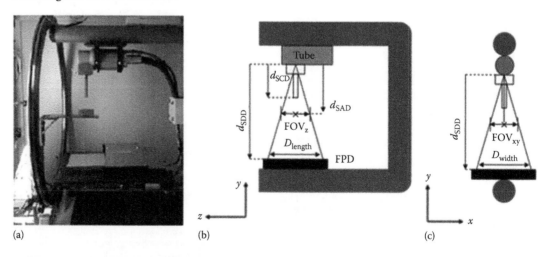

(a)   (b)   (c)

FIGURE 18.7 (a) Interior overview of the Princess Margaret Cancer Centre image-guided microirradiator. Visible in the picture are the x-ray tube and flat-panel detector (FPD) mounted on a C-arm gantry with front-mounted stabilizing ring, three-axis animal positioning stage with solid water phantom setup for illustrative purposes, and interchangable collimator for dose delivery. (b, c) Overall geometry of the system along side (b) and front (c) views with $d_{SAD} = 30.7$ cm, $d_{SDD} = 64.5$ cm, and $d_{SCD} = 23$ cm. The imaging field of view at isocenter is $FOV_z = FOV_{xy} = 9.7$ cm with the detector size of $D_{length} = D_{width} = 20.4$ cm.

Initial imaging characterization of the system was performed with a 10 cm-long, 2.5 cm-diameter water phantom for CT number uniformity and noise assessment, and a micro-CT performance evaluation phantom (Shelley Model vmCT 610, Shelley Medical Imaging Technologies, London, ON, Canada) for linearity, geometry, modulation transfer function, imaging resolution, and CT number accuracy quantification (Clarkson et al. 2011). With a dose of ~1 cGy in water (delivered at 40 kVp, 30 mA·s), the CT numbers were linear ($R^2 \geq 0.998$) and average noise across four 10 mm-diameter by 1 mm-long cylindrical regions of interest (ROIs) in the homogeneous water phantom was 30 HU. The presampled modulation transfer function, as determined by a slanted edge test, was 0.64 and 1.35 mm^{-1} at the 50% and 10% levels, respectively. Linear attenuation coefficients measured in 17.0 mm^3 ROIs across eight different inserts mimicking common anatomical radiodensities (from −1000 to 2960 HU) reached a maximum standard deviation of 0.02 cm^{-1} for both 40 kVp, 30 mA·s and 100 kVp, 6 mA·s image acquisition settings. End-to-end 3D targeting accuracy, assessed with seven measurements of image-guided targeting of a ball bearing affixed to pieces of radiochromic film, was 0.20 ± 0.09 mm (average ± SD). This targeting accuracy benefited from the implementation of dynamic stage corrections to compensate for minor gantry flex as the gantry rotated (Clarkson et al. 2011).

Typical research uses of the device range from complex tumor targeting to broad irradiations typical of conventional preclinical studies. By the end of 2013, 105 users had acquired over 11,000 CBCT images and delivered a similar number of irradiations using the two X-RAD 225Cx units installed at the Princess Margaret Cancer Centre. Complex applications use multiple beams and soft-tissue localization (Zheng et al. 2012), such as in the irradiation of orthotopic cervix models (Hill et al. 2012) or U87 brain tumor models (Chung et al. 2013). Other targeted applications include irradiations in window chamber models of spinal (Figley et al. 2013), subcutaneous (Maeda et al. 2012), and cranial (Burrell et al. 2012) structures for tumor and normal tissue radiation response quantification.

Current research and development efforts with the Princess Margaret Cancer Centre small animal micro-irradiator are focused on optimizing and delivering complex radiation plans and improved dose calculations for kV energies and relatively small field sizes (as low as 1 mm in diameter) typically employed for preclinical investigations. Initial work on developing a framework for delivering arbitrary 2D dose distributions with the fixed collimation of the system has focused on step-and-shoot stage motions using a dose calculation engine based on empirical measurements of the beam profile. Optimization of two challenging dose distributions, outlined in Figure 18.8,

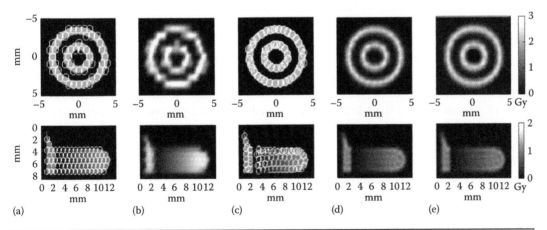

(a) (b) (c) (d) (e)

FIGURE 18.8 Planar dose optimization and delivery results for two illustrative dose distributions: a bullseye dose distribution with 1 mm-wide rings alternating between 0 and 2 Gy (top row) and a "sock" dose distribution containing an exponentially decaying 1 Gy region, a 1.5 × 4 mm 1 Gy rectangle, a 7.5 × 4 mm rectangular region with a linear dose increase from 0 to 1 Gy over 7.5 mm, and a constant 1 Gy semicircular region with 2 mm radius (bottom row). (a) Initial beam positions overlaid on desired dose distribution. Each elliptical region denotes the region of the beam that is at least half the maximum beam value. (b) Calculated dose for initial beam placements and weights. (c) Optimized beam positions delineated on desired dose. (d) Calculated dose for optimized beam positions and weights. (e) Dose delivery validation. Beam positions were mapped to automated stage motions for delivery of the optimized dose. Each dose distribution is the average of five deliveries measured with radiochromic film.

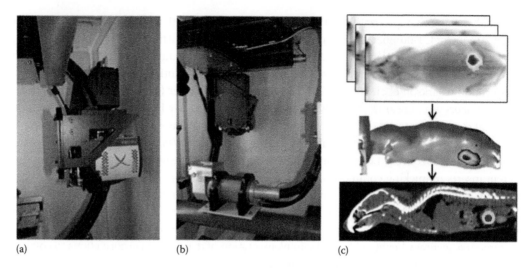

(a)   (b)   (c)

FIGURE 18.9 Bioluminescence camera integration. (a) Camera (iXon Ultra 897, Andor Technology plc., Belfast, United Kingdom) mounted on gantry ring. (b) Wider field image showing camera position with respect to x-ray tube and flat panel x-ray detector. Software-controlled filter wheel is visible at the front of the camera. (c) Example workflow of 3D source reconstruction. Bioluminescence measurements of a 3-week orthotopic prostate model made at multiple projections before mapping the bioluminescence signal to the surface mesh of the animal determined using cone beam computed tomography. Source is then reconstructed using diffuse optical tomography techniques. (Image courtesy of Dr. Robert Weersink.)

demonstrated an ability to deliver distributions with millimeter-scale heterogeneity with submillimeter targeting accuracy for plan optimization times of approximately 5 min (Stewart et al. 2013). Film validation of the delivered optimized dose distributions revealed a dosimetric delivery error of 3.9% and a mean absolute delivery error of 0.01 Gy across a 0–1 Gy linear dose gradient over 7.5 mm using the 1 mm circular collimator. Related work has focused on improving the accuracy of 3D dose calculations with Monte Carlo-based methods using phase space files derived analytically through the focal spot intensity, predetermined x-ray spectrum, and collimator geometry (Granton and Verhaegen 2013). Such an approach yielded a 19- to 1200-fold calculation time improvement over Monte Carlo-derived phase space files for circular collimators ranging from 25 to 1 mm in diameter. The dosimetric results agreed within 1% for all collimators except for the 1 and 2 mm collimators, which agreed within 7% and 4%, respectively, as a result of focal spot occlusion.

Complementing the ongoing dosimetric research, investigations are under way to integrate bioluminescence imaging for targeting structures not readily visible in CBCT images and longitudinal treatment response assessment (Weersink et al. 2011). The bioluminescence integration, described in Figure 18.9, consists of an iXon Ultra 897 electron-multiplying charge-coupled device camera (Andor Technology plc., Belfast, United Kingdom) mounted to the gantry ring of the micro-irradiator to facilitate 360° image acquisitions. The geometry of the system provides a 10 × 10 cm field of view with a 0.2 mm resolution at the isocenter. A software-controlled filter wheel enables detection of multiple wavelengths (from 560 to 680 nm) to aid in determining the depth of the bioluminescence source. Initial experiments have demonstrated the feasibility of recovering and registering the 2D bioluminescence projections to the 3D CBCT image surface, tomographic bioluminescence source reconstruction based on diffuse optical tomography, and 2D targeting based on the recovered bioluminescence signal. Preliminary results suggest a 1.3 mm error in reconstructing the location of a point bioluminescence source and a 2D irradiation targeting accuracy using bioluminescence guidance of 0.3 mm.

18.5 COMMISSIONING SMALL ANIMAL RADIATION PLATFORMS

This section will provide a brief overview of the commissioning procedures needed to characterize the mechanical and dosimetric behavior of small animal radiation research platforms. Only a brief summary will be given here; for technical details, the original papers or manufacturer's guidelines are to be consulted. Additional commissioning may be needed to provide input data for dose calculations; this is covered in Chapter 20.

Several recent papers have dealt at length with commissioning of small animal radiation research platforms, which is essential to guarantee their targeting and dosimetric accuracy as well as the image quality of the onboard imager (Rodriguez et al. 2009; Tryggestad et al. 2009; Zhou et al. 2010; Clarkson et al. 2011; Newton et al. 2011; Pidikiti et al. 2011; Lindsay et al. 2013). Point dose measurements for determining the absolute machine output for a large field or sometimes also for relative output factors for different field sizes may be done with small ionization chambers, provided the field is large compared to the sensitive volume of the ionization chamber. Half-value layers may also be measured with an ionization chamber in large fields. The dosimetry protocol for low- and medium-energy x-rays TG61 (Ma et al. 2001) may provide guidance to derive these quantities.

The dosimeter of choice for measuring the 2D dose distribution (depth dose and lateral dose profiles) from small fields is the self-developing radiochromic film (Rodriguez et al. 2009; Tryggestad et al. 2009; Pidikiti et al. 2011). This is mainly due to the high spatial resolution (better than 100 μm) of film of the types EBT, EBT2, and EBT3 (Ashland Specialty Ingredients, Wayne, NJ). Film readout is usually done with a high-quality document scanner. Prescanning of the film can be avoided by using the three-color read-out procedure (van Hoof et al. 2012). Care has to be taken to calibrate this film accurately, to position it accurately during the measurement, and to read out the film at the correct time, in accordance with the read-out procedure used during the calibration procedure. When used correctly, this type of dosimeter can achieve accuracies of a few percent. Three-dimensional dosimetry in small animal irradiators has also been reported (Newton et al. 2011) using PRESAGE plastic dosimeters. This system allows recording 3D dose distributions with a resolution of 200 μm in a quasi-tissue-equivalent medium by using an optical CT readout system. Because of optical artifacts at the dosimeter's surface, no reliable data could be obtained at depths shallower than 4 mm. Figure 18.10 shows the dose distribution in three planes for a combined irradiation with five beam sizes.

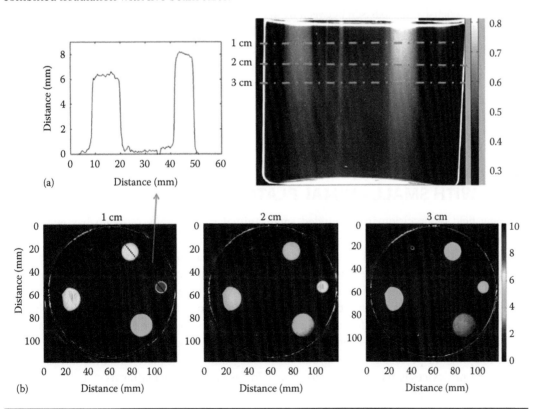

FIGURE 18.10   PRESAGE dosimeter irradiated with five circular fields (diameter 20, 15, 10, 2.5, and 1 mm) incident on the top surface of the cylinder in (a). Three axial planes are shown in (b) along with a line profile through the center of the 15 and 10 mm fields. (From Newton, J. et al., *Med. Phys.*, 38(12), 6754, 2011. With permission.)

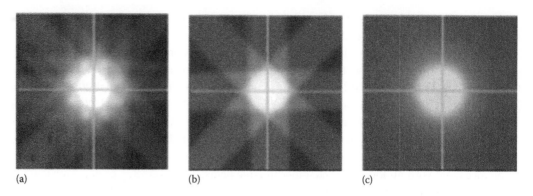

(a)　　　　　　　　　　　　(b)　　　　　　　　　　　　(c)

FIGURE 18.11 Evaluation of mechanical flex compensation system of X-RAD 225Cx system (other systems exhibit similar behavior). In (a–c), radiochromic film is irradiated with beams parallel to the film. A star-shot (eight-beam) irradiation, with flex correction (a) off and (b) on. (c) 360° arc with flex correction on. (From Clarkson, R. et al., *Med. Phys.*, 38, 845, 2011. With permission.)

Imaging and targeting capabilities of the systems have also received much attention (Zhou et al. 2010; Clarkson et al. 2011). Dedicated micro-CT phantoms exist for quantification of the CT performance (e.g., Shelley model vmCT 610, Shelley Medical Imaging Technologies, London, ON, Canada). These allow quantifying many imaging characteristics and identifying the performance limits of the devices. A study (Clarkson et al. 2011) found CT noise to be about 30 HU in a uniform water phantom at an imaging dose of 1 cGy. Targeting accuracy of the system was determined to have a mean displacement error of better than 0.12 mm with a standard deviation of better than 0.17 mm. Figure 18.11 shows a test for the mechanical flex correction of a system, demonstrating that this correction is essential to achieve accurate dose distributions (Clarkson et al. 2011).

A comprehensive comparison between the mechanical and dosimetric commissioning of three similar X-RAD 225Cx devices was published recently (Lindsay et al. 2013), showing that differences in radiation output of the machines may occur. For large fields, the absolute radiation output was found to differ by 20%, while for the smallest field in the study (2.5 mm diameter) the output differed by more than 50%. The latter is caused by small differences in collimator alignment and electron focal spot shape. The mechanical flex of the heavy systems during rotation were found to be unique, requiring individual corrections.

18.6 NEW DEVELOPMENTS AND RESEARCH WITH SMALL ANIMAL PLATFORMS

The new field of precision image-guided small animal radiotherapy research has engendered much enthusiasm among researchers in radiotherapy. An estimated 30 institutes now have operational research platforms, and many more are considering acquiring this new technology. This clearly shows that the radiotherapy community has fully embraced the urgent need for this type of research. Much effort went into developing the new technology and assessing its performance. Many papers were published in recent years on the development and commissioning of the systems. There is still much room for further development of these platforms by adding various forms of imaging such as bioluminescent imaging, fluorescent imaging, photoacoustic imaging, ultrasound imaging, CT perfusion imaging, and so on. Integration of the platforms with various standalone imaging devices is also a field where more effort is needed. The platforms could be equipped with more complex animal stages, allowing noncoplanar arcs to be delivered. Collimation systems could have variable apertures (e.g., the Stanford system mentioned in this chapter). Some work has been done on simultaneous stage motion and beam delivery, enabling true intensity-modulated radiotherapy in small animals. This would then allow the comparison of several dose painting approaches to derive optimal human therapy dose distributions. Motion compensation with gating signals (respiratory, cardiac, and others) is another technology that is already well established in small animal MRI imaging, which could potentially be adapted to precision irradiation. Extension of the technology toward ion beam therapy would also be highly desirable (Narayanan et al. 2013; Sandison et al. 2013). The onboard imager in some systems could be used as a dosimeter (Granton et al. 2012) (see Chapter 20). Treatment planning for small animals needs to become a mature field; only a few initial efforts have been presented (see Chapter 20) and this has, so far, prevented complex radiation studies. For instance, inverse planning, which is now the standard in human radiotherapy, needs to be developed.

These advanced irradiation/imaging systems are so new that not many biology studies have been published yet, and papers are scattered widely over journals and research fields. A good example of recent work that is enabled by the new generation of research platforms is a study in which an orthotopic lung tumor was implanted in a rat (Song et al. 2010). A combination of x-ray imaging and bioluminescent imaging was used to target the tumor accurately and to spare the heart and liver (Figure 18.12). The latter imaging mode can also be used to study the response of the tumor to irradiation (Rehemtulla et al. 2000; Luker and Luker 2008). In their preliminary results, the authors mention that in untreated tumors the bioluminescent signal increased over time, whereas in irradiated animals the signal remained constant. The knowledge gained may be beneficial for stereotactic body radiotherapy treatments of lung cancer. The limiting factor in lung radiotherapy is often damage to the remaining healthy lung volume; the same group (Cho et al. 2010) also studied radiation-induced toxicity to healthy mouse lung. These studies are excellent examples of how the new platforms may help develop an understanding of the radiobiology of hypofractionated radiotherapy, which may ultimately aid in optimizing therapy.

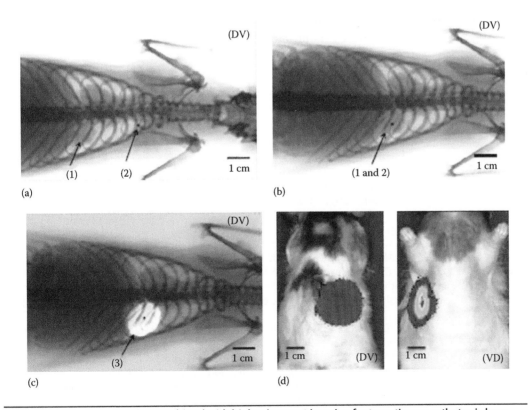

FIGURE 18.12 X-ray imaging combined with bioluminescent imaging for targeting an orthotopic lung tumor in a rat. (a) Prelocalization x-ray image with (1) the tumor center of mass and (2) the centre of the radiation beam. (b) Postlocalization x-ray image with beam coinciding with tumor. (c) X-ray image superimposed on beam image. (d) Bioluminescent image in dorsal–ventral (DV) and ventral–dorsal (VD) orientation. (From Song, K. et al., *Phys. Med. Biol.*, 55, 7345, 2010. With permission.)

Another application for this new technology is the radiation bystander or "abscopal" effect (Mothersill and Seymour 2001; Kaminski et al. 2005) and the related clinical translation of spatially fractionated or "grid" radiotherapy (Penagaricano et al. 2010). The technology provides the capability to irradiate small subregions of tissue, potentially with sharp beam edges, and is therefore ideal for studying these phenomena in vivo.

Clearly, a vast range of biological research topics can be addressed. An example is the use of gold nanoparticles to enhance both CT image contrast and therapeutic dose deposition (24, 74). Other examples include the effect of radiation on novel transgenic tumor models, the synergy of radiation with either radiosensitizers (for tumors) or radioprotectors (for normal tissue), the exploration of novel fractionation schedules, or RBE studies. Synergistic studies with other treatment modalities are also expected to become popular on these platforms. The combination of radiation and other agents to prevent inhibition of apoptosis in pancreatic cancer has been reported (73). The biggest challenge here lies in the development of realistic animal models that allow translation of the experimental findings to human radiotherapy.

Also, noncancer research benefits from the new technology. In a study of treatment of Huntington's disease with selective serotonin reuptake inhibitors (Duan et al. 2008), precision irradiation of the mouse brain was done. The radiation blocks neurogenesis, which was found in this experiment to be a requirement for the drug to be successful. This indicated that neurogenesis-enhancing drugs might be beneficial for the disease investigated in this work. Radiotherapy is already used in the successful treatment of noncancer diseases especially in neurology. But such therapy is also expected in the near future to play an important role mainly in the treatment of other noncancer diseases where it is currently not being used, for example, by acting as a radiosurgical scalpel for treating cardiac disease or renal denervation for curing hypertension (Bert et al. 2012). Also in this field, small animal research platforms may provide ideal tools for fundamental exploration and human translation.

This nonexhaustive discussion clearly shows the immense research potential of small animal research platforms. Much work lies ahead to develop versatile research platforms and to formulate research questions and protocols from which we may draw important conclusions concerning the mechanisms and treatment of cancer and noncancer human disease.

REFERENCES

Attix, F. (2004). Charged-particle interactions in matter. In *Introduction to Radiological Physics and Radiation Dosimetry*. John Wiley & Sons, New York, pp. 160–202.

Bazalova, M., L. Beaulieu et al. (2007). Correction of CT artifacts and its influence on Monte Carlo dose calculations. *Med. Phys.* **34**(6): 2119–2132.

Bazalova, M., J. F. Carrier et al. (2008a). Dual-energy CT-based material extraction for tissue segmentation in Monte Carlo dose calculations. *Phys. Med. Biol.* **53**(9): 2439–2456.

Bazalova, M., J. F. Carrier et al. (2008b). Tissue segmentation in Monte Carlo treatment planning: A simulation study using dual-energy CT images. *Radiother. Oncol.* **86**(1): 93–98.

Bazalova, M. and E. E. Graves. (2011). The importance of tissue segmentation for dose calculations for kilovoltage radiation therapy. *Med. Phys.* **38**(6): 3039–3049.

Bazalova, M., H. Zhou et al. (2009). Kilovoltage beam Monte Carlo dose calculations in submillimeter voxels for small animal radiotherapy. *Med. Phys.* **36**(11): 4991–4999.

Bert, C., R. Engenhart-Cabillic et al. (2012). Particle therapy for noncancer diseases. *Med. Phys.* **39**(4): 1716–1727.

Bijl, H. P., P. van Luijk et al. (2002). Dose-volume effects in the rat cervical spinal cord after proton irradiation. *Int. J. Radiat. Oncol. Biol. Phys.* **52**(1): 205–211.

Bijl, H. P., P. van Luijk et al. (2003). Unexpected changes of rat cervical spinal cord tolerance caused by inhomogeneous dose distributions. *Int. J. Radiat. Oncol. Biol. Phys.* **57**(1): 274–281.

Bijl, H. P., P. van Luijk et al. (2006). Influence of adjacent low-dose fields on tolerance to high doses of protons in rat cervical spinal cord. *Int. J. Radiat. Oncol. Biol. Phys.* **64**(4): 1204–1210.

Burrell, K., R. P. Hill et al. (2012). High-resolution in-vivo analysis of normal brain response to cranial irradiation. *PloS One* **7**(6): e38366.

Cho, J., R. Kodym et al. (2010). High dose–per-fraction irradiation of limited lung volumes using an image-guided, highly focused irradiator: Simulating stereotactic body radiotherapy regimens in a small-animal model. *Int. J. Radiat. Oncol. Biol. Phys.* **77**: 895–902.

Chung, C., S. Jalali et al. (2013). Imaging biomarker dynamics in an intracranial murine glioma study of radiation and antiangiogenic therapy. *Int. J. Radiat. Oncol. Biol. Phys.* **85**(3): 805–812.

Clarkson, R., P. Lindsay et al. (2011). Characterization of image quality and image guidance performance of a pre-clinical micro-irradiator. *Med. Phys.* **38**: 845–856.

Crosbie, J. C., R. L. Anderson et al. (2010). Tumor cell response to synchrotron microbeam radiation therapy differs markedly from cells in normal tissues. *Int. J. Radiat. Oncol. Biol. Phys.* **77**(3): 886–894.

Debus, J., M. Scholz et al. (2003). Radiation tolerance of the rat spinal cord after single and split doses of photons and carbon ions. *Radiat. Res.* **160**(5): 536–542.

DesRosiers, C., M. S. Mendonca et al. (2003). Use of the Leksell Gamma Knife for localized small field lens irradiation in rodents. *Technol. Cancer Res. Treat.* **2**(5): 449–454.

Duan, W., Q. Peng et al. (2008). Sertraline slows disease progression and increases neurogenesis in N171–82Q mouse model of Huntington's disease. *Neurobiol. Dis.* **30**: 312–322.

Fahrig, R., M. Moreau et al. (1997). Three-dimensional computed tomographic reconstruction using a C-arm mounted XRII: Correction of image intensifier distortion. *Med. Phys.* **24**(7): 1097–1106.

Figley, S. A., Y. Chen et al. (2013). A spinal cord window chamber model for in vivo longitudinal multimodal optical and acoustic imaging in a murine model. *PloS One* **8**(3): e58081.

Ford, N. L., M. M. Thornton et al. (2003). Fundamental image quality limits for microcomputed tomography in small animals. *Med. Phys.* **30**(11): 2869–2877.

Granton, P. V., M. Podesta et al. (2012). A combined dose calculation and verification method for a small animal precision irradiator based on onboard imaging. *Med. Phys.* **39**(7): 4155–4166.

Granton, P. V. and F. Verhaegen. (2013). On the use of an analytic source model for dose calculations in precision image-guided small animal radiotherapy. *Phys. Med. Biol.* **58**(10): 3377–3395.

Graves, E. E., H. Zhou et al. (2007). Design and evaluation of a variable aperture collimator for conformal radiotherapy of small animals using a microCT scanner. *Med. Phys.* **34**(11): 4359–4367.

Hamstra, D. A., D. J. Rice et al. (1999). Combined radiation and enzyme/prodrug treatment for head and neck cancer in an orthotopic animal model. *Radiat. Res.* **152**(5): 499–507.

Hill, R., N. Chaudary et al. (2012). Studies of Radiochemotherapy in an Orthotopic Cervix Cancer Model. Symposium on novel targeting drugs and radiotherapy: From the bench to the clinic. In *ESTRO Meeting*, Toulouse, France.

Huda, W., H. A. Kissi et al. (2002). Optimizing the x-ray photon energy for digital radiographic imaging systems. *Proc. SPIE* **4682**: 633–644 .

Jaffray, D. A., J. H. Siewerdsen et al. (2002). Flat-panel cone-beam computed tomography for image-guided radiation therapy. *Int. J. Radiat. Oncol. Biol. Phys.* **53**(5): 1337–1349.

Kagadis, G., G. Loudos et al. (2010). In vivo small animal imaging: Current status and future prospects. *Med. Phys.* **37**: 6421–6442.

Kalender, W. (2005). *Computed Tomography: Fundamentals, System Technology, Image Quality, Applications.* Publicis Corporate Publishing, Erlangen, Germany.

Kaminski, J. M., E. Shinohara et al. (2005). The controversial abscopal effect. *Cancer Treat. Rev.* **31**(3): 159–172.

Khan, F. (1994). *The Physics of Radiation Therapy.* Lippincott Williams & Wilkins, Baltimore, MD.

Kiehl, E. L., S. Stojadinovic et al. (2008). Feasibility of small animal cranial irradiation with the microRT system. *Med. Phys.* **35**(10): 4735–4743.

Kirkby, C., C. Field et al. (2007). A Monte Carlo study of the variation of electron fluence in water from a 6 MV photon beam outside of the field. *Phys. Med. Biol.* **52**(12): 3563–3578.

Laissue, J. A., H. Blattmann et al. (2007). Prospects for microbeam radiation therapy of brain tumours in children to reduce neurological sequelae. *Dev. Med. Child Neurol.* **49**(8): 577–581.

Lambin, P., S. F. Petit et al. (2010). The ESTRO Breur Lecture 2009. From population to voxel-based radiotherapy: Exploiting intra-tumour and intra-organ heterogeneity for advanced treatment of non-small cell lung cancer. *Radiother. Oncol.* **96**(2): 145–152.

Lechtman, E., N. Chattopadhyay et al. (2011). Implications on clinical scenario of gold nanoparticle radiosensitization in regards to photon energy, nanoparticle size, concentration and location. *Phys. Med. Biol.* **56**(15): 4631–4647.

Lechtman, E., S. Mashouf et al. (2013). A Monte Carlo-based model of gold nanoparticle radiosensitization accounting for increased radiobiological effectiveness. *Phys. Med. Biol.* **58**(10): 3075–3087.

Lindsay, P., P. Granton et al. (2013). Multi-institutional dosimetric and geometric commissioning of image-guided small animal irradiators. *Med. Phys.* **41**(3): 031714.

Luker, G. D. and K. E. Luker. (2008). Optical imaging: Current applications and future directions. *J. Nucl. Med.* **49**(1): 1–4.

Ma, C. M., C. W. Coffey et al. (2001). AAPM protocol for 40–300 kV x-ray beam dosimetry in radiotherapy and radiobiology. *Med. Phys.* **28**(6): 868–893.

Maeda, A., M. K. Leung et al. (2012). In vivo optical imaging of tumor and microvascular response to ionizing radiation. *PLoS One* **7**(8): e42133.

Maeda, M., M. Tomita et al. (2010). Bystander cell death is modified by sites of energy deposition within cells irradiated with a synchrotron x-ray microbeam. *Radiat. Res.* **174**(1): 37–45.

Martinez-Rovira, I., J. Sempau et al. (2012). Monte Carlo-based treatment planning system calculation engine for microbeam radiation therapy. *Med. Phys.* **39**(5): 2829–2838.

Matinfar, M., E. Ford et al. (2009). Image-guided small animal radiation research platform: Calibration of treatment beam alignment. *Phys. Med. Biol.* **54**(4): 891–905.

Moiseenko, V., C. Duzenli et al. (2007). In vitro study of cell survival following dynamic MLC intensity-modulated radiation therapy dose delivery. *Med. Phys.* **34**(4): 1514–1520.

Mothersill, C. and C. Seymour. (2001). Radiation-induced bystander effects: Past history and future directions. *Radiat. Res.* **155**(6): 759–767.

Narayanan, M., C. Mochizuki et al. (2013). Monte Carlo simulation of a precision proton radiotherapy platform designed for small-animal experiments (abstract). In *AAPM 55*, Indianapolis, IN.

Newton, J., M. Oldham et al. (2011). Commissioning a small-field biological irradiator using point, 2D, and 3D dosimetry techniques. *Med. Phys.* **38**(12): 6754–6762.

Nikjoo, H. and L. Lindborg. (2010). RBE of low energy electrons and photons. *Phys. Med. Biol.* **55**(10): R65–R109.

Obenaus, A. and A. Smith. (2004). Radiation dose in rodent tissues during micro-CT imaging. *J. X-Ray Sci. Technol.* **12**(4): 241–249.

Paulus, M. J., S. S. Gleason et al. (2001). A review of high-resolution x-ray computed tomography and other imaging modalities for small animal research. *Lab. Anim. (NY)* **30**(3): 36–45.

Penagaricano, J. A., E. G. Moros et al. (2010). Evaluation of spatially fractionated radiotherapy (GRID) and definitive chemoradiotherapy with curative intent for locally advanced squamous cell carcinoma of the head and neck: Initial response rates and toxicity. *Int. J. Radiat. Oncol. Biol. Phys.* **76**(5): 1369–1375.

Pidikiti, R., S. Stojadinovic et al. (2011). Dosimetric characterization of an image-guided stereotactic small animal irradiator. *Phys. Med. Biol.* **56**(8): 2585–2599.

Raabe, A., H. P. Beck-Bornholdt et al. (2001). Impact of pulmonary metastases of the R1H-tumour on radiation tolerance of rat lung. *Int. J. Radiat. Biol.* **77**(9): 947–954.

Rehemtulla, A., L. D. Stegman et al. (2000). Rapid and quantitative assessment of cancer treatment response using in vivo bioluminescence imaging. *Neoplasia* **2**(6): 491–495.

Ritman, E. L. (2002). Molecular imaging in small animals—Roles for micro-CT. *J. Cell Biochem. Suppl.* **39**: 116–124.

Rodriguez, M. and R. Jeraj. (2008). Design of a radiation facility for very small specimens used in radiobiology studies. *Phys. Med. Biol.* **53**(11): 2953–2970.

Rodriguez, M., H. Zhou et al. (2009). Commissioning of a novel microCT/RT system for small animal conformal radiotherapy. *Phys. Med. Biol.* **54**(12): 3727–3740.

Sandison, G., E. Ford et al. (2013). A small animal proton IGRT facility for comparative studies of tumor and normal tissue responses. In *First Symposium on Precision Image-Guided Small Animal RadioTherapy*, Maastricht, the Netherlands.

Shahine, B., M. Al-Ghazi et al. (1999). Experimental evaluation of interface doses in the presence of air cavities compared with treatment planning algorithms. *Med. Phys.* **26**: 350–355.

Siewerdsen, J. H. and D. A. Jaffray. (1999). A ghost story: Spatio-temporal response characteristics of an indirect-detection flat-panel imager. *Med. Phys.* **26**(8): 1624–1641.

Slatkin, D. N., P. Spanne et al. (1992). Microbeam radiation therapy. *Med. Phys.* **19**(6): 1395–1400.

Song, K., R. Pidikiti et al. (2010). An x-ray image guidance system for small animal stereotactic irradiation. *Phys. Med. Biol.* **55**: 7345–7362.

Stewart, J. M., P. E. Lindsay et al. (2013). Two-dimensional inverse planning and delivery with a preclinical image guided microirradiator. *Med. Phys.* **40**(10): 101709.

Stojadinovic, S., D. A. Low et al. (2006). Progress toward a microradiation therapy small animal conformal irradiator. *Med. Phys.* **33**(10): 3834–3845.

Stojadinovic, S., D. A. Low et al. (2007). MicroRT-small animal conformal irradiator. *Med. Phys.* **34**(12): 4706–4716.

Syme, A., C. Kirkby et al. (2009). Relative biological damage and electron fluence in and out of a 6 MV photon field. *Phys. Med. Biol.* **54**(21): 6623–6633.

Tillner, F., P. Thute et al. (2012). Bildgeführtes Präzisionsbestrahlungsgerät für Kleintiere (SAIGRT)—Vom Konzept zur praktischen Umsetzung. *Jahrestagung der DGMP* **43**: 647–650.

Tryggestad, E., M. Armour et al. (2009). A comprehensive system for dosimetric commissioning and Monte Carlo validation for the small animal radiation research platform. *Phys. Med. Biol.* **54**(17): 5341–5357.

van Eerde, M. R., H. H. Kampinga et al. (2001). Comparison of three rat strains for development of radiation-induced lung injury after hemithoracic irradiation. *Radiother. Oncol.* **58**(3): 313–316.

van Hoof, S. J., P. V. Granton et al. (2012). Evaluation of a novel triple-channel radiochromic film analysis procedure using EBT2. *Phys. Med. Biol.* **57**(13): 4353–4368.

van Luijk, P., H. P. Bijl et al. (2005). Data on dose-volume effects in the rat spinal cord do not support existing NTCP models. *Int. J. Radiat. Oncol. Biol. Phys.* **61**(3): 892–900.

Verhaegen, F., P. Granton et al. (2011). Small animal radiotherapy research platforms. *Phys. Med. Biol.* **56**(12): R55–R83.

Verhaegen, F., B. Reniers et al. (2005). Dosimetric and microdosimetric study of contrast-enhanced radiotherapy with kilovolt x-rays. *Phys. Med. Biol.* **50**(15): 3555–3569.

Wang, G. and M. Vannier. (2001). Micro-CT scanners for biomedical applications: An overview. *Adv. Imaging* **16**: 18–27.

Weersink, R., P. Lindsay et al. (2011). Integration of bioluminescence imaging and cone-beam CT for image-guided small animal irradiation. *Med. Phys.* **38**(6): 3440.

Wiegman, E. M., H. Meertens et al. (2003). Loco-regional differences in pulmonary function and density after partial rat lung irradiation. *Radiother. Oncol.* **69**(1): 11–19.

Willekens, I., N. Buls et al. (2010). Evaluation of the radiation dose in micro-CT with optimization of the scan protocol. *Contrast Media Mol. Imaging* **5**(4): 201–207.

Wong, J., E. Armour et al. (2008). High-resolution, small animal radiation research platform with x-ray tomographic guidance capabilities. *Int. J. Radiat. Oncol. Biol. Phys.* **71**(5): 1591–1599.

Workman, P., E. O. Aboagye et al. (2010). Guidelines for the welfare and use of animals in cancer research. *Br. J. Cancer* **102**(11): 1555–1577.

Zheng, J., P. Lindsay et al. (2012). Use of small animal MRI and cone-beam CT for image-guided radiotherapy of orthotopic tumors in mice. In *103rd Annual Meeting of the American Association for Cancer Research. Cancer Res.* **72**(8 Suppl): Abstract nr 4060, Chicago, IL. AACR, Philadelphia, PA.

Zhou, H., M. Rodriguez et al. (2010). Development of a micro-computed tomography-based image-guided conformal radiotherapy system for small animals. *Int. J. Radiat. Oncol. Biol. Phys.* **78**(1): 297–305.

19

Dosimetry of Ionizing Radiation in Small Animal Imaging

Michael G. Stabin

19.1 FUNDAMENTALS OF RADIATION DOSIMETRY

The effects of ionizing radiation on living organisms are evaluated using the quantity "absorbed dose," which is the amount of energy deposited per unit mass of any material, in this case living tissue. A formal definition is given by the International Commission on Radiation Units and Measurements (1993):

$$D = \frac{d\bar{\varepsilon}}{dm} \qquad (19.1)$$

This depicts the mean energy $d\bar{\varepsilon}$ imparted by ionizing radiation in matter of mass dm. In most applications in radiation protection and nuclear medicine, dose is averaged over whole organs or tissues (e.g., liver, brain, active marrow). There are specialized applications in which three-dimensional dose distributions (Kolbert et al. 1997) or microscopic dose distributions (Rossi 1968) are of interest, but they are mostly related to research application, not daily practical dosimetry. The SI unit for quantifying absorbed dose is the gray (Gy), which is equal to 1 J of energy absorbed in 1 kg of matter (J/kg). In humans, another quantity, "equivalent dose," is defined as $H_{T,R} = w_R D_{T,R}$. Here, the quantity H is the equivalent dose (for tissue T and radiation type R), and w_R is a "radiation weighting factor," which is related to a quantity called the relative biological effectiveness (RBE); this expresses the ratio of doses of different radiation types needed to produce a given biological endpoint (e.g., 63% cell killing in an in vitro cell survival study).

In humans, another quantity is defined, called "effective dose." Certain organs and tissues in the body are assigned dimensionless "tissue weighting factors," w_T, which are proportional to their radiosensitivities for expressing fatal cancers or genetic effects. Current values recommended by the International Commission on Radiological Protection (2007) are provided in Table 19.1.

The sum of the weighted equivalent doses for a given exposure is the "effective dose" (E):

$$E = \sum_T H_T \times w_T \tag{19.2}$$

The effective dose is meant to represent the equivalent dose that, if received uniformly by the whole body, would result in the same total risk as that actually incurred by a given actual nonuniform irradiation. This is an example calculation for a theoretical nonuniform exposure to several organs and the resulting value of effective dose:

Organ	Weighting Factor	Equivalent Dose (mSv)	Weighted Dose Equivalent (mSv)
Liver	0.05	0.25	0.0125
Kidneys	0.005	0.31	0.00155
Ovaries	0.08	0.10	0.008
Red marrow	0.12	0.45	0.054
Bone surfaces	0.01	0.55	0.0055
Thyroid	0.05	0.17	0.0085
Total (effective dose)			0.090

The actual doses received by the organs are given in the third column. The effective dose of 0.09 mSv is a uniform whole-body dose, which would have theoretically resulted in the same overall risk as the actual nonuniform dose pattern that actually occurred. Use of effective dose allows simplification of dose reporting when different doses are received by different organs and the direct comparison of, for example, two different nuclear medicine cardiac imaging agents, or a nuclear medicine scan and a computed tomography (CT) examination. The weighting factors have substantial inherent uncertainties, and the overall calculation, of course, embodies all of these combined uncertainties, so its use should be made understanding these uncertainties. Also, effective dose should only be used to discuss risks to *populations*, not *individuals*, and it should not be used to discuss doses to any nonhuman species. Note that in the above calculation not all of the organ weighting factors were used; if one has a complete list of organ doses, all of the weighting factors may be used, but if only some organ doses are known, and the other doses are relatively small, this partial calculation gives a very reasonable value for E.

TABLE 19.1
Tissue Weighting Factors Recommended by the ICRP for the Calculation of "Effective Dose"

Organ or Tissue	Recommended Tissue Weighting Factor, w_T
Gonads	0.08
Red marrow	0.12
Colon	0.12
Lungs	0.12
Stomach	0.12
Bladder	0.04
Breasts	0.12
Liver	0.04
Esophagus	0.04
Thyroid	0.04
Skin	0.01
Bone surfaces	0.01
Brain	0.01
Salivary glands	0.01
Remainder	0.12

19.2 BIOLOGICAL EFFECTS FROM RADIATION EXPOSURES: TYPES OF EFFECTS, RELATION TO DOSE LEVELS, EXPERIENCE

The effects of ionizing radiation on living tissue are thought to be related to damage to DNA in the cell nucleus. Ionizing radiation may cause DNA damage via direct interaction with components of DNA ("direct" effects); however, it is known that far more effects are caused by ionization and excitation of water molecules, and subsequent interactions of formed free radicals with DNA ("indirect effects"). The most important species in indirect effects of radiation is the OH radical (OH•). Biological effects from ionizing radiation are broadly grouped into two categories, "nonstochastic effects" and "stochastic effects." Nonstochastic effects are generally observed relatively soon after exposure to radiation, compared to stochastic effects. The major identifying characteristics of nonstochastic effects are (Stabin 2007) that (1) there is a threshold of dose below which the effects will not be observed, (2) above this threshold, the *severity* of the effect increases with dose, and (3) the effects are clearly associated with the radiation exposure. Examples of nonstochastic effects include reddening of the skin (erythema), hair loss (epilation), depression of bone marrow cell division, physical deformities to the unborn child, or directly observable cell killing and damage to any major organ (e.g., ablation of the thyroid). If a person is exposed to a large dose of radiation in a uniform manner, there may be an immediate wave of nausea, vomiting, and diarrhea (the "NVD" syndrome), then a period of days to weeks when no symptoms are observed (the "latent period"), and then a period of manifestation of symptoms. A quantity known as the $LD_{50/30}$, which is the dose that has a 50% chance of resulting in lethality in a population within 30 days of exposure to radiation, is known to be around 3–4 Gy in humans. Damage to the bone marrow, with subsequent decreases in cell division and production of blood elements, may occur at doses as low as 0.1 Gy, but higher levels such as 3–4 Gy may result in death of the organism. At higher doses, damage to the gastrointestinal tract or central nervous system will occur, and the higher the dose, the shorter will be the latent periods and periods of manifest illness. Other species have different $LD_{50/30}$ values, for example, dogs and primates have values like humans, around 3–4 Gy, while rats, gerbils, and rabbits have higher values, between 7 and 10 Gy (Stabin 2007). Human skin may have a threshold of around 3 Gy for epilation and 6 Gy for erythema. Most human organs will begin to manifest observable damage with doses of 10–40 Gy.

For stochastic effects, the major characteristics are that (1) a threshold may not be observed, (2) the *probability* of the effect increases with dose, and (3) one cannot definitively associate the effect with the radiation exposure. Two important stochastic effects are cancer and hereditary effects (the latter, however, never have been clearly demonstrated in any human population, but have been demonstrated in some animal species, e.g., fruit flies and mice). The induction of cancer by ionizing radiation is one of the best studied effects of radiation, but much controversy remains about what effects, if any, radiation may cause at low doses and dose rates, such as are common in radiation protection and diagnostic medical imaging applications. Several kinds of cancer are known to be caused by high levels of radiation, most cataloged in the survivors of the atomic bombings in Japan in the 1940s. The latest report of the Committee on the Biological Effects of Ionizing Radiation (BEIR) of the National Academies of the Sciences (National Research Council 2006) states that for all cancers except leukemia, the data are "not inconsistent with" a linear, no-threshold (LNT) model, that is, a model that is a simple linear function down to zero dose, thus suggesting that any dose of radiation, no matter how small, carries some small risk of cancer induction (or promotion). By contrast, the French Academy of Sciences and the National Academy of Medicine expressed "doubts on the validity of using LNT for evaluating the carcinogenic risk of low doses (100 mSv) and even more for very low doses (10 mSv). The LNT concept can be a useful pragmatic tool for assessing rules in radioprotection for doses above 10 mSv; however since it is not based on biological concepts of our current knowledge, it should not be used without precaution for assessing by extrapolation the risks associated with low and even more so, with very low doses (10 mSv) …" (Aurengo et al. 2005). Some (Luckey 2008) even argue that low levels of radiation may be beneficial, as they may stimulate repair mechanisms within cells (sometimes referred to as "hormesis").

19.3 RADIATION DOSIMETRY FOR NUCLEAR MEDICINE IMAGING (SPECT, PET)

To estimate absorbed dose in a given organ of the body from a radiopharmaceutical administration, we will calculate the average amount of energy deposited per unit mass of any organ or tissue. A generic equation for the *absorbed dose rate* in an organ may be shown as

$$\dot{D} = \frac{kA \sum_i n_i E_i \phi_i}{m} \tag{19.3}$$

where
  $\dot{D}$ is the absorbed dose rate (Gy/s)
  A is the activity (MBq) in a "source" region (a region with a significant uptake of the compound)
  n is the number of radiations with energy E emitted per nuclear transition
  E is the energy per radiation (MeV)
  ϕ is the fraction of energy emitted from a source that is absorbed in the target
  m is the mass of target region (kg)
  k is a proportionality constant that incorporates appropriate unit conversions (Gy kg/MBq s MeV)

Integrating Equation 19.3 to obtain the total dose received is not difficult, as usually the only term that changes with time is the activity A. Activity is most often characterized by one or more exponential terms, as activity in biological systems is often cleared by first-order functions. Radioactive decay is, of course, an exponential function; the activity at any time can be given as

$$A(t) = A_0 e^{-\lambda t} \tag{19.4}$$

Here, λ is the "radioactive decay constant," which is the ratio of the natural logarithm of 2 over the physical half-life of the radioactive substance ($\lambda = \ln(2)/T_{1/2} = 0.693/T_{1/2}$). Many materials are also cleared from the body in a first-order manner; the quantity of material X in a system as a function of time may be shown as

$$X(t) = X_0 e^{-\lambda_b t} \tag{19.5}$$

where
   $X(t)$ is the amount of the *nonradioactive* substance at time t
   X_0 is the initial amount of substance X
   λ_b is the *biological* disappearance constant $= 0.693/T_b$
   T_b is the *biological* half-time for removal

The two combined effects are summed in an "effective disappearance constant":

$$\lambda_e = \lambda_b + \lambda_p \tag{19.6}$$

And it is easy to show that an "effective half-time" is given as

$$T_e = \frac{T_b \times T_p}{T_b + T_p} \tag{19.7}$$

The effective half-time is always shorter than the smaller of the two half-lives: for example,

$$T_b = 10 \text{ days} \quad T_p = 20 \text{ days} \quad T_{eff} = \frac{20 \times 10}{20 + 10} = 6.67 \text{ days}$$

$$T_b = 10 \text{ days} \quad T_p = 10 \text{ days} \quad T_{eff} = \frac{10 \times 10}{10 + 10} = 5 \text{ days}$$

Now, to integrate activity over time, assuming that activity is characterized by one exponential term, we have

$$\tilde{A} = \int_0^\infty A(t)dt = \int_0^\infty A_0 e^{-\lambda_e t}\, dt = \frac{A_0}{\lambda_e} = 1.443 A_0 T_e \tag{19.8}$$

If there are two or more exponential terms, it is easy to expand Equation 19.7 to have more than one term. Then, the integral of Equation 19.3 is

$$\dot{D} = \frac{k\tilde{A} \sum_i n_i E_i \phi_i}{m} \tag{19.9}$$

The units of $\tilde{A}$ are, for example Bq s, which are the number of disintegrations that have occurred in an organ with significant uptake of activity (1 Bq = 1 disintegration per second of activity). This term has been called "cumulated activity" by some (Loevinger et al. 1988). The term ϕ gives the fraction of energy emitted in a source region that is absorbed in that region or any other target region of interest.

Values of ϕ are calculated using mathematical models of the human body (also sometimes called "anthropomorphic phantoms"). Older models were based on combining simple geometric shapes to represent the various organs, the whole body, the skeleton, and so on (Cristy and Eckerman 1987). More recently, models have been derived based on medical images of real humans and animals. The RAdiation Dose Assessment Resource (RADAR) task group of the Society of Nuclear Medicine (SNM) has developed 15 human and 8 animal models for use in dose calculations (Shi et al. 2008; Keenan et al. 2010; Stabin et al. 2012). In the images, the models are represented as groups of "voxels" (small rectangular "volume elements"), and points in each organ are randomly sampled as starting locations for photons or electrons of various energies, and the particles' transport within the body are simulated using "Monte Carlo" methods. When all of a particle's energy is spent, locations where interactions occurred are noted, and after simulating many particles, the average values of the fraction of starting energy deposited in all of the body structures can be calculated; these are the values of ϕ that are needed for the calculation of dose. Then, all of the terms except the cumulated activity are often combined into "dose factors" (Stabin and Siegel 2003).

For a numerical example, if we administer 500 MBq of a substance labeled with ^{131}I to an adult female patient, 20% is taken up by the liver, and has an effective half-time of 15 h, the cumulated activity is

$$\tilde{A} = 1.443 \times 500 \text{ MBq} \times 0.2 \times 15 \text{ h} \times \frac{3600 \text{ s}}{\text{h}} = 7.8 \times 10^6 \text{ MBq s} \qquad (19.10)$$

The dose to the liver itself, the marrow, and ovaries of the patient are estimated by looking up the appropriate dose factors (these were taken from the OLINDA/EXM software (Stabin et al. 2005)):

$$D_{Liver} = 7.8 \times 10^6 \text{ MBq s} \times 2.86 \times 10^{-5} \frac{\text{mGy}}{\text{MBq s}} = 223 \text{ mGy}$$

$$D_{Marrow} = 7.8 \times 10^6 \text{ MBq s} \times 2.98 \times 10^{-7} \frac{\text{mGy}}{\text{MBq s}} = 2.32 \text{ mGy} \qquad (19.11)$$

$$D_{Ovaries} = 7.8 \times 10^6 \text{ MBq s} \times 1.58 \times 10^{-7} \frac{\text{mGy}}{\text{MBq s}} = 1.23 \text{ mGy}$$

Doses to other organs from the 20% in the liver and from the other 80% of the activity would be added to these to form a complete calculation for this administration. Effective doses for a number of common nuclear medicine studies are shown in Table 19.2.

TABLE 19.2
Values of Effective Dose for Some Nuclear Medicine Studies

Type of Study	Typical Activity Administered (MBq)	Radiopharmaceutical	Effective Dose (mSv)
Brain	740	^{99m}Tc-DTPA	3.6[a]
Bone	740	^{99m}Tc-MDP	4.2[a]
Lung	185	^{99m}Tc-MAA	2.0[a]
Lung	370	[133]Xe	0.26[a]
Kidney	740	^{99m}Tc-MAG3	5.2[a]
Heart	1100	^{99m}Tc-MIBI	9.9[a]
Heart	74	^{201}Tl-chloride	10[b]
Tumor, others	370	^{18}F-FDG	7.0[b]

[a] ICRP Publication 80 (Valentin 1998).
[b] ICRP Publication 106 (International Commission on Radiological Protection 2008).

19.4 RADIATION DOSIMETRY FOR CT IMAGING

Radiation doses in CT are somewhat more complicated to characterize, as CT systems pass along a predetermined length of the body, irradiating organs, while imaging systems with single detector rows (SDCT) or multiple detector rows (MDCT) rotate around the body, sometimes not irradiating whole organs uniformly. The rotation may be in an axial or helical manner; in the latter case, sequential image slices may have various values of "pitch," which is defined as the ratio of the travel of the imaging table per x-ray tube rotation to the width of the imaging slice (The American Association of Physicists in Medicine 2008). Currently, commercial MDCT systems can acquire up to 64 channels of data simultaneously. In either MDCT or SDCT systems, the tube current (given in milliampere-seconds or mAs) and exposure time per rotation determine the number of x-ray photons emitted by the imaging system. The radiation dose per mAs, however, varies between different CT scanner types, models, and kilovoltage potential (kVp) settings. The principal dose quantity in CT dose calculations is the "computed tomography dose index" (CTDI), given as

$$CTDI = \frac{1}{NT} \int_{-\infty}^{\infty} D(z)dz \qquad (19.12)$$

Here

N is the number of tomographic sections imaged in a single axial scan
T is the width of the tomographic section along the axis of rotation imaged by one data channel
$D(z)$ is the radiation dose profile along the axis of rotation (z)

The CTDI gives the average absorbed dose along the z-axis from a series of contiguous irradiations. The CTDI given over a 100 mm scan (the length of many commercially available "pencil" ionization chambers) is often cited:

$$CTDI_{100} = \frac{1}{NT} \int_{-50\,mm}^{50\,mm} D(z)dz \qquad (19.13)$$

A factor f, which is 0.94 rad/R for tissue, can relate the reading from a measured exposure in R:

$$CTDI_{100}\,(\text{rad}) = \frac{C \times f\,(\text{rad/R}) \times 100\,\text{mm ion chamber reading (R)}}{NT\,(\text{mm})} \qquad (19.14)$$

Here, C is the unitless conversion factor for the measuring chamber, which may often be very close to unity. The integrated dose along the length of a given scan is given by another quantity called the "dose–length product" (DLP):

$$DLP = \frac{N \times T \times \left[\frac{1}{3} CTDI_{100,center} + \frac{2}{3} CTDI_{100,edge} \right]}{I} \times \text{scan length (cm)} \qquad (19.15)$$

The two CTDI measurements are made in the center and edge of standard acrylic phantoms, and I is the table increment distance per axial scan (mm). Note that the pitch is $I/(N \times T)$. DLP is given in units of mGy, as for any general representation of absorbed dose, and represents the total energy

TABLE 19.3

Conversion Factors Recommended by the AAPM (the American Association of Physicists in Medicine 2008) to Convert DLP Values to Effective Dose Values for Adults of Standard Physique and Pediatric Patients of Various Ages

Region of Body	Newborn	1-Year-Old	5-Year-Old	10-Year-Old	Adult
Head/neck	0.013	0.0085	0.0057	0.0042	0.0031
Head	0.011	0.0067	0.004	0.0032	0.0021
Neck	0.017	0.012	0.011	0.0079	0.0059
Chest	0.039	0.026	0.018	0.013	0.014
Abdomen/pelvis	0.049	0.03	0.02	0.015	0.015
Trunk	0.044	0.028	0.019	0.014	0.015

TABLE 19.4
Typical Effective Dose Values for CT Exams

Exam Type	Effective Dose (mSv)
Head	1–2
Chest	5–7
Abdomen	5–7
Pelvic	3–4
Abdomen/pelvis	8–14
Coronary artery, calcium	1–3
Coronary angiography	5–15

Source: The American Association of Physicists in Medicine, The Measurement, Reporting, and Management of Radiation Dose in CT Report of AAPM Task Group 23 of the Diagnostic Imaging Council CT Committee, College Park, MD, 2008.

absorbed from a complete scan, which can be used to evaluate potential biological effects, as discussed earlier (The American Association of Physicists in Medicine 2008). However, the AAPM also provided factors that allowed conversion of DLP values to effective dose (Table 19.3). The AAPM also gave typical effective dose values for typical CT exams (Table 19.4).

19.5 COMBINED DOSIMETRY FOR PET/CT AND SPECT/CT IMAGING

Very commonly, images for a positron emission tomography (PET) or single photon emission computed tomography (SPECT) study are combined with a CT image, so the patient will receive dose from the radiopharmaceutical as well as the CT imaging session. Effective doses may be added directly, as the quantity represents the same concept for either an external or internal source. In both cases, there are considerable uncertainties, as noted earlier, but the convenience of the quantity becomes very clear in this case. So if a chest CT of perhaps 5 mSv is combined with a heart study using ^{99m}Tc-MIBI (see Tables 19.2 and 19.4), with an effective dose of about 10 mSv, the total effective dose is ~15 mSv. A number of recent events involving high CT exposures to patients, some of whom were children, have received widespread attention (Bogdanich 2009; Goldstein 2009). Some bodies have expressed interest in tracking of patient dose information over time and in reducing patient radiation doses as much as possible (Strauss et al. 2009). Data that show the risks of the stochastic effects discussed include doses over 100 mSv; such doses will be rare in common diagnostic imaging. However, for complicated cases involving multiple PET/CT or SPECT/CT studies over a short time frame (e.g., 5 or 10 PET/CTs over a few months), doses at this level could be encountered. We know that the direct benefits to the patient population significantly outweigh any theoretical cancer risks (Zanzonico and Stabin 2014). Nonetheless, these theoretical risks are thought to be proportional to a person's cumulative radiation exposure (National Research Council 2006), so it is always prudent to keep radiation exposures as low as is reasonably achievable (ALARA), for example, using lower mAs settings for smaller and thinner patients, and scaling radiopharmaceutical dosages based on patient weight. On the other hand, one of the worst things for the patient is to administer *too little radiation*, so that no diagnostic benefit is received and the study has to be repeated. Physicians and physicists should work together to ensure they are using reasonably accurate dosimetry information in making decisions about patient care and imaging, and make reasoned decisions about the possible risks and benefits for a given study in a given patient. Patients should be as knowledgeable as possible about all aspects of their medical care, including radiation dose, and should participate actively in the decision-making process. One should not be flippant about radiation doses, with the benefits so clearly outweighing the risks, but some have an unhealthy fear of some uses of radiation because of the sometimes hysterical information about radiation dose and risk distributed in popular media. Neither extreme attitude is helpful. We have very advanced and reasonably accurate techniques for calculating doses from radiopharmaceuticals and CT exams, and we should confidently rely on this information for the choices that we make, understanding the uncertainties inherent in the numbers and their application to any given patient.

19.6 CT AND PHARMACEUTICAL DOSIMETRY FOR SMALL ANIMALS

Small animal imaging focuses on testing of new imaging methods and pharmaceuticals to ultimately provide new technology for healthcare delivery to humans. Radiation doses to the animals themselves are of interest as well. Radiation doses in CT imaging are significantly higher than those for human CT imaging, as more photons are needed to obtain sufficient counts in the much smaller voxels into which the data are reconstructed. Doses to small animals for radiopharmaceuticals will be much higher than those for humans per MBq of administered activity, as the animal organ masses are so much smaller (see Tables 19.5 and 19.6; Keenan et al. 2010), based on the "MOBY" and "ROBY" models of Segars (Segars and Tsui 2007)), and dose is energy absorbed per unit mass, but of course much smaller amounts of activity will be administered to a small animal than to a human.

TABLE 19.5
Organ Masses in the Three Mouse Models in the RADAR Series

	Organ Mass (g)		
	25 g	**30 g**	**35 g**
Brain	0.466	0.568	0.666
Heart	0.235	0.291	0.342
Stomach	0.055	0.069	0.082
Small intestine	1.74	2.12	2.49
Large intestine	0.583	0.709	0.830
Kidneys	0.302	0.374	0.432
Liver	1.74	2.15	2.57
Lungs	0.087	0.107	0.131
Pancreas	0.305	0.378	0.450
Skeleton	2.18	2.61	3.01
Spleen	0.111	0.136	0.157
Testes	0.160	0.197	0.228
Thyroid	0.014	0.016	0.020
Bladder	0.060	0.075	0.088
Body	24.11	29.80	35.27

Source: Keenan, M.A. et al., *J. Nucl. Med.*, 51(3), 471, 2010.

TABLE 19.6				
Organ Masses in the Five Rat Models in the RADAR Series				
	Organ Mass (g)				
	200 g	**300 g**	**400 g**	**500 g**	**600 g**
Brain	1.57	2.32	3.16	3.93	4.54
Heart	1.80	2.64	3.55	4.39	5.28
Stomach	0.941	1.40	1.89	2.37	2.86
Small intestine	10.6	15.5	20.8	25.6	30.8
Large intestine	7.86	11.5	15.5	19.2	23.1
Kidneys	2.06	3.03	4.09	5.06	6.09
Liver	7.55	11.2	15.2	18.8	22.8
Lungs	0.594	0.884	1.21	1.50	1.82
Pancreas	0.368	0.535	0.732	0.908	1.10
Skeleton	15.3	22.0	29.2	35.2	38.7
Spleen	0.607	0.884	1.18	1.45	1.74
Testes	0.174	0.245	0.321	0.386	0.460
Thyroid	0.191	0.275	0.368	0.457	0.549
Bladder	0.475	0.682	0.916	1.12	1.34
Body	226	335	443	547	643

Source: Keenan, M.A. et al., *J. Nucl. Med.*, 51(3), 471, 2010.

We can repeat our earlier example for the 400 g rat model. The animal is about 400/70,000 times smaller than a human, or about 0.6%. If we simply adjust the administered activity by body mass, we would administer just about 3 MBq of a substance labeled with ^{131}I to the animal, and the cumulated activity and dose would be

$$\tilde{A} = 1.443 \times 3 \text{ MBq} \times 0.2 \times 15 \text{ h} \times \frac{3600 \text{ s}}{\text{h}} = 4.6 \times 10^4 \text{ MBq s}$$

$$D_{Liver} = 4.6 \times 10^4 \text{ MBq s} \times 2.06 \times 10^{-3} \frac{\text{mGy}}{\text{MBq s}} = 96 \text{ mGy} \tag{19.16}$$

Other investigators have developed animal phantoms using geometric shapes, similar to Cristy and Eckerman (1987) human phantoms; an overview of these models is presented by Keenan et al. (2010). The newer and more realistic image-based models are clearly more suitable for dosimetry as they better represent real organ geometries and overlap, so we will limit our discussion to results presented with this technology. Quantities such as DLP and CTDI are not usually quoted for animal studies, and the quantity "effective dose" is not applicable, as the tissue weighting factors are for expression of human cancers. Mostly, researchers have presented average organ doses for either CT or pharmaceutical dose calculations.

Figueroa et al. (2008) used thermoluminescent dosimeters (TLDs) to measure doses from microCT imaging, and estimated the average doses of about 80 mGy to most organs of mice. Visvikis et al. (2006) estimated doses 90 mGy to mice imaged with a microCT imaging system (70 kVp, 90 mAs) and doses between 1–5 mGy/MBq administered for ^{18}FDG, except for the bladder wall, which appeared to receive 50 mGy/MBq (they did not cite the actual activities administered). Taschereau and Chatziioannou (2007) measured distributions of ^{18}FDG, ^{18}FLT, and ^{18}F fluoride ion in mice and developed four different digital mouse phantoms based on the MOBY mouse phantom of Segars and Tsui (2007). They presented doses per an assumed 7.4 MBq injection of the compounds; the urinary bladder received the highest doses around 2–4 Gy, while other organs received between ~20 and 200 mGy, although for fluoride ion skeletal structures received up to 500 mGy. El Ali et al. (2012) evaluated the dosimetry of ^{124}I as sodium iodide, with administration of between 0.7 and 22 kBq to rats between 350 and 600 g. They calculated dose factors using Monte Carlo methods with the same MOBY model as Segars and Tsui (2007) but using thyroid volumes measured with microCT. Their thyroid doses per unit of administered activity were estimated to be between 5,000 and 10,000 mGy/MBq, resulting in actual thyroid doses of between 5 and 220 Gy, for the range of activities administered. Reasonable images of the animal thyroids were obtained with the lowest levels of activity administered. Padilla et al. (2008) presented a model of a "female hound cross" (~25 kg), which was imaged by CT and used to develop an anatomical model for dosimetry. Absorbed fractions for internal sources of photons and electrons were calculated; no actual dose estimates were provided. Johnson et al. (2011) measured distribution and retention of ^{99m}Tc-MDP in mice and estimated skeletal doses of 590 mGy from administration of 120 MBq of the agent using the Keenan et al. (2010) MOBY models. Most other organs received between 30 and 70 mGy, but the urinary bladder dose was reported at 3.7 Gy.

Clearly, more time and experience are needed in calculating doses to small animals and in relating these radiation doses to any observed effects. Experience with possible therapy agents would be most likely to result in situations in which actual nonstochastic effects might be observed. Of particular interest would be studies with alpha emitters or other high linear energy transfer radiation types to evaluate how to interpret these effects in the context of radiation (not tissue) weighting factors.

REFERENCES

Aurengo, A., D. Averbeck et al. (2005). Dose-effect relationships and estimation of the carcinogenic effects of low doses of ionizing radiation. *Acad. Sci.—Natl. Acad. Med.*

Bogdanich, W. (2009). Radiation overdoses point up dangers of CT scans. *New York Times.*

Cristy, M. and K. Eckerman. (1987). *Specific Absorbed Fractions of Energy at Various Ages from Internal Photons Sources.* Oak Ridge National Laboratory, Oak Ridge, TN.

El-Ali, H. H., M. Eckerwall et al. (2012). The combination of in vivo 124I-PET and CT small animal imaging for evaluation of thyroid physiology and dosimetry. *Diagnostics* **2**(2): 10–22.

Figueroa, S. D., C. T. Winkelmann et al. (2008). TLD assessment of mouse dosimetry during microCT imaging. *Med. Phys.* **35**(9): 3866–3874.

Goldstein, J. (2009). Hospital mistake gives patients radiation overdose. *Wall Street Journal.*

International Commission on Radiation Units and Measurements. (1993). Quantities and units in radiation protection dosimetry. ICRU Report No. 51, Bethesda, MD.

International Commission on Radiological Protection. (2007). Recommendations of the ICRP, ICRP publication 103. *Ann. ICRP* **37**.

International Commission on Radiological Protection. (2008). Radiation dose to patients from radiopharmaceuticals—Addendum 3 to ICRP publication 53. ICRP publication 106. *Ann. ICRP* **38**.

Johnson, L. C., R. W. Johnson et al. (2011). Radiation dose-based comparison of PET and SPECT for preclinical bone imaging. *Nucl. Sci. Symp. Med. Imaging Conf. Rec.* **2011**: 2845–2850.

Keenan, M. A., M. G. Stabin et al. (2010). RADAR realistic animal model series for dose assessment. *J. Nucl. Med.* **51**(3): 471–476.

Kolbert, K. S., G. Sgouros et al. (1997). Implementation and evaluation of patient-specific three-dimensional internal dosimetry. *J. Nucl. Med.* **38**(2): 301–308.

Loevinger, R., T. F. Budinger et al. (1988). *MIRD Primer for Absorbed Dose Calculations.* Society of Nuclear Medicine, New York.

Luckey, T. D. (2008). Atomic bomb health benefits. *Dose Response* **6**(4): 369–382.

National Research Council. (2006). *Health Risks from Exposure to Low Levels of Ionizing Radiation: BEIR VII Phase 2.* The National Academies Press, Washington, DC.

Padilla, L., C. Lee et al. (2008). Canine anatomic phantom for preclinical dosimetry in internal emitter therapy. *J. Nucl. Med.* **49**(3): 446–452.

Rossi, H. H. (1968). Microdosimetric energy distribution in irradiated matter. In F. H. Attix and W. C. Roesch (eds.), *Radiation Dosimetry.* Vol. I, *Fundamentals.* Academic Press, New York.

Segars, W. and B. Tsui. (2007). 4D MOBY and NCAT phantoms for medical imaging simulation of mice and men. *J. Nucl. Med. Meeting Abstracts* **48**(Suppl. 2): 203.

Shi, C. Y., X. G. Xu et al. (2008). SAF values for internal photon emitters calculated for the RPI-P pregnant-female models using Monte Carlo methods. *Med. Phys.* **35**(7): 3215–3224.

Stabin, M. G. (2007). *Radiation Protection and Dosimetry: An Introduction to Health Physics.* Springer, New York.

Stabin, M. G., M. A. Emmons et al. (2012). Realistic reference adult and paediatric phantom series for internal and external dosimetry. *Radiat. Protect. Dosim.* **149**(1): 56–59.

Stabin, M. G. and J. A. Siegel. (2003). Physical models and dose factors for use in internal dose assessment. *Health Phys.* **85**(3): 294–310.

Stabin, M. G., R. B. Sparks et al. (2005). OLINDA/EXM: The second-generation personal computer software for internal dose assessment in nuclear medicine. *J. Nucl. Med.* **46**(6): 1023–1027.

Strauss, K. J., M. J. Goske et al. (2009). Image Gently Vendor Summit: Working together for better estimates of pediatric radiation dose from CT. *AJR Am. J. Roentgenol.* **192**(5): 1169–1175.

Taschereau, R. and A. F. Chatziioannou. (2007). Monte Carlo simulations of absorbed dose in a mouse phantom from 18-fluorine compounds. *Med. Phys.* **34**(3): 1026–1036.

The American Association of Physicists in Medicine. (2008). The measurement, reporting, and management of radiation dose in CT Report of AAPM Task Group 23 of the Diagnostic Imaging Council CT Committee, College Park, MD.

Valentin, J. (1998). Radiation dose to patients from radiopharmaceuticals: (Addendum 2 to ICRP Publication 53) ICRP Publication 80 Approved by the Commission in September 1997. *Ann. ICRP* **28**.

Visvikis, D., M. Bardies et al. (2006). Use of the GATE Monte Carlo package for dosimetry applications. *Nucl. Instrum. Methods Phys. Res. A* **569**(2): 335–340.

Zanzonico, P. and M. Stabin. (2014) Quantitative benefit-risk analysis of medical radiation exposures. *Semin. Nucl. Med.* **44**(3): 210–214.

Zanzonico, P. and M. G. Stabin. Benefits of medical radiation exposures. http://hps.org/hpspublications/articles/Benefitsofmedradexposures.html (accessed September 10, 2015).

20

Treatment Planning
for Small Animals

Frank Verhaegen

20.1 INTRODUCTION

In Chapter 18, we discussed why small animal precision irradiation with photons is preferably done at kilovolt (kV) instead of megavolt (MV) energies. This is mostly to avoid extensive dose buildup and dose re-buildup regions near medium interfaces and to avoid large beam penumbras (see Chapter 18). Some animal studies were done in the past with MV beams, often without treatment planning or dosimetry. In many cases, no imaging was available. When using a treatment planning system (TPS) to calculate the complex 3D dose distribution for human radiotherapy, several problems are encountered. These TPSs use calculation models that are not intended or commissioned for small beams (usually not below 3 cm field size). Forcing them to calculate doses for small fields may lead to very wrong absolute machine radiation output. Among the few animal studies using small beams of high-energy photon beams accurately is that of DesRosiers et al. (2003), who irradiated rats with small beams from a GammaKnife device employing multiple ^{60}Co sources (1.25 MeV) with small collimators. They used a dedicated TPS for this device. A second problem arises because radiotherapy TPSs perform calculations in voxelized geometries derived from computed tomography (CT) images. In case where a human CT scanner is used to image a small animal, the voxel resolution may be too coarse (no better than 1 mm) to capture the details of the animal's anatomy; in case an animal CT scanner is used, the voxel resolution may be much better, but human radiotherapy TPSs are not equipped to handle very small voxels or very large numbers of voxels. Similar problems may arise with positron emission tomography (PET) or magnetic resonance imaging (MRI) when other forms of anatomical and/or functional imaging are required. A third problem is encountered when the animal irradiator employs kV photon beams. There is currently no clinical TPS that handles kV beams accurately. So, using a clinical TPS may lead to large uncertainties in small animal dose calculations.

This overview of inadequacies of radiotherapy TPS to calculate dose for small animals makes it clear that dedicated calculation techniques are needed. Recently, several methods have been proposed from relatively simple pencil beam type calculations to Monte Carlo photon transport simulations. The latter is arguably the most accurate dose calculation technique available under all conditions (Seco and Verhaegen 2013), but it also suffers from some sensitivities to, for example, the tissue composition. In the following sections, we will first highlight some issues related to small animal dose calculation, followed by an overview of some methods to calculate the dose. Next, a section will be devoted to tissue segmentation. Then, a topic intimately related to treatment planning will be discussed: dose verification. Finally, some considerations on future developments will be mentioned. This chapter focuses on methodologies and not on, for example, specific treatment plans for the study of certain radiation effects.

20.2 SPECIFIC ISSUES FOR SMALL ANIMAL RADIATION DOSE CALCULATION

Significant differences exist between treatment planning for radiotherapy research in small animals and clinical radiotherapy for human patients. Targeting small structures in animals (tumors, or subregions in them) requires often small beams, and we have already seen that kV x-rays provide sharper beams and a higher treatment accuracy over MV beams. In the commercially available devices, typically a broad x-ray spectrum of about 225 kV is used for radiotherapy studies with collimators of a fixed beam size between 0.5 mm and several centimeters, but other beam geometries and photon energies are also possible (see Chapter 18). In commercial devices, the x-ray tube is mounted on a movable support to enable irradiation of the specimen from different sides. The same x-ray tube, but with a lower kV setting and a broader field, typically encompassing a whole mouse or rat is used to create a cone beam CT image (CBCT) of the specimen. The CBCT imaging panel typically has a high resolution of about 10.0 μm (much better than human scanners), resulting in 3D CBCT images with a large number of voxels easily ranging in the tens of millions, imposing heavy demands on computer memory.

In the MV energy range of photons, there is basically only one type of interaction that is important to model: Compton scatter (for more background on photon interactions see, e.g., Attix 2004). In this process, the energy of the scattered photon and its flight direction can be significantly altered with respect to the incoming photon. A secondary electron is set in motion during the process, which can have a significant range, thereby depositing its energy at some distance from its starting point. In the kV energy range, Compton interactions are still important, but the range of the secondary electrons is much smaller. More importantly, a second process, photoelectric effect, becomes significant at low-photon energies. In this process, the incoming photon is absorbed by an atom, and its energy is transferred to an orbital electron (mostly from the K shell) minus its shell binding energy. In the process, one or more characteristic photons and low-energy Auger electrons may also be emitted. The probability of the photoelectric effect, almost negligible at high MV energies, becomes highly significant at low photon energies. For example, in water, for photon energies below about 25 keV, the photoelectric effect is the dominant process, whereas for (human) cortical bone, this is already the case for about 55 keV. This means that in kV photons the photoelectric effect cannot be ignored without losing dose calculation accuracy.

Figure 20.1b shows that the mass energy absorption coefficients for various human tissues differ significantly from that of water. This quantity is closely related to the absorbed radiation dose. Cortical bone and skeletal muscle differ by more than a factor of six in their energy absorption around 30 keV (for the mass attenuation coefficients, a measure for photon attenuation, the differences are slightly smaller). Assuming that these human data also apply to small animals, it is clear that assigning all media to water may lead to significant dose calculation errors due to ignoring the photoelectric effect whose cross section increases with increasing atomic number and decreasing photon energy, both to a power between 3 and 4. Animal tissue compositions are very hard to find in the literature; a rare old example can be found for dogs (Moran et al. 1992). Tissue analysis for

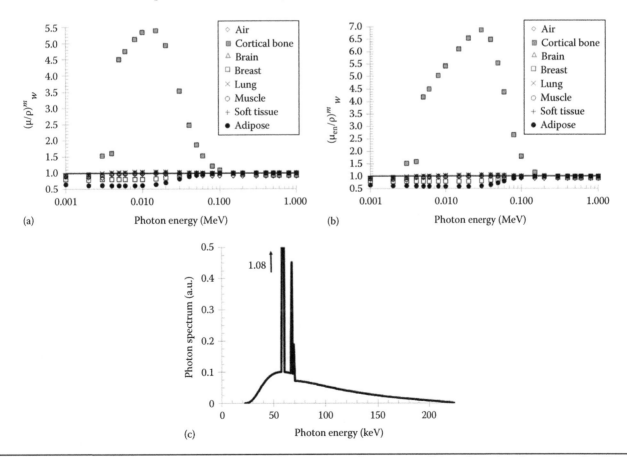

FIGURE 20.1   (a) Photon mass attenuation coefficients and (b) mass energy absorption coefficients for various (human) media of interest. (c) Typical 225 kVp x-ray spectrum used for small animal studies (area normalized to unity).

animals (or humans) with modern techniques such as inductively coupled plasma mass spectrometry does not seem to be present in the literature. A systematic determination of the chemical composition of various normal and cancerous tissues for animals (and humans) would probably benefit treatment planning, among other applications.

Even muscles may differ from water in their energy absorption by more than 3% at ~40 keV. In contrast, MV dose calculations are far less sensitive to the tissue composition. The largest differences for photons from ^{60}Co up to 24 MV are around 12% for cortical bone compared to water (Siebers et al. 2000). For all other media, differences with water are nearly negligible for practical purposes. It is therefore clear that to avoid excessive problems with heterogeneous dose distributions due to tissue absorption, photon energies above 100 keV should be used preferably for small animal radiotherapy research. Several irradiation devices that are now commercially available have opted for 225 kV as the x-ray tube potential (see Figure 20.1 and Chapter 18).

Due to the sensitivity of energy absorption in different tissues, it is, therefore, important that dose calculation for small animal irradiators in the kV range be implemented with great care. It should be emphasized that large dose enhancements in, for example, bone in kV beams do not correspond to the situation in MV beam treatments, which may invalidate some preclinical studies. Conversely, kV irradiators may be used for studies of differential tissue absorption.

Another issue closely related to the use of kV photons for small animal radiotherapy research is the relative biological effectiveness (RBE) of low-energy photons. It is known from the literature that low-energy photons (say, <50 keV) are more potent in causing various forms of biological damage in cells or living organisms than higher energy photons (Nikjoo and Lindborg 2010). In other words, the dose required to cause the same level of damage by low-energy photons is smaller than the dose needed to see the same level of damage for a higher energy photon quality. RBE for low-energy photons can easily reach values of 2 or more, whereas for MV photons all RBE values are essentially unity. A state-of-the-art treatment planning system should be able to provide information on the photon spectrum in various animal tissues, thereby allowing estimates of RBE values. All the issues discussed in this section are obviously of importance when translating findings from animal studies into human radiotherapy practice, so it is essential to be aware of them.

20.3 IMAGING INFORMATION NEEDED FOR TREATMENT PLANNING

20.3.1 CT IMAGING

Since accurate dose calculation in small animals for kV beams requires tissue knowledge, a 3D high-resolution image of the animal's anatomy is needed. The required spatial resolution should typically be of the order of 100–200 µm to ensure a sufficient volumetric accuracy to image anatomical structures, tumors, or subparts thereof. Modern commercial small animal irradiators come equipped with CBCT x-ray imaging equipment. Figure 20.2 shows a typical CBCT image of the same mouse alive and after postmortem plastination. CBCT images usually have somewhat lower imaging quality and resolution than a small animal CT scanner. Images may be affected by photon scatter, resulting in reduced imaging contrast, but this is less an issue in small animals than in human subjects where the larger scattering volume may lead to significant image degradation. The images in Figure 20.2 allow identification of many structures and are relatively free from CT artifacts.

A typical workflow for treatment planning involves the acquisition of a 3D CT or 3D CBCT image, which will consist of many small voxels. To perform a dose calculation, at the very least the tissue density needs to be assigned to every voxel. This is typically done by acquiring an image of a calibration phantom with known densities, and constructing a Hounsfield unit (HU) versus mass density (ρ) calibration curve. Figure 20.3 shows a CT image of a possible calibration phantom geometry with several inserts of known materials (Du et al. 2007).

Figure 20.4 depicts a typical HU versus ρ curve obtained with only three calibration materials in a phantom. A piecewise linear curve was drawn through these points. This curve serves to assign

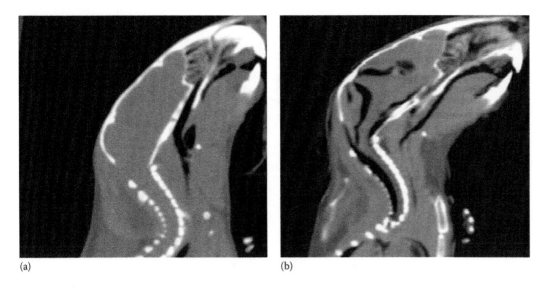

(a)   (b)

FIGURE 20.2   Sagittal plane reconstruction in a CBCT scan of the head section of the same mouse (a) alive and (b) after postmortem plastination. The plastination process preserves many anatomical details.

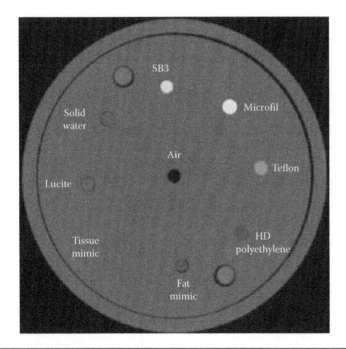

FIGURE 20.3   CT imaging of a typical calibration phantom for a small animal irradiator with eight different material inserts.

the appropriate density to each voxel with known HU. One should be aware that the calibration curve depends on the kV–filter combination (i.e., the radiation quality). For many types of dose calculation algorithms, for example, Monte Carlo radiation transport simulations, additional information is needed. These algorithms require knowledge of the tissue type. The traditional method to assign tissues to voxels is to subdivide the whole HU range into intervals that belong to the same tissue (Figure 20.4). The assignment into different tissue types is a topic of research to which we will return later in this chapter.

In a recent study (Yang et al. 2013), dual-energy CT imaging (Bazalova et al. 2008a,b; Landry et al. 2011a, 2011b) was used to optimally segment tissues in small animals while minimizing the imaging radiation dose, which is important for longitudinal studies that may involve repeated imaging with high doses of x-rays. This form of imaging offers the advantage over single-energy CT

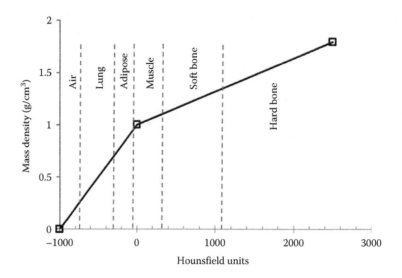

FIGURE 20.4 Typical HU versus mass density calibration curve. Only three calibration media were used (air at −1000 HU, water at 0 HU, and bone equivalent material at 2500 HU; the latter value depends on the material and scanner used).

imaging of an added material characteristic (often the atomic number Z) that can be extracted in addition to density. The method requires two consecutive interleaved or simultaneous exposures with two different x-ray spectra, which usually involves an increased dose compared to single energy CT. The (ρ, Z) information may be used to improve treatment planning for small animals. A recent study (Yang et al. 2013) reported that the dual-energy CT method outperforms single-energy CT for tissue segmentation. In addition, as long as proper kV combinations are used (e.g., 50–80 kV), radiation dose can be limited.

20.3.2 Coordinate Systems

If the onboard CT imager is used that is available in commercial small animal irradiators, then the CT image and the irradiation space are in the same coordinate system. In case a separate CT imager is used or another imaging device is employed, as discussed in the next section, a procedure to ensure a common system of coordinates is needed. This may be done, for example, through fiducial markers attached to the animal or to its carrier. This is a nontrivial issue, which will determine the irradiation accuracy to a large extent and therefore deserves attention. In radiotherapy clinics, this is taken care of by laser alignment systems on the imaging devices and accelerators and through the use of robust fiducial systems (markers or frames attached to the patient). For animal radiation research laboratories, this problem often needs to be solved for each institute. Radiotherapy images are also highly standardized via the DICOM (Digital Imaging and Communications in Medicine) imaging standard, which largely is absent in the small animal research community.

20.3.3 Alternative Imaging Methods

Other imaging methods may be required to extract information for treatment planning. Single-photon emission computed tomography (SPECT) or PET imaging may be used to distinguish metabolically highly active regions by using the ^{18}F-deoxyglucose (FDG) tracer. The same imaging modality may be used to detect hypoxic (low in oxygen) regions in tumors with tracers such as ^{18}F-fluoroazomycin-arabinoside (FAZA). Such information may serve to assess subregions of tumors that, for example, could be the subject of radiation boost studies.

Also, MRI imaging could be used in treatment planning in the new field of precision radiation research in small animals. Recently, an MRI-based workflow for irradiation of brain tumors was proposed by researchers from the University of Ghent, Belgium (Vanhove et al. 2013). This involved

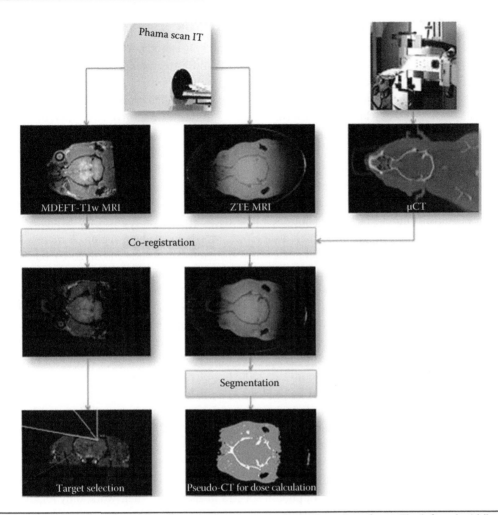

FIGURE 20.5 MRI-based workflow for assigning tissues to the brain region of a rat. The left and middle panels refer to different MRI image sequences, whereas the right panel pertains to the CBCT procedure on a small animal irradiator. (Courtesy of C. Vanhove, Ghent University, Ghent, Belgium.)

special MR sequences to generate pseudo-CT images with a limited number of tissue classes. Figure 20.5 shows their workflow; animal studies offer the opportunity to explore different imaging workflows compared to human radiotherapy.

Ultrasound imaging may also be used for the visualization of soft tissue structures for small animal treatment planning, but, to our knowledge, this modality has not yet been reported in combination with precision irradiators. There have been some recent reports on the use of specialized ultrasound techniques for the characterization of tissue damage after irradiation (Vlad et al. 2009, 2011). Specifically, for small animal studies also optical or photoacoustic imaging may be useful to visualize certain processes in the tumor or healthy tissue. Other chapters in this volume discuss various forms of small animal imaging, most of which may play a role in guiding treatment planning for preclinical studies or in assessing treatment outcome.

20.4 DOSE CALCULATION MODELS FOR SMALL ANIMALS

To perform accurate kV photon dose calculations in the many small voxels that constitute the small animal voxel phantom, in principle the only suitable technique is Monte Carlo simulation (Seco and Verhaegen 2013). Other techniques, with less accuracy, can be based, for example, on ray-tracing, superposition convolution, or other models. We will first describe some of the simpler methods, followed by Monte Carlo simulation.

20.4.1 ANALYTICAL METHOD FOR ^{192}Ir IRRADIATOR

In Chapter 18, the small animal irradiator based on the ^{192}Ir isotope at Washington University is discussed. It has a software interface for stage control and a treatment planning system (micro-RTP) that is derived from CERR, a MATLAB®-based in-house research treatment planning system (Deasy et al. 2003). The authors developed a fast analytical dose calculation scheme (based on results calculated with Monte Carlo techniques) that models a uniform water geometry, thereby ignoring tissue heterogeneities. The model uses a multiplication of factors that can be derived from measurements or Monte Carlo simulations, not unlike the older dose calculation models once commonly employed in radiotherapy. The energy of photons emitted from the ^{192}Ir isotope ranges from a few keV to 885 keV, with an average energy of 380 keV. Therefore, the differences in absorbed dose in the different mouse tissues are not as large as for kV x-rays (Figure 20.1), but still their dose calculation accuracy could be improved by taking nonwater heterogeneities into account. They estimate that the water-only assumption causes only small dose errors, mostly in the lung region. A more accurate approach taking into account heterogeneities was considered (Stojadinovic et al. 2007), but it is not a simple matter to adapt their parameterized model to include tissue heterogeneities. The fast algorithm was reported to result in dose errors limited to 10% in water, near beam edges and close to the entrance region for small fields (<10 mm). These researchers (Stojadinovic et al. 2006) reported a build-up region of less than 1.5 mm, and their papers indicate that the beam penumbra extends up to 3 mm. The Washington University researchers have also worked on developing an inline tandem combination of a micro cone beam CT scanner and a high-power 320 kV x-ray tube capable of providing a dose rate of 20 Gy/min (Hope et al. 2006; Izaguirre et al. 2009, 2010), for which probably another dose calculation approach may be needed.

A somewhat similar simple dose calculation model specifically designed for fast calculation of small arc segments has been reported (Marco-Rius et al. 2013) for kV irradiators. Tissue heterogeneities were approximately taken into account in this work by using estimates of radiological pathlengths of photons through tissues. These simplified approaches are inaccurate in non-water geometries and in cases where photon scatter is relevant. These methods only provide estimates of radiation dose and not, for example, of photon spectra in tissues. In their work, only relative dose distributions were demonstrated, whereas it has been shown that the absolute output of small radiation beams is much harder to model accurately (Granton et al. 2012; Granton and Verhaegen 2013).

20.4.2 SUPERPOSITION–CONVOLUTION DOSE CALCULATIONS

The Johns Hopkins group in Baltimore has developed an algorithm for dose calculation in small animals, based on their previous work for the implementation of a superposition–convolution algorithm for calculation for linear accelerators (Jacques et al. 2010, 2011). This method is centered on the calculation of terma, that is, the total energy released to matter, which is obtained by multiplying the local photon energy fluence (energy times fluence) by the energy absorption coefficient. The local energy fluence is derived from the primary energy fluence emanating from the photon source, corrected for inverse-square falloff with distance and photon attenuation. The method is essentially a ray-tracing approach, which also models first-order Compton scatter, which may not be a bad approximation in small animal irradiators. An assumption in the Johns Hopkins model is that the kernels are generated for different densities of water-like materials. Electron transport is included in the dose deposition kernel. For 225 keV electrons (the maximum electron energy for the commercial x-ray irradiators), for example, the electron range in most materials is limited to ~0.3 mm, but in lung the range can be a few millimeters, which can exceed several times the voxel dimensions.

The superposition–convolution algorithm can handle various grid resolutions, from coarse to fine. Their forward planning approach accepts a photon spectrum as input or phase space files derived from Monte Carlo simulations. Dose reporting is done as dose to water-in-medium, but can be converted to dose to medium-in-medium by multiplying the former by ratios of photon energy absorption coefficients of medium and water (Siebers et al. 2000; Landry et al. 2011c). The group also performed experimental validation by using multislab phantoms of plastic, cork, aluminum, and carbon with a radiochromic film (RCF) sandwiched in between. From these comparisons, it appeared that for some materials (mostly with higher atomic numbers than encountered in animals) correction factors for the superposition–convolution approach are needed, based on Monte Carlo simulations. It had already been reported that the superposition–convolution approach for kV photons may lead to problems in nonwater media where dose discontinuities arise (e.g., bone, lung) (Verhaegen et al. 2004).

The Johns Hopkins group recently described in a conference proceeding (Cho and Kazanzides 2012) the use of 3D Slicer to create an interface for the TPS and their SARRP (Small Animal Radiation Research Platform) system (see Chapter 24). The 3D Slicer is a popular freeware for medical image visualization and analysis (Pieper et al. 2004, 2012). Such a user-friendly interface is essential to allow a broad group of users (not necessarily experts in treatment planning) to efficiently produce radiation plans for small animals. The Baltimore group implemented the dose calculation on graphical processing units (GPUs) in the CUDA language for fast processing (subminute dose calculations). This dose calculation method is now commercially available (Xstrahl Ltd., Camberley, United Kingdom).

20.4.3   Monte Carlo Simulation

For more than 60 years, Monte Carlo particle simulation has been employed to faithfully reproduce real-life particle interactions. Since the late 1970s, the approach has become popular in radiotherapy as well (Andreo 1991; Verhaegen and Seuntjens 2003; Seco and Verhaegen 2013). The Monte Carlo technique models particle transport, interactions with other particles and fields, and production of secondary particles based on cross sections and transport theories. The latter nowadays offer such a degree of accuracy for the types of particles and the energy ranges encountered in medical physics that dose calculations can attain an accuracy of better than 1%. The Monte Carlo technique is often considered the best possible method to calculate dose distributions and related quantities such as photon spectra. It does not suffer from assumptions such as local electron energy deposition as was discussed in the previous section on the superposition–convolution method. The Monte Carlo method has been used to model a wide variety of radiation sources such as linear accelerators, brachytherapy sources, and proton beam lines, and it has also been applied successfully to model x-ray units (Verhaegen et al. 1999; Ding et al. 2008; Ding and Coffey 2009; Downes et al. 2009). Recently, the technique has also been employed to model small animal radiation research platforms (Tryggestad et al. 2009; Verhaegen et al. 2011; Granton et al. 2012). Note that the superposition–convolution method described earlier uses Monte Carlo simulations to determine the photon energy fluence emanating from the SARRP device (Tryggestad et al. 2009). The frequently heard complaint that Monte Carlo simulations are slow is now becoming less relevant with the advent of fast computers and parallel processing techniques and corresponding high-performance centers. Even GPU can now be used to run simulations very efficiently (Hissoiny et al. 2011a,b). This is only going to improve in the future, and therefore Monte Carlo simulations are now a serious option for many fields of applications.

A group in Toronto (Princess Margaret Hospital) has developed a graphical user interface, DOSCTP (Chow and Leung 2007, 2008), enabling basic treatment planning of kV beams of 0.5–5 cm diameter. The Monte Carlo code DOSXYZnrc (Walters et al. 2009) is employed within DOSCTP. The basic TPS is capable of creating 3D conformal plans; Figure 20.6 shows an example of an arc beam delivery. The results may be exported to a commercial TPS for more detailed analysis. DOSCTP was validated against a commercial treatment planning system for MV beams (Chow and Leung 2007) and later also for kV beams (Chow et al. 2010). The same authors studied the dose increase in mouse bone due to kV x-rays with Monte Carlo simulations in a slab phantom (Chow 2010). They found dose enhancements of over fivefold for 100 kV x-rays, at the entrance side of the beam. Proximity of lung was also found to have an effect on the bone dose. The authors later extended their study to real mouse geometries where hot spots in bone were reported, but very little effect due to the presence of lungs was noted for kV irradiations (Chow et al. 2010).

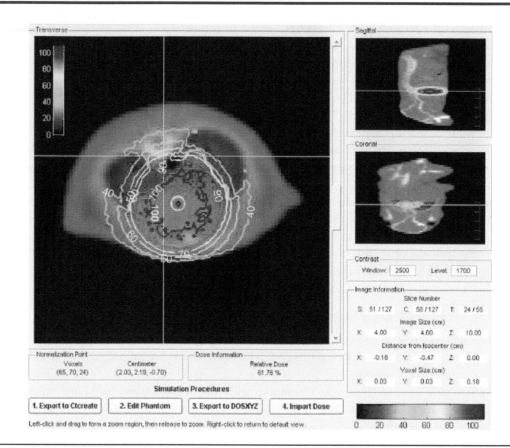

FIGURE 20.6 Arc irradiation of a small target in a mouse chest with 225 kVp x-rays, planned on a cone beam CT image. (Reproduced from Chow, J.C. and Leung, M.K., *Med. Phys.*, 34(12), 4810, 2007. With permission.)

The group at Johns Hopkins University has explored 3D kV Monte Carlo treatment planning for their SARRP platform. Their solution was to couple dose computation using DOSXYZnrc (Walters et al. 2009) with beam definition and dose display performed in the commercial TPS Pinnacle (Philips, Eindhoven, the Netherlands). An in-house developed C-code, interactively interfacing with the user via the Pinnacle user-scripting environment, is used to automatically generate the input files required for DOSXYZnrc: beam angle geometry and beam weighting, and the CT phantom creation. Once the dose is computed by DOSXYZnrc scripts, C-codes are used to automatically convert the output file into a Pinnacle-format dose image for display. The system uses a set of precomputed phase space files for each of the available SARRP nozzle collimators generated with BEAMnrc (Tryggestad et al. 2009). A basic case was shown as early as 2008 (Wong et al. 2008).

Around the same time, the Stanford group (Chapter 24) also developed a Monte Carlo model for their micro-irradiator, designed around a modified 120 kV micro-CT scanner. They provide information on how to optimize the variance reduction techniques in Monte Carlo particle transport (essentially, speed-up techniques) (Bazalova et al. 2009). Simulation times of up to about 100 h were reported to achieve a 1% statistical uncertainty. It is conceivable that larger uncertainties are acceptable for small animal radiotherapy. The Stanford group also used Monte Carlo simulations to investigate the dose conformality to small spherical and ellipsoidal targets in a phantom delivered by discrete beams spread out in an arc (Motomura et al. 2010). They studied centered and off-centered targets in a phantom, and also performed planning for a tumor in a mouse lung. They showed dose distributions that were close to human lung tumor cases. To compare dose distributions, often dose volume histograms are employed, a tool borrowed from human radiotherapy. These display in a cumulative manner the volume that at least received a certain dose. Figure 20.7 shows an example for a target and a nontarget region. Dose volume histograms have great value in optimizing and comparing dose plans.

Recently, a group in Maastricht (the Netherlands) developed a stand-alone TPS SmART-Plan (Small Animal RadioTherapy planning system) (van Hoof et al. 2013). It is a MATLAB program

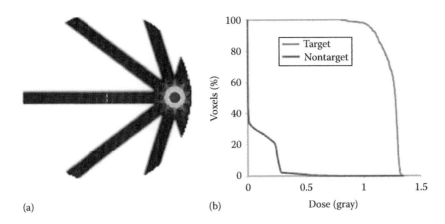

FIGURE 20.7   (a) Dose distribution in an axial plane of a cylindrical phantom from five small beams centered on a target in a discrete arc configuration. The high dose region (red) is the target. (b) Dose volume histogram for the target (blue, upper curve) and the area outside the target (nontarget, red, lower curve). (Reproduced from Motomura, A.R. et al., *Med. Phys.*, 37(2), 590, 2010. With permission.)

that follows the same workflow as a clinical TPS but is much simplified compared to the latter. Figure 20.8 shows the workflow of SmART-Plan. The workflow is kept very simple and robust; the tasks on the left have to be performed sequentially with the option of retracing to an earlier step to make corrections. First, the DICOM-compliant images of an animal of an imaging modality (e.g., CT) are uploaded, possibly resampled on another grid, and then they are processed to assign densities and materials to each voxel in the CT to density and material (CT2MD) step. The end result of this step is a mathematical phantom that can be used for Monte Carlo simulations. Next, structures such as tumors, healthy organs, etc., are delineated, and the structure contours are stored, for example, for subsequent dose analysis. Then the radiation beams are defined, which can be static or arcs, resulting in a treatment plan and the input file for the simulations. Next the dose calculation is performed with the DOSXYZnrc Monte Carlo code, where parallel processing techniques are used to run the calculations efficiently on a multicore workstation. Upon completion of the calculation, dose analysis can be performed in terms of, for example, dose volume histograms. Finally a radiation device control file is written (.ini file, at the moment of writing specifically for the PXI X-RAD 225Cx device), which allows to move the animal stage, position, and the beams and irradiate the specimen. SmART-Plan is commercially available (PXI Inc., North Branford, CT).

Figure 20.9 shows a validation case for SmART-Plan. A plastified mouse (PlastiMouse™) was sliced into two, and a piece of RCF was inserted in between the two parts. A certain dose was planned with SmART-Plan, delivered with an X-RAD 225Cx device, and compared to the measured film dose. This case demonstrates that, for a realistic mouse phantom, the agreement between calculated and measured dose is good, with significant dose differences present only in the small beam penumbra region for a 5 mm collimator. For voxels with doses between 80% and 100% of the maximum dose in the slice containing the RCF, no voxels exceeded a dose difference of more than 10% with respect to the maximum dose in that slice.

Another aspect of the work of the Maastricht group is the detailed Monte Carlo model they built for their irradiator, including a sophisticated model for the electron focal spot on the x-ray target (Granton and Verhaegen 2013). Figure 20.10 shows on the left side the detailed Monte Carlo simulation model with a 2D intensity distribution for the electron source. This was derived from focal spot size measurements, with a pinhole camera recording the emitted photons from the target. This kind of level of detailed modeling is required to reproduce the radiator output for especially the smallest fields of 0.5 to a few millimeters, where the focal spot size is of a size comparable to that of the beam aperture. It is well known that modeling such small beams presents many challenges (Aspradakis et al. 2010).

In contrast to Monte Carlo dose calculations in animals, which require only a few minutes nowadays, simulations of full treatment devices are still time consuming. In the same paper, Granton and

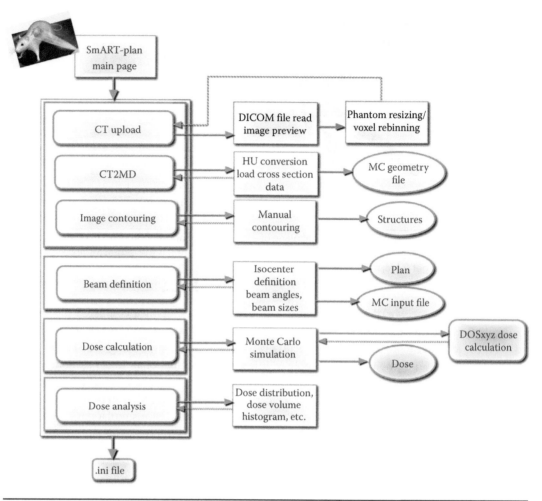

FIGURE 20.8 Flowchart of SmART-Plan. The tasks depicted on the left-hand side are to be performed sequentially top/down, from CT upload to Dose analysis and writing the device control (.ini) file. (Reproduced from van Hoof, S.J. et al., *Radiother. Oncol.*, 109(3), 361, 2013. With permission.)

Verhaegen (2013) presented an analytical photon source model (Figure 20.10b) that allows speeding up of the generation of an output phase space file. This file contains millions of photons normally generated by a full Monte Carlo simulation of the radiation device, and is to be used for the specimen dose calculation. By replacing the device Monte Carlo simulation with an analytical method, speedups of more than 1000 can be achieved.

Finally, the efforts to build a Monte Carlo simulation platform based on the Geant4 Monte Carlo code for animal dose calculation from the Dresden consortium (Tillner et al. 2012) deserves mention. They interfaced the Geant4 code with an in-house system to perform contouring and setting up beams, and this is also an example of a TPS developed for a specific type of animal irradiator (see Chapter 18). So far, no animal TPS is capable of handling various types of irradiators.

20.5 ISSUES RELATED TO TREATMENT PLANNING FOR SMALL ANIMALS

20.5.1 SPECIMEN PHOTON SCATTER

Photon scatter in large geometries, such as radiotherapy patients for MV photon beams, is a well-understood phenomenon. Also, in kV photon beams and the smaller geometries encountered in small animals, photon scatter may be an issue, but it is not a well-studied phenomenon, especially for the new

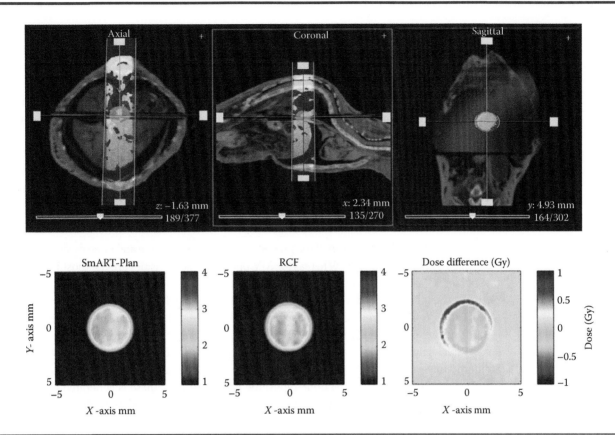

FIGURE 20.9   2D CT slices containing target location and dose distribution through the head region of a PlastiMouse phantom (top row). The bottom row shows the calculated (SmART-Plan) and measured (RCF, radiochromic film) dose distributions in the sagittal plane and their difference (calculated – measured).

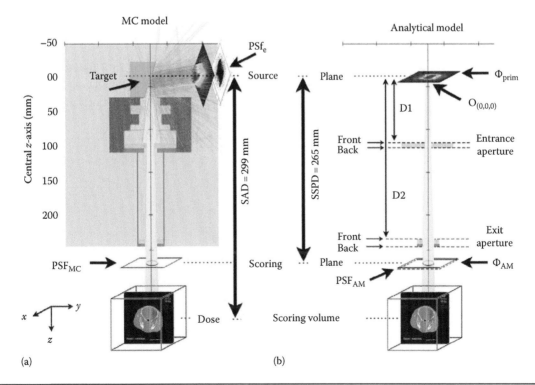

FIGURE 20.10   (a) Full Monte Carlo model of an x-ray animal irradiator including detailed model for primary electron source. (b) Analytical model to derive phase space file (PSF) with photons for animal dose calculation. (Reproduced from Granton, P.V. and Verhaegen, F., *Phys. Med. Biol.*, 58(10), 3377, 2013. With permission.)

small field irradiators. The scatter may prevent accurate dose calculations from being performed, other than by Monte Carlo methods, and they may have a deleterious effect on the imaging contrast for the onboard CT imagers. A recent study (Noblet et al. 2013) showed that for a large field animal irradiator (160 kVp; Faxitron CP-160, Tucson, AZ), the surface dose to a small 2.8 cm-diameter phantom could be overestimated by 15% when the conventional x-ray dosimetry protocol AAPM TG-61 (Ma et al. 2001) is used when compared to GATE Monte Carlo simulations of the same setup. At the exit side of the phantom, the dose overestimation rose to more than 30%. The authors attributed this to the significantly different photon scatter conditions in small and large geometries (for which TG-61 is valid). It has to be pointed out that this study mostly focused on the lack of scatter for larger beams irradiating a very small phantom. For the small field precision irradiators, no studies have been published yet, to our knowledge.

20.5.2   TISSUE SEGMENTATION FOR SMALL ANIMALS

In Section 20.3.1, we briefly outlined the procedure to assign soft tissue and bone media to voxel geometries, based on mass densities. Whereas for MV photons voxel densities are the most important information, for kV photons, also the effective atomic numbers of the tissues must be known to take the photoelectric interactions into account, which largely dominate the energy deposition in the kV energy range. Large dosimetry errors can be made if, especially, bone and adipose are misassigned (Figure 20.1).

For Monte Carlo simulations in MV photons, commonly only four media are assigned in most studies (air, lung, soft tissue, and a single bone type, usually cortical bone). In recent work from Stanford University (Zhou et al. 2009) focused mostly on human geometries, the researchers advocated using 47 different artificial bone types; the composition of these is generated by realizing, for example, that the relative weight fraction of Ca and P correlates well with bone density in bone tissues (Figure 20.11). The 47 bone types were needed to limit errors in calculated terma to 2%. In a more recent study from the same group (Bazalova and Graves 2011), the influence of tissue assignment on Monte Carlo dose calculations was studied in detail for different radiation sources: kV x-ray units of 120 kV (2.5 mm Al), 225 kV (4 mm Al), 225 kV (0.5 mm Cu), 320 kV (1.5 mm Pb, 5 mm Sn, 1 mm Cu, 4 mm Al), and a brachytherapy ^{192}Ir source (the information in brackets is the added filtration). The authors point out that there is no clear relationship between absorbed dose in tissue and mass density of the tissues for photon energies ≤ 225 kV. In other words, also information on effective atomic number is needed to achieve accurate dose calculations.

They compared a simple 4-tissue assignment scheme to an 8-tissue scheme and a 39-tissue scheme. They interpolated the compositions of the all-human ICRU-44 tissue database (1989) to make artificial new media, thereby also adjusting the weight fractions of Ca and H to match the

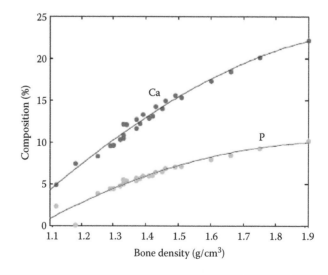

FIGURE 20.11   Relationship between relative weight fraction of Ca and P and bone density in human bone tissues. (Reproduced from Zhou, H. et al., *Med. Phys.*, 36(3), 1008, 2009. With permission.)

effective atomic numbers. They reported large dose errors when the simpler tissue assignment schemes were used in a cylindrical phantom with simple tissue inserts. A dose discrepancy of up to 27% was found for a 120 kV x-ray unit irradiating soft tissue. For a treatment plan of a mouse lung irradiation, they also noticed large local dose errors when using simple tissue assignment strategies, but the target dose volume histogram (in the lung) was fairly insensitive to tissue misassignment in the surrounding regions, whereas larger errors might occur in the organs at risk. The largest dose errors were obtained when adipose was incorrectly defined as muscle or cartilage. These studies clearly show the importance of correct tissue recognition. One should be aware that all studies published so far assigned human tissues to animals, for lack of information on animal tissue compositions. A better approach may be dual-energy CT imaging, where effective atomic numbers or even chemical compositions to some extent can be determined directly (Landry et al. 2011a, 2011b, 2013).

20.5.3 Dose Reporting

A matter related to the tissue assignment discussed in the previous section is dose reporting. Accurate dose calculation for kV photons requires that the specific tissues are taken into account when photons are transported through them. Modeling the whole geometry as water, even with the proper densities, would lead in many cases to unacceptable dose errors. Dose reporting is a different matter, though. The most natural way for Monte Carlo simulations to report dose is in terms of dose to medium-in-medium ($D_{m,m}$). This means that the particles are transported in the proper medium, and when it comes to scoring their dose contribution in a voxel, the material interaction coefficients of the medium are used. For other dose calculation algorithms mentioned in this chapter, dose to water-in medium ($D_{w,m}$) may be the quantity scored. To obtain $D_{w,m}$, the photon interaction coefficients of water are used. One should realize that $D_{m,m}$ and $D_{w,m}$ are two entirely different quantities, although they are both absorbed doses. This issue also exists for MV photon beams but there the numerical difference between the two dose quantities is limited to about 12% in cortical bone and much smaller in most other tissues. Conversion of one of the dose quantities into the other involves multiplication by ratios of mass stopping powers for water and the medium for high-energy photons (Siebers et al. 2000), or by multiplication by ratios of energy absorption coefficients for low-energy photons (Landry et al. 2011). More complex conversions are also possible depending on the so-called cavity theory one invokes (Tedgren and Carlsson 2013). In the case of low-energy photons, the differences between $D_{m,m}$ and $D_{w,m}$ can be very large, as we have already seen in Figure 20.1.

A reason why these two, potentially confusing, dose quantities exist next to each other (Beaulieu et al. 2012; Enger et al. 2012; Tedgren and Carlsson 2013) is that one or the other is thought to correlate better with biological radiation effects, but that it is currently unknown which one correlates best. Another reason is that, although $D_{m,m}$ may be a more natural quantity to calculate, especially for Monte Carlo methods, $D_{w,m}$ is easier to measure since much of the knowledge about dose measurements is based on the medium (water). One should also be aware that conversion of one quantity into the other adds uncertainty. This discussion is ongoing in, for example, brachytherapy (Beaulieu et al. 2012). For small animal dosimetry, one should strive for clarity when reporting dose distributions to allow for meaningful comparisons.

20.5.4 Commissioning TPS

Commissioning of small animal irradiators and their dedicated TPS may require special measurement procedures. We refer to the literature (Tryggestad et al. 2009; Lindsay et al. 2013; Wack et al. 2013) and manufacturer's instructions for this specialized subject.

20.6 DOSE VERIFICATION

Photon dose calculations in the kV range are quite difficult to perform and sensitive to tissue composition. The small beams used in precision irradiation studies also may lead to differences between calculated and real dose distributions. In human radiotherapy, some techniques exist to verify the delivered dose distribution. An increasingly popular technique is based on portal dosimetry using the imaging panel opposite the linear accelerator beam exit from the patient. The MV photons penetrating the patient interact with the imaging panel to form a portal image. The photons contributing to the image contain information on the absorbed dose in the patient. Methods have been published to convert the portal images into dose distributions into the plane of the portal imager and also into 3D dose distributions inside the patient (van Elmpt et al. 2008, 2009). The Maastricht University group has developed a method to compare 2D portal images with the onboard imager in their X-RAD

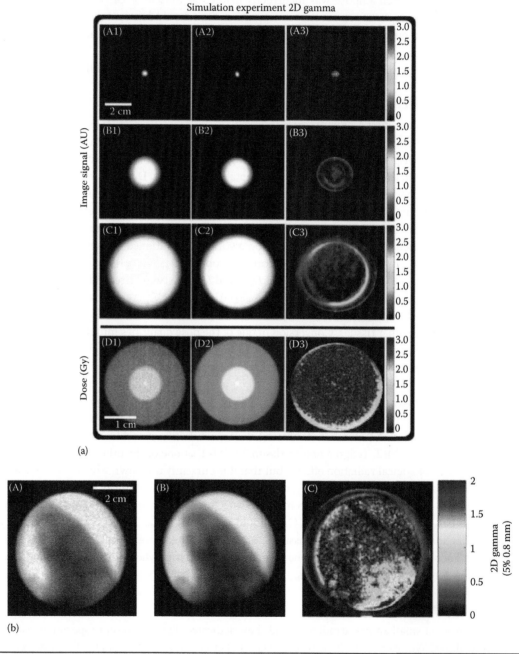

FIGURE 20.12 (a) Simulated and measured images at the level of the onboard imager for three collimator sizes (1, 4, and 6 mm). Last column shows gamma distributions with acceptance criteria (5%; 0.8 mm) to measure dose and distance-to-agreement differences. The last row shows results for the compound field consisting of all three circular fields. In this case, only small differences in the beam penumbras are visible. (b) Simulated onboard image (A) of a mouse irradiation (nose on top left, shoulder and front paw at bottom) with a 25 mm-diameter beam at the isocenter (the beam was chosen quite large to also show the mouse anatomy in the imaging field) and (B) measured onboard image. Also the gamma (5%; 0.8 mm) difference image is shown. In this case, the animal shows a slight motion of the shoulder region between planning and irradiation time points, leading to dose differences. (Reproduced from Granton, P.V. et al., *Med. Phys.*, 39(7), 4155, 2012. With permission.)

225Cx to precalculated images. Figure 20.12 illustrates this for a compound field existing of three circular fields without specimen in the beam (a), and for a mouse irradiation with a 25 mm-radius beam (b). Their study shows the feasibility of onboard dosimetry in small animal research platforms, for which conversion of the portal images to dose distributions is required. Full 3D dose reconstruction for small animals based on onboard imaging should in principle also be possible.

20.7 FUTURE DEVELOPMENTS

It is clear that precision image-guided small animal radiation research has a bright future. Many discoveries potentially benefiting large groups of patients is what is being hoped for. Precise and accurate radiation planning software and versatile radiation delivery platforms need to be developed to achieve the full potential of these research platforms. These may then empower animal studies, which may be translated successfully to human radiotherapy. Treatment planning systems are only now becoming available that can handle the complexities of the combination of small beams, low-energy photons, and small geometric structures in animals. Also, other beam modalities than low-energy photons may appear on the scene of precision irradiation animal studies, such as low-energy protons (Narayanan et al. 2013; Sandison et al. 2013). Clearly, intensive development of planning and dose calculation methods for various types of radiation beams is required. Equally, there is a need for beam delivery verification techniques, for example, based on onboard imagers, especially due to the difficulties in low-energy photon dose calculations.

To advance treatment planning for small animals, many techniques can be borrowed from the mature field of treatment planning for human radiotherapy. To conclude this chapter, we will briefly discuss an application in inverse treatment planning for animals, a technique that is commonly applied in its human counterpart. Researchers at the Princess Margaret Cancer Centre (Toronto,

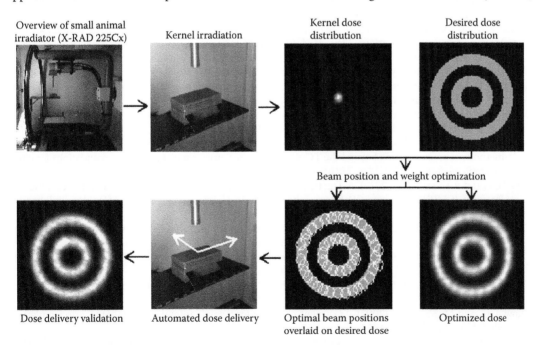

FIGURE 20.13 Overview of the inverse optimization method. Clockwise from top left: Self-shielded small animal irradiator with x-ray tube and flat-panel detector mounted on C-arm gantry, solid water phantom on animal positioning stage, and interchangeable collimator. Setup for empirically measured dose kernel, with radiochromic film placed at depth in a solid water phantom. Based on the kernel dose distribution and the desired dose distribution, the framework determined an optimal set of beam positions and weights that minimized the difference between the optimized and desired dose distributions. The beam positions and weights were then converted to a set of automated stage motions for delivery. Finally, delivery was validated with radiochromic film measurements. (Courtesy of James Stewart, Princess Margaret Cancer Center, Toronto, Ontario, Canada.)

Canada) have developed a 2D dose optimization framework based on the RCF measurements of the interchangeable collimator X-RAD 225Cx small animal irradiator. The technique determined a set of beam positions and weights (or beam-on times) to optimally deliver arbitrary planar dose distributions at a constant depth. During the optimization process, dose was calculated in a linear superposition manner using a convolution of the beam positions, multiplied by their respective weights, with the measured dose kernel. The optimization (Stewart 2013) minimized the difference between the calculated and desired dose summed across the entire distribution. The optimal beam positions and weights that minimized this difference were then converted to a set of animal stage motions, which linearly moved the stage through the set of all beam positions and pausing at each position according to the respective beam weight.

Figure 20.13 shows the inverse planning framework and its validation for a dose distribution from a 5 mm-radius bulls-eye dose distribution. Optimization of the distribution required 2.1 min, and automated delivery of five repeated irradiations showed a 0.3 mm targeting accuracy with dosimetric delivery error of 3.9%.

These and many other developments in the field of small animal treatment planning will be available to the research community in the near future. Planning and radiation dose delivery to small animal models of unprecedented accuracy will be available at a comparable level to human radiotherapy.

REFERENCES

ICRU-44. (1989). Tissue substitutes in radiation dosimetry and measurement. ICRU Report 44, Bethesda, MD.

Andreo, P. (1991). Monte Carlo techniques in medical radiation physics. *Phys. Med. Biol.* **36**(7): 861–920.

Aspradakis, M. M., J. Byrne et al. (2010). Small field MV photon dosimetry. IPEM Report 103, York, U.K.

Attix, F. (2004). Gamma and x-ray interactions in matter. In *Introduction to Radiological Physics and Radiation Dosimetry*. John Wiley & Sons, New York, pp. 124–159.

Bazalova, M., J. F. Carrier et al. (2008a). Dual-energy CT-based material extraction for tissue segmentation in Monte Carlo dose calculations. *Phys. Med. Biol.* **53**(9): 2439–2456.

Bazalova, M., J. F. Carrier et al. (2008b). Tissue segmentation in Monte Carlo treatment planning: A simulation study using dual-energy CT images. *Radiother. Oncol.* **86**(1): 93–98.

Bazalova, M. and E. E. Graves. (2011). The importance of tissue segmentation for dose calculations for kilovoltage radiation therapy. *Med. Phys.* **38**(6): 3039–3049.

Bazalova, M., H. Zhou et al. (2009). Kilovoltage beam Monte Carlo dose calculations in submillimeter voxels for small animal radiotherapy. *Med. Phys.* **36**(11): 4991–4999.

Beaulieu, L., A. Carlsson Tedgren et al. (2012). Report of the Task Group 186 on model-based dose calculation methods in brachytherapy beyond the TG-43 formalism: Current status and recommendations for clinical implementation. *Med. Phys.* **39**(10): 6208–6236.

Cho, N. P. and P. Kazanzides. (2012). A Treatment Planning System for the Small Animal Radiation Research Platform (SARRP) based on 3D Slicer. Johns Hopkins University, Baltimore, MD. http://hdl.handle.net/10380/3364.

Chow, J. (2010). Depth dose dependence of the mouse bone using kilovoltage photon beams: A Monte Carlo study for small-animal irradiation. *Radiat. Phys. Chem.* **79**(5): 567–574.

Chow, J. and M. Leung. (2008). A graphical user interface for calculation of 3D dose distribution using Monte Carlo simulations. *J. Phys. Conf. Ser.* **102**: 012003.

Chow, J., M. Leung et al. (2010). Dosimetric variation due to the photon beam energy in the small animal irradiation: A Monte Carlo study. *Med. Phys.* **37**(10): 5322–4329.

Chow, J. C. and M. K. Leung. (2007). Treatment planning for a small animal using Monte Carlo simulation. *Med. Phys.* **34**(12): 4810–4817.

Deasy, J. O., A. I. Blanco et al. (2003). CERR: A computational environment for radiotherapy research. *Med. Phys.* **30**(5): 979–985.

DesRosiers, C., M. S. Mendonca et al. (2003). Use of the Leksell Gamma Knife for localized small field lens irradiation in rodents. *Technol. Cancer Res. Treat.* **2**(5): 449–454.

Ding, G. X. and C. W. Coffey. (2009). Radiation dose from kilovoltage cone beam computed tomography in an image-guided radiotherapy procedure. *Int. J. Radiat. Oncol. Biol. Phys.* **73**(2): 610–617.

Ding, G. X., D. M. Duggan et al. (2008). Accurate patient dosimetry of kilovoltage cone-beam CT in radiation therapy. *Med. Phys.* **35**(3): 1135–1144.

Downes, P., R. Jarvis et al. (2009). Monte Carlo simulation and patient dosimetry for a kilovoltage cone-beam CT unit. *Med. Phys.* **36**(9): 4156–4167.

Du, L. Y., J. Umoh et al. (2007). A quality assurance phantom for the performance evaluation of volumetric micro-CT systems. *Phys. Med. Biol.* **52**(23): 7087–7108.

Enger, S. A., A. Ahnesjo et al. (2012). Dose to tissue medium or water cavities as surrogate for the dose to cell nuclei at brachytherapy photon energies. *Phys. Med. Biol.* **57**(14): 4489–4500.

Granton, P. V., M. Podesta et al. (2012). A combined dose calculation and verification method for a small animal precision irradiator based on onboard imaging. *Med. Phys.* **39**(7): 4155–4166.

Granton, P. V. and F. Verhaegen. (2013). On the use of an analytic source model for dose calculations in precision image-guided small animal radiotherapy. *Phys. Med. Biol.* **58**(10): 3377–3395.

Hissoiny, S., B. Ozell et al. (2011a). GPUMCD: A new GPU-oriented Monte Carlo dose calculation platform. *Med. Phys.* **38**(2): 754–764.

Hissoiny, S., B. Ozell et al. (2011b). Validation of GPUMCD for low-energy brachytherapy seed dosimetry. *Med. Phys.* **38**(7): 4101–4107.

Hope, A., S. Stojadinovic et al. (2006). A prototype rotational immobilization system for a proposed static-gantry microRT device with tomographic capabilities (abstract). *Med. Phys.* **33**: 2272.

Izaguirre, E., H. Chen et al. (2010). Implementation of a small animal image guided microirradiator: The microIGRT (abstract). *Med. Phys.* **37**: 3457.

Izaguirre, E., B. Kassebaum et al. (2009). Preclinical image guided microirradiators: Concepts, design and implementation (abstract). *Med. Phys.* **36**: 2720.

Jacques, R., R. Taylor et al. (2010). Towards real-time radiation therapy: GPU accelerated superposition/convolution. *Comput. Meth. Programs Biomed.* **98**(3): 285–292.

Jacques, R., J. Wong et al. (2011). Real-time dose computation: GPU-accelerated source modeling and superposition/convolution. *Med. Phys.* **38**(1): 294–305.

Landry, G., P. V. Granton et al. (2011a). Simulation study on potential accuracy gains from dual energy CT tissue segmentation for low-energy brachytherapy Monte Carlo dose calculations. *Phys. Med. Biol.* **56**(19): 6257–6278.

Landry, G., K. Parodi et al. (2013). Deriving concentrations of oxygen and carbon in human tissues using single- and dual-energy CT for ion therapy applications. *Phys. Med. Biol.* **58**(15): 5029–5048.

Landry, G., B. Reniers et al. (2011b). Extracting atomic numbers and electron densities from a dual source dual energy CT scanner: Experiments and a simulation model. *Radiother. Oncol.* **100**(3): 375–379.

Landry, G., B. Reniers et al. (2011c). The difference of scoring dose to water or tissues in Monte Carlo dose calculations for low energy brachytherapy photon sources. *Med. Phys.* **38**(3): 1526–1533.

Lindsay, P. E., P. V. Granton et al. (2013). Multi-institutional dosimetric and geometric commissioning of image-guided small animal irradiators. *Med. Phys.* **41**(3): 031714.

Ma, C. M., C. W. Coffey et al. (2001). AAPM protocol for 40–300 kV x-ray beam dosimetry in radiotherapy and radiobiology. *Med. Phys.* **28**(6): 868–893.

Marco-Rius, I., L. Wack et al. (2013). A fast analytic dose calculation method for arc treatments for kilovoltage small animal irradiators. *Phys. Med.* **29**(5): 426–435.

Moran, J. M., D. W. Nigg et al. (1992). Macroscopic geometric heterogeneity effects in radiation dose distribution analysis for boron neutron capture therapy. *Med. Phys.* **19**(3): 723–732.

Motomura, A. R., M. Bazalova et al. (2010). Investigation of the effects of treatment planning variables in small animal radiotherapy dose distributions. *Med. Phys.* **37**(2): 590–599.

Narayanan, M., C. Mochizuki et al. (2013). Monte Carlo simulation of a precision proton radiotherapy platform designed for small-animal experiments (abstract). In *AAPM 55*, Indianapolis, IN.

Nikjoo, H. and L. Lindborg. (2010). RBE of low energy electrons and photons. *Phys. Med. Biol.* **55**(10): R65–109.

Noblet, C., S. Chiavassa et al. (2013). Underestimation of dose delivery in preclinical irradiation due to scattering conditions. *Phys. Med.* **30**(1): 63–68.

Pieper, S., M. Halle et al. (2004). 3D slicer. In *First IEEE International Symposium on Biomedical Imaging: From Nano to Macro*, Boston, MA.

Pinter, C., A. Lasso et al. (2012). SlicerRT: Radiation therapy research toolkit for 3D slicer. *Med. Phys.* **39**(10): 6332–6338.

Sandison, G., E. Ford et al. (2013). A small animal proton IGRT facility for comparative studies of tumor and normal tissue responses. In *First Symposium on Precision Image-Guided Small Animal RadioTherapy*, Maastricht, the Netherlands.

Seco, J. and F. Verhaegen. (eds.). (2013). *Monte Carlo Techniques in Radiation Therapy.* Taylor & Francis, CRC Press, Boca Raton, FL.

Siebers, J. V., P. J. Keall et al. (2000). Converting absorbed dose to medium to absorbed dose to water for Monte Carlo based photon beam dose calculations. *Phys. Med. Biol.* **45**(4): 983–995.

Stewart, J. (2013). Two-dimensional inverse planning and delivery for precision preclinical radiobiological investigations. In *First Symposium on Precision Image-Guided Small Animal RadioTherapy*, Maastricht, the Netherlands.

Stojadinovic, S., D. A. Low et al. (2006). Progress toward a microradiation therapy small animal conformal irradiator. *Med. Phys.* **33**(10): 3834–3845.

Stojadinovic, S., D. A. Low et al. (2007). MicroRT-small animal conformal irradiator. *Med. Phys.* **34**(12): 4706–4716.

Tedgren, A. C. and G. A. Carlsson. (2013). Specification of absorbed dose to water using model-based dose calculation algorithms for treatment planning in brachytherapy. *Phys. Med. Biol.* **58**(8): 2561–2579.

Tillner, F., P. Thute et al. (2012). Bildgeführtes Präzisionsbestrahlungsgerät für Kleintiere (SAIGRT)—Vom Konzept zur praktischen Umsetzung. *Jahrestagung der DGMP* **43**: 647–650.

Tryggestad, E., M. Armour et al. (2009). A comprehensive system for dosimetric commissioning and Monte Carlo validation for the small animal radiation research platform. *Phys. Med. Biol.* **54**(17): 5341–5357.

van Elmpt, W., L. McDermott et al. (2008). A literature review of electronic portal imaging for radiotherapy dosimetry. *Radiother. Oncol.* **88**(3): 289–309.

van Elmpt, W., S. Nijsten et al. (2009). 3D in vivo dosimetry using megavoltage cone-beam CT and EPID dosimetry. *Int. J. Radiat. Oncol. Biol. Phys.* **73**(5): 1580–1587.

van Hoof, S. J., P. V. Granton et al. (2013). Development and validation of a treatment planning system for small animal radiotherapy: SmART-Plan. *Radiother. Oncol.* **109**(3): 361–366.

Vanhove, C., B. Descamps et al. (2013). MRI-based workflow for radiotherapy planning on a small animal radiation research platform. Society of Nuclear Medicine and Molecular Imaging, Vancouver, British Columbia, Canada.

Verhaegen, F., P. Granton et al. (2011). Small animal radiotherapy research platforms. *Phys. Med. Biol.* **56**(12): R55–R83.

Verhaegen, F., A. E. Nahum et al. (1999). Monte Carlo modelling of radiotherapy kV x-ray units. *Phys. Med. Biol.* **44**(7): 1767–1789.

Verhaegen, F., C. Schulze et al. (2004). Heterogeneity corrected convolution dose calculation for kV x-rays versus Monte Carlo simulation. In *46th Annual Meeting of the American Association of Physicists in Medicine*, Pittsburgh, PA.

Verhaegen, F. and J. Seuntjens. (2003). Monte Carlo modelling of external radiotherapy photon beams. *Phys. Med. Biol.* **48**(21): R107–R164.

Vlad, R. M., S. Brand et al. (2009). Quantitative ultrasound characterization of responses to radiotherapy in cancer mouse models. *Clin. Cancer Res.* **15**(6): 2067–2075.

Vlad, R. M., M. C. Kolios et al. (2011). Ultrasound imaging of apoptosis: Spectroscopic detection of DNA-damage effects at high and low frequencies. *Meth. Mol. Biol.* **682**: 165–187.

Wack, L., W. Ngwa et al. (2013). High throughput film dosimetry in homogeneous and heterogeneous media for a small animal irradiator. *Phys. Med.* **30**(1): 36–46.

Walters, B., I. Kawrakow et al. (2009). DOSXYZnrc users manual. NRCC Report PIRS-794revB. Ottawa Ionizing Radiation Standards National Research Council of Canada, Ottawa, Ontario, Canada.

Wong, J., E. Armour et al. (2008). High-resolution, small animal radiation research platform with x-ray tomographic guidance capabilities. *Int. J. Radiat. Oncol. Biol. Phys.* **71**(5): 1591–1599.

Yang, C.-C., J.-A. Yu et al. (2013). Optimization of the scan protocols for CT-based material extraction in small animal PET/CTstudies. *Nucl. Instrum. Methods Phys. Res. A* **731**: 299–304.

Zhou, H., P. J. Keall et al. (2009). A bone composition model for Monte Carlo x-ray transport simulations. *Med. Phys.* **36**(3): 1008–1018.

21

Radiolabeled Agents for Molecular Imaging and/or Therapy

Dimitrios Psimadas and Eirini A. Fragogeorgi

maging agents hold much promise for diagnosing disease, monitoring disease progression, tracking therapeutic response, and enhancing our knowledge of physiology and pathophysiology. It is an indisputable fact that unique criteria direct the identification of opportunities for the discovery and the development of new single-photon emission computed tomography (SPECT) and positron emission tomography (PET) agents. Although the procedure parallels the common drug development pathway, special requirements need to be fulfilled during the process.

Radiolabeled agent development has the flexibility to pursue functional or nonfunctional targets as long as they play a role in the specific disease or mechanism of interest and meet image ability requirements. However, it is true that their innovation is tempered by relatively small diagnostic imaging agent markets, intellectual property challenges, radiolabeling constraints, and adequate target concentrations for imaging. At the same time, preclinical imaging is becoming a key translational tool for proof of mechanism and concept studies.

Experimental molecular imaging of small animal models is rapidly expanding as it provides an efficient tool to study biological processes noninvasively. Radiolabeled tracer development and application can provide nondestructive imaging information, allowing time-related phenomena to be repeatedly studied in a single animal. Recently, there has been great progress in related technologies and a number of dedicated imaging systems have overcome the resolution limitations associated with imaging small animals.

Nuclear medicine is the most widely spread molecular imaging medical specialty, where radioisotope-based tracers are used. The main steps in molecular imaging with radioisotopes are as follows: (1) design of molecules that can target specific receptors, (2) attachment of radioisotopes that emit photons (for SPECT) or positrons (for PET) without altering the biological properties of the target molecules, and (3) *in vivo* administration and imaging, with the use of high resolution and sensitivity devices.

Central for the quality of any imaging strategy is the specificity and sensitivity of the method. Important for high specificity is the absence of cross-reactivity with nontarget molecular structures and lack of accumulation in nonspecific compartments in order to avoid false-positive results. High sensitivity is determined by the contrast, which, in turn, is dependent on a high ratio between radioactivity concentrations in the tumor and normal tissue in order to avoid false negatives. An optimal tracer molecule for the imaging of solid tumors should thus accumulate with high specificity in tumor tissues and not in healthy tissues. The nonbound tracer should furthermore be eliminated rapidly from the blood pool and other nontarget compartments, while being retained for an extended period of time in the target site to achieve maximal contrast. Just recently, a mechanistic model for predicting the relationship between the molecular size and affinity of a tracer molecule and tumor uptake has been developed (Schmidt and Wittrup 2009). It has been shown that intermediate-sized targeting agents (~25 kDa), such as single-chain variable fragments (scFvs), have the lowest tumor uptake and would thus not be suitable as imaging agents, while higher tumor uptakes may be reached with both smaller and larger tracer molecules. Furthermore, a high affinity for the target is important for high retention of smaller proteins in the tumor, and also small peptides and proteins reach their maximum tumor uptake considerably faster than large macromolecules. Interestingly, the model also predicts, consistent with experimental observations, that for large molecules (e.g., nanoparticles and liposomes), affinity to the target has only little influence on tumor uptake, and that tumor accumulation may be attributed mainly to enhanced permeability and retention effects (Kirpotin et al. 2006). Since imaging on the same day of injection is generally considered favorable, both for patient comfort and in matching with the short half-life of many of the radionuclides (e.g., PET isotopes), the time to reach maximum contrast is also considered to be an important factor.

A number of different tracer formats, ranging in size from small molecules and peptides to large macromolecules, have been investigated and are currently under development in preclinical and clinical phases. Some of the radionuclide-based molecular imaging agents that are already being used in nuclear medicine practice today are summarized in Table 21.1.

In this chapter, radiolabeled molecules that are currently used in the biomedical imaging research field are discussed. Latest developments in monoclonal antibodies (mAbs) and small

TABLE 21.1
SPECT/PET Molecular Imaging Agents Used in Clinical Practice for Cancer Management

SPECT/PET Imaging Agent	Trade Name	Target	Indication
[^{18}F]FDG	MetaTrace FDG®	Glucose transporters	Tumor cell proliferation and metabolism
[^{111}In]Pentetreotide	OctreoScan™	Somatostatin receptors	Neuroendocrine tumors
[^{111}In]Ibritumomab tiuxetan	Zevalin®	CD20	Non-Hodgkin's lymphoma
[^{111}In]Satumomab pendetide	OncoScint®	TAG-72	Colorectal, ovarian carcinoma
[^{111}In]Capromab pendetide	ProstaScint®	PSMA	Prostate carcinoma
[^{99m}Tc]Nofetumomab	Verluma®	EpCAM	Small-cell lung cancer
[^{99m}Tc]Acritumomab	CEA-Scan®	CEA	Colorectal carcinoma

regulatory peptides radiolabeled with a variety of SPECT and PET radioisotopes are summarized here under the prism of small animal imaging application for *in vivo* preclinical evaluation.

21.1 RADIOLABELING

The development of a tracer for radionuclide-based molecular imaging is based on the choice of a suitable radioisotope as well as the labeling method or conjugating moiety, which can all have substantial influence on the *in vivo* performance of a tracer (Tolmachev and Orlova 2010). It is important that the physical half-life of the radioisotope matches the biological half-life of the tracer, and additionally it is long enough to allow time for the accumulation of the tracer in the target tissue. However, radioisotopes with unnecessarily long half-lives should be avoided in order to limit the dose received by the patient. Large macromolecules with long circulation times, like antibodies, are usually labeled with relatively long-lived radioisotopes like ^{111}In (T1/2 = 67.2 h), while more short-lived radioisotopes, like ^{18}F (T1/2 = 110 min) and ^{99m}Tc (T1/2 = 6 h), are more suitable for the labeling of rapidly cleared peptides.

Another important issue is that the radioisotope of choice should be stably attached to the ligand substance for at least the duration of the imaging session and that it does not dissociate *in vivo*. The low stability of the radioactive complex could lead to high unspecific uptake of radioactivity in healthy tissue and low levels of accumulated radioactivity at the target site, which thereby reduces image quality. The attachment of the radioisotope to the molecule is preferably site-specifically directed at defined residues but can also be randomly distributed within the molecule's structure. Other important factors to consider are the speed of the labeling procedure, especially when short-lived PET radioisotopes are used, and the conditions under which the labeling is carried out which cannot be very harsh (high temperatures, extreme pH, etc.) because of the risk of destroying the secondary structure, and thereby binding activity, of the ligand substance. Radioisotopes with suitable characteristics for diagnostic purposes are presented in Table 21.2.

TABLE 21.2
"Diagnostic" Radioisotopes

Radionuclide	Production/Availability	Decay	E_γ/E_β^+ (keV)
^{18}F	Cyclotron	β^+	633.5
^{64}Cu	Cyclotron	Electron capture (EC), β^-, and β^+	γ, 1345.8; β^-, 578; β^+, 651
^{67}Ga	Cyclotron	EC	γ, 93.3, 184.6, and 300.2
^{68}Ga	Cyclotron	EC and β^+	γ, 1077.3; β^+, 1899
^{89}Zr	Cyclotron	EC and β^+	γ, 909.2; β^+, 395.5
^{99m}Tc	^{99}Mo/^{99m}Tc generator	Isomeric transition (IT)	γ, 142.7
^{111}In	Cyclotron	EC/Auger	γ, 171.3 and 245.4
^{123}I	Cyclotron	EC	γ, 127, 159

TABLE 21.3
"Therapeutic" Radioisotopes

Radionuclide	Production/Availability	Decay	E_β^-/E_α (MeV)
⁹⁰Y	⁹⁰Sr/⁹⁰Y generator	β⁻	β⁻, 2.27
¹⁵³Sm	Nuclear reactor	β⁻ and γ	β⁻, 0.69, 0.64; γ, 0.1032, 0.0697
¹⁶⁶Ho	Nuclear reactor	β⁻ and γ	β⁻, 1.855, 1.773; γ, 0.0806, 1.3794
¹⁷⁷Lu	Nuclear reactor	β⁻ and γ	β⁻, 0.497; γ, 0.2084, 0.1129
¹⁸⁸Re	¹⁸⁸W/¹⁸⁸Re generator	β⁻ and γ	β⁻, 2.118, 1.962; γ, 0.155
¹⁸⁶Re	Nuclear reactor	β⁻ and γ	β⁻, 1.071, 0.933; γ, 0.1372
²¹¹At	Accelerator	EC, α, and γ	γ, 0.687, 0.6696; α, 5.868
²¹²Bi	²²⁴Ra/²¹²Bi generator	β⁻, γ, and α	β⁻, 2.251; γ, 0.7273; α, 6.051

The concept of specific cell targeting by systemic delivery of cytotoxic radiation via a targeting agent is an attractive way to increase the delivered dose to the tumor cells and decrease adverse effects in nontarget tissues. This is a critical issue in cancer treatment. Depending on the size and the physiology of the targeted tumors, radioisotopes with different modes of decay may be suitable. The concept is that they should emit particle radiation that can destroy tumor cells either from a relative distance of some millimeters (β particles) or by entering the nucleus (α particles, Auger electrons). The most common "therapeutic" radioisotopes are presented in Table 21.3.

21.2   MONOCLONAL ANTIBODIES

Antibodies are proteins produced by the B cells. They have two ends: one end sticks to proteins on the outside of white blood cells and the other end sticks to an antigen (e.g., the germ or a damaged cell) and helps to kill it. The end of the antibody that sticks to the white blood cell is always the same and it is called the constant end. The end of the antibody that recognizes antigens varies depending on the cell it is designed to recognize and it is called the variable end (Figure 21.1). Each B cell makes antibodies with a different variable end. In the concept that cancer cells are not normal cells, there are antibodies that recognize them by specifically targeting antigens expressed on the cancer cell membrane or ligands.

MAbs represent a lot of copies of one type of antibody and are designed to recognize and attach to specific antigens on the cell surface. Each mAb recognizes one particular antigen, so they work in different ways depending on the antigen they are targeting. For tumor targeting, the major advantages of these targeted molecules are their tumor specificity and low toxicity profile. Development of the hybridoma technique (Kohler and Milstein 1975) allowed the isolation of large quantities of antibodies with predefined specificity. With the identification of the tumor-associated target antigens,

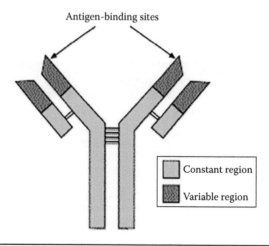

FIGURE 21.1   Typical structure of an antibody.

real progress has been made in developing treatment and/or diagnostic strategies using mAbs. To date, several mAbs have been approved by pharmaceutical agencies around the world for the treatment of cancer, while some newer types are still in clinical trials.

For therapy applications and in order to destroy tumor cells by mAbs, several mechanisms are available: via effector cells or complement-dependent cytotoxicity or through conjugation of the mAb to toxins, drugs, or radioisotopes. Since antigen expression within tumors is heterogeneous, tumor cells that do not express the antigen may evade tumor cell lysis by effector cell- or complement-mediated cytotoxicity, which may eventually lead to tumor recurrence. The same applies to radiolabeled mAbs, since internalization of a mAb conjugated to a radioisotope is required to mediate cell death (Oosterwijk et al. 2003).

Radiolabeling of antibodies was developed in 1950, when Eisen observed that proteins could be labeled with ^{131}I without altering their immunological properties and specificity (Eisen and Keston 1949). Besides ^{131}I, other radioisotopes (e.g., ^{90}Y, ^{177}Lu, ^{153}Sm, ^{186}Re, and ^{188}Re) have since been investigated in order to induce tumor cell death for therapy purposes. One major advantage of radiolabeled antibodies is that it is not necessary to bind to every tumor cell to induce cytotoxicity, since the radioisotopes are β-emitters, and thus can be effective for several cell diameters. This is called "crossfire effect" and can overcome the drawback of heterogeneity of antigen expression, as the radiation destroys the antigen-negative cells as well. The disadvantage of this effect is that normal cells are also exposed to radiation. This is more crucial for the radiosensitive organs such as the bone marrow.

Target expression levels in tumors can be also determined *in vivo* using immuno-PET and immuno-SPECT. For both molecular imaging techniques, a mAb is labeled with a radioisotope (positron or γ-emitter) and injected intravenously, and the 3D distribution of the radiolabeled mAb is subsequently visualized with a SPECT or PET scanner. With these molecular imaging techniques, several limitations of existing assays (e.g., immunohistochemistry) can be circumvented. Advantages include noninvasive screening and visualization of specific membranous target expression. Moreover, these whole body scans enable target visualization in the whole tumor, in multiple lesions simultaneously and take target accessibility into account (Heskamp et al. 2013).

A wide array of mAbs against tumor-associated antigens has been produced, for example, mAbs against carcino-embryonic antigen (CEA) (mainly expressed in colorectal and medullary thyroid carcinomas), MUC-1 (mainly ovarian and breast cancer), CD-20 (non-Hodgkin's lymphoma [NHL]), and TAG72 (mainly ovarian and colorectal cancer). The safety and efficacy of these newly developed mAbs after radiolabeling with various radioisotopes have been investigated in clinical trials (Larson 1987). NHL tumors are relatively radiosensitive and mAbs have shown good access to the tumor and high antitumor activity. The outcome of this research was the first registered treatment with radiolabeled mAbs directed against the surface antigen CD-20 expressed on B-cell NHL: ^{90}Y-labeled anti-CD20 mAb ibritumomab tiuxetan (Zevalin; Biogen Idec, Boston, MA, and Schering, Berlin, Germany) and ^{131}I-labeled anti-CD20 mAb tositumomab (Bexxar®; GSK, Philadelphia, PA).

21.2.1 RADIOLABELED mAbs FOR HER IMAGING

HER is a class of receptors that play a vital role in the development of various cancers. There have been mAbs developed that target HER1 and HER2. Trastuzumab, pertuzumab, cetuximab, and panitumumab are mAbs that target HER1 and/or HER2 and are widely used to determine the expression levels of those receptors *in vivo* prior to therapy. ^{111}In-trastuzumab SPECT and $^{64}Cu/^{89}Zr$-trastuzumab PET can identify HER2-positive lesions in patients with primary and metastatic breast cancer (Perik et al. 2006; Dijkers et al. 2010; Tamura et al. 2013). Although the exact predictive value of these imaging modalities for trastuzumab response remains to be investigated, they have great potential. In mice, ^{111}In-pertuzumab SPECT (a HER2 dimerization inhibitor) sensitively imaged HER2 downregulation during trastuzumab treatment in an *in vivo* human breast cancer model, illustrating the feasibility of monitoring HER2 expression levels during HER2-mediated treatment (McLarty et al. 2009).

Nayak et al. performed SPECT/CT and MR imaging studies in mice bearing subcutaneous and orthotopic malignant pleural mesothelioma tumors by using HER1- and HER2-targeted [111]In- and ^{125}I-labeled panitumumab and trastuzumab, respectively (Nayak et al. 2013). The mean tumor uptake of ^{111}In-panitumumab was significantly greater than for ^{111}In-trastuzumab and ^{125}I-panitumumab, as was clearly shown from biodistribution and small animal imaging studies. The combination of HER assessment by using radiolabeled antibodies and tumor monitoring by using MR imaging proved to be an attractive diagnostic and prognostic tool for patient treatment, so they can potentially be used for therapeutic interventions with HER-targeted immunotherapies alongside radioimmunotherapy (RIT) (Figure 21.2).

Promising results for visualization of HER1 *in vivo* with molecular imaging techniques were obtained with ^{111}In-cetuximab (SPECT) and ^{64}Cu-cetuximab and [89]Zr-cetuximab (PET) *in vivo* cancer models (Aerts et al. 2009; Hoeben et al. 2011; Corcoran and Hanson 2014). The results obtained have led to a phase I trial, investigating the predictive value of [89]Zr-cetuximab PET to cetuximab therapy in stage IV cancer patients.

Promising preclinical imaging results for visualization of tumor-associated HER expression obtained with ^{111}In-cetuximab (SPECT) and ^{64}Cu or [89]Zr-cetuximab (PET) (Aerts et al. 2009; Hoeben et al. 2011; Corcoran and Hanson 2014) have led to a phase I trial, investigating the predictive value of the PET tracer [89]Zr-cetuximab, to monitor cetuximab therapy outcome in stage-IV cancer patients.

21.2.2   RADIOLABELED mAbs AGAINST VEGF

Vascular endothelial growth factor-A (VEGF-A) plays a central role in inducing the formation of new blood vessels during physiological and pathological angiogenesis. Bevacizumab and ranibizumab are antibodies directed against VEGF-A. The visualization of tumor VEGF expression using molecular imaging techniques was the aim of many research studies. ^{111}In-bevacizumab (SPECT) can image VEGF expression in various malignancies (Nagengast et al. 2011). There was, however, no direct correlation between radiolabeled bevacizumab uptake and VEGF-A expression in tumor as determined by conventional methods suggesting that other factors may play a role in bevacizumab targeting (e.g., vascular volume and permeability). Nevertheless, there is an ongoing effort for noninvasive measurement of VEGF levels in tumors.

This could provide essential information in the evaluation of anti-angiogenic therapy and could lead to better understanding and patient-tailored therapy. This could lead to a better understanding of the biology of these tumors and consequently could provide a more efficient evaluation of the anti-angiogenic treatment effects and new patient-tailored therapy options. In a recent study, [166]Ho-bevacizumab was used for biodistribution evaluation and to acquire dosimetric aspects of the radiolabeled antibody in mice using SPECT (Khorami-Moghadam et al. 2013). [166]Ho-bevacizumab imaging in wild-type rats showed distinct accumulation of the radiotracer in the chest region. Accumulation of the radiolabeled antibody in liver, spleen, kidney, bone, and other tissues as well as sufficient tumor uptake demonstrates a similar pattern to the majority of radiolabeled anti-VEGF immunoconjugates. [166]Ho-bevacizumab can become a potential compound for diagnosis and treatment studies and follow-up of VEGF expression in oncology.

A study by Nagengast et al. (2011) used two radioisotopes for the development of radiolabeled bevacizumab: (1) the PET radioisotope [89]Zr, which has good characteristics for antibody imaging (long half-life that matches the pharmacokinetic of antibodies and relative low average positron energy) and (2) the widely available and long-lived radioisotope, ^{111}In, for SPECT and potential therapy. *Ex vivo* biodistribution results for ^{111}In-bevacizumab were similar to those of [89]Zr-bevacizumab. Images were obtained by micro-PET at 24, 72, and 168 h after injection of [89]Zr-bevacizumab in nude mice with human SKOV-3 ovarian tumor xenografts. Both ^{111}In-bevacizumab and [89]Zr-bevacizumab can be used clinically for noninvasive *in vivo* VEGF imaging.

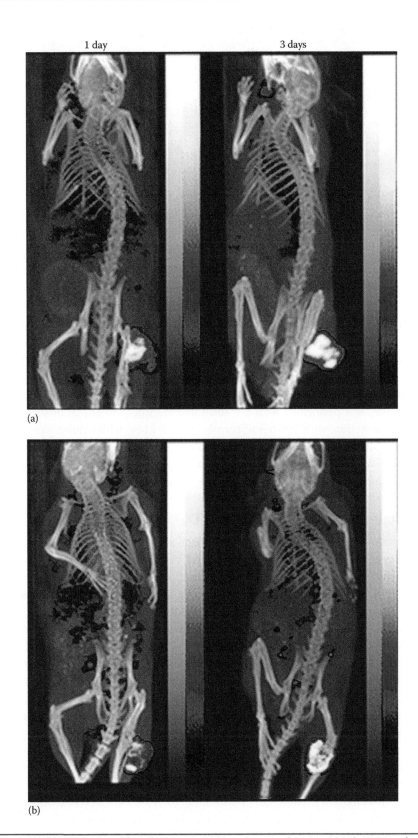

FIGURE 21.2 SPECT/CT images of female athymic (NCr) nu/nu mice bearing (a) NCI-H226 and (b) MSTO-211H subcutaneous tumor xenografts intravenously injected via the tail vein with ^{111}In-CHX-A0-DTPA–panitumumab on days 1 and 3 p.i. (Reprinted by Nayak, T.K. et al., *Radiology*, 267(1), 173, 2013. With permission from RSNA.)

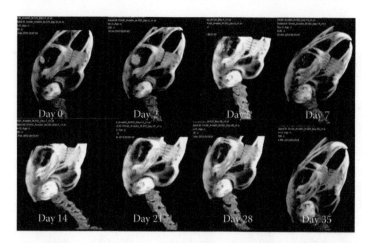

FIGURE 21.3 Image presentation of the clearance patterns of ^{124}I-bevacizumab within a rabbit's vitreous cavity showing no tracer detection on day 28. (Reprinted by Christoforidis, J.B. et al., *Invest. Ophthalmol. Vis. Sci.*, 52(8), 5899, 2011. With permission from ARVO/IOVS.)

Putting cancer aside, intravitreal injection of anti-VEGF agents is the most commonly performed procedure for retina treatment. A clinical question that often arises is whether anti-VEGF agents actually remain within the vitreous cavity after intravitreal placement. A study by Christoforidis et al. tried to answer this question by determining whether radiolabeled bevacizumab and ranibizumab remain confined within the vitreous cavity after intravitreal injection (Christoforidis et al. 2011). Radiolabeling was performed with ^{124}I and three anesthetized Dutch-belted rabbits underwent intravitreal injection with ^{124}I-bevacizumab (Figure 21.3), while three were injected with ^{124}I-ranibizumab. All rabbits were imaged with a micro-PET/CT scanner on various days after injection. Results showed no significant escape of radiolabeled bevacizumab and ranibizumab from the vitreous cavity and both imaging agents remained detectable until 28 and 21 days, respectively.

21.2.3 RADIOLABELED mAbs AGAINST CD20

The first therapeutic mAb approved for the treatment of lymphoma was rituximab, that targets CD20, which is a surface antigen found on all mature B cells including all types of B-cell NHL.

Rituximab has been radiolabeled with various radioisotopes for either imaging or RIT. ^{131}I-rituximab SPECT visualized CD20-positive cells in NHL patients, although these studies primarily focused on hematologic toxicity due to subsequent rituximab-RIT.

^{131}I-rituximab SPECT not only visualized CD20-positive cells in NHL patients but also had therapeutic applications, and its treatment had been further assessed either for blood or bone marrow toxicity (Boucek and Turner 2005; Leahy et al. 2006). ^{123}I-rituximab SPECT revealed very low tumor accumulation up to 48 h postinjection in four patients with highly CD20-positive primary central nervous system lymphomas, most likely because rituximab cannot pass the blood brain barrier; thus these tumors are unlikely to respond to rituximab treatment despite apparent CD20 expression (Dietlein et al. 2005). These findings stress the importance not only of target expression levels but also of target accessibility. Clinical imaging results so far indicate that molecular imaging approaches could be more predictive of mAb therapy response than conventional methods.

The purpose of another study by Muldoon et al. was to determine whether blood–brain barrier disruption (BBBD)-enhanced delivery of ^{90}Y-rituximab increased efficacy in female athymic nude rats with intracerebral human MC116 B-lymphoma cells. ^{90}Y-rituximab was effective at decreasing tumor volume and improving survival and was not affected by combination with methotrexate or by BBBD (Muldoon et al. 2011).

Two radiolabeled mAbs already approved by pharmaceutical agencies are ^{90}Y-ibritumomab tiuxetan (Zevalin) and ^{131}I-Tositumomab (Bexxar). Unlike rituximab, these are murine antibodies and conjugated to the therapeutic β-emitting radionuclides ^{90}Y and ^{131}I for RIT (Nowakowski and Witzig 2006).

Because ^{90}Y is a pure therapeutic β-emitter, it requires a surrogate radioisotope (γ-emitter), such as ^{111}In, for imaging. Thus, for Zevalin imaging, ibritumomab tiuxetan can be labeled with ^{111}In, while for Bexxar, ^{131}I can be used for both imaging and therapy since it emits both β and γ radiation. Alternatively, PET imaging with ^{89}Zr-ibritumomab tiuxetan can be used to monitor the biodistribution of ^{90}Y-ibritumomab tiuxetan as shown in mice and can thus be used to predict biodistribution and dose-limiting organ during therapy (Rizvi et al. 2012). Finally, because Zevalin and Bexxar are murine antibodies that might induce human anti-mouse antibodies, these agents are not normally administered repeatedly.

21.2.4   RADIOLABELED mAbs AGAINST PSMA

Prostate-specific membrane antigen (PSMA) has several optimal characteristics for targeting by antibodies as it is a highly expressed prostate-restricted nonsecreted protein anchored to the plasma membrane and its expression increases as tumor grade increases (Osborne et al. 2013).

The first clinical agent for targeting PSMA in prostate cancer was ^{111}In-capromab, which consists of a murine antibody radiolabeled with ^{111}In (Elsasser-Beile et al. 2006). An important fact that limits its use as a good imaging agent is that only nonviable cells with damaged cell membranes bind the mAb.

A promising next-generation antibody (J591) that targets the extracellular domain of PSMA may provide significant benefits to prostate cancer imaging. Although J591 can be administered unlabeled for regular immunotherapy, the current opinion is that conjugation of J591 with toxins or radionuclides is a more promising approach (Morris et al. 2005). For RIT purposes, J591 has been labeled with ^{90}Y or ^{177}Lu in phase I studies in metastatic castration-resistant prostate cancer patients with promising results (Vallabhajosula et al. 2005; Tagawa et al. 2013).

^{177}Lu-J591 can be used for both imaging and therapy because it emits both β and γ radiation, while for ^{90}Y-J591 imaging, a surrogate radioisotope is required.

In order to improve slow pharmacokinetics of J591, a diabody based on scFv fragments of J591 was produced and site-specifically radiolabeled with $[^{99m}Tc(CO)_3]^+$ via the C-terminal His tag and evaluated in a subcutaneous DU145/DU145-PSMA prostate carcinoma xenograft model (Kampmeier et al. 2014). J591C diabody binds to PSMA-expressing cells with low nanomolar affinity and subsequent SPECT studies allowed imaging of tumor xenografts with high contrast from 4 h postinjection. The radiolabeled diabody presented favorable properties required for further development for antibody-based imaging of PSMA expression in prostate cancer.

21.2.5   RADIOLABELED mAbs AGAINST CARBONIC ANHYDRASE IX

Girentuximab is a mAb that binds to carbonic anhydrase IX (CAIX), which is a heat-sensitive transmembranous glycoprotein, and it is developed for the treatment of renal cell carcinoma (RCC). Previous studies reported almost ubiquitous expression (>90%) of CAIX in clear cell RCC (ccRCC). Girentuximab has been successfully labeled to ^{111}In, ^{131}I, and ^{124}I, though intrapatient comparisons revealed higher uptake of residualizing ^{111}In-girentuximab preparations (Brouwers et al. 2005; Pryma et al. 2011).

BALB/c nude mice bearing xenografts of human RCC subcutaneously were imaged by ^{124}I-girentuximab PET/CT, and excellent localization of the imaging agent in the tumor was demonstrated (Lawrentschuk et al. 2011). ^{124}I-girentuximab targets RCC with correlation between uptake on noninvasive PET–CT studies and traditional biodistribution studies opening the possibility of using PET/CT in future studies.

Stillebroer et al. made the hypothesis that labeling girentuximab with the residualizing positron emitter ^{89}Zr would overcome rapid excretion from the tumor cells after internalization of ^{124}I-girentuximab and thus lead to higher tumor uptake and more sensitive detection of ccRCC lesions (Stillebroer et al. 2013). Nude mice with CAIX-expressing ccRCC xenografts were injected with ^{89}Zr-girentuximab or ^{124}I-girentuximab. The xenografts were clearly visualized and tumor uptake of ^{89}Zr-girentuximab was significantly higher compared to ^{124}I-girentuximab. This indicates the potential of girentuximab immune-PET in the diagnosis and staging of ccRCC and confirms that PET imaging of ccRCC tumors with the ^{89}Zr-labeled mAb could be more sensitive than the ^{124}I-labeled mAb.

21.2.6 OTHER RADIOLABELED mAbs

The CD44 antigen is a cell-surface glycoprotein involved in cell–cell interactions, cell adhesion, and migration, playing a crucial role in cancer development and metastasis. RG 7356 is a humanized antibody targeting the constant region of CD44, which has been radiolabeled with ^{89}Zr by Vuqts et al. for preclinical evaluation in tumor-bearing mice and normal cynomolgus monkeys (Vugts et al. 2014). Studies with ^{89}Zr-RG7356 were performed in mice bearing tumor xenografts that differ in the level of CD44 expression and RG7356 responsiveness, and immuno-PET whole body biodistribution studies were performed in normal cynomolgus monkeys to determine normal organ uptake after administration of a single dose. ^{89}Zr-RG7356 selectively targets CD44(+) responsive and nonresponsive tumors in mice and CD44(+) tissues in monkeys indicating the importance of accurate antibody dosing in humans to obtain optimal tumor targeting and showing, moreover, that efficient binding of RG7356 to CD44(+) tumors may not be sufficient in itself to drive an antitumor response.

The L-type amino acid transporter-1 (LAT1, SLC7A5) is upregulated in a wide range of human cancers, positively correlated with the biological aggressiveness of tumors, and is a promising target for both imaging and therapy. A recent study by Ikotun et al. describes the development and biological evaluation of a novel ^{89}Zr labeled antibody, ^{89}Zr-DFO-Ab2, which targets the extracellular domain of LAT1 in a preclinical model of colorectal cancer (Ikotun et al. 2013). This tracer demonstrated specificity for LAT1 *in vitro* and *in vivo* with excellent tumor imaging properties in mice with tumor xenografts. PET imaging studies showed high tumor uptake, with optimal tumor-to-nontarget contrast achieved at 7 days postinjection, demonstrating the potential of immune-PET agents for imaging specific amino acid transporters.

The CD138 antigen is strongly expressed on myeloma cells in 100% of patients. In a study by Cherel et al., intravenous injection of mouse myeloma cells into a syngeneic mouse strain resulted in a rapid invasion of the marrow and limb paralysis, and radioimmunotherapy was performed 10 days after cell engraftment with an intravenous injection of an anti-mouse CD138 antibody radiolabeled with the α-emitter ^{213}Bi (Cherel et al. 2013). The short range of alpha-particles enables localized irradiation of tumor cells within the bone marrow and a cytotoxic effect on isolated cells due to the high linear energy transfer of α-particles. This study demonstrates promising therapeutic efficacy of ^{213}Bi-CD138 for the treatment of residual disease in the case of multiple myeloma, with only moderate and transient toxicity.

21.3 SMALL REGULATORY PEPTIDES

Radiolabeled antibodies are a potentially promising class of molecular imaging agents and good results have been published up to now. Nevertheless, they also show disadvantages, which are mainly attributed to their high molecular weight resulting in sequestration by reticuloendothelial cells and liver Kupffer cells, leaving only a small amount of antibody available for binding to the target and leading to reduced specificity and diagnostic accuracy (Signore et al. 2010). Another important issue that limits the use of antibodies is, in most cases, the need for radioisotopes with long half-lives, since antibodies generally have a long elimination half-life and usually need several days to reach optimal target-to-background levels.

A strategy that can bypass the drawbacks of radiolabeled antibodies is the radiolabeling of peptides. In general, peptides are small amino acid sequences that have low molecular weight, consist of fewer than 50 amino acids, and regulate many metabolic processes (Reubi 2003). These small peptides are rapidly taken up and retained in the target tissues by coupling to specific receptors, functioning either as agonists or antagonists in specific biological processes. Most agonists exert their function by binding to G-coupled proteins, which results in internalization of the peptide–receptor complex, whereas the majority of antagonists remain on the external side of the cellular membrane (Krohn 2001). These characteristics, together with their rapid plasma clearance due to renal excretion, make them very suitable for imaging. Up to now, their clinical use has been limited to tumor imaging and treatment (mostly neuroendocrine tumors) and to some extent to the diagnosis of infectious and inflammatory diseases. In several tumor types, and to a lesser extent in some inflammatory diseases, many of the peptide receptors are overexpressed, thereby forming a potential target for molecular imaging and therapy (Reubi 2003). The peptide synthesis and development is relatively cheap and can be performed quickly with automated synthesis units. Peptides are not immunogenic, and they usually demonstrate good tumor penetration and present fast tumor uptake, having low bone marrow uptake (Eberle and Mild 2009).

The principle of the development of a radiopeptide for successful receptor targeting can be summarized as follows:

1. Identification of the molecular target (receptor) using receptor autoradiography or immunohistochemistry. Receptor density, homogeneity, and incidence are important factors in predicting successful *in vivo* targeting.
2. Design and synthesis of a peptide analog based on the structure of the endogenous ligand (natural peptide) that exhibits very high affinity for the corresponding receptor by various ways such as cyclization, oligomerization, insertion of nonnatural or D-amino acid residues or other nonpeptide residues, substitution of peptide bonds, N-methylation, C-terminal amidation, and N-terminal acylation.
3. For radiolabeling, the peptide is either covalently coupled, often via a spacer, to a chelator that can complex radionuclides (i.e., ^{111}In, ^{99m}Tc, ^{68}Ga, ^{64}Cu), or it carries a prosthetic group that can be labeled with other radiometals such as radioiodine (*I) including ^{123}I, ^{124}I, ^{125}I, and ^{131}I radionuclides or ^{18}F (Figure 21.4). Labeling protocols should allow very high labeling yield, radiochemical purity, and specific activity and the radiopeptides should retain the affinity for the receptor.
4. *In vitro* radiopeptide binding studies in cells or cell membrane preparations and other molecular biology studies, which allow their screening and evaluation, regarding their affinity for the receptor, their internalization rate, dissociation from the tumor cells, etc.

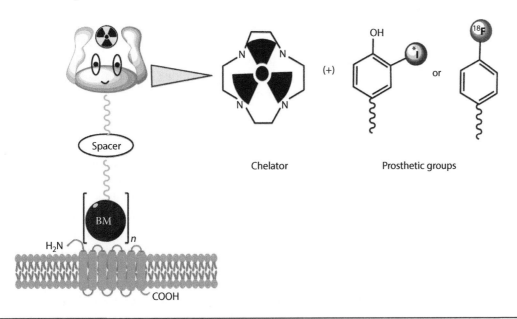

FIGURE 21.4   Illustration of the main strategies (via a chelator or a prosthetic group) applied for the radiolabeling of a targeting biomolecule (BM) (i.e., peptide) for receptor binding. A spacer is used to bridge the radiometal complex and the targeting biomolecule (BM).

TABLE 21.4
Regulatory Peptides Suitable for Imaging Probes

Regulatory Peptide	Receptor	Indication	Radiopeptide
Somatostatin (SST)	sstr$_2$	Neuroendocrine tumors (NET)	Pentetreotide
			DOTA-TOC/TATE/NOC
			Depreotide
Cholecystokinin (CCK)/gastrin	CCK$_2$	Medullary thyroid carcinoma (MTC)	DTPA/DOTA-minigastrin
			DTPA/DOTA-CCK$_8$
			Demogastrin
Bombesin (BN)	GRP-R	Prostate and breast cancer	AOC-BN
			DTPA/DOTA-panbombesin
			Demobesin
			DOTA-PESIN
			BAY 86-4367
GLP-1	GLP-1 R	Insulinomas	DTPA/DOTA-exendin-4
α-MSH	MCr	Melanomas	CCMSH
RGD peptides	αvβ$_3$	Various	Galacto-cRGDfK
			NC100692

5. *In vivo* evaluation studies by using biodistribution and imaging techniques with suitable animal models, in order to evaluate the pharmacological behavior and pharmacokinetics of the radiopeptides. Many aspects should be taken into consideration for further development, such as the accumulation in target and nontarget tissues, the clearance from the body, the excretory pathway and the *in vivo* stability of the radiopeptide.

6. Finally, radiopeptides, which successfully go through all tests, after toxicological studies and established preparation, may enter clinical studies in humans (de Visser et al. 2008).

Table 21.4 presents a small debrief of the currently used radiopeptides in preclinical research and in clinical use.

21.3.1 Somatostatin Analogs

Somatostatin (SST) is a cyclic peptide hormone that is expressed in the central and peripheral nervous systems and is present in two forms: SST-14 consisting of 14 amino acids and SST-28 consisting of 28 amino acids. SST inhibits the release of hormones such as growth hormone, glucagon, and insulin by binding to five different G-coupled SST cell membrane receptors (sstr1–5), which recognize the ligand and generate a transmembrane signal. The resulting hormone–receptor complexes have the ability to be internalized. Neuroendocrine tumors frequently express a high density of sstr, with the sstr2 subtype generally showing the highest density (Weckbecker et al. 2003).

In order to develop metabolically stable peptide analogs for molecular imaging and given the fact that SST have a plasma half-life of 3 min, introduction of D-amino acids and shortening of the molecule to the bioactive core sequence resulted in eight amino acid–containing SST analogs, such as octreotide (OC). The first commercially available agent was ^{111}In-pentetreotide for sstr scintigraphy (OctreoScan®). However, this radiopeptide has moderate binding affinity to sstr2 and is not a suitable chelator for β-emitters complexation. The modified OC analogs *N,N',N'',N'''*-tetraacetic acid (DOTA)-TOC and DOTA-TATE, labeled with ^{111}In and ^{68}Ga for SPECT and PET imaging respectively, and with ^{90}Y and ^{177}Lu for targeted radionuclide therapy, are routinely used in many hospitals. [99m]Tc, ^{18}F, and ^{64}Cu-labeled OC analogs have also been developed and are promising for SPECT and PET imaging clinical application (Fani and Maecke 2012).

All of the analogs mentioned here have high affinity for sstr2. However, radiolabeled SST-based probes with a broader receptor subtype affinity profile may target a broader spectrum of tumors but also they may increase the net tumor uptake, given the presence of several receptor subtypes on the same tumor cell. Modifications in current OC analogs gave rise to such imaging tracers with encouraging results. Pan-somatostatin radiopeptides with high affinity for all receptor subtypes have also been recently developed (Fani et al. 2012).

All somatostatin analogs mentioned earlier are agonists that induce receptor internalization. However, high-affinity sstr2- and sstr3-selective SST antagonists can perform as well as or even better than agonists in terms of *in vivo* uptake in corresponding tumor xenografts, despite their poor internalization (Ginj et al. 2006).

21.3.2 BOMBESIN ANALOGS

Bombesin (BN) is a small neuropeptide of 14 amino acids originally isolated from frog skin. One of its mammalian counterparts is the 27-amino acid gastrin-releasing peptide (GRP). Both peptides show similar biological behavior in humans. GRP and bombesin can bind to GRP receptors (GRPR), which are G-protein coupled receptors (GPCRs) and are present on cells of many human cancer types, such as prostate, breast, and pancreatic cancer and small-cell lung carcinoma, making radio-labeled BN analogs attractive candidates for targeting GRPR (Fani and Maecke 2012). The half-life of natural BN is about 2–3 min and, therefore, more resistant analogs have been developed as molecular imaging tracers. The binding site of BN consists of eight amino acid residues, and in the literature peptides that incorporate these eight peptides are usually termed BN. Radiolabeled BN analogs are usually based on the eight amino acid sequence peptide analogs that have modified amino acid residues in order to improve their stability and allow conjugation of chelators for radiolabeling.

[99m]Tc-labeling of an agonist (Demobesin-4) and an antagonist (Demobesin-1) with similar affinity for GRPR showed superiority of the antagonist in terms of high and prolonged tumor retention in GRPR-positive xenografts and rapid background clearance, even though the agonist has already entered clinical trials (http://clinicaltrials.gov/show/NCT00989105) (Cescato et al. 2008).

A number of studies have investigated the effect of the spacer between the chelator and the pharmacophoric group in a series of BN peptidic derivatives radiolabeled with ^{99m}Tc or ^{111}In, after *in vivo* biodistribution and imaging studies in small animals. Introduction of hydrophilic carbohydrate linker moieties and polar linkers with different charges significantly influenced the biodistribution profile (Garcia Garayoa et al. 2007).

BN derivatives bearing the chelators 4,8,11-tetraazabicyclo[6.6.2]hexadecane-4,11-diacetic acid (CB-TE2A) and 1,4,7,10-tetraazacyclodoadecane-DOTA have been radiolabeled with ^{64}Cu, which is a radioisotope that has clinical potential for application in both diagnostic imaging and radionuclide therapy (Garrison et al. 2007). The direct comparison of the ^{64}Cu-CB-TE2A-aminooctanoic acid (AOC)-BN and ^{64}Cu-DOTA-AOC-BN analogs has been made in a prostate cancer xenograft severely compromised immune-deficient mouse model bearing PC-3 prostate cancer xenografts. The pharmacokinetic and small-animal PET/CT studies demonstrate significantly improved nontarget tissue clearance for the ^{64}Cu-CB-TE2A8-AOC-BN, which is attributed to the improved *in vivo* stability of the ^{64}Cu-CB-TE2A chelate complex as compared with the ^{64}Cu-DOTA chelate complex (Garrison et al. 2007) (Figure 21.5).

Slightly modified analogs of the universal BN ligand (Pradhan et al. 1998) bind with high affinity to all bombesin receptor subtypes and were coupled to diethylenetriaminepentaacetic acid (DTPA) and DOTA using γ-aminobutyric acid as a spacer. The resulting panbombesin analogs radiolabeled with ^{111}In, ^{177}Lu, or ^{90}Y exhibited high and specific uptake in GRPR-positive tissue (pancreas) and tumor (Zhang et al. 2004).

Recently, BN conjugated to DOTA via 4-polyethylene glycol (PEG4) spacer (DOTA-PESIN) was used for therapeutic purposes after radiolabeling with ^{177}Lu versus the α emitter [213]Bi in a prostate carcinoma xenografts model (alpha- versus beta-particle radiopeptide therapy) (Wild et al. 2011), where it was shown that α therapy was more efficacious than β therapy.

Finally, after radiolabeling with ^{18}F of various BN analogs with different charged spacers, it has been shown that a negatively charged BN derivative had increased tumor uptake compared to the corresponding positively charged analogs. Among all analogs, the ^{18}F-BAY 86-4367 containing two L-cysteic moieties has been studied in two prostate cancer xenograft models (PC-3 and LNCaP) and has showed better specific and effective GRPR-based targeting *in vivo* than the benchmark radio-tracers ^{18}F-fluoroethylcholine and ^{18}F-FDG (Honer et al. 2011).

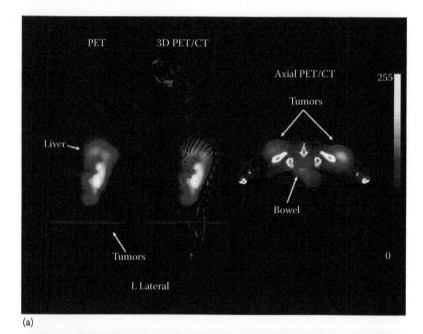

(a)

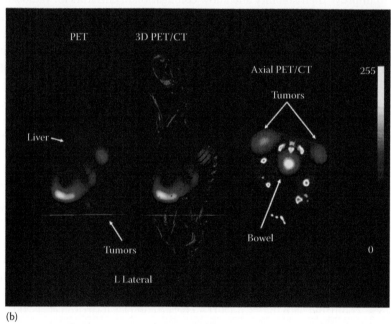

(b)

FIGURE 21.5 Small-animal PET, fused small-animal PET/CT and axial images of (a) 64Cu-DOTA-8-AOC-BN and (b) 64Cu-CB-TE2A8-AOC-BN, in PC-3 tumor-bearing SCID mice at 20 h p.i. (From Garrison, J.C. et al., *J. Nucl. Med.*, 48(8), 1327, 2007. This research was originally published in JNM. Figure 6. Copyright by the Society of Nuclear Medicine and Molecular Imaging, Inc. With permission.)

In addition, rapid tumor targeting and fast renal and hepatobiliary excretion were identified, whereas PET studies provided clear and specific visualization of the PC-3 tumor xenografts. The favorable preclinical data shown by 18F-BAY 86-4367 makes it a considerable candidate for further clinical evaluation.

21.3.3 CHOLECYSTOKININ/GASTRIN ANALOGS

Another hormone that is suitable as a molecular target is cholecystokinin (CCK), which is structurally and functionally related to gastrin. Both hormones function in the gastrointestinal tract and in the CNS and exert their function via CCK/gastrin GPCRs. In tumor tissues, the most important

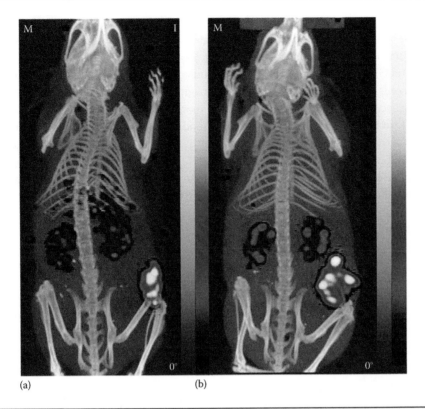

FIGURE 21.6 Small-animal SPECT/CT images showing the biodistribution of the (a) dimer compound ^{111}In-MGD5 and (b) the monomer compound ^{111}In-APH070 at 4 h p.i. (From Sosabowski, J.K. et al., *J. Nucl. Med.*, 50(12), 2082, 2009. This research was originally published in JNM. Figure 7. Copyright by the Society of Nuclear Medicine and Molecular Imaging, Inc. With permission.)

receptor subtypes are CCK1 and CCK2, the latter having high affinity for gastrin and CCK in humans (Reubi et al. 1997). Several CCK2-selective peptide analogs are in development for colonic, gastric, brain, and medullary thyroid cancer imaging (Reubi et al. 1997).

The radiolabeled CCK/gastrin analogs are mostly based on the C-terminal octapeptide CCK8 and on minigastrin (MG). CCK8 can be in nonsulfated form (CCK8) or in sulfated form (sCCK8). MG is a C-terminal truncated nonsulfated form having 13 amino acid residues. Studies in both CCK1 and CCK2 receptor-expressing tumors in mice have shown that the uptake of [99mTc]-labeled sCCK8 was significantly higher than that of CCK8 (Laverman et al. 2004).

Radioiodinated as well as ^{111}In-labeled analogs of MG and CCK8 have been studied indicating that MG analogs had greater uptake in receptor-positive tissues but also extremely high kidney uptake (Fani and Maecke 2012). Slight modifications in the structure of the peptidic derivatives have been performed (e.g., reducing the number of glutamate residues and insertion of histidine tags) and results showed improved binding affinity, significantly reduced kidney uptake, and increased tumor to kidney ratio (Mather et al. 2007).

N_4-derivatized MG analogs (Demogastrin) were labeled with [99mTc] showing the most encouraging results compared to other [99mTc]- and ^{111}In-labeled analogs in order to be used as a diagnostic tool in preliminary clinical studies (Froberg et al. 2009). Further improvement of the *in vivo* performance and metabolic stability was accomplished mainly by dimerization and not by cyclization of analogs, which after conjugation with DOTA were radiolabeled with ^{111}In and showed an increase in binding affinity and tumor uptake *in vivo* (Sosabowski et al. 2009) (Figure 21.6).

21.3.4 Glucagon-Like Peptide 1/Exendin Analogs

The receptors for glucagon-like peptide 1 (GLP-1) also belong to the G-coupled protein receptor group. The function of these GLP-1 receptors (GLP-1R) is not yet completely understood, but receptor stimulation seems to result in the secretion of insulin and calcitonin and increased proliferation of tumor cells with inhibition of apoptosis (Korner et al. 2007). Abnormal receptor expression has been shown in endocrine, neural, and embryonic tumors (Korner et al. 2007).

The natural GLP-1 ligand has 30 amino acids and a half-life of less than 2 min, so modification is essential in order to be suitable as a molecular imaging agent. Exendin-4 is a metabolically resistant naturally occurring peptide identified in Gila monster, which shares high homology with the human GLP-1. *In vivo* evaluation of ^{111}In-DTPA-exendin-4 in tumor-bearing transgenic Rip1Tag2 mice (tumors of the pancreatic β cells with extremely high density of GLP-1R) indicated high tumor uptake and good tumor visualization (Wild et al. 2006). Based on the Auger e- of ^{111}In, ^{111}In-DTPA exendin-4 was additionally evaluated for its therapeutic efficiency in the same tumor model, showing a dose-dependent reduction of the tumor volume without significant acute toxicity (Wicki et al. 2007). Recently, ^{111}In-DTPA exendin-4 has been applied in humans for the imaging of insulinomas showing promising results (Christ et al. 2009).

[99mTc]- and [68Ga]-labeled analogs have been developed and evaluated, also showing encouraging results having similar behavior and tumor uptake, with the [99mTc]-labeled analog being currently under clinical evaluation (Fani et al. 2012).

Another field of research involving GLP-1/exendin analogs is the monitoring of β cell mass (BCM) in pancreas during the course of diabetes development and antidiabetic treatment due to the overexpression of GLP-1R on native pancreatic β cells. Furthermore, the method might be used for monitoring of islet cell graft survival after transplantation. It has been shown that the uptake of ^{111}In-DTPA exendin-3 correlates with the BCM in a linear manner in rats with alloxan-induced diabetes (Brom et al. 2014).

Finally, PET imaging has also been performed using GLP-1/exendin analogs. In a proof-of-principle study, a new vinyl sulfone-DOTA-exendin-4 analog radiolabeled with [64Cu] detected transplanted human islets in the liver of mice (Wu et al. 2011). Additionally, a conformationally constrained GLP-1 analog via lactam bridges, namely EM3106B, has been developed and radiolabeled with [18F] showing specific and high uptake in GLP-1R-positive tumors and low uptake in tumors with low GLP-1R expression (Gao et al. 2011).

21.3.5   RGD (Arg-Gly-Asp) Peptides

Imaging of angiogenesis has become increasingly important with the rising use of targeted antiangiogenic therapies like bevacizumab for the response assessment of such therapies. One promising approach of noninvasive angiogenesis imaging is the imaging of specific molecular markers of the angiogenic cascade like integrins. Integrins are a family of heterodimeric transmembrane glycoproteins that are involved in a wide range of cell extracellular matrix (ECM) interactions and cell–cell interactions. At least 24 distinct integrins are formed by a combination of 18 α and 8 β subunits. Among them $\alpha\nu\beta_3$ is highly expressed on activated and proliferating endothelial cells during tumor angiogenesis and metastasis and has a very important role in the regulation of tumor growth and metastasis (Desgrosellier and Cheresh 2010). The $\alpha\nu\beta_3$ integrin binds to Arg-Gly-Asp (RGD)-containing components of the ECM like vitronectin and fibronectin. For molecular imaging of integrin expression, the use of radiolabeled RGD peptides is the only approach that has been successfully translated into the clinic (Gaertner et al. 2012).

The first RGD peptides that were radiolabeled were cyclic pentapeptides (RGDfV, RGDfK). *In vivo* evaluation of radioiodinated RGD peptides in xenografts showed receptor-specific tumor uptake, but also unfavorable hepatobiliary elimination (Haubner et al. 1999). Various RGD peptide derivatives were radiolabeled with [99mTc] via appropriate chelators and technetium cores, showing encouraging results (Janssen et al. 2002; Psimadas et al. 2012). Already a [99mTc]-labeled RGD analog containing a double-bridged peptide, [99mTc]-NC100692, introduced by GE Healthcare, has been evaluated in breast cancer patients (Bach-Gansmo et al. 2006). Glycosylation of the RGD derivatives showed improvement in the pharmacokinetics and led to [18F]-Galacto-RGD, which alongside [18F]-fluciclatide have entered clinical studies (Haubner et al. 2005; Kenny et al. 2008).

DOTA, TE2A, and other macrocyclic chelators have been conjugated with RGD peptide derivatives giving rise to analogs that can be radiolabeled with ^{111}In, ^{64}Cu, and ^{68}Ga, some of them having advantageous properties for visualization of $\alpha\nu\beta_3$ integrin-expressing tumors (Dumont et al. 2011).

The pharmacokinetics of RGD-based probes were further improved by PEGylation (conjugation of polyethylene glycol) (Chen et al. 2004), whereas their binding affinity for the $\alpha\nu\beta3$ receptor was further improved by multimerization (more than one RGD moiety in the same targeting molecule) (Janssen et al. 2002; Dijkgraaf et al. 2011). Even though multimerization does not guarantee improvement on tumor to background tissue ratios, a favorable balance between the number of binding RGD moieties and the overall size of the molecule as well as the use of an appropriate linker (e.g., PEG) may significantly improve the overall performance of the RGD peptide tracer.

21.3.6 OTHER PEPTIDE ANALOGS

The neuropeptide vasoactive intestinal peptide (VIP) is a secretin-like peptide, which regulates a broad spectrum of biological activities, including vasodilation, stimulating secretion of various hormones, immunomodulation, and promotion of cell proliferation. Its receptors (VPAC1 and VPAC2) function through two distinct GPCR subtypes that can also be internalized (Shetzline et al. 2002). Only the VPAC1 receptor is found on tumor cells of neuroendocrine origin and neuroblastomas. The proteolytic degradation of VIP *in vivo* as well as high VPAC receptor expression in normal tissues hampers the applicability of this target for the imaging of neuroendocrine tumors. Recently, a more stable ^{64}Cu-labeled VIP analog has been developed for neuroendocrine and human breast tumors. *In vivo* studies in athymic BALB/c nude mice bearing subcutaneous VIP expressing estrogen-dependent human T47D breast tumors showed a higher tumor uptake than the ^{99m}Tc-labeled VIP analog (Thakur et al. 2004). Recently, a ^{18}F-labeled VIP analog has been developed and evaluated with micro-PET imaging study in C26 colon carcinoma–bearing mice, confirming high tumor specificity and good tumor/muscle radioactivity uptake ratio (Cheng et al. 2013).

Neurotensin (NT) is a peptide that acts as a neurotransmitter in the CNS but also as a hormone in the ileum, inducing physiological effects, including hypotension, analgesia and gut contraction, and increasing vascular permeability. Three different G-coupled receptor subtypes of NT have been found, which overexpress in many tumor types (Reubi et al. 1998). Several NT analogs have been developed and radiolabeled with various radioisotopes, including ^{99m}Tc, ^{111}In, and [18]F (Fani et al. 2012).

The α-melanocyte-stimulating hormones (α-MSH) are produced in the pituitary gland and regulate skin pigmentation (Miao and Quinn 2007). The G-coupled receptors for α-MSH have been found in 80% of metastases from melanomas. Many α-MSH analogs have been developed for melanoma imaging, but unfortunately their *in vivo* tumor targeting properties are not satisfactory. New modified derivatives are currently under development and evaluation in melanoma-bearing mice, including [18]F-, [64]Cu-, and ^{177}Lu-labeled α-MSH analogs (Yang et al. 2009; Gao et al. 2011; Lim et al. 2012).

21.3.7 OTHER SMALL VITAMIN-BASED ANALOGS

The application of small organic targeting moieties is favorable to surmount the drawbacks of long circulation times as well as possible immunogenicity encountered with antibodies. In this respect, vitamins dispose several advantages: small size, cost-effectiveness, easy chemical modification, and nonimmunogenicity. During cancer cell proliferation, there is an increased demand for certain vitamins such as folates, vitamin B12, biotin, and riboflavin that act as co-enzymes of several biochemical pathways like aminoacid and DNA and RNA synthesis (Russell-Jones et al. 2004).

Folic acid (Figure 21.7a) is the most widely studied water-soluble vitamin of the B-complex group that binds with high affinity to folate receptors (FRs) (KD < 10^{-9} M), which are glycoproteins existing in several isoforms (Antony 1996). While being compared with other targeting moieties (e.g., peptides), folic acid is relatively stable in solution over a wide range of pH values and at elevated temperatures and is nontoxic to healthy organs or tissues (Low et al. 2008). In normal organs and tissues, FR expression is highly restricted to only a few sites where it is located on the apical side of polarized epithelia in the lung, the placenta, the choroid plexus of the brain, and the proximal tubule

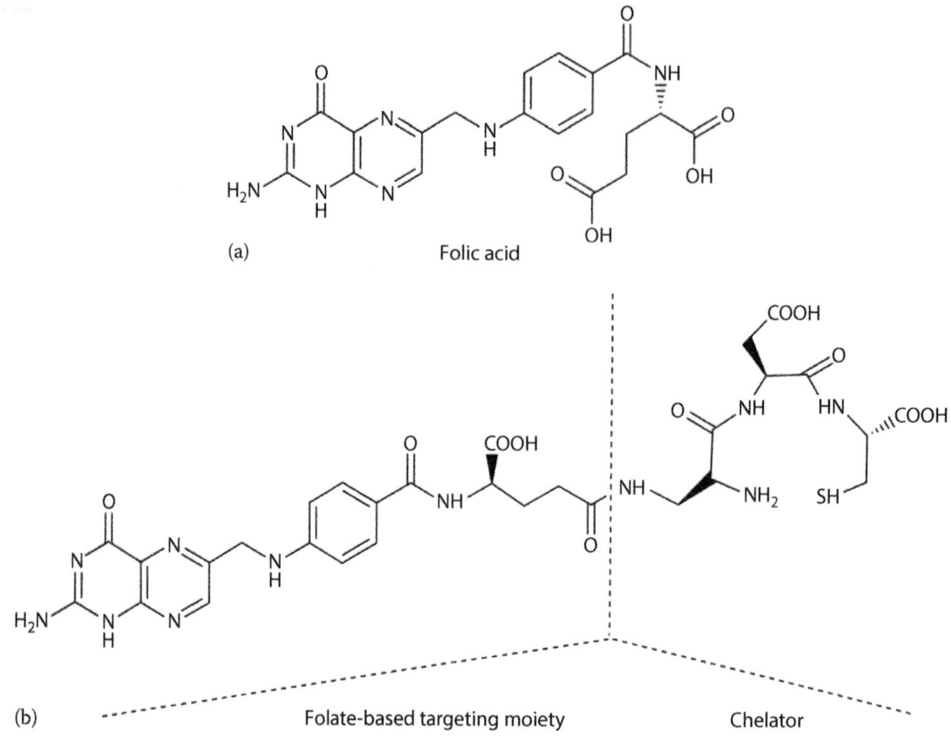

FIGURE 21.7 Structure of (a) folic acid and (b) EC20.

TABLE 21.5
Folate Receptor-Targeted Radionuclide SPECT/PET Imaging and Therapeutic Agents

Radionuclide	Folate-Based Conjugates	References
99mTc	DTPA	Ke et al. (2004), Muller (2013), and Muller and Schibli (2013)
	EC20	
	HYNIC	
	PnAO	
111In	DTPA	
66Ga, 67Ga, and 68Ga	DF	
	NODAGA	
	DOTA	
	DO3A	
64Cu and 67Cu	BTSC	
	CYCLAM	
18F	FDG	
	Glucose	
	FPCH	
	AZA	
	Click	
	FBCH	
	FBA	
177Lu	DOTA	
	DOTA-Bz-EDA	

Abbreviations: DTPA, pentetic acid or di-ethylene tri-amine penta-acetic acid; EC20, a folate-containing peptide consisting of the amino acid sequence Pte-D-Glu-β-Dpr-Asp-Cys; HYNIC, 6-hydrazino-pyridine-3-carboxylic acid; PnAO, propylene amine oxime; DF, deferoxamine; NODAGA, 1,4,7-triaza-1-glutaric acid-4,7-diacetic acid; DOTA, 1,4,7,10-tetraaza-cyclo-dodecane-1,4,7,10-tetraacetic acid; DO3A, 1,4,7,10-tetraaza-cyclo-dodecane-1,4,7-triacetic acid; BTSC, bis(thiosemicarbazones); CYCLAM, 1,4,8,11-tetra-azacyclo-tetradecane; FDG, fluoro-deoxy-glucose; FPCH, fluoro-propyl-choline; AZA, azomycin-arabinoside; Click, 6-[18F]-fluoro-1-hexyne as a prosthetic group in the presence of Cu(I) as a catalyst was reacted with an azide-derivatized folate precursor; FBCH, fluoro-benzene-carbohydrazide; FBA, fluoro-benzyl-amine; DOTA-Bz-EDA, DOTA-benzyl-ethylenediamine.

cells of the kidneys (Weitman et al. 1992; Parker et al. 2005). The FR has been identified as a target associated with a variety of frequent tumor types (e.g., ovarian, lung, brain, renal, and colorectal) (Parker et al. 2005). Most of the folate-based radioconjugates have been developed for SPECT and PET imaging (Table 21.5) (Ke et al. 2004; Muller and Schibli 2011).

Only two of these folate radiotracers, ^{111}In-diethylenetriaminepentaacetic acid (DTPA)-folate and ^{99m}Tc-EC20 (Etarfolatide[TM]; Endocyte Inc.) (Figure 21.7b), have been tested in human patients (Siegel et al. 2003; Fisher et al. 2008). Regarding PET imaging, extensive research has been dedicated to the preclinical evaluation of folate-based radiotracers but none of them have been tested in clinical trials yet (Muller 2013). In the view of therapeutic applications of folate-based radioconjugates based on particle-emitting radioisotopes, several studies have been undertaken. However, a therapeutic concept with folate radioconjugates has not yet been envisaged for clinical application. The reason is the generally high hydrophilic character of folate radioconjugates and the undesired uptake in kidneys with low tumor-to-kidney ratios (<0.2) that may result in damage to the renal tissue (Muller and Schibli 2013a). In an attempt to improve the tissue distribution profile, a clinically used antifolate (pemetrexed) was administered shortly before the radiofolate ^{111}In-DTPA-folate and resulted in a significant reduction of kidney uptake with no increase of tumor accumulation being observed (Muller et al. 2006). Radiolabeling studies of vitamin B12 (cobalamin), dating back almost 40 years, have demonstrated not only the visualization of tumor tissues but also the undesired high accumulation of radioactivity in normal tissues (liver, spleen, and especially in the kidneys). Vitamin B12 is bound to soluble transport proteins in circulation, namely transcobalamin I (TCI), intrinsic factor, and transcobalamin II (TCII) and the corresponding cellular receptors (Muller and Schibli 2013). Findings based on [99m]Tc(CO)$_3$-PAMA-C$_4$-cobalamin have demonstrated that the transport of B12 derivatives into malignant tissues is mediated via TCI rather than TCII and that the spacer's length could significantly affect the overall tissue distribution (Waibel et al. 2008).

Transporters or receptors of vitamins other than folic acid and vitamin B12 could be used for tumor targeted radioimaging or even radiotherapeutic purposes. Such a candidate might be vitamin C, which could be used for brain targeting as was shown by ascorbate-conjugated nanoparticles that were successfully delivered to brain presumably via the sodium-dependent ascorbic acid transporter SVCT2 (Salmaso et al. 2009).

REFERENCES

Aerts, H. J., L. Dubois et al. (2009). Disparity between in vivo EGFR expression and 89Zr-labeled cetuximab uptake assessed with PET. *J. Nucl. Med.* **50**(1): 123–131.

Antony, A. C. (1996). Folate receptors. *Annu. Rev. Nutr.* **16**: 501–521.

Bach-Gansmo, T., R. Danielsson et al. (2006). Integrin receptor imaging of breast cancer: A proof-of-concept study to evaluate ^{99m}Tc-NC100692. *J. Nucl. Med.* **47**(9): 1434–1439.

Boucek, J. A. and J. H. Turner. (2005). Validation of prospective whole-body bone marrow dosimetry by SPECT/CT multimodality imaging in (131)I-anti-CD20 rituximab radioimmunotherapy of non-Hodgkin's lymphoma. *Eur. J. Nucl. Med. Mol. Imaging* **32**(4): 458–469.

Brom, M., W. Woliner-van der Weg et al. (2014). Non-invasive quantification of the beta cell mass by SPECT with (1)(1)(1)In-labelled exendin. *Diabetologia* **57**(5): 950–959.

Brouwers, A. H., P. F. Mulders et al. (2005). Lack of efficacy of two consecutive treatments of radioimmunotherapy with 131I-cG250 in patients with metastasized clear cell renal cell carcinoma. *J. Clin. Oncol.* **23**(27): 6540–6548.

Cescato, R., T. Maina et al. (2008). Bombesin receptor antagonists may be preferable to agonists for tumor targeting. *J. Nucl. Med.* **49**(2): 318–326.

Chen, X., R. Park et al. (2004). MicroPET imaging of brain tumor angiogenesis with 18F-labeled PEGylated RGD peptide. *Eur. J. Nucl. Med. Mol. Imaging* **31**(8): 1081–1089.

Cheng, D., Y. Liu et al. (2013). F-18 labeled vasoactive intestinal peptide analogue in the PET imaging of colon carcinoma in nude mice. *Biomed. Res. Int.* **2013**: 420480.

Cherel, M., S. Gouard et al. (2013). 213Bi radioimmunotherapy with an anti-mCD138 monoclonal antibody in a murine model of multiple myeloma. *J. Nucl. Med.* **54**(9): 1597–1604.

Christ, E., D. Wild et al. (2009). Glucagon-like peptide-1 receptor imaging for localization of insulinomas. *J. Clin. Endocrinol. Metab.* **94**(11): 4398–4405.

Christoforidis, J. B., M. M. Carlton et al. (2011). PET/CT imaging of I-124-radiolabeled bevacizumab and ranibizumab after intravitreal injection in a rabbit model. *Invest. Ophthalmol. Vis. Sci.* **52**(8): 5899–5903.

Corcoran, E. B. and R. N. Hanson. (2014). Imaging EGFR and HER2 by PET and SPECT: A review. *Med. Res. Rev.* **34**(3): 596–643.

de Visser, M., S. M. Verwijnen et al. (2008). Update: Improvement strategies for peptide receptor scintigraphy and radionuclide therapy. *Cancer Biother. Radiopharm.* **23**(2): 137–157.

Desgrosellier, J. S. and D. A. Cheresh. (2010). Integrins in cancer: Biological implications and therapeutic opportunities. *Nat. Rev. Cancer* **10**(1): 9–22.

Dietlein, M., H. Pels et al. (2005). Imaging of central nervous system lymphomas with iodine-123 labeled rituximab. *Eur. J. Haematol.* **74**(4): 348–352.

Dijkers, E. C., T. H. Oude Munnink et al. (2010). Biodistribution of 89Zr-trastuzumab and PET imaging of HER2-positive lesions in patients with metastatic breast cancer. *Clin. Pharmacol. Ther.* **87**(5): 586–592.

Dijkgraaf, I., C. B. Yim et al. (2011). PET imaging of alphavbeta(3) integrin expression in tumours with (6)(8)Ga-labelled mono-, di- and tetrameric RGD peptides. *Eur. J. Nucl. Med. Mol. Imaging* **38**(1): 128–137.

Dumont, R. A., F. Deininger et al. (2011). Novel (64)Cu- and (68)Ga-labeled RGD conjugates show improved PET imaging of alpha(nu)beta(3) integrin expression and facile radiosynthesis. *J. Nucl. Med.* **52**(8): 1276–1284.

Eberle, A. N. and G. Mild. (2009). Receptor-mediated tumor targeting with radiopeptides. Part 1. General principles and methods. *J. Recept. Signal Transduct. Res.* **29**(1): 1–37.

Eisen, H. N. and A. S. Keston. (1949). The immunologic reactivity of bovine serum albumin labeled with trace-amounts of radioactive iodine (I131). *J. Immunol.* **63**(1): 71–80.

Elsasser-Beile, U., P. Wolf et al. (2006). A new generation of monoclonal and recombinant antibodies against cell-adherent prostate specific membrane antigen for diagnostic and therapeutic targeting of prostate cancer. *Prostate* **66**(13): 1359–1370.

Fani, M. and H. R. Maecke. (2012). Radiopharmaceutical development of radiolabelled peptides. *Eur. J. Nucl. Med. Mol. Imaging* **39**(Suppl. 1): S11–S30.

Fani, M., H. R. Maecke et al. (2012). Radiolabeled peptides: Valuable tools for the detection and treatment of cancer. *Theranostics* **2**(5): 481–501.

Fisher, R. E., B. A. Siegel et al. (2008). Exploratory study of ^{99m}Tc-EC20 imaging for identifying patients with folate receptor-positive solid tumors. *J. Nucl. Med.* **49**(6): 899–906.

Froberg, A. C., M. de Jong et al. (2009). Comparison of three radiolabelled peptide analogues for CCK-2 receptor scintigraphy in medullary thyroid carcinoma. *Eur. J. Nucl. Med. Mol. Imaging* **36**(8): 1265–1272.

Gaertner, F. C., H. Kessler et al. (2012). Radiolabelled RGD peptides for imaging and therapy. *Eur. J. Nucl. Med. Mol. Imaging* **39**(Suppl. 1): S126–S138.

Gao, H., G. Niu et al. (2011). PET of insulinoma using (1)(8)F-FBEM-EM3106B, a new GLP-1 analogue. *Mol. Pharm.* **8**(5): 1775–1782.

Garcia Garayoa, E., C. Schweinsberg et al. (2007). New [^{99m}Tc]bombesin analogues with improved biodistribution for targeting gastrin releasing-peptide receptor-positive tumors. *Q. J. Nucl. Med. Mol. Imaging* **51**(1): 42–50.

Garrison, J. C., T. L. Rold et al. (2007). In vivo evaluation and small-animal PET/CT of a prostate cancer mouse model using 64Cu bombesin analogs: Side-by-side comparison of the CB-TE2A and DOTA chelation systems. *J. Nucl. Med.* **48**(8): 1327–1337.

Ginj, M., H. Zhang et al. (2006). Radiolabeled somatostatin receptor antagonists are preferable to agonists for in vivo peptide receptor targeting of tumors. *Proc. Natl. Acad. Sci. USA* **103**(44): 16436–16441.

Haubner, R., W. A. Weber et al. (2005). Noninvasive visualization of the activated alphavbeta3 integrin in cancer patients by positron emission tomography and [18F]Galacto-RGD. *PLoS Med.* **2**(3): e70.

Haubner, R., H. J. Wester et al. (1999). Radiolabeled alpha(v)beta3 integrin antagonists: A new class of tracers for tumor targeting. *J. Nucl. Med.* **40**(6): 1061–1071.

Heskamp, S., H. W. van Laarhoven et al. (2013). Tumor-receptor imaging in breast cancer: A tool for patient selection and response monitoring. *Curr. Mol. Med.* **13**(10): 1506–1522.

Hoeben, B. A., J. D. Molkenboer-Kuenen et al. (2011). Radiolabeled cetuximab: Dose optimization for epidermal growth factor receptor imaging in a head-and-neck squamous cell carcinoma model. *Int. J. Cancer* **129**(4): 870–878.

Honer, M., L. Mu et al. (2011). 18F-labeled bombesin analog for specific and effective targeting of prostate tumors expressing gastrin-releasing peptide receptors. *J. Nucl. Med.* **52**(2): 270–278.

Ikotun, O. F., B. V. Marquez et al. (2013). Imaging the L-type amino acid transporter-1 (LAT1) with Zr-89 immunoPET. *PLoS One* **8**(10): e77476.

Janssen, M. L., W. J. Oyen et al. (2002). Tumor targeting with radiolabeled alpha(v)beta(3) integrin binding peptides in a nude mouse model. *Cancer Res.* **62**(21): 6146–6151.

Kampmeier, F., J. D. Williams et al. (2014). Design and preclinical evaluation of a ^{99m}Tc-labelled diabody of mAb J591 for SPECT imaging of prostate-specific membrane antigen (PSMA). *EJNMMI Res.* **4**(1): 13.

Ke, C. Y., C. J. Mathias et al. (2004). Folate-receptor-targeted radionuclide imaging agents. *Adv. Drug Deliv. Rev.* **56**(8): 1143–1160.

Kenny, L. M., R. C. Coombes et al. (2008). Phase I trial of the positron-emitting Arg-Gly-Asp (RGD) peptide radioligand 18F-AH111585 in breast cancer patients. *J. Nucl. Med.* **49**(6): 879–886.

Khorami-Moghadam, A., B. Bolouri et al. (2013). Preclinical evaluation of holmium-166 labeled anti-VEGF-A(Bevacizumab). *J. Labelled Comp. Radiopharm.* **56**(8): 365–369.

Kirpotin, D. B., D. C. Drummond et al. (2006). Antibody targeting of long-circulating lipidic nanoparticles does not increase tumor localization but does increase internalization in animal models. *Cancer Res.* **66**(13): 6732–6740.

Kohler, G. and C. Milstein. (1975). Continuous cultures of fused cells secreting antibody of predefined specificity. *Nature* **256**(5517): 495–497.

Korner, M., M. Stockli et al. (2007). GLP-1 receptor expression in human tumors and human normal tissues: Potential for in vivo targeting. *J. Nucl. Med.* **48**(5): 736–743.

Krohn, K. A. (2001). The physical chemistry of ligand-receptor binding identifies some limitations to the analysis of receptor images. *Nucl. Med. Biol.* **28**(5): 477–483.

Larson, S. M. (1987). Lymphoma, melanoma, colon cancer: Diagnosis and treatment with radiolabeled monoclonal antibodies. The 1986 Eugene P. Pendergrass New Horizons Lecture. *Radiology* **165**(2): 297–304.

Laverman, P., M. Behe et al. (2004). Two technetium-99m-labeled cholecystokinin-8 (CCK8) peptides for scintigraphic imaging of CCK receptors. *Bioconjug. Chem.* **15**(3): 561–568.

Lawrentschuk, N., F. T. Lee et al. (2011). Investigation of hypoxia and carbonic anhydrase IX expression in a renal cell carcinoma xenograft model with oxygen tension measurements and (1)(2)(4)I-cG250 PET/CT. *Urol. Oncol.* **29**(4): 411–420.

Leahy, M. F., J. F. Seymour et al. (2006). Multicenter phase II clinical study of iodine-131-rituximab radioimmunotherapy in relapsed or refractory indolent non-Hodgkin's lymphoma. *J. Clin. Oncol.* **24**(27): 4418–4425.

Lim, J. C., Y. D. Hong et al. (2012). Synthesis and biological evaluation of a novel (177)Lu-DOTA-[Gly(3)-cyclized(Dap(4), (d)-Phe(7), Asp(10))-Arg(11)]alpha-MSH(3–13) analogue for melanocortin-1 receptor-positive tumor targeting. *Cancer Biother. Radiopharm.* **27**(8): 464–472.

Low, P. S., W. A. Henne et al. (2008). Discovery and development of folic-acid-based receptor targeting for imaging and therapy of cancer and inflammatory diseases. *Acc. Chem. Res.* **41**(1): 120–129.

Mather, S. J., A. J. McKenzie et al. (2007). Selection of radiolabeled gastrin analogs for peptide receptor-targeted radionuclide therapy. *J. Nucl. Med.* **48**(4): 615–622.

McLarty, K., B. Cornelissen et al. (2009). Micro-SPECT/CT with 111In-DTPA-pertuzumab sensitively detects trastuzumab-mediated HER2 downregulation and tumor response in athymic mice bearing MDA-MB-361 human breast cancer xenografts. *J. Nucl. Med.* **50**(8): 1340–1348.

Miao, Y. and T. P. Quinn. (2007). Alpha-melanocyte stimulating hormone peptide-targeted melanoma imaging. *Front. Biosci.* **12**: 4514–4524.

Morris, M. J., C. R. Divgi et al. (2005). Pilot trial of unlabeled and indium-111-labeled anti-prostate-specific membrane antigen antibody J591 for castrate metastatic prostate cancer. *Clin. Cancer Res.* **11**(20): 7454–7461.

Muldoon, L. L., S. J. Lewin et al. (2011). Imaging and therapy with rituximab anti-CD20 immunotherapy in an animal model of central nervous system lymphoma. *Clin. Cancer Res.* **17**(8): 2207–2215.

Muller, C. (2013). Folate-based radiotracers for PET imaging—Update and perspectives. *Molecules* **18**(5): 5005–5031.

Muller, C., M. Bruhlmeier et al. (2006). Effects of antifolate drugs on the cellular uptake of radiofolates in vitro and in vivo. *J. Nucl. Med.* **47**(12): 2057–2064.

Muller, C. and R. Schibli. (2011). Folic acid conjugates for nuclear imaging of folate receptor-positive cancer. *J. Nucl. Med.* **52**(1): 1–4.

Muller, C. and R. Schibli. (2013a). Prospects in folate receptor-targeted radionuclide therapy. *Front. Oncol.* **3**: 249.

Muller, C. and R. Schibli. (2013b). Single photon emission computed tomography tracer. *Recent. Results Cancer Res.* **187**: 65–105.

Nagengast, W. B., M. N. Hooge et al. (2011). VEGF-SPECT with (1)(1)(1)In-bevacizumab in stage III/IV melanoma patients. *Eur. J. Cancer* **47**(10): 1595–1602.

Nayak, T. K., M. Bernardo et al. (2013). Orthotopic pleural mesothelioma in mice: SPECT/CT and MR imaging with HER1- and HER2-targeted radiolabeled antibodies. *Radiology* **267**(1): 173–182.

Nowakowski, G. S. and T. E. Witzig. (2006). Radioimmunotherapy for B-cell non-Hodgkin lymphoma. *Clin. Adv. Hematol. Oncol.* **4**(3): 225–231.

Oosterwijk, E., C. R. Divgi et al. (2003). Monoclonal antibody-based therapy for renal cell carcinoma. *Urol. Clin. North Am.* **30**(3): 623–631.

Osborne, J. R., N. H. Akhtar et al. (2013). Prostate-specific membrane antigen-based imaging. *Urol. Oncol.* **31**(2): 144–154.

Parker, N., M. J. Turk et al. (2005). Folate receptor expression in carcinomas and normal tissues determined by a quantitative radioligand binding assay. *Anal. Biochem.* **338**(2): 284–293.

Perik, P. J., M. N. Lub-De Hooge et al. (2006). Indium-111-labeled trastuzumab scintigraphy in patients with human epidermal growth factor receptor 2-positive metastatic breast cancer. *J. Clin. Oncol.* **24**(15): 2276–2282.

Pradhan, T. K., T. Katsuno et al. (1998). Identification of a unique ligand which has high affinity for all four bombesin receptor subtypes. *Eur. J. Pharmacol.* **343**(2–3): 275–287.

Pryma, D. A., J. A. O'Donoghue et al. (2011). Correlation of in vivo and in vitro measures of carbonic anhydrase IX antigen expression in renal masses using antibody 124I-cG250. *J. Nucl. Med.* **52**(4): 535–540.

Psimadas, D., M. Fani et al. (2012). Synthesis and comparative assessment of a labeled RGD peptide bearing two different (9)(9)mTc-tricarbonyl chelators for potential use as targeted radiopharmaceutical. *Bioorg. Med. Chem.* **20**(8): 2549–2557.

Reubi, J. C. (2003). Peptide receptors as molecular targets for cancer diagnosis and therapy. *Endocr. Rev.* **24**(4): 389–427.

Reubi, J. C., J. C. Schaer et al. (1997). Cholecystokinin(CCK)-A and CCK-B/gastrin receptors in human tumors. *Cancer Res.* **57**(7): 1377–1386.

Reubi, J. C., B. Waser et al. (1998). Neurotensin receptors: A new marker for human ductal pancreatic adenocarcinoma. *Gut* **42**(4): 546–550.

Rizvi, S. N., O. J. Visser et al. (2012). Biodistribution, radiation dosimetry and scouting of 90Y-ibritumomab tiuxetan therapy in patients with relapsed B-cell non-Hodgkin's lymphoma using 89Zr-ibritumomab tiuxetan and PET. *Eur. J. Nucl. Med. Mol. Imaging* **39**(3): 512–520.

Russell-Jones, G., K. McTavish et al. (2004). Vitamin-mediated targeting as a potential mechanism to increase drug uptake by tumours. *J. Inorg. Biochem.* **98**(10): 1625–1633.

Salmaso, S., J. S. Pappalardo et al. (2009). Targeting glioma cells in vitro with ascorbate-conjugated pharmaceutical nanocarriers. *Bioconjug. Chem.* **20**(12): 2348–2355.

Schmidt, M. M. and K. D. Wittrup. (2009). A modeling analysis of the effects of molecular size and binding affinity on tumor targeting. *Mol. Cancer Ther.* **8**(10): 2861–2871.

Shetzline, M. A., J. K. Walker et al. (2002). Vasoactive intestinal polypeptide type-1 receptor regulation. Desensitization, phosphorylation, and sequestration. *J. Biol. Chem.* **277**(28): 25519–25526.

Siegel, B. A., F. Dehdashti et al. (2003). Evaluation of 111In-DTPA-folate as a receptor-targeted diagnostic agent for ovarian cancer: Initial clinical results. *J. Nucl. Med.* **44**(5): 700–707.

Signore, A., S. J. Mather et al. (2010). Molecular imaging of inflammation/infection: Nuclear medicine and optical imaging agents and methods. *Chem. Rev.* **110**(5): 3112–3145.

Sosabowski, J. K., T. Matzow et al. (2009). Targeting of CCK-2 receptor-expressing tumors using a radiolabeled divalent gastrin peptide. *J. Nucl. Med.* **50**(12): 2082–2089.

Stillebroer, A. B., G. M. Franssen et al. (2013). ImmunoPET imaging of renal cell carcinoma with (124)I- and (89) Zr-labeled anti-CAIX monoclonal antibody cG250 in mice. *Cancer Biother. Radiopharm.* **28**(7): 510–515.

Tagawa, S. T., M. I. Milowsky et al. (2013). Phase II study of Lutetium-177-labeled anti-prostate-specific membrane antigen monoclonal antibody J591 for metastatic castration-resistant prostate cancer. *Clin. Cancer Res.* **19**(18): 5182–5191.

Tamura, K., H. Kurihara et al. (2013). 64Cu-DOTA-trastuzumab PET imaging in patients with HER2-positive breast cancer. *J. Nucl. Med.* **54**(11): 1869–1875.

Thakur, M. L., M. R. Aruva et al. (2004). PET imaging of oncogene overexpression using 64Cu-vasoactive intestinal peptide (VIP) analog: Comparison with 99mTc-VIP analog. *J. Nucl. Med.* **45**(8): 1381–1389.

Tolmachev, V. and A. Orlova. (2010). Influence of labelling methods on biodistribution and imaging properties of radiolabelled peptides for visualisation of molecular therapeutic targets. *Curr. Med. Chem.* **17**(24): 2636–2655.

Vallabhajosula, S., S. J. Goldsmith et al. (2005). Prediction of myelotoxicity based on bone marrow radiation-absorbed dose: Radioimmunotherapy studies using 90Y- and 177Lu-labeled J591 antibodies specific for prostate-specific membrane antigen. *J. Nucl. Med.* **46**(5): 850–858.

Vugts, D. J., D. A. Heuveling et al. (2014). Preclinical evaluation of 89Zr-labeled anti-CD44 monoclonal antibody RG7356 in mice and cynomolgus monkeys: Prelude to Phase 1 clinical studies. *MAbs* **6**(2): 567–575.

Waibel, R., H. Treichler et al. (2008). New derivatives of vitamin B12 show preferential targeting of tumors. *Cancer Res.* **68**(8): 2904–2911.

Weckbecker, G., I. Lewis et al. (2003). Opportunities in somatostatin research: Biological, chemical and therapeutic aspects. *Nat. Rev. Drug Discov.* **2**(12): 999–1017.

Weitman, S. D., R. H. Lark et al. (1992). Distribution of the folate receptor GP38 in normal and malignant cell lines and tissues. *Cancer Res.* **52**(12): 3396–3401.

Wicki, A., D. Wild et al. (2007). [Lys40(Ahx-DTPA-111In)NH2]-Exendin-4 is a highly efficient radiotherapeutic for glucagon-like peptide-1 receptor-targeted therapy for insulinoma. *Clin. Cancer Res.* **13**(12): 3696–3705.

Wild, D., M. Behe et al. (2006). [Lys40(Ahx-DTPA-111In)NH2]exendin-4, a very promising ligand for glucagon-like peptide-1 (GLP-1) receptor targeting. *J. Nucl. Med.* **47**(12): 2025–2033.

Wild, D., M. Frischknecht et al. (2011). Alpha- versus beta-particle radiopeptide therapy in a human prostate cancer model (213Bi-DOTA-PESIN and 213Bi-AMBA versus 177Lu-DOTA-PESIN). *Cancer Res.* **71**(3): 1009–1018.

Wu, Z., I. Todorov et al. (2011). In vivo imaging of transplanted islets with 64Cu-DO3A-VS-Cys40-Exendin-4 by targeting GLP-1 receptor. *Bioconjug. Chem.* **22**(8): 1587–1594.

Yang, J., H. Guo et al. (2009). Evaluation of a novel Arg-Gly-Asp-conjugated alpha-melanocyte stimulating hormone hybrid peptide for potential melanoma therapy. *Bioconjug. Chem.* **20**(8): 1634–1642.

Zhang, H., J. Chen et al. (2004). Synthesis and evaluation of bombesin derivatives on the basis of pan-bombesin peptides labeled with indium-111, lutetium-177, and yttrium-90 for targeting bombesin receptor-expressing tumors. *Cancer Res.* **64**(18): 6707–6715.

Section VII
Image Quantification

22

Quantification in Nuclear Preclinical Imaging

Istvan Szanda

22.1 INTRODUCTION

Preclinical imaging is a powerful tool to enhance the ways in which biological processes are studied. Different imaging technologies (modalities) provide different and usually complementary information about the tissues. All modalities convert the acquired information to images, regardless of how abstract (i.e., relaxation of the spins in MRI) or substantially different (i.e., sound waves in ultrasound [US]) the underlying physical phenomena providing the information are. In order to draw valid conclusions from the images regarding the underlying biochemical processes, understanding each modality and their limitations is necessary.

In general, quantification covers a set of approaches that aim to remove, or at least minimize variability in the observation process. Variability has three main sources. The first source is the animal biology; for instance, studies involving the comparison of drug effect on multiple groups of animals, variability arising from the different states of the animals has to be minimized. The second is bias from the scanning process. As an example, for longitudinal studies where the same animal is imaged multiple times, it is important for the scanner to be well-calibrated for each imaging session. The third is subjectivity in image analysis. For instance, for studies when different people analyze the images due to the large number of animals or multicenter trials, the variability introduced by the observer may alter the false results. The aim of this chapter is to discuss these sources and summarize strategies aiming to achieve quantification.

In this chapter, quantification is referred to as a numerical expression of concentration. Optical and nuclear modalities (single photon emission computerized tomography [SPECT] and positron emission tomography [PET]) provide primarily functional information; in these techniques, a tracer or imaging agent is injected into the animal and the regional concentration is investigated. For modalities, providing primarily anatomical information, characteristics of the tissue can be measured during the image acquisition process without the injection of any substrate. In computed tomography (CT), magnetic resonance imaging (MRI), and ultrasound (US), the object is exposed to external radiation (X-rays, electromagnetic waves, and US waves, respectively), and the interaction between the tissue and the external radiation is utilized for imaging purposes. In CT, interaction is related to the linear attenuation (as a function of the density of a certain tissue) of the material. In US, besides attenuation, reflection and backscatter may also contribute significantly to the imaging process. In MRI, interaction is related to the concentration of the magnetically active nuclei and their relaxation properties. Similar to the primarily functional modalities, in CT, MRI, and US, it is also possible to inject an external substrate in the form of contrast agents, which changes the measured characteristics of the tissue. In CT, for instance, gold nanoparticles can be injected to enhance contrast due to their higher attenuation properties. In US, injected microbubbles can enhance scatter due to their nonlinear characteristics, resulting in a higher contrast compared to blood. In MRI, gadolinium contrast agents or hyperpolarized nuclei (i.e., carbon) can significantly enhance signal-to-noise ratio.

All modalities produce organized numbers in the form of images. However, modalities can be characterized in terms of how quantitative they are, which is determined by the nature of the underlying physical phenomena and the acquisition process. Among the primarily functional modalities, optical techniques are on the lower end of the quantitative spectrum due to the limitation due to optical density. For the primarily anatomic modalities, similarly, ultrasound can offer limited quantification due to high attenuation, and also the nonlinear propagation of the sound waves. Other techniques, like MRI, can be considered as more quantitative (James and Gambhir 2012). However, for MRI, the underlying physical processes are very complex allowing only an indirect measurement of the concentration of the magnetically active nuclei; moreover, the acquisition procedure is complex, having a direct impact on the response of the examined system. MRI is characterized by very low sensitivity; this results in the requirement of large volumes of contrast agents that may perturb the system subject of examination, and in some cases cause toxicity (Hasebroock and Serkova 2009). CT is also a more quantitative modality, as the X-ray attenuation of the tissue is well defined and in good correlation with the density of the normal tissue. However, density is not a straightforward consequence of any biological function and concerns, for some contrast agents are similar to MRI (Hildebrandt et al. 2008; Taylor et al. 2012).

Due to their potential of being highly quantitative, this chapter focuses on nuclear imaging modalities. The underlying physical principle of the measurement is simple: every molecule labeled with a radioactive isotope emits some form of radiation. The emitted radiation may be detected directly in SPECT or indirectly via an intermediate process (annihilation) in PET. In nuclear imaging modalities, the acquisition process has no impact on the underlying processes of radioactive decay. Moreover, these isotopes are unlikely to be found in the body prior to injection, and gamma photons are likely to penetrate the object. During the scanning process, regions containing more radioactivity produce proportionally more photons, which may be detected by the scanner. So with these modalities, collected counts from a region used to produce images may be a linear function of the concentration of the tracer in the tissue, which greatly contributes to being quantitative. Moreover, these techniques are very sensitive compared to MRI and CT. From an imaging agent viewpoint, there is approximately 7–10 orders of magnitude difference between MRI and PET/SPECT sensitivity: while in the case of MRI 10^{-3}–10^{-5} M concentration is necessary for detection, for nuclear techniques it is only 10^{-10}–10^{-12} M (James and Gambhir 2012). As a result, only a small amount of tracer needs to be injected, which is unlikely to perturb the system.

22.2 BIOLOGICAL ASPECTS OF QUANTIFICATION

The impact of the animal state on the results can be significant; researchers have reported a tenfold difference in the results due to environmental conditions (Fueger et al. 2006). In this section, issues regarding variability arising from the response of the animal's response to environmental conditions are briefly reviewed.

22.2.1 ANESTHESIA

One of the main differences between clinical and preclinical imaging is the requirement for animal anesthesia in most cases. The impact of anesthesia on the animal physiology is profound, and the uptake of radiotracer into different organs can be affected through changes in systematic physiology and local perfusion (Balaban and Hampshire 2001; Constantinides et al. 2011). These differences can be especially important in studies where transgenic animals are compared to wild-type animals, as transgenic animals can be more susceptible to anesthesia (Balaban and Hampshire 2001; Roth et al. 2002). In order to control undesirable impacts of anesthesia, monitoring of physiological functions may be necessary. Depth of anesthesia can be approximated by monitoring respiration rate and rectal body temperature. Blood sampling, while invasive, can provide arterial blood gas concentration; however, repeated sampling is problematic due to the difficulties in cannulation of the animal and the small total blood volume in the mouse.

Anesthesia is typically administered via intravenous injection (such as propofol, pentobarbital, or ketamine) or inhalation (such as isoflurane). One such example of their impact on metabolism, anesthesia may cause hyperglycemia (Fueger et al. 2006), which may have a profound impact on 2-deoxy-2-(18F)fluoro-D-glucose (FDG) studies, widely used in PET imaging. For other tracers, investigating the impact of anesthesia on tracer uptake may be required.

22.2.2 BODY TEMPERATURE

Adequate warmth is especially important in small animals, because their body surface to body weight ratio, and their metabolic rates, which is nearly tenfold higher than that for larger mammalian species (Balaban and Hampshire 2001). One of the practical consequences is that the loss of fluids during the scan procedure can be significant, causing fluid imbalances in their metabolism and altering physiological parameters. For mice, the temperature range, where no active processes in the body are necessary to maintain body temperature, lies way above room temperature (Ambrosini et al. 2009). Below this temperature, there may be increased activity in brown adipose tissue and muscles, altering uptake in the muscles; for example, changing the tumor-to-muscle ratio in tracer uptake.

22.2.3 METABOLIC CYCLES AND FASTING

Eating and sleeping are part of the natural cycles and the physiological state highly depends on what time of the day animals are imaged. For instance, rodents are naturally more active during the evening than in the morning, so animals studied in the morning compared to the ones studied later during the day can produce different tracer uptake (Hildebrandt et al. 2008). One of the strategies to handle this is to image the rodents at the same time of the day. Eating is another factor, which has a significant impact on the physiological state. For instance, for FDG PET studies, high blood glucose levels can reduce the tracer uptake in the organs, as glucose and the radiolabeled analogue compete with each other for the same receptors. Fasting of the animals is recommended to have stabilized FDG uptake (Ambrosini et al. 2009; Kreissl et al. 2011).

22.2.4 TRACER ADMINISTRATION

Tracer administration may have a significant impact on the physiology of the animal. In order to produce a sufficient number of counts for imaging, injection of relatively large volumes (as a function of the specificity of the radiotracer) may be necessary, which may have a profound impact on the fluid balance of the animal.

Besides volume, radiation dose received by the animal is desired to be minimal (Funk et al. 2004); this is especially true for combining CT scans, which also deliver radiation dosage, alongside the nuclear imaging acquisitions. The received dose can result in profound radiotherapy, making it difficult to decide whether the observed changes (for instance, in tumor size) are the results of the treatment response or the scanning procedure in longitudinal studies (Schambach et al. 2010).

22.2.5 MOTION AND OTHER FACTORS

Respiratory motion may have a profound effect on quantification in chest and abdominal regions, resulting in additional blurring in the images (Segars et al. 2004). Various other factors can influence the state of the animal, which are previously discussed in Chapter 3. Examples are stressors like noise, holding conditions and animal handling by staff (Balcombe et al. 2004). Uptake period may also have a significant impact on the results, as well as profound factors like inflammation and infection.

As environmental factors may have a significant impact on animal studies, the effects of each environmental stimulus are advised to be separately assessed. However, there is an ongoing debate about whether a more rigorous environmental standardization will enhance the otherwise generally poor reproducibility of animal studies, or whether standardization itself introduces bias to an experiment. Although it has been suggested that varying environmental conditions might eventually lead to more robust results (Richter et al. 2009; Martic-Kehl et al. 2012), they have to be adjusted in a controlled manner.

22.3 TECHNOLOGICAL ASPECTS OF QUANTIFICATION

In this section, three aspects of instrumentalization impacting quantification are covered. The first corresponds to the corrections that are implemented in the scanners. The second, briefly reviews reconstruction as an important factor, and the third covers some aspects of the CT technology widely used in preclinical scanners to correct for photon losses.

22.3.1 Corrections Applied in Nuclear Modalities

In an ideal situation, all generated photons should be detected by the scanner with the characteristics (i.e., energy and momentum) they have emitted with. However, this is rarely the case. Corrections are necessary to take into account the underlying physical and technological processes to compensate for inaccuracies of the acquisition process. Some corrections are necessary due to physical phenomena (for instance, attenuation and scatter), while others are necessary because characteristics of the scanner are not ideal (for instance, at high count rates, the sensitivity decreases). From a technological viewpoint, robust and effective corrections are vital for quantitative studies. The following sections briefly review different phenomena, their direct impact on quantification and the main strategies for their correction.

22.3.2 Attenuation and Scatter Correction

Photons may travel different distances within the object prior to reaching the detector's surface. Photon–tissue interactions can be characterized on the macroscopic level as attenuation and scatter. When photons have lost their energy completely and cannot reach the detector, the process is called attenuation. When photons reach the detector but with reduced energy and the direction of travel may have changed, the process is called scatter. Although this is macroscopic differentiation, underlying microscopic interactions are very similar in both cases.

The impact of attenuation on quantification is straightforward: the lost photons decrease the perceived number of counts from a region, proportionally underestimating its activity concentration. As this process depends on how deep an organ is located within the body, regions close to the skin (for instance, subcutaneous tumors) appear to contain more activity compared to an organ with the same amount of activity located deeper within the animal. If scattered photons reached the detector but their energy is lower than a threshold value, often called the lower energy discriminator, they are also considered by the scanner as lost photons as in the case of attenuation. Scattered photons that have high enough energy when detected by the scanner are not considered as lost counts. However, as their direction may have changed whilst traveling through the object, and therefore may be mispositioned, resulting in additional blurring and background signal in the image, which can lead to errors in quantification of a specific region.

The cross section of gamma photons changes with the energy, meaning that photons with a higher energy may attenuate or scatter with lower likelihood than photons carrying lower energies. As photon energies in SPECT studies are lower than in PET studies, it suggests that attenuation is a more significant problem for SPECT. However, in PET both annihilation photons are attenuated, attenuation factors are reported to be higher in clinical studies (Bailey 1998). Moreover, scatter fraction depends on the energy resolution of the detector, which is usually higher for SPECT than for PET. In the case of multiple-isotope SPECT imaging, correction of scatter is especially important; as a result of scatter, higher-energy photons from isotope "A" may lose their energy and be detected in the lower energy window setup for isotope "B." This phenomenon is usually referred to as cross-talk (Hutton et al. 2011).

In the first preclinical scanners, attenuation properties were determined by scanning the animal using an external radioactive source with a known activity level. By calculating the fraction of the photons penetrating the object from different projection angles, attenuation maps were generated. This method is called the transmission method. The main advantages are that radiation sources used for generating attenuation maps are similar (i.e., Cs-137 used for SPECT images) or even identical (i.e., Ge-68 used for PET images) to the nuclear isotopes injected; while in addition, attenuation maps may not suffer from co-registration issues. On the other hand, generating and attenuation map may take a long time and generated images might be very noisy. The clinical success of combining nuclear modalities with CT, preclinically triggered the usage of a similar technology; in state-of-the-art scanners today, CT may be used to generate attenuation maps. In these cases, scaling the linear attenuation factor from the CT energy range to the nuclear modality range is necessary (Kinahan et al. 1998). Attenuation maps generated by the CT are usually applied during reconstruction (in the case of iterative methods). Due to utilizing CT to generate attenuation maps, accuracy of CT quantification is linked to quantification of PET and SPECT. Further discussion on this matter is found in Section 22.3.8.

22.3.3 Detector Uniformity Correction

For SPECT and PET, detection of gamma photons is performed with a scintillator crystal, which can be either continuous or pixelated. This crystal converts the energy of gamma into visible photons. The visible light is then converted into electrons and forms an electronic pulse. The most widely used technology to complete the second process is the application of photomultiplier tubes (PMT). By default, no detector gives homogeneous response to homogeneous irradiation. Large differences can be handled by adjusting the hardware appropriately, for instance, by adjustment of high voltage individually for each PMT to compensate for their different amplification. However, in order to correct for smaller inhomogeneities in detector sensitivity, corrections are necessary on an image level as well. For SPECT and PET, this correction is usually referred to as uniformity calibration and normalization.

Differences in detector efficiency often result in various artifacts within the images that can reduce the efficiency of the image analysis. For instance, by showing uptake where there is no uptake or decrease the perceived uptake because of artifacts.

For SPECT, uniformity correction usually involves removing the collimators and homogeneously irradiating the detectors with a point source. After the collection of a considerable number of counts, a pixel-by-pixel correction factor is calculated based on the variation in the "flood" images. These correction factors are stored as uniformity tables for application to future image acquisitions.

Although in PET the principle of the method is similar, coincidence detection introduces technological complications. As one annihilation event is detected by two detectors simultaneously, the aim of normalization is to correct different sensitivities in "lines of responses" (LORs) joining the pairs of detectors. Apart from the variation in individual detector efficiencies similarly to SPECT, in PET geometric factors may also play a significant part due to coincidence detection. The most straightforward way of obtaining a full set of normalization coefficients would be to perform a scan where every LOR is illuminated by the same source; however, it is not possible due to the large number of LORs. In practice, the coefficients are modeled as products of geometric factors and intrinsic crystal efficiencies. If the geometric factors are known, intrinsic crystal efficiencies can be calculated (Badawi et al. 1998).

22.3.4 Randoms Correction (PET)

In PET scanners, detected photonpairs meeting temporal (detected within a short period of time, usually a few nanoseconds) and spatial criteria (detected approximately in the opposite direction) are counted as coincidence events originating from the same annihilation event. However, scatter, attenuation, and geometrical factors cause many photons to be detected unpaired. Random coincidences occur when two unpaired photons are detected by the scanner meeting the requirement of coincidence detection. Probability of this occurrence can be calculated from the number of detected events (singles) per detector and the coincidence time window. At higher count levels, the random rate is significantly higher; it is proportional to the square of the singles count rate level (Brasse et al. 2005).

A high random rate has an impact on both image quality and the number of counts. As randoms have no specific distribution, they create a flat background, similar to scatter events. This background signal can compromise the ratio when a low uptake region is compared to a high uptake region (for instance, tumor-to-muscle ratio) at high activities.

The most common randoms correction method is based on the subtraction of random coincidences. In this method, two independent coincidence circuits test whether the detected event fulfils the coincidence criteria. The first circuit tests in the standard time window (usually a few nanoseconds) to determine the prompt coincidence events, which can be true events, scattered events, or randoms. The second circuit tests the events in a significantly delayed window in which only random events are detected. As the numbers of randoms are equal in both windows, delayed window measurement can be subtracted from the standard time window measurement, which will then contain only true and scattered events (Brasse et al. 2005).

22.3.5 Deadtime Correction

PET and SPECT scanners can be modeled as a series of subsystems, where each of them needs a minimum time to reset to a standby state between two consecutive events to record them as separate. Deadtime corresponds to the period after the recording of a particle or pulse when a detector is unable to record another one. At high count rates, the probability of radioactive decay occurring within unresponsive time slot can be significantly high, resulting in count losses compromising the scanner's linear response. There are different sources of deadtime depending on the electronics. The first is the so-called integration time during which electrons from PMTs are acquired. The second one is the reset time, during which the electronics cannot take new events. The third one is the coincidence circuit, which only occurs in PET systems due to temporal sorting introducing a need for complicated coincidence circuits. While in SPECT, only a relatively small number of events reach the detector due to the collimation technology, in PET events are sorted after detection. Due to the higher count rates and the time necessary for coincidence sorting, deadtime may have a greater impact on PET than SPECT scanners.

As a result of deadtime, count rates detected by the scanner are not a linear function of the activity in the field of view (FOV). Comparing data acquired at low count levels to data acquired at high count levels may not be valid if deadtime is not adequately corrected. Where absolute quantification in MBq is necessary, as may be in multicenter trials, deadtime correction is extremely important.

The most straightforward method is to calculate a correction factor corresponding to the nonlinearity of the scanner. Data can be multiplied with this factor to correct for the deadtime effect. This correction, however, does not take into account the spatial distribution of the tracer in the animal. Other methods measure the "live time" of the different electronic blocks or by fitting an analytic model to the system parameters (Bailey 2005).

22.3.6 Finite Resolution Effects

Spatial resolution is a key factor in preclinical imaging as organs can be very small. Spatial resolution is limited by different factors depending on the modality: in case of SPECT, these are technological; while in PET, technological factors as well as physical phenomena need to be considered.

As a complex activity distribution may be considered a superposition of several point sources, technological factors in SPECT can be studied by measuring the collimator-detector response (CDR) by imaging a point source. Main components of the CDR are intrinsic response, geometric response, septal penetration, and septal scatter (Rahmim and Zaidi 2008). Intrinsic response covers noncollimator related parameters like statistical variation in the PMT output and inter-crystal scatter. As shape of the collimators defines the angle of acceptance, the geometric response gets wider with an increasing distance from the detector surface (Shokouhi et al. 2009). Septal penetration refers to the photons not absorbed by the collimators; septal scatter corresponds to the photons that are scattered by the collimators but detected in the accepted energy window.

In PET, due to the lack of collimators, technological factors are related to the detector itself; the finite stopping power of the crystal may result in scattering and crystal penetration. In these cases, more than one crystal produces light as a response to the incident gamma photon, resulting in an uncertainty in the spatial localization of the annihilation event. A special consequence of the finite stopping power is the so-called parallax effect, where photons coming from oblique angles may only be absorbed by the neighborhood crystals not the one that was initially reached by the photon. Parallax effect results in worsening resolution with increasing radial distance from the center (Wang et al. 2006). Although small crystals may be attractive, as the size of the crystal may have a fundamental impact on the resolution of the PET scanner (Moses 2011), decreasing the crystals' size results in a decreased stopping power.

Other factors in PET include noncolinearity of the annihilation gamma photons and the positron range. The former is due to the original nonzero momentum of the positron or electron at the moment of annihilation; the amount of blurring due to noncolinearity is given roughly by 0.022 multiplied by the detector separation. This effect may only be significant in clinical scanners where the diameter of the scanner is larger. Positron range is defined as the distance that a positron travels before annihilation due to its nonzero kinetic energy. This energy is highly dependent on the radionuclide and also results in a degraded resolution depending on the radioisotope.

The tissue fraction effect has consequences on images generated by both modalities and relates to sampling of an image. Voxel and anatomical borders rarely do not match, which means that one voxel can contain information about more than one anatomical tissue. The simplest strategy to minimize this effect is to decrease the voxel size, provided the resulting noise increase is acceptable.

Finite resolution leads to a phenomenon called the partial volume effect. Figure 22.1 is an illustration of this: if the object is small, the observed activity concentration may be largely dependent on the size of the object. In other words, small regions may look "colder" than they are (in extreme cases remaining below the threshold of detectability), whilst activity concentration observed in large regions may be closer to the true values. An example of a consequence of this bias is a small but highly proliferating tumor (in case of high tracer uptake) that appears less aggressive than it is. There are different strategies to correct for finite resolution in SPECT and PET. In some cases, similar to the parallax effect, measuring a more precise detector response can be applied (Wang et al. 2006). However, in most of the cases, researchers attempt to include *a priori* information known about the scanner (for instance, detector response) or the object

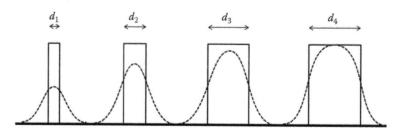

FIGURE 22.1 Schematic, two-dimensional illustration of the partial volume effect. The solid line corresponds to the true activities in regions with increasing diameters ($d_1 - d_4$) while dashed lines represent the observed activity concentration. In small objects, due to the finite resolution of the scanner resulting in a blurring effect, uptake is significantly underestimated.

(positron range) into an iterative reconstruction. Partial volume correction is a vast, complicated field being one of the highly complex inverse problems in physics. Soret et al. (2007) and more recently Erlandsson et al. (2012) have reviewed several different approaches used in PET and SPECT imaging.

22.3.7 RECONSTRUCTION METHODS

Reconstruction is a complex part of the imaging procedure requiring research and a specific attention from a quantification viewpoint. There are two main types of reconstruction methods available: analytical and statistical. The most widely used analytical method is filtered back-projection, which is a discretized form of a mathematically well-defined X-ray transformation. The main advantages are its linearity and speed of calculation. Statistical methods, like maximum likelihood expectation maximization (MLEM), require more computational power and may be nonlinear. However, *a priori* knowledge of physics and technological aspects can be incorporated into statistical methods, which may have a dramatic impact on the quality of the images and the observed resolution of the scanner. For example, these aspects can include the geometrical model of the scanner, as well as detector response, positron range, intercrystal scatter, attenuation, and scatter effects. One of the drawbacks of the statistical methods is that they are sensitive to inaccuracies in the scanner model, so minimizing these inaccuracies is important. Moreover, as the results might be sensitive to different reconstruction parameters, like the number of iterations, choosing these parameters for each specific study may be important for robust quantification.

22.3.8 CT QUANTIFICATION

If the attenuation correction is based on CT-generated attenuation maps, CT quantification has a direct impact on the quantification of small animal emission tomography. There are two main challenges in preclinical CT from an attenuation correction viewpoint. First is beam hardening, which is related to the phenomenon that the linear attenuation coefficient is dependent on the energy of the X-ray. As no X-ray source is monochromatic, it contains a range of lower and higher energies. Lower energy rays have higher absorption rate, so the relative ratio of the higher energy rays increases while penetrating through an object. In other words, "the beam is hardening," resulting in a similar effect as attenuation in emission tomography: the CT numbers are lower in the center of a homogeneous object than closer to the edges. The other challenge is the relatively high noise coming from the cone-beam technology. Compared to the fan-beam CT widely used clinically, with the higher number of scattered photons noise increases in images (Scarfe and Farman 2008). Applying beam hardening correction and appropriate noise-reduction filtering in CT reconstruction may improve precision of the attenuation correction.

22.3.9 IS SPECT OR PET MORE ADVANTAGEOUS FROM A QUANTIFICATION VIEWPOINT?

There is an ongoing debate whether clinical PET or SPECT is the prime nuclear imaging technique (Mariani et al. 2008; Hicks and Hofman 2012). However, it can be argued that both preclinical nuclear modalities offer specific advantages making them complementary research tools.

The main advantages of preclinical SPECT are the following. First of all, in case of the clinical systems, the resolution of preclinical SPECT is significantly higher than the resolution of preclinical PET. This is due to the pinhole technology primarily used in preclinical SPECT compared to the parallel-hole collimator technology in clinical SPECT; as well as the lack of limitation coming from a positron range in PET. SPECT offers this very important advantage of higher spatial resolution, which may result in a lower partial volume effect. From an application viewpoint, PET imaging is possible with only one tracer at a time due to the energy of the annihilation photons being independent of the isotope, while in SPECT, multiple tracers can be concurrently injected for imaging.

Due to their different gamma energies, isotopes can be separated allowing the targeting of different biological functions at the same time. Moreover, due to the longer half-lives of SPECT tracers, this modality may be extremely useful if biological processes need to be imaged on a time scale of days, which may not be practical with commonly used PET isotopes.

On the other hand, the sensitivity of PET is greater by approximately two orders of magnitude than SPECT due to the lack of collimation. Moreover, scan times with PET systems are shorter due to the lack of rotation, which makes dynamic imaging highly feasible and can provide valuable information about tracer kinetics. Unlike SPECT, where there is always a compromise between resolution and sensitivity: the small pinholes may result in higher spatial resolution but lower sensitivity while the large ones may result in lower spatial resolution but higher sensitivity (Peterson and Shokouhi 2012), in 3D PET, high sensitivity and resolution can be achieved simultaneously. Due to the magnification effect of the pinholes in SPECT, the highest spatial resolution can be achieved while limiting the axial FOV; due to the lack of magnification effect in PET scanners, axial FOV of the scanner is usually much larger (Peterson and Shokouhi 2012). PET usually allows the imaging of a whole mouse or even a rat in one bed position, which provides dynamic biodistribution information about every organ taking full advantage of the noninvasive nature of imaging.

Most technological issues regarding the quantification of SPECT and PET (i.e., attenuation and detector response), as well as correction strategies (i.e., CT-based attenuation map and modeling detector response in the reconstruction), may be similar. As the available tracers for these modalities may be different, both preclinical SPECT and PET are valuable research tools providing great advantages to research.

22.4 IMAGE ANALYSIS ASPECTS OF QUANTIFICATION

Beyond the animal handling and the scanning, image analysis is the next step in the workflow of imaging experiments. This section discusses subjectivity introduced by the person performing the image analysis and a number of strategies to handle this.

22.4.1 LEGACY OF *EX VIVO* BIODISTRIBUTION

While in clinical practice, imaging introduced a substantially new way of diagnosing patients; biodistribution in animals was studied *ex vivo* long before preclinical imaging. With this technique, tracer is injected into the animals, and after a certain uptake period, organs are separated *ex vivo* into vials and measured in a gamma counter. The uptake of an organ is often expressed as percentage injected dose per gram; to calculate an uptake specific to the tissue, the activity measured in a calibrated gamma counter is divided by the injected dose and the weight of the organ. As preclinical imaging is a relatively new technique, *ex vivo* biodistribution and imaging are often used side-by-side, evoking the need for comparison of the results determined by the two techniques.

22.4.2 MAIN COMPONENT OF IMAGE ANALYSIS: REGIONS OF INTERESTS

The main component of image analysis is to segment the images by defining regions of interest (ROI). It has been reported previously that even in the case of the most widely used tracer (FDG) in clinical PET, the definition of ROI may have a profound impact on quantification accuracy, estimated to result in 50% change in quantification values (Boellaard 2009).

There are two components of the ROI definition: the first is to define the outline and the second is to report a number within the outline. In principle, there are five main strategies to define ROI in order to report uptake values. The integral method involves drawing a large ROI around the organ, which may collect the activity detected outside of the organ boundaries due to the partial volume effect. For integral method, the sum of counts may be reported. In automatic methods, predefined algorithms are used to determine the contour of the organ (Prieto et al. 2012). It can be, for instance, the pixels that are above a threshold value (often defined as percentage of the maximum in the organ or as an absolute counts level) that are included in the ROI. For automatic methods, the

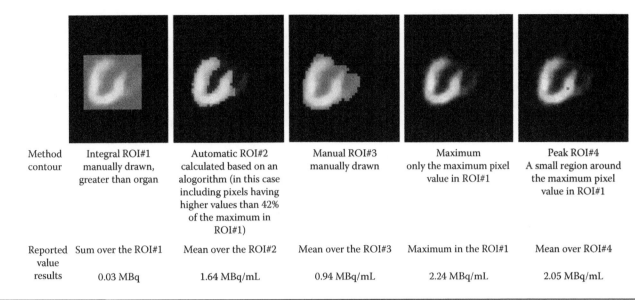

Method contour	Integral ROI#1 manually drawn, greater than organ	Automatic ROI#2 calculated based on an alogorithm (in this case including pixels having higher values than 42% of the maximum in ROI#1)	Manual ROI#3 manually drawn	Maximum only the maximum pixel value in ROI#1	Peak ROI#4 A small region around the maximum pixel value in ROI#1
Reported value results	Sum over the ROI#1	Mean over the ROI#2	Mean over the ROI#3	Maximum in the ROI#1	Mean over ROI#4
	0.03 MBq	1.64 MBq/mL	0.94 MBq/mL	2.24 MBq/mL	2.05 MBq/mL

FIGURE 22.2 Illustration of the five methods of ROI definition in the same mouse heart. Different methods produce a variety of ROIs resulting in different quantification numbers.

mean value calculated over the voxels in the ROI may be reported. Another method is the manual definition of the ROI around the organ, where the mean may be taken from. The next method is the maximum method, where a contour is defined similar to the integral method, and from the maximum pixel value is taken. Finally, the so-called peak method is similar to the previous one, but instead of taking only one pixel value, the mean of the voxels in a small region around the maximum is taken. Figure 22.2 illustrates the different methods for ROI definition. Besides the aforementioned methods, researchers have developed more sophisticated techniques for image segmentation (Maroy et al. 2008; Zaidi and El Naqa 2010).

Every method has advantages and drawbacks. In order to decide which method is the most adequate for a particular study, both the study design and the fundamental research question have to be taken into account. For instance, in a study where only an intra-animal comparison is needed, like evaluation of the uptake in lymph nodes on the inflamed side compared to the non-inflamed side, relative numbers may be sufficient. This is different for multicenter trials or kinetic studies, where uptake values may have to be expressed in activity concentration (MBq/mL) for comparison of data from different centers. One of the approaches for overcoming the need for MBq/mL in multicenter trials is the definition of tumor-to-blood or tumor-to-muscle ratio for a tracer. However, it may increase sensitivity to environmental parameters, which may already be subjected to greater variability in multicenter trials.

The need for comparison of image-based uptakes to *ex vivo* biodistribution data defines different priorities. In *ex vivo* studies, dissected organs can be weighed and the sums of counts in organs are usually measured in a gamma counter method, which is not affected by partial volume effect. For comparison, image-based quantification needs to provide a volume of an organ, which then can be converted to weight. Simultaneously, partial volume effect is desired to be minimal. Different methods fulfill these requirements to a different extent. For instance, while maximum method is the least affected by the partial volume effect (and most affected by the noise), taking only a maximum pixel value in the organ and comparing it with a sum of counts in an *ex vivo* dissected organ may not be relevant. As integral method produces a sum of counts over a region, it may lead to more comparable uptake values than other methods. However, the large ROIs around an organ can contain other tissues (for instance, vascularization around the heart and ureter or adrenal glands around the kidneys) resulting in overestimated counts and weights compared to *ex vivo* biodistribution. Manual segmentation may provide more adequate data for both volumes and counts. However, if the manual method aims for a precise volume, uptake values may be underestimated due to the partial volume effect.

Another consideration is the extent of functional homogeneity of the investigated organ. Using maximum and peak methods may introduce bias for large organs, as they consist of different types of tissues resulting in heterogeneous uptake. For instance, different sections of the heart may have different uptake of a tracer and also contains blood, the kidney contains urine, which may be significantly hotter than the rest of the tissue, and the stomach wall and its contents will differ as well. The same problem exists for tumors as they may not be homogeneous (Bedard et al. 2013).

While in larger organs the inhomogeneity can cause significant bias, for smaller organs it may be a result of the partial volume effect. For smaller organs, the integral method can be advantageous as it is not likely to be affected by the partial volume effect, whilst manual segmentation may lead to an underestimated uptake. The maximum method might be a beneficial choice as it is the least sensitive to partial volume effect. However, maximum value in an organ may only be a measure of noise rather than the biologically relevant uptake. Therefore, the peak method might be a good compromise.

Another aspect that needs to be considered is the intra- and inter-observer variability of a method. Typically, the maximum and peak methods are the most robust, as the rules of selection are straightforward. Integral method may be more affected by the subjectivity of the researcher as it involves a manual definition of the ROIs. For automatic methods, ROI definition might not be subjective. An example of subjectivity in automatic methods is setting a threshold value manually, which is unlikely to be the same for every organ, species, scanner, and count level and noise level is threshold is defined based on the maximum pixel. Manual segmentation involves high levels of subjectivity due to, for instance, the different window levels for image visualization and also intra- and inter-observer variability, especially in multicenter trials (Boellaard 2011). However, one of the main advantages of the manually defined ROI is that an experienced and trained researcher can adapt the method for the specific needs of a study and image characteristics. Incorporating these factors into a computer program proves to be very difficult. For this reason, manual segmentation may still be considered as the gold standard in clinical imaging (Zaidi and El Naqa 2010).

The last concern is whether estimation of weights is necessary. As mentioned earlier, while for *ex vivo*, biodistribution measuring the weight is profound, in image analysis, measuring the volume accurately can be difficult as nuclear modalities usually provide only poor anatomical reference. A profound way to improve the accuracy of the drawn outlines would be the use of an anatomical modality to guide segmentation. However, in most cases, the soft tissue contrast of preclinical CT is generally too low for adequate segmentation. With the emerging technology of combining nuclear modalities with MR, this issue can be overcome. Comparison of the soft tissue contrast in CT and MRI images of the same mouse is illustrated in Figure 22.3. Alternatively, anatomical atlases can aid brain and more recently whole body segmentation; however, for these approaches, precise image registration may be difficult to achieve (Gutierrez and Zaidi 2012).

22.4.3 *Ex Vivo* Biodistribution or Image-Based Quantification?

Although *ex vivo* biodistribution and image-based quantification exist simultaneously, they are fundamentally different: while *ex vivo* biodistribution is invasive, imaging is a noninvasive technique. With *ex vivo* biodistribution, temporal information is only possible when using multiple animals

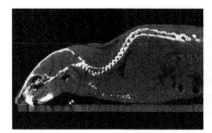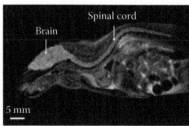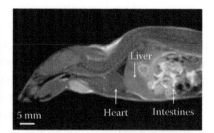

FIGURE 22.3 CT (45 kVp tube voltage), T2 and T1 weighted MRI images of the same animal. Soft tissue contrast of MRI is superior compared to the CT. (Courtesy of Kalman Nagy, Karolinska Institute, Stockholm, Sweden. Images were acquired by a nanoScan, Copyright PET/CT and PET/MRI, respectively.)

introducing inter-animal variability. On the other hand, imaging provides temporal information of tracer kinetics within the same animal. This feature not only provides focus on the standard uptake value but also (especially in PET) an opportunity to study parameters such as the glucose metabolic rate and the pharmacokinetic rate constants by compartmental modeling analysis in dynamic studies (Vriens et al. 2010).

As discussed earlier in this chapter, two main sources of bias in imaging analysis are partial-volume effect and ROI definition. On the other hand, errors during the dissection procedure are very difficult to estimate. Main sources are the excess/loss of tissue during organ excision, especially in the case of small organs like the thyroid, which is close to the trachea, salivary glands, and thymus. Similarly, in some cases, the amount of body fluids changes during the dissection. For instance, the heart may lose an excess amount of bloody fluids while being excised from the body (Cheng et al. 2011).

Although the debate on the preference of one method over the other is still ongoing, imaging has a significant potential to be exploited due to its noninvasive nature.

22.5 CONCLUSION AND FINAL REMARKS

This chapter summarizes the different aspects of quantification in preclinical PET and SPECT imaging. Three aspects covering biological, technological, and image analysis aspects were investigated. The study of these aspects suggest that for successful preclinical studies aiming for reproducibility and objectivity, general understanding of the impact of different parameters and environmental factors is necessary. The largest variability may come, first, from the biological models and their response to environmental effects due to the complexity of physiological functions and, second, from the image analysis due to its subjectivity. Unfortunately, these are the hardest to control. Technological factors may be easier to monitor; however, rigorous quality control of the scanner is essential to achieve reliable results.

ACKNOWLEDGMENTS

I am grateful to Dr. David Lewis and Dr. Greg Mullen for their comments on the biological aspects; to Prof. Paul Marsden, Dr. Lefteris Livieratos, and Dr. Peter Major for their comments on the technological aspects and the image analysis section; to Dr. Kris Thielemans for his comments on the reconstruction section; and to Julia Wykrota for her views on structure and consistency.

REFERENCES

Ambrosini, V., C. Quarta et al. (2009). Small animal PET in oncology: The road from bench to bedside. *Cancer Biother. Radiopharm.* **24**(2): 277–285.

Badawi, R. D., M. A. Lodge et al. (1998). Algorithms for calculating detector efficiency normalization coefficients for true coincidences in 3D PET. *Phys. Med. Biol.* **43**(1): 189–205.

Bailey, D. L. (1998). Transmission scanning in emission tomography. *Eur. J. Nucl. Med.* **25**(7): 774–787.

Bailey, D. L. (2005). *Positron Emission Tomography: Basic Sciences.* Springer, New York.

Balaban, R. S. and V. A. Hampshire. (2001). Challenges in small animal noninvasive imaging. *ILAR J.* **42**(3): 248–262.

Balcombe, J. P., N. D. Barnard et al. (2004). Laboratory routines cause animal stress. *Contemp. Top. Lab. Anim. Sci.* **43**(6): 42–51.

Bedard, P. L., A. R. Hansen et al. (2013). Tumour heterogeneity in the clinic. *Nature* **501**(7467): 355–364.

Boellaard, R. (2009). Standards for PET image acquisition and quantitative data analysis. *J. Nucl. Med.* **50**(Suppl. 1): 11S–20S.

Boellaard, R. (2011). Methodological aspects of multicenter studies with quantitative PET. *Methods Mol. Biol.* **727**: 335–349.

Brasse, D., P. E. Kinahan et al. (2005). Correction methods for random coincidences in fully 3D whole-body PET: Impact on data and image quality. *J. Nucl. Med.* **46**(5): 859–867.

Cheng, D., M. Rusckowski et al. (2011). Improving the quantitation accuracy in noninvasive small animal single photon emission computed tomography imaging. *Nucl. Med. Biol.* **38**(6): 843–848.

Constantinides, C., R. Mean et al. (2011). Effects of isoflurane anesthesia on the cardiovascular function of the C57BL/6 mouse. *ILAR J.* **52**: e21–e31.

Erlandsson, K., I. Buvat et al. (2012). A review of partial volume correction techniques for emission tomography and their applications in neurology, cardiology and oncology. *Phys. Med. Biol.* **57**(21): R119–R159.

Fueger, B. J., J. Czernin et al. (2006). Impact of animal handling on the results of 18F-FDG PET studies in mice. *J. Nucl. Med.* **47**(6): 999–1006.

Funk, T., M. Sun et al. (2004). Radiation dose estimate in small animal SPECT and PET. *Med. Phys.* **31**(9): 2680–2686.

Gutierrez, D. F. and H. Zaidi. (2012). Automated analysis of small animal PET studies through deformable registration to an atlas. *Eur. J. Nucl. Med. Mol. Imaging* **39**(11): 1807–1820.

Hasebroock, K. M. and N. J. Serkova. (2009). Toxicity of MRI and CT contrast agents. *Expert. Opin. Drug Metab. Toxicol.* **5**(4): 403–416.

Hicks, R. J. and M. S. Hofman. (2012). Is there still a role for SPECT-CT in oncology in the PET-CT era? *Nat. Rev. Clin. Oncol.* **9**(12): 712–720.

Hildebrandt, I. J., H. Su et al. (2008). Anesthesia and other considerations for in vivo imaging of small animals. *ILAR J.* **49**(1): 17–26.

Hutton, B. F., I. Buvat et al. (2011). Review and current status of SPECT scatter correction. *Phys. Med. Biol.* **56**(14): R85–R112.

James, M. L. and S. S. Gambhir. (2012). A molecular imaging primer: Modalities, imaging agents, and applications. *Physiol. Rev.* **92**(2): 897–965.

Kinahan, P. E., D. W. Townsend et al. (1998). Attenuation correction for a combined 3D PET/CT scanner. *Med. Phys.* **25**(10): 2046–2053.

Kreissl, M. C., D. B. Stout et al. (2011). Influence of dietary state and insulin on myocardial, skeletal muscle and brain [F]-fluorodeoxyglucose kinetics in mice. *EJNMMI Res.* **1**: 8.

Mariani, G., L. Bruselli et al. (2008). Is PET always an advantage versus planar and SPECT imaging? *Eur. J. Nucl. Med. Mol. Imaging* **35**(8): 1560–1565.

Maroy, R., R. Boisgard et al. (2008). Segmentation of rodent whole-body dynamic PET images: An unsupervised method based on voxel dynamics. *IEEE Trans. Med. Imaging* **27**(3): 342–354.

Martic-Kehl, M. I., S. M. Ametamey et al. (2012). Impact of inherent variability and experimental parameters on the reliability of small animal PET data. *EJNMMI Res.* **2**(1): 26.

Moses, W. W. (2011). Fundamental limits of spatial resolution in PET. *Nucl. Instrum. Methods Phys. Res. A* **648**(Suppl. 1): S236–S240.

Peterson, T. E. and S. Shokouhi. (2012). Advances in preclinical SPECT instrumentation. *J. Nucl. Med.* **53**(6): 841–844.

Prieto, E., P. Lecumberri et al. (2012). Twelve automated thresholding methods for segmentation of PET images: A phantom study. *Phys. Med. Biol.* **57**(12): 3963–3980.

Rahmim, A. and H. Zaidi. (2008). PET versus SPECT: Strengths, limitations and challenges. *Nucl. Med. Commun.* **29**(3): 193–207.

Richter, S. H., J. P. Garner et al. (2009). Environmental standardization: Cure or cause of poor reproducibility in animal experiments? *Nat. Methods* **6**(4): 257–261.

Roth, D. M., J. S. Swaney et al. (2002). Impact of anesthesia on cardiac function during echocardiography in mice. *Am. J. Physiol. Heart Circ. Physiol.* **282**(6): H2134–H2140.

Scarfe, W. C. and A. G. Farman. (2008). What is cone-beam CT and how does it work? *Dent. Clin. North Am.* **52**(4): 707–730, v.

Schambach, S. J., S. Bag et al. (2010). Application of micro-CT in small animal imaging. *Methods* **50**(1): 2–13.

Segars, W. P., B. M. Tsui et al. (2004). Development of a 4-D digital mouse phantom for molecular imaging research. *Mol. Imaging Biol.* **6**(3): 149–159.

Shokouhi, S., S. D. Metzler et al. (2009). Multi-pinhole collimator design for small-object imaging with SiliSPECT: A high-resolution SPECT. *Phys. Med. Biol.* **54**(2): 207–225.

Soret, M., S. L. Bacharach et al. (2007). Partial-volume effect in PET tumor imaging. *J. Nucl. Med.* **48**(6): 932–945.

Taylor, U., A. Barchanski et al. (2012). Toxicity of gold nanoparticles on somatic and reproductive cells. *Adv. Exp. Med. Biol.* **733**: 125–133.

Vriens, D., E. P. Visser et al. (2010). Methodological considerations in quantification of oncological FDG PET studies. *Eur. J. Nucl. Med. Mol. Imaging* **37**(7): 1408–1425.

Wang, Y., J. Seidel et al. (2006). Performance evaluation of the GE healthcare eXplore VISTA dual-ring small-animal PET scanner. *J. Nucl. Med.* **47**(11): 1891–1900.

Zaidi, H. and I. El Naqa. (2010). PET-guided delineation of radiation therapy treatment volumes: A survey of image segmentation techniques. *Eur. J. Nucl. Med. Mol. Imaging* **37**(11): 2165–2187.

23

Performance Assessment of Small Animal Imaging Systems and Common Standards

Nancy L. Ford

23.1   INTRODUCTION

All imaging equipment produces a pictorial representation of the object that was imaged. Obtaining quantitative information from these images provides the true power behind the technology, taking the important step from a pretty picture to measured data that can be compared across study groups, between research studies, and across imaging platforms. Ensuring that the data obtained are reliable and accurate requires good image quality. In this chapter, image quality is described by various objective metrics that can be measured in most imaging platforms. To obtain the best images possible will require tailoring the imaging protocol for different studies; however, practical limitations may reduce the flexibility of the optimization process, particularly for *in vivo* studies. To ensure reliable results from the imaging equipment over time, quality assurance programs and industry standards are needed to identify when individual systems go out of calibration or are functionally unstable. Regular assessment of system performance will enable long-term studies to be performed without concerns about the validity of any results obtained from the images.

23.2   IMAGE QUALITY METRICS

23.2.1   SIGNAL AND NOISE

For all imaging modalities, what is measured in the image should be representative of the subject. The image receptor will collect the signal from the imaging chain, which includes the signal from the subject and additional background signal, which we refer to as noise. How the signal is created depends on the imaging modality. The signal may be a map of the transmission through the subject, as is the case for x-ray imaging and computed tomography (CT) or a map of the distribution of radioactive tracers, as in single-photon emission computed tomography (SPECT) and positron emission tomography (PET) imaging. In ultrasound imaging, the signal is an echo of the ultrasonic pulse that is reflected at the tissue boundaries, whereas in magnetic resonance imaging (MRI), the signal is related to the interaction of a radio frequency pulse and the static magnetic field with the tissue of interest. In all cases, the signal that represents the subject and is measured by the image receptor contains the important information about the subject that is the focal point of the imaging study.

However, there are other sources of signal that do not originate in the subject that may also be recorded by the image receptor. These signals could be from objects that were in the field of view, such as supporting structures (cradles, immobilization devices, etc.) and supportive care instrumentation (catheters, anesthesia delivery hoses, heating units, ECG monitoring, etc.). Alternatively, the unwanted signals could be added by the electronics within the detector, or be related to the stochastic nature of the imaging modalities (photon counting statistics, scattered signals, attenuation, etc.). All of these unwanted signals are considered to be noise and degrade the information contained within the image.

To produce images containing useful information about the tissue or animal under investigation, the intensity of the signal originating from the subject must be greater than that of the background noise. By looking at the image, the observer must be able to differentiate between the tissue or animal and the background noise. If the signal and the noise have similar intensities within the image, it is very difficult to delineate where the tissue of interest begins and surrounding background tissues end. The ratio between the intensities of the signal and the background are referred to as the signal-to-noise ratio (SNR). Detectability of a signal within a noisy background depends on a number of criteria including the size of the region producing the signal, the relative intensities of the signal and background, and the observer. Albert Rose was the first person to attempt to evaluate the detectability of a signal in medical images based on human visual perception. In his seminal paper (Rose 1948), he suggested an SNR threshold value of 5 was required for detection, which is now known as the Rose criterion. Subsequent studies of observer acuity have shown that the Rose Criterion is limited to high-contrast signals in a noise-limited image. To fully assess the detection limits requires more complex modeling of each specific imaging scenario (Burgess 1999).

To ensure that the observer will be able to clearly visualize the boundaries between adjacent tissues or detect anomalies within a tissue, the Rose criterion must be achieved, although a larger difference between the signal and the noise is desirable. To achieve the detection threshold, the intensity of the signal can be increased or the intensity of the noise can be decreased or potentially both. Increasing the signal may not always be practical, as the signal depends on interactions within the tissues; increasing the signal may require increased doses of x-rays or radiotracers or extended imaging times, which may prove to be detrimental to the animal or tissue that is being imaged. However, introducing a contrast agent that will artificially increase or decrease the signal originating from tissues in which the contrast agent has accumulated can serve to improve the signal detectability in the image. Decreasing the noise also has practical limitations, as the noise can result from the imaging receptor, electronics, scattered signal, or be related to the physics of the imaging modality (photon counting statistics, acoustic reflection in tissues, magnetic susceptibility, etc.). Eliminating the noise from the images entirely is unrealistic, but by ensuring that the signal of interest exceeds the background levels in accordance with the Rose criterion, the images produced will be useful for detecting and measuring the tissues of interest.

23.2.2 IMAGE ACCURACY

For quantitative measurements, it is imperative that the images accurately represent the object that was imaged. In some cases, accuracy errors are easy to detect; if the image does not look like the object, then errors have undoubtedly occurred. Obvious errors would include stretching or skewing the image such that circular objects do not appear to be round or linear objects are not continuous along the full length. However, subtle errors are not as easy to detect. In the more subtle cases, the image appears to be correct, with no obvious distortions or omissions. In small animal imaging, there are biological variations between animals, even for genetically identical littermates, that make it difficult to assess if the image is stretched or misaligned. Any misalignment or distortions may alter the measurements obtained, leading to systematic errors in the measured values.

Correction factors can be applied if the distortion is well characterized; corrections are often applied in CT imaging to account for gravitational forces acting on the gantry as it rotates causing the projection images to be slightly misaligned. Similarly, corrections can be applied to counteract inhomogeneity in the magnetic field for MRI studies. To determine the correction factors, the misalignment or distortion must be well understood with phantom images.

To ensure accurate measurements are obtained, regular quality assurance testing is advised. To perform accuracy testing, a phantom containing a known distribution of objects can be imaged. Typically, the objects are high-contrast linear objects, like thin wires or narrow channels, or spherical objects, like beads or voids. The test objects are oriented within the image with a known orientation and spacing. The dimensions of these objects are also known (length, width, height). Measuring the object itself and the spacing between adjacent objects, and comparing the measured values with the known object dimensions, will provide the geometric accuracy in all three directions.

23.2.3 CONTRAST AND CONTRAST RESOLUTION

Contrast is a very important requirement for any imaging system. Contrast refers to the ability to distinguish between different objects in the image or to pick a feature out of the background of the image. The contrast mechanisms are different for each imaging modality and are outlined in more detail in the modality specific chapters of this book. Understanding the contrast mechanisms of the different imaging modalities will assist in the selection of the most appropriate imaging tool for the study.

In general, the contrast can be described as the difference between the average signal intensity of the object and the average signal intensity of the background, where a larger difference would correspond to an image where the object is easily located within the background. Some objects will be readily identified within the image even if the image is not considered to have the best quality; these objects are considered to be high-contrast objects that are readily observed due to their physical characteristics (composition, attenuation, and acoustic or magnetic properties). Often, the high-contrast objects are different tissue types, such as bone, lungs, fat, or organs. However, when the tissue properties are similar, the contrast observed in the images will be reduced and the target tissue will be harder to identify against the background tissue. Low-contrast objects could be different tissues or organs with similar physical characteristics, or could be a diseased region within the healthy background tissue. In the case of low-contrast objects, obtaining the best image possible will greatly improve the probability of detecting and visualizing the low-contrast objects.

To assess the contrast achieved in an image, a comprehensive set of image quality assessment tools are needed to assess the high-contrast case and the low-contrast case as both of these cases are important for different imaging studies. Phantoms with a series of different target materials, representing the different tissue types and including both low-contrast and high-contrast objects are available. Typically, the test objects have a range of known sizes for each contrast level. When imaged, the tool can provide a single image with a range of contrast levels in a single image acquisition, allowing the contrast to be determined for all of the tissue types included. The different sizes allow the observer to assess the smallest object size for each contrast since the high-contrast objects will be easier to locate and, therefore, would likely be seen even for smallest object size in the phantom compared with a low-contrast object.

Assessing the contrast as a function of size is also known as the contrast resolution. The contrast resolution test describes the smallest target object that can be detected for a given contrast level. An observer grades the image to identify the smallest object that is detected, requiring complete delineation of the periphery of the object, for each contrast level. To be considered visualized, the entire periphery of the object must be clearly distinguished from the background; a patchy appearance or the inability to follow the edge of the object would be unacceptable for this test. Results of this image quality test are usually quoted as the minimum detectable target size for each contrast level, expressed as a percent difference from the background. Typically, the low-contrast objects are more difficult to see than the high-contrast objects, especially for the smaller sizes. This implies that small, low-contrast objects will be the most difficult to detect in any imaging system.

To improve the contrast of an image, a couple of approaches are possible. By improving the detector technology, the sensitivity of the detector can be increased to ensure that the signal intensity representing the important target organ is sufficient to allow the observer to easily visualize the organ of interest. However, changing the detector is not always a feasible solution; to improve the contrast of an organ in an individual research study, contrast-enhancing media can be used. Contrast agents can be introduced that accumulate at the desired location and alter the contrast between the target organ and the background tissues. Typically, contrast agents have a known residence time and clearance pathway and are introduced intravenously. In some cases, the contrast material is introduced along with the disease model of interest, such as introducing labeled cells in a tumor model. The contrast media has different physical properties than the background tissue, resulting in a brighter or darker appearance in the images compared to non-contrast-enhanced regions.

23.2.4 Spatial Resolution

The spatial resolution of an imaging system describes how closely spaced two objects can be positioned and be identified as separate entities in the image. This idea is also known as the limiting spatial resolution. The limiting resolution can be assessed subjectively with a series of phantom objects consisting of high-contrast objects spaced a known distance apart. The observer can then decide which objects appear as separate entities and estimate the limiting resolution in lines pairs per millimeter (lp/mm). The phantom object is often a bar pattern for x-ray imaging, comprised of metal foil embedded in plastic spacers, or a star pattern for nuclear imaging, consisting of the radioisotope injected into narrow channels of a background material.

A more objective method of determining the spatial resolution involves estimating the response of the imaging system to a sharp change in the signal intensity due to a sharp edge or a point source of a high-contrast material compared with the background of the image. The ratio of the contrast between the high-contrast material and the background measured in the image to that of the object being imaged results in the modulation transfer function (MTF), which is a quantitative metric used to describe the resolution of an imaging system. The modulation transfer function is graphed with the spatial frequency along the x-axis and the MTF value (ranging from 0 to 1) along the y-axis. The curve is shaped like a sinc() function and reaches the first zero at the Nyquist frequency, which corresponds to the pixel or voxel size in the ideal case. MTF is related to the subjective measurement of the limiting resolution described earlier; for the average viewer, the limiting spatial resolution obtained with the high-contrast repeating pattern of known separation will correspond with the spatial frequency at which the MTF curve drops to 10% of the maximum value.

Many manufacturer's specifications and scientific papers often refer to the resolution of their images as the image pixel size (or voxel size for 3D images). This practice can be misleading, as the pixel size on the computer screen may not be related to the limiting resolution of the imaging system. In digital images, the image pixels can be broken into smaller units and displayed with a "higher resolution" or smaller pixel size. By artificially resampling the data in software post-acquisition, the data can be displayed using a finer grid, but the inherent resolution of the image is unchanged.

23.2.5 ARTIFACTS

Artifacts have a number of different causes and appearances but are undesirable signals that degrade the diagnostic quality of the images. In this section, some of the most commonly occurring artifacts are described; however, there may be additional artifacts that are not listed here and the frequency and severity of the listed artifacts may change with the introduction of new technologies.

In the case of ghosting artifacts, objects appear in the images that were not in the field of view during the acquisition. These ghost images are typically due to lag or afterglow in the detector. Some of the signal from the previous image remains in the detector and is added to the signal obtained during the new acquisition. The resulting image is a superposition of the current object and the previous object. The previous object will have much lower signal intensity, giving it a ghost-like appearance. However, the ghost object may have sufficient intensity to overshadow the real object or obscure important information in the image. One solution to this artifact is to read the detector out multiple times, ensuring that the image is stored, and any residual signal is flushed from the detector prior to acquiring the next image. Typically, ghosting is a known issue and taken care of by the manufacturer of commercially available equipment.

Electromagnetic interference can also cause a variety of artifacts, typically in magnetic resonance imaging. These artifacts present in a number of different ways, including a striped or herringbone appearance. These characteristic patterns consist of alternating bright and dark bands superimposed on the object. Different appearances occur for different imaging sequences as a result of an inhomogeneous magnetic field across the field of view. The inhomogeneity could be transient during the image, resulting from power fluctuations, or persist throughout the image acquisition. Improved shielding and uniformity of the magnetic field will reduce the occurrence of these artifacts.

Cross-talk describes the sharing of signal between adjacent pixels in the detector. In this case, the signal that represents an object is a summation of the signal from multiple pixels. Sharing the signal across 2 or 3 pixels in each direction can lead to a reduction in the spatial resolution of the imaging system. Instead of representing an important feature with a single pixel or a small patch of pixels, cross-talk would imply that the apparent size can be 2 or more pixels larger than the real size in all directions. Quantitative analysis of an affected image would result in systematic overestimation of any linear or volumetric measurement. To overcome this issue, detectors are typically built with a small buffer between pixels. Although this solution does eliminate the cross-talk problem, it also implies a reduced sensitive area (or fill factor) for each pixel, thereby reducing the measured signal.

Detector saturation results in a washed out appearance of the image, with reduced contrast and dynamic range. Saturation occurs when the signal received is more than an individual pixel in the detector can accommodate. The result is either the affected pixel is pinned at the highest intensity value or the signal wraps around and the pixel value is the difference between the true signal and the maximum range for the pixel. Either of these cases would make the signal intensity either extremely bright or extremely dark relative to the unaffected neighboring pixels. Saturation can occur in isolation with a single pixel or small group of pixels displaying this behavior. In the case of isolated pixels exhibiting saturation, the culprit is generally the electronics controlling that section of the detector. If the affected area is small relative to the detector and the behavior is predictable, a correction map could be used to eliminate the saturation and remap the affected pixels to values that more closely approximate the correct signal intensities. In some cases, the saturation affects a large region of the detector, or even the entire field of view. For large areas of saturation, the problem is often related to the integration time of the pixels and the intensity of the signal reaching the detector. The user can often alter imaging parameters to compensate for cases where the entire detector is saturated.

Aliasing usually appears as a Moire interference pattern superimposed on the image. Aliasing means that the imaging system has undersampled the object and the image representation is not accurate as a result, with higher frequency signals being presented as low-frequency signals. All frequencies above the Nyquist frequency, defined as half the sampling frequency, will be aliased resulting in reduced fine detail in the image. Unfortunately, aliasing is unlikely to be completely eliminated in any digital imaging application; however, increasing the sampling frequency can reduce the effects of aliasing. For spatially aliased images, the solution is to increase the number of pixels in the image that represent the object. In addition, using a field of view that is smaller than the object can cause aliasing, so increasing the size of the field of view may also improve the images. For images that are obtained from a series of projections at different angles around the subject, such as CT, SPECT, and PET, increasing the number of angles used to obtain the data can reduce angular aliasing. Finally, for studies that are monitoring dynamic processes, temporal aliasing can occur. Reducing the temporal aliasing will require faster electronics to obtain each image faster, thereby increasing the sampling frequency.

The tissues and tissue boundaries within the animal can cause artifacts as well. For some modalities, the signal is weak or nonexistent from certain tissues, such as bone in MRI or air in ultrasound, leaving signal voids in the image. In addition, the boundary between two tissues may not be easily delineated due to boundary effects. Boundary effects are caused by interference of the signal as it passes through the boundary between two tissue types or between a tissue and an implanted device or foreign body. At the interface, the signal from one tissue cancels or partially cancels the signal intensity of the other tissue and vice versa. Tissue-related artifacts result in images appearing inhomogeneous across a tissue and may be addressed through altered imaging parameters. However, some tissue-related artifacts are difficult to eliminate completely, as in the case of implanted devices. For the implanted devices, the signal intensity from the surrounding tissue and the implant are sufficiently different that the resulting artifact is often very stark in the image. Post-processing algorithms can be applied to try and reduce the impact of the implant, but in reality, these regions of the image may not have sufficient information to produce a faithful reproduction of the subject.

For *in vivo* imaging, motion artifacts are often observed. The appearance of this artifact is a blurring most noticeable at the edges of tissues that are in motion. Motion artifacts make it difficult to clearly see the edges of a tissue or a diseased region or to detect very small lesions and can be thought of as a net loss in resolution. Gross anatomical motion is easily avoided by anesthetizing and restraining the animal for the duration of the imaging session. Normal respiratory and cardiac function cause a lot of tissue movement through the thorax and into the abdomen. In addition, peristaltic motion may cause motion artifacts to be observed in the intestinal area as food and gases move along the intestinal tract. For very short image acquisition times, the motion observed might not degrade the image sufficiently to warrant intervention, particularly for anatomical regions that are not impacted by physiological motion. In cases where the features of interest in the imaging session are moving throughout the scan or immediately adjacent to a moving organ, a motion artifact will undoubtedly degrade the image obtained. For most imaging modalities, some method of gating the image acquisition with the physiological signal (respiratory or cardiac) is available either by implementing specialized hardware to synchronize the imaging with the physiological signals or with software corrections to address and minimize the impact of the motion on the final image.

23.3 PRACTICAL CONSIDERATIONS FOR PROTOCOL OPTIMIZATION

Many researchers tend to use the preset vendor-supplied protocols, but for many preclinical imaging systems, the parameters can be altered to optimize the images obtained. The main reason to adjust the parameters is to improve the image quality for a specific sample type. For specimen imaging, the optimization can really improve the images and resulting quantitative analysis. However, for *in vivo* studies, there are a few practical limitations on the settings that can be used, including the radiation dose for modalities that use ionizing radiation, and the total time the animal spends under general anesthesia.

23.3.1 IONIZING RADIATION DOSE

For some imaging and therapy procedures in small animals, ionizing radiation may become an important concern. For high-resolution imaging, the amount of radiation required to obtain an adequate image may be much higher than what is used in similar clinical imaging situations (Ford et al. 2003). From observing patients that have been exposed to ionizing radiation, there is sufficient evidence of radiation damage in high-dose imaging studies, or patients that have undergone numerous imaging sessions, to suggest that exposure to ionizing radiation in medical imaging and therapy carries inherent risk of damage to cells and DNA. There have not been many studies about the effects of ionizing radiation from preclinical imaging or therapeutic studies; however, the tissues are similar enough to human tissues that the risk of damage should be considered.

Many ethics panels may not be concerned about the radiation dose received by the animals, since imaging is less invasive than other procedures the animals will be subjected to. However, radiation received by the animal may be an important issue from a scientific standpoint that could be overlooked. For many experiments, the imaging session is immediately followed by euthanasia, meaning the rodents will be euthanized long before the adverse effects of the radiation can be observed, so the impact of the radiation is of no scientific consequence. However, as the scan times are reduced and the technologies improved to become more sensitive to the presence of disease, longitudinal imaging studies will become more prevalent. Furthermore, some of these imaging platforms are being developed for monitoring the drug distribution within the animals and how the disease responds to the treatment over a period of days, weeks, or months. With the increased number of imaging sessions, and the extended survival time post-irradiation, attention should be given to ensuring that the radiation received during the imaging session does not adversely impact the scientific outcome of the study.

There are a few precautions that can be used to reduce the impact of ionizing radiation on the animals. Optimizing the imaging parameters in PET and SPECT to identify the minimum amount of radioactive material that can be introduced for adequate detection of the agent can reduce the volume and activity of tracers used, thereby managing the radiation burden to the animal. For x-ray imaging applications, optimizing the imaging parameters to reduce the total exposure time or the intensity of the x-ray beam during the imaging session can reduce the ionizing dose to the animal. Reducing the field of view in x-ray imaging can also be used to reduce the amount of tissue exposed to ionizing radiation, potentially sparing particularly sensitive organs. Careful planning to ensure that the acquired images contain the organ of interest is crucial to avoid repeated exposures and can be easily achieved by using a low-dose scout view to verify positioning or a precisely aligned animal holder that will guarantee consistent positioning.

23.3.2 MAXIMUM SCAN TIMES

To produce higher quality images, increasing the scan time can enable higher resolution or improved SNR in many imaging modalities. For specimen imaging, the improvement in image quality may be a good investment of time. However, for *in vivo* small animal imaging, there are practical limitations on imaging session duration. During *in vivo* imaging sessions, the animals must be anesthetized by either inhaled or injectable agents. For the injectable agents, each dose will only last for a short period of time (up to 20 or 30 min), whereas the inhaled agents can be delivered at a low concentration continuously. In both cases, there are limitations on how long an animal can be reliably anesthetized without adverse events. For injectable agents, the limitation is mostly imposed by the drugs administered, how well the dose is tolerated, and whether the anesthesia period can be extended by giving additional doses of the drug without reaching the lethal limit. For inhaled agents, the concentration can be continuously monitored and adjusted as needed. These limitations are species and drug specific, but if handled well, the animals can remain anesthetized for a lengthy experiment if needed.

To ensure that the anesthesia is not impacting the imaging study, it is important to carefully select the anesthesia method, as the drugs used will interact with the physiology in very different ways (altering the respiratory rates, cardiac rates, body temperature, etc.). Changes in the physiology may impact the images obtained, particularly for functional imaging sessions. The combination of drugs used, supportive care while under anesthesia, and careful planning of procedures to minimize the time required will ensure the data obtained are more consistent between animals and therefore more scientifically valid.

23.4 DEVELOPING COMMON STANDARDS

To ensure that preclinical imaging equipment performs well, it is important to assess the image quality and accuracy routinely. Similarly, for preclinical therapeutic systems, the accuracy of targeting a specific region and the dose delivery needs to be carefully calibrated. This section gives a brief overview of the phantoms used to assess imaging and therapeutic equipment, quality assurance programs that have been developed, and progress on development of industry standards for preclinical equipment.

23.4.1 Phantoms

Phantoms have been developed to assess the imaging capabilities of preclinical systems. Some phantoms are custom built in a specific laboratory, but there are also some commercially available phantoms. Image quality phantoms designed to objectively assess micro-CT imaging systems have been developed. A quality assurance phantom was designed and tested by Du et al. (2007) to assess the performance of a high-speed volumetric micro-CT scanner with respect to spatial resolution, geometric accuracy, noise, uniformity, linearity, and CT number accuracy. This phantom has made the transition from a laboratory tool to a commercially available product.

Custom phantoms to assess specific imaging tasks have also been reported. A phantom to assess the voxel scaling accuracy in micro-CT has been described and tested for a variety of micro-CT systems covering the range of image resolutions (Waring et al. 2012). To quantify the mineral density of different tissue types in micro-CT images, calibration phantoms for dentine (K_2HPO_4) and bone (hydroxyapatite) have been fabricated and compared (Zou et al. 2009). Zou et al. found that the K_2HPO_4 solutions provided better uniformity than the hydroxyapatite phantoms and exhibited similar composition to dentine. A novel hydroxyapatite phantom with an expanded range of concentrations has been described by Deuerling et al. (2010). In this study, Deuerling et al. demonstrated that calibrating bone images using their phantom resulted in more accurate measurements of bone mineral density and tissue mineral density. A calibration phantom to measure the trabecular spacing, trabecular thickness, and structure model index has been described as a PMMA cylinder containing aluminum wires, meshes, and spheres (Perilli et al. 2006).

A phantom to enable improved spherical lesion detection in micro-SPECT and micro-PET has been developed by Di Filippo et al. (2010). In this phantom, the fillable hollow spheres have a smaller diameter more suited to micro-PET and micro-SPECT imaging tasks. In addition, the phantom boasts a symmetric, in-place rotation design and uses the principles of superposition to eliminate the wall effect.

In the field of small animal ultrasound imaging, measuring blood flow in the small vessels is an important challenge, which can be achieved with high frequency ultrasound. Qian et al. (2010) have extended ultrasound particle image velocimetry (Echo-PIV) to include high-frequency ultrasound. Their study also described a straight small vessel mimic phantom and a stenotic small vessel phantom for validating the micro-EPIV technique.

To assess the imaging parameters in situations that are more consistent with small animal imaging, anatomically correct rodent phantoms have been developed. Some of these phantoms are simulated models, whereas some are physical phantoms that can be placed and imaged in the scanning devices. Two main purposes of the simulation phantoms are to provide a co-registration tool to an "average" animal and to enable delineation of low-contrast organs within the image based on the relative location in the simulated phantom. Zhang et al. (2009) have described a simulated anatomical mouse model based on contrast-enhanced micro-CT imaging. To develop the 3D voxel-based model, each organ was digitally extracted from the micro-CT image of a BALB/c mouse, which is a very commonly used strain in research. Wang et al. (2012) have described a statistical mouse atlas developed using contrast-enhanced micro-CT images of 45 mice from a variety of sizes and strains. In this study, the statistical model outperformed competing single subject models in registration accuracy. A 4D digital mouse phantom (MOBY) has been described using MRI imaging to provide anatomical detail, and includes realistic motion due to the cardiac and respiratory cycles (Segars et al. 2004). Dogdas et al. (2007) used micro-CT images along with cryosectioning to identify different organs in a single mouse postmortem and build the Digimouse atlas, which is suitable for co-registration of small animal images. The existing whole body atlases have been adapted by creating articulating atlases to eliminate the errors due to postural variability and positioning differences (Baiker et al. 2010; Khmelinskii et al. 2011).

Alternatively, the simulated rodent phantoms can also be used to quantify the tissue composition of rodents and determine a more accurate estimation of the dose received during micro-CT, PET, SPECT imaging, or therapeutic irradiation. Radiation assessment dose resource (RADAR) animal model series has been developed using nonuniform rational B-spline (NURBS) to facilitate dose calculations in preclinical studies (Keenan et al. 2010). Dosimetry calculations have been performed to determine the absorbed dose to subcutaneous tumors in radionuclide studies (Larsson et al. 2011). S-values have been calculated for a range of radioisotopes, which provides a database for assessing radiation dose in mice (Xie and Zaidi 2013). Dosimetry estimates for individual organs of a mouse subjected to micro-CT have also been reported using a realistic model (Taschereau et al. 2006), with the highest doses found in bony structures. The bony structures also exhibited a more heterogeneous dose distribution than the soft tissues. Mauxion et al. (2013) caution that the segmentation of organs should be done with care, and model information reported meticulously when performing Monte Carlo–based dosimetric evaluations in preclinical studies.

23.4.2 PERFORMANCE EVALUATION AND QUALITY ASSURANCE PROGRAMS

Evaluating the performance of imaging equipment is especially important when introducing a new modality or when implementing improvements to hardware design. Assessing all aspects of performance is critical to ensure that any improvements gained are not at the expense of a different imaging parameter. Although industry standards and commercially available phantoms are not available for all modalities, and those that are available are newly developed, assessing performance can be done on a case-by-case basis using custom-built phantom objects. As an example, new PET scanners have been assessed and compared across generations to demonstrate technological advances (Knoess et al. 2003; Tai et al. 2005).

 Calibration of imaging systems is also important, particularly if quantitative measurements are to be obtained from the images. Calibration is also needed for many reconstruction algorithms to ensure that the signal observed in the images is related to the composition of the object under investigation. Patel et al. described a method of using the acquired projection data obtained in a rotating-object micro-CT scanner for geometric calibration in the reconstruction. The projections can be aligned and corrected for geometrical distortions on an individual basis prior to reconstruction, resulting in a self-calibrating system that does not require additional scans of calibration objects (Patel et al. 2009). Cone beam micro-CT is also used in image-guided micro-irradiators for visualizing the target region and improving the accuracy of radiation delivery. Clarkson et al. (2011) have characterized the image quality achieved in their micro-irradiator over a period of many months and measured the targeting accuracy that they could achieve for image-guided radiation therapy. Eloot et al. (2010) have developed a phantom for micro-CT that includes both image quality and dosimetry information in a single scan. This custom-built phantom can be used for a routine quality assurance program to evaluate the stability of micro-CT systems and could be used for the optimization of imaging protocols for specific imaging tasks. For specimen micro-CT scanners, a quality control program and phantoms have been proposed to assess the noise, uniformity, and accuracy in the images (Stoico et al. 2010). The program begins with establishing the baseline values during acceptance testing followed by monthly scans of the phantom and comparing the measured values to the baseline values. Any measurements that are out of the acceptable range trigger maintenance and calibration of the micro-CT system to ensure that long-term studies are not impacted by fluctuations in the micro-CT equipment.

23.4.3 REGULATORY AND/OR INDUSTRY STANDARDS

Preclinical imaging modalities have not been subjected to the same regulations and standards as their clinical counterparts. Aside from compliance with national and regional safety regulations, such as radiation and electrical safety, preclinical devices have been sold or built largely without regulatory oversight. As more preclinical devices enter the commercial marketplace, a set of standards will become more important for comparing the capabilities of different products and characterizing the improvements achieved with new technologies.

The National Electrical Manufacturers Association (NEMA) has developed some standards for some medical imaging devices, including small animal positron emission tomography machines (NEMA 2008). To date, no other NEMA standards devoted to preclinical imaging exist. The NEMA standards define a process or product by developing a description of the product (dimensions, composition, etc.), operating specifications and their tolerances, and safety information. In addition, the NEMA standards include a set of tests that can be performed to assess the performance of the device with respect to its desired functionality. In the preclinical PET standard, there is a phantom described to assess the image quality achieved with a preclinical PET scanner (NEMA 2008). With this phantom, measurements of the spatial resolution, scatter fraction, count rate, and sensitivity can be performed. The phantom also enables image quality analysis, including assessing uniformity, attenuation corrections, and scatter corrections.

The NEMA NU 4-2008 standard has been applied to characterize preclinical PET imaging systems. Goertzen et al. (2012) performed image quality analysis on 11 different models of commercially available PET scanners manufactured since 2000. Using the NEMA standards, the authors were able to show improvements in the system performance of the newer scanners compared with the first-generation designs. However, the NEMA standard does not account for the differences in scanner designs and reconstruction algorithms, and the authors of the study caution against direct comparisons between the models of different vendors based on the NEMA standards alone (Goertzen et al. 2012). The specific imaging task should be considered in addition to the results of the NEMA testing.

NEMA standards have been used to characterize laboratory-built preclinical PET scanners (Wong et al. 2012). In this study, the NEMA image quality measurements, along with the results from mouse-like and rat-like phantoms, were used to demonstrate that the laboratory-built scanner had excellent imaging performance, including spatial resolution, count rate, and sensitivity (Wong et al. 2012). Additional phantom studies were performed in a Micro Deluxe phantom and an Ultra-Micro Hot Spot phantom, both made by Data Spectrum Corp. *In vivo* imaging in healthy mice was also performed to augment the assessment of the scanner during a more relevant imaging task. In the absence of NEMA standards for preclinical SPECT imaging systems, Harteveld et al. have also used the NEMA NU-4 protocol to assess the image quality for a multi-pinhole SPECT system under mouse imaging conditions (Harteveld et al. 2011). In the future, the NEMA NU-4 standard could be altered to address the specific needs of the multi-pinhole SPECT scanners.

REFERENCES

Baiker, M., J. Milles et al. (2010). Atlas-based whole-body segmentation of mice from low-contrast Micro-CT data. *Med. Image Anal.* **14**(6): 723–737.

Burgess, A. E. (1999). The Rose model, revisited. *J. Opt. Soc. Am. A* **16**(3): 633–646.

Clarkson, R., P. E. Lindsay et al. (2011). Characterization of image quality and image-guidance performance of a preclinical microirradiator. *Med. Phys.* **38**(2): 845–856.

Deuerling, J. M., D. J. Rudy et al. (2010). Improved accuracy of cortical bone mineralization measured by polychromatic microcomputed tomography using a novel high mineral density composite calibration phantom. *Med. Phys.* **37**(9): 5138–5145.

Difilippo, F. P., S. L. Gallo et al. (2010). A fillable micro-hollow sphere lesion detection phantom using superposition. *Phys. Med. Biol.* **55**(18): 5363–5381.

Dogdas, B., D. Stout et al. (2007). Digimouse: A 3D whole body mouse atlas from CT and cryosection data. *Phys. Med. Biol.* **52**(3): 577–587.

Du, L. Y., J. Umoh et al. (2007). A quality assurance phantom for the performance evaluation of volumetric micro-CT systems. *Phys. Med. Biol.* **52**(23): 7087–7108.

Eloot, L., N. Buls et al. (2010). Quality control of micro-computed tomography systems. *Radiat. Prot. Dosimetry* **139**(1–3): 463–467.

Ford, N. L., M. M. Thornton et al. (2003). Fundamental image quality limits for microcomputed tomography in small animals. *Med. Phys.* **30**(11): 2869–2877.

Goertzen, A. L., Q. Bao et al. (2012). NEMA NU 4-2008 comparison of preclinical PET imaging systems. *J. Nucl. Med.* **53**(8): 1300–1309.

Harteveld, A. A., A. P. Meeuwis et al. (2011). Using the NEMA NU 4 PET image quality phantom in multipinhole small-animal SPECT. *J. Nucl. Med.* **52**(10): 1646–1653.

Keenan, M. A., M. G. Stabin et al. (2010). RADAR realistic animal model series for dose assessment. *J. Nucl. Med.* **51**(3): 471–476.

Khmelinskii, A., M. Baiker et al. (2011). Articulated whole-body atlases for small animal image analysis: Construction and applications. *Mol. Imaging Biol.* **13**(5): 898–910.

Knoess, C., S. Siegel et al. (2003). Performance evaluation of the microPET R4 PET scanner for rodents. *Eur. J. Nucl. Med. Mol. Imaging* **30**(5): 737–747.

Larsson, E., M. Ljungberg et al. (2011). Monte Carlo calculations of absorbed doses in tumours using a modified MOBY mouse phantom for pre-clinical dosimetry studies. *Acta Oncol.* **50**(6): 973–980.

Mauxion, T., J. Barbet et al. (2013). Improved realism of hybrid mouse models may not be sufficient to generate reference dosimetric data. *Med. Phys.* **40**(5): 052501.

National Electrical Manufacturers Association (NEMA). (2008). Performance measurements for small animal positron emission tomographs (PETs). NEMA Standards Publication NU 4-2008. NEMA, Rosslyn, VA.

Patel, V., R. N. Chityala et al. (2009). Self-calibration of a cone-beam micro-CT system. *Med. Phys.* **36**(1): 48–58.

Perilli, E., F. Baruffaldi et al. (2006). A physical phantom for the calibration of three-dimensional x-ray microtomography examination. *J. Microsc.* **222**(Pt 2): 124–134.

Qian, M., L. Niu et al. (2010). Measurement of flow velocity fields in small vessel-mimic phantoms and vessels of small animals using micro ultrasonic particle image velocimetry (micro-EPIV). *Phys. Med. Biol.* **55**(20): 6069–6088.

Rose, A. (1948). The sensitivity performance of the human eye on an absolute scale. *J. Opt. Soc. Am.* **38**(2): 196–208.

Segars, W. P., B. M. W. Tsui et al. (2004). Development of a 4-D digital mouse phantom for molecular imaging research. *Mol. Imaging Biol.* **6**(3): 149–159.

Stoico, R., S. Tassani et al. (2010). Quality control protocol for in vitro micro-computed tomography. *J. Microsc.* **238**(2): 162–172.

Tai, Y. C., A. Ruangma et al. (2005). Performance evaluation of the microPET focus: A third-generation microPET scanner dedicated to animal imaging. *J. Nucl. Med.* **46**(3): 455–463.

Taschereau, R., P. L. Chow et al. (2006). Monte Carlo simulations of dose from microCT imaging procedures in a realistic mouse phantom. *Med. Phys.* **33**(1): 216–224.

Wang, H., D. B. Stout et al. (2012). Estimation of mouse organ locations through registration of a statistical mouse atlas with micro-CT images. *IEEE Trans. Med. Imaging* **31**(1): 88–102.

Waring, C. S., J. S. Bax et al. (2012). Traceable micro-CT scaling accuracy phantom for applications requiring exact measurement of distances or volumes. *Med. Phys.* **39**(10): 6022–6027.

Wong, W. H., H. Li et al. (2012). Engineering and performance (NEMA and animal) of a lower-cost higher-resolution animal PET/CT scanner using photomultiplier-quadrant-sharing detectors. *J. Nucl. Med.* **53**(11): 1786–1793.

Xie, T. and H. Zaidi. (2013). Monte Carlo-based evaluation of S-values in mouse models for positron-emitting radionuclides. *Phys. Med. Biol.* **58**(1): 169–182.

Zhang, X., J. Tian et al. (2009). An anatomical mouse model for multimodal molecular imaging. *Conf. Proc. IEEE Eng. Med. Biol. Soc.* **2009**: 5817–5820.

Zou, W., J. Gao et al. (2009). Characterization of a novel calibration method for mineral density determination of dentine by x-ray micro-tomography. *Analyst* **134**(1): 72–79.

24

Monte Carlo Simulations in Imaging and Therapy

Panagiotis Papadimitroulas, George C. Kagadis, and George K. Loudos

24.1   MONTE CARLO SIMULATIONS

24.1.1   INTRODUCTION

In the literature, most of the scientific studies in the field of medical physics have used Monte Carlo (MC) techniques for the evaluation and verification of methods and algorithms, the design of new systems, and the quantification of new tracers in imaging or in dosimetry. The basis of MC numerical techniques can be described as statistical methods where random number generators are used to perform more realistic simulations for specified situations. Hence, the importance of such simulation programs is necessary in understanding the underlying physical processes that are taking place in a clinical procedure. The most important aspect of the application of MC simulations in the field of medical physics is the simplicity of the calculations of the physical interactions that are taking place (e.g., very-low-energy photon interactions, production of secondary charged particles, photon elastic scattering, electron ionization, and attenuated particles). Alongside, a critical aspect regarding the daily application of MC methods in hospital and institutional workstations is the consumption of time and resources for the acquisition of such procedures. Realistic simulations for clinical practice still require large computing resources. Thus, over the last 20 years, computer science has been rapidly evolving by enabling high-performance computers (HPC), clusters and grids of central processing units, and lately graphical physical units (GPUs) that are extensively used for the acquisition of MC simulations.

Nowadays, clinical calculations are becoming more personalized, which tends to bring MC methods into daily clinical practice. Thus, there is an increasing interest over the last decade on the MC applications both in *nuclear imaging and in radiotherapy/dosimetry*. Alongside, *in vivo* imaging studies have emerged as a critical component of preclinical biomedical research. As small animal imaging provides a noninvasive way of assaying biological information and function, the number of scientific studies rapidly increases. According to scopus (www.scopus.com), relevant documents on "small animal" and "Monte Carlo" total almost 1220 in the last 40 years, as shown in Figure 24.1. Oncological studies, new imaging modalities, novel radiotracers, nanomedicine, dose assessment, and biological impact of irradiation are just some of the scientific interests in small animal applications that MC techniques contribute in daily practice.

24.1.2   MONTE CARLO CODES

Since the early 1990s, there are several codes that have been extensively used for the simulation of radiation physics. Electron gamma shower (EGS) (Nelson et al. 1985) and integrated TIGER series (ITS) (Halbleib et al. 1992) were the standard tools for the calculation of the transport of

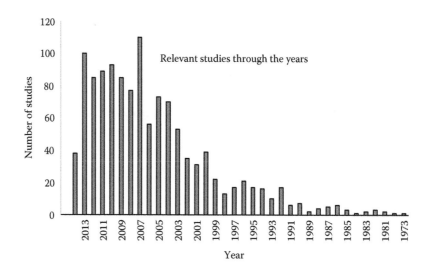

FIGURE 24.1   Number of documents relevant to "small animal" and "Monte Carlo" cited in Scopus in the period of 1973–2014.

electron and photon beams mainly in the field of radiotherapy (RT) and for dosimetric studies. ITS consists of three independent codes, namely TIGER, CYLTRAN, and ACCEPT, simulating photon and electron interactions down to 1 keV. MCNP (Monte Carlo N-particle transport) (Brown 2003) was also a standard code for radiation physics, which was extensively applied in the field of medicine in the late 1990s. Initially, due to the association of MCNP in simulating neutrons, it was widely used for military purposes. Lately, after the involvement of electron transportation and high-energy photons and low-energy photon simulations, MCNP was established as "gold standard" in the field of diagnostic and nuclear and therapeutic medicine. ETRAN (Seltzer 1991) was also one of the first widely used MC programs using the stopping powers of ICRU 37 (ICRU 1984). Most of the data in ICRU Report 35 were calculated based on ETRAN code, as it provided very accurate electron transport. In 2001, PENELOPE was published, which is a code system for MC simulations of coupled electron and photon transport in arbitrary materials and complex quadric geometries (Sempau et al. 1997, 2003). Two years later, CERN made available in the public domain the Geant4 simulation toolkit, which is a generic MC code, handling the physical processes of particles interacting with matter (http://geant4.cern.ch/). Its areas of application include high energy, nuclear and accelerator physics, as well as studies in medical and space science (Agostinelli et al. 2003). Geant4 managed to be one of the "strongest" reference MC codes in the field of medical physics and for clinical and preclinical simulations. This statement can be understood by the future widely used MC codes that were developed based on the core of Geant4. Geant4 application for tomographic emission (GATE) (Jan et al. 2004), Geant4-based architecture for medicine-oriented simulations (GAMOS) (Arce et al. 2008), TOPAS (Tool for PArticle Simulation) (Perl et al. 2012), and PTSIM (Particle Therapy Simulation) (Akagi et al. 2011) are typical examples of such codes. PTSIM and TOPAS are specified for RT and particle therapy studies. GAMOS is a user-friendly framework to implement GEANT4 simulations for nuclear medicine (NM) applications, which has been recently validated in small animal positron emission tomography (PET) imaging (Canadas et al. 2011). Finally, GATE is an advanced toolkit developed by the international OpenGATE collaboration (http://www.opengatecollaboration.org/home) and dedicated to numerical simulations in medical imaging and RT (PET, single-photon emission tomography [SPECT], computed tomography [CT], RT, dosimetry).

During the 1990s, more MC codes were developed, dedicated for specific medical applications. The SimSET package (Harrison et al. 1993) (http://depts.washington.edu/simset/html/simset_main.html) stands for "Simulation System for Emission Tomography" and models the appropriate physical processes and instrumentation that are used in nuclear medical imaging (NMI). It is designed for modeling both single-photon and positron emitters for SPECT (including scintigraphy) and PET applications, respectively. In its latest release, SimSET is based on the photon history generator, which generates and tracks the photons within the field of view (FOV) of the modeled system. Typical simulations proceed decay by decay, tracking the photons through the media of the phantom, the collimator module, and then through the detector module. Finally, a binning module is responsible for archiving the detected photons or the coincident events into an output formatted array.

Among the dedicated codes that were developed for SPECT applications is the SIMIND (Ljungberg and Strand 1989), SimSPECT (derived from MCNP) (Yanch et al. 1992), and MCMATV (Smith 1993), while a series of codes were dedicated for PET simulations, such as PETSIM (Thompson et al. 1992), EIDOLON (Zaidi et al. 1998), MC code (Reilhac et al. 1999), PET–EGS (Castiglioni et al. 1999), and PET SORTEO (Reilhac et al. 2004).

In Table 24.1, several MC codes are summarized according to categories, divided into two main groups: the generic codes and the dedicated ones. All MC codes are based on some common features such as the random number generator, the probability distribution functions (pdfs), sampling rules, and scoring (or tallying). However, there are also important differences between the MC codes, which mainly regard the accuracy of the physical processes modeling, flexibility, efficiency, execution time, and whether they are user friendly. Generally, even if the dedicated MC codes are faster in executing simulations than the generic, they usually suffer in terms of validation, accuracy, and support (Buvat and Castiglioni 2002).

TABLE 24.1
Main Monte Carlo Codes Categorized in "Generic" and "Dedicated," Including Their Main Characteristics

Name	Description
Generic codes	
EGS4	Radiation dosimetry
	Transport of electrons–photons (programming in MORTRAN)
ITS	High-energy physics
	Includes TIGER–CYLTRAN–ACCEPT
	Transport of electrons–photons
	Accurate electron cross sections to 1 keV
	NIST database (programming in FORTRAN)
MCNP	Radiation dosimetry
	Continuous energy
	Transport of neutrons–photons–electrons (programming in FORTRAN)
GEANT	High-energy physics
	Transport of electrons–photons (programming in C++)
ETRAN	Radiation dosimetry
	Transport of electrons–photons
Dedicated codes	
SIMIND	SPECT
SimSPECT	SPECT (derived from MCNP)
MCMATV	SPECT
PETSIM	PET
EIDOLON	PET
Reilhac et al.	PET
PET-EGS	PET
PET-SORTEO	PET
SIMSET	SPECT–PET
PTSIM	Particle therapy–RT–dosimetry (based in Geant4)
TOPAS	Particle therapy–RT–dosimetry (based in Geant4)
GAMOS	SPECT–PET–RT–dosimetry (based in Geant4)
GATE	SPECT–PET–RT–dosimetry (based in Geant4)

24.2 COMPUTATIONAL SMALL ANIMAL MODELS

24.2.1 DIGITAL PHANTOMS IN PRECLINICAL SIMULATIONS

Over the past 40 years, science has evolved, and imaging and RT techniques have become more and more individualized thus increasing the need for rapid and accurate diagnostic tools and therapeutic schemes. MC calculations provide such a tool for accurate computations in terms of physical processes. Although for the completeness of a clinical and/or a preclinical study, MC calculations are not enough. Thus, in radiation protection, radiological imaging, and RT research, a lot of effort is put into modeling the anatomy of the specimen.

For example, in order to estimate dose deposition, taking into account the inhomogeneity within the body, and to determine the amount of the energy deposited in various parts of the organs by external or internal ionizing sources, a pattern of the anatomy is needed. Physical and computational models serve as mimicking the interior and exterior anatomical features of the human or animal body.

Computational phantoms have been integrated with MC simulations since the 1960s and are used to simulate radiation transport inside the human body for modeling as realistically as possible the clinical situations determining the radiation interactions within the body. In the field of medical physics, the relevant physical processes include photons/electrons with energies up to 20 MeV, protons up to 250 MeV, and in some case of radiation protection, neutron sources are found in nuclear reactors with energies ranging from a few keV to 10 MeV. The probability of an interaction occurring within an organ or tissue is determined by "nuclear cross sections" that are associated with the tissue electron density, the tissue chemical composition, and the radiation energy (Xu 2010).

In the past decades, the rapid evolvement in genetics and molecular imaging, combined with the development of techniques for genetically engineering small animals, has led to increased interest in *in vivo* small animal research. Mice and other rodents are widely used for experimental modeling of preclinical studies. With the rise of small animal imaging, new instrumentation, data acquisition strategies, image processing, and reconstruction techniques are being developed and evaluated. A major challenge that still remains is how to evaluate all these experimental results and generate generalized computational phantoms as reference animal models. MC simulations can provide a ground truth tool for the evaluation and improvement of molecular imaging systems. Currently, there is a lack of realistic digital phantoms modeling the mouse anatomy and physiological functions for reference use in molecular imaging research.

For many decades, stylized and mathematical animal models were used for radiation dosimetry studies. The replacement of these phantoms came by the use of CT image data for the creation of realistic whole body models (Keenan et al. 2010). Research in the field of small animal imaging was based on the simple definition of the early human phantoms, which were modeled using geometrical shapes, for individual organs. In 1994, Hui et al. (1994) developed such a model, based on 10 athymic mice, including major organs, namely, liver, spleen, kidneys, lungs, heart, stomach, small intestine, large intestine, thyroid, pancreas, bone, marrow, and rest of the body (remainder tissues), for calculating cross-organ beta doses from ^{90}Y. At the beginning of the new millennium, Flynn et al. (2001) created another mouse model for the dosimetry of ^{131}I and ^{90}Y beta emitters, taking into account the heterogeneity within the tumor and kidneys. Few years later, Konijnenberg et al. introduced a stylized representation of Wistar rats (Figure 24.2) incorporating it in MCNP-4C MC simulator for calculating several dose factors (Konijnenberg et al. 2004).

The models that followed the stylized forms of mice were based on voxelized formats. Hindorf et al. (2004) developed another mouse model using voxel-based geometrical shapes for the definition of 10 organs. In 2003, MR images of a female athymic mouse were used by Kolbert et al. (2003) in order to realistically model the liver, kidneys, and spleen, with a 3D representation. CT data of a transgenic mouse (body mass 27 g) and a Sprague-Dawley rat (body mass 248 g) were used to define voxel-based digital phantoms by Stabin et al. (2006). X-ray CT, PET, and cryosection data of a normal nude male mouse were used by Dogdas et al. in order to create a 3D atlas of a whole body mouse

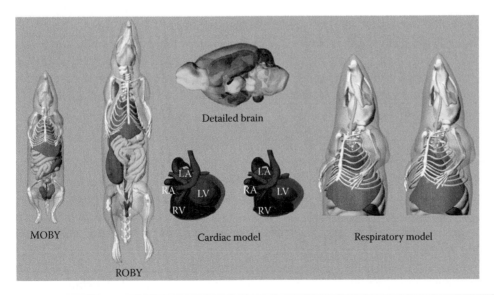

FIGURE 24.2 Representation of the MOBY and ROBY computational models, developed by Segars et al. (Reprinted from Segars, W.P. et al., *Mol. Imaging Biol.*, 6(3), 149, 2004. With permission.)

(Dogdas et al. 2007). The potential applications of the proposed model include the automated labeling of anatomical structures and computer studies phantom to simulate small animal PET, SPECT, CT, or MRI imaging systems.

The range of applications in small animal studies, incorporating MC simulations, is continuously increasing, with more and more realistic digital phantoms and new techniques for further applications. In 2008, in Japan, Sakae Kinase presented a voxelized frog phantom based on segmented images of cryosection data for internal dose evaluation purposes (Kinase 2008). Recently, the IT'IS foundation has been developing a series of animal models including mice and rats (male, female, and pregnant) as well as a pig voxelized model based on MR data, for several applications (http://www.ITIS.ethz.ch/). More details on the IT'IS animal models will be described in Section 24.2.3.

24.2.2   MOBY AND ROBY MODELS

Segars and Tsui (2009) developed a 4D series of hybrid, computational, anatomical phantoms based on nonuniform rational B-splines (NURBS) surfaces (Piegl 1991). Among these digital models, there are two small animal phantoms capable of providing a wealth of realistic multimodality imaging data from subjects of various anatomies and motions: a mouse whole body and a rat whole body model, namely, MOBY and ROBY respectively (Segars et al. 2004).

The 4D MOBY phantom was created based on a $256 \times 256 \times 1024$ 3D magnetic resonance microscopy dataset of a normal 16-week-old male C57BL/6 mouse, with a spatial resolution of $110 \, \mu m$ along the whole body (Segars and Tsui 2010). Such a detailed resolution allows the creation of a fine realistic phantom. The major enhancement of the 4D MOBY phantom compared to previous models is the incorporation of motions and the flexibility of the NURBS surfaces. NURBS provides the user with the ability to easily control the size and the structures of several organs within the body, according to the user needs. Furthermore, the important ability of manipulating cardiac and respiratory motion of the mouse model provides an enormous potential to study the effects of anatomical, physiological, physical, and instrumentation factors on small animal research.

In addition, a 4D digital rat computational model, namely, ROBY, was developed by the same group. According to the methodology that MOBY was created, MR rat segmented images (obtained from the U.S. Air Force Research Laboratory at Brooks Air Force Base in Texas) were used for the extraction of the surfaces. The phantom is a realistic and flexible representation of the anatomy of a rat, based on NURBS. This model also incorporates cardiac and respiratory motion with all the basic anatomical variations, appropriate for executing realistic molecular imaging simulations. The motion models were derived by dynamic time curves extracted by datasets of a healthy rat. The flexibility of ROBY including the variations of the involuntary motions provides a unique, accurate tool in the field of small animal molecular imaging and dosimetry. The known anatomy (attenuation—activity maps) and the controlled mechanisms of motion can provide the ground truth in MC platforms for the evaluation, validation, and optimization of several preclinical imaging and dosimetric protocols.

24.2.3   IT'IS ANIMAL MODELS

IT'IS foundation developed a series of high-resolution animal models (rats and mice of different gender and age) in cad format, so as to perform detailed dosimetry for *in vivo* exposure systems. They extended their models to include a male pig phantom, based on MRI data, as a dosimetric tool used in research and development. The pig phantom is a domestic pig weighing 35 kg and includes 103 types of tissue.

The other models include two male Sprague-Dawley rats (one big and one small), one female rat with three mammary tumors, and a pregnant rat and a rat pup with undefined sex. The two male rats can be modified according to the needs, in terms of changing the posture of the spine, tail, and legs.

Furthermore, four mouse models are included in their dataset; a male PIM1 mouse, which can be modified by the user, a male and a female OF1 mouse, and a pregnant B6F3C1 mouse. All the models are based on MR segmented data, and each defined tissue parameter is analytically described. The included tissue parameters can be assigned for various physical processes and can be used for a variety of applications, for example, electric conductivity, permittivity and permeability for electromagnetic simulations, or thermal conductivity, heat generation rate, heat transfer rate, and heat capacity for thermal simulations. Density and dielectric properties are also included in the materials database provided by IT'IS (www.ITIS.ethz.ch/database).

24.3 PRECLINICAL SYSTEMS

24.3.1 INSTRUMENTATION

The dramatic advances of molecular research over the past 20 years have increased the need for molecular imaging instrumentation and the optimization of imaging system design. The advantage of the noninvasive techniques of SPECT, PET, and CT imaging resulted in the establishment of these modalities in the early diagnosis of several diseases including oncology. This growing interest for *in vivo* imaging techniques activated the research of small animal instrumentation using MC tools, including the design of new medical imaging systems, prototypes, and evaluation and optimization of the small animal devices. Detectors, collimators, and shielding design are some of the most common parts of the system that are usually investigated. In addition, correction methods (e.g., scatter–random–attenuation–motion corrections, reconstruction algorithms, and partial volume effect) are studied for the improvement of image quantification (Zaidi 1999).

The tomographs in small animal NMI should have higher spatial resolution and sensitivity than clinical systems, and so the research community has put a big effort in using MC methods for the investigation of high count rate systems with increased spatial resolution (Khalil et al. 2011). Chung et al. (2004) investigated the effects of a dual-layer phoswich detector (lutetium oxyorthosilicate [LSO] and lutetium-yttrium aluminum perovskite [LuYAP]) as a function of crystal order and length on the depth of interaction (DOI) capability using GATE MC simulations. According to their simulated results, the sensitivity was increasing according to LSO crystal lengths regardless of the order of the layers. Furthermore, the effects of the crystal order of the detector on DOI measurements and image quality were investigated and characterized as a function of the frontal layer length. Zhang et al. (2011) performed several MC simulations in order to decide the system design and parameters of the system performance so as to create a commercial bench top low-cost preclinical PET scanner for satisfying typical resolution and sensitivity requirements needed for biodistribution and organ uptake studies.

Quantitative measurements always have been a challenge for SPECT, but they are of vital importance in small animal imaging applications. The requirements for accurate and quantitative images are complicated by limitations in the hardware components as well as in the imaging processes. Photons interacting within the tissue can be absorbed and scattered before detection by the scintillator. Hwang et al. (Hwang and Hasegawa 2005; Hwang et al. 2008) used GATE MC toolkit to perform studies for estimating several scatter and attenuation parameters in small animal SPECT. In their study, they suggest that for many isotopes, scatter may contribute approximately 20%–25% to the total counts, and in the case of ^{99m}Tc, less than 10% of all photons are probably scattered in rodent-sized phantom. Furthermore, they showed that photon attenuation can reduce the measured radioactivity by approximately 50% in case of ^{125}I. Partial volume errors are significant but are highly dependent on the geometry and the size of the specimen.

24.3.2 COMMERCIAL PRECLINICAL SYSTEMS

The advantages of MC techniques in small animal research led to investigations using validated simulated commercial preclinical systems. Many groups with installed preclinical systems in their laboratories extended their studies using MC simulators. The performance and the distribution of new radiotracers are usually tested in simulated imaging systems in order to test their diagnostic value as well as several dosimetric studies have been acquired in MC platforms, using validated commercial CT beams.

Merherb et al. (2007) modeled in GATE the performance of the Philips MOSAIC™ animal PET system, taking into account the ring PET geometry, the detectors, shielding, electronic processing, and the dead times. The MOSAIC is the first commercial preclinical PET system based on Anger logic detectors (Surti et al. 2003). The measured and simulated sensitivities were compared resulting in differences less than 3%, while scatter fractions agreed within 8%. In addition the simulated volumetric spatial resolution at the center of the FOV differed less than 6% compared to the experimental resolution values. Rey et al. (2007) have already studied the performance of the Lausanne ClearPET ring scanner comparing their experimental measurements with the GATE MC simulated data. Energy spectra, count rate performance, and noise equivalent count rate curves were simulated showing good agreement to the experimental curves. All the simulations were calculated using LSO and low-density LuYAP crystals as they are used in clear PET. In 2005, the PET part of the optical-PET system was characterized using GATE simulation toolkit. Specifically in the study of Rannou et al. (2004), the absolute sensitivity, DOI, and the spatial resolution were investigated for a variety of geometric configurations.

Jan et al. (2005) validated in GATE the modeling of the microPET FOCUS system in terms of spatial resolution, count rate, and image quantification, implementing realistic simulations with the use of a voxelized rat brain phantom. Correction approaches were presented in the same study for the positron range and the gamma accolinearity effects in rat brain dopaminergic studies. Realistic mouse studies have been also performed using two small animal PET scanners by Siemens in the PET-SORTEO MC code by Lartizien et al. (2007). In their study, they validated the performance of PET-SORTEO in small animal imaging by comparing simulated and experimental results of the MicroPET R4 and the Focus 220 PET scanners. The simulated energy spectra were in good agreement with the experimental spectra when using an aluminum cylinder to model the attenuating back compartment of the PMTs in both systems. Regarding the spatial resolution, a very good agreement was obtained between the modeled and real systems, while scatter fraction discrepancies were observed in the range of 250 and 300 keV, which was reported to be an effect of the intrinsic natural radioactivity of the LSO crystal. This was a similar study to the one presented by Jan et al. (2005) simulating the Focus 220 system in GATE achieving almost equivalent results in spatial resolution and count rate measurements with much less computational needs. Kim et al. (2007) carried out a comparative evaluation study of three microPET systems from the Siemens series using the GATE toolkit. Actually in the present study, the sensitivity and the scatter fraction of the Inveon, the R4 and the Focus120 microPET (Kim et al. 2007) systems were compared. SimSET MC toolkit was also used by Shokouhi et al. (2005) in order to model the rat conscious animal PET (RatCAP) and investigate the effects of the ring's geometry in image quality.

Several small animal SPECT systems have been modeled, evaluated, and validated using MC simulations. The Inveon small animal SPECT scanner with a single pinhole collimator was modeled by Konik et al. using GATE package (Konik et al. 2012). Following the NEMA-rat phantoms, the scatter fractions were calculated in the range of 10%–15% for ^{99m}Tc, 17%–23% and 8%–12% for the two energy peaks of ^{111}In (171 and 245 keV), and 32%–40% for ^{125}I. The Triumph X-SPECT preclinical scanner was also simulated comparing the simulated performance to the specifications provided by the manufacturer, Gamma Medica (Yu et al. 2013). The results were well matched in terms of sensitivity and spatial resolution at radius of rotation (ROR) ranging from 20 up to 35 mm. Lee et al in 2013 developed and validated a complete GATE model of the Inveon Trimodal Imaging Platform by Siemens including the SPECT, PET, and CT scanners. The validation of the models was based on both experimental and published data. Regarding SPECT and CT, the agreement of the simulated results to the empirical data was approximately 5%, while the PET simulation was validated following the NEMA NU-4 standard.

24.3.3 PROTOTYPE PRECLINICAL SYSTEMS

The investigation of small animal scanners is increasing by several laboratories that develop novel prototype systems in order to optimize either sensitivity or resolution. There is also a big effort on the development of low cost and small bench top systems. Sakellios et al. (2006) simulated in GATE two prototype systems for SPECT and PET using voxelized phantoms and rotating-head detectors. These are two dedicated small FOV scanners commercially available, and the simulations have been validated for their spatial resolution and sensitivities. MOBY was used in order to produce realistic bone scan images both for SPECT and PET. More recently, a simulation of a new prototype scanner "HyperSPECT" was presented by Tibbelin et al. (2012), which is a high-resolution small animal system with in-line x-ray optics. The full 3D simulation of the system was performed in two separate parts including ray-tracing, prism-array lens, pre- and post-collimators, and scintillator-based detector based on the MANTIS MC platform (Badano and Sempau 2006). Another CsI(Tl) based, prototype animal camera was validated in GATE platform by the members of OpenGateCollaboration, with excellent agreement of the simulated and the experimental data (Lazaro et al. 2004). The validation was implemented in terms of spatial resolution, sensitivity, scatter fraction, and energy spectra.

A main challenge in the development of new detectors in the field of animal imaging is the achievement of a satisfactory comprehension of the instrument behavior regarding the CT scanners. Brunner et al. (2007) simulated an innovative microCT scanner, the PIXSCAN, so as to understand and characterize its performance and the behavior of the contrast for soft tissues. The GATE toolkit was used to model a prototype of the scanner (PIXSCAN-XPAD2) so as to achieve a proof of principle of the system. Spatial resolution, contrast measurement, and dose reduction were estimated for the ultimate PIXSCAN performance. Furthermore, the MCNP4C code was validated in the work of Ay and Zaidi simulating a cone beam small animal CT scanner, the SkyScan 1076 (Ay and Zaidi 2005). The results were compared to measured data as well as to already simulated profiles by the accelerated MC simulator (AMCS) by Colijn et al. (2004), showing good agreement. Discrepancies were observed between the AMCS and the MCNP4C due to the underestimation of the scattered component and the overestimation of primary component in AMCS. The SIERRA MC code (Boone et al. 2000) was used to evaluate the radiation dose to small animals as a result of x-ray exposure (Boone et al. 2004). The x-ray spectra of a typical mouse CT system was modeled in SIERRA (Oxford x-ray tube, tungsten anode, 0.04 mm beryllium window) and the results were compared with reasonable accuracy to measured data.

24.4 MONTE CARLO APPLICATIONS

24.4.1 IMAGING

MC techniques are widely used in nuclear imaging for evaluation purposes and to tackle a variety of difficulties that are met by experimental or analytical approaches. Several studies have been implemented using MC methods for the optimization of (1) the detector design (crystal, geometry, collimator) and (2) the correction algorithms (scatter, motion, attenuation, denoising, normalization, partial volume effect) so as to perform quantitatively and qualitatively accurate imaging with the best possible diagnostic information in preclinical studies.

Deloar et al. (2003) used the MCNP code for the evaluation of the penetration and scattering parameters in a conventional pinhole SPECT system, modeling several types of collimators and using dedicated phantoms filled with ^{123}I, ^{131}I, ^{99m}Tc, and ^{201}Tl. In the study, it is reported that for a 1 mm pinhole aperture, the penetration levels were 78%, 28%, 23%, and <4% for each isotope, respectively, while for a 2 mm pinhole aperture, the respective levels were 65%, 16%, 12%, and <4%. The results showed that for the ^{201}Tl solutions, scatter correction due to the phantom is appropriate while the penetration effect can be neglected. Instead collimator scatter correction can be ignored in the cases of ^{99m}Tc and ^{123}I, while penetration and object scatter correction are essential for quantitative applications.

Another study used MC techniques for the investigation of the optimal number of pinholes in multi-pinhole SPECT dedicated to mouse brain imaging applications (Cao et al. 2005). The studied pinholes had double-knife edges with opening angles (120°) covering a large FOV in the mouse brain region. The main purpose was to reduce the injected dose to the animal. For this reason, 37 MBq injected activity was used, whereas the simulated results indicated that this was sufficient for quantitative neuroreceptor binding studies using a multi-pinhole collimator. No optimal number of pinholes are reported in the study, although they vary according to the collimator configurations.

In PET acquisitions, the scatter fraction has been earlier calculated in the range of 8% for mice and ~20% for rats (Kuntner and Stout 2014). Several labs are imaging—two to eight mice simultaneously (Aide et al. 2010; Rominger et al. 2010), although in a recent simulation study using a MOBY phantom and LabPET™-8 scanner in Geant4, it was shown that the scatter fraction can increase by 25% and 64% when imaging simultaneously three or five mice, respectively (Prasad and Zaidi 2014).

MC simulations can also play a crucial role in imaging by studying the optical properties of bioluminescence tomography. Alexandrakis et al. (2006) explored the effects of several magnitude optical property errors on the acquired images of a combined optical-PET system.

24.4.2 Diagnostic and Therapeutic Dosimetry

The field of preclinical targeted radionuclide therapy (TRT) has raised the interest of the scientific community recently in individualized and accurate dosimetric schemes, which are of high priority and of great importance. The relationship between the small animal absorbed dose assessment and the biological effects due to radiation toxicity can be translated from preclinical results to clinical practice. Dose estimations are appropriate when new radiopharmaceuticals are tested, and thus MC simulations combined with voxelized computational models can provide an adequate tool for ground truth. Radiation dose assessment is of crucial interest both in diagnostic NM and CT imaging, as well as in radio-immunotherapy applications. To date, the dosimetric calculations for several radionuclides used in diagnosis and therapy are based on the medical internal radiation dose (MIRD) (Loevinger et al. 1991) formalism following the expression of absorbed dose as

$$D_k = \Sigma_h A_h \cdot S_{(k \leftarrow h)} \qquad (24.1)$$

where

D_k is the mean absorbed dose in Grays (Gy) from the source emitter h to the target k
A_h is the cumulated activity in source h in Becquerel times seconds (Bq·s)
$S_{(k \leftarrow h)}$ is the mean absorbed dose in target k per unit of cumulated activity in the source h in Gy/(Bq·s)

Radiation absorbed doses in SPECT and PET imaging of small animals are still not very well established. Funk et al. (2004) estimated the whole-body radiation dose to rodents (mice and rats) for several isotopes, namely, ^{18}F, ^{99m}Tc, ^{201}Tl, ^{111}In, ^{123}I, and [125]I. The MCNP code was used with mouse- and rat-sized ellipsoids (30 and 300 g respectively), assuming a uniform source distribution in order to extract the S-values of the different isotopes. In the field of PET diagnosis, Taschereau and Chatziioannou (2007) calculated internal absorbed dose distribution in mice for applications with fluorine-18-labeled compounds, such as the [18]FDG, [18]FLT, and fluoride ion. GATE simulation toolkit was used for the extraction of the absorbed doses in several organs (Sarrut et al. 2014), as well as the PENELOPE electromagnetic process was used for the transport of electrons and positrons. Figure 24.3 represents an absorbed dose map of MOBY mouse, created in GATE using the "dose actors" tool. The MOBY mouse phantom, with low-resolution voxel size (400 μm), was used in order to represent a 16-week-old male C57B1/6 mouse of weight 33 g. According to the authors, depending on the tracer, the most critical organs of relatively high doses are the bladder wall, heart, bone marrow, bones, and kidneys. Also it is stated that absorbed doses from microPET can be higher than the absorbed dose from microCT, depending on the injected activity, the tracer, and the organ.

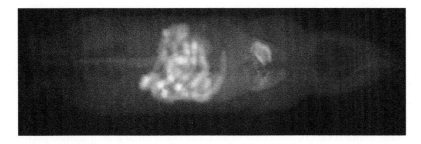

FIGURE 24.3 Absorbed dose map created in GATE, using a whole body normal biodistribution of ^{99m}Tc-Sestamibi in MOBY mouse model.

Peixoto et al. (2008) modeled in MCNP a tomographic rat phantom based on CT images of an adult male Wistar rat, calculating the absorbed fractions (AFs) for monoenergetic photon and electron energies. The same code was used by Zhang et al. (2012) incorporating a newly developed voxel mouse model based on cryosection images of a normal nude mouse for the absorbed dose assessment in photon beam RT. The organ dose conversion coefficients were obtained for 22 different external monoenergetic photon beams in the energy range of 10 keV to 10 MeV. The effects of the weight and the body size of the mouse were also investigated. The accuracy of the dosimetric calculations in preclinical dedicated studies is of high importance. Konijnenberg et al. (2004) studied the development of a generalized computational model for internal dosimetry in a rat. The absorbed organ dosimetry was investigated for ^{90}Y, ^{111}In, and ^{177}Lu radionuclide therapy. A spherical tumor of 0.25 g was also incorporated into the model in the right femur. According to the MCNP dosimetric simulations, the dose volume histograms showed that ^{111}In and ^{177}Lu are likely to have a higher threshold for renal damage compared to ^{90}Y.

24.4.3 S-VALUES

S-value (or elsewhere found as *S*-factor) is defined as the product of the emitted energy per disintegration and the AFs for the given combination of source and target regions, for the type of radiation emitted, divided by the mass of the target region.

$$S_{target \leftarrow source} = \frac{\sum_i n_i E_i \varphi_i (target \leftarrow source)}{m} \tag{24.2}$$

where n_i, E_i, and φ_i are the number of disintegrations emitted per transformation and the energy per emitted radiation and the AFs, respectively. *S*-values are, therefore, strongly dependent on the geometry of both the source and target regions and on the quality of the emitted radiation. *S*-values are generally calculated for anatomic models with the MC method for both animals and humans (Hindorf et al. 2004).

Many authors have used several simulation codes for developing human, animal, mathematical, and voxel-based models. More specifically, Bitar et al. (2007) simulated mono-energetic photon and electron sources using the MCNP4c2 MC code and *S*-values were calculated for 16 radionuclides and for a large number of target and source combinations for a 30 g mouse. In the study of Kolbert et al. (2003), a list of *S*-values was calculated on a model of an athymic mouse from MR data including spleen, liver, and kidneys. Hindorf et al. (2004) studied the parameters that influence *S*-value calculation in mouse models. Starting from a simple mathematical model, they have calculated *S*-value in different cases. According to their study, organ mass and shape, as well as source–target distance, have a noticeable impact on dose estimation. Mohhamadi and Kinase (2011) also calculated AFs and *S*-values for ^{90}Y, ^{131}I, ^{153}Sm, and ^{188}Re in several organs using the Digimouse voxelized model. In a similar study (Xie et al. 2010), the dosimetric characteristics on the skeleton of a Sprague–Dawley rat model were calculated for ^{32}P, ^{89}Sr, ^{90}Y, ^{143}Pr, and ^{169}Er. Boutaleb et al. (2009) calculated *S*-values with the MCNPX transport code and produced two datasets of S-values in order to compare the results and assess the impact of mouse models for preclinical dosimetry in TRT. The first dataset was developed using a voxel-based model of a 30 g female nude mouse (Bitar et al. 2007a,b) and the second one originated using the Digimouse model (Dogdas et al. 2007), which is based on a 28 g normal nude male mouse. On the series of Radiation Dose Assessment Resource (RADAR) (Keenan et al. 2010) for small animal studies, several scaled models of ROBY and MOBY were used for the AFs calculation of monoenergetic electron/photon sources and for the S-values generation of several radionuclides. More recently, Mauxion et al. (2013) generated a 30 g mouse phantom based on the realistic hybrid model MOBY. Dosimetric assessment of S-values was performed with two MC codes (MCNPX v2.7a and GATE v6.1) for ^{18}F, demonstrating that the comparison between two "similar" realistic digital mice leads to different S-values. In this study, the authors conclude that even with the use of the same software and incorporating realistic computational whole-body mouse models, there could be large variations on the resultant dosimetric characteristics. Thus, it is not obvious to refer to such models as "reference" models, as slight changes in the geometry may have great impact on the dosimetric values. Finally, in a similar work (Xie and Zaidi 2013), the variation of the absorbed dose coefficients was investigated as a function of the whole body mass. Furthermore, variations were observed in self S-values in different sizes of mice, which were not related to the organ mass variations of each model.

24.5   DISCUSSION

MC simulations have become ubiquitous in the medical physics field over the last 40 years. Small animal preclinical studies have taken advantage from these techniques over the last ~20 years. The number of groups that are using MC codes for their preclinical applications is continuously increasing. The tools that are available to scientists, in terms of (1) accuracy in physical processes modeling (even to low energies), (2) computational resources due to IT development (e.g., GPU, clusters, HPC), and (3) high resolution, hybrid, realistic animal models with accurate anatomical description and motion incorporation, can further their research.

There are still issues highly debated, as research in preclinical application becomes more and more specialized and with finer accuracy. To date, MC codes serve as a standard tool for the investigation of the instrumentation in the development of the imaging systems (SPECT, PET, CT, optical systems), as well as for the evaluation and the optimization of several pre- and post-processing techniques. Reconstruction, filtering, de-noising, and correction methods are well validated using simulated data before standardizing the algorithms.

In several preclinical applications, such as the evaluation of imaging protocols, the investigation of biokinetics of new tracers, and the study of biodistribution and organ uptake, it is of high interest to understand and quantify related parameters. The known anatomy in the computerized rodent models combined with the knowledge of the history of each particle in a simulated "experiment" provides the ground truth to the researcher. Thus, combined studies can be applied in a specific model, in a multi-tasking study, for example, studying the dosimetric characteristics in parallel to an imaging acquisition. NM has always been interested in the biological aspects of particle transportation within the tissues. MC methods played a crucial role in radiobiology, estimating and predicting the biological response caused by the ionizing deposited energy.

Dosimetry has been a debated field for many years, as it is quite difficult to be applied experimentally, and there is always a need for a standard reference model. Research today has become increasingly specialized and personalized, and thus there is a need to investigate individualized parameters at a very high level of accuracy and detail. Many studies have investigated the absorbed dose per organ for a variety of radionuclides for both imaging and therapy applications. The AFs and S-values for different organs in a plethora of mouse models were calculated using several MC codes. Currently, there is no reference computational animal model reported, as variations have been reported in most of the studies.

REFERENCES

Agostinelli, S., J. Allison et al. (2003). Geant4—A simulation toolkit. *Nucl. Instrum. Methods Phys. Res. A* **506**(3): 250–303.

Aide, N., C. Desmonts et al. (2010). High-throughput small animal PET imaging in cancer research: Evaluation of the capability of the Inveon scanner to image four mice simultaneously. *Nucl. Med. Commun.* **31**(10): 851–858.

Akagi, T., T. Aso et al. (2011). The PTSim and TOPAS projects, Bringing Geant4 to the particle therapy clinic. *Prog. Nucl. Sci. Technol.* **2**: 912–917.

Alexandrakis, G., F. R. Rannou et al. (2006). Effect of optical property estimation accuracy on tomographic bioluminescence imaging: Simulation of a combined optical-PET (OPET) system. *Phys. Med. Biol.* **51**(8): 2045–2053.

Arce, P., P. Rato et al. (2008). GAMOS: A Geant4-based easy and flexible framework for nuclear medicine applications. In *IEEE Nuclear Science Symposium Conference Record*, Dresden, Germany.

Ay, M. R. and H. Zaidi. (2005). Development and validation of MCNP4C-based Monte Carlo simulator for fan- and cone-beam x-ray CT. *Phys. Med. Biol.* **50**(20): 4863–4885.

Badano, A. and J. Sempau. (2006). MANTIS: Combined x-ray, electron and optical Monte Carlo simulations of indirect radiation imaging systems. *Phys. Med. Biol.* **51**(6): 1545–1561.

Bitar, A., A. Lisbona et al. (2007a). S-factor calculations for mouse models using Monte-Carlo simulations. *Q. J. Nucl. Med. Mol. Imaging* **51**(4): 343–351.

Bitar, A., A. Lisbona et al. (2007b). A voxel-based mouse for internal dose calculations using Monte Carlo simulations (MCNP). *Phys. Med. Biol.* **52**(4): 1013–1025.

Boone, J. M., M. H. Buonocore et al. (2000). Monte Carlo validation in diagnostic radiological imaging. *Med. Phys.* **27**(6): 1294–1304.

Boone, J. M., O. Velazquez et al. (2004). Small-animal x-ray dose from micro-CT. *Mol. Imaging* **3**(3): 149–158.

Boutaleb, S., J. P. Pouget et al. (2009). Impact of mouse model on preclinical dosimetry in targeted radionuclide therapy. *Proc. IEEE* **97**(12): 2076–2085.

Brown, F. B. (2003). MCNP—A general Monte Carlo N-particle transport code. v. R. LA-UR-03-1987. Los Alamos National Laboratory, Los Alamos, NM.

Brunner, F. C., R. Khoury et al. (2007). Simulation of PIXSCAN, a photon counting micro-CT for small animal imaging. In *Fourth International Conference on Imaging Technologies in Biomedical Sciences from Medical Images to Clinical Information—Bridging the Gap*, IOP Publishing Ltd. and SISSA, Milos Island, Greece.

Buvat, I. and I. Castiglioni. (2002). Monte Carlo simulations in SPET and PET. *Q. J. Nucl. Med.* **46**(1): 48–61.

Canadas, M., P. Arce et al. (2011). Validation of a small-animal PET simulation using GAMOS: A GEANT4-based framework. *Phys. Med. Biol.* **56**(1): 273–288.

Cao, Z., G. Bal et al. (2005). Optimal number of pinholes in multi-pinhole SPECT for mouse brain imaging—A simulation study. *Phys. Med. Biol.* **50**(19): 4609–4624.

Castiglioni, I., O. Cremonesi et al. (1999). Scatter correction techniques in 3D PET: A Monte Carlo evaluation. *IEEE Trans. Nucl. Sci.* **46**(6): 2053–2058.

Chung, Y. H., Y. Choi et al. (2004). Characterization of dual layer phoswich detector performance for small animal PET using Monte Carlo simulation. *Phys. Med. Biol.* **49**(13): 2881–2890.

Colijn, A. P., W. Zbijewski et al. (2004). Experimental validation of a rapid Monte Carlo based micro-CT simulator. *Phys. Med. Biol.* **49**(18): 4321–4333.

Deloar, H. M., H. Watabe et al. (2003). Evaluation of penetration and scattering components in conventional pinhole SPECT: Phantom studies using Monte Carlo simulation. *Phys. Med. Biol.* **48**(8): 995–1008.

Dogdas, B., D. Stout et al. (2007). Digimouse: A 3D whole body mouse atlas from CT and cryosection data. *Phys. Med. Biol.* **52**(3): 577–587.

Flynn, A. A., A. J. Green et al. (2001). A mouse model for calculating the absorbed beta-particle dose from (131) I- and (90)Y-labeled immunoconjugates, including a method for dealing with heterogeneity in kidney and tumor. *Radiat. Res.* **156**(1): 28–35.

Funk, T., M. Sun et al. (2004). Radiation dose estimate in small animal SPECT and PET. *Med. Phys.* **31**(9): 2680–2686.

Halbleib, J. A., R. P. Kensek et al. (1992). ITS version 3.0: The integrated TIGER series of coupled electron/photon Monte Carlo transport codes. Sandia National Laboratories, Albuquerue, NM.

Harrison, R., S. Vannoy et al. (1993). Preliminary experience with the photon history generator module of a public-domain simulation system for emission tomography. In *IEEE Nuclear Science Symposium and Medical Imaging Conference*, San Francisco, CA.

Hindorf, C., M. Ljungberg et al. (2004). Evaluation of parameters influencing S values in mouse dosimetry. *J. Nucl. Med.* **45**(11): 1960–1965.

Hui, T. E., D. R. Fisher et al. (1994). A mouse model for calculating cross-organ beta doses from yttrium-90-labeled immunoconjugates. *Cancer* **73**(Suppl. 3): 951–957.

Hwang, A. B., B. L. Franc et al. (2008). Assessment of the sources of error affecting the quantitative accuracy of SPECT imaging in small animals. *Phys. Med. Biol.* **53**(9): 2233–2252.

Hwang, A. B. and B. H. Hasegawa. (2005). Attenuation correction for small animal SPECT imaging using x-ray CT data. *Med. Phys.* **32**(9): 2799–2804.

ICRU (1984). ICRU Report 37: Stopping powers for electrons and positrons. ICRU Reports, Bethesda, MD.

Jan, S., A. Desbree et al. (2005). Monte Carlo simulation of the microPET FOCUS system for small rodents imaging applications. In *Nuclear Science Symposium Conference Record*, Fajardo, Puerto Rico.

Jan, S., G. Santin et al. (2004). GATE: A simulation toolkit for PET and SPECT. *Phys. Med. Biol.* **49**(19): 4543–4561.

Keenan, M. A., M. G. Stabin et al. (2010). RADAR realistic animal model series for dose assessment. *J. Nucl. Med.* **51**(3): 471–476.

Khalil, M. M., J. L. Tremoleda et al. (2011). Molecular SPECT imaging: An overview. *Int. J. Mol. Imaging* **2011**: 796025.

Kim, J. S., J. S. Lee et al. (2007a). Performance measurement of the microPET focus 120 scanner. *J. Nucl. Med.* **48**(9): 1527–1535.

Kim, J. S., J. S. Lee et al. (2007b). Comparative evaluation of three microPET series systems using Monte Carlo simulation: Sensitivity and scatter fraction. In *Nuclear Science Symposium, IEEE*, Honolulu, HI.

Kinase, S. (2008). Voxel-based frog phantom for internal dose evaluation. *J. Nucl. Sci. Technol.* **45**: 1049–1052.

Kolbert, K. S., T. Watson et al. (2003). Murine S factors for liver, spleen, and kidney. *J. Nucl. Med.* **44**(5): 784–791.

Konijnenberg, M. W., M. Bijster et al. (2004). A stylized computational model of the rat for organ dosimetry in support of preclinical evaluations of peptide receptor radionuclide therapy with (90)Y, (111)In, or (177) Lu. *J. Nucl. Med.* **45**(7): 1260–1269.

Konik, A., M. T. Madsen et al. (2012). GATE simulations of small animal SPECT for determination of scatter fraction as a function of object size. *IEEE Trans. Nucl. Sci.* **59**(5): 1887–1891.

Kuntner, C. and D. B. Stout. (2014). Quantitative preclinical PET imaging: Opportunities and challenges. *Front. Phys.* **2**: 1–12.

Lartizien, C., C. Kuntner et al. (2007). Validation of PET-SORTEO Monte Carlo simulations for the geometries of the MicroPET R4 and Focus 220 PET scanners. *Phys. Med. Biol.* **52**(16): 4845–4862.

Lazaro, D., I. Buvat et al. (2004). Validation of the GATE Monte Carlo simulation platform for modelling a CsI(Tl) scintillation camera dedicated to small-animal imaging. *Phys. Med. Biol.* **49**(2): 271–285.

Lee, S., J. Gregor et al. (2013). Development and validation of a complete GATE model of the Siemens Inveon trimodal imaging platform. *Mol. Imaging* **12**(7): 1–13.

Ljungberg, M. and S. E. Strand. (1989). A Monte Carlo program for the simulation of scintillation camera characteristics. *Comput. Methods Programs Biomed.* **29**(4): 257–272.

Loevinger, R., T. F. Budinger et al. (1991). *Mird Primer for Absorbed Dose Calculations.* Society of Nuclear Medicine, New York.

Mauxion, T., J. Barbet et al. (2013). Improved realism of hybrid mouse models may not be sufficient to generate reference dosimetric data. *Med. Phys.* **40**(5): 052501.

Merheb, C., Y. Petegnief et al. (2007). Full modelling of the MOSAIC animal PET system based on the GATE Monte Carlo simulation code. *Phys. Med. Biol.* **52**(3): 563–576.

Mohammadi, A. and S. Kinase. (2011). Influence of voxel size on specific absorbed fractions and S-values in a mouse voxel phantom. *Radiat. Prot. Dosimetry* **143**(2–4): 258–263.

Nelson, W. R., H. Hirayama et al. (1985). The EGS4 code system Stanford. Stanford Linear Accelerator Center, Menlo Park, CA.

Peixoto, P. H., J. W. Vieira et al. (2008). Photon and electron absorbed fractions calculated from a new tomographic rat model. *Phys. Med. Biol.* **53**(19): 5343–5355.

Perl, J., J. Shin et al. (2012). TOPAS: An innovative proton Monte Carlo platform for research and clinical applications. *Med. Phys.* **39**(11): 6818–6837.

Piegl, L. (1991). On NURBS: A survey. *IEEE Comput. Graph. Appl.* **11**(1): 55–71.

Prasad, R. and H. Zaidi. (2014). Scatter characterization and correction for simultaneous multiple small-animal PET imaging. *Mol. Imaging Biol.* **16**(2): 199–209.

Rannou, F. R., V. Kohli et al. (2004). Investigation of OPET performance using GATE, a Geant4-based simulation software. *IEEE Trans. Nucl. Sci.* **51**(5): 2713–2717.

Reilhac, A., M. C. Gregoire et al. (1999). A PET Monte Carlo simulator from numerical phantom: Validation against the EXACT ECAT HR ± scanner. In *IEEE Nuclear Science Symposium*, Seattle, WA.

Reilhac, A., C. Lartizien et al. (2004). PET-SORTEO: A Monte Carlo-based simulator with high count rate capabilities. *IEEE Trans. Nucl. Sci.* **51**(1): 46–52.

Rey, M., S. Jan et al. (2007). Count rate performance study of the Lausanne ClearPET scanner demonstrator. *Nucl. Instrum. Methods Phys. Res. A* **571**: 207–210.

Rominger, A., E. Mille et al. (2010). Validation of the octamouse for simultaneous 18F-fallypride small-animal PET recordings from 8 mice. *J. Nucl. Med.* **51**(10): 1576–1583.

Sakellios, N., J. L. Rubio et al. (2006). GATE simulations for small animal SPECT/PET using voxelized phantoms and rotating-head detectors. In *IEEE Nuclear Science Symposium*, San Diego, CA.

Sarrut, D., M. Bardies et al. (2014). A review of the use and potential of the GATE Monte Carlo simulation code for radiation therapy and dosimetry applications. *Med. Phys.* **41**(6): 064301.

Segars, W. P., B. M. Tsui et al. (2004). Development of a 4-D digital mouse phantom for molecular imaging research. *Mol. Imaging Biol.* **6**(3): 149–159.

Segars, W. P. and B. M. W. Tsui. (2009). MCAT to XCAT: The evolution of 4-D computerized phantoms for imaging research. *Proc. IEEE* **97**(12): 1954–1968.

Segars, W. P. and B. M. W. Tsui. (2010). The MCAT, NCAT, XCAT, and MOBY computational human and mouse phantoms. In X. G. Xu and K. F. Eckerman (eds.), *Handbook of Anatomical Models for Radiation Dosimetry*. CRC Press, Boca Raton, FL.

Seltzer, S. M. (1991). Electron-photon Monte Carlo calculations: The ETRAN code. *Int. J. Radiat. Appl. Instrum. A* **42**(10): 917–941.

Sempau, J., E. Acosta et al. (1997). An algorithm for Monte Carlo simulation of coupled electron-photon transport. *Nucl. Instrum. Methods Phys. Res. B* **132**(3): 377–390.

Sempau, J., J. M. Fernández-Varea et al. (2003). Experimental benchmarks of the Monte Carlo code penelope. *Nucl. Instrum. Methods Phys. Res. B* **207**(2): 107–123.

Shokouhi, S., P. Vaska et al. (2005). System performance Simulations of the RatCAP awake rat brain scanner. *IEEE Trans. Nucl. Sci.* **52**(5): 1305–1310.

Smith, M. F. (1993). Modelling photon transport in non-uniform media for SPECT with a vectorized Monte Carlo code. *Phys. Med. Biol.* **38**(10): 1459–1474.

Stabin, M. G., T. E. Peterson et al. (2006). Voxel-based mouse and rat models for internal dose calculations. *J. Nucl. Med.* **47**(4): 655–659.

Surti, S., J. S. Karp et al. (2003). Design evaluation of A-PET: A high sensitivity animal PET camera. *IEEE Trans. Nucl. Sci.* **50**(5): 1357–1363.

Taschereau, R. and A. F. Chatziioannou. (2007). Monte Carlo simulations of absorbed dose in a mouse phantom from 18-fluorine compounds. *Med. Phys.* **34**(3): 1026–1036.

Thompson, C. J., J. Moreno-Cantu et al. (1992). PETSIM: Monte Carlo simulation of all sensitivity and resolution parameters of cylindrical positron imaging systems. *Phys. Med. Biol.* **37**(3): 731–749.

Tibbelin, S., P. Nillius et al. (2012). Simulation of HyperSPECT: A high-resolution small-animal system with in-line x-ray optics. *Phys. Med. Biol.* **57**(6): 1617–1629.

Xie, T., D. Han et al. (2010). Skeletal dosimetry in a voxel-based rat phantom for internal exposures to photons and electrons. *Med. Phys.* **37**(5): 2167–2178.

Xie, T. and H. Zaidi. (2013). Monte Carlo-based evaluation of S-values in mouse models for positron-emitting radionuclides. *Phys. Med. Biol.* **58**(1): 169–182.

Xu, X. G. (2010). Computational phantoms for radiation dosimetry: A 40-year history of evolution. In X. G. Xu and K. F. Eckerman (eds.), *Handbook of Anatomical Models for Radiation Dosimetry*. CRC Press, Boca Raton, FL.

Yanch, J. C., A. B. Dobrzeniecki et al. (1992). Physically realistic Monte Carlo simulation of source, collimator and tomographic data acquisition for emission computed tomography. *Phys. Med. Biol.* **37**(4): 853–870.

Yu, A. R., S. Park et al. (eds.). (2013). Experiment and simulation for performance assessment of Triumph X-SPECT preclinical scanner. *J. Nucl. Med.* **54**(Suppl. 2): 2162.

Zaidi, H. (1999). Relevance of accurate Monte Carlo modeling in nuclear medical imaging. *Med. Phys.* **26**(4): 574–608.

Zaidi, H., C. Labbe et al. (1998). Implementation of an environment for Monte Carlo simulation of fully 3-D positron tomography on a high-performance parallel platform. *Parallel Comput.* **24**: 1523–1536.

Zhang, H., Q. Bao et al. (2011). Performance evaluation of PETbox: A low cost bench top preclinical PET scanner. *Mol. Imaging Biol.* **13**(5): 949–961.

Zhang, X., X. Xie et al. (2012). Organ dose conversion coefficients based on a voxel mouse model and MCNP code for external photon irradiation. *Radiat. Prot. Dosimetry* **148**(1): 9–19.

Section VIII
Applications

25

Small Animal Imaging and Therapy
How They Affect Patient Care

Lawrence W. Dobrucki

Over the past few decades, advances in life sciences, such as molecular biology, genomics, and proteomics, and the recent growth in the instrumentation and computational sciences, have stimulated the development of novel strategies focused on early detection and treating disease based on an individual's unique profile, an approach called "individualized medicine."

The growth of individualized medicine, which includes the development of personalized treatment regimes, will be aided by research efforts that provide better understanding of molecular events associated with normal and pathological processes. This research will contribute to the overall knowledge of the biochemical mechanisms that initiate the disease process, will allow for the identification of disease subtypes and will aid in predicting patient's response to the therapy. However, the described process of advancing patient care through individualized medicine is quite complex and slow. Through recent developments in imaging technology and advances in genomic medicine, it became evident that molecular imaging approaches including preclinical research bring a promise to accelerate, simplify, and reduce the costs of delivering improved health care to patients and has a great potential to facilitate the implementation of individualized medicine in clinics.

During recent decades, the translation of preclinical imaging approaches to clinical trials and patient use provided strong evidence that current clinical imaging applications have been proven to be effective to diagnose diseases such as cancer, neurological disorders, and cardiovascular disease in their initial stages, which permitted effective treatment associated with reduced morbidity and mortality. In addition, it was demonstrated that noninvasive imaging of therapeutic response reduced patients' exposure to toxic and ineffective treatments and allowed for early inclusion of alternative therapeutic interventions. Finally, these imaging approaches allowed the development of molecularly targeted treatments of cancer and certain endocrine disorders.

Critical reviews of current clinical applications of molecular imaging and their challenges have aided to identify emerging opportunities in molecular imaging, which should be addressed in the relatively near future. These opportunities include understanding the relationship between brain chemistry and behavior, understanding the metabolism and pharmacology of new drugs, assessment of the efficacy of new therapeutics and other forms of treatment allowing for quick introduction to clinical practice, and employment of targeted imaging or theranostic approaches for clinical use. In addition, these current challenges include the development of new technology platforms such as integrated microfluidic chips that would accelerate and reduce the costs associated with the synthesis and evaluation of novel molecular imaging probes and the development of higher resolution, higher sensitivity imaging instruments to detect and quantify disease processes faster and more accurately.

In spite of these exciting possibilities to improve patient health care through the translation of preclinical imaging approaches, the deteriorating infrastructure, loss of federal research support, and long and inefficient approval process for new imaging probes are jeopardizing the advancement of molecular imaging approaches. Understanding these challenges and careful systemic planning on how to address them is critical to revitalize the field to realize its potential.

25.1 SMALL ANIMAL IMAGING IN THE DEVELOPMENT OF A NEW GENERATION OF THERANOSTICS

25.1.1 CLINICAL IMAGING MODALITIES

New diagnostic and therapeutic agents are typically complex compounds for which biodistribution and pharmacokinetic profiles *in vivo* are difficult to assess. Therefore, multiple imaging approaches have been proposed to model targeted uptake and predict pharmacokinetics *in vivo*. These noninvasive imaging strategies have become essential tools used in basic and applied research. Typical workflow allows for dynamic imaging to assess the biodistribution of a studied probe or therapeutic agent in the same animal at different time points or stages of disease. Performing imaging studies in the same animal is cost-effective and minimizes variations between individuals, which results in enhanced reproducibility and repeatability as compared to traditional methods based on postmortem investigations.

Small animal imaging provides noninvasive means to assess biodistribution and pharmacokinetics of new experimental drugs and imaging probes in physiologically relevant environments *in vivo*. Intense development of new instrumentation and novel imaging strategies resulted in the availability of small animal dedicated imaging systems, which can be used to produce information about anatomical structures and physiological function. Novel trends in hybridization characterized by the combination of two or more imaging modalities led to the design and construction of multimodal imaging systems, which provided the most accurate information on the function of the studied agents with excellent anatomical references (Dobrucki and Sinusas 2005a,b,c).

Over the past three decades, multiple new imaging modalities have been developed but only few are currently used in both preclinical and clinical researches involving molecular imaging agents, including nuclear techniques (positron emission tomography [PET] and single photon emission computed tomography [SPECT]), x-ray computed tomography, magnetic resonance imaging, and optical imaging. The optimal choice of the most suitable imaging application for a certain study depends on the availability of both the instrumentation and imaging probe and prioritization of requested features. Detailed description of each imaging modality with the focus on their strengths and weaknesses is beyond the scope of this chapter; therefore, the readers are directed to the relevant chapters of this book or a few excellent recently published reviews (Bengel 2009; Dobrucki and Sinusas 2010; Mitsos et al. 2012).

25.1.1.1   Nuclear Imaging Modalities

Nuclear imaging includes SPECT and PET that belongs to clinical imaging modalities, which have been proved to provide excellent sensitivity (in the picomolar range), good temporal resolution (from seconds to minutes), and reasonable spatial resolution (one to few millimeters) (Basu et al. 2011).

In PET imaging, positrons emitted from an unstable atomic nucleus undergo an annihilation process with electrons, which results in the production of two gamma photons of 511 keV energy, separated by 180° and detected by a ring of gamma detectors located around the imaged object. The information about the location of the annihilation process derived from the detection of two 511 keV photons is then used to measure radioactive tracer accumulation in the tissue of interest or its consumption over time.

The major strength of PET imaging is that radionuclides such as carbon-11, nitrogen-13, or oxygen-15 can be incorporated in the molecule with minimal interference to the function of pharmaceuticals. This allows for developing radiolabeled probes, which are chemically almost identical to the parent compounds. This strategy has been successfully employed to develop PET tracers, which can pass the blood–brain barrier and can be used for imaging brain function (Judenhofer et al. 2008; Mariani et al. 2010).

PET imaging is also advantageous when considering its superior sensitivity, practically limitless penetration depth, and its ability for high-temporal resolution dynamic imaging which can provide useful information on pharmacokinetic parameters and can be used in compartmental modeling of probe's cellular uptake and washout rates.

One important disadvantage of PET is its fundamental limitation in achieving very high-spatial resolution, which is associated with the fact that the localization of positron emission is not the same as the place of annihilation and strongly depends on both the energy of emitted positron and the electron density of the tissue.

The labeling of PET pharmaceuticals can be complicated and due to the short half-life of PET isotopes, the presence of an on-site production cyclotron is required. Among the clinically available radioisotopes, only fluorine-18 has an adequate half-life (~2 h) that allows for required delivery times to be hours, not minutes. Other radioisotopes such as copper-64 (~12.7 h half-life) and zirconium-89 (~3.3 days half-life) are currently evaluated in preclinical models with the promise to be translated to clinical use in the near future (Wadas et al. 2007; De Silva et al. 2012; Zeng et al. 2012).

SPECT imaging is historically an older technique than PET and is based on the detection of gamma radiation emitted from an unstable atomic nucleus. Typically, an animal or patient injected with a SPECT radiopharmaceutical is imaged from several angles (projections), which enables

3D reconstruction of an image and further image processing and analysis. Radioisotopes used in SPECT imaging are characterized by their specific emission spectra, which allow for simultaneous multiple isotope imaging. Also, their relatively long half-lives (hours to days) allow for longitudinal imaging studies for up to several weeks with single administration of the studied radiopharmaceutical. Since different radioisotopes have different physicochemical properties, labeling the molecules or bioactive agents requires access to the radiochemistry resources and the knowledge of traditional chemistry. Also, many SPECT isotopes (technetium-99m, indium-111, or radioactive iodine isotopes such as iodine-125 and iodine-123) need to be chelated before labeling the parent molecule. This process can increase the molecular weight of the target molecule, may change the overall charge, and finally cause steric hindrances that may affect the pharmacokinetic properties of the studied molecule (Mariani et al. 2010).

25.1.1.2   X-Ray Computed Tomography

The general principle of how x-rays are used to produce 3D diagnostic images has not changed since the discovery of x-rays by Dr. Wilhelm Roentgen in 1895. In principle, x-rays produced by high-voltage x-ray tube pass through the patient, and are detected by an oppositely located detector. In 3D computed tomography (CT) imaging, both x-ray source and detector rotate around the animal or the patient acquiring projection images, which are further reconstructed in 3D space and processed. In contrast to nuclear imaging techniques where radiation originates from the imaged object, in x-ray CT, the ionizing radiation is produced by the imaging instrument and detected. The image contrast depends on the linear attenuation coefficient and since differences in linear attenuation coefficients for soft tissues are small ($\mu_{water} = 0.21$ cm^{-1}, $\mu_{fat} = 0.18$ cm^{-1}, $\mu_{muscle} = 0.20$ cm^{-1}, $\mu_{bone} = 0.38$ cm^{-1}), x-ray based CT techniques are not optimal to image soft tissues.

Despite these limitations, x-ray CT has been proven useful to provide anatomical information in localizing the radiolabeled molecules in both biodistribution and dynamic modeling studies. Current hybrid systems providing combined PET-CT or SPECT-CT modalities are available not only for clinical use but also for preclinical research using small animal dedicated instruments. These multimodal imaging systems allow for simultaneous or sequential imaging with various modalities, which can be coregistered without moving the imaged object (Figure 25.1) (Sinusas et al. 2008; Nahrendorf et al. 2009).

25.1.1.3   Magnetic Resonance

In contrast to x-ray CT, magnetic resonance imaging (MRI) provides the best soft tissue determination and is traditionally used for high-resolution anatomic and functional imaging. Isotopes with intrinsic magnetic moment and angular momentum such as carbon-13, hydrogen-2, nitrogen-15, and phosphorus-31 are typically used in MRI. When an isotope is placed in strong magnetic field, the nucleus of the atom is aligned with magnetic field. A short radio frequency (RF) signal is then used to perturb this alignment. After a pulse, the nucleus will realign and return to its natural state at a certain rate, called the relaxation time, emitting an RF signal that is recorded, analyzed, and used to produce MR images. Similarly, magnetic resonance spectroscopy allows for measuring relative concentrations of molecules in the target tissue (Wang, et al. 2009).

Although traditionally MRI was used for fine-structure imaging studies, contrast agents containing gadolinium, manganese, or ferric compounds are commonly used for local biodistribution investigations. Ultra-small paramagnetic iron oxide is a nanoparticle-based contrast agent, which has been successfully used in cardiovascular applications (i.e., atherosclerotic carotid plaque imaging), oncological studies (i.e., tumor vascular morphology or lymph node metastasis imaging), and neuroimaging (i.e., diffusion in brain disorders). Despite excellent spatial resolution and lack of ionizing radiation, MRI suffers from relatively low sensitivity, which hindered the development of MRI-compatible targeted molecular imaging agents for clinical translation. In contrast to nuclear imaging techniques, where picomolar concentrations of an imaging probe are needed for reliable imaging results, MRI requires millimolar levels of paramagnetic contrast agents. This limitation and the potential nephrotoxicity of gadolinium-based

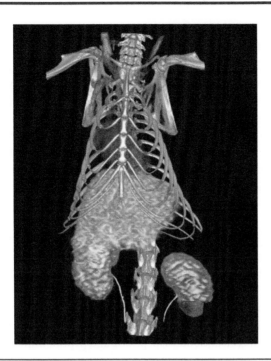

FIGURE 25.1   A representative PET-CT image of a healthy Lewis rat. The rat was injected with an iodine-based x-ray contrast agent prior to CT imaging to visualize the heart and adjacent veins and arteries. The biodistribution profile of copper-64 labeled cRGDyK peptide targeted at the avb3 integrin expressed in angiogenic vessels was studied 60 min after i.v. injection. ^{64}Cu-cRGDyK accumulates mainly in the liver and kidneys (green accumulation). The image also visualizes the elimination of ^{64}Cu-cRGDyK through kidney filtration and excretion to the bladder.

contrast agents should be taken under consideration during attempts to translate MRI-based animal imaging strategies to clinical practice.

25.1.1.4   Optical Imaging

In optical imaging, two phenomena (fluorescence and bioluminescence) are used to produce light originating from tissues, which is then monitored by a common CCD camera and recorded for processing. Typically, molecules are labeled with fluorescent moieties (i.e., cyanine-based Cy3 or Cy5 dyes), injected to the biological system, and their biodistribution is followed *in vivo* using 2D or 3D optical system scanners. Another method uses transfection of cells with a gene (i.e., green fluorescence protein [GFP] gene), which can be used to study cell function *in vivo*. Despite the clear advantages of fluorescence imaging, such as relatively easy usability, high throughput, inexpensive, and lack of ionizing radiation, its clinical translation is limited by reduced depth of penetration (in millimeters), surface reflectance, absorption (i.e., by hemoglobin), scattering, and autofluorescence. On the other hand, bioluminescence is more sensitive and is not affected by surface reflectance or scattering, although it still suffers from limited penetration depth. All of these disadvantages practically excluded optical techniques from clinical whole-body 3D imaging and limited their use to only local investigations including image-guided surgery applications or postmortem tissue analyses (Lee et al. 2012).

25.1.2   Traditional Imaging Agents

Small animal imaging methods are ideal to study new generation diagnostic and therapeutic agents. Since high-sensitivity techniques such as nuclear imaging approaches or optical imaging require only nano- or picomolar levels of labeled compounds, preclinical imaging strategies have proven very effective in obtaining preliminary results of biodistribution, accumulation, pharmacokinetics, and metabolic excretion routes of diagnostic or therapeutic agents. In contrast to traditional *ex vivo* or postmortem techniques, small animal imaging allows the serial assessment of these parameters *in vivo* achieved over time without sacrificing the animal.

Most of the commercially available pharmaceuticals are small molecules whose mechanism of action is typically based on direct interactions as an agonist or antagonist with the target tissue or cells. When labeled with radioactive metals, paramagnetic moieties, or optical molecules, small pharmaceutical compounds become powerful tools to assess the pathophysiology of the disease process, which allows not only the early detection of pathological processes but also the monitoring of the response to the therapeutic intervention. For example, a glucose analogue radiolabeled with fluorine-18 fluorodeoxyglucose is commonly used to measure tissue glucose consumption associated with enhanced metabolism. This phenomenon has been used in several PET imaging applications in neuroscience (activation of certain brain areas), cardiology (myocardial metabolism), and oncology (detection of primary tumors and metastases) (Rudd et al. 2002; Ogawa et al. 2004; Tawakol et al. 2006; Mullani et al. 2008).

Past few decades saw a period of extensive research focused on the development and characterization of molecularly targeted imaging agents for *in vivo* characterization and measurement of biological processes at the cellular, whole organ, or body level. Historically, the earliest molecular imaging approaches involved native monoclonal antibodies conjugated with the labels, which provided analytical signal. Despite their superior sensitivity and selectivity, monoclonal antibodies suffered from long circulation times and slow excretion, which significantly contributed to poor image quality and strong background signal. To overcome this limitation, monoclonal antibodies were later replaced by engineered antibodies, peptides, and peptidomimetic probes for imaging cell-specific antigens. More recently, nanoparticle-based constructs were proposed that not only allow molecular imaging based on direct interactions between target and the ligand but also serve as novel delivery technologies that enable sustained and controlled release of therapeutic agents (Criscione et al. 2011).

25.1.3 Theranostic Approaches

Nanoparticle-based constructs have traditionally been designed as carriers of therapeutic agents, which are natively unstable *in vivo*. More recent studies led to the introduction of theranostic approaches by integrating therapy with molecular imaging in a single material to provide diagnostic information through a variety of *in vivo* imaging modalities. Over the last decade, multiple nanoparticle formulations have been proposed that integrate imaging probes containing targeting moieties (such as antibodies, peptides, or peptidomimetics) and targeted controlled delivery of therapy.

Recent interest in theranostic strategies has led to the development of multifunctional nanocarrier platforms, which include combinations of drug–polymer conjugates, paramagnetic nanoparticles, protein constructs, solid lipid nanoparticles, dendritic macromolecules, liposomes, micelles, metal nanoparticles and nanocarbons, which are typically conjugated to a targeting moiety, and are briefly summarized in Table 25.1 (Muthu et al. 2014).

Drug–polymer conjugates including protein conjugates and small-molecule drug conjugates have been successfully used in studies focused on imaging and therapy of primary and metastatic

TABLE 25.1
Selected Theranostic Nanoparticle Platforms for Simultaneous Diagnostic Imaging and Therapy

Types	Therapeutic Agents	Diagnostic Agents	Size (nm)	Targeting Moieties
Drug–polymer conjugates	Copper-64	Copper-64 (PET imaging)	N/A	cRGD peptide
Polymeric nanoparticles	Docetaxel	Quantum dots (fluorescence)	~250	Folic acid
Solid lipid nanoparticles	Paclitaxel siRNA	Quantum dots (fluorescence)	~130	cRGD peptide
Dendrimers	Phthalocyanines	Phthalocyanines (fluorescence)	~62	LHRH
Liposomes	Doxorubicin	Technetium-99m (SPECT imaging)	100–200	DNA aptamer for nucleonin
Micelles	Iron oxide nanoparticles	Iron oxide (MRI)	~178	Passive targeting
Metal-core nanoparticles	Doxorubicin	Gold (x-ray CT)	~55	CPLGLAGG peptide

tumors (Vasey et al. 1999). Solid lipid nanoparticles are a safe and effective alternative to polymeric nanoparticles and are typically synthesized with a solid biocompatible hydrophobic core (i.e., triglyceride core) containing a therapeutic agent (Muller et al. 2000). These nanoparticles are characterized by high drug-loading efficiency and small size, which allow for effective passive targeting and the capacity for stable, controlled drug release for long periods of time. Dendrimers, highly branched spherical polymers, contain multiple sites available for functionalization with both therapeutic and diagnostic agents via encapsulation, entrapment, or covalent linkage. Phthalocyanine-functionalized dendrimers have been recently used for targeted delivery of antitumor paclitexel for both fluorescence imaging and efficient photodynamic therapy (Baker 2009). Recent studies suggest that dendrimer constructs have the potential for translation to clinical practice. Unfortunately, unprotected dendrimers suffer from *in vivo* toxicity; therefore, other safer nanoparticle-based agents have been developed. Liposomes were found to be an efficient alternative for drug delivery and imaging (Xing et al. 2013). They are characterized by low toxicity, good biocompatibility, and capacity for loading drugs and diagnostic agents such as iron oxide (for MRI), gold nanoparticles (for CT), quantum dots (for optical imaging), and radioisotopes (for PET or SPECT imaging). Recent studies demonstrated that by using different liposome formulations and surface modifications it is possible to achieve strong bioimaging contrast and optimal drug delivery for extended durations (Saad et al. 2008). Despite these clear advantages, multimodal nanoparticles often suffer from reduced chemical and biological stability and targeting specificity. In the design of multimodal nanoparticles for potential translation to human applications, both biocompatibility and long-term toxicity of multimodal constructs should be carefully evaluated.

25.2 DEVELOPMENT OF MOLECULAR IMAGING PROBES USED IN THE CLINIC

Traditionally, drugs have been developed by screening natural products and observing their physiological effects. In the twentieth century, this drug development model underwent a revolution by incorporating processes focused on chemical modifications and the use of analytical chemistry techniques to purify and identify biologically active compounds. Despite these advancements, the assessment of physiological activities was still done by direct observations *in vivo*. As genomics and proteomics evolved, the process of new drug development shifted to chemical organic synthesis and high throughput analysis techniques, which allowed for the characterization of new pharmaceuticals based on their efficient binding to the target. Only a small fraction of screened compounds have the potential to become a clinically effective drug; therefore, there is a tremendous need to develop new techniques, which can better characterize the drug–target interactions. Molecular imaging brings the promise to noninvasively assess physiological effects of the therapy in small animals, which when translated to patients can aid in developing new pharmaceuticals with increased clinical efficacy. Due to the complexity of drug–ligand interactions, the use of laboratory animals is instrumental to our understanding of complex *in vivo* effects on biochemical pathways and physiology. The question remains what can be learned from small animal studies that will accelerate the use of molecular imaging probes in the clinic.

25.2.1 USE OF ANIMAL MODELS IN TARGET VALIDATION

The prime example of recent advances of genomic and proteomic approaches in the development of novel drugs and molecular imaging probes is the use of transgenic animals. With the introduction of knockout animals in the late 1980s, it became possible to identify and validate proposed targets of new therapeutic or diagnostic compounds. Although the use of transgenic animals in which genes can be up- or downregulated *in vivo* has been instrumental in predicting targeting, their use in assessing drug efficacy is somewhat limited. It is partially due to the recent shift in paradigm from an emphasis on the complete disease characterization to a focus on a single target. This approach is a clear departure from the early testing of physiological effects of nature-derived therapeutics when often the target was unknown.

Despite this limitation, use of knockout animals is currently the method of choice in the development and characterization of new chemical entities and validation of target binding. In fact, there is an ongoing effort to produce knockout animals for all targets for which new pharmaceuticals or imaging probes can be potentially developed. In addition, transgenic animals have been used as animal models for certain diseases, but when therapeutic effectiveness is considered, there are valid concerns if results obtained in knockout animals can be extrapolated to humans. It is partially due to the fact that most drugs are not targeting single-gene disease but are rather targeting a key biochemical step controlled by multiple gene interactions in the progression of the disease. Furthermore, the human physiology is maintained by actions of various feedback and compensatory mechanisms, which are particularly challenging to model using laboratory animals including transgenics. Because of these compensatory mechanisms, it is very difficult to associate therapeutic effect with a single gene or gene expression product which was a cause for the failure of several therapeutics to show clinical efficacy in patients despite promising results obtained in laboratory animals (Sinusas 2004; Eckelman et al. 2005).

These concerning observations constitute a reason for a recent decline in successful U.S.-based Food and Drug Administration (FDA) approvals for both new therapeutics and molecular imaging agents over the last 20 years. Despite recent initiatives and published FDA guidance on medical imaging indications for drugs and biologics (Harapanhalli 2010), this decrease in successful FDA filings provides a lesson to be learned for the new generation of clinical scientists and basic science researchers focused on the development, characterization, and clinical translation of molecular imaging probes.

25.2.2 Practical Aspects of Development of New Molecular Imaging Probes

The paradigm shift toward combinatorial chemistry and the availability of high throughput screening approaches resulted in target-based drug and imaging probe development that affect a single target. This implies that a disease and its progression is dependent on a single protein expression product, which is in clear contrast to the approaches using physiology as an end point that can be affected by multiple targets. For new drug or imaging agent development, this could represent an oversimplification leading to a genetic reductionism and inability to develop optimal animal models to simulate a clinical situation. For example, this oversimplification can reduce a disease to a single genetic abnormality modeled with transgenic animals but it may not represent a clinical multifactorial disease. As a result, fully appropriate animal models for many human diseases are not yet available which makes a translation of target efficacy to disease efficacy less certain. Recent recommendation for development of drug and molecular imaging agents is to combine a rational target-based approach with strong physiology and disease focus (Sams-Dodd 2005).

The discovery and development of molecular imaging probes has dramatically evolved by incorporation of postgenomic techniques including molecular biology, nanomedicine, proteomics, protein–protein interactions, and reporter genes. As a result, multiple targeted molecular probes have been synthesized and characterized in *in vitro* models and *in vivo* using animals. The question addressed here is whether the development of imaging probes has been predominantly oriented toward target validation or toward impacting a particular disease.

For the molecular imaging probe to have diagnostic value, there must be a well-characterized target; therefore, target validation is a critical step in the development process. On the other hand, without a clear potential to impact patient care through noninvasive monitoring of disease progression or efficacy of therapeutic intervention, many molecular imaging probes will not reach their full clinical potential nor they will be approved by the regulatory agencies. To paraphrase the FDA recommendations: "the ability to produce beautiful images is not enough for successful regulatory approval, there must be a clear benefit for patient outcome." Other factors should be also considered during the development of target-based imaging agents. For example, what kind of targets should be chosen: those being early and specific indicators of the disease or being unique to assess the therapeutic effect? To address these concerns, an instructive systematic approach for the development of molecular imaging agents has been recently proposed (Eckelman et al. 2005). This approach involves three steps including target validation, sensitivity assessment, and model validation. First step involves studies focused on proving that the imaging agent targets the proposed binding protein. It is typically done by competitive binding studies performed *in vitro* or *in vivo*. A number of studies were also based on utilizing knockout animals to validate targeting of the imaging agents. Indeed, transgenic animals have been particularly useful in accelerating the target validation process using molecular imaging techniques. The next level of validation involves the assessment of probe's sensitivity. Because in targeted studies the outcome is often a measure of target density, the analytical signal originating from imaging probe should therefore change with the target density. These correlation studies are typically performed in systems where the target density gradually changes such as in wild-type, knockouts and animals overexpressing the target of interest. Finally, the last but most important step involves model validation. The difficulty to properly identify the optimal animal model arises from the fact that human disease is multifactorial and is almost impossible to model using biological *in vivo* systems where the relationship between the target expression and pathology is linear.

To address the limitations associated with using laboratory animals in the model validation phase of the development of molecular imaging agents, nonlinear biological systems called complex adaptive systems (CASs) have been recently utilized (Eckelman et al. 2005). These systems are typically described by complex quadratic equations, which allows for the presence of more than one equilibrium point (such as homeostasis managed by feedback and compensatory mechanisms), and their solution depends on the initial conditions of the biological system. A practical example of CASs involves the second-order reaction of ligand–receptor binding where the concentration of ligand–receptor complex increases with increased ligand levels before reaching linear region and a plateau value. It has been proposed that such complex interactions between facilitating and inhibitory mechanisms can be best studied using more than a single molecular imaging probe or with a single imaging agent targeting multiple receptors. SPECT imaging and its ability for simultaneous multiple isotope imaging or multiplexed optical imaging are the method of choice where a series of single-target tracers are used. In contrast, when a single multiple-target imaging agent should be used, in this regard, multimodal nanoparticle-based imaging constructs present distinct advantages over other approaches.

25.3   FUTURE DIRECTIONS

Small animal imaging has a great potential to substantially accelerate, simplify, and reduce the cost of delivering and improving health care, and in result significantly augmenting the quality of life. To realize this promise, the experts in the field from both academic institutions and industrial partners need to focus their research on the development of new imaging probes to improve the understanding of the biology and organ function, the development of hybrid imaging instruments enabling multimodal imaging approaches, introduction of theranostic approaches into clinical practice, the use of molecular imaging as a tool in the discovery and evaluation of new drugs, and finally, the translation of research from bench to bedside, including investment in infrastructure and training of clinical scientists.

REFERENCES

Baker, J. R., Jr. (2009). Dendrimer-based nanoparticles for cancer therapy. *Hematol. Am. Soc. Hematol. Educ. Program* **2009**: 708–719.

Basu, S., T. C. Kwee et al. (2011). Fundamentals of PET and PET/CT imaging. *Ann. N Y Acad. Sci.* **1228**: 1–18.

Bengel, F. M. (2009). Clinical cardiovascular molecular imaging. *J. Nucl. Med.* **50**(6): 837–840.

Criscione, J. M., L. W. Dobrucki et al. (2011). Development and application of a multimodal contrast agent for SPECT/CT hybrid imaging. *Bioconjug. Chem.* **22**(9): 1784–1792.

De Silva, R. A., S. Jain et al. (2012). Copper-64 radiolabeling and biological evaluation of bifunctional chelators for radiopharmaceutical development. *Nucl. Med. Biol.* **39**(8): 1099–1104.

Dobrucki, L. W. and A. J. Sinusas. (2005a). Cardiovascular molecular imaging. *Semin. Nucl. Med.* **35**(1): 73–81.

Dobrucki, L. W. and A. J. Sinusas. (2005b). Molecular cardiovascular imaging. *Curr. Cardiol. Rep.* **7**(2): 130–135.

Dobrucki, L. W. and A. J. Sinusas. (2005c). Molecular imaging—A new approach to nuclear cardiology. *Q. J. Nucl. Med. Mol. Imaging* **49**(1): 106–115.

Dobrucki, L. W. and A. J. Sinusas. (2010). PET and SPECT in cardiovascular molecular imaging. *Nat. Rev. Cardiol.* **7**(1): 38–47.

Eckelman, W. C., S. Rohatagi et al. (2005). Are there lessons to be learned from drug development that will accelerate the use of molecular imaging probes in the clinic? *Nucl. Med. Biol.* **32**(7): 657–662.

Harapanhalli, R. S. (2010). Food and Drug Administration requirements for testing and approval of new radio-pharmaceuticals. *Semin. Nucl. Med.* **40**(5): 364–384.

Judenhofer, M. S., H. F. Wehrl et al. (2008). Simultaneous PET-MRI: A new approach for functional and morphological imaging. *Nat. Med.* **14**(4): 459–465.

Lee, J. H., G. Park et al. (2012). Design considerations for targeted optical contrast agents. *Quant. Imaging Med. Surg.* **2**(4): 266–273.

Mariani, G., L. Bruselli et al. (2010). A review on the clinical uses of SPECT/CT. *Eur. J. Nucl. Med. Mol. Imaging* **37**(10): 1959–1985.

Mitsos, S., K. Katsanos et al. (2012). Therapeutic angiogenesis for myocardial ischemia revisited: Basic biological concepts and focus on latest clinical trials. *Angiogenesis* **15**(1): 1–22.

Mullani, N. A., R. S. Herbst et al. (2008). Tumor blood flow measured by PET dynamic imaging of first-pass 18F-FDG uptake: A comparison with 15O-labeled water-measured blood flow. *J. Nucl. Med.* **49**(4): 517–523.

Muller, R. H., K. Mader et al. (2000). Solid lipid nanoparticles (SLN) for controlled drug delivery—A review of the state of the art. *Eur. J. Pharm. Biopharm.* **50**(1): 161–177.

Muthu, M. S., D. T. Leong et al. (2014). Nanotheranostics—Application and further development of nanomedicine strategies for advanced theranostics. *Theranostics* **4**(6): 660–677.

Nahrendorf, M., D. Sosnovik et al. (2009). Multimodality cardiovascular molecular imaging, Part II. *Circ. Cardiovasc. Imaging* **2**(1): 56–70.

Ogawa, M., S. Ishino et al. (2004). (18)F-FDG accumulation in atherosclerotic plaques: Immunohistochemical and PET imaging study. *J. Nucl. Med.* **45**(7): 1245–1250.

Rudd, J. H., E. A. Warburton et al. (2002). Imaging atherosclerotic plaque inflammation with [18F]-fluorodeoxyglucose positron emission tomography. *Circulation* **105**(23): 2708–2711.

Saad, M., O. B. Garbuzenko et al. (2008). Receptor targeted polymers, dendrimers, liposomes: Which nanocarrier is the most efficient for tumor-specific treatment and imaging? *J. Control. Release* **130**(2): 107–114.

Sams-Dodd, F. (2005). Target-based drug discovery: Is something wrong? *Drug Discov. Today* **10**(2): 139–147.

Sinusas, A. (2004). Imaging of angiogenesis. *J. Nucl. Cardiol.* **11**(5): 617–633.

Sinusas, A., F. Bengel et al. (2008). Multimodality cardiovascular molecular imaging: Part I. *Circ. Cardiovasc. Imaging* **1**(3): 244–256.

Tawakol, A., R. Q. Migrino et al. (2006). In vivo 18F-fluorodeoxyglucose positron emission tomography imaging provides a noninvasive measure of carotid plaque inflammation in patients. *J. Am. Coll. Cardiol.* **48**(9): 1818–1824.

Vasey, P. A., S. B. Kaye et al. (1999). Phase I clinical and pharmacokinetic study of PK1 [N-(2-hydroxypropyl) methacrylamide copolymer doxorubicin]: First member of a new class of chemotherapeutic agents-drug-polymer conjugates. Cancer Research Campaign Phase I/II Committee. *Clin. Cancer Res.* **5**(1): 83–94.

Wadas, T. J., E. H. Wong et al. (2007). Copper chelation chemistry and its role in copper radiopharmaceuticals. *Curr. Pharm. Des.* **13**(1): 3–16.

Wang, Y., L. de Rochefort et al. (2009). Magnetic source MRI: A new quantitative imaging of magnetic biomarkers. *Conf. Proc. IEEE Eng. Med. Biol. Soc.* **1**: 53–56.

Xing, H., L. Tang et al. (2013). Selective delivery of an anticancer drug with aptamer-functionalized liposomes to breast cancer cells in vitro and in vivo. *J. Mater. Chem. B Mater. Biol. Med.* **1**(39): 5288–5297.

Zeng, D., N. S. Lee et al. (2012). 64Cu Core-labeled nanoparticles with high specific activity via metal-free click chemistry. *ACS Nano* **6**(6): 5209–5219.

26

Applications for Drug Development

Jessica Kalra, Donald T. Yapp, Murray Webb, and Marcel B. Bally

26.1   A HISTORY OF ANIMAL MODELS USED IN DRUG DEVELOPMENT

For a large part of human history, drugs derived from plants, animals, and minerals were discovered through experimentation and observation in both humans and animals. Scientists such as Edward Jenner (1749–1823) were even able to test some of their ideas in humans. For example, Jenner's observation that milkmaids showed immunity to small pox led to his hypothesis that the fluid from a woman with a cow pox blister may provide protection to others. In order to validate this theory, Jenner inoculated the suppurate material into a young boy and challenged the subject with two rounds of small pox (Willis 1997; Tan 2004). By doing this, Jenner demonstrated the concept of passive immunity, and laid the foundations for contemporary immunology. It was not until the 1800s, when human testing came under great scrutiny, that a larger emphasis was placed on tests in animal models. Starting with the discovery of diphtheria toxin and antitoxin in the late 1880s (Holmes 2000), animal models have become standard in the scientific assessment of new chemical agents and a means through which scientists can gain a broader understanding of complex disease processes. One of the most significant examples of the utility of animal models in drug discovery was by the renowned scientist Louis Pasteur (1822–1895). His study of infectious disease in animal models led to the development of vaccines against both cholera (Smith 2012) and rabies (Horsley 1889; Smith 2012). As discussed by Devita et al. in a detailed review on the history of cancer chemotherapy, Paul Ehrlich (1854–1915), the German physician and scientist who coined the term "chemotherapy," was the first to use animals to screen a small number of chemicals for activity against disease (DeVita and Chu 2008). In 1921, Dr. Frederick Banting and Dr. Charles Best established the first animal model of diabetes by removing the pancreas from previously nondiabetic dogs. More importantly, by grinding up excised pancreatic tissue and extracting specific substances, the researchers identified the hormone insulin and demonstrated that the injection of purified insulin in diabetic dogs was able to control blood sugar levels (Banting and Best 1990; Rafuse 1996). George Clowes (1877–1958) of the Roswell Park Memorial Institute (RPMI) established the first transplantable tumor systems in rodents (DeVita and Chu 2008), which led to the testing of a large number of chemicals for antitumor activity (DeVita and Chu 2008). From the early 1900s onwards, research efforts have focused on identifying the ideal animal model for drug testing, an effort that continues to this day.

Animal models have been pivotal in determining the function of the many compounds discovered or synthesized *de novo* by chemists for over eight decades. In many cases, such studies led to the discovery and development of compounds that are now used clinically. Examples include lithium as an anticonvulsant and treatment for bipolar disease, and halothane as a general anesthetic. Animal testing has made treatments possible for serious illness such as hypertension, cancer, and AIDS, and the eradication of diseases such as diphtheria. Despite the clear societal benefits of testing drugs in animals, the use of animals for research remains controversial. There are those who believe that testing in animals is simply unethical. Those who accept the necessity of preclinical testing are trying to better regulate the humane use of animals for research by defining guidelines to balance the procedures against the information obtained. In Canada, the Canadian Council of Animal Care is "responsible for setting and maintaining the standards for the ethical use and care of animals in science" (Canadian Council of Animal Care 2014). Comparable organizations in the United States (e.g., Association for Assessment and Accreditation of Laboratory Animal Care International), Europe (e.g., ETS 123), and Asia (e.g., Japanese Association for Laboratory Animal Science), emphasize the three "Rs" of animal research: Replace, Reduce, and Refine. Each of the aforementioned organizations recognizes that animal models are an essential tool in the development of new pharmaceutical products. Further, the data from such studies are required by various regulatory bodies that must assess the safety of new products prior to initiation of human clinical trials and prior to final approval. It is argued here that small animal imaging has the potential to refine how animals are used in the context of drug development programs, and that noninvasive imaging methods will result in a net reduction in the number of animals used in preclinical studies. Over the next decade, imaging will form the basis of a high content screening (HCS) platform for animal research in drug discovery.

26.2   THE DRUG DISCOVERY PROCESS AND SMALL ANIMAL IMAGING

The drug discovery process is long and challenging. Only 1 out of every 10,000 candidate compounds make it through to marketing approval by regulatory bodies. The pharmaceutical industry estimates that it can cost between 800 million and 1 billion dollars to develop an approved drug. This figure incorporates all the costs of drug development programs that do not progress to market approval. Even if these are not considered, independent estimates place the costs to develop an approved drug at approximately 100 million dollars (Morgan et al. 2011). Perhaps more important to those suffering from life-threatening diseases, it typically will take 10–15 years to see a therapeutic agent go from the discovery phase through to approval. The process is now fully regulated by government agencies, which mandate a development program that consists of well-recognized steps: basic research, discovery, preclinical development, and clinical testing (Figure 26.1). The basic research leading up to discovery work involves obtaining a greater understanding of the disease state. At this time, researchers attempt to identify the key elements of the diseased state that are different from the normal state. These studies can be done in vitro, in vivo, and even in silico. Discovery research begins with the identification of targets for treating or preventing a disease. The challenge is to discover which targets are relevant and to determine their role in the disease or disease process. The next step is to validate that targeting a particular molecule is helpful in the management of a condition. Identifying chemical compounds that modify a target can be done through high-throughput screening or computer-based modeling of chemicals and targets that will predict whether chemicals have the capacity to bind and modulate the activity of the target in a way that will alter the disease state. Once a chemical has been identified, the chemical is manipulated to achieve optimal binding or processing of the target. The drug candidate is then used in preclinical studies to (1) examine the efficacy of this chemical in vivo, (2) define what dose can be safely administered and provide information on potential toxicities, and (3) determine the metabolism and pharmacokinetic properties of the drug. At this point, successful agents can move on to clinical trials. Prior to entering Phase I studies, the manufacturing of candidate drugs must be scaled up by laboratories that are certified in good manufacturing practices. In Phase I clinical trials, healthy volunteers (or in the case of cancer, patients that have failed previous treatment options) are treated with the drug candidate to determine the safety and pharmacokinetics of the drug in humans. In Phase II clinical trials, between 100 and 250 patients with disease are treated with the drug in order to evaluate the efficacy and optimal dose in humans. Additionally, safety is retested, this time in the disease state. In Phase III clinical trials, a

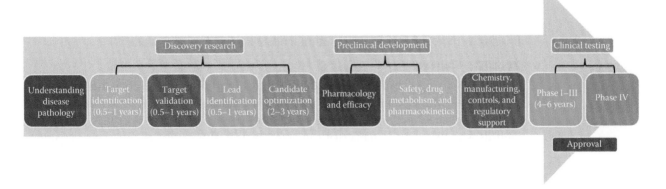

FIGURE 26.1   Key steps in the drug discovery and development process. Drug discovery and development research is conducted through four major steps prior to marketing approval: (1) basic research, (2) discovery research, (3) preclinical development, and (4) clinical testing.

larger cohort of patients, between 1000 and 3000 individuals, is recruited. This trial is conducted in order to confirm the effectiveness of the new drug across a broad population range, to monitor side effects, and to compare the agent to established treatments. Finally, Phase IV clinical trials are sometimes conducted after the drug has been on the market in order to collect additional information including the treatment risks, benefits, and optimal use. In recent decades, animal models have been used extensively in the preclinical stages of the drug discovery and development process, but the use of small animal imaging has generally been confined to basic research. It is anticipated that this will soon change for three reasons: (1) preclinical studies using imaging methods will be validated in clinical studies; (2) the tools that are used clinically are being reconfigured for use in small animals, facilitating the translation between the preclinical and clinical settings; and (3) regulatory bodies will accept preclinical safety studies defined using small animal imaging, and the quality management support for small animal imaging will meet the current Good Laboratory Practices (cGLP) guidelines. With the selection of appropriate animal models and small animal imaging, animals can be used at nearly every stage from basic research to drug discovery and development. It is reasonable to expect that using preclinical models and imaging early on will help make the drug development process more efficient and potentially enhance the success rate of compounds transitioning from candidate drugs to approved drugs. In the "Critical path opportunities" publication, the U.S. Department of Health and Human Services and the Food and Drug Administration highlight the importance of improving the predictive value of disease models as well as advancing the use of imaging both in the preclinical and clinical settings. Specifically, this document stresses the need for animal models that are better able to translate drug efficacy, pharmacokinetic, pharmacodynamic, and toxicity data, as well as the need for advancements in biomarker imaging and noninvasive imaging for monitoring disease progression and therapeutic efficacy (U.S. Department of Health and Human Services and Food and Drug Administration 2006).

Currently, there are five major imaging techniques that are widely available and commonly used in drug discovery. These are (1) x-ray and computed tomography (CT), (2) ultrasound (US) imaging, (3) optical imaging (OI), (4) magnetic resonance imaging (MRI), and (5) radionuclide-based imaging (SPECT and PET). The technical aspects of each of these modalities have been thoroughly described in previous chapters. Here we focus on the utility of each modality as it pertains to the relevant stages in the drug discovery and development process, animal modeling of disease, target discovery and validation, efficacy, safety, and toxicity profiling.

26.2.1 Stage 1: Basic Research—Using Animal Models to Understand the Pathophysiology of Disease

A robust understanding of the molecular pathogenesis of disease is vital in the discovery process because it provides the backdrop for target identification and validation of the effectiveness of treatment candidates. Animal models are commonly used to understand the pathophysiology of cancer and inflammatory, infectious, and neurodegenerative diseases. Often, the same disease models are used to evaluate drug candidates in the drug development pipeline. The following section describes some of the standard animal models used in these research areas in the context of small animal imaging. A thorough review of preclinical models can be found in Chapter 4.

26.2.1.1 Cancer

Animal models have been used since the 1950s to study tumors in situ, whether examining basic biological concepts of cancer development or therapeutic response. Over the past six decades, the diversity of models used in oncology research has increased, from cell lines injected into animals, animals with reporting systems that can assess biological function, to transgenic organisms that develop spontaneous disease. In each case, small animal imaging can play roles in monitoring disease progression, localizing disease, and evaluating changes in pathology as a result of intervention in the same animal over time.

Tumor xenografts arise when a small number of human tumor cells are transplanted into immune-compromised, and now humanized, syngeneic animals, typically mice. Over a period of weeks to months, the cells are able to establish a solid tumor, which can then be used to examine biological or pharmacological phenomena in situ. The first generation of xenografts was developed using murine tumor cell lines that could be inoculated into immune competent, syngeneic mice. The advantage of such a system was simply speed (tumors developed in 8–12 days) and convenience (mice were readily available and relatively inexpensive). These models were used to identify a large number of the cytotoxic drugs that currently form the basis of standard of care therapies for many different cancers. However, these models were highly criticized. First, the tumor cell lines had fast doubling times, which were unrepresentative of most human tumors. Second, the tumors were of mouse, rather than human, origin. To address this, investigators started to explore the use of human tumor cell lines injected subcutaneously. Again, the advantage of these models was convenience. Solid surface-localized tumors are easily accessible and can be measured directly using tools such as calipers. However, these tumors grow only in immune-compromised mice, and it is well recognized that interactions with immune cells is important in terms of cell growth, intra and intercellular signaling, and response to therapy. Still, the subcutaneous model provided investigators with a robust system to test the effects of therapeutic agents that went on to become very successful drugs such as doxorubicin (Di Marco et al. 1969) and, more recently, trastuzumab (Baselga et al. 1998; Vogel et al. 2001). Similar to the murine tumor models, human tumor cell line xenograft assessments were primarily based on evaluations of tumor growth rate and associated long-term survival, where survival was defined by a humane endpoint of tumor size (typically >10% of the body weight of the mouse) or animal health status. It is interesting to note that more frequently investigators are now turning to Response Evaluation Criteria In Solid Tumor (RECIST), which are being used in the majority of clinical trials (Eisenhauer et al. 2009). Here, complete response (CR) is defined as the disappearance of all target lesions (measurable disease); partial response (PR) is defined by a 30% decrease in the sum of the longest diameter measurable disease; stable disease (SD) is defined as insufficient reduction to define either PR or progressive disease (PD); and PD is at least a 20% increase in the sum of the longest diameter measurable disease. These criteria are frequently acquired through the use of imaging studies in patients, and these same imaging tools can be used in animals. Importantly, little information is available on how tumors respond immediately following initiation of treatment, and, again, animal imaging may have an important role in this arena as more researchers begin to explore early time points in drug efficacy studies.

It is widely agreed that subcutaneous tumor models are unable to recapitulate the complexity of the disease with respect to tumor heterogeneity, metastasis (site of tumor growth), or the effects of the local microenvironment, all of which are influenced temporally following treatment. At the most simplistic level, small animal imaging enables the use of more representative animal models such as orthotopic disease (arising in the tissue of origin) or metastatic disease, or use of genetically engineered models (GEMs). Orthotopic tumors were developed in part to better replicate the morphology and behavior of the original tumor within its organ of origin. In an orthotopic model, tumor cells are injected into, or in some cases implanted (stitched) onto, organs from which the tumors originated. In many cases, these organs lie deep within the body, and inspecting the tumor by palpation is not possible. At the same time, tumor progression creates severe effects on the health status of the animal necessitating euthanasia. Similarly, as an attempt to model metastatic disease, tumor cells are introduced directly into the blood stream either using a tail vein or an intracardiac injection (Conley 1979; Drake et al. 2005; van der Horst et al. 2011; Zhou and Zhao 2014). Alternatively, tumor cells are selected for their propensity to establish metastatic disease from solid tumors grown orthotopically (Guerin et al. 2013). Tumors that form systemically are difficult to locate without necropsy, and again there are significant animal welfare issues that arise because of our inability to adequately monitor disease progression in these models. Orthotopic and metastatic models are frequently developed using human cancer cell lines, so the results obtained when using these models is imperfect because the immune system of the mouse must be compromised to allow tumor growth. The GEMs try to address this limitation by genetically manipulating animals in a manner that allow tumors to arise in distinct sites (organs) in an immune-competent setting. This requires alterations (deletions, mutations, or overexpression) in one or several genes involved in malignant transformation or tumor progression. As a result, the animal develops spontaneous tumors that can be evaluated in the context of a drug discovery/development program. In each of the cases described, namely, orthotopic, metastatic, and spontaneous tumors, locating early tumor growth or monitoring disease burden in a manner that facilitates RECIST assessment is not trivial. Most frequently, animals need to be euthanized at selected time points during a time course of anticipated disease development in order to assess disease presence or disease burden. To achieve statistical relevance, such studies must use a large number of animals per selected time point. Use of noninvasive small animal imaging has the potential to provide time-course information using fewer animals and facilitate assessments of disease progression at early stages of tumor development and within sites of metastasis. Imaging will allow preclinical scientists to adopt RECIST evaluation methods, which, in turn, will help identify PD at a time point before tumor development impacts animal welfare. It is important to recognize that, when first establishing the value of small animal imaging methods, more rather than fewer animals are needed. This is simply a consequence of a need to validate data collected using established conventional methods compared to data collected using small animal imaging. Once this is done for a particular model, imaging should result in a significant reduction in the number of animals required to monitor tumor development and progression and for studies assessing the therapeutic potential of a selected drug candidate. In some cases, noninvasive small animal imaging will also enable the collection of physiologically relevant data such as vascular perfusion and metabolism within the tissue being imaged. Figure 26.2 provides a few examples of how small

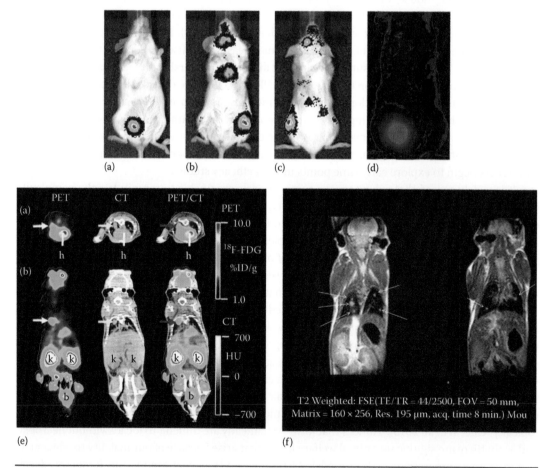

FIGURE 26.2 Small animal imaging of cancer models. Optical imaging modalities include bioluminescence imaging (a–c) and fluorescence imaging (d). (e) Detection of pulmonary metastasis with ^{18}F-FDG PET/CT (e). (From Deroose, C. et al., *J. Nucl. Med.*, 48, 295. Copyright 2007. With permission.) (f) Transgenic mouse model (EML4-ALK) in which tumors spontaneously develop imaged using T2-weighted MRI. (Images courtesy of Simonetta Geninatti, University of Turin, Torino, Italy.)

animal imaging may be used to visualize tumors in models of cancer. OI modalities include bioluminescence imaging (Figure 26.2a through c) and fluorescence imaging (Figure 26.2d). Figure 26.2a illustrates bioluminescent breast cancer cells inoculated in the mammary fat pad of a mouse, while Figure 26.2b and c represent the ventral (Figure 26.2b) and dorsal (Figure 26.2c) view of an animal that received an intracardiac inoculation of the same BL human breast cancer cell line. Figure 26.2d shows an orthotopic breast tumor developed from a human breast cancer cell line transfected with the fluorescent protein mKate. Figure 26.2e illustrates the detection of pulmonary metastasis with ^{18}F-FDG PET/CT (Figure 26.2e). Transverse (Figure 26.2a) and coronal (Figure 26.2b) sections of co-registered ^{18}F-FDG PET and CT images of a SCID mouse 45 days after tail-vein injection of A375M-Fluc melanoma cells are also shown. White arrows indicate a hypermetabolic area seen on the ^{18}F-FDG PET image. Yellow arrows show the water-density nodule in the dorsal apex of the right lung in the CT image. The PET/CT fusion image confirms registration of the hypermetabolic image on anatomic reference (red arrow). Physiologic tracer uptake in the heart (h), kidneys (k), and bladder (b) is marked. In Figure 26.2f, a transgenic mouse model (EML4-ALK) in which tumors spontaneously develop is imaged using T2-weighted MRI (FSE(TE/TR) = 44/2500, field of view = 50 mm, matrix = 160 × 256, resolution 195 µm, acquisition time = 8 min) (Figure 26.2f). Although the lungs are not usually visible with this type of imaging, lesions developing in the lungs are clearly seen; lesions were found to be 0.4–1.0 mm.

As discussed in detail in Chapter 10, tumor cells engineered to express fluorescent proteins or luciferase can be easily generated in the lab, and many well-characterized cell lines are also commercially available. Small numbers of labeled cells can be implanted subcutaneously, orthotopically, or by other routes as indicated previously, and OI can then be used to monitor disease progression. As an example, one of our studies used luciferase-positive breast cancer cells to establish orthotopic ascites and even metastatic breast cancer in animals. Bioluminescent imaging (BLI) enabled us to follow disease from inoculation to the end of the study. As shown in Figure 26.3, bioluminescent human breast cancer cells were inoculated orthotopically (Figure 26.3a) via intracardiac injection (Figure 26.3b whole body and Figure 26.3c organs) or intraperitoneally (Figure 26.3d). BLI was used to monitor tumor growth over time. Images shown were acquired on days 0, 1, 7, 14, 21, and 28 (Kalra et al. 2011). Minimal disease burden can be visualized as early as day 1 depending on the site of inoculation. BLI studies performed in the late 1990s showed that as few as 1×10^3 luciferase-positive human cervical carcinoma cells (Hela-Luc) could be visualized using BLI following injection (Edinger et al. 1999). BLI can also be used to monitor tumor progression in transgenic animals. For example, Hsieh et al. (2005) developed a PSA-Luc mouse model that would enable researchers to monitor the development of prostate tumors. The same group developed the TRAMP-luc transgenic mouse, which progresses to form prostate metastases in bone and other soft tissues that can be visualized using BLI (Hsieh et al. 2007).

Although generally less expensive and easier to perform than other imaging modalities, OI is used in only a small fraction of oncology studies. PET has been a dominant imaging tool used to evaluate tumor progression in vivo since the 1980s. Early studies show that radiolabeled fluorodeoxy glucose (FDG) uptake is high in metabolically active cells such as tumor cells, and this phenotype could be exploited as a tool to image tumors. Today, PET imaging enables more than just localization of metabolically active tumor cells. Imaging probes have been developed to evaluate other functional and metabolic properties of the disease. For example, Bradbury et al. (2008) used dynamic small animal PET imaging to evaluate tumor proliferation with radiolabeled 3′-deoxy-3′-18F-fluorothymidine (a compound that is incorporated into the DNA of actively dividing cells) in a genetically engineered mouse model of high-grade glioma. In another example, Koglin et al. (2011) used a small-molecule ^{18}F labeled tracer to specifically image early tumorigenesis by assessing expression of the cysteine/glutamate exchanger, a marker of oxidative stress. In our own institution, ^{18}F-EF5 has been used to examine tumor hypoxia in an androgen-sensitive model for prostate cancer (Yapp et al. 2007). EF5 is reduced in a hypoxic environment, and the reduced form is retained within the cell. Other examples

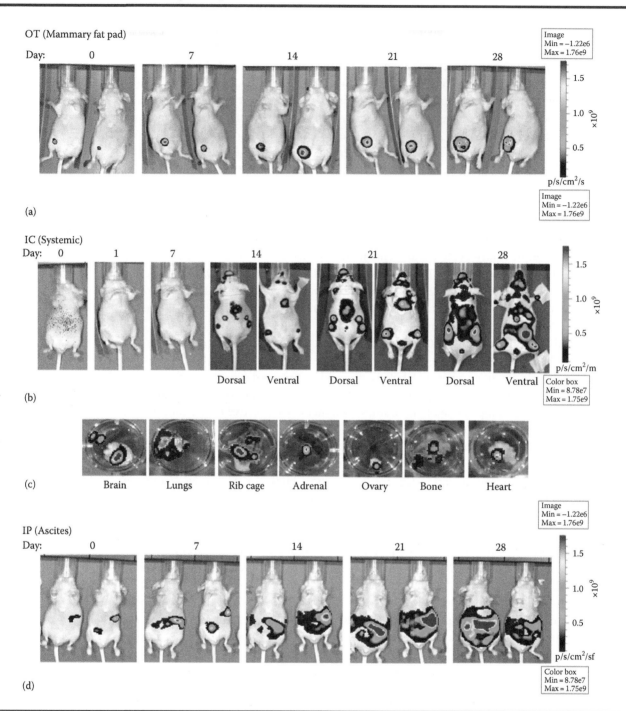

FIGURE 26.3 Growth of LCC6[WT-Luc] cells in female NCr nude mice following inoculation orthotopically (o.t.) (mammary fat pad), intracardiacally (i.c.) (left ventricle), or intraperitoneally (i.p.). Bioluminescent human breast cancer cells were inoculated o.t. (a), via intracardiac injection (b, whole body and c, organs), or i.p. (d). Bioluminescence imaging was used to monitor tumor growth over time.

where PET imaging has been useful in evaluating physiological functions in tumors include changes in tumor vasculature (Yapp et al. 2013) and markers of oxidative stress (Webster et al. 2014).

Ultrasonography (US) is truly a noninvasive approach in small animal imaging. Traditionally used for anatomical assessments of the cardiovascular system (ECG), US has been a valuable tool in cardiovascular disease models; however, US can also be used to indirectly detect the presence of proliferative diseases such as cancer or inflammation where cellular infiltrations cause thickening of mucosal linings. As shown by Plengsuriyakarn et al. (2012), US can be used to detect the thickening of visceral walls, which is indicative of cholangiosarcoma (CCA). This group was able to show that US imaging of sediments in the gallbladder, thickening of gallbladder wall, and hypoechogenicity of the liver correlated with the histopathological presence of CCA. In this way, US may be used as a noninvasive method of monitoring the development and progression of CCA in preclinical drug efficacy studies (Plengsuriyakarn et al. 2012).

26.2.1.2 Inflammation

Inflammatory disease is marked by changes in tissue architecture, infiltration of immune cells, and alterations in vascular permeability. Modeling inflammatory pathologies in animals has been challenging; however, robust preclinical models of asthma (Bates et al. 2009), rheumatoid arthritis (RA) (Anthony and Haqqi 1999; Joe and Wilder 1999; Hu et al. 2013), diabetes (Akash et al. 2013), and inflammatory bowel disease (IBD) (Valatas et al. 2013; Webb 2014) are widely used. In each of these models, inflammation is induced and physiological endpoints are used to monitor disease. For example, in a murine model of asthma, upon inhalation of ovalbumin (OVA), animals develop airway hyper-responsiveness (AHR), cellular infiltration into the lungs, and increases in eosinophil numbers, which are trademark features of human asthma (Bates et al. 2009). Readouts for assessing endpoints of asthma are invasive, typically requiring animal sacrifice. These readouts include plethysmographs to monitor AHR, and examination of bronchial lavage fluid for cellularity and protein levels. Small animal imaging modalities are noninvasive and have been used to measure asthma endpoints including AHR. Similar to the tumor models described previously, imaging allows investigators to get a sense of temporal changes using fewer animals. Brown et al. have suggested that high-resolution CT is one of the best tools available to measure changes in airway size, lung volume, and bronchial contraction and relaxation in a living animal (Brown and Mitzner 2003). More recently, inducible fluorescent probes have become available, which allow the imaging of cellular and subcellular processes involved in pulmonary inflammation (Haller et al. 2008). In a study published by Haller et al., fluorescence molecular tomography was used to visualize and quantify the pulmonary response to lipopolysaccharide (LPS)-induced airway inflammation over time. The investigators use activated cysteine proteases as molecular markers for acute inflammation, and a cathepsin-sensitive activatable fluorescence probe to visualize lung inflammation over time (Haller et al. 2008; Ntziachristos 2009).

The RA mouse model is produced through collagen peptide injections, and the readouts are ankle and paw swelling, redness, and stiffness. This model is considered extremely invasive. Long-term assessments are the norm, by which time the animal may experience difficultly in managing pain. Early signs of disease cannot be adequately examined without imaging. In 2008, Lee et al. illustrated that anatomical changes observed using small animal MRI and small animal CT could be correlated with histological changes in a rat model of RA (Lee et al. 2008). As shown in Figure 26.4, this group was able to use MRI to capture joint effusion, soft tissue swelling, bone erosion, and bone marrow changes. CT evaluation of bone mass density also enabled earlier less invasive examination of disease progression compared to histology (Lee et al. 2008). On day 8, bone marrow in the distal tibia (Figure 26.4a and b) showed patches of low signal intensity on T1-weighted images and high signal intensity on inversion recovery (yellow arrows), with focal areas of bone erosion (thick white arrows). The soft tissue swelling and edema are shown to have spread to the ankle joint (white arrows). Decreased bone density can be seen in the sagittal micro-CT image of the distal tibia (yellow arrowheads) (Figure 26.4c). Histology shows inflammatory infiltrate in the bone marrow and surrounding soft tissue (black arrowheads) (Figure 26.4d). Tibiotalar joint surfaces are intact on all modalities. Put et al. demonstrate that ^{99m}Tc-labeled nanobodies targeting the macrophage mannose receptor and SPECT-CT could be used to monitor joint inflammation in an RA mouse model (Put et al. 2013). More recently, Fuchs et al. demonstrated that 3′-deoxy-3′-18F-fluorothymidine PET enables detection of cellular proliferation in an animal model of RA, a primary indication for inflammation (Fuchs et al. 2013).

Under circumstances where inflammation is the result of pathology as opposed to the etiology of the disease, small animal imaging is still useful in monitoring disease outcomes. In one example, the apolipoprotein E knock-out mouse (apoE$^{-/-}$) was used as a model for atherosclerosis. The homozygous knockout leads to hypercholesterolemia and atherosclerosis. Micro-PET studies with ^{18}F-FDG proved useful in examining atherosclerotic plaques and inflammation within lesions.

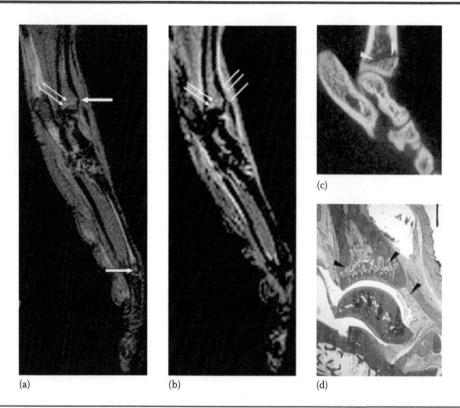

(a)   (b)   (c)   (d)

FIGURE 26.4 Early findings of small animal MRI and small animal CT correlated with histological changes in a rat model of rheumatoid arthritis. (a and b) T1-weighted MR images of bone marrow in an RA mouse model. (c) Bone density using sagittal micro-CT imaging of the distal tibia. (d) Histology of the bone marrow and surrounding soft tissue. (From Lee, S.W. et al., *NMR Biomed.*, 21, 527, 2008. Copyright 2007. With permission.)

Studies such as the one done by Ogawa et al. (2004) have been able to correlate the FDG uptake with macrophage infiltration of the plaque (Ogawa et al. 2004; Haq et al. 2013).

26.2.1.3 Infectious Disease

In many animal models of infectious disease, rodents are challenged with a pathogen, and signs of disease and treatment efficacy are evaluated through examination of temperature, behavioral changes, and quantitation of microbial load at defined time points after treatment. Mouse models for parasitic, fungal, viral, and bacterial infections are numerous, and have been invaluable in understanding the pathogenesis of disease, the associated immune response, as well as identification of early biomarkers and effective treatments. Noninvasive small animal imaging has now become an important research tool for preclinical studies of infectious diseases, as longitudinal studies of the same animals can be performed during the infection and treatment periods. In a simple but elegant example, Zhao et al. used GFP-expressing bacteria to resolve the spatiotemporal behavior of bacteria in vivo and the physiological response of the animal (Figure 26.5) (Zhao et al. 2001). Specifically, this group was able to follow *Escherichia coli*-GFP through the mouse gastrointestinal tract after kanamycin treatment (Zhao et al. 2001). In a more complex example, the malaria mouse model was established by inoculating mice with *Plasmodium berghei* Antwerpen-Kasapa (ANKA), which is capable of producing cerebral disease in animals. Several investigators have used T1- and T2-weighted MRI and MR angiography or flow-alternating arterial inversion spin-labeling MRI studies and single-voxel proton spectroscopy to demonstrate decreased cerebral blood flow, as well as neuronal and axonal injury in mice infected with *P. berghei*-ANKA (Penet et al. 2005, 2007). In a recent publication, Bagci et al. demonstrated, using multimodal PET-CT imaging of animals, that pulmonary infections could be tracked over time and correlated with infection severity (Bagci et al. 2013). Their study used an automated method for measuring areas of abnormal uptake of ^{18}F-FDG on PET images and an interactive method for analyzing the corresponding anatomical structures on CT images (Bagci et al. 2013).

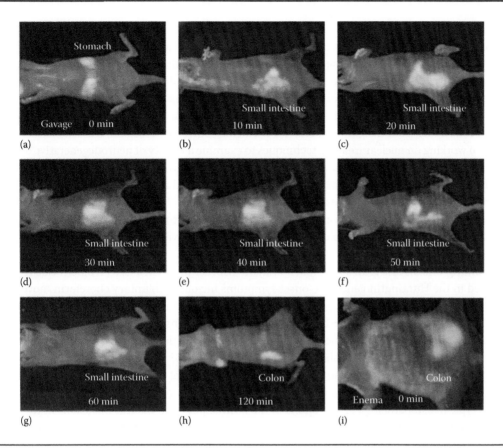

FIGURE 26.5 Whole-body imaging of *E. coli*-GFP infection in various organs. (a) *E. coli*-GFP infection in the stomach immediately after gavage of 1011 *E. coli*-GFP. (b) *E. coli*-GFP infection in the small intestine 10 min after gavage. (c) *E. coli*-GFP infection in the small intestine 20 min after gavage. (d) *E. coli*-GFP infection in the small intestine 30 min after gavage. (e) *E. coli*-GFP infection in the small intestine 40 min after gavage. (f) *E. coli*-GFP infection in the small intestine 50 min after gavage. (g) *E. coli*-GFP infection in the small intestine 60 min after gavage. (h) *E. coli*-GFP infection in the colon 120 min after gavage. (i) *E. coli*-GFP infection in the colon immediately after enema of 1011 *E. coli*-GFP. (From Zhao, M. et al., *Proc. Natl. Acad. Sci. USA*, 98, 9814. Copyright 2001. With permission.)

26.2.1.4 Neurodegenerative Disease

For major neurodegenerative diseases including Parkinson's, Huntington's, and Alzheimer's disease, there are well-established animal models that have been vetted for over five decades. Some of the key behavioral features of neurodegenerative diseases are difficult to establish in rodent models, while histopathological analyses are more definitive but require ex vivo analyses of cortical tissues and, therefore, studies involving a large number of animals. Neuroimaging tools such as MRI and PET can be used to complement ex vivo studies and enable researchers to monitor disease progression noninvasively and longitudinally. Research has shown that micro-PET and SPECT are sensitive enough imaging tools, capable of detecting picomolar concentrations of neuroreceptors, transporters, and neurotransmitters with the appropriate radioligands.

The etiology of Parkinson's disease remains under investigation. Both genetic (Parkin, Alpha synuclein) and environmental factors (pesticides, head trauma) have been implicated in this disease. The major clinical features of Parkinson's disease include tremor, muscle rigidity, bradykinesia, postural instability, and cognitive deficits. Postmortem examinations of cortical tissues show degeneration of dopamine neurons in the substantia nigra. Experimental rodent models of Parkinson's disease can be successfully induced using chemicals (6-hydroxydopamine [6-OHDA], 1-methyl-4-phenyl-1,2,3,6-tetrahydropyridine, rotenone, paraquat, or epoxomicin) or genetically (α-synuclein knock-in mice or DJ-1 knock-out mice) (Garcia-Alloza and Bacskai 2004). For each model, only some neuropathological and behavioral phenotypes may be present, and no single model is able to recapitulate all of the features of Parkinson's disease. Nevertheless, MRI has been successfully used to study anatomical or morphological changes in vivo, while PET and SPECT have provided valuable insights into the mechanisms of nigrostriatal degeneration in animal models of Parkinson's disease. For example, in 2007, Pellegrino et al. used PET to investigate dopamine transporter status using (11)C-2β-carbomethoxy-3β-(4-fluorophenyl)-tropane as well as dopamine receptor modulation using (11)C-raclopride in a 6-OHDA model of Parkinson's disease (Pellegrino et al. 2007). A group led by Prof. Muller from the Clinic of Nuclear Medicine, University Hospital, Düsseldorf, Germany, has been working on nuclear imaging techniques to examine a variety of neurodegenerative diseases including Parkinson's disease. In several publications, Muller's group describes the molecular imaging of the dopaminergic synapse in vivo. In one paper published in 2011 (Nikolaus et al. 2011), this group used a series of radionuclides to examine DA transporter (DAT) and/or Dopamine (2) receptor binding with SPECT or PET in a 6-OHDA rat model. Interestingly, their studies were conducted before and after treatment with haloperidol, L-DOPA, and methylphenidate, the gold standard therapies for Parkinson's disease.

Huntington's disease is an autosomal dominant disorder in which a CAG trinucleotide repeat is observed in the Huntingtin gene. Late onset symptoms include involuntary choreform movements and cognitive impairment. Histologically, patients have some neuronal loss in the striatum. Animal models of Huntington's disease (HD) are induced by neurotoxins such as quinolinic acid, ibotenic acid, or 3-nitropropionic acid, which destroy or impair the corticostriatal projection. A rat transgenic HD model has been established using a cDNA carrying 51 CAG trinucleotide repeats under the control of the Huntingtin gene promoter. The transgenic rat exhibits adult-onset neurological phenotypes with reduced anxiety, cognitive impairments, and progressive motor dysfunction, as well as typical histopathological alterations in the form of neuronal nuclear inclusions in the brain. Von Horsten et al. have been able to use MRI to image striatal shrinkage and ventricular enlargement in transgenic animals. This group has also used ^{18}F-FDG PET to image brain glucose metabolism and shown a reduction similar to that seen in patients with HD (von Horsten et al. 2003). In 2000, Araujo et al. used micro-PET and [3-(29–18F]fluoroethyl)spiperone to image dopamine receptor binding as well as ^{18}F-FDG to image glucose metabolism in a quinolinic-induced rat model of HD (Araujo et al. 2000).

Patients with Alzheimer's disease (AD) present with signs characteristic of dementia including memory loss, cognitive decline, and alterations in personal care. Morphologically, there can be significant cortical atrophy and enlarged ventricles, while histologically AD has two features: β-amyloid plaques and neurofibrillary tangles. The most common rodent models of AD are the tau and amyloid β (Aβ) transgenic mouse and rat models. Aβ is one of the histological markers that has been the main focus of imaging studies. Poduslo et al. used a putrescine and gadolinium-labeled Aβ peptide (Aβ1-40) for imaging Aβ plaques in vivo using MRI (Poduslo et al. 2002). Jack et al. (2007) showed for the first time that T2-weighted MRI can be used to identify plaques as small as 50 μm in vivo (Jack et al. 2007). More recently, Grand'maison et al. used anatomical MRI scans to correlate cortical thickness with predisposition to Aβ plaque deposition in mutant amyloid precursor protein (APP) transgenic mice. They suggest that this technology may be used as a predictive tool for disease progression (Grand'maison et al. 2013).

The previous sections have illustrated how small animal imaging can be used to enhance the knowledge gleaned from in vitro studies or OMICs (genomics, proteomics, etc.) approaches. A preclinical study using animal models and imaging can complete the bottom-up picture. In vitro work begins with a molecular understanding of disease, followed by cellular evaluation. Imaging-based studies of the whole animal sheds light on tissues, organs, and the physiological perspective of the disease. In this way, more endpoints can be explored, and confirmation of in vitro data can be accomplished before moving on to stage 2 of the drug discovery process, where the focus is on the drug target. Once established, a validated animal model will be used to test how target manipulation impacts the disease state. Therefore, these very same animal models will continue to be useful throughout the target discovery, validation, and drug efficacy steps.

26.2.2 STAGE 2: TARGET DISCOVERY, TARGET VALIDATION, AND DRUG DISCOVERY—DEFINING THE MOLECULAR AND GENETIC DIFFERENCES BETWEEN THE DISEASE STATE AND NORMAL STATE IN ORDER TO IDENTIFY AND VALIDATE NOVEL DRUG TARGETS

Traditionally, research performed in vitro has been the mainstay for discovery science. In recent years, OMICs profiling has provided investigators with a means for the rapid and robust determination of molecular differences in diseased cells versus their normal counterparts. OMICs data often generate a significant number of putative molecular and genetic targets for disease modification (Ho et al. 2010; Deyati et al. 2013; Russell et al. 2013; Berg 2014). Methods for narrowing down the number of targets to the top candidates require the use of animal studies, and in this context noninvasive imaging is proving to be an indispensable tool. Part of the discovery process is confirming that the target identified in vitro maintains its integrity as a candidate in vivo and exploring how that target behaves under physiological conditions. In vivo studies that help in the discovery process include transgenic knock-in and knock-out animals, as well as conditional expression models. In the case where the gene or molecule of interest is missing, an investigator may query the result of target input. To this end, recombinant genes that act as reporters may be used to track protein expression in vivo.

Knock-in models can be valuable in understanding the consequences of overexpressing a target. For example, Ishikawa et al. produced a luciferase-positive COX-2 knock-in mouse model to be used in the study of acute sepsis and inflammation (Ishikawa et al. 2006). Specifically, the COX-2 gene was modified to include the coding sequence for firefly luciferase under the same promoter. In this way, whenever and wherever the COX-2 gene was activated, luciferase was transcribed and translated as well. This construct enabled the visualization of COX-2-mediated processes such as inflammation using noninvasive BLI (Ishikawa et al. 2006). The utility of such a system is boundless in the discovery of novel anti-inflammatory agents that target the COX-2 pathway. New agents can be tested, producing a tangible noninvasive BL readout. In another example, Gu et al. used a conditional knock-in to understand the regulation and function of multidrug resistance protein 1 (MDR-1) in vivo. MDR-1 is part of a family of glycoproteins that mediate drug efflux from cells, which in turn plays a role in drug resistance. Rather than directly labeling MDR-1, Gu et al. constructed a cre-lox MDR-1–luciferase mouse model to study MDR-1 induction in vivo (Gu et al. 2009). Agents that target MDR-1 may be used to sensitize disease to cytotoxic therapy. The cre-lox MDR-1–luciferase mouse model and BLI may be used to evaluate these candidate drugs.

At the moment, targets overexpressed in disease states are the main focus of research, where treatment can be achieved through target inhibition or ablation. Small animal imaging can be used in existing animal models of disease to evaluate the physiological outcome of knocking out a gene of interest, so long as the knock-out is not embryonically lethal. To exemplify this, Berr et al. used MRI to examine the stomach lining of Tff-1 knock-out animals as compared with control (Berr et al. 2003). The Tff-1 gene produces a protein that is part of the trefoil family, which is protective to the mucosal layer of the GI tract. Tff-1$^{-/-}$ animals have a high propensity toward the development of spontaneous gastric tumors, while Tff-1 overexpression is protective. The researchers use MRI imaging to illustrate that Tff-1$^{-/-}$ animals have thickened gastric walls, which exhibit an irregular surface and nodular structures that are bright within the lumen, indicative of gastric carcinoma. Imaging provides more evidence that Tff-1 should be considered as a target for gene therapy of gastric carcinoma (Berr et al. 2003). As another example, researchers at Dongguk University College of Medicine, Goyang, Korea, are using a combination of imaging modalities to examine inflammation in the apoE$^{-/-}$ mouse. As discussed earlier, apo-lipoprotein E mediates cholesterol metabolism in peripheral tissues. Certain polymorphisms are associated with dysfunction in receptor binding and poor lipid transport and a higher risk for atherosclerosis. As part of this disease process, matrix metalloproteinase (MMP) activity is increased, as is the infiltration of tissue with macrophage. Shin et al. used a combined near-infrared imaging of labeled MMP and macrophage tracking using BLI to study inflammation in these animals (Deguchi et al. 2006; Shin et al. 2013). In each of the models discussed, important target-related biological information may be obtained through knock-in or knock-out animal models and in vivo imaging, but, more importantly, once established, these knock-out models can be used to validate drug candidates that target the molecule of interest.

There is a subtle yet important difference between the target discovery and the target validation. In the former, investigators use in vitro and in vivo studies to explore disease versus normal states in order to identify differences in genetic and molecular processes. Once a genetic/molecular difference has been thoroughly vetted in the disease state, the validation program begins. Validation focuses on understanding the physiological consequences of target manipulation, which could involve target inhibition/ablation or target induction/replacement. This validation process may involve the use of "tool" compounds to modulate target activity, or, alternatively (as described in the previous section), genetic methods to knock in or knock out the target. "Tool" compounds may be identified through existing literature, through high-throughput in vitro screens, or via molecular modeling. Perhaps more importantly, these "tool" compounds can define the scaffold from which a medicinal chemistry program can be developed to identify drug candidates potentially useful in the management of a defined disease process.

As an example of how small animal imaging may be used in the validation process, we turn to a series of imageable models developed to examine targets within the inflammatory pathway that are induced by LPS. IkappaBalpha is involved in the pro-inflammatory response, which makes it an important target in inflammatory diseases such as type B gastritis as well as cancer. Zhang et al. have produced an IkappaBalpha-luc transgenic mouse model for IkappaBalpha-promoter-driven luciferase activity. Using this model, they were able to demonstrate a rapid and systemic induction of IkappaBalpha expression in the transgenic mice following inoculation with LPS (Zhang et al. 2005). When animals were treated with bortezomib, a potent proteasome inhibitor, a significant suppression of LPS-induced luciferase activity was observed, suggesting that bortezomb is able to target IkappaBalpha; more importantly, it validated IkappaBalpha as a target in reducing the physiological signs of inflammation (Zhang et al. 2005). NF-kappaB has been identified as a target in cancer as well as inflammatory and autoimmune diseases. NF-kappaB luciferase reporter mice have been used to test anti-inflammatory agents in vivo using noninvasive BLI. For example, Austenaa et al. (Austenaa et al. 2009) showed that retinoic acid successfully targets NF-kappaB in vivo. Administration of retinoic acid repressed LPS-induced whole-body luminescence and decreased the physiological response to LPS. In another study, the NF-kappaB luciferase reporter mouse was used to illustrate the anti-inflammatory activity of the mushroom *Antrodia camphorata* (Hseu et al. 2010), providing more validation for targeting NF-kappaB in inflammatory conditions. Small animal imaging can play a role in the target discovery, target validation, and the drug discovery program, and, if used appropriately, can provide valuable information regarding the mechanism of action and physiological response to targeting a specific pathway.

26.2.3 STAGE 3: PRECLINICAL DEVELOPMENT: EFFICACY—DETERMINING HOW WELL A NOVEL THERAPEUTIC CANDIDATE PERFORMS

Efficacy studies using small animal imaging come in many forms, and many examples can be found in the literature. In the simplest form, these studies rely on animal models of disease, and treatment-induced changes in disease initiation or progression are followed. As an example of the conventional approach to an efficacy study, as well as how small animal imaging may play a role, Andreu et al. at Imperial College, London, United Kingdom, created a model of *Mycobacterium tuberculosis* carrying a red-shifted firefly luciferase gene (Andreu et al. 2013). The organism was then used to infect mice, and disease progression was monitored using BLI before and after treatment with isoniazid. This animal model can now be used as an elegant tool to examine the efficacy of novel therapeutics for tuberculosis (Andreu et al. 2013). In a recent study, Myburgh et al. modified the GVR35 strain of *Trypanosoma brucei brucei* to express luciferase, and used this luciferase-positive parasite to rapidly evaluate compounds for treatment efficacy using serial, noninvasive BLI in mice (Myburgh et al. 2013). Our own laboratory used BLI to evaluate the efficacy of a novel small-molecule inhibitor against integrin-linked kinase (ILK) in the treatment of an orthotopic model of breast cancer (Kalra et al. 2009). For this study, luciferase-positive breast cancer cells were inoculated into the mammary fat pad of mice and allowed to develop over 2 weeks. At this time, one group of animals was treated with the novel ILK-targeting agent, and tumor growth in the treated and untreated animals was monitored using BLI. We were able to show that the small-molecule inhibitor was successful in suppressing tumor growth using this method of imaging (Kalra et al. 2009). In 2011, Verreault et al. demonstrated that mKate2-transduced glioblastoma cells can be used to monitor tumor progression using noninvasive fluorescence imaging. Furthermore, this study examined the efficacy of temozolimide in animals with orthotopic implants of the mKate2-expressing cells. Figure 26.6 shows the results of this study (Verreault et al. 2011). The sensitivity of low mKate2- and medium mCherry-expressing U251MG cells to irinotecan (CPT-11) (Figure 26.6a) and temozolomide (TMZ) (Figure 26.6b) was assessed in vitro by MTT assay and compared with that of the parental line. MKate2-expressing U251MG line was implanted orthotopically, and animals were left untreated or treated with TMZ. Representative images of control and TMZ-treated mKate2-expressing tumor-bearing animals are shown in Figure 26.6c. MKate2 fluorescent signal (scaled counts/s) was measured before treatment initiation (day 21) and at the end of the study (day 42), which showed significant treatment-induced reduction in fluorescence signal intensity (Figure 26.6d). Even less conventional

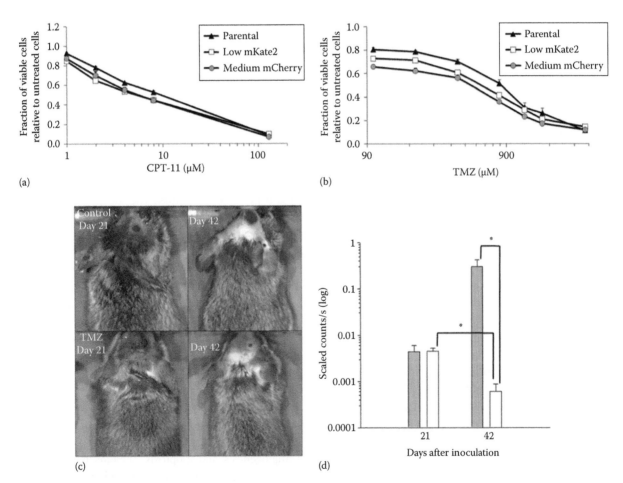

FIGURE 26.6 *In vivo* sensitivity of mKate2-expressing tumor cells to chemotherapeutic drugs. The sensitivity of low mKate2- and medium mCherry-expressing U251MG cells to irinotecan (CPT-11) and temozolomide (TMZ) was assessed in vitro (a and b) and in vivo (c and d).

imaging modalities such as US can be used in a drug efficacy study. For example, Chen et al. used US echogenicity as a measure of tissue inflammation in order to compare the anti-inflammatory agents indomethacin, aspirin, and COX-2 inhibitors in bone fracture mouse models. This group correlated US images to histological markers of inflammation to confirm that their noninvasive ultrasound method was capable of monitoring post-fracture tissue inflammation and response to drug treatment (Chen et al. 2014).

As articulated by Gwyther et al. in their 2008 review, small animal imaging can be used to monitor disease-specific physiological processes affected by drugs that do not appear when using conventional assessments (Gwyther and Schwartz 2008). The importance of this point is illustrated in cases where a drug target does not directly impact disease burden. As an example, many novel targets in the treatment of cancer have cytostatic effects as opposed to the cytotoxic effects of previous generations of drugs. In this way, target ablation may not have any impact on the size of the tumor; rather, changes in the physiology of the disease are of more importance. Gwyther posits that MRI and CT studies, for example, can be used to evaluate efficacy end points such as altered tissue integrity, vascular function, and angiogenesis, while PET can be used to examine altered metabolism. As an example of the latter, PET-FDG was used to show that the tyrosine kinase inhibitor imatinib mesylate was able to reduce the metabolic activity within metastatic liver lesions before changes in tumor volume could be seen in gastrointestinal tumors (Choi et al. 2007). This study established the proof of concept that FDG imaging by PET can be used to supply early indications of therapeutic activity in patients treated with this drug, and provides an excellent example of how a preclinical imaging study can guide the use of imaging in the context of a clinical trial. PET and SPECT are able to deliver information about physiological functions at the molecular level and therefore can give readouts for glucose metabolism, blood flow and perfusion, receptor–ligand binding rates, apoptosis, and hypoxia. To demonstrate the latter, Yapp et al. used ^{18}F-EF5 and PET to evaluate tumor hypoxia in the Shionogi tumor model for prostate cancer (Yapp et al. 2007). This marker of hypoxia can be useful in drug efficacy studies, where a targeted treatment impacts tumor vasculature and hypoxia. Outside of PET, Doppler imaging and microbubble technology is increasing the utility of US in functional imaging. A 2011 study from the University of San Francisco showed that noninvasive, microbubble, contrast-enhanced ultrasound imaging can be used to monitor blood flow and blood vessel perfusion in sunitinib-treated pancreatic tumor mouse models (Olson et al. 2011).

In vivo cell and molecular imaging is quickly becoming a very important part of a drug efficacy program. Molecular imaging makes it possible to evaluate the mechanism of drug action in vivo. OI is perhaps one of the easiest to implement and least expensive modalities in molecular imaging and has been useful as a first-line approach in understanding cellular and molecular pathogenesis and drug efficacy. Despite its lack of clinical translatability, OI is widely used in vivo to track single cells or molecules in the whole body without the use of ionizing radiation. Reporter gene assays have been used to yield fundamental biological information on transcriptional regulation, signal transduction, protein–protein interactions, cell trafficking, and targeted drug action in vivo at the cellular and molecular levels. For example, Viola et al. inoculated mice with breast carcinoma cells transfected with an HIF-1α luciferase reporter construct (Viola et al. 2008). HIF-1α is a key player in the stress response elicited by low levels of oxygen. Animals were treated with cyclophosphamide or paclitaxel, which caused an increase in HIF-1α protein levels as quantified using BLI (Viola et al. 2008).

Labeled cell lines and molecules are now commercially available, as are novel systems that may be used to track spontaneous disease and treatment effects. Perkin Elmer offers a wide range of luciferase-positive cell lines and fluorescent biosensors that actively target and bind to specific biomarkers for inflammation, osteogenesis, angiogenesis, and apoptosis. The biosensors available come in three types: (1) activatable agents that are optically silent upon injection but are activated in vivo through cleavage by specific protease biomarkers of disease (cathepsin, matrix metalloproteinases, elastases); (2) targeted agents that bind to specific biomarkers (Her2/neu, hydroxyapatite, annexin, integrin, lectin); and (3) physiological sensors that allow for in vivo imaging of highly stable and localized molecules to enable imaging of disease physiology, vasculature, vascular permeability, and angiogenesis in response to targeted treatment. The California-based nanobiotechnology company Zymera has established a self-illuminating red to near-infrared (NIR) emitting quantum dot technology for preclinical in vivo imaging. The product is a Q-dot nanocrystal conjugate known as a bioluminescence resonance energy transfer-Q-dot (BRET-Q-dot), which is able to convert biochemical energy to detectable photon energy through the activity of luciferase. The BRET-Q-dot probes are associated with streptavidin, and are designed as biosensors for biotinylated antibodies bound in vivo. The development and application of bioconjugated quantum dots and multifunctional nanoparticles as a tool in molecular imaging is discussed in a thorough review by Rhyner et al. (2006).

26.2.4 STAGE 3: PRECLINICAL DEVELOPMENT: TOXICITY, AND SAFETY—IS A NOVEL THERAPEUTIC CANDIDATE GOING TO CAUSE MORE HARM THAN GOOD?

Once a drug has been identified and validated, the candidate agent is tested for a variety of characteristics: (1) how long does the agent stay in the blood stream (pharmacokinetics), (2) where does the agent distribute in the body (biodistribution), (3) is the agent metabolized and if so what is it metabolized to and where does metabolism occur, (4) how is the agent eliminated from the body, and (5) does the agent exert toxic effects on normal tissues, organs, and/or functions? Small animal imaging is proving useful as a tool to address many of these questions, but will likely not replace the formal toxicity tests to be conducted under current Good Laboratory Practice (cGLP) guidelines. Examples of how small animal imaging may be used in the assessment of toxicity and safety are discussed below.

PET, and SPECT can be used to measure the pharmacokinetics, target organ localization, molecular target expression and function, and biodistribution of drugs with contrast enhancing properties. In a comprehensive review, van der Veldt et al. examined the use of PET as a method for measuring drug delivery to tumors in vivo. The authors indicate that radiolabeled drugs can be used to monitor drug pharmacokinetics noninvasively, and the uptake of radiopharmaceuticals can be used to make predictions on treatment outcome (van der Veldt et al. 2013). To exemplify this, a group led by Takashima tested the use of ^{18}F-FDG and PET to evaluate gastrointestinal absorption in rats (Hume et al. 2013) and humans (Shingaki et al. 2012). They were able to estimate the rate of gastric emptying and intestinal absorption of ^{18}F-FDG using time profiles of radioactivity in the stomach and small intestine. In an elegant study published in 2013, Hume et al. synthesized ^{11}C-labeled metformin and used this radiolabeled drug with micro-PET to examine the effect of the toxin extrusion transporter (MATE1 and 2) inhibitor pyrimethamine on ^{11}C metformin bioavailability. The researchers were able to use small animal PET to follow ^{11}C metformin transport through the liver, kidney, and bladder. Animals treated with pyrimethamine had a significant increase in ^{11}C metformin PET signal from the liver, indicating that pyrimethamine treatment could prolong the bioavailability of metformin (Hume et al. 2013). Use of micro-PET in pharmacokinetic distribution and biodistribution studies is a relatively new and promising approach because it is highly sensitive and noninvasive. However, the biggest problem with using PET to examine biodistribution and/or pharmacokinetics is synthesizing the drug with a PET isotope that has the right properties and maintains the chemical composition and structure of the drug. Additionally, the radioisotope ^{11}C used in the Hume study did not interfere with metformin structure or function; however, it is important to note that ^{11}C has a half-life of 22 min. The short half-life of the radioligand places a constraint on PET scanning time. Isotopes such as ^{64}Cu or ^{68}Ga have longer half-lives, but these isotopes have a higher likelihood of changing the pharmacokinetic and biodistribution profile of the drug.

PET is not the only imaging modality being used in the exploration of biodistribution. BLI has also been used in preclinical assessments of drug delivery and bioavailability. As an example of the utility of BLI in the study of bioavailability, luciferin transporter conjugates have been used as tools for real-time determination of drug uptake into cells and tissues in vivo (Jones et al. 2006; Wender et al. 2007). In 2010, researchers at the Washington University School of Medicine developed a transgenic mouse model that expressed an inducible Gal4-LUC to evaluate molecule-specific pharmacodynamics of rapamycin in vivo (Pan et al. 2010). In preliminary research done in our own lab, we have been using BLI and FLI to image nanoparticle formulations of drugs in vivo, comparing the enhanced permeability and retention effect of primary to metastatic tumors. Luciferase-positive cell lines were used to establish orthotopic breast cancer or disseminated disease, which can be imaged using BLI. As illustrated in Figure 26.7, lipid nanoparticles prepared using fluorescent lipids can be imaged in vivo using FLI. Colocalization of fluorescent and luminescent signals allow the comparison of drug distribution to different tumors over time. Bioluminescent imaging of breast cancer cells inoculated in the mammary fat pad of a mouse on day 42 post inoculation is shown in Figure 26.7a. On day 42, animals were subject to i.v. injection of fluorescent-labeled liposomes. These animals were imaged 1, 2, 4, 8, and 24 h after injection (Figure 26.7b). Fluorescent imaging shows that labeled liposomes are cleared systemically after 24 h, but are retained within the tumor tissue at this time. Tumors were excised and stained with a vasculature marker (blue) and imaged using fluorescence microscopy (Figure 26.7c). Fluorescent liposomes (magenta) were shown to localize around blood vessels with increasing concentrations over 24 h. The concentration of labeled liposomes was quantified (Figure 26.7d) using a fluorimeter to confirm that higher concentrations occur at later time points. OI was also employed by Cekanova et al., who used fluorescence endoscopy to visualize the uptake of fluorocoxib, a novel cyclooxygenase-2-targeted OI agent, in a canine model of colon cancer (Cekanova et al. 2012). In 2008, Baker et al. used ^{19}F magnetic resonance spectroscopy to monitor the delivery of 5-fluorouracil (5-FU) to tumor tissue in vivo (Baker et al. 2008). Viglianti et al. showed that T1-weighted MRI can be used to measure tumor drug concentration in vivo following the systemic administration of liposomes containing both the drug (doxorubicin [DOX]) and the contrast agent (manganese [Mn]) (Viglianti et al. 2004, 2006).

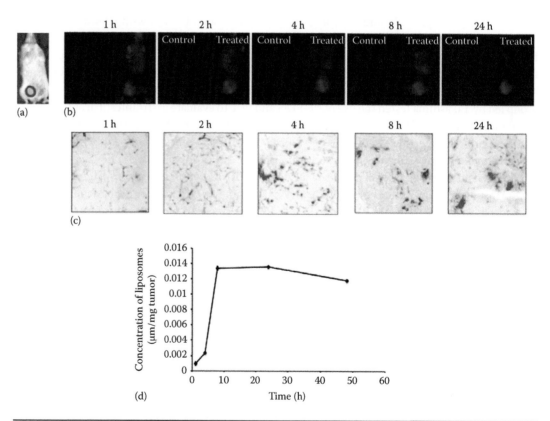

FIGURE 26.7   Small animal imaging of the enhanced permeability and retention effect. Bioluminescent imaging of breast cancer cells inoculated in the mammary fat pad of a mouse (a) is compared to fluorescent imaging of labeled liposomes after i.v. injection (b). Tumors were excised and stained with a vasculature marker (blue) and imaged using fluorescence microscopy (c). The concentration of labeled liposomes was quantified (d) using a fluorimeter.

Radiolabeled drugs can be used to determine whether drugs appropriately target, bind to, or are taken up through receptor interactions with the targeted cells. As an example, it is known that some breast cancer patients who have Her2/neu-positive cancers are intrinsically insensitive to trastuzumab, the therapeutic monoclonal antibody used in combination with other broad-spectrum anticancer drugs to treat patients with Her2/neu-positive cancers (Figueroa-Magalhaes et al. 2014). To query the mechanism of this resistance, researchers asked whether these patients may be insensitive to the drug because it fails to bind to Her2/neu-positive cells in situ. To answer this question, Ferreira et al. developed a ^{64}Cu radiolabeled trastuzumab that can be used with PET imaging to evaluate biodistribution of trastuzumab in vivo (Ferreira et al. 2010). This technology also has the potential to be rapidly translated into clinical applications.

To date, very little has been done using small animal imaging to assess toxicity; however, as imaging modalities become widely available, these tools may also be useful in the evaluation of drug safety. Already, Bell et al. have used optical coherence tomography (OCT) to visualize drug-induced changes in vaginal tissue morphology following the application of topical microbicides. Micro-CT systems employed in small animal imaging are excellent methods for imaging soft tissue structures, skeletal abnormalities, and tumors, and can therefore be used to evaluate tissue-specific toxicities or metabolic dysfunction in response to treatment where this can be predicted. As an example, in 2013, Jeyabalan et al. used micro-CT and bone histomorphometry studies to evaluate the effect of the antidiabetic drug metformin on bone mass and fracture healing in a type 2 diabetes mouse model. They were able to show no significant differences in cortical and trabecular bone architecture in metformin-treated rodents and no effect on bone resorption or significant differences of healing between the groups, indicating that metformin has no adverse effects on bone mass and fracture healing in a type 2 diabetes setting (Jeyabalan et al. 2013). In an effort to determine the long-term physiological consequences of using therapeutics targeting PI3 kinase and mTOR, Smith et al. used micro-CT to evaluate bone volume and bone strength. They were able to show that long-term use of these targeted agents can be detrimental to bone health (Smith et al. 2013). The best use of imaging in toxicity and safety assessments will likely be in early stages of toxicity, where, for example, markers of liver dysfunction (alanine transaminase [ALT], or aspartate transaminase [AST]) or renal failure (creatinine, blood urea nitrogen [BUN]) may be evaluated noninvasively. However, before this can be done, strict guidelines and standard operating procedures that consider animal welfare will need to be put in place.

It is important to mention that validated animal models and imaging modalities may also provide key insights in the optimization process of drug development research. Optimization in this context refers to the modification of a drug to make it more effective and/or safer by changing its chemical structure. These drug analogs are then tested in vitro and in vivo. The final result of this stage is an optimized drug candidate that has the best chance of succeeding in the latter phases of drug development and clinical trials. Properly designed in vivo studies incorporating the use of small animal imaging may be helpful in making the comparisons between drug candidates and different formulations. Accordingly, Cirstou-Hapca et al. used BLI to examine the differential delivery of polymeric immuno-nanoparticles consisting of paclitaxel-loaded nanoparticles coated with anti-Her2/neu monoclonal antibodies (trastuzumab), compared with free drugs. They found that in a disseminated ovarian cancer model, the antibody-coated nanoparticle showed improved activity compared to the free drug (Cirstoiu-Hapca et al. 2010).

As already suggested, the correct use of an animal model and noninvasive imaging can answer very specific and physiologically relevant questions that can impact clinical trial design and potentially clinical applications.

26.3 CONCLUSIONS AND FUTURE PERSPECTIVES

Small animal imaging has arisen from the utility of imaging methods in the context of human disease detection and health monitoring. It is very reasonable to assume that, if comparable imaging methods are used in preclinical models, the results obtained in such models would be more predictive of human disease. Using cancer as an example and clinically applicable RECIST criteria, it is interesting to ask the question of how many experimental drugs actually induce a partial or complete response in relevant animal models as detected through imaging methods. Effective use of conventional methods such as survival studies and ex vivo examination of tumor tissue using immunohistochemistry, integrated with small animal imaging, will be key to identify drug candidates that have a greater potential of working in patients. Further, imaging tools have already allowed researchers to pursue larger and more robust studies examining multiple end points in situ. The power of these technologies is only now being realized in this era of high content data. Creative scientists are using imaging in medium- to high-throughput studies, combining modalities and labeling medicines to gain insight into disease while at the same time addressing disease management. Multimodal imaging and theranostics are two innovative methodologies that exemplify this ingenuity.

26.3.1 MULTIMODAL IMAGING

No single imaging modality is currently able to provide the required sensitivity, specificity, and resolution needed for drug development purposes. For example, CT has high spatial resolution and provides a good anatomical reference; however, it is of limited value for detecting molecular processes in vivo. The use of multiple imaging modalities concurrently, such as PET-CT and SPECT-CT, has already demonstrated utility in drug discovery and development studies. Furthermore, the integration of multimodal imaging information with bioinformatics and pharmacodynamic, pharmacokinetic, and pharmacogenetic profiles will be able to create a greater understanding of drug targets, efficacy, and utility, making preclinical studies more predictive and therefore more meaningful. Several groups are already on the leading edge of this type of combinatorial research. Deroose et al. illustrated how PET-CT and BLI can be used to monitor tumor melanoma growth and metastasis (Deroose et al. 2007), while Chan et al. used dual BLI and PET-CT to image molecular alterations as well as metabolic and anatomical alterations in tumor-bearing animals after treatment with novel small-molecule inhibitors of HSP90 (Chan et al. 2012). In a review by Jaffer et al., the utility of OI in addition to noninvasive tomography in the context of vascular disease is comprehensively discussed, illustrating the importance of a multimodal approach (Jaffer et al. 2009).

26.3.2 THERANOSTICS

Theranostics is an emerging branch of personalized medicine that combines diagnostic, prognostic, and therapeutic capabilities into a single agent. Theranostics aims to produce therapeutic protocols that are specific to individual patients and in addition have the ability for monitoring patient response following treatment. Imaging is the key to the diagnostic strategies employed in theranostics. Small animal imaging studies continue to provide the proof of concept for the utility, accessibility, and cost effectiveness of pursuing theranostics. Several recent reviews highlight theranostic research that has been successful to date (Xie et al. 2010; Kelkar and Reineke 2011; Choi et al. 2012). In 2011, Kalber et al. published an intriguing paper in which they describe a gadolinium-labeled derivative of the tubulin binding agent cholchicinic acid. This group was able to use MR approaches to image cancer in animals bearing subcutaneous ovarian xenografts, while at the same time eliciting cholchicine-mediated tumor cell death (Kalber et al. 2011). Kenny et al. have constructed MR-sensitive liposomal nanoparticles that also act as siRNA delivery vehicles. They were able to silence Survin expression using siRNA, and showed that this treatment was able to slow tumor growth, while at the same time using the nanoparticle to visualize tumors in vivo using MR (Kenny et al. 2011).

26.4 CONCLUSIONS

The use of small animal imaging is starting to have an impact on drug development programs; however, for this methodology to replace existing approaches used to define efficacy and toxicity, there is a need for more model development, more experience with existing models, additional validation studies, and application of standard operating methods that will be auditable through defined quality management systems. The effort is worthwhile because, in the end, noninvasive imaging will result in a reduction in the number of animals used in studies, it will provide more refined (content rich) information collected in longitudinal manner, and it will replace current methods that are certainly not highly predictive of activity (toxicity or efficacy) of drug candidates in humans. Although speculative, it is anticipated that information collected using small animal imaging will be comparable to data collected using similar imaging equipment used to detect and monitor disease development in humans. In the end, one must recognize that there are limitations to using any small animal imaging modality as part of a drug discovery endeavor. These include cost, complexity of study design, and at the moment little in the way of validated study guidelines. It should also be noted that there is no perfect model, and for this reason it is best to collect data using multiple models. These data when taken in aggregate will provide the most robust information to help improve our decision-making capabilities related to three key drug development milestones: (1) Is the therapeutic approach effective as determined in preclinical models? (2) Is the compound suitable for clinical development? (3) Will the candidate drug be safe at doses that have the potential to be effective in patients?

REFERENCES

Akash, M. S., K. Rehman, and S. Chen. 2013. An overview of valuable scientific models for diabetes mellitus. *Curr. Diabetes Rev.* **9**: 286–293.

Andreu, N., A. Zelmer et al. 2013. Rapid in vivo assessment of drug efficacy against Mycobacterium tuberculosis using an improved firefly luciferase. *J. Antimicrob. Chemother.* **68**: 2118–2127.

Anthony, D. D. and T. M. Haqqi. 1999. Collagen-induced arthritis in mice: An animal model to study the pathogenesis of rheumatoid arthritis. *Clin. Exp. Rheumatol.* **17**: 240–244.

Araujo, D. M., S. R. Cherry, K. J. Tatsukawa, T. Toyokuni, and H. I. Kornblum. 2000. Deficits in striatal dopamine D(2) receptors and energy metabolism detected by in vivo microPET imaging in a rat model of Huntington's disease. *Exp. Neurol.* **166**: 287–297.

Austenaa, L. M., H. Carlsen, K. Hollung, H. K. Blomhoff, and R. Blomhoff. 2009. Retinoic acid dampens LPS-induced NF-kappaB activity: Results from human monoblasts and in vivo imaging of NF-kappaB reporter mice. *J. Nutr. Biochem.* **20**: 726–734.

Bagci, U., B. Foster et al. 2013. A computational pipeline for quantification of pulmonary infections in small animal models using serial PET-CT imaging. *EJNMMI Res.* **3**: 55.

Baker, J. H., J. Lam et al. 2008. Irinophore C, a novel nanoformulation of irinotecan, alters tumor vascular function and enhances the distribution of 5-fluorouracil and doxorubicin. *Clin. Cancer Res.* **14**: 7260–7271.

Banting, F. G. and C. H. Best. 1990. Pancreatic extracts. 1922. *J. Lab. Clin. Med.* **115**: 254–272.

Baselga, J., L. Norton, J. Albanell, Y. M. Kim, and J. Mendelsohn. 1998. Recombinant humanized anti-HER2 antibody (Herceptin) enhances the antitumor activity of paclitaxel and doxorubicin against HER2/neu overexpressing human breast cancer xenografts. *Cancer Res.* **58**: 2825–2831.

Bates, J. H., M. Rincon, and C. G. Irvin. 2009. Animal models of asthma. *Am. J. Physiol. Lung Cell Mol. Physiol.* **297**: L401–L410.

Berg, E. L. 2014. Systems biology in drug discovery and development. *Drug Discov. Today* **19**: 113–125.

Berr, S. S., J. K. Roche, W. El-Rifai, M. F. Smith Jr., and S. M. Powell. 2003. Magnetic resonance imaging of gastric cancer in Tff1 knock-out mice. *Magn. Reson. Med.* **49**: 1033–1036.

Bradbury, M. S., D. Hambardzumyan et al. 2008. Dynamic small-animal PET imaging of tumor proliferation with 3′-deoxy-3′-18F-fluorothymidine in a genetically engineered mouse model of high-grade gliomas. *J. Nucl. Med.* **49**: 422–429.

Brown, R. H. and W. Mitzner. 2003. Understanding airway pathophysiology with computed tomography. *J. Appl. Physiol.* **95**: 854–862.

Canadian Council of Animal Care. 2014. http://www.ccac.ca/en_/standards/guidelines.

Cekanova, M., M. J. Uddin et al. 2012. Single-dose safety and pharmacokinetic evaluation of fluorocoxib A: Pilot study of novel cyclooxygenase-2-targeted optical imaging agent in a canine model. *J. Biomed. Opt.* **17**: 116002.

Chan, C. T., R. E. Reeves et al. 2012. Discovery and validation of small-molecule heat-shock protein 90 inhibitors through multimodality molecular imaging in living subjects. *Proc. Natl. Acad. Sci. USA* **109**: E2476–E2485.

Chen, Y. C., Y. H. Lin, S. H. Wang, S. P. Lin, K. K. Shung, and C. C. Wu. 2014. Monitoring tissue inflammation and responses to drug treatments in early stages of mice bone fracture using 50 MHz ultrasound. *Ultrasonics* **54**: 177–186.

Choi, H., C. Charnsangavej et al. 2007. Correlation of computed tomography and positron emission tomography in patients with metastatic gastrointestinal stromal tumor treated at a single institution with imatinib mesylate: Proposal of new computed tomography response criteria. *J. Clin. Oncol.* **25**: 1753–1759.

Choi, K. Y., G. Liu, S. Lee, and X. Chen. 2012. Theranostic nanoplatforms for simultaneous cancer imaging and therapy: Current approaches and future perspectives. *Nanoscale* **4**: 330–342.

Cirstoiu-Hapca, A., F. Buchegger, N. Lange, L. Bossy, R. Gurny, and F. Delie. 2010. Benefit of anti-HER2-coated paclitaxel-loaded immuno-nanoparticles in the treatment of disseminated ovarian cancer: Therapeutic efficacy and biodistribution in mice. *J. Control. Release* **144**: 324–331.

Conley, F. K. 1979. Development of a metastatic brain tumor model in mice. *Cancer Res.* **39**: 1001–1007.

Deguchi, J. O., M. Aikawa et al. 2006. Inflammation in atherosclerosis: Visualizing matrix metalloproteinase action in macrophages in vivo. *Circulation* **114**: 55–62.

Deroose, C. M., A. De et al. 2007. Multimodality imaging of tumor xenografts and metastases in mice with combined small-animal PET, small-animal CT, and bioluminescence imaging. *J. Nucl. Med.* **48**: 295–303.

DeVita, V. T., Jr. and E. Chu. 2008. A history of cancer chemotherapy. *Cancer Res.* **68**: 8643–8653.

Deyati, A., E. Younesi, M. Hofmann-Apitius, and N. Novac. 2013. Challenges and opportunities for oncology biomarker discovery. *Drug Discov. Today* **18**: 614–624.

Di Marco, A., M. Gaetani, and B. Scarpinato. 1969. Adriamycin (NSC-123,127): A new antibiotic with antitumor activity. *Cancer Chemother. Rep.* **53**: 33–37.

Drake, J. M., C. L. Gabriel, and M. D. Henry. 2005. Assessing tumor growth and distribution in a model of prostate cancer metastasis using bioluminescence imaging. *Clin. Exp. Metastasis* **22**: 674–684.

Edinger, M., T. J. Sweeney, A. A. Tucker, A. B. Olomu, R. S. Negrin, and C. H. Contag. 1999. Noninvasive assessment of tumor cell proliferation in animal models. *Neoplasia* **1**: 303–310.

Eisenhauer, E. A., P. Therasse et al. 2009. New response evaluation criteria in solid tumours: Revised RECIST guideline (version 1.1). *Eur. J. Cancer* **45**: 228–247.

Ferreira, C. L., D. T. Yapp et al. 2010. Comparison of bifunctional chelates for (64)Cu antibody imaging. *Eur. J. Nucl. Med. Mol. Imaging* **37**: 2117–2126.

Figueroa-Magalhaes, M. C., D. Jelovac, R. M. Connolly, and A. C. Wolff. 2014. Treatment of HER2-positive breast cancer. *Breast* **23**: 128–136.

Fuchs, K., U. Kohlhofer et al. 2013. In vivo imaging of cell proliferation enables the detection of the extent of experimental rheumatoid arthritis by 3′-deoxy-3′-18f-fluorothymidine and small-animal PET. *J. Nucl. Med.* **54**: 151–158.

Garcia-Alloza, M. and B. J. Bacskai. 2004. Techniques for brain imaging in vivo. *Neuromol. Med.* **6**: 65–78.

Grand'maison, M., S. P. Zehntner et al. 2013. Early cortical thickness changes predict beta-amyloid deposition in a mouse model of Alzheimer's disease. *Neurobiol. Dis.* **54**: 59–67.

Gu, L., W. M. Tsark, D. A. Brown, S. Blanchard, T. W. Synold, and S. E. Kane. 2009. A new model for studying tissue-specific mdr1a gene expression in vivo by live imaging. *Proc. Natl. Acad. Sci. USA* **106**: 5394–5399.

Guerin, E., S. Man, P. Xu, and R. S. Kerbel. 2013. A model of postsurgical advanced metastatic breast cancer more accurately replicates the clinical efficacy of antiangiogenic drugs. *Cancer Res.* **73**: 2743–2748.

Gwyther, S. J. and L. H. Schwartz. 2008. How to assess anti-tumour efficacy by imaging techniques. *Eur. J. Cancer* **44**: 39–45.

Hag, A. M., R. S. Ripa, S. F. Pedersen, R. P. Bodholdt, and A. Kjaer. 2013. Small animal positron emission tomography imaging and in vivo studies of atherosclerosis. *Clin. Physiol. Funct. Imaging* **33**: 173–185.

Haller, J., D. Hyde, N. Deliolanis, R. de Kleine, M. Niedre, and V. Ntziachristos. 2008. Visualization of pulmonary inflammation using noninvasive fluorescence molecular imaging. *J. Appl. Physiol.* **104**: 795–802.

Ho, E. A., E. Ramsay et al. 2010. Characterization of cationic liposome formulations designed to exhibit extended plasma residence times and tumor vasculature targeting properties. *J. Pharm. Sci.* **99**: 2839–2853.

Holmes, R. K. 2000. Biology and molecular epidemiology of diphtheria toxin and the tox gene. *J. Infect. Dis.* **181**(Suppl 1): S156–S167.

Horsley, V. 1889. On rabies: Its treatment by M. Pasteur, and on the means of detecting it in suspected cases. *Br. Med. J.* **1**: 342–344.

Hseu, Y. C., H. C. Huang, and C. Y. Hsiang. 2010. Antrodia camphorata suppresses lipopolysaccharide-induced nuclear factor-kappaB activation in transgenic mice evaluated by bioluminescence imaging. *Food Chem. Toxicol.* **48**: 2319–2325.

Hsieh, C. L., Z. Xie et al. 2005. A luciferase transgenic mouse model: Visualization of prostate development and its androgen responsiveness in live animals. *J. Mol. Endocrinol.* **35**: 293–304.

Hsieh, C. L., Z. Xie et al. 2007. Non-invasive bioluminescent detection of prostate cancer growth and metastasis in a bigenic transgenic mouse model. *Prostate* **67**: 685–691.

Hu, Y., W. Cheng, W. Cai, Y. Yue, J. Li, and P. Zhang. 2013. Advances in research on animal models of rheumatoid arthritis. *Clin. Rheumatol.* **32**: 161–165.

Hume, W. E., T. Shingaki et al. 2013. The synthesis and biodistribution of [(11)C]metformin as a PET probe to study hepatobiliary transport mediated by the multi-drug and toxin extrusion transporter 1 (MATE1) in vivo. *Bioorg. Med. Chem.* **21**: 7584–7590.

Ishikawa, T. O., N. K. Jain, M. M. Taketo, and H. R. Herschman. 2006. Imaging cyclooxygenase-2 (Cox-2) gene expression in living animals with a luciferase knock-in reporter gene. *Mol. Imaging Biol.* **8**: 171–187.

Jack, C. R., Jr., M. Marjanska et al. 2007. Magnetic resonance imaging of Alzheimer's pathology in the brains of living transgenic mice: A new tool in Alzheimer's disease research. *Neuroscientist* **13**: 38–48.

Jaffer, F. A., P. Libby, and R. Weissleder. 2009. Optical and multimodality molecular imaging: Insights into atherosclerosis. *Arterioscler. Thromb. Vasc. Biol.* **29**: 1017–1024.

Jeyabalan, J., B. Viollet et al. 2013. The anti-diabetic drug metformin does not affect bone mass in vivo or fracture healing. *Osteoporos Int.* **24**: 2659–2670.

Joe, B. and R. L. Wilder. 1999. Animal models of rheumatoid arthritis. *Mol. Med. Today* **5**: 367–369.

Jones, L. R., E. A. Goun, R. Shinde, J. B. Rothbard, C. H. Contag, and P. A. Wender. 2006. Releasable luciferin-transporter conjugates: Tools for the real-time analysis of cellular uptake and release. *J. Am. Chem. Soc.* **128**: 6526–6527.

Kalber, T. L., N. Kamaly et al. 2011. Synthesis and characterization of a theranostic vascular disrupting agent for in vivo MR imaging. *Bioconjug. Chem.* **22**: 879–886.

Kalra, J., M. Anantha et al. 2011. Validating the use of a luciferase labeled breast cancer cell line, MDA435LCC6, as a means to monitor tumor progression and to assess the therapeutic activity of an established anticancer drug, docetaxel (Dt) alone or in combination with the ILK inhibitor, QLT0267. *Cancer Biol. Ther.* **11**: 826–838.

Kalra, J., C. Warburton et al. 2009. QLT0267, a small molecule inhibitor targeting integrin-linked kinase (ILK), and docetaxel can combine to produce synergistic interactions linked to enhanced cytotoxicity, reductions in P-AKT levels, altered F-actin architecture and improved treatment outcomes in an orthotopic breast cancer model. *Breast Cancer Res.* **11**: R25.

Kelkar, S. S. and T. M. Reineke. 2011. Theranostics: Combining imaging and therapy. *Bioconjug. Chem.* **22**: 1879–1903.

Kenny, G. D., N. Kamaly et al. 2011. Novel multifunctional nanoparticle mediates siRNA tumour delivery, visualisation and therapeutic tumour reduction in vivo. *J. Control. Release* **149**: 111–116.

Koglin, N., A. Mueller et al. 2011. Specific PET imaging of xC-transporter activity using a (1)(8)F-labeled glutamate derivative reveals a dominant pathway in tumor metabolism. *Clin. Cancer Res.* **17**: 6000–6011.

Lee, S. W., J. M. Greve et al. 2008. Early findings of small-animal MRI and small-animal computed tomography correlate with histological changes in a rat model of rheumatoid arthritis. *NMR Biomed.* **21**: 527–536.

Morgan, S., P. Grootendorst, J. Lexchin, C. Cunningham, and D. Greyson. 2011. The cost of drug development: A systematic review. *Health Policy* **100**: 4–17.

Myburgh, E., J. A. Coles et al. 2013. In vivo imaging of trypanosome-brain interactions and development of a rapid screening test for drugs against CNS stage trypanosomiasis. *PLoS Negl. Trop. Dis.* **7**: e2384.

Nikolaus, S., R. Larisch et al. 2011. Pharmacological challenge and synaptic response—Assessing dopaminergic function in the rat striatum with small animal single-photon emission computed tomography (SPECT) and positron emission tomography (PET). *Rev. Neurosci.* **22**: 625–645.

Ntziachristos, V. 2009. Optical imaging of molecular signatures in pulmonary inflammation. *Proc. Am. Thorac. Soc.* **6**: 416–418.

Ogawa, M., S. Ishino et al. 2004. (18)F-FDG accumulation in atherosclerotic plaques: Immunohistochemical and PET imaging study. *J. Nucl. Med.* **45**: 1245–1250.

Olson, P., G. C. Chu, S. R. Perry, O. Nolan-Stevaux, and D. Hanahan. 2011. Imaging guided trials of the angiogenesis inhibitor sunitinib in mouse models predict efficacy in pancreatic neuroendocrine but not ductal carcinoma. *Proc. Natl. Acad. Sci. USA* **108**: E1275–E1284.

Pan, D., S. D. Caruthers et al. 2010. Nanomedicine strategies for molecular targets with MRI and optical imaging. *Future Med. Chem.* **2**: 471–490.

Pellegrino, D., F. Cicchetti et al. 2007. Modulation of dopaminergic and glutamatergic brain function: PET studies on parkinsonian rats. *J. Nucl. Med.* **48**: 1147–1153.

Penet, M. F., F. Kober et al. 2007. Magnetic resonance spectroscopy reveals an impaired brain metabolic profile in mice resistant to cerebral malaria infected with *Plasmodium berghei* ANKA. *J. Biol. Chem.* **282**: 14505–14514.

Penet, M. F., A. Viola et al. 2005. Imaging experimental cerebral malaria in vivo: Significant role of ischemic brain edema. *J. Neurosci.* **25**: 7352–7358.

Plengsuriyakarn, T., V. Eursitthichai et al. 2012. Ultrasonography as a tool for monitoring the development and progression of cholangiocarcinoma in Opisthorchis viverrini/dimethylnitrosamine-induced hamsters. *Asian Pac. J. Cancer Prev.* **13**: 87–90.

Poduslo, J. F., T. M. Wengenack et al. 2002. Molecular targeting of Alzheimer's amyloid plaques for contrast-enhanced magnetic resonance imaging. *Neurobiol. Dis.* **11**: 315–329.

Put, S., S. Schoonooghe et al. 2013. SPECT imaging of joint inflammation with nanobodies targeting the macrophage mannose receptor in a mouse model for rheumatoid arthritis. *J. Nucl. Med.* **54**: 807–814.

Rafuse, J. 1996. Seventy-five years later, insulin remains Canada's major medical-research coup. *CMAJ* **155**: 1306–1308.

Rhyner, M. N., A. M. Smith, X. Gao, H. Mao, L. Yang, and S. Nie. 2006. Quantum dots and multifunctional nanoparticles: New contrast agents for tumor imaging. *Nanomedicine (Lond)* **1**: 209–217.

Russell, C., A. Rahman, and A. R. Mohammed. 2013. Application of genomics, proteomics and metabolomics in drug discovery, development and clinic. *Ther. Deliv.* **4**: 395–413.

Shin, I. J., S. M. Shon et al. 2013. Characterization of partial ligation-induced carotid atherosclerosis model using dual-modality molecular imaging in ApoE knock-out mice. *PLoS One* **8**: e73451.

Shingaki, T., T. Takashima et al. 2012. Imaging of gastrointestinal absorption and biodistribution of an orally administered probe using positron emission tomography in humans. *Clin. Pharmacol. Ther.* **91**: 653–659.

Smith, G. C., W. K. Ong et al. 2013. Extended treatment with selective phosphatidylinositol 3-kinase and mTOR inhibitors has effects on metabolism, growth, behaviour and bone strength. *FEBS J.* **280**: 5337–5349.

Smith, K. A. 2012. Louis pasteur, the father of immunology? *Front. Immunol.* **3**: 68.

Tan, S. Y. 2004. Edward Jenner (1749–1823): Conqueror of smallpox. *Singapore Med. J.* **45**: 507–508.

U.S. Department of Health and Human Services and Food and Drug Administration. 2006. Critical paths opportunities list. Silver Spring, MD.

Valatas, V., M. Vakas, and G. Kolios. 2013. The value of experimental models of colitis in predicting efficacy of biological therapies for inflammatory bowel diseases. *Am. J. Physiol. Gastrointest. Liver Physiol.* **305**: G763–G785.

van der Horst, G., J. J. van Asten et al. 2011. Real-time cancer cell tracking by bioluminescence in a preclinical model of human bladder cancer growth and metastasis. *Eur. Urol.* **60**: 337–343.

van der Veldt, A. A., E. F. Smit, and A. A. Lammertsma. 2013. Positron emission tomography as a method for measuring drug delivery to tumors in vivo: The example of [(11)C]docetaxel. *Front. Oncol.* **3**: 208.

Verreault, M., D. Strutt, D. Masin, D. Fink, R. Gill, and M. B. Bally. 2011. Development of glioblastoma cell lines expressing red fluorescence for non-invasive live imaging of intracranial tumors. *Anticancer Res.* **31**: 2161–2171.

Viglianti, B. L., S. A. Abraham et al. 2004. In vivo monitoring of tissue pharmacokinetics of liposome/drug using MRI: Illustration of targeted delivery. *Magn. Reson. Med.* **51**: 1153–1162.

Viglianti, B. L., A. M. Ponce et al. 2006. Chemodosimetry of in vivo tumor liposomal drug concentration using MRI. *Magn. Reson. Med.* **56**: 1011–1018.

Viola, R. J., J. M. Provenzale et al. 2008. In vivo bioluminescence imaging monitoring of hypoxia-inducible factor 1alpha, a promoter that protects cells, in response to chemotherapy. *AJR Am. J. Roentgenol.* **191**: 1779–1784.

Vogel, C. L., M. A. Cobleigh et al. 2001. First-line Herceptin monotherapy in metastatic breast cancer. *Oncology* **61**(Suppl 2): 37–42.

von Horsten, S., I. Schmitt et al. 2003. Transgenic rat model of Huntington's disease. *Hum. Mol. Genet.* **12**: 617–624.

Webb, D. R. 2014. Animal models of human disease: Inflammation. *Biochem. Pharmacol.* **87**: 121–130.

Webster, J. M., C. A. Morton et al. 2014. Functional imaging of oxidative stress with a novel PET imaging agent, 18F-5-fluoro-L-aminosuberic acid. *J. Nucl. Med.* **55**: 657–664.

Wender, P. A., E. A. Goun et al. 2007. Real-time analysis of uptake and bioactivatable cleavage of luciferin-transporter conjugates in transgenic reporter mice. *Proc. Natl. Acad. Sci. USA* **104**: 10340–10345.

Willis, N. J. 1997. Edward Jenner and the eradication of smallpox. *Scott. Med. J.* **42**: 118–121.

Xie, J., S. Lee, and X. Chen. 2010. Nanoparticle-based theranostic agents. *Adv. Drug. Deliv. Rev.* **62**: 1064–1079.

Yapp, D. T., C. L. Ferreira et al. 2013. Imaging tumor vasculature noninvasively with positron emission tomography and RGD peptides labeled with copper 64 using the bifunctonal chelates DOTA, oxo-DO3a. and PCTA. *Mol. Imaging* **12**: 263–272.

Yapp, D. T., J. Woo et al. 2007. Non-invasive evaluation of tumour hypoxia in the Shionogi tumour model for prostate cancer with 18F-EF5 and positron emission tomography. *BJU Int.* **99**: 1154–1160.

Zhang, N., M. H. Ahsan, L. Zhu, L. C. Sambucetti, A. F. Purchio, and D. B. West. 2005. Regulation of IkappaBalpha expression involves both NF-kappaB and the MAP kinase signaling pathways. *J. Inflamm. (Lond)* **2**: 10.

Zhao, M., M. Yang et al. 2001. Spatial-temporal imaging of bacterial infection and antibiotic response in intact animals. *Proc. Natl. Acad. Sci. USA* **98**: 9814–9818.

Zhou, H., and D. Zhao. 2014. Ultrasound imaging-guided intracardiac injection to develop a mouse model of breast cancer brain metastases followed by longitudinal MRI. *J. Vis. Exp.* **85**(Mar 6).

27

Imaging of Intracellular Targets

Veerle Kersemans

27.1 INTRODUCTION TO INTRACELLULAR TARGETS FOR MOLECULAR IMAGING

27.1.1 BACKGROUND

Noninvasive imaging has evolved over the past decades into an important tool for diagnosing, understanding, and monitoring disease. Molecular imaging has distinguished itself as an interdisciplinary field that enables in vivo visualization, characterization, and quantification of biologic processes at the cellular and subcellular level with high sensitivity, specificity, spatial, and temporal resolution. Since measurements are made in the intact organism, longitudinal studies using identical or alternative biological imaging assays at different time points can be performed and multiple molecular events can be monitored near-simultaneously. This approach allows researchers not only to identify critical molecular and/or cellular processes associated with initiation of disease but also to facilitate the development of novel pharmaceuticals. As a result, much earlier detection of disease than when using traditional imaging techniques or relying on clinical symptoms can be made, evaluation of therapeutic efficacy has become possible and personalized therapeutic regimens can be implemented (Weissleder 1999; Massoud and Gambhir 2003; Chun et al. 2008; Weissleder and Pittet 2008; Morse and Gillies 2010; Skotland 2012).

Molecular imaging relies on the use of suitable reporter agents, the exception being magnetic resonance imaging or spectroscopy (MRI or MRS), where the imaging signal often emanates from endogenous molecules. Imaging probes in general must have the ability to reach the intended target at sufficient concentration and for a sufficient length of time to be detectable in vivo. Rapid excretion, nonspecific binding, metabolism, and delivery barriers all counteract this process and must be overcome. The basic underlying difficulty is to design agents with high target-to-background ratios or in other words, a high degree of specificity for its molecular target. The ideal marker will be expressed in the target tissue of a given pathology, but not in any other normal, unaffected tissue to avoid false-positive and confounding imaging results. To achieve this goal, different strategies, from small molecules to nanoparticles, have been pursued of which a summary is visualized in Figure 27.1. However, more

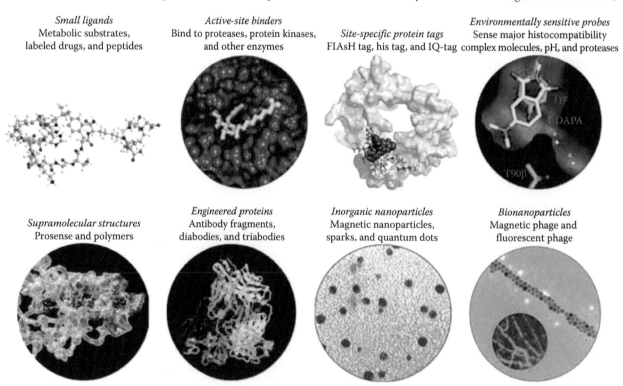

Small ligands
Metabolic substrates, labeled drugs, and peptides

Active-site binders
Bind to proteases, protein kinases, and other enzymes

Site-specific protein tags
FlAsH tag, his tag, and IQ-tag

Environmentally sensitive probes
Sense major histocompatibility complex molecules, pH, and proteases

Supramolecular structures
Prosense and polymers

Engineered proteins
Antibody fragments, diabodies, and triabodies

Inorganic nanoparticles
Magnetic nanoparticles, sparks, and quantum dots

Bionanoparticles
Magnetic phage and fluorescent phage

FIGURE 27.1 High-affinity imaging agents with appropriate pharmacokinetics are essential for imaging at the molecular level. (Reprinted from Weissleder, R. and Pittet, M.J., *Nature*, 452, 580, 2008. With permission.)

often than not, attaining high specificity is not sufficient as typical target concentrations are in the pico- or nanomolar range. As a result, signal amplification strategies, chemical and/or biological, are often harnessed to reach high levels of imaging signal per unit level of target. Biological amplification refers to targets with inherent signal amplification such as overexpression of receptors/enzymes or trapping, accumulation of target following internalization, enzymatic conversion, and pretargeting. Some examples of chemical amplification strategies include using multiple copies of a label in a single molecule, pretargeting (Barbet et al. 1998), pairing of deactivators and fluorochromes (Weissleder et al. 1999), and releasing caged compounds (Sharkey et al. 2005; Jaffer et al. 2007; Weissleder and Pittet 2008). All of the injectable imaging agents rely on reporter molecules such as fluorochromes (indocyanines, quantum dots), radiotracers (^{18}F, ^{11}C, ^{111}In, ^{99m}Tc), or magnetic labels (Gd-chelates, magnetic nanoparticles, hyperpolarized molecules) to be detected.

To date, many of the molecular imaging approaches have focused on extracellular targets, exploiting receptors, ion channels, transporters, antigens and enzymes at the cell surface or extracellular environment. They are often regarded as the ideal molecular imaging structure as these targets are relatively easy to access in vivo. As a result, both the chance of getting enough substrate to reach the target tissue and the chance of obtaining successful imaging of a disease when the target is upregulated are very good. Moreover, since the target is situated at the cell surface, background level may be acceptable, especially if the targeting molecule shows rapid renal excretion. Although these imaging markers still need to overcome many physiological barriers when administered in living subjects (Figure 27.2), they represent a less stringent pharmacological challenge in terms of hydrophilicity, chemical nature, and size. In case of agonists, receptor binding may cause internalization of the imaging probe, but as long as they are not degraded in lysosomes, receptor cycling can contribute to signal amplification. Conversely, some applications require lysosomal degradation in order to switch-on the signal, as described by Ogawa et al. (2009). Additionally, it needs to be noted that the selectivity of the latter techniques still is based upon binding to and selectivity for the extracellular receptors. To date, [^{18}F]-labeled fluorodeoxyglucose (FDG) is still the most used extracellular molecular imaging marker, both clinically and preclinically. Its working mechanism relies on the overexpression of the extracellular receptor Glut-1. Although the assessment of glucose

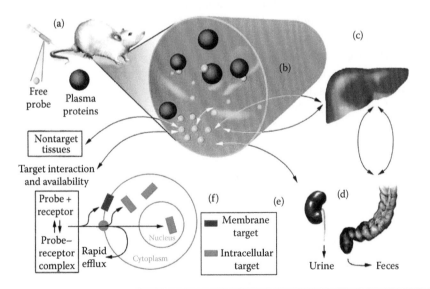

FIGURE 27.2 Pharmacokinetics of molecular imaging probes. Molecular imaging probes need to overcome many biological barriers when administered to living subjects. These probes are subject to all the pharmacokinetic rules and constraints that govern the concentration of "drugs" in plasma, including absorption/delivery (a), distribution (b), metabolism (c), excretion/reabsorption in the enterohepatic circulation (d), urinary excretion (e), and other factors within the vascular compartment (b; e.g., plasma half-life, protein binding). Rapid excretion, nonspecific binding/trapping in nontarget tissues, metabolism, and delivery barriers are all important obstacles to be overcome before availability to target(s) for interaction (f). (Reprinted from Massoud, T.F. and Gambhir, S.S., *Genes Dev.*, 17, 545, 2003. With permission.)

metabolism is used in many disease areas, including oncology, cardiology, and neurology, FDG is not a "magic bullet," and over the years a host of new molecular imaging agents have been developed that target more specific aspects of disease initiation and progression. A few examples of these include antibody-based imaging exploiting disease-specific cell surface antigens, perfusion imaging, peptide-based imaging targeting cell surface receptors, and amino acid– or nucleotide transporter–based imaging (Massoud and Gambhir 2003; Pysz et al. 2010; Zhu et al. 2010; Jacobs et al. 2012).

Although imaging extracellular targets is relatively easy and straightforward and thus most often preferred, many more opportunities arise when intracellular targets can be exploited. Not only are there more available targets, many of these are the focus of drug discovery efforts. Although the latter presents a great challenge, as targets are difficult to reach, it is also of great importance for understanding intracellular events and elucidating various biological phenomena. However, background levels may prove a problem, as excretion can be slow when no specific cellular export mechanisms exist for these imaging probes. This chapter will focus on the different aspects of preclinical imaging of intracellular targets, including the mechanisms of targeting, cell labeling and tracking, reporter genes, and theranostics.

27.1.2 Bypassing Physiological and Biological Barriers

Targeted imaging probes that interact with intracellular components or events are becoming increasingly important. The intracellular compartment presents many different processes that can potentially be targeted, some more challenging than others. These encompass both cytosolic targets and nuclear targets. Hence, no matter what the end destination is, the intracellular targeted imaging probe has to bypass many biological barriers, avoiding elimination and/or degradation at each step. Indeed, biological barriers such as the immune system, liver, kidneys, reticuloendothelial system (RES), and blood–brain barrier are intrinsic to the body's defense mechanism and they are very efficient at removing foreign materials and thus also in preventing contrast agents to reach their target destinations. Although these mechanisms are designed to protect us from viruses, bacteria, and other toxins, they impose great challenges on imaging probe design. These barriers can restrict contrast agents by blocking their movement, causing physical changes, or by inducing a negative host response using biochemical signaling (Lee et al. 2007). On the other hand, systemic distribution or off-target accumulation of contrast agents can increase the pharmacological dose to normal organs leading to increased toxicity and side effects (Kievit and Zhang 2011).

As a result, a high-specificity agent is useless if it cannot reach its target and much effort must be geared toward optimal pharmacodynamic and pharmacokinetic properties, which are largely related to physicochemical properties, including morphology, hydrodynamic size, charge, and other surface properties (Chouly et al. 1996; Dobrovolskaia et al. 2008). Adjusting these physicochemical properties can help to overcome the challenge of physiological barriers. The first barrier encountered is the blood of which its high ionic strength can cause contrast agents to be sequestered or clustered. Moreover, the probes can bind nonspecifically to plasma proteins, which not only alter their biodistribution but also potentially prevent them of reaching their target tissue as opsonization can lead to recognition by the reticuloendothelial system, followed by removal from circulation. The latter is especially true for hydrophobic contrast agents, which results in a short circulation time. However, uptake by the RES can be effectively avoided by using surface modification with polyethylene glycol (PEG) to increase circulatory half-life from minutes to many hours or days (Klibanov et al. 1991; Park 2002). Therefore, PEGylated contrast agents are commonly regarded to as "stealth" agents because they are not readily recognized by the RES (Kresse et al. 1998; Harris and Chess 2003). In addition to coping with the vascular environment, contrast agents must also avoid breakdown and/or chemical modification in liver and kidneys.

Once the imaging agent survives the physiological barriers, specific targeting can begin. In oncologic applications, the first targeting step is usually achieved by enhanced perfusion and retention (EPR) where the imaging probes will extravasate into the tumor stroma through the fenestrations of the angiogenic vasculature. Once in the extracellular space, they can target specific features of the tumor cell. However, this is further complicated in brain malignancies where the tight junctions between epithelial cells of the blood–brain barrier prevent EPR. Nevertheless, EPR enhancement strategies, such as the administration of Bradykinin, have been reported and mathematical formulations have recently become available to optimize EPR properties of imaging probes (Wu et al. 1998; Dellian et al. 2000; Decuzzi et al. 2005; Ferrari 2005).

While at this point the end station is reached for contrast agents interacting with extracellular targets, this is not the case for those aimed at intracellular targets. After the contrast agent has reached its target and/or extravasated from the blood, it must be taken up by the targeted cells and overcome the cell membrane, a negatively charged phospholipid layer that separates the inside of the cell from the extracellular space. Although almost every molecule will diffuse over time through the lipid bilayer down its own concentration, the rate of diffusion shows enormous variation depending on the size and hydrophobicity of the molecule. Small, nonpolar molecules will traverse it readily whilst hydrophilic compounds are prevented entry efficiently. Besides this passive transport, eukaryotic cells have the potential to use a number of different endocytosis mechanisms, or active transport, to translocate molecules across the cell membrane. These processes will be discussed later. However, imaging probes entering the cell through any form of endocytosis will face yet another problem: lysosomal degradation. Indeed, many endosomes are translocated into lysosomes where hydrolytic and enzymatic reactions can metabolize imaging probes, rendering them ineffective upon cellular processing. To date, many strategies have been proposed to facilitate endosomal escape, reviewed by Varkouhi et al. (2011), allowing the imaging probe to avoid degradation and subsequent interaction with its intended target (Belting et al. 2005). However, the journey of imaging probes targeting intranuclear structures does not end here as one more barrier stands in their way: the nuclear envelope. This membrane consists of a double lipid bilayer where the intermembrane space is continuous with the endoplasmic reticulum. This is again a very efficient barrier and solutions to cross this obstacle are discussed in the following texts.

27.1.3 Different Imaging Modalities

In addition to target selection and probe design, one must also take into account the choice of imaging modality. As reviewed by others, the various existing imaging technologies differ in five main aspects: spatial and temporal resolution, depth penetration, energy expended for image generation (ionizing or nonionizing, depending on which component of the electromagnetic radiation spectrum is exploited for image generation), availability of injectable/biocompatible molecular probes, and the respective detection threshold of probes for a given technology (Massoud and Gambhir 2003; Condeelis and Weissleder 2010). All imaging modalities have their intrinsic advantages and disadvantages and their individual characteristics are discussed elsewhere in this handbook. However, it might be useful to draw your attention to the following aspects.

Selection of the imaging modality will depend on the required sensitivity and resolution, target density, amplification method, probe design, chemistry compatibility, etc. As intracellular target density is often low, probe design to image intracellular targets is often geared toward nuclear medicine and optical imaging modalities. As a result, sensitivity plays a key role and is prioritized over resolution. Moreover, given the vast number of ligands and enzymatic precursors that can be radiolabeled, nuclear modalities are well suited for imaging of molecular events. One of the additional advantages of positron emission tomography (PET) is that imaging probes or existing molecules known to interact with a specific target can be modified with a radiolabel while minimally perturbing the parent molecule. For optical and MRI approaches, this is usually not possible because the signaling portion is itself a large molecule (e.g., a fluorochrome) or a bulky atom (e.g., gadolinium).

Dosing and related toxicity issues must be considered as well. Whereas most radiolabeled molecular imaging probes are given in low doses (nonpharmacological, nanogram levels), molecular imaging probes for MRI and optical techniques are usually given in mass levels (typically micrograms to milligrams). However, both the problem of bulk and mass can be solved by clever probe design. Examples include liposomes, micelles, or nanoparticles that can contain large amounts of contrast agent as their cargo.

27.2 MECHANISMS TO TARGET INTRACELLULAR MOIETIES AND THEIR APPLICATIONS

As mentioned earlier, once the imaging probes have conquered the physiological barriers, they need to overcome the biological barriers of which the cell membrane is the first obstacle. Luckily for us, membranes are not impenetrable walls, the more because nutrients must enter the cell and waste products have to leave in order for cells to survive. The same principles can be applied to transfer molecular imaging agents across the cell membrane. In general, three major ways can be distinguished for getting in and out of cells: diffusion (passive), protein-mediated transport (passive or active), and endocytosis/exocytosis (active). The movement of compounds across the membrane can be either "passive," occurring without the input of cellular energy, or "active," requiring the cell to expend energy in transporting it. In addition to these naturally occurring transport mechanisms, imaging probes can be forced across the cell membrane through pharmacological and/or physical intervention. This section will discuss the aforementioned strategies as well as ways to reach intranuclear targets.

27.2.1 DIFFUSION

Diffusion is one of the several transport phenomena that occur in nature and, according to Fick's laws, the diffusion flux is reversely proportional to the gradient of concentrations (Philibert 2005). The rate at which a molecule diffuses across a membrane depends on its size and its degree of hydrophobicity. As a result, only low-molecular-weight molecules that are lipophilic can slip between the lipids in the bilayer and cross from one side to the other easily whilst this is not the case for larger and/or polar molecules, unless they are very small and uncharged (e.g., water and ethanol). All in all, relatively few endogenous molecules are capable to cross membranes by diffusion, and one should keep in mind that this process is not target specific. In other words, diffusion of a lipophilic imaging probe will take place through membranes of all cells, including those we do not want to image. As a result, high background levels can be expected. Because of the high ionic strength of blood, these compounds also tend to stick to serum proteins such as albumin, prolonging their circulation times, which will again contribute to higher background levels.

Noninvasive imaging of hypoxia using radiopharmaceuticals is one example where imaging probes enter the cell through diffusion. Identifying hypoxic tissue has therapeutic implications for multiple disease states as healthy cells deprived of oxygen will invariably die (e.g., stroke, myocardial ischemia, and diabetes). On the other hand, tumor cells are gradually exposed to chronic hypoxia, resulting in an adaptation mechanism leading to resistance to therapy (Rajendran and Krohn 2005). Radiolabeled 2-nitroimidazoles, predominantly [18F]-fluoromisonidazole (FMISO), have been studied the most extensively in the context of hypoxia (Nunn et al. 1995). Their octanol/water partition coefficient is close to 1 so that they are freely diffusible and only less than 5% is protein bound allowing efficient transport from blood into tissues (Grunbaum et al. 1987; Rasey et al. 1999). The mechanism by which 2-nitroimidazoles are reduced and retained in hypoxic tissues is well understood and displayed in Figure 27.3. In short, specificity of the compounds toward intracellular low oxygen levels is obtained through differential efflux under hypoxic conditions. These compounds are reduced into reactive intermediary metabolites by intracellular reductases in a process, which is directly related to the level of oxygenation/hypoxia. This causes a gradient, which is favorable for the detection of hypoxic cells. Subsequently, these intermediary metabolites covalently bind to thiol groups of intracellular proteins and thereby are trapped within viable hypoxic cells. Since FMISO was proposed as a PET tracer for noninvasive imaging of tumor hypoxia, several other tracers also have been evaluated for this purpose (e.g., [18F]-fluoroerythronitroimidazole [FETNIM], [18F]-fluoroazomycin-arabinofuranoside [FAZA], and 2-(2-nitro-1H-imidazol-1-yl)-N-(2,2,3,3,3-pentafluoropropyl)-acetamide [[18F]-EF5]) with the goal of a possibly more optimal partition coefficient and therefore faster clearance properties, less nonspecific retention, or fewer metabolites (Ballinger 2001).

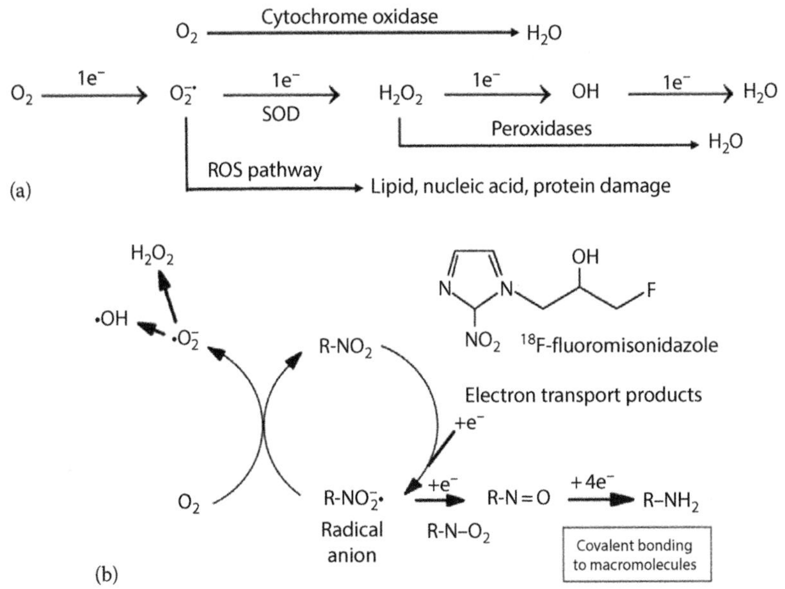

FIGURE 27.3 Mechanism of reduction and intracellular retention of nitroimidazoles. (a) Molecular oxygen is a terminal electron acceptor in the mitochondrial respiratory chain. Electrons mostly react with O_2 to form H_2O in a process catalyzed by cytochrome c oxidase, complex IV. Some electrons escape and combine with O_2 to form superoxide radical anions that react with superoxide dismutase (SOD), which is abundant in anaerobic organisms; alternatively, reactive oxygen species (ROS) oxidize critical biomolecules, leading to cellular dysfunction and death. H_2O_2 is substrate for catalase and peroxidase enzymes and eventually produces more H_2O. (b) The mechanism of reduction and intracellular retention of nitroimidazoles involves buildup of steady-state level of 1 electron radical anion, which reacts preferentially with O_2 if it is present to return tracer to its original form. If O_2 is not present, it accepts another electron to produce an intermediate that is sequentially reduced to a highly efficient alkylating agent, RNH_2, resulting in cellular retention of the labeled tracer. (Reprinted from Krohn, K.A. et al., *J. Nucl. Med.*, 49(Suppl. 2), 129S, 2008. With permission.)

27.2.2 PROTEIN-MEDIATED TRANSPORT: CARRIERS AND CHANNELS

In order to cross the hydrophobic interior of the cell membrane, hydrophilic and large molecules need the help of membrane transport proteins. These integral membrane proteins provide a continuous protein-lined pathway through the bilayer. Two major classes can be distinguished: channel proteins, which form a narrow pore through which ions can pass (passive transport), and carrier proteins, which translocate specific molecules across (passive or active transport). However, these transport mechanisms are less relevant to imaging of intracellular targets. Indeed, ion channels are important in muscle and neuronal excitation, and a well-known example of a carrier protein that carries out passive transport is the glucose transporter, exploited by the most used extracellular molecular imaging marker [^{18}F]-FDG. Active transporters include ATP-driven ion pumps, coupled transporters (or secondary active transporters), and ABC transporters. This final type of transporter is specialized in pumping small molecules, including drugs and other toxins, out of the cells and is not only involved in multidrug resistance in cancers but also contributes to the blood–brain barrier.

27.2.3 ENDOCYTOSIS-MEDIATED UPTAKE

27.2.3.1 General

Large and hydrophilic molecules can enter the cell through different types of endocytosis, a process by which cells absorb molecules by invagination of the plasma membrane. Substances may be taken up by endocytosis after binding to a receptor, by interacting with other structures on the cell surface, or just by being engulfed together with fluid surrounding the cell when an endocytic vesicle is formed. There are multiple types of endocytic pathways of which the clathrin-mediated endocytosis pathway is best known and by far the best studied. Additionally, several clathrin-independent endocytosis mechanisms have been described (Sandvig et al. 2008; Howes et al. 2010; Kumari et al. 2010), including dynamin-dependent mechanisms and dynamin-independent mechanisms. It should be noted that receptor-mediated endocytosis can involve several of the aforementioned mechanisms, dependent on the type of receptor being used.

Whatever the mechanism of endocytosis, its cargo will become trapped in the endosomes, will not enter the cytosol, and are degraded by specific enzymes in the lysosomes unless specific transporters in the endosomal membrane are available. However, some exceptions exist as retrograde transport has been reported for some toxins (Gallazzi et al. 2003; Sandvig and van Deurs 2005) and some viruses and bacteria are also able to enter the cytosol (Mercer et al. 2010; Stavru et al. 2011). Thus, one must not only devise a strategy to translocate the imaging probes across the cell membrane but also design the agents in such a way that endosomal escape can take place. A comprehensive review on endosomal escape pathways for delivery of biological has been published by Varkouhi et al. (2011).

27.2.3.2 Receptor-Mediated Transport

Cell surface receptors have been exploited mainly for imaging extracellular targets. However, one can benefit from the same approach to study intracellular targets, a route that is often preferred when bispecific antibodies or multifunctional nanoparticles are involved. Although many applications of the latter exist for therapy (Flavell et al. 1992; Cao and Lam 2003; Oh et al. 2009; Govindan and Goldenberg 2012), only a few are applied for imaging, let alone imaging of intracellular targets (Capello et al. 2004).

Tian et al. used bispecific peptide conjugates that linked peptide nucleic acids (PNAs) directed against CCND1, myc, or K-RAS mRNA to d(CSKC)c (CSKC stands for Cys-Ser-Lys-Cys), a circular peptide that binds to insulin-like growth factor-1 receptor (IGF-1R). The d(CSKC)c peptide binds to IGF-1R, overexpressed in breast cancer tumors. Following the internalization of the peptide–receptor complex, the PNAs will bind their respective mRNAs, and the imaging label (in this case Copper-64 for PET) will allow external imaging of the expression of these genes (Tian et al. 2005).

Another example was reported by Cornelissen et al. (2009b). They used indium-111-labeled radioimmunoconjugates directed against the cyclin-dependent kinase inhibitor p21$^{\text{WAF-1/Cip-1}}$ which is overexpressed in breast cancer following trastuzumab treatment. They designed immunoconjugates composed of epidermal growth factor (EGF) and an anti-p27Kip1 antibody that recognized the extracellular EGF receptor on breast cancer cells and the intranuclearly located p27$^{\text{Kip1}}$, respectively. EGF, which was used to target a specific cell type, was found to be relatively resilient to cross-linking with macromolecules. No major decrease in affinity for its receptor was noted following the conjugation of EGF to not only the anti-p27Kip1 antibody but also to nuclear localizing sequences (NLS) which routed the construct from the cytoplasm into the nucleus following EGF-receptor binding and internalization. Once in the nucleus, the anti-p27Kip1 antibody could interact with its target, p27$^{\text{Kip-1}}$. This modular approach could be applied to synthesize other bispecific immunoconjugates that use alternative receptors for tumor targeting and internalization or that probe different intracellular protein epitopes.

The results presented recently by Dong et al. show the same principle for a nanoparticle approach (Dong et al. 2012). They designed a novel multifunctional SnO_2 nanoprobe, which contains a cell-targeting moiety (folic acid targeting the folate receptor), as well as a conjugated gene probe to specifically recognize a target sequence (molecular beacon). Visualization of the delivery and intracellular response is made possible through fluorescence of the SnO_2. However, this approach has only been used and tested in vitro.

27.2.3.3 Cell Penetrating Peptides

A very popular approach to translocate moieties across the cell membrane barrier, both for therapy and imaging applications, is to make use of cell penetrating peptides (CPPs) (Kersemans et al. 2008; Kersemans and Cornelissen 2010; Choi et al. 2011). The in vivo potential of CPPs to unlock intracellular targets became clear when Schwarze et al. first demonstrated that the fusion of beta-galactosidase protein to the trans-activating transcriptional activator (TAT) from human immunodeficiency virus 1 resulted in the delivery of the biologically active fusion protein to all tissues in mice, including the brain (Schwarze et al. 1999).

CPPs are short polypeptides that contain several positively charged amino acids (polycationic) or have sequences that contain a pattern of alternating polar and nonpolar, hydrophobic amino acids (amphipathic) (Madani et al. 2011). Proteoglycans play an important role in the initial contact between CPPs and the cell surface, after which endocytosis is initiated. However, no general uptake mechanism has been elucidated, and depending on many parameters including the nature of the CPP, its cargo, and reporter group, a different endocytosis uptake route for internalization is preferred (Mayor and Pagano 2007). Once in the cell, little is known about their intracellular trafficking, but since endocytosis is involved, one can assume either routing toward the lysosome and subsequent degradation by proteases or endosomal escape and subsequent interaction with its intracellular target. This highlights not only the necessity to better understand the endolysosomal pathway and the structure–activity relationships needed for the cargo to reach the cytoplasm but also to take into account endosomal escape when developing the imaging probe.

A comprehensive review, highlighting some of the applications of CPPs in molecular imaging and molecular radiotherapy was published by Kersemans and Cornelissen (2010). They organized the use of CPPs in molecular imaging by different categories: (1) direct application, (2) CPP constructs with an intracellular or intranuclear target, (3) CPP constructs with an extracellular target, (4) activatable CPPs, and (5) the use of CPP to track prelabeled cells. The latter is visualized in Figure 27.4. As is apparent from these publications, CPPs hold great potential as in vivo delivery vectors of contrast agents for imaging intracellular targets. Although CPP-mediated intracellular cargo delivery lacks cell specificity, it should not be a show-stopper as demonstrated by the Reilly RM group. They conjugated TAT to various radiolabeled immunoglobulins to promote their cellular internalization into human breast cancer cells and the attachment of the radiolabel indium-111 allowed for in vivo SPECT imaging (Hu et al. 2007; Cornelissen et al. 2009). Although the TAT-peptide routes the immunoconjugates to the intracellular space of each cell, interaction of the immunoconjugates with their intracellular target will result in specific retention, which will generate the imaging contrast needed. Indeed, the TAT peptide not only promotes cellular internalization but it also contains specific amino acids that stimulate active export when not retained, that is, bound to its target, in the cell. In other words, imaging contrast is achieved through differential retention, rather than uptake (Rayne et al. 2010).

An elegant way to overcome the problem of tissue nonspecificity of CPPs was presented by Nobel laurate Roger Tsien in 2004 (Jiang et al. 2004; Kersemans and Cornelissen 2010). They developed the so-called "activatable" CPPs or constructs that only generate a signal in tissues that contain a specific cleaving enzyme that releases the CPP. Following its release, the CPP construct could enter the surrounding cells resulting in both a slower blood clearance when compared to its "plain"

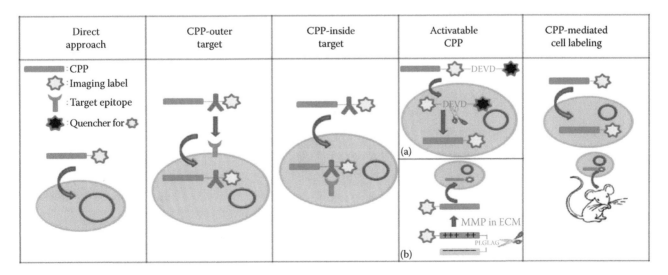

FIGURE 27.4 A schematic overview of the various CPP-construct approaches for molecular imaging. Activatable constructs generate a signal either (a) inside or (b) outside the cells using a specific cleaving enzyme. (Reprinted from Kersemans, V. and Cornelissen, B., *Pharmaceuticals*, 3, 600, 2010. With permission.)

alternative and in excellent target to nontarget imaging contrast. Although they used this approach for imaging extracellular targets, the same principle can be applied to study intracellular targets. This is illustrated by Johnson et al. who demonstrated the utility of caspase-activatable CPPs for high throughput and high-resolution live cell microscopy imaging of apoptosis (Johnson et al. 2012).

27.2.3.4 Liposomes and Micelles

Liposomes and micelles are artificially prepared bilayers or monolayers, respectively, which can be used to deliver imaging probes to their sites of action. The lipid bilayer/monolayer can fuse with other bilayers such as the cell membrane, thus depositing the liposome contents into the cytoplasm. A well-known application is the transfection of DNA (lipofection) into a host cell.

Because of their structural versatility, these macromolecules have found their way into nano-medicine as drug carrier systems where they are predominantly used to enhance in vivo efficiency of anticancer drugs. Soon thereafter, it was also recognized that liposomes when labeled with gamma radiation–emitting radionuclides might serve as radiopharmaceuticals to visualize pathological processes (Boerman et al. 2000; Torchilin 2007). The same principle can be applied for imaging intracellular targets to increase the local spatial accumulation of contrast agents. However, examples of the latter have not been published yet. Liposomal-mediated contrast agents can be divided into two main categories: liposomes encapsulating the contrast agent within the core and liposomes in which the contrast agent is dispersed within the lipid bilayer. Figure 27.5 provides an overview of the different strategies associated with liposomes for the delivery of contrast agents and/or therapeutics (Mody et al. 2009).

Liposomes and micelles are taken up in the pathological area through the EPR effect, which allows large molecules to extravasate and passively accumulate in the interstitial tumor space based on the cut-off size of the leaky vasculature (Yuan et al. 1995; Torchilin 2007). Furthermore, the surface of liposomes and micelles can be functionalized to target specific cells (antibody modification), increase intracellular delivery (CPPs), avoid the RES (PEG), and control the location of contrast agent release (pH-sensitive moieties). However, the problem still remains that any molecule entering the cell through endocytosis

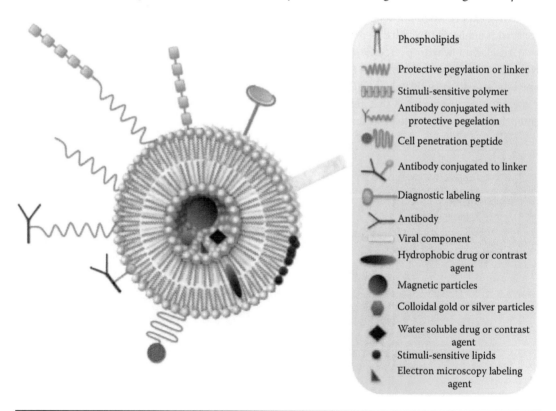

FIGURE 27.5 Different strategies associated with liposomes to be used as a carrier system for various types of drugs and contrast agents. (Reprinted from Mody, V.V. et al., *Adv. Drug Deliv. Rev.*, 61(10), 795, 2009. With permission.)

will eventually be directed toward the lysosomes and hence only a small fraction of the cargo will reach the cytoplasm if no effective endosomal escape mechanism takes place. The latter has been achieved by including pH-sensitive (Torchilin et al. 1993; Shalaev and Steponkus 1999) or membrane-destabilizing (Yessine and Leroux 2004) components into the liposome structure. Dioleoylphosphatidylethanolamine (DOPE), a fusogenic lipid, is such an example and is frequently used in pH-sensitive liposomal formulations. Incorporation of DOPE into cationic liposomes significantly improves the endosomal escape by membrane fusion and destabilization. Its fusogenicity comes from its strong tendency to form a non-lamellar structure/phase (non-bilayer or hexagonal phase) due to its cone-shape geometry and its weak acidity, which stabilizes its lamellar, bilayer phase at neutral pH but becomes partially protonated at low pH, resulting in destabilization of the lamellar phase (Zhu and Torchilin 2012).

27.2.4   Physical Intervention

Apart from the methods described earlier that use the inherent cellular uptake machinery, contrast agents can also be internalized into the cells by direct physical manipulation. This approach has been used abundantly for in vitro applications (Delehanty et al. 2009), but fewer examples exist for in vivo settings, let alone for preclinical imaging of intracellular targets.

Electroporation in combination with hypothermia has been used in vivo for intracellular delivery of DNA/RNA (Aihara and Miyazaki 1998; Boutin et al. 2008), which could be translated to contrast agent delivery as well. Indeed, Leroy-Willig applied electroporation of rat tibialis muscle after local intramuscular injection of a plasmid coding for an MRI contrast agent. The amount of trapped contrast agent was determined for a range of electric field intensity (Leroy-Willig et al. 2005). They used a protocol proposed by Mir et al. which was developed for therapeutic gene transfer in mouse muscle and showed minimal adverse effects (Mir et al. 1999). A few years later, Holm et al. reported on its use for facilitated intracellular delivery of MRI contrast agent following systemic i.v. administration (Holm 2009). This study suggests that the contrast agent remains trapped for periods up to 2 months and this in concentrations high enough to allow for intracellular MRI imaging throughout this period.

Drugs, or other cargo molecules such as contrast agents, can also be released ultrasonically from liposomes and micelles that circulate in the blood and retain their cargo of drugs until they enter an insonated volume of tissue. Gas-filled microbubbles in the circulatory system cavitate upon insonation and disrupt surrounding cells and membranes, thus allowing the passage of drugs into the targeted tissue. Although a recently published retrospective study reported that efficient intracellular uptake of molecules into cells in vitro is rare (Liu et al. 2012), US-mediated destruction of microbubbles has been shown to be successful in vivo. An example was published by Bekeredjian et al. who infused rats with albumin microbubbles with a plasmid encoding a luciferase reporter through the jugular vein for 20 min (Bekeredjian et al. 2003). During this time, four acoustic power bursts were applied to episodically destroy the microbubbles in the myocardium. Using optical imaging, they showed this method can deliver plasmids to the heart cells, achieving transient transgene expression with high tissue specificity. Another example includes the use of US-sensitive (sonosensitive) liposomes to image drug delivery or gene expression using MRI (Jung et al. 2012).

27.2.5   Targeting the Nucleus: An Extra Barrier to Overcome

Intracellular targets also include those situated in the cell nucleus and, in order to reach these, another barrier needs to be overcome: the nuclear envelope, a double membrane that is continuous with the endoplasmic reticulum. All transport across this membrane exclusively occurs via the nuclear pore complexes, which are embedded in the nuclear envelope. Although passive diffusion through these complexes is possible for molecules up to 45 kDa (ions, metabolites, and smaller proteins), active transport is the predominant mechanism to transfer molecules to and from the nucleus (Gorlich and Kutay 1999; van der Aa et al. 2006). This nucleocytoplasmic transport is described well by Conti and Izaurralde (2001).

As a result, access to the nucleus is a highly restricted process and NLSs are required to actively transport molecules across the nuclear membrane. A visual overview of NLS-mediated transport is presented in Figure 27.6. One of the best understood and most abundantly used NLSs is the NLS

(a)   (b)   (c)

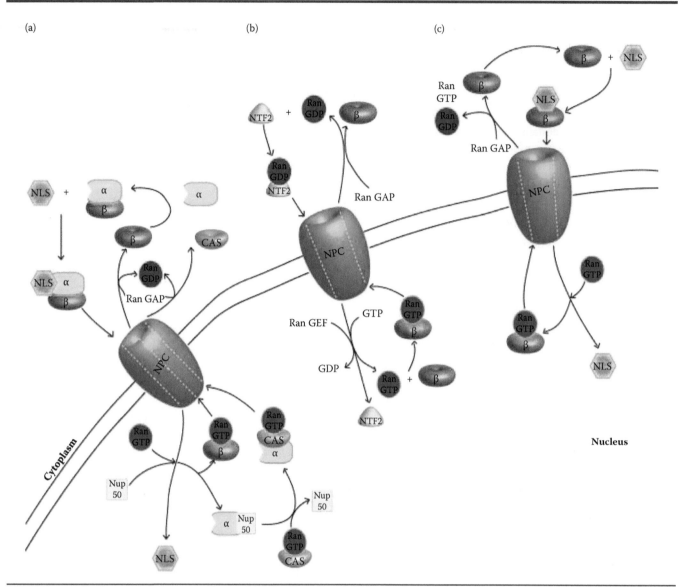

FIGURE 27.6   Nuclear protein import pathways. (a) IMP α/β mediated, (b) Ran cycle, and (c) IMP β mediated. (Reprinted from Pouton, C.W. et al., *Adv. Drug Deliv. Rev.*, 59(8), 698, 2007. With permission.)

peptide derived from the SV-40 large T antigen, a single stretch of basic amino acids (PKKKRKV[132]) (Kalderon et al. 1984). Additionally, the cell-penetrating peptide TAT (mentioned in Section 27.2.3.3) is not only able to translocate cargos to the cytoplasm but it also harbors a NLS sequence to deliver moieties to the nucleus. The amount of NLSs conjugated to the cargo differs considerably between different papers and no definitive conclusion can be drawn concerning the amount of NLS peptide needed to obtain efficient nuclear import. However, it is most likely that it is dependent on the 3D structure of the cargo used, and for each application, the amount of NLS peptides conjugated to the cargo needs to be optimized. However, the selectivity of the nuclear pore not only comes down to the presence of a NLS but it is also dependent on the size of the cargo. The maximum load for nuclear pore complex import was determined by Pante and Kann using NLS-coated gold beads (Pante and Kann 2002) and they showed that transit is restricted to a size/structure approximately below 40–60 nm limit. This was further illustrated by the fact that hepatitis B viral capsids of 36 nm could cross the NPC intact (Pante and Kann 2002). As visualized in Figure 27.6, molecules are not only actively imported but also exported from the nucleus through the nuclear pore complexes. However, through the specific interaction of the imaging agents with its intranuclear target, specific retention is obtained, resulting in a shift in steady state toward the nuclear compartment and hence the generation of imaging if target density is high enough.

An example of intranuclear imaging is presented by Cornelissen et al. (2011) who imaged DNA damage in vivo using immunoconjugates. Signal amplification was achieved by focusing on the phosphorylated histone H2A-variant H2AX, γH2AX, an early and almost universal feature of the eukaryotic response to DNA double strand breaks. Hundreds of copies of γH2AX accumulate in foci at DNA double strand breaks. The estimated number of γH2AX formed following doses of 2–10 Gy irradiation would be 2.4×10^5 to 1.2×10^6 which is comparable to the number of targets needed for extracellular molecular imaging. They attached the TAT-peptide covalently to a fluorophore- or ^{111}In-labeled anti-γH2AX antibody to reach the nucleoplasm. Tat/IgG conjugation ratio was 5:1 as determined by radioiodination of the TAT-peptide. As already mentioned, TAT is not only able to route the antibody to the cytoplasm but it also contains a NLS sequence to import its cargo to the nucleus where it is able to specifically bind to its target and image DNA double strand breaks.

27.3 CELL LABELING AND TRACKING

A widely used application for preclinical imaging of intracellular contrast agents is the tracking of cells in vivo. Cell tracking is used to study the fate of injected cells during cell-based therapy. However, other medical applications, such as understanding how metastatic tumors spread, benefit from this technique. Frangioni and Hajjar described the desirable characteristics for a cell tracking imaging agent (Frangioni and Hajjar 2004) and are as follows: (1) it should be biocompatible, (2) should be safe for the injected cells and should not interfere with their function, (3) should be highly sensitive (capable of detecting single cells), (4) should not dilute with cell division, (5) should not transfer to other cells, and (6) should be capable of being imaged for months to years after initial labeling.

The majority of cell tracking studies can be regarded as "indirect" as cells are preloaded with contrast agents before they are injected into an organism and followed over time. The first preclinical studies and the concept of MRI cell tracking were introduced in the early 1990s but it was not until the emergence of stem cell therapy or cancer vaccine therapy that these methods became robust (Bulte et al. 1993).

First, cells need to be rendered visible for this approach to work. Superparamagnetic iron oxide nanoparticles (SPIONs) for MRI have been most extensively explored for this purpose, and more recently, quantum dots and gold nanoparticles as contrast agents for fluorescence and optoacoustic imaging, respectively, have entered the arena (McCormack et al. 2011; Ricles et al. 2011). Numerous methods have been employed for cellular labeling, including nonspecific or spontaneous cellular internalization by phagocytotic cells following the addition of the contrast agents to the cell culture medium (Bulte 2009). This uptake is dependent on the cell type, particle size, and particle coating, and a higher concentration and exposure time usually resulted in a higher labeling efficiency. However, the uptake efficiency is difficult to predict and, as a result, other approaches such as surface modification of nanoparticles using conjugation to CPPs, complexation with transfection agents, and magnetoelectroporation were devised (Moore et al. 2002; Arbab et al. 2004; Rogers et al. 2006). As discussed in previous paragraphs, the intracellular fate of the contrast agent plays an important role for cell tracking to be successful. Whereas SPIONs mainly accumulate in endosomal compartments resembling lysosomes, CPP-conjugated nanoparticles have been found to be homogeneously distributed within the cell, including the nucleus (Taylor et al. 2012). Delivery of the contrast agents to the lysosomes can have a huge impact with respect to their metabolism and long-term stability. Additionally, when iron particles are concerned, toxicity following dissolution in the lysosomes needs to be taken into account as well (Bulte 2009). Even though iron is an essential mineral involved in various biological processes, its overload can be toxic to cells (Soenen and De Cuyper 2010). Once cells are labeled with the imaging probes, their migration can be tracked with the imaging modality of choice following an injection intravenously or at the site of interest. However, as soon as the cells enter the body, extra challenges might arise that eventually can result in the loss of imaging contrast. As labeled cells proliferate in vivo, the nanoparticles are divided between the daughter cells, which can migrate to other regions of the tissue or body, hence weakening the signal or resulting in false-positive results. Additionally, it is not possible to discriminate live from dead cells and once cells die, the SPIONs remain in and around dead cells until cleared away, primarily by cells of the macrophage lineages, again transferring the label to host cells. Finally, SPIONs generate negative contrast, which might interfere with endogenous blood derivatives such as methemoglobin and makes the in vivo visualization of SPIONs difficult.

Although most studies used MRI contrast agents, some applications using the radioisotope indium-111 have made it into clinic as well. One example has been described by de Vries et al. who studied the migratory capacity of both immature and mature dendritic cells for anti-tumor vaccinations (e.g., anti-melanoma vaccination) (De Vries et al. 2003). This study relied on earlier preclinical work performed by the same group who studied the biodistribution on ^{111}In-labeled dendritic cells in the murine B16 melanoma tumor model (Eggert et al. 1999). For this purpose, dendritic cells were labeled with [^{111}In]-oxinate following its addition to the cell culture medium for 20 min. Once the labeling efficiency was determined, the cells were injected subcutaneously (thighs or abdomen), intraperitoneally, or intraveneously in the lateral tail vein of mice and cell migration was tracked using planar SPECT imaging. Not only did they show that cell tracking using ^{111}In-labeled cells was feasible, but the method also allowed for semi-quantitative analysis of the amount of labeled cells that accumulate in tissues. By doing so, they were able to prove that the distribution of dendritic cells to lymphoid tissues is dependent on the route of vaccination and that mature, but not immature, dendritic cells migrated specifically into T-cell areas of secondary lymphoid organs, a necessity for the therapy to work.

More recently, methods for "direct" monitoring and tracking of endogenous cells in vivo have been developed. Again, MRI seems to be the imaging modality of choice and endogenous cell labeling has been achieved peripherally where iron oxide MRI contrast agents were phagocytosed by circulating macrophages (Wiart et al. 2007; Panizzo et al. 2009). Also, neuronal stem and progenitor cells are able to internalize iron oxide beads following an injection near the subventricular zone of the lateral ventrical (Sumner et al. 2009). The latter was crucial as other injection sites in the lateral ventricle did not lead to efficient labeling of progenitor cells. Using this approach, the authors were not only able to label 30% of the migrating progenitor cells in vivo but also showed using MRI that within 2 weeks post-injection these labeled cells populated the olfactory bulb. A similar study makes use of the commercially available Endorem, a ferumoxide MRI contrast agent composed of dextran-coated SPIONs (Panizzo et al. 2009). As with the previous study, they were able to visualize endogenous migrating neuroblasts. However, their method allowed a smaller concentration of contrast agent and the use of growth factors to promote cell labeling.

A different approach is published recently by Miao et al. (2012). Although they have not performed the in vivo proof of principle yet, it might be an elegant and promising alternative to the aforementioned strategies. They designed a radioactive [^{125}I]-labeled nanoparticle that is cleaved in vivo by furin, an intracellular protein that is overexpressed in many diseases, including cancer and Alzheimer. Once the nanoparticle entered the cell, presumably following endocytosis, an active intermediate is formed following furin cleavage, yielding an amphiphilic dimer. The latter has a hydrophobic macrocyclic core, which lends itself for the self-assembly of the ^{125}I-nanoparticle via π–π stacking. Not only will this self-assembly result in signal amplification but it will also prevent the cells to have the nanoparticle pumped out because of its big size and hydrophobic core.

27.4 REPORTER GENES IN PRECLINICAL IMAGING

Reporter genes have been used for several decades to study intracellular molecular events in vitro. However, it is much more recent that this approach was adopted by the molecular imaging community to visualize intracellular targets by using extracellular probes. This indirect imaging strategy involves the use of a reporter transgene placed under the control of upstream promoter elements and specific probes to produce an image that reflects reporter gene expression. The reporter transgene usually encodes for an enzyme (e.g., HSV1-tk) that selectively metabolizes the imaging probe and results in its entrapment and accumulation in the transduced cell. Alternatively, a reporter gene can

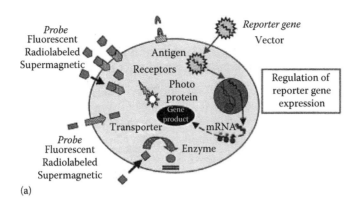

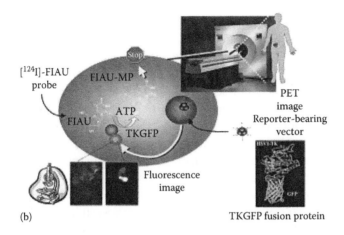

FIGURE 27.7 Schematic of different reporter genes and gene products and dual modality imaging of HSV1-tk/GFP reporter gene expression. Panel (a) illustrates that different reporter genes have been developed and the reporter gene products include enzymes (HSV1-tk, HSV1-sr39tk, and firefly and Renilla luciferases), transporters (hNIS), receptors (hSSTR2, hD2R), cell surface antigens and proteins, and fluorescent proteins (eGFP, mRFP). Different reporter probes corresponding to specific reporter genes have also been developed and used for nuclear (PET), magnetic resonance (MR), and optical (fluorescence and biolumi-nescence) imaging as described and referenced in the text. Panel (b) illustrates the steps involved for dual modality imaging of HSV1-tk/GFP reporter gene expression. (Reprinted from Blasberg, R.G., *Nucl. Med. Biol.*, 30(8), 879, 2003. With permission.)

encode for an extracellular or intracellular receptor that irreversibly binds or transports the imaging probe. The promoter element can be either constitutive or "induced" following the activation by specific endogenous transcription factors (e.g., hypoxia responsive element). Finally, three imaging modalities or any combination of them are used to image these reporter gene constructs: optical, MRI, and nuclear medicine (PET, SPECT) (Blasberg 2003; Gilad et al. 2007; Kang and Chung 2008). An example for each modality will be discussed briefly in the following and a diagram explaining reporter gene imaging is presented in Figure 27.7.

PET imaging has been used initially to study reporter gene expression. To date, three main application categories can be distinguished: enzyme type (e.g., HSV1-tk), transporter type (e.g., sodium/iodide symporter [NIS]), and receptor type (e.g., dopamine, norepinephrine, and somatostatine) (Kang and Chung 2008). The specificity of the method is achieved by the retention of the reporter probe, that is, radiotracer, in the tissue expressing the reporter gene. This condition can be met by (1) the conversion of a positron-labeled enzyme substrate to a "trapped" metabolic product or (2) the binding of a radiolabeled ligand to its receptor. The herpes simplex virus 1 thymidine kinase reporter gene (HSV1-tk) has been explored extensively by several groups and it has the advantage of relaxed substrate specificity when compared to its mammalian form (Tjuvajev et al. 1996; Haberkorn et al. 1997; Gambhir et al. 1998). In HSV1-tk-transfected cells, the *HSV1-tk* gene is transcribed to HSV1-tk mRNA and translated to TK, which then phosphorylates its substrate. HSV1-tk can phosphorylate acycloguanosine and uracil derivatives and substrates such as ^{14}C-, ^{131}I-, or ^{124}I-labeled 5-iodo-2′-fluoro-2′-deoxy-1-β-D-arabinofuranosyl-5′-iodouracil (FIAU), 8-[^{18}F]fluoroganciclovir (FGCV), 9-[4-[^{18}F]fluoro-3-(hydroxymethyl)butyl]guanine (^{18}F-FHBG), and 9-[3-[^{18}F]fluoro-1-(hydroxy-2-propoxy)methyl]guanine have been investigated (Gambhir et al. 1999). Once these substrates are phosphorylated, they can no longer pass the cell membrane and thus remain trapped within the target cells. Moreover, because the amount of radioactivity reflects the levels of gene expression, quantitative in vivo assays can be performed. The same principle can be applied for the other reporter gene categories for PET but instead of being an enzyme substrate, the imaging probe will bind to a receptor or transporter instead. Apart from PET, applications for SPECT have been introduced as well as they usually are easier and cheaper to synthesize. Examples for SPECT include ^{131}I-labeled metaiodobenzylguanidine (^{131}I-MIBG), ^{111}In-octreotide, and ^{99m}Tc-pertechnetate, which target norepinephrine transporter, somatostatin receptor, and NIS, respectively (Rogers et al. 1999; Haberkorn 2001; Altmann et al. 2003).

Two major approaches are pursued for optical imaging: bioluminescence reporter gene imaging using a construct encoding for Luciferase from the firefly *Photinus pyralis* (FLuc) and fluorescence imaging constructs using green fluorescent protein (GFP) (Misteli and Spector 1997; Gilad et al. 2007; Close et al. 2011). Bioluminescence imaging of FLuc requires the injection, usually intraperitoneally, of D-luciferin. The subsequent enzymatic turnover produces light, which allows the readout and semi-quantitation of the reporter gene activity. Luciferase is considered as an excellent marker for gene expression because of its lack of posttranslational modifications and a relatively short in vivo half-life of approximately 3 h. An example of such an approach is presented by Qayum et al. who used a luciferase reporter construct driven by three hypoxia-responsive element-binding sequences for hypoxia-inducible factor. This provided them with a quantifiable optical read-out of hypoxia. Following in vitro experiments, they used the same cell line in vivo to generate tumors subcutaneously in mice. Treatment of these tumors with signaling inhibitors of the RAS-PI3K-AKT pathway resulted in a decreased bioluminescence signal when compared to untreated animals, indicating a reduction in tumor hypoxia (Qayum et al. 2009). In contrast to bioluminescence imaging, animal fluorescence imaging is hampered by low signal-to-background ratios and autofluorescence (Troy et al. 2004). As a result, GFP fluorescence has generally been visualized in organs such as the lung or liver only during postmortem examinations. In the past years, a number of groups have developed several near-infrared fluorescent proteins that avoid overlapping emissions from tissue or organic compounds and enable real-time imaging to take place without interference from autofluorescence events. An interesting approach, taking advantage of both bioluminescence and fluorescence at the same time, was presented by Iglesias and Costoya (2009). They developed a genetically encoded biosensor that is induced by the presence of hypoxia. It comprised a regulatory module that contains a transcriptional enhancer able to bind the alpha subunit of the HIF-1 transcription factor, and a dual tracer formed by a fusion protein of a near-infrared fluorophore (mCherry) and FLuc. By fusing a fluorescent to a bioluminescent protein, they obtained a bioluminescence resonance energy transfer phenomenon, turning this fusion protein into a new class of hypoxia-sensing genetically encoded biosensor.

Efforts toward the development of MRI reporter genes have been made for at least 15 years. Nevertheless, the field is still in its early developmental stage, a major factor being the inherent low sensitivity of MRI. Indeed, in addition to the general requirements for a reporter gene, a suitable MR reporter gene needs to generate a translated protein that can produce detectable contrast with a sufficient signal-to-noise ratio. Gilad et al. classified the MRI reporter genes as (1) enzyme based (e.g., β-galactosidase), (2) metalloprotein based (e.g., transferrin, tyrosinase, ferritin) and (3) based on engineered surface reporters as ligands for targeted contrast agents (e.g., antigen–antibody systems) (Gilad et al. 2007). Of these systems, the β-galactosidase, encoded by the lacZ gene, and the iron-binding metalloproteins are the most explored ones. An example of the former is presented by Moats et al. who developed a gadolinium-based substrate for the β-galactosidase reporter (Moats et al. 1997). Following the activation of the lacZ gene, β-galactosidase is transcribed, which will cleave the galactose group that hinders the inner sphere relaxation enhancement effect of the paramagnetic chelated gadolinium. This, in turn, will result in increased imaging contrast. Tyrosinase and ferritin are examples of iron-binding proteins and of the latter approach. Meir et al. used ferritin as an endogenous MRI reporter protein, which would directly change the MR signal, in analogy to fluorescent proteins in optical imaging, without the need to administer additional contrast materials (Cohen et al. 2005). The overexpression of ferritin resulted in a redistribution of iron and in a significant increase in net intracellular iron content. These phenomena caused a change in relaxation rate that was significant and with high enough sensitivity for detection by MRI.

Today, multimodality imaging approaches are gaining more and more interest. As a result, dual and triple reporter genes that enable us to combine PET, optical, and MRI are being developed. This has the huge advantage that more than one intracellular process can be monitored at the same time (Blasberg 2003; Gilad et al. 2007). An example of such dual reporter is published by Doubrovin et al. (Doubrovin et al. 2001; Blasberg 2003). They showed that p53-dependent gene expression can be imaged in vivo with PET and by in situ fluorescence. For this purpose, a retroviral vector was generated by placing the HSV1-tk and eGFP fusion gene (HSV1-tk-eGFP, a dual-reporter gene) under the control of a p53-specific response element. Tumor-bearing mice, inoculated with the transfected cell lines, were treated with N,N'-bis(2-chloroethyl)-N-nitrosourea (BCNU), etoposide, or UV radiation. Following PET imaging using ^{18}F-FIAU and GFP fluorescence imaging, the upregulation of p53 transcriptional activity was demonstrated and correlated with the expression of p53-dependent downstream genes, including p21. A trifusion multimodality reporter harboring a bioluminescence reporter gene (synthetic Renilla luciferase), a fluorescence reporter gene (monomeric red fluorescence protein), and a PET reporter gene (truncated version of HSV1-tk) was described by Ray et al. (2004). Not only does it combine the advantages of each single imaging modality, but it also allows moving between imaging technologies without having to use a different reporter gene for each application. Using this approach, they were able to follow metastatic tumor growth in living mice. Indeed, the bioluminescence signal is easily detectable from superficial metastases from any region of the body. On the other hand, PET revealed metastases from deep inside the body, but signals from metastases in the abdomen/pelvis are somewhat obscured by the nonspecific signal in the gastrointestinal and urinary tract due to tracer clearance. Although autofluorescence of biological molecules limits the detection of metastases by in vivo fluorescence imaging, it was a useful tool to visualize metastases in sacrificed animals.

27.5 THERANOSTICS

Theranostics aim to combine imaging and therapy, which will allow physicians to monitor therapy in real time. As a result, dosing and drug regimens can be adjusted on a patient-to-patient basis avoiding both overtreatment that would result in harmful side effects and undertreatment that would lead to incomplete disease eradication. The development of theranostics has evolved hand in hand with the design of nanoparticles and, although such nanoparticles are more complex to develop, their advantages are well worth the effort. Indeed, they can be engineered to (1) overcome the biological barriers, (2) accumulate in target cells more specifically, (3) reach higher concentrations of contrast agent at the target site, (4) improve endosomal escape and ensure the desired intracellular trafficking, (5) combine both therapy and imaging, (6) avoid expulsion by multidrug resistance proteins and efflux pumps, and (7) contain different reporter groups to either visualize several targets at the same time or use different imaging modalities. Thanks to their versatility, lipid-based nanoparticles are most widely used to combine imaging and therapy in one single compound. Moreover, the encapsulation of the imaging and/or therapeutic probe has huge advantages as it will shield them from opsonization and enzymatic degradation.

Many theranostic nanoparticle formulations have been developed for extracellular targets (Cole et al. 2011; Yu et al. 2012). Fortunately, intracellular targeting is catching up, especially with regards to the intracellular delivery of DNA and small interfering RNA. An example of such an application is presented by Kievit et al. (2010). They investigated targeted gene delivery to C6 glioma cells in a xenograft mouse model using chlorotoxin (CTX)-labeled nanoparticles. The developed nanovector consisted of an iron oxide nanoparticle core of which the superparamagnetic property provides MRI contrast for imaging in vivo. The core was coated with a copolymer of chitosan and PEG to avoid the RES and bypass the blood–brain barrier and with polyethylenimine (PEI) for endosomal escape through the proton sponge effect. The DNA was loaded into nanoparticle through the electrostatic interaction between negatively charged DNA and positively charged coating and contained a GFP-encoding sequence to monitor gene expression using optical fluorescence imaging. Finally, CTX was conjugated through a PEG linker to ensure that CTX was free to interact with the target cells and thus provide the target specificity of the nanoparticle. Kumar et al. synthesized a breast tumor-targeted nanodrug designed to specifically shuttle siRNA to human breast cancer while simultaneously allowing for the noninvasive monitoring of the siRNA delivery process (Kumar et al. 2010). This theranostic consisted of SPIONs for MRI monitoring, Cy5.5 fluorescence dye for near-infrared optical imaging, and siRNA to target the tumor-specific antiapoptotic gene BIRC5. Specific breast cancer tumor targeting was achieved by decorating the nanoparticle with EPPT (Glu-Pro-Pro-Thr) synthetic peptides that bind to the overexpressed uMUC-1 antigen.

Finally, radionuclide therapies using Auger-electron emitters can be considered as another class of theranostics. The advantage of Auger electrons–emitting isotopes is that a single isotope can be used for both imaging and therapy (e.g., ^{111}In, ^{123}I, ^{125}I, and ^{99m}Tc). However, their track length is very short (in nanometer range), which means that all the energy from the Auger electrons is deposited in a very small sphere around the decaying radionuclide. Hence, for them to be effective as a therapy tool, they need to be in close proximity of the cell's DNA. A comprehensive review on this topic was published by Cornelissen and Vallis (2010), and recently, the same group exploited this approach even further (Cornelissen et al. 2012). The same ^{111}In-labeled immunoconjugate construct as mentioned in Section 27.2.5 of this chapter was used but at a higher specific activity (6 MBq/μg). Following intravenous injection, ^{111}In-DTPA-anti-γH2AX-Tat is translocated to the cell nucleus where it binds to γH2AX and so is retained at sites of DNA damage. They demonstrated that, besides a SPECT imaging agent, the construct could also be used to deliver a significant amount of therapeutic radioisotope directly to sites of preexisting DNA strand breaks. In other words, the Auger electron–emitting radionuclide that accumulates at the sites of DNA double strand breaks, through association with γH2AX, increased the DNA damage in cells exposed to a genotoxic insult. Using this approach, simple double strand breaks were converted into more complex damage, reducing the probability of successful repair of the DNA double strand break and thus increasing the likelihood of cell kill.

27.6 SUMMARY

Imaging intracellular targets presents numerous challenges and many barriers, both physiological and biological, that need to be overcome. This chapter described these challenges and offered approaches to overcome these difficulties. In short, multifunctional proteins have been constructed for targeting contrast agents to the cytoplasm or nucleus of specific cells. Such proteins will have moieties to (1) avoid nonspecific organ uptake, RES scavenging and metabolization by the excretion organs, (2) target-specific cells, (3) ensure intracellular translocation, (4) facilitate endosomal escape, (5) provide imaging signal and, potentially, (6) to traffic intracellular signals to the cell nucleus. Additionally, applications using preclinical imaging of intracellular targets are discussed, including cell labeling and tracking, reporter genes, and theranostics.

REFERENCES

Aihara, H. and J. Miyazaki. (1998). Gene transfer into muscle by electroporation in vivo. *Nat. Biotechnol.* **16**(9): 867–870.

Altmann, A., M. Kissel et al. (2003). Increased MIBG uptake after transfer of the human norepinephrine transporter gene in rat hepatoma. *J. Nucl. Med.* **44**(6): 973–980.

Arbab, A. S., G. T. Yocum et al. (2004). Efficient magnetic cell labeling with protamine sulfate complexed to ferumoxides for cellular MRI. *Blood* **104**(4): 1217–1223.

Ballinger, J. R. (2001). Imaging hypoxia in tumors. *Semin. Nucl. Med.* **31**(4): 321–329.

Barbet, J., P. Peltier et al. (1998). Radioimmunodetection of medullary thyroid carcinoma using indium-111 bivalent hapten and anti-CEA x anti-DTPA-indium bispecific antibody. *J. Nucl. Med.* **39**(7): 1172–1178.

Bekeredjian, R., S. Chen et al. (2003). Ultrasound-targeted microbubble destruction can repeatedly direct highly specific plasmid expression to the heart. *Circulation* **108**(8): 1022–1026.

Belting, M., S. Sandgren et al. (2005). Nuclear delivery of macromolecules: Barriers and carriers. *Adv. Drug Deliv. Rev.* **57**(4): 505–527.

Blasberg, R. G. (2003). In vivo molecular-genetic imaging: Multi-modality nuclear and optical combinations. *Nucl. Med. Biol.* **30**(8): 879–888.

Blasberg, R. G. (2003). Molecular imaging and cancer. *Mol. Cancer Ther.* **2**(3): 335–343.

Boerman, O. C., P. Laverman et al. (2000). Radiolabeled liposomes for scintigraphic imaging. *Prog. Lipid Res.* **39**(5): 461–475.

Boutin, C., S. Diestel et al. (2008). Efficient in vivo electroporation of the postnatal rodent forebrain. *PLoS One* **3**(4): e1883.

Bulte, J. W. (2009). In vivo MRI cell tracking: Clinical studies. *AJR Am. J. Roentgenol.* **193**(2): 314–325.

Bulte, J. W., L. D. Ma et al. (1993). Selective MR imaging of labeled human peripheral blood mononuclear cells by liposome mediated incorporation of dextran-magnetite particles. *Magn. Reson. Med.* **29**(1): 32–37.

Cao, Y. and L. Lam. (2003). Bispecific antibody conjugates in therapeutics. *Adv. Drug. Deliv. Rev.* **55**(2): 171–197.

Capello, A., E. P. Krenning et al. (2004). Increased cell death after therapy with an Arg-Gly-Asp-linked somatostatin analog. *J. Nucl. Med.* **45**(10): 1716–1720.

Choi, Y. S., J. Y. Lee et al. (2011). Cell penetrating peptides for tumor targeting. *Curr. Pharm. Biotechnol.* **12**(8): 1166–1182.

Chouly, C., D. Pouliquen et al. (1996). Development of superparamagnetic nanoparticles for MRI: Effect of particle size, charge and surface nature on biodistribution. *J. Microencapsul.* **13**(3): 245–255.

Chun, H. J., J. Narula et al. (2008). Intracellular and extracellular targets of molecular imaging in the myocardium. *Nat. Clin. Pract. Cardiovasc. Med.* **5**(Suppl. 2): S33–S41.

Close, D. M., T. Xu et al. (2011). In vivo bioluminescent imaging (BLI): Noninvasive visualization and interrogation of biological processes in living animals. *Sensors (Basel)* **11**(1): 180–206.

Cohen, B., H. Dafni et al. (2005). Ferritin as an endogenous MRI reporter for noninvasive imaging of gene expression in C6 glioma tumors. *Neoplasia* **7**(2): 109–117.

Cole, A. J., V. C. Yang et al. (2011). Cancer theranostics: The rise of targeted magnetic nanoparticles. *Trends Biotechnol.* **29**(7): 323–332.

Condeelis, J. and R. Weissleder. (2010). In vivo imaging in cancer. *Cold Spring Harb. Perspect. Biol.* **2**(12): a003848.

Conti, E. and E. Izaurralde. (2001). Nucleocytoplasmic transport enters the atomic age. *Curr. Opin. Cell Biol.* **13**(3): 310–319.

Cornelissen, B., S. Darbar et al. (2012). Amplification of DNA damage by a gammaH2AX-targeted radiopharmaceutical. *Nucl. Med. Biol.* **39**(8): 1142–1151.

Cornelissen, B., V. Kersemans et al. (2009a). 111In-labeled immunoconjugates (ICs) bispecific for the epidermal growth factor receptor (EGFR) and cyclin-dependent kinase inhibitor, p27Kip1. *Cancer Biother. Radiopharm.* **24**(2): 163–173.

Cornelissen, B., V. Kersemans et al. (2009b). In vivo monitoring of intranuclear p27(kip1) protein expression in breast cancer cells during trastuzumab (Herceptin) therapy. *Nucl. Med. Biol.* **36**(7): 811–819.

Cornelissen, B., V. Kersemans et al. (2011). Imaging DNA damage in vivo using gammaH2AX-targeted immunoconjugates. *Cancer Res.* **71**(13): 4539–4549.

Cornelissen, B. and K. A. Vallis. (2010). Targeting the nucleus: An overview of Auger-electron radionuclide therapy. *Curr. Drug Discov. Technol.* **7**(4): 263–279.

De Vries, I. J., D. J. Krooshoop et al. (2003). Effective migration of antigen-pulsed dendritic cells to lymph nodes in melanoma patients is determined by their maturation state. *Cancer Res.* **63**(1): 12–17.

Decuzzi, P., S. Lee et al. (2005). A theoretical model for the margination of particles within blood vessels. *Ann. Biomed. Eng.* **33**(2): 179–190.

Delehanty, J. B., H. Mattoussi et al. (2009). Delivering quantum dots into cells: Strategies, progress and remaining issues. *Anal. Bioanal. Chem.* **393**(4): 1091–1105.

Dellian, M., F. Yuan et al. (2000). Vascular permeability in a human tumour xenograft: Molecular charge dependence. *Br. J. Cancer* **82**(9): 1513–1518.

Dobrovolskaia, M. A., P. Aggarwal et al. (2008). Preclinical studies to understand nanoparticle interaction with the immune system and its potential effects on nanoparticle biodistribution. *Mol. Pharm.* **5**(4): 487–495.

Dong, H., J. Lei et al. (2012). Target-cell-specific delivery, imaging, and detection of intracellular microRNA with a multifunctional SnO_2 nanoprobe. *Angew. Chem. Int. Ed. Engl.* **51**(19): 4607–4612.

Doubrovin, M., V. Ponomarev et al. (2001). Imaging transcriptional regulation of p53-dependent genes with positron emission tomography in vivo. *Proc. Natl. Acad. Sci. USA* **98**(16): 9300–9305.

Eggert, A. A., M. W. Schreurs et al. (1999). Biodistribution and vaccine efficiency of murine dendritic cells are dependent on the route of administration. *Cancer Res.* **59**(14): 3340–3345.

Ferrari, M. (2005). Cancer nanotechnology: Opportunities and challenges. *Nat. Rev. Cancer* **5**(3): 161–171.

Flavell, D. J., S. Cooper et al. (1992). Effectiveness of combinations of bispecific antibodies for delivering saporin to human acute T-cell lymphoblastic leukaemia cell lines via CD7 and CD38 as cellular target molecules. *Br. J. Cancer* **65**(4): 545–551.

Frangioni, J. V. and R. J. Hajjar. (2004). In vivo tracking of stem cells for clinical trials in cardiovascular disease. *Circulation* **110**(21): 3378–3383.

Gallazzi, F., Y. Wang et al. (2003). Synthesis of radiometal-labeled and fluorescent cell-permeating Peptide—PNA conjugates for targeting the bcl-2 proto-oncogene. *Bioconjug. Chem.* **14**(6): 1083–1095.

Gambhir, S. S., J. R. Barrio et al. (1998). Imaging of adenoviral-directed herpes simplex virus type 1 thymidine kinase reporter gene expression in mice with radiolabeled ganciclovir. *J. Nucl. Med.* **39**(11): 2003–2011.

Gambhir, S. S., J. R. Barrio et al. (1999). Imaging adenoviral-directed reporter gene expression in living animals with positron emission tomography. *Proc. Natl. Acad. Sci. USA* **96**(5): 2333–2338.

Gilad, A. A., P. T. Winnard, Jr. et al. (2007). Developing MR reporter genes: Promises and pitfalls. *NMR Biomed.* **20**(3): 275–290.

Gorlich, D. and U. Kutay. (1999). Transport between the cell nucleus and the cytoplasm. *Annu. Rev. Cell Dev. Biol.* **15**: 607–660.

Govindan, S. V. and D. M. Goldenberg. (2012). Designing immunoconjugates for cancer therapy. *Expert Opin. Biol. Ther.* **12**(7): 873–890.

Grunbaum, Z., S. J. Freauff et al. (1987). Synthesis and characterization of congeners of misonidazole for imaging hypoxia. *J. Nucl. Med.* **28**(1): 68–75.

Haberkorn, U. (2001). Gene therapy with sodium/iodide symporter in hepatocarcinoma. *Exp. Clin. Endocrinol. Diabetes* **109**(1): 60–62.

Haberkorn, U., A. Altmann et al. (1997). Multitracer studies during gene therapy of hepatoma cells with herpes simplex virus thymidine kinase and ganciclovir. *J. Nucl. Med.* **38**(7): 1048–1054.

Harris, J. M. and R. B. Chess. (2003). Effect of pegylation on pharmaceuticals. *Nat. Rev. Drug Discov.* **2**(3): 214–221.

Holm, D. A. (2009). *Electroporation Facilitated Intracellular Delivery of Contrast Agent to Probe Sub-Cellular Metabolite Compartmentalization in Vivo.* ISMRM, Honolulu, HI.

Howes, M. T., S. Mayor et al. (2010). Molecules, mechanisms, and cellular roles of clathrin-independent endocytosis. *Curr. Opin. Cell Biol.* **22**(4): 519–527.

Hu, M., P. Chen et al. (2007). 123I-labeled HIV-1 tat peptide radioimmunoconjugates are imported into the nucleus of human breast cancer cells and functionally interact in vitro and in vivo with the cyclin-dependent kinase inhibitor, p21(WAF-1/Cip-1). *Eur. J. Nucl. Med. Mol. Imaging* **34**(3): 368–377.

Iglesias, P. and J. A. Costoya. (2009). A novel BRET-based genetically encoded biosensor for functional imaging of hypoxia. *Biosens. Bioelectron.* **24**(10): 3126–3130.

Jacobs, A. H., B. Tavitian et al. (2012). Noninvasive molecular imaging of neuroinflammation. *J. Cereb. Blood Flow Metab.* **32**(7): 1393–1415.

Jaffer, F. A., P. Libby et al. (2007). Molecular imaging of cardiovascular disease. *Circulation* **116**(9): 1052–1061.

Jiang, T., E. S. Olson et al. (2004). Tumor imaging by means of proteolytic activation of cell-penetrating peptides. *Proc. Natl. Acad. Sci USA* **101**(51): 17867–17872.

Johnson, J. R., B. Kocher et al. (2012). Caspase-activated cell-penetrating peptides reveal temporal coupling between endosomal release and apoptosis in an RGC-5 cell model. *Bioconjug. Chem.* **23**(9): 1783–1793.

Jung, S. H., K. Na et al. (2012). Gd(inverted question mark)-DOTA-modified sonosensitive liposomes for ultrasound-triggered release and MR imaging. *Nanoscale Res. Lett.* **7**(1): 462.

Kalderon, D., B. L. Roberts et al. (1984). A short amino acid sequence able to specify nuclear location. *Cell* **39**(3 Pt 2): 499–509.

Kang, J. H. and J. K. Chung. (2008). Molecular-genetic imaging based on reporter gene expression. *J. Nucl. Med.* **49**(Suppl. 2): 164S–179S.

Kersemans, V. and B. Cornelissen. (2010). Targeting the tumour: Cell penetrating peptides for molecular imaging and radiotherapy. *Pharmaceuticals* **3**: 600–620.

Kersemans, V., K. Kersemans et al. (2008). Cell penetrating peptides for in vivo molecular imaging applications. *Curr. Pharm. Des.* **14**(24): 2415–2447.

Kievit, F. M., O. Veiseh et al. (2010). Chlorotoxin labeled magnetic nanovectors for targeted gene delivery to glioma. *ACS Nano* **4**(8): 4587–4594.

Kievit, F. M. and M. Zhang. (2011). Cancer nanotheranostics: Improving imaging and therapy by targeted delivery across biological barriers. *Adv. Mater.* **23**(36): H217–H247.

Klibanov, A., K. Maruyama et al. (1991). Activity of amphipathic poly(ethylene glycol) 5000 to prolong the circulation time of liposomes depends on the liposome size and is unfavorable for immunoliposome binding to target. *Biochim. Biophys. Acta* **1062**(2): 142–148.

Kresse, M., S. Wagner et al. (1998). Targeting of ultrasmall superparamagnetic iron oxide (USPIO) particles to tumor cells in vivo by using transferrin receptor pathways. *Magn. Reson. Med.* **40**(2): 236–242.

Krohn, K. A., J. M. Link et al. (2008). Molecular imaging of hypoxia. *J. Nucl. Med.* **49**(Suppl. 2): 129S–148S.

Kumar, M., M. Yigit et al. (2010). Image-guided breast tumor therapy using a small interfering RNA nanodrug. *Cancer Res.* **70**(19): 7553–7561.

Kumari, S., S. Mg et al. (2010). Endocytosis unplugged: Multiple ways to enter the cell. *Cell Res.* **20**(3): 256–275.

Lee, J. H., Y. M. Huh et al. (2007). Artificially engineered magnetic nanoparticles for ultra-sensitive molecular imaging. *Nat. Med.* **13**(1): 95–99.

Leroy-Willig, A., M. F. Bureau et al. (2005). In vivo NMR imaging evaluation of efficiency and toxicity of gene electrotransfer in rat muscle. *Gene Ther.* **12**(19): 1434–1443.

Liu, Y., J. Yan et al. (2012). Can ultrasound enable efficient intracellular uptake of molecules? A retrospective literature review and analysis. *Ultrasound Med. Biol.* **38**(5): 876–888.

Madani, F., S. Lindberg et al. (2011). Mechanisms of cellular uptake of cell-penetrating peptides. *J. Biophys.* **2011**: 414729.

Massoud, T. F. and S. S. Gambhir. (2003). Molecular imaging in living subjects: Seeing fundamental biological processes in a new light. *Genes Dev.* **17**(5): 545–580.

Mayor, S. and R. E. Pagano. (2007). Pathways of clathrin-independent endocytosis. *Nat. Rev. Mol. Cell Biol.* **8**(8): 603–612.

McCormack, D. R., K. Bhattacharyya et al. (2011). Enhanced photoacoustic detection of melanoma cells using gold nanoparticles. *Lasers Surg. Med.* **43**(4): 333–338.

Mercer, J., M. Schelhaas et al. (2010). Virus entry by endocytosis. *Annu. Rev. Biochem.* **79**: 803–833.

Miao, Q., X. Bai et al. (2012). Intracellular self-assembly of nanoparticles for enhancing cell uptake. *Chem. Commun. (Camb.)* **48**(78): 9738–9740.

Mir, L. M., M. F. Bureau et al. (1999). High-efficiency gene transfer into skeletal muscle mediated by electric pulses. *Proc. Natl. Acad. Sci. USA* **96**(8): 4262–4267.

Misteli, T. and D. L. Spector. (1997). Applications of the green fluorescent protein in cell biology and biotechnology. *Nat. Biotechnol.* **15**(10): 961–964.

Moats, R. A., S. E. Fraser et al. (1997). A "smart" magnetic resonance imaging agent that reports on specific enzymatic activity. *Angew. Chem. Int. Ed. Engl.* **36**(7): 726–728.

Mody, V. V., M. I. Nounou et al. (2009). Novel nanomedicine-based MRI contrast agents for gynecological malignancies. *Adv. Drug Deliv. Rev.* **61**(10): 795–807.

Moore, A., P. Z. Sun et al. (2002). MRI of insulitis in autoimmune diabetes. *Magn. Reson. Med.* **47**(4): 751–758.

Morse, D. L. and R. J. Gillies. (2010). Molecular imaging and targeted therapies. *Biochem. Pharmacol.* **80**(5): 731–738.

Nunn, A., K. Linder et al. (1995). Nitroimidazoles and imaging hypoxia. *Eur. J. Nucl. Med.* **22**(3): 265–280.

Ogawa, M., N. Kosaka et al. (2009). In vivo molecular imaging of cancer with a quenching near-infrared fluorescent probe using conjugates of monoclonal antibodies and indocyanine green. *Cancer Res.* **69**(4): 1268–1272.

Oh, S., B. J. Stish et al. (2009). A novel reduced immunogenicity bispecific targeted toxin simultaneously recognizing human epidermal growth factor and interleukin-4 receptors in a mouse model of metastatic breast carcinoma. *Clin. Cancer Res.* **15**(19): 6137–6147.

Panizzo, R. A., P. G. Kyrtatos et al. (2009). In vivo magnetic resonance imaging of endogenous neuroblasts labelled with a ferumoxide-polycation complex. *Neuroimage* **44**(4): 1239–1246.

Pante, N. and M. Kann. (2002). Nuclear pore complex is able to transport macromolecules with diameters of about 39 nm. *Mol. Biol. Cell* **13**(2): 425–434.

Park, J. W. (2002). Liposome-based drug delivery in breast cancer treatment. *Breast Cancer Res.* **4**(3): 95–99.

Philibert, J. (2005). One and a half century of diffusion: Fick, Einstein, before and beyond. *Diff. Fundam.* **2**: 1.1–1.10.

Pouton, C. W., K. M. Wagstaff et al. (2007). Targeted delivery to the nucleus. *Adv. Drug Deliv. Rev.* **59**(8): 698.

Pysz, M. A., S. S. Gambhir et al. (2010). Molecular imaging: Current status and emerging strategies. *Clin. Radiol.* **65**(7): 500–516.

Qayum, N., R. J. Muschel et al. (2009). Tumor vascular changes mediated by inhibition of oncogenic signaling. *Cancer Res.* **69**(15): 6347–6354.

Rajendran, J. G. and K. A. Krohn. (2005). Imaging hypoxia and angiogenesis in tumors. *Radiol. Clin. North Am.* **43**(1): 169–187.

Rasey, J., G. Martin et al. (1999). Quantifying hypoxia with radiolabeled fluoromisonidazole: Pre-clinical and clinical studies. In H.-J. Machulla (ed.), *The Imaging of Hypoxia*. Kluwer Academic Publishers, Dordrecht, the Netherlands, pp. 85–117.

Ray, P., A. De et al. (2004). Imaging tri-fusion multimodality reporter gene expression in living subjects. *Cancer Res.* **64**(4): 1323–1330.

Rayne, F., S. Debaisieux et al. (2010). Phosphatidylinositol-(4,5)-bisphosphate enables efficient secretion of HIV-1 Tat by infected T-cells. *EMBO J.* **29**(8): 1348–1362.

Ricles, L. M., S. Y. Nam et al. (2011). Function of mesenchymal stem cells following loading of gold nanotracers. *Int. J. Nanomed.* **6**: 407–416.

Rogers, B. E., S. F. McLean et al. (1999). In vivo localization of [(111)In]-DTPA-D-Phe1-octreotide to human ovarian tumor xenografts induced to express the somatostatin receptor subtype 2 using an adenoviral vector. *Clin. Cancer Res.* **5**(2): 383–393.

Rogers, W. J., C. H. Meyer et al. (2006). Technology insight: In vivo cell tracking by use of MRI. *Nat. Clin. Pract. Cardiovasc. Med.* **3**(10): 554–562.

Sandvig, K., M. L. Torgersen et al. (2008). Clathrin-independent endocytosis: From nonexisting to an extreme degree of complexity. *Histochem. Cell Biol.* **129**(3): 267–276.

Sandvig, K. and B. van Deurs. (2005). Delivery into cells: Lessons learned from plant and bacterial toxins. *Gene Therapy* **12**(11): 865–872.

Schwarze, S. R., A. Ho et al. (1999). In vivo protein transduction: Delivery of a biologically active protein into the mouse. *Science* **285**(5433): 1569–1572.

Shalaev, E. Y. and P. L. Steponkus. (1999). Phase diagram of 1,2-dioleoylphosphatidylethanolamine (DOPE):water system at subzero temperatures and at low water contents. *Biochim. Biophys. Acta* **1419**(2): 229–247.

Sharkey, R. M., T. M. Cardillo et al. (2005). Signal amplification in molecular imaging by pretargeting a multivalent, bispecific antibody. *Nat. Med.* **11**(11): 1250–1255.

Skotland, T. (2012). Molecular imaging: Challenges of bringing imaging of intracellular targets into common clinical use. *Contrast Media Mol. Imaging* **7**(1): 1–6.

Soenen, S. J. and M. De Cuyper. (2010). Assessing iron oxide nanoparticle toxicity in vitro: Current status and future prospects. *Nanomedicine (London)* **5**(8): 1261–1275.

Stavru, F., C. Archambaud et al. (2011). Cell biology and immunology of *Listeria monocytogenes* infections: Novel insights. *Immunol. Rev.* **240**(1): 160–184.

Sumner, J. P., E. M. Shapiro et al. (2009). In vivo labeling of adult neural progenitors for MRI with micron sized particles of iron oxide: Quantification of labeled cell phenotype. *Neuroimage* **44**(3): 671–678.

Taylor, A., K. M. Wilson et al. (2012). Long-term tracking of cells using inorganic nanoparticles as contrast agents: Are we there yet? *Chem. Soc. Rev.* **41**(7): 2707–2717.

Tian, X., A. Chakrabarti et al. (2005). External imaging of CCND1, MYC, and KRAS oncogene mRNAs with tumor-targeted radionuclide-PNA-peptide chimeras. *Ann. NY Acad. Sci.* **1059**: 106–144.

Tjuvajev, J. G., R. Finn et al. (1996). Noninvasive imaging of herpes virus thymidine kinase gene transfer and expression: A potential method for monitoring clinical gene therapy. *Cancer Res.* **56**(18): 4087–4095.

Torchilin, V. P. (2007). Targeted pharmaceutical nanocarriers for cancer therapy and imaging. *AAPS J.* **9**(2): E128–E147.

Torchilin, V. P., F. Zhou et al. (1993). pH-sensitive liposomes. *J. Liposome Res.* **3**: 201–255.

Troy, T., D. Jekic-McMullen et al. (2004). Quantitative comparison of the sensitivity of detection of fluorescent and bioluminescent reporters in animal models. *Mol. Imaging* **3**(1): 9–23.

van der Aa, M. A., E. Mastrobattista et al. (2006). The nuclear pore complex: The gateway to successful nonviral gene delivery. *Pharm. Res.* **23**(3): 447–459.

Varkouhi, A. K., M. Scholte et al. (2011). Endosomal escape pathways for delivery of biologicals. *J. Control. Release* **151**(3): 220–228.

Weissleder, R. (1999). Molecular imaging: Exploring the next frontier. *Radiology* **212**(3): 609–614.

Weissleder, R. and M. J. Pittet. (2008). Imaging in the era of molecular oncology. *Nature* **452**(7187): 580–589.

Weissleder, R., C. H. Tung et al. (1999). In vivo imaging of tumors with protease-activated near-infrared fluorescent probes. *Nat. Biotechnol.* **17**(4): 375–378.

Wiart, M., N. Davoust et al. (2007). MRI monitoring of neuroinflammation in mouse focal ischemia. *Stroke* **38**(1): 131–137.

Wu, J., T. Akaike et al. (1998). Modulation of enhanced vascular permeability in tumors by a bradykinin antagonist, a cyclooxygenase inhibitor, and a nitric oxide scavenger. *Cancer Res.* **58**(1): 159–165.

Yessine, M. A. and J. C. Leroux. (2004). Membrane-destabilizing polyanions: Interaction with lipid bilayers and endosomal escape of biomacromolecules. *Adv. Drug Deliv. Rev.* **56**(7): 999–1021.

Yu, M. K., J. Park et al. (2012). Targeting strategies for multifunctional nanoparticles in cancer imaging and therapy. *Theranostics* **2**(1): 3–44.

Yuan, F., M. Dellian et al. (1995). Vascular permeability in a human tumor xenograft: Molecular size dependence and cutoff size. *Cancer Res.* **55**(17): 3752–3756.

Zhu, L., G. Niu et al. (2010). Preclinical molecular imaging of tumor angiogenesis. *Q. J. Nucl. Med. Mol. Imaging* **54**(3): 291–308.

Zhu, L. and V. P. Torchilin. (2013). Stimulus-responsive nanopreparations for tumor targeting. *Integr. Biol. (Camb.)* **5**(1): 96–107.

28

Imaging of Cell Trafficking and Cell Tissue Homing

Veerle Kersemans

28.1 INTRODUCTION TO IMAGING OF CELL TRAFFICKING AND CELL TISSUE HOMING

28.1.1 BACKGROUND

The idea of using our body's own resources to regenerate or replace diseased tissue or to fight disease is very captivating, and, as a result, the concept of cellular therapy has been around since 1667 when Jean-Baptiste Denys documented the first blood transfusion (Aarntzen et al. 2012). Today, the field of cellular therapy is thriving and offers hope for the treatment of cancer and cardiovascular and neurological diseases. Transplanted therapeutic cells have the potential to regenerate tissue, replace the function of lost cells, and correct aberrant processes (Aarntzen et al. 2012). Currently, nearly 26,000 clinical trials are listed worldwide, which involve some sort of cellular therapy (http://www.clinicaltrials.gov) (Srinivas et al. 2010). However, the underlying mechanism for treatment failure or success is unknown, which makes translation of preclinical studies into the clinic still challenging. One of the main problems is the difficulty of detecting, localizing, and examining the cells in vivo upon systemic readministration at both the cellular and molecular level in order to determine their ultimate fate, functionality, and differentiation. Additionally, safety and efficacy of the cellular therapy needs to be proven and, for this, one needs to be able to follow the cells in vivo. Indeed, one has to be certain that the transplanted cells end up alive in the target tissue and do not cause any toxicity.

These requirements, together with the noninvasive nature of imaging, have led to a growing interest in techniques that allow in vivo localization of transplanted cells. Indeed, imaging plays an important role in cellular therapeutics because of its clear advantages. It not only allows longitudinal investigations and immediate translation into human practice but also has the ability to provide specific information about cell numbers, their viability, and functionality (Srinivas et al. 2013). The concept of labeling cells has been around for nearly 40 years following the introduction of Indium-111 labeled platelets and leucocytes for the detection of abscesses and inflammation (Thakur et al. 1976, 1977).

Broadly speaking, the same principles as described in Chapter 27 for the detection of intracellular targets can be applied for cell labeling. The ideal imaging label should be inert and/or biocompatible, highly specific to target cells, and detected at the single (or few) cell level. At present, however, no single probe or label exists that meets these criteria in combination with a single imaging technique.

Two major categories for cell labeling can be distinguished, which will be discussed in this chapter: direct cell labeling, which does not involve genetic modification, and indirect cell labeling, which uses reporter genes. However, cell labeling techniques not only rely on cell-penetrating peptides and receptor-mediated uptake but also still use electroporation, diffusion, and transfection agents to a great extent. Since 1976, many direct and indirect cell labeling procedures have been developed for use in different imaging modalities. As with visualizing intracellular targets, the choice of labeling procedure and modality will depend on the cellular processes that need to be studied as well as the readouts that are most desirable.

No matter which strategy is used, the detectability of the cell-tracking agents must be tuned to the time frame of the study. For example, if the therapeutic action of the transplanted cells occurs within a short period of time, then it is of no use loading a cell with potentially harmful longer lasting compounds. Conversely, if long-term viability or long-term cell location is ambitioned, then a short-term label will not provide the complete picture of cell distribution.

28.1.2 IMAGING MODALITIES

Imaging can play a pivotal role in the widespread acceptance of cellular therapies, as it can offer longitudinal and kinetic data. Providing the availability of suitable contrast agents, long-term tissue function and survival can be studied and information on cell numbers, functionality, and localization can be obtained. The ideal imaging setup would bring together high sensitivity, high resolution, and low toxicity. Unfortunately, such a system does not exist (yet) and each imaging modality has its own

TABLE 28.1

Advantages and Disadvantages of Different Imaging Modalities with Respect to Cellular Imaging

Modality	Spectrum Used	Spatial Resolution	Temporal Resolution	Molecular Sensitivity	Tissue Penetration	Quantification Capability	Cost	Used Clinically to Track Cells?
SPECT	γ-rays	0.5–2 mm (preclinical) 7–15 mm (clinical)	Seconds to minutes	10^{-10}–10^{-11}	No limit	High	High	Yes
PET	Annihilation photons	1–4 mm (preclinical) 6–10 mm (clinical)	Seconds to minutes	10^{-11}–10^{-12}	No limit	High	Medium high	Yes
Ultrasound	High frequency sound waves	50–500 μm (preclinical) 0.1–1 mm (clinical)	Seconds to minutes	[a]	1–200 mm	Low	Low	No
MRI	Radiowaves	25–100 μm (preclinical) 0.2–1 mm (clinical)	Minutes to hours	10^{-3}–10^{-5}	No limit	Low ^{19}F-MRI: high	High	Yes
CT	X-rays	30–500 μm (preclinical) 0.5–1 mm (clinical)	Minutes	[a]	No limit	Low	Medium	No
Fluorescence	Visible to infrared light	3–10 mm	Seconds to minutes	10^{-9}–10^{-11}	1–20 mm	Medium	Low	No
Bioluminescence	Visible to infrared light	3–10 mm	Seconds to minutes	10^{-13}–10^{-16}	1–20 mm	Medium	Low	No

[a] Not well characterized.

advantages and disadvantages with respect to cellular imaging, which are summarized in Table 28.1. Techniques for in vivo detection of grafted cells include optical imaging, nuclear medicine imaging, magnetic resonance imaging, computed tomography (CT) imaging, and ultrasound imaging.

Nuclear medicine techniques such as single-photon emission computed tomography (SPECT) and positron emission tomography (PET) require the use of radioactive labels. Although they cannot reach the powerful resolution of MRI, they are highly sensitive and can detect contrast agents in the picomolar range (10^{-12} mol/L). As a result, radioactive contrast agents can be used in such minute amounts that biological reactions are not evoked. Different isotopes are used for different applications, and their physical half-life, together with the biological half-life of the contrast agent, will determine how long the cells can be monitored. Generally speaking, isotopes with a long half-life, such as indium-111 for SPECT and copper-64 or iodine-124 for PET, are used for direct cell tracking, as cells can be followed for up to 14 days (Adonai et al. 2002; Chin et al. 2003; Daldrup-Link et al. 2005; Agger et al. 2007; Pittet et al. 2007; Schachinger et al. 2008; Vrtovec et al. 2013). On the other hand, shorter lived isotopes such as technetium-99m for SPECT and fluorine-18 for PET are mostly applied for indirect cell labeling using reporter genes and allow cell tracking for 2–8 h after transplantation (Koehne et al. 2003; Kim et al. 2005; Kang et al. 2006; Doubrovin et al. 2007; Terrovitis et al. 2008).

MRI has exquisite soft tissue contrast, and its high spatial resolution lends itself to tracking transplanted cells. Also, the absence of radiation makes MRI an attractive alternative to nuclear imaging. It does, however, have lower sensitivity compared to nuclear and optical imaging modalities, which was illustrated by Kraitchman et al. (2005). In his dual-modality SPECT/MRI study, mesenchymal stem cells were labeled with both indium-111 oxide and iron oxides. Upon intravenous injection, the cells could only be tracked dynamically by SPECT, as MRI lacked in sensitivity. MRI contrast agents for cell tracking can be divided into (1) paramagnetic agents (gadolinium- or manganese-based) for T_1-based imaging (Granot et al. 2007; Gilad et al. 2008a; Tseng et al. 2010); (2) superparamagnetic agents (iron oxide-based) for T_2- and T_2^*-based imaging (Bulte et al. 1993; Zhu et al. 2006; Bulte 2009; Long and Bulte 2009; Muja and Bulte 2009; Naumova et al. 2010; Xie et al. 2010; Cromer Berman et al. 2011; Vandsburger et al. 2013b; Azene et al. 2014); (3) chemical exchange saturation transfer (CEST) agents (Ferrauto et al. 2013); and (4) fluorine-19 labeled agents for nonproton MRI (Srinivas et al. 2007, 2010; Barnett et al. 2011). While T_2/T_2^* agents result in hypointensity, T_1 agents create hyperintensity hot spots, which makes identification easier. Another concern

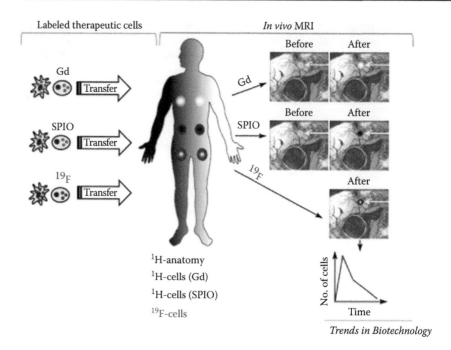

FIGURE 28.1   Cell tracking using MRI with contrast agents and ^{19}F labels. Therapeutic cells (e.g., dendritic or stem cells) are labeled using contrast agents (Gd or SPIO) or a ^{19}F label. Conversely, typical anatomical MRI utilizes the ^{1}H from H_2O in tissues. (From Srinivas, M., Heerschap, A. et al.: F-19 MRI for quantitative in vivo cell tracking. *Trends Biotechnol.* 2010. 28(7). 363–370. Copyright Wiley-VCH Verlag GmbH & Co. KGaA. Reproduced with permission.)

for T_2/T_2^* agents is the ability to discriminate their hypointensities from other causes such as air, calcium, and hemorrhage. Additionally, both classes of contrast agents are not detected directly but instead through their effects on the intrinsic ^{1}H signal, which makes their quantification difficult. Fluorine-19 tracers create positive contrast and quantification of cell numbers can be carried out directly from the image. These characteristics are illustrated in Figure 28.1 (Srinivas et al. 2010). CEST agents form a relatively new class of contrast agents for cell labeling. They resonate at unique frequencies away from water resonance, and the contrast generated by CEST agents can be selectively "turned on." As a result, multiple cell populations that are labeled with distinct CEST agents can be visualized in a multispectral manner (Vandsburger 2014).

Optical imaging techniques can provide high sensitivity for cell tracking with a reported detectability of 10^{-9}–10^{-12} and 10^{-15}–10^{-17} M for fluorescence and bioluminescence, respectively (Zhang and Wu 2007). Fluorescent imaging is hampered by tissue autofluorescence and limited tissue penetration. Nevertheless, near-infrared dyes have been used for cell tracking preclinically (Swirski et al. 2007; Foster et al. 2008; Cova et al. 2013). Some of the problems presented by the use of conventional fluorescent dyes can be resolved by quantum dots. These nanoparticles can be functionalized, provide multicolor cell labeling, and are highly resistant to photobleaching and chemical degradation. As a result, they are well suited for repeated imaging (Slotkin et al. 2007; Noh et al. 2008). However, as in the case of all fluorescent dyes, they still suffer from a limited spatial resolution due to the light scattering in tissue and they cannot provide information about the functional status of the transplanted cells (Solanki et al. 2008). Bioluminescence, on the other hand, uses endogenous reporters, which will provide signals only when the cells are alive. Moreover, there is a direct relationship between the number of cells and the bioluminescent signal, and the total absence of background allows data to be acquired semiquantitatively (Sutton et al. 2008). Optical imaging based on reporter genes, such as the green fluorescence protein for fluorescence and luciferase for bioluminescence, has been extensively used as an alternative to the direct labeling approaches (Huang et al. 2012; Mezzanotte et al. 2013).

Tracking transplanted cells using ultrasound can be done by using microbubble contrast agents, acoustically active liposomes, gold nanoparticles, and perfluorocarbon nanoparticles (Weller et al. 2003; Cui et al. 2008; Herbst et al. 2010). These agents can be functionalized to recognize a variety of cell epitopes, but, thus far, tracking the transplanted cells in vivo using ultrasound has not been reported. Moreover, ultrasound imaging is not a whole-body imaging technique, has poor spatial resolution, and depth penetration is limited.

CT is the most recent addition to the assortment of imaging modalities to track cells in vivo. It offers the possibility to quantify the number of transplanted cells and visualize their distribution with high resolution (Torrente et al. 2006; Villa et al. 2010). However, one has barely begun to explore CT for cell tracking and homing, and not many reports have been published. So far, the approaches have focused on the use of bismuth or barium (Barnett et al. 2006; Azene et al. 2014), gadolinium labeled gold particles (Arifin et al. 2011), and perfluorooctylbromide nano-emulsions (Barnett et al. 2011).

Unfortunately, no single imaging modality is able to provide us with both anatomical and functional information at both high spatial and temporal resolution. It is, therefore, compelling to combine several synergistic modalities, as the strengths of different imaging modalities can be maximized. Multimodal imaging can be performed using a single label that is visible using different modalities or a combination of imaging labels (Tran et al. 2007). However, combining more than one method would also imply increased cost and time and the need for more trained personnel. That said, the multitude of preclinical studies indicates the necessity for multimodal imaging, not only to gain complementary information but also to validate a more experimental imaging method with an established one (Nahrendorf et al. 2009; Arifin et al. 2011; Barnett et al. 2011; Srinivas et al. 2013).

28.1.3 TOXICITY OF CELL LABELS

One major concern for cell imaging technologies is the potential effect of the contrast media on cell biology and survival. For direct cell labeling, toxicity can be due to the route of administration, the hydrodynamic size, the surface charge, the type of coating, and the concentration of the label or nanoparticle. Also, some labels are inherently toxic, such as heavy metals in quantum dots for optical imaging (Su et al. 2010).

Iron oxide nanoparticles for magnetic resonance imaging contain iron and, although it is a natural and essential mineral, an overdose can be toxic to cells (Soenen et al. 2011). Moreover, aggregation of these nanoparticles in the presence of a magnetic field can lead to embolization (Gupta et al. 2007). Several studies have looked into the adverse effects of iron oxide particles on cells, but the reports have been contradictory. Importantly, however, the variety of the applied methods and model systems to assess cytotoxicity plays a critical role and can contribute significantly to these contradictory findings (Soenen and De Cuyper 2010). As a result, many highlighted the need for a more standardized procedure for the assessment of nanoparticle cytotoxicity. In general, the toxicity associated with iron oxide nanoparticles, or their coatings, appears to be closely related to the release of ions and the generation of free radicals following passage through the cell's lysosomes (Soenen and De Cuyper 2010; Taylor et al. 2012). On the positive side, the uptake and toxicity of these nanoparticles can be manipulated by modifications in the surface chemistry and selecting the right size and biocompatible coating of the nanomaterial (Alkilany and Murphy 2010).

Radiotoxicity must be considered when using cell labels for PET and SPECT imaging, particularly when long-lived isotopes such as indium-111 or copper-64 are used (Brenner et al. 2004; Gholamrezanezhad et al. 2009). Significant impairment of proliferation and function of [In-111] oxine-labeled progenitor cells was observed by Brenner et al. (2004). However, Jin et al., who compared indium-111 incubation activity and cell viability, suggest that 100% viability can be achieved when the incubated activity is kept below 0.9 MBq (Jin et al. 2005). All in all, the overall cytotoxicity of radionuclides is considered to be fairly low if proper titration of the radiation dose is performed (Adonai et al. 2002; Bindslev et al. 2006; Zanzonico et al. 2006; Ritchie et al. 2007; Kircher et al. 2008; Patel et al. 2010).

28.2   METHODS FOR CELL LABELING

28.2.1   OVERVIEW

Irrespective of the imaging modality, in vivo cell tracking requires the labeling of cells with the contrast agent of choice. In general, ex vivo cell labeling strategies can be divided into two classes: direct and indirect. Direct labeling involves the internalization of detectable probes. This technique is relatively simple and cheap, and high-contrast cell-to-tissue ratios can be achieved. However, not only does the label get diluted with each cell cycle, it also can be distributed unevenly between the progeny cells or lost because of cell death or marker decay (Zhou et al. 2005; Grimm et al. 2006). As a result, the time that is available for observing the directly labeled cells is limited, which makes detection of the transplanted cells relatively short-lived. Direct cell labeling is thus best for obtaining confirmation of cell delivery success and to answer the question: "Where do the cells go?"

Indirect labeling procedures rely on the detection of a reporter gene. For these applications, a reporter gene is introduced into the cell, which is transformed into a nonnative enzyme, receptor, or fluorescent or bioluminescent protein that accumulates and can be detected (Weissleder et al. 2000; Ponomarev et al. 2004). If the reporter gene expression is stable, the cells can be monitored over their entire lifetime and information about cell proliferation, activation, or death can be extracted. Although the cell label is passed on to the next generation without dilution, the reporter gene can be silenced, which would suppress the signal (Krishnan et al. 2006). Potential disadvantages include costs, cellular dysfunction or death, immunogenicity of gene products, and potential risk of uncontrolled growth and malignancy. As a result, indirect labeling is possible only in preclinical settings, as these aspects preclude clinical application in patients at this time.

Both direct and indirect labeling are useful methods for in vivo cell tracking, but it was not until 2012 that their advantages and disadvantages were comparatively investigated. Park et al. described a radiolabel approach using [^{64}Cu]-copper-pyruvaldehyde-bis(N-4-methylthiosemicarbazone) (^{64}Cu-PTSM) and [^{124}I]-2′-fluoro-2′-deoxy-1-β-D-arabinofuranosyl-5-iodouracil (^{124}I-FIAU) to track chronic myelogenous leukemia cells and investigated cell labeling efficiency, label retention, and cell viability (Park et al. 2012). They concluded that for this specific application, ^{64}Cu-PTSM was preferred for in vivo cell tracking and that both approaches had little effect on cell viability.

In addition to direct and indirect cell labeling, one can also consider in vivo or in situ cell labeling and microencapsulation. The former technique can be used for both direct and indirect cell labeling and is based on systemic or local injection of the imaging probe into the organism. Following injection, the imaging probe is specifically taken up by the target cells by, for instance, receptor-specific accumulation. This strategy allows repetitive imaging of the targeted cell population but requires the development of individual agents for each unique target. Microencapsulation involves encapsulating the transplanted cells together with a label (Arifin et al. 2013). Microencapsulation of therapeutic cells has been widely used to achieve cellular immunoprotection following transplantation. More recently, contrast agents have been added to these microcapsules to allow visualization by MRI, CT and US. This way, large amounts of contrast agents can be incorporated with negligible toxicity and at the same time overcome the sensitivity limits of MRI.

28.2.2   DIRECT CELL LABELING

For direct cell labeling, cells are harvested, labeled ex vivo with probes such as radiotracers, fluorophores, or paramagnetic nanoparticles, and then re-infused, allowing them to be visualized by nuclear, optical, and MR imaging, respectively. Several strategies exist to introduce the label into cells, but many are still based upon nonspecific spontaneous uptake, or phagocytosis (Jasmin et al. 2012). As a result, phagocytic cells such as macrophages, microglia, and dendritic cells are widely used for cell labeling, as they readily take up the cell label. Additionally, these cells are migratory cells, and thus are very relevant for cell-tracking studies (Schulze et al. 1995; Weissleder et al. 1997; Franklin et al. 1999; Zelivyanskaya et al. 2003). Unfortunately, uptake through phagocytosis is highly unpredictable, and factors such as cell type, incubation time, concentration, and the physicochemical properties of the label all can interfere with this process (Weissleder et al. 1997; Politi et al. 2007; Mailander et al. 2008). In contrast to the label concentration and incubation time, which are easiest to control and which allow for systematic studies of their influence on phagocytosis, the effects of the physicochemical parameters are much more difficult to determine and interpretation of these results is often problematic (Levy et al. 2010; Cho et al. 2011; Taylor et al. 2012).

Fortunately, other strategies can be employed to not only overcome the problems associated with phagocytosis and to enhance the internalization of the cell label but also to label nonphagocytic cells. These methods make use of (1) external agents or the so-called forced-entry approaches such as transfection agents (Frank et al. 2002, 2003; Rudelius et al. 2003; Bulte et al. 2004; Bulte and Kraitchman 2004; Modo et al. 2005; Cova et al. 2013), electroporation (Walczak et al. 2005; Suzuki et al. 2007; Kim et al. 2011b), magnetofection (Scherer et al. 2002; Plank et al. 2003), or sonoporation (Mo et al. 2008); or (2) specific targeting strategies such as functionalization with specific ligands or cell-penetrating peptides and the use of specific cell membrane transporters (Bulte et al. 1999, 2004; Josephson et al. 1999; Ahrens et al. 2003; Doyle et al. 2007; Park et al. 2012). Examples of transfection agents are poly-L-lysine (Frank et al. 2003), cationic liposome (van den Bos et al. 2003), or protamine sulfate (Arbab et al. 2004), which form complexes with labeled nanoparticles and greatly improve their efficiency of endocytosis in cells that cannot avidly phagocytize. On the other hand, specific targeting can be achieved by using monoclonal antibodies (Weissleder et al. 1997; Bulte et al. 1999; Ahrens et al. 2003), specific receptor ligands such as transferrin (Qian et al. 2002), or specific trapping systems for 2-[^{18}F]-fluoro-2-deoxy-d-glucose (FDG) or ^{64}Cu-PTSM (Koike et al. 1997; Adonai et al. 2002; Hofmann et al. 2005; Huang et al. 2008).

As Taylor et al. pointed out, it is also worth noting that, as of yet, most strategies for labeling cells for tracking are still based on nonspecific uptake or "forced" entry. This is in contrast to methods used in molecular imaging, as described in the previous chapter, which rely on the use of specific targeting strategies employing ligands that bind to specific receptors in or on the cell (Taylor et al. 2012). Anyway, direct cell labeling does not involve extensive manipulation of the cells as compared to indirect labeling strategies and, therefore, is preferred for clinical implementation.

28.2.3 INDIRECT CELL LABELING

While the imaging strategies discussed in the previous section provide solutions to short-term visualization of transplanted cells, they do not offer a solution to assess the biology and longitudinal viability of cells. For preclinical applications, the reporter gene approach can be used as an alternative to address these problems. The basic principle is to transfer a genetically engineered gene, the reporter gene, into the genome of a cell ex vivo. This gene is not normally expressed in the target cell, and usually encodes an enzyme, receptor, or transporter. The gene transfer can be achieved via viral or nonviral methods, using lentiviruses (De et al. 2003) or nanoparticles, polymers, electroporation, and transfection agents (Lam and Dean 2010), respectively. The choice of the promotor to drive the reporter gene depends on the question to be answered. It can be either a constitutive promotor to monitor cell location, migration, targeting, proliferation, or viability, or an inducible promotor to study the activation of a specific cell function (Blasberg 2003; Dimayuga and Rodriguez-Porcel 2011). Next, cells can be visualized at specified time points after intravenous injection of an imaging tracer that is specific for the reporter gene product. Because the accumulation of the imaging tracer is dependent on both the expression of the reporter gene and the activity of the gene product, the imaging signal will be dependent on the viability of the transplanted cells (Bengel et al. 2005). Additionally, stable transfection will ensure that the reporter will not be diluted upon cell division (Cao et al. 2006).

Since early proof-of-principle studies by Wu et al. (2003), several reporter genes have been developed for cell tracking, of which the transferrin receptor, ferritin receptor, herpes simplex virus type 1 thymidine kinase (HSV1), and luciferase are the most popular ones. An overview of the large number of reporter genes for MR imaging can be found in a recent paper by Vandsburger et al. (2013). Imaging the trafficking of cells using reporter genes has been performed using optical, nuclear, and magnetic resonance imaging or a combination thereof (Zhou et al. 2006). However, each modality tends to use its own set of reporter genes.

The most commonly used reporter gene for radionuclide imaging is the enzyme-based system herpes simplex virus type 1 thymidine kinase (HSV1), which can be found intracellularly (Min and Gambhir 2008). Thymidine kinases phosphorylate, and thereby trap, radiolabeled nucleoside analogs such as FIAU, fluoropenciclovir, and 9-[4-fluoro-3-(hydrommethyl)butyl] guanine (FHBG) into the cells (Thakur et al. 1976; Koehne et al. 2003; Wu et al. 2003; Cao et al. 2006). As a result, when used in nonpharmacological tracer doses, these substrates are nontoxic and serve as PET or SPECT targeted reporter probes by their accumulation in only cells expressing the HSV1 gene. Other applications for cell tracking using nuclear imaging involve the use of receptor- and transporter-based systems such as the dopamine-2 and somatostatin receptors and the sodium-iodine transporter, respectively (Liang et al. 2001; Miyagawa et al. 2005; Kircher et al. 2011).

Two optical imaging techniques, namely luminescence and fluorescence, have been used for in vivo cell tracking, but their application is limited because of the nontomographic character of the techniques and the limited tissue penetration of the generated light. Fluorescent imaging is mainly based on the transfection of cells with green fluorescent protein or its variants, while bioluminescence frequently uses the luciferase reporter genes (Tang et al. 2003; Borovjagin et al. 2010; Dimayuga and Rodriguez-Porcel 2011).

Over the past decade, MRI reporter gene strategies have also been developed, and three distinct categories have emerged (Kim et al. 2005; Vandsburger et al. 2013). The first one is based on the overexpression of iron-regulatory proteins such as ferritin and the transferrin receptor, with the ultimate goal of accumulating as much iron as possible inside the cells (Gilad et al. 2007; Liu et al. 2009; Naumova et al. 2010; Vandsburger et al. 2013). The second strategy utilizes reporter genes that encode a plasma-membrane-bound protein or antigen. These are subsequently targeted by intravenously administered antibody-conjugated iron oxide nanoparticles (Chung et al. 2011). Finally, the last broad strategy generates CEST contrast through either constitutive expression of CEST reporter genes or through the expression of transgenic kinases for targeted CEST contrast agents (Vandsburger 2014).

Finally, some efforts are geared toward the development of multimodality reporter genes as cell tracking, in nature, demands both high sensitivity and spatial resolution. Especially the use of fusion proteins that enable optical and nuclear imaging have become attractive. An example of such a system was reported by Ray et al., who combined red fluorescent protein, Renilla luciferase, and HSV1 for in vitro fluorescent microscopy or cell sorting, in vivo high-sensitivity bioluminescence imaging, and in vivo quantitative tomographic PET/SPECT imaging, respectively (Ray et al. 2004).

28.2.4 In Situ Labeling

In contrast to ex vivo cell labeling, in vivo or in situ cell labeling consists of systemic or local injection of the contrast agent, which is then either taken up nonspecifically by the relevant (phagocytic) cells or targeted specifically to the relevant cell type (Nieman et al. 2010; Vreys et al. 2010). This strategy omits the ex vivo cell handling, purification, and labeling and is thus a lot easier, less labor intensive, and cheaper to implement. Again, only viable cells will be able to take up the label effectively, and longitudinal studies, even with short-lived radiotracers, are feasible. However, broad application of in situ labeling for cell tracking is often hampered by insufficient accumulation of the contrast agent in the target cells, nonspecific uptake, the need for higher doses to generate sufficient signal, and clearance of the label (Srinivas et al. 2010; Vande Velde et al. 2012).

Micrometer-sized iron oxide particles (MPIO) and superparamagnetic iron oxide (SPIO) have been proven to be useful agents for in situ cell labeling. These can be injected as such or in combination with a transfection agent to reduce particle spread and thus increase labeling efficiency. The latter approach has been successfully used in neurogenesis studies (Vreys et al. 2010). Unmodified MPIOs and SPIOs are taken up by the phagocytic cells, including circulating blood monocytes and tissue macrophages (Long et al. 2009). Additionally, these particles have been conjugated to optical labels, which created dual-modality imaging agents, or functionalized with antibodies or antigens to allow specific cell targeting.

28.2.5 MICROENCAPSULATION

Microencapsulation has been used widely to protect therapeutic cells from attack by the immune system after transplantation. Microcapsules are typically prepared from hydrogels and consist of a porous coating, usually alginate, which allows flow of nutrients and metabolites but restricts larger entities such as antibodies or antigen-presenting cells (Lim and Sun 1980; Arifin et al. 2013). Interestingly, microcapsules can be made visible by loading them with contrast agents for CT, US, and MRI, which makes them a valuable tool for cell-tracking and homing studies. An overview of the different contrast agent microcapsules and their applications in terms of cell tracking is presented by Arafin et al. (2002). A recent further addition to the field is the capsule-in-capsules, which, as the name predicts, consist of two separate capsules: the inner one contains the contrast agent, while the cells are enclosed in an outer one (Kim et al. 2011a). A schematic of the capsule-in-capsules is presented in Figure 28.2. The advantage of this approach is that the contrast agents are physically separated from the cells and toxicity issues can be avoided. As a result, the amount

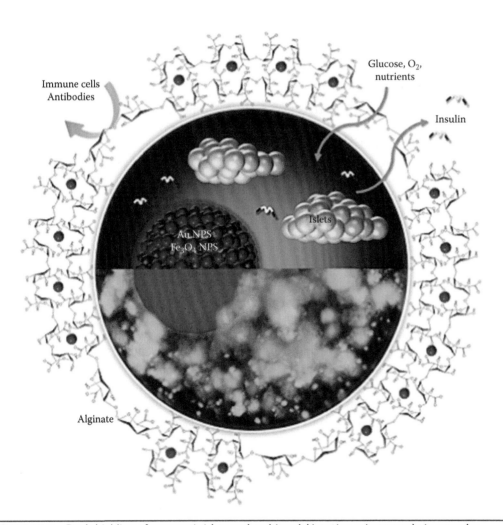

FIGURE 28.2 Dual shielding of pancreatic islets and multimodal imaging using capsule-in-capsules (CICs). The semipermeable outer alginate membrane blocks the penetration of immune cells and antibodies, but allows unhindered diffusion of nutrients, glucose, oxygen, and insulin produced by islets. The inner capsule, containing iron oxide and gold nanoparticle contrast agents, which enable concurrent MRI, computed tomography (CT), and ultrasound (US) imaging, prevents direct exposure of nanoparticles to cells. Islets within CICs exhibit improved insulin secretion compared with single capsules that contain the two types of nanoparticles and islets altogether. NP, nanoparticle. (From Arifin, D.R., Kedziorek, D.A. et al.: Microencapsulated cell tracking. *NMR Biomed.* 2013. 26(7). 850–859. Copyright Wiley-VCH Verlag GmbH & Co. KGaA. Reproduced with permission.)

of contrast agent can be increased to enhance sensitivity without inducing cytotoxicity. However, it should be noted that labeled, microencapsulated cells, similar to directly labeled cells, do not allow assessment of cell viability and function.

28.3 DIFFICULTIES

The list of properties for an ideal cell-tracking imaging agent seems endless. It should be biocompatible, nontoxic, highly sensitive, quantifiable, and noninterfering with the cell's function, have migratory capacity, be detectable throughout the body at any tissue depth, be nondegradable in the cell, not be diluted or lost upon cell division, not be transferred to other cells in vivo, and, last, be imageable for months to years after the initial labeling. Consequently, as one can imagine, combining all these properties into one compound is a big challenge and, to date, impossible. Indeed, while many of the issues have been solved, some are difficult to address because of their contradictory nature. For example, biocompatibility can often not be combined with the need for retention of the contrast agent for longitudinal studies. Most of these obstacles are related to direct cell labeling, although indirect labeling using reporter genes has its specific hurdles as well, which will be discussed in this section.

28.3.1 LABEL DEGRADATION

Once it has been established that a tracking agent is biocompatible and nontoxic to the cells, they should incorporate rapidly into cells and be fully retained. However, many agents fail in the latter requirement and are degraded and recycled in the lysosomal bodies of the cells and/or effluxed, which makes interpretation of the data problematic. Indeed, if the eliminated agent, be it a metabolite or the intact compound, is not rapidly excreted, it has the potential to redistribute or be taken up by other cells, which will result in a false-positive secondary signal. Additionally, degradation of the contrast agent leads to a continuous drop in sensitivity and thus long-term tracking of cells becomes impossible. However, these days, it is possible to gain information on lysosomal degradation in an in vitro setting using citrate-containing buffers at low pH (Levy et al. 2010; Soenen et al. 2010). At least for SPION particles and quantum dots, such studies have revealed that their dissolution is dependent on their surface coating (Hu and Gao 2010; Soenen et al. 2010). On the other hand, as pointed out by Taylor et al., accumulation of the cell-tracking agent in cell organelles such as endosomes and lysosomes can also be a positive aspect of the labeling process (Taylor et al. 2012). For example, SPIONs inherently absorb to cell proteins, which alters their conformation and thus their biological stability and activity (Mahmoudi et al. 2011). Accumulation of these particles in cell organelles would reduce their exposure to cytosolic and nuclear proteins, and thus minimize the disruption of protein function. Thompson et al. reported another advantage when they showed that clustering of MR labels in dense vacuoles resulted in increased local contrast enhancement as compared to cytosolic distribution (Thompson et al. 2005).

28.3.2 LABEL DILUTION

Another common issue for agents used in direct cell labeling is the dilution of the contrast agent during cell proliferation, which results in loss of detectability or errors in quantification of cell numbers (Walczak et al. 2007). Label dilution is even more of an issue for fast dividing cells and/or if daughter cells migrate to other regions of the body. However, this problem can be overcome by using indirect cell labeling strategies or, for slower proliferating cells, if a high cell loading of the contrast agent can be achieved. Additionally, in some cases one can exploit the presence of endogenous proteins such as bacterial thymidine kinase to track the presence of bacterial cells in vivo using 1-(2′-deoxy-2′-fluoro-β-D-arabinofuranosyl)-5-[^{125}I] iodouracil ([^{125}I]-FIAU) (Diaz et al. 2007). Such a strategy allows repetitive targeting of the positive cell population in vivo but requires the development of individual agents for each unique target.

28.3.3 Transfer of Cell Label

Typically, in vivo imaging modalities detect just the label regardless of whether it is still attached to the targeting vector, contained in the relevant cells, effluxed and excreted or transferred to other cells. Therefore, transfer of cell label to other cells can become a big problem if the fate of the cell label is not characterized carefully beforehand by, for instance, histology or flow cytometry.

Cell label transfer usually occurs when the transplanted cells die. As a result, the label can be taken up by resident phagocytes in an effort to clear the cell debris from the body by neighboring cells or by inflammatory cells. These processes lead to false positives if the cell label is not cleared rapidly from the site. Indeed, movement of label in the images suggests that cells are alive and migrating. However, one cannot distinguish between healthy cells and dead cells phagocytozed by macrophages; nor can one detect whether the cell is leaking and the label is taken up by neighboring cells (Winter et al. 2010). Several studies have documented this issue, but an elegant one was reported by Berman et al., who transplanted neural stem cells in both immunocompetent and immunodeficient mice and showed a remarkable difference in signal intensity (Berman et al. 2011).

More recently, Luciani et al. reported another process by which the cell can release its label (Luciani et al. 2010). They showed that, once the magnetic nanoparticles are internalized by monocytes or macrophages, they can be released by the cells upon stress through the formation of microvessicles. These vesicles are then taken up by naïve macrophages, which results in an intercellular transfer of the magnetic label.

Once again, the problem of cell label transfer can be avoided when indirect cell labeling is used. Additionally, bifunctional imaging agents could be designed such that one of the targets is known to be cell-specific (Srinivas et al. 2013).

28.3.4 Sensitivity and Cell Quantification

Quantification of imaging data begins with quantifying the cell labeling efficiency. Therefore, cell loading needs to be precise and reproducible so that the average amount of label per cell can be determined. For this, it is important to take into account possible aggregation and sedimentation of nanoparticles if long incubation times are needed. Aggregates can remain after cell harvesting, which results in an overestimation of the labeling efficiency (Ge et al. 2009). Therefore, it is self-evident to use stringent methods to eliminate unbound particles from the culture dish if accurate quantification is desired. However, determining the cell labeling efficiency is the more easy part of the process, as the in vivo system (human or animal) can throw up many problems that impede image quantification. These include cell label dilution and transfer, which were discussed previously.

Lastly, quantitation will also depend on the imaging modality used, as not all modalities exhibit a linear relationship between signal intensity and the cell label concentration. This is the case for ^{1}H-MRI, where the relationship between iron concentrations and signal intensity may not be monoexponential and strictly depends on the acquisition parameters (Politi 2007). Although many obstacles need to be taken into account, quantitative cell tracking is possible as was reported by Srinivas et al. for ^{19}F-MRI (Srinivas et al. 2007, 2009) and by Ponomarev and Su for PET (Su et al. 2004; Ponomarev 2009).

28.3.5 Limits of Reporter Genes

Although many of the difficulties with cell labeling and subsequent cell tracking are related to directly labeled cells, reporter-gene-based imaging can have its limitations as well. One such problem was described by Gilad et al., who reported cytosolic, nuclear, and membrane accumulation of reporter gene products when they are overexpressed in the cells (Gilad et al. 2007). Also, reporter gene products have the potential to interfere with the normal biology of transduced cells and their signaling pathways (Gilad et al. 2007, 2008b).

28.4 SUMMARY

Cellular therapy has become an interesting avenue for the treatment of many medical conditions in disease areas such as cardiology, neurology, oncology, and immunology. Transplanted cells can be used to repair damaged tissue, replace lost cells, or correct for aberrant processes. However, a major hurdle for widespread clinical acceptance of cellular therapy is the fact that the mechanisms underlying therapy success, or failure, are still poorly understood. In order to clarify these, cell labeling was introduced to follow the transplanted cells in an in vivo setting and to derive information on tissue function or survival, transplanted cell numbers, localization, or functionality.

Cell can be labeled ex vivo by internalizing contrast agents or relying on reporter genes. Although this approach allows assessment of label toxicity before cells are transplanted into the living organism, labels with relatively long lifetimes are needed to allow longitudinal imaging. In addition, the lifetime of the label is restricted by leakage from the transplanted cells, dilution following proliferation, degradation in the lysosomes, and radioactive decay. On the other hand, cells can be labeled in situ, allowing repetitive assessment using short probe lifetimes. However, disadvantages include the possible nonspecific uptake in nontargeted tissues and often insufficient accumulation of the contrast agent in the target cells.

Lastly, the choice of the imaging modality plays an important role, and each modality has its own advantages and disadvantages; to date, no single imaging modality offers high sensitivity, high resolution, and low toxicity. Therefore, it is important to consider multimodality imaging so that the strengths of individual systems can be combined. Although multimodality imaging often presents logistical and financial challenges, both in a preclinical and clinical setting, it will provide additional information so that information gathering can be maximized ion one single imaging session.

REFERENCES

Aarntzen, E. H., M. Srinivas et al. (2012). In vivo tracking techniques for cellular regeneration, replacement, and redirection. *J. Nucl. Med.* **53**(12): 1825–1828.

Adonai, N., K. N. Nguyen et al. (2002). Ex vivo cell labeling with ^{64}Cu-pyruvaldehyde-bis(N^4-methylthiosemicarbazone) for imaging cell trafficking in mice with positron-emission tomography. *Proc. Natl. Acad. Sci. USA* **99**(5): 3030–3035.

Agger, R., M. S. Petersen et al. (2007). T cell homing to tumors detected by 3D-coordinated positron emission tomography and magnetic resonance imaging. *J. Immunother.* **30**(1): 29–39.

Ahrens, E. T., M. Feili-Hariri et al. (2003). Receptor-mediated endocytosis of iron-oxide particles provides efficient labeling of dendritic cells for in vivo MR imaging. *Magn. Reson. Med.* **49**(6): 1006–1013.

Alkilany, A. M. and C. J. Murphy (2010). Toxicity and cellular uptake of gold nanoparticles: What we have learned so far? *J. Nanopart. Res.* **12**(7): 2313–2333.

Arbab, A. S., G. T. Yocum et al. (2004). Efficient magnetic cell labeling with protamine sulfate complexed to ferumoxides for cellular MRI. *Blood* **104**(4): 1217–1223.

Arifin, D. R., D. A. Kedziorek et al. (2013). Microencapsulated cell tracking. *NMR Biomed.* **26**(7): 850–859.

Arifin, D. R., C. M. Long et al. (2011). Trimodal gadolinium-gold microcapsules containing pancreatic islet cells restore normoglycemia in diabetic mice and can be tracked by using US, CT, and positive-contrast MR imaging. *Radiology* **260**(3): 790–798.

Azene, N., Y. Fu et al. (2014). Tracking of stem cells in vivo for cardiovascular applications. *J. Cardiovasc. Magn. Reson.* **16**(1): 7.

Barnett, B. P., D. L. Kraitchman et al. (2006). Radiopaque alginate microcapsules for x-ray visualization and immunoprotection of cellular therapeutics. *Mol. Pharm.* **3**(5): 531–538.

Barnett, B. P., J. Ruiz-Cabello et al. (2011). Fluorocapsules for improved function, immunoprotection, and visualization of cellular therapeutics with MR, US, and CT imaging. *Radiology* **258**(1): 182–191.

Barnett, B. P., J. Ruiz-Cabello et al. (2011). Use of perfluorocarbon nanoparticles for non-invasive multimodal cell tracking of human pancreatic islets. *Contrast Media Mol. Imaging* **6**(4): 251–259.

Bengel, F. M., V. Schachinger et al. (2005). Cell-based therapies and imaging in cardiology. *Eur. J. Nucl. Med. Mol. Imaging* **32**(Suppl. 2): S404–S416.

Berman, S. C., C. Galpoththawela et al. (2011). Long-term MR cell tracking of neural stem cells grafted in immunocompetent versus immunodeficient mice reveals distinct differences in contrast between live and dead cells. *Magn. Reson. Med.* **65**(2): 564–574.

Bindslev, L., M. Haack-Sorensen et al. (2006). Labelling of human mesenchymal stem cells with indium-111 for SPECT imaging: Effect on cell proliferation and differentiation. *Eur. J. Nucl. Med. Mol. Imaging* **33**(10): 1171–1177.

Blasberg, R. G. (2003). In vivo molecular-genetic imaging: Multi-modality nuclear and optical combinations. *Nucl. Med. Biol.* **30**(8): 879–888.

Borovjagin, A. V., L. R. McNally et al. (2010). Noninvasive monitoring of mRFP1- and mCherry-labeled oncolytic adenoviruses in an orthotopic breast cancer model by spectral imaging. *Mol. Imaging* **9**(2): 59–75.

Brenner, W., A. Aicher et al. (2004). ^{111}In-labeled CD34+ hematopoietic progenitor cells in a rat myocardial infarction model. *J. Nucl. Med.* **45**(3): 512–518.

Bulte, J. W. (2009). In vivo MRI cell tracking: Clinical studies. *AJR Am. J. Roentgenol.* **193**(2): 314–325.

Bulte, J. W., A. S. Arbab et al. (2004). Preparation of magnetically labeled cells for cell tracking by magnetic resonance imaging. *Methods Enzymol.* **386**: 275–299.

Bulte, J. W. and D. L. Kraitchman (2004). Iron oxide MR contrast agents for molecular and cellular imaging. *NMR Biomed.* **17**(7): 484–499.

Bulte, J. W., L. D. Ma et al. (1993). Selective MR imaging of labeled human peripheral blood mononuclear cells by liposome mediated incorporation of dextran-magnetite particles. *Magn. Reson. Med.* **29**(1): 32–37.

Bulte, J. W., S. Zhang et al. (1999). Neurotransplantation of magnetically labeled oligodendrocyte progenitors: Magnetic resonance tracking of cell migration and myelination. *Proc. Natl. Acad. Sci. USA* **96**(26): 15256–15261.

Cao, F., S. Lin et al. (2006). In vivo visualization of embryonic stem cell survival, proliferation, and migration after cardiac delivery. *Circulation* **113**(7): 1005–1014.

Chin, B. B., Y. Nakamoto et al. (2003). 111In oxine labelled mesenchymal stem cell SPECT after intravenous administration in myocardial infarction. *Nucl. Med. Commun.* **24**(11): 1149–1154.

Cho, E. C., Q. Zhang et al. (2011). The effect of sedimentation and diffusion on cellular uptake of gold nanoparticles. *Nat. Nanotechnol.* **6**(6): 385–391.

Chung, J., K. Kee et al. (2011). In vivo molecular MRI of cell survival and teratoma formation following embryonic stem cell transplantation into the injured murine myocardium. *Magn. Reson. Med.* **66**(5): 1374–1381.

Cova, L., P. Bigini et al. (2013). Biocompatible fluorescent nanoparticles for in vivo stem cell tracking. *Nanotechnology* **24**(24): 245603.

Cromer Berman, S. M., P. Walczak et al. (2011). Tracking stem cells using magnetic nanoparticles. *Wiley Interdiscip Rev. Nanomed. Nanobiotechnol.* **3**(4): 343–355.

Cui, W., J. Wang et al. (2008). Stem cell tracking using ultrasound contrast agents. *Circulation* **118**: S642.

Daldrup-Link, H. E., R. Meier et al. (2005). In vivo tracking of genetically engineered, anti-HER2/neu directed natural killer cells to HER2/neu positive mammary tumors with magnetic resonance imaging. *Eur. Radiol.* **15**(1): 4–13.

De, A., X. Z. Lewis et al. (2003). Noninvasive imaging of lentiviral-mediated reporter gene expression in living mice. *Mol. Ther.* **7**(5 Pt. 1): 681–691.

Diaz, L. A., Jr., C. A. Foss et al. (2007). Imaging of musculoskeletal bacterial infections by [124I]FIAU-PET/CT. *PLoS One* **2**(10): e1007.

Dimayuga, V. M. and M. Rodriguez-Porcel (2011). Molecular imaging of cell therapy for gastroenterologic applications. *Pancreatology* **11**(4): 414–427.

Doubrovin, M. M., E. S. Doubrovina et al. (2007). In vivo imaging and quantitation of adoptively transferred human antigen-specific T cells transduced to express a human norepinephrine transporter gene. *Cancer Res.* **67**(24): 11959–11969.

Doyle, B., B. J. Kemp et al. (2007). Dynamic tracking during intracoronary injection of 18F-FDG-labeled progenitor cell therapy for acute myocardial infarction. *J. Nucl. Med.* **48**(10): 1708–1714.

Ferrauto, G., D. Delli Castelli et al. (2013). In vivo MRI visualization of different cell populations labeled with PARACEST agents. *Magn. Reson. Med.* **69**(6): 1703–1711.

Foster, A. E., S. Kwon et al. (2008). In vivo fluorescent optical imaging of cytotoxic T lymphocyte migration using IRDye800CW near-infrared dye. *Appl. Opt.* **47**(31): 5944–5952.

Frank, J. A., B. R. Miller et al. (2003). Clinically applicable labeling of mammalian and stem cells by combining superparamagnetic iron oxides and transfection agents. *Radiology* **228**(2): 480–487.

Frank, J. A., H. Zywicke et al. (2002). Magnetic intracellular labeling of mammalian cells by combining (FDA-approved) superparamagnetic iron oxide MR contrast agents and commonly used transfection agents. *Acad. Radiol.* **9**(Suppl. 2): S484–S487.

Franklin, R. J. M., K. L. Blaschuk et al. (1999). Magnetic resonance imaging of transplanted oligodendrocyte precursors in the rat brain. *Neuroreport* **10**(18): 3961–3965.

Ge, Y., Y. Zhang et al. (2009). Effect of surface charge and agglomerate degree of magnetic iron oxide nanoparticles on KB cellular uptake in vitro. *Colloids Surf. B Biointerfaces* **73**(2): 294–301.

Gholamrezanezhad, A., S. Mirpour et al. (2009). Cytotoxicity of 111In-oxine on mesenchymal stem cells: A time-dependent adverse effect. *Nucl. Med. Commun.* **30**(3): 210–216.

Gilad, A. A., P. Walczak et al. (2008a). MR tracking of transplanted cells with "positive contrast" using manganese oxide nanoparticles. *Magn. Reson. Med.* **60**(1): 1–7.

Gilad, A. A., P. T. Winnard, Jr. et al. (2007). Developing MR reporter genes: Promises and pitfalls. *NMR Biomed.* **20**(3): 275–290.

Gilad, A. A., K. Ziv et al. (2008b). MRI reporter genes. *J. Nucl. Med.* **49**(12): 1905–1908.

Granot, D., Y. Addadi et al. (2007). In vivo imaging of the systemic recruitment of fibroblasts to the angiogenic rim of ovarian carcinoma tumors. *Cancer Res.* **67**(19): 9180–9189.

Grimm, J., F. K. Swirski et al. (2006). A nanoparticle-based cell labelling agent for cell tracking with SPECT/CT. *Mol. Imaging* **5**: 364.

Gupta, A. K., R. R. Naregalkar et al. (2007). Recent advances on surface engineering of magnetic iron oxide nanoparticles and their biomedical applications. *Nanomedicine (Lond.)* **2**(1): 23–39.

Herbst, S. M., M. E. Klegerman et al. (2010). Delivery of stem cells to porcine arterial wall with echogenic liposomes conjugated to antibodies against CD34 and intercellular adhesion molecule-1. *Mol. Pharm.* **7**(1): 3–11.

Hofmann, M., K. C. Wollert et al. (2005). Monitoring of bone marrow cell homing into the infarcted human myocardium. *Circulation* **111**(17): 2198–2202.

Hu, X. and X. Gao (2010). Silica-polymer dual layer-encapsulated quantum dots with remarkable stability. *ACS Nano* **4**(10): 6080–6086.

Huang, J., C. C. Lee et al. (2008). Radiolabeling rhesus monkey CD34+ hematopoietic and mesenchymal stem cells with ^{64}Cu-pyruvaldehyde-bis(N^4-methylthiosemicarbazone) for microPET imaging. *Mol. Imaging* **7**(1): 1–11.

Huang, N. F., J. Okogbaa et al. (2012). Bioluminescence imaging of stem cell-based therapeutics for vascular regeneration. *Theranostics* **2**(4): 346–354.

Jasmin, L., A. Jelicks et al. (2012). Mesenchymal bone marrow cell therapy in a mouse model of chagas disease. Where do the cells go? *PLoS Negl. Trop. Dis.* **6**(12): e1971.

Jin, Y., H. Kong et al. (2005). Determining the minimum number of detectable cardiac-transplanted 111In-tropolone-labelled bone-marrow-derived mesenchymal stem cells by SPECT. *Phys. Med. Biol.* **50**(19): 4445–4455.

Josephson, L., C. H. Tung et al. (1999). High-efficiency intracellular magnetic labeling with novel superparamagnetic-Tat peptide conjugates. *Bioconjug. Chem.* **10**(2): 186–191.

Kang, W. J., H. J. Kang et al. (2006). Tissue distribution of 18F-FDG-labeled peripheral hematopoietic stem cells after intracoronary administration in patients with myocardial infarction. *J. Nucl. Med.* **47**(8): 1295–1301.

Kim, J., D. R. Arifin et al. (2011a). Multifunctional capsule-in-capsules for immunoprotection and trimodal imaging. *Angew. Chem. Int. Ed. Engl.* **50**(10): 2317–2321.

Kim, T., E. Momin et al. (2011b). Mesoporous silica-coated hollow manganese oxide nanoparticles as positive T1 contrast agents for labeling and MRI tracking of adipose-derived mesenchymal stem cells. *J. Am. Chem. Soc.* **133**(9): 2955–2961.

Kim, Y. H., D. S. Lee et al. (2005). Reversing the silencing of reporter sodium/iodide symporter transgene for stem cell tracking. *J. Nucl. Med.* **46**(2): 305–311.

Kircher, M. F., S. S. Gambhir et al. (2011). Noninvasive cell-tracking methods. *Nat. Rev. Clin. Oncol.* **8**(11): 677–688.

Kircher, M. F., J. Grimm et al. (2008). Noninvasive in vivo imaging of monocyte trafficking to atherosclerotic lesions. *Circulation* **117**(3): 388–395.

Koehne, G., M. Doubrovin et al. (2003). Serial in vivo imaging of the targeted migration of human HSV-TK-transduced antigen-specific lymphocytes. *Nat. Biotechnol.* **21**(4): 405–413.

Koike, C., M. Watanabe et al. (1997). Tumor cells with organ-specific metastatic ability show distinctive trafficking in vivo: Analyses by positron emission tomography and bioimaging. *Cancer Res.* **57**(16): 3612–3619.

Kraitchman, D. L., M. Tatsumi et al. (2005). Dynamic imaging of allogeneic mesenchymal stem cells trafficking to myocardial infarction. *Circulation* **112**(10): 1451–1461.

Krishnan, M., J. M. Park et al. (2006). Effects of epigenetic modulation on reporter gene expression: Implications for stem cell imaging. *FASEB J.* **20**(1): 106–108.

Lam, A. P. and D. A. Dean (2010). Progress and prospects: Nuclear import of nonviral vectors. *Gene Ther.* **17**(4): 439–447.

Levy, M., F. Lagarde et al. (2010). Degradability of superparamagnetic nanoparticles in a model of intracellular environment: Follow-up of magnetic, structural and chemical properties. *Nanotechnology* **21**(39): 395103.

Levy, R., U. Shaheen et al. (2010). Gold nanoparticles delivery in mammalian live cells: A critical review. *Nano Rev* **1**, 4889. doi: 10.3402/nano.v1i0.4889.

Liang, Q., N. Satyamurthy et al. (2001). Noninvasive, quantitative imaging in living animals of a mutant dopamine D2 receptor reporter gene in which ligand binding is uncoupled from signal transduction. *Gene Ther.* **8**(19): 1490–1498.

Lim, F. and A. M. Sun (1980). Microencapsulated islets as bioartificial endocrine pancreas. *Science* **210**(4472): 908–910.

Liu, J., E. C. Cheng et al. (2009). Noninvasive monitoring of embryonic stem cells in vivo with MRI transgene reporter. *Tissue Eng. Part C Methods* **15**(4): 739–747.

Long, C. M. and J. W. Bulte (2009). In vivo tracking of cellular therapeutics using magnetic resonance imaging. *Expert Opin. Biol. Ther.* **9**(3): 293–306.

Long, C. M., H. W. van Laarhoven et al. (2009). Magnetovaccination as a novel method to assess and quantify dendritic cell tumor antigen capture and delivery to lymph nodes. *Cancer Res.* **69**(7): 3180–3187.

Luciani, N., C. Wilhelm et al. (2010). The role of cell-released microvesicles in the intercellular transfer of magnetic nanoparticles in the monocyte/macrophage system. *Biomaterials* **31**(27): 7061–7069.

Mahmoudi, M., M. A. Shokrgozar et al. (2011). Irreversible changes in protein conformation due to interaction with superparamagnetic iron oxide nanoparticles. *Nanoscale* **3**(3): 1127–1138.

Mailander, V., M. R. Lorenz et al. (2008). Carboxylated superparamagnetic iron oxide particles label cells intracellularly without transfection agents. *Mol. Imaging Biol.* **10**(3): 138–146.

Mezzanotte, L., M. Aswendt et al. (2013). Evaluating reporter genes of different luciferases for optimized in vivo bioluminescence imaging of transplanted neural stem cells in the brain. *Contrast Media Mol. Imaging* **8**(6): 505–513.

Min, J. J. and S. S. Gambhir (2008). Molecular imaging of PET reporter gene expression. *Handb. Exp. Pharmacol.* (185 Pt 2): 277–303.

Miyagawa, M., M. Beyer et al. (2005). Cardiac reporter gene imaging using the human sodium/iodide symporter gene. *Cardiovasc. Res.* **65**(1): 195–202.

Mo, R., S. Lin et al. (2008). Preliminary in vitro study of ultrasound sonoporation cell labeling with superparamagnetic iron oxide particles for MRI cell tracking. *Conf. Proc. IEEE Eng. Med. Biol. Soc.* **2008**: 367–370.

Modo, M., M. Hoehn et al. (2005). Cellular MR imaging. *Mol. Imaging* **4**(3): 143–164.

Muja, N. and J. W. Bulte (2009). Magnetic resonance imaging of cells in experimental disease models. *Prog. Nucl. Magn. Reson. Spectrosc.* **55**(1): 61–77.

Nahrendorf, M., D. E. Sosnovik et al. (2009). Multimodality cardiovascular molecular imaging, Part II. *Circ. Cardiovasc. Imaging* **2**(1): 56–70.

Naumova, A. V., H. Reinecke et al. (2010). Ferritin overexpression for noninvasive magnetic resonance imaging-based tracking of stem cells transplanted into the heart. *Mol. Imaging* **9**(4): 201–210.

Nieman, B. J., J. Y. Shyu et al. (2010). In vivo MRI of neural cell migration dynamics in the mouse brain. *Neuroimage* **50**(2): 456–464.

Noh, Y. W., Y. T. Lim et al. (2008). Noninvasive imaging of dendritic cell migration into lymph nodes using near-infrared fluorescent semiconductor nanocrystals. *FASEB J.* **22**(11): 3908–3918.

Park, J. J., T. S. Lee et al. (2012). Comparison of cell-labeling methods with (1)(2)(4)I-FIAU and (6)(4)Cu-PTSM for cell tracking using chronic myelogenous leukemia cells expressing HSV1-tk and firefly luciferase. *Cancer Biother. Radiopharm.* **27**(10): 719–728.

Patel, D., A. Kell et al. (2010). Cu^{2+}-labeled, SPION loaded porous silica nanoparticles for cell labeling and multifunctional imaging probes. *Biomaterials* **31**(10): 2866–2873.

Pittet, M. J., J. Grimm et al. (2007). In vivo imaging of T cell delivery to tumors after adoptive transfer therapy. *Proc. Natl. Acad. Sci. USA* **104**(30): 12457–12461.

Plank, C., F. Scherer et al. (2003). Magnetofection: Enhancing and targeting gene delivery with superparamagnetic nanoparticles and magnetic fields. *J. Liposome Res.* **13**(1): 29–32.

Politi, L. S. (2007). MR-based imaging of neural stem cells. *Neuroradiology* **49**(6): 523–534.

Politi, L. S., M. Bacigaluppi et al. (2007). Magnetic resonance-based tracking and quantification of intravenously injected neural stem cell accumulation in the brains of mice with experimental multiple sclerosis. *Stem Cells* **25**(10): 2583–2592.

Ponomarev, V. (2009). Nuclear imaging of cancer cell therapies. *J. Nucl. Med.* **50**(7): 1013–1016.

Ponomarev, V., M. Doubrovin et al. (2004). A novel triple-modality reporter gene for whole-body fluorescent, bioluminescent, and nuclear noninvasive imaging. *Eur. J. Nucl. Med. Mol. Imaging* **31**(5): 740–751.

Qian, Z. M., H. Li et al. (2002). Targeted drug delivery via the transferrin receptor-mediated endocytosis pathway. *Pharmacol. Rev.* **54**(4): 561–587.

Ray, P., A. De et al. (2004). Imaging tri-fusion multimodality reporter gene expression in living subjects. *Cancer Res.* **64**(4): 1323–1330.

Ritchie, D., L. Mileshkin et al. (2007). In vivo tracking of macrophage activated killer cells to sites of metastatic ovarian carcinoma. *Cancer Immunol. Immunother.* **56**(2): 155–163.

Rudelius, M., H. E. Daldrup-Link et al. (2003). Highly efficient paramagnetic labelling of embryonic and neuronal stem cells. *Eur. J. Nucl. Med. Mol. Imaging* **30**(7): 1038–1044.

Schachinger, V., A. Aicher et al. (2008). Pilot trial on determinants of progenitor cell recruitment to the infarcted human myocardium. *Circulation* **118**(14): 1425–1432.

Scherer, F., M. Anton et al. (2002). Magnetofection: Enhancing and targeting gene delivery by magnetic force in vitro and in vivo. *Gene Ther.* **9**(2): 102–109.

Schulze, E., J. T. Ferrucci et al. (1995). Cellular uptake and trafficking of a prototypical magnetic iron-oxide label in-vitro. *Invest. Radiol.* **30**(10): 604–610.

Slotkin, J. R., L. Chakrabarti et al. (2007). In vivo quantum dot labeling of mammalian stem and progenitor cells. *Dev. Dyn.* **236**(12): 3393–3401.

Soenen, S. J. and M. De Cuyper (2010). Assessing iron oxide nanoparticle toxicity in vitro: Current status and future prospects. *Nanomedicine (Lond.)* **5**(8): 1261–1275.

Soenen, S. J., U. Himmelreich et al. (2010). Intracellular nanoparticle coating stability determines nanoparticle diagnostics efficacy and cell functionality. *Small* **6**(19): 2136–2145.

Soenen, S. J., P. Rivera-Gil et al. (2011). Cellular toxicity of inorganic nanoparticles: Common aspects and guidelines for improved nanotoxicity evaluation. *Nano Today* **6**(5): 446–465.

Solanki, A., J. D. Kim et al. (2008). Nanotechnology for regenerative medicine: Nanomaterials for stem cell imaging. *Nanomedicine* **3**(4): 567–578.

Srinivas, M., E. H. Aarntzen et al. (2010). Imaging of cellular therapies. *Adv. Drug Deliv. Rev.* **62**(11): 1080–1093.

Srinivas, M., A. Heerschap et al. (2010). F-19 MRI for quantitative in vivo cell tracking. *Trends Biotechnol.* **28**(7): 363–370.

Srinivas, M., I. Melero et al. (2013). Cell tracking using multimodal imaging. *Contrast Media Mol. Imaging* **8**(6): 432–438.

Srinivas, M., P. A. Morel et al. (2007). Fluorine-19 MRI for visualization and quantification of cell migration in a diabetes model. *Magn. Reson. Med.* **58**(4): 725–734.

Srinivas, M., M. S. Turner et al. (2009). In vivo cytometry of antigen-specific t cells using 19F MRI. *Magn. Reson. Med.* **62**(3): 747–753.

Su, H., A. Forbes et al. (2004). Quantitation of cell number by a positron emission tomography reporter gene strategy. *Mol. Imaging Biol.* **6**(3): 139–148.

Su, Y. Y., M. Hu et al. (2010). The cytotoxicity of CdTe quantum dots and the relative contributions from released cadmium ions and nanoparticle properties. *Biomaterials* **31**(18): 4829–4834.

Sutton, E. J., T. D. Henning et al. (2008). Cell tracking with optical imaging. *Eur. Radiol.* **18**(10): 2021–2032.

Suzuki, Y., S. Zhang et al. (2007). In vitro comparison of the biological effects of three transfection methods for magnetically labeling mouse embryonic stem cells with ferumoxides. *Magn. Reson. Med.* **57**(6): 1173–1179.

Swirski, F. K., C. R. Berger et al. (2007). A near-infrared cell tracker reagent for multiscopic in vivo imaging and quantification of leukocyte immune responses. *PLoS One* **2**(10): e1075.

Tang, Y., K. Shah et al. (2003). In vivo tracking of neural progenitor cell migration to glioblastomas. *Hum. Gene Ther.* **14**(13): 1247–1254.

Taylor, A., K. M. Wilson et al. (2012). Long-term tracking of cells using inorganic nanoparticles as contrast agents: Are we there yet? *Chem. Soc. Rev.* **41**(7): 2707–2717.

Terrovitis, J., K. F. Kwok et al. (2008). Ectopic expression of the sodium-iodide symporter enables imaging of transplanted cardiac stem cells in vivo by single-photon emission computed tomography or positron emission tomography. *J. Am. Coll. Cardiol.* **52**(20): 1652–1660.

Thakur, M. L., R. E. Coleman et al. (1977). Indium-111-labeled leukocytes for the localization of abscesses: Preparation, analysis, tissue distribution, and comparison with gallium-67 citrate in dogs. *J. Lab. Clin. Med.* **89**(1): 217–228.

Thakur, M. L., M. J. Welch et al. (1976). Indium-LLL labeled platelets: Studies on preparation and evaluation of in vitro and in vivo functions. *Thromb. Res.* **9**(4): 345–357.

Thompson, M., D. M. Wall et al. (2005). In vivo tracking for cell therapies. *Q. J. Nucl. Med. Mol. Imaging* **49**(4): 339–348.

Torrente, Y., M. Gavina et al. (2006). High-resolution x-ray microtomography for three-dimensional visualization of human stem cell muscle homing. *FEBS Lett.* **580**(24): 5759–5764.

Tran, N., P. R. Franken et al. (2007). Intramyocardial Implantation of bone marrow-derived stem cells enhances perfusion in chronic myocardial infarction: Dependency on initial perfusion depth and follow-up assessed by gated pinhole SPECT. *J. Nucl. Med.* **48**(3): 405–412.

Tseng, C. L., I. L. Shih et al. (2010). Gadolinium hexanedione nanoparticles for stem cell labeling and tracking via magnetic resonance imaging. *Biomaterials* **31**(20): 5427–5435.

van den Bos, E. J., A. Wagner et al. (2003). Improved efficacy of stem cell labeling for magnetic resonance imaging studies by the use of cationic liposomes. *Cell Transplant.* **12**(7): 743–756.

Vande Velde, G., S. Couillard-Despres et al. (2012). In situ labeling and imaging of endogenous neural stem cell proliferation and migration. *Wiley Interdiscip. Rev. Nanomed. Nanobiotechnol.* **4**(6): 663–679.

Vandsburger, M. (2014). Cardiac cell tracking with MRI reporter genes: Welcoming a new field. *Curr. Cardiovasc. Imaging Rep.* **7**: 9250.

Vandsburger, M. H., M. Radoul et al. (2013a). Ovarian carcinoma: Quantitative biexponential MR imaging relaxometry reveals the dynamic recruitment of ferritin-expressing fibroblasts to the angiogenic rim of tumors. *Radiology* **268**(3): 790–801.

Vandsburger, M. H., M. Radoul et al. (2013b). MRI reporter genes: Applications for imaging of cell survival, proliferation, migration and differentiation. *NMR Biomed.* **26**(7): 872–884.

Villa, C., S. Erratico et al. (2010). Stem cell tracking by nanotechnologies. *Int. J. Mol. Sci.* **11**(3): 1070–1081.

Vreys, R., G. Vande Velde et al. (2010). MRI visualization of endogenous neural progenitor cell migration along the RMS in the adult mouse brain: Validation of various MPIO labeling strategies. *Neuroimage* **49**(3): 2094–2103.

Vrtovec, B., G. Poglajen et al. (2013). Effects of intracoronary CD34(+) stem cell transplantation in nonischemic dilated cardiomyopathy patients 5-year follow-up. *Circ. Res.* **112**(1): 165–173.

Walczak, P., D. A. Kedziorek et al. (2005). Instant MR labeling of stem cells using magnetoelectroporation. *Magn. Reson. Med.* **54**(4): 769–774.

Walczak, P., D. A. Kedziorek et al. (2007). Applicability and limitations of MR tracking of neural stem cells with asymmetric cell division and rapid turnover: The case of the shiverer dysmyelinated mouse brain. *Magn. Reson. Med.* **58**(2): 261–269.

Weissleder, R., H. C. Cheng et al. (1997). Magnetically labeled cells can be detected by MR imaging. *J. Magn. Reson. Imaging* **7**(1): 258–263.

Weissleder, R., A. Moore et al. (2000). In vivo magnetic resonance imaging of transgene expression. *Nat. Med.* **6**(3): 351–355.

Weller, G. E., E. Lu et al. (2003). Ultrasound imaging of acute cardiac transplant rejection with microbubbles targeted to intercellular adhesion molecule-1. *Circulation* **108**(2): 218–224.

Winter, E. M., B. Hogers et al. (2010). Cell tracking using iron oxide fails to distinguish dead from living transplanted cells in the infarcted heart. *Magn. Reson. Med.* **63**(3): 817–821.

Wu, J. C., I. Y. Chen et al. (2003). Molecular imaging of cardiac cell transplantation in living animals using optical bioluminescence and positron emission tomography. *Circulation* **108**(11): 1302–1305.

Xie, J., J. H. Wang et al. (2010). Human serum albumin coated iron oxide nanoparticles for efficient cell labeling. *Chem. Commun.* **46**(3): 433–435.

Zanzonico, P., G. Koehne et al. (2006). [131I]FIAU labeling of genetically transduced, tumor-reactive lymphocytes: Cell-level dosimetry and dose-dependent toxicity. *Eur. J. Nucl. Med. Mol. Imaging* **33**(9): 988–997.

Zelivyanskaya, M. L., J. A. Nelson et al. (2003). Tracking superparamagnetic iron oxide labeled monocytes in brain by high-field magnetic resonance imaging. *J. Neurosci. Res.* **73**(3): 284–295.

Zhang, S. J. and J. C. Wu (2007). Comparison of imaging techniques for tracking cardiac stem cell therapy. *J. Nucl. Med.* **48**(12): 1916–1919.

Zhou, R., P. D. Acton et al. (2006). Imaging stem cells implanted in infarcted myocardium. *J. Am. Coll. Cardiol.* **48**(10): 2094–2106.

Zhou, R., D. H. Thomas et al. (2005). In vivo detection of stem cells grafted in infarcted rat myocardium. *J. Nucl. Med.* **46**(5): 816–822.

Zhu, J. H., L. F. Zhou et al. (2006). Tracking neural stem cells in patients with brain trauma. *New Engl. J. Med.* **355**(22): 2376–2378.

29

Imaging of Cardiovascular Disease

Aleksandra Kalinowska and Lawrence W. Dobrucki

29.1 INTRODUCTION

In the late 1990s, predictions were made that by the year 2020 cardiovascular disease would become a dominant health problem and the major cause of mortality (Murray and Lopez 1997). The prophecy has already been fulfilled almost a decade earlier—cardiovascular disease is a prevalent problem in today's aging American society. It affects over 70% of people above 60 years of age and already about 12% of young citizens only in their 20s (Roger 2011).

Molecular imaging has the potential to significantly impact preclinical research and future clinical cardiovascular care. The introduction of novel imaging technologies allows noninvasive diagnosis and risk assessment. It plays a pivotal role in designing individually tailored treatment, as well as in observing the development of cardiovascular pathologies and in monitoring the efficacy of complex therapies.

Although some applications of molecular imaging are well established, other clinical applications are still under development. With recent trends in medical practice, the focus of research on cardiovascular imaging is shifting from post-event treatment to effective prevention. For instance, one of the lately emerging applications of molecular imaging is early detection of unstable plaques in atherosclerosis. However, in both vascular and myocardial applications, the ability to reliably prognosticate the onset of a cardiovascular disease in its early stage requires high-resolution technologies that allow visualization of diminutive changes.

This chapter will discuss the application of previously described molecular imaging modalities, such as single-photon emission computed tomography (SPECT), positron-emission tomography (PET), computed tomography (CT), magnetic resonance (MRI), and fluorescence imaging, which have the potential to be clinically significant in the treatment of patients with cardiovascular diseases. It also discusses about hybrid imaging methods, which give the most effective combination of certainty, sensitivity, specificity, and spatial resolution and, therefore, are currently showing the biggest promise of successful application in clinical settings.

Patients with conditions, such as heart failure, acute myocardial infarction, myocarditis, or deep venous thrombosis (DVT), have the potential to benefit from developments in molecular imaging. Each cardiovascular disease can be linked to certain pathophysiological processes of the cardiovascular system, such as atherosclerosis, stenosis, ventricular remodeling, thrombosis, inflammation, and/or apoptosis. Many diseases, when considered at all stages of advancement, can be linked to more than one of these research areas.

It is worth noting that molecular signals commonly overlap between different biological processes, which make precise imaging even more challenging. For most processes, key molecular events or signaling proteins have been identified and serve as imaging targets for tracer-enhanced imaging. In other cases, characteristic abnormalities in perfusion as well as changes in vessel-wall or myocardium fiber architecture have been defined and provide a basis for other imaging modalities. All these will be the topic of further discussion in this chapter.

29.2 CAUSES OF CARDIOVASCULAR DISEASE

Certainly, the most common cardiovascular pathologies that are prevalent in today's society include hypertension and atherosclerosis. These in turn lead to further complications of the cardiovascular system. The essential hypertension that comes naturally with old age and affects 95% of all hypertensive patients does not have a precisely identified source, being caused by a range of environmental and genetic factors. However, its diagnosis is relatively easy, making it possible for the clinician to introduce appropriate treatment, such as pharmacological therapy.

Atherosclerosis, on the other hand, is a chronic disease, that is significantly harder to diagnose noninvasively and can benefit greatly from the continuous advancements in molecular imaging. It is generally characterized as a condition, in which high-risk fatty plaques accumulate in the arterial wall and is considered a systemic disease of the vessel wall as it can involve the aorta as well as coronary, carotid, and peripheral arteries. Stable atherosclerotic lesions may remain asymptomatic for decades (Sanz and Fayad 2008). However, with the progression of atherosclerotic diseases, many symptoms, such as stenosis, inflammation, and revascularization, can be observed and used for targeted imaging approaches (Figure 29.1).

In all related studies, atherosclerosis has been noted to induce inflammatory response, which is fundamental to the formation, progression, and rupture of atherosclerotic plaques. These processes include monocyte adhesion, cytokine secretion, as well as reactive oxygen species activation, and are key biomarkers to the early diagnosis of atherosclerosis.

If treatment is not administered, the inflammatory response progresses significantly and causes further complications. The process is complex—monocytes in the arterial wall differentiate into macrophages, which then internalize previously modified lipoproteins and settle as foam cells in the arterial tunica intima. If the organism's anti-inflammatory signals do not counterbalance

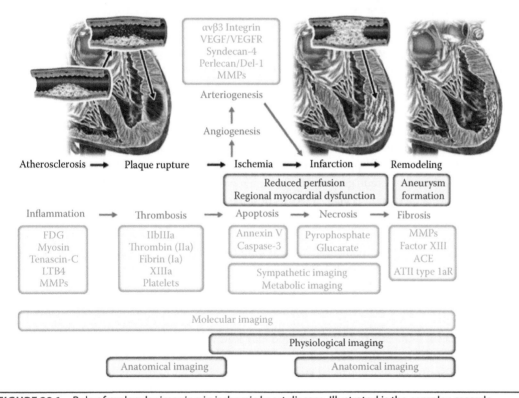

FIGURE 29.1 Role of molecular imaging in ischemic heart disease. Illustrated is the complex cascade of molecular and cellular events associated with the progression from early to advanced atherosclerosis, ischemic injury, and subsequent angiogenesis, arteriogenesis, and post-infarction remodeling. Highlighted within the green boxes are some of the potential molecular targets for imaging these molecular processes, in temporal relation to conventional imaging indexes (blue boxes) using standard imaging approaches that focus on the evaluation of physiology and changes in anatomic structure. Molecular imaging provides a noninvasive evaluation of molecular events that precede the manifestation of the pathophysiological or anatomic changes associated with ischemic heart disease. *Abbreviations:* ACE, angiotensin-converting enzyme; ATII type 2aR, angiotensin II type 2 adrenergic receptor; LTB4, leukotriene B4; MMP, matrix metalloproteinase; VEGF, vascular endothelial growth factor; VEGFR, vascular endothelial growth factor receptor. (Reprinted from *JACC Cardiovasc. Imaging*, 4, Sinusas, A.J., Thomas, J.D., and Mills, G., The future of molecular imaging, 799–806. Copyright 2011, with permission from Elsevier.)

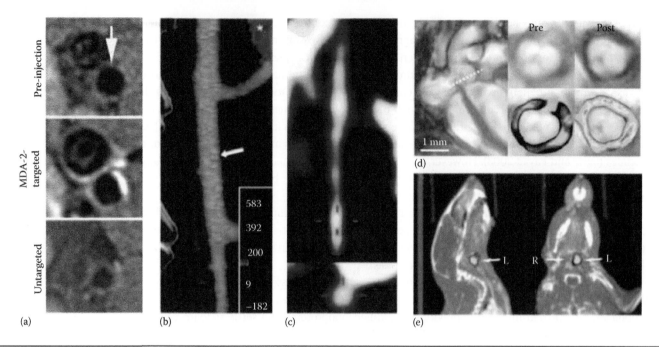

FIGURE 29.2 Atherosclerosis imaging. (a) Imaging of inflammatory atherosclerosis in rabbits using Gd-loaded oxLDL-targeted micelles. (b) Macrophage-targeted nanoparticles enable CT imaging of inflammatory atherosclerosis in a rabbit model. (c) FDG PET imaging in a rabbit model of atherosclerosis. (d) MRI of VCAM-1 in the aortic root of apoE$^{-/-}$ mice. After injection of VCAM-1-targeted nanoparticles, a signal decrease was observed on T2*-weighted gradient MRI. (e) RP782 MMP-targeted microSPECT-CT 3 weeks after carotid injury in the apoE$^{-/-}$ mouse. Arrows point to the injured left (L) and noninjured right (R) carotid arteries. (Reprinted from Nahrendorf, M. et al., *Circ. Cardiovasc. Imaging*, 2, 56, 2009. With permission.)

atherogenic pathways, plaques undergo destabilization. Destabilization may eventually result in ruptures and thrombi formation. A thromboembolic event at the site of or downstream from the plaque rupture is a quite common clinical consequence of atherosclerosis. Imaging of thrombosis has therefore been proven a successful approach in helping confirm the diagnosis of atherosclerotic changes.

Similarly, when the plaque grows, arterial lumen narrows and subsequently blood perfusion is reduced. Abnormal perfusion is also the focus of some imaging strategies and will be discussed in later parts of this chapter. As a parallel process, local ischemia caused by a lipid-laden plaque leads to myocardial hypoxia. In the organism's attempt to restore proper blood perfusion, revascularization is stimulated and, therefore, biomarkers of angiogenic and arteriogenic processes, such as integrin avb3 or vascular endothelial growth factor, appear in areas of present atherosclerotic plaques. These processes are very similar to those of new vessel formation, known as angiogenesis, and thus one can refer to the next chapter for a thorough description of applicable imaging strategies.

At a late stage of development, if coronary arteries are affected, chronic ischemia in the heart muscle may occur—the atherosclerotic plaque may lead to myocardial infarction (MI), stroke, or even sudden death (Hansson 2005). Until recently, atherosclerosis was detectable only at a very late stage of the disease when arterial stenosis or reduced organ perfusion would become clearly visible on low-sensitivity imaging scans. Novel targeted molecular imaging approaches permit its diagnosis at an early stage when characteristic but still miniscule changes begin appearing in vessel morphology. Radiotracer-based imaging enables the identification of lesions in vascular beds of the coronary and carotid arteries prior to significant vessel occlusion and gives hope of lowering the occurrence rate of damaging myocardial ischemia (Figure 29.2).

29.3 IMAGING OF INFLAMMATION

One of the most significant pathological processes in the body associated with cardiovascular diseases, such as atherosclerosis or aortic aneurysms, are inflammatory changes localized in and around the spot of interest. An inflammatory response involves the interaction of multiple immigrant cells, such as lymphocytes, neutrophils, and macrophages, which are part of the immunological system and appear in high concentration around an inflammatory site. Imaging approaches focus on either the metabolic activity or membrane markers of inflammatory cells and thus make radiotracer-based imaging techniques most suitable for application. Modalities such as PET are ideal because they allow specific noninvasive detection of changes in the region of interest. Novel hybrid scanners combine PET with MRI and CT and thus allow better localization of the pathological changes in situ, combining PET's high specificity with the high-spatial resolution of MR and CT.

Initial research targeted cells such as technetium-99m (^{99m}Tc) or indium-111 (^{111}In)-labeled macrophages and lymphocytes as biomarkers for molecular imaging. Other groups chose a different approach and focused on increased metabolic activity by analyzing gallium-67 (^{67}Ga) citrate and fluorine-18 (^{18}F)-fluorodeoxyglucose (^{18}FDG) as potential enhancers for PET scanners (Figure 29.3).

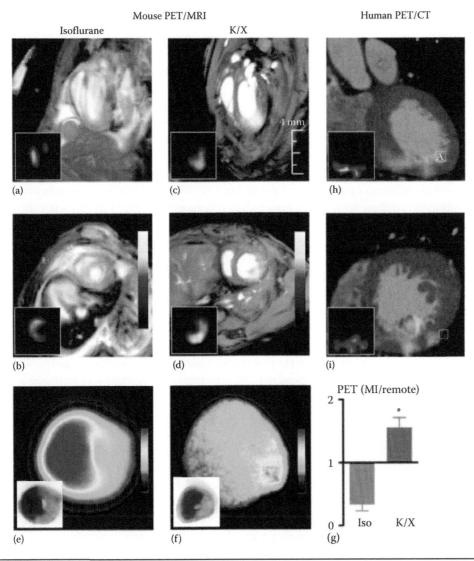

FIGURE 29.3 ^{18}FDG PET/GdDTPA MRI (a–d) ^{18}FDG PET/MRI short- and long-axis views acquired in mice on day 5 after MI using different anesthesia (K/X: ketamine/xylazine). Insets show PET signal. MRI used delayed enhancement cine gradient echo. (e, f) Autoradiography of short-axis rings in mice with isoflurane (e) versus K/X (f). Insets depict the infarct as unstained pale tissue on TTC. (g) Infarct to remote myocardium SUV ratio in ^{18}FDG scans with respective anesthesia. Mean ± SEM. *P < 0.01. (h, i) ^{18}FDG PET/CT in a patient 5 days after right coronary artery occlusion shows increased PET signal in the injured inferior LV wall. (Reprinted from *J. Am. Coll. Cardiol.*, 59(2), Lee, W.W., Marinelli, B., van der Laan, A.M., Sena, B.F., Gorbatov, R., Leuschner, F., Dutta, P. et al., PET/MRI of inflammation in myocardial infarction, 153–163. Copyright 2012, with permission from Elsevier.)

With time, a number of various biomarkers have been taken into consideration and investigated—these include choline metabolism, translocator protein (TSPO), somatostatin receptor (SSTR), vascular adhesion protein (VAP-1), matrix metalloproteinases (MMPs), integrin receptors, and VCAM-1 (Lee et al. 2012). Out of all these, FDG currently remains the most used radiopharmaceutical agent both in research and in clinical practice (Basu et al. 2009).

29.3.1 MACROPHAGES

Macrophages are a key inflammatory factor and are present in large amounts in regions where inflammatory changes had taken place. They are metabolically active with an increased glucose uptake and, as phagocytic cells, they internalize various carbohydrate and polyol-coated nanoparticles, which can be injected and can thus serve as characteristic markers for imaging.

To take advantage of the increased metabolic rate in active inflammatory cells compared to peripheral nonimmigrant cells, a glucose analog, ^{18}FDG, has been used to show an enhanced signal in PET images. The accumulated tagged glucose is a positron emitter, making it a perfect biomarker for sensitive PET scans. Because of the intensified glucose transport, there is a high concentration of sugars in the inflammatory cells. Hexokinase-mediated intracellular phosphorylation makes the concentration stable enough to give reliable images.

Issues have been raised that the beating heart already has a high uptake of glucose due to smooth muscle activity and it has been shown that ^{18}FDG PET detection of macrophage-rich coronary plaques suffers from substantial background uptake by a metabolically active myocardium, especially in the left ventricle. However, myocardial ^{18}FDG signal can be suppressed by a rich-in-fat low-on-glucose diet and β-adrenergic blockers applied prior to imaging and lowered to a level where it has almost no influence on image contrast, making ^{18}FDG suitable for imaging many cardiovascular conditions, including for instance myocarditis (Cheng et al. 2010). ^{18}FDG does not accumulate in normal vascular structures, rendering it applicable for imaging areas with a high metabolic rate, such as lipid-laden plaques (Tawakol et al. 2006). However, although it remains the most commonly applied tracer in the imaging of inflammation in clinical settings, not all conditions can be detected with ^{18}FDG-PET. A study done on chronic inflammation in the wall of asymptomatic abdominal aortic aneurysms showed an insufficient signal increase in the area of interest due to locally higher metabolism (Tegler et al. 2012).

Alternative strategies have thus been considered for macrophage imaging—recent research has been done on the use of nanoparticle-enhanced MRI. Superparamagnetic iron oxides are considered pioneer agents in this approach because these magnetic nanoparticles with 3 nm superparamagnetic iron oxide cores were the first to be noticed to reduce the transverse relaxation of hydrogen atoms on T2- and T2*-weighted images (Jaffer et al. 2007). Such particles, or similar ultrasmall paramagnetic iron oxides (USPIO), are engulfed by macrophages *in vivo* and result in a detectable decrease of MRI signal, allowing for targeted imaging.

Studies on nanoparticle-enhanced MRI of macrophages have already advanced to human testing. In a recent study, signal intensity was shown to weaken in proportion to the magnitude of inflammation at sites where an atherosclerotic plaque had formed (Trivedi et al. 2006). USPIO-enhanced carotid MRI was also applied during a clinical trial for statin treatment efficacy and showed significant results in the detection of the suppression of macrophage accumulation in carotid atheroma (Tang et al. 2009). These results show great promise for introducing the technology into clinical settings.

As a whole, however, no imaging modality is ideal. Magnetic resonance studies provide unparalleled versatility and soft tissue contrast but are not highly sensitive, which makes injections of relatively large amounts of nanoparticles (2–20 mg Fe/kg) necessary for sufficient signal enhancement. Lower concentrations of nanoparticles can be used by nuclear techniques, such as PET, which are incomparably more sensitive. In a recent study, the detection threshold of a nanoparticle in an imaging phantom was ≈5 μg Fe/mL using T2-weighted MRI and 0.1 μg Fe/mL for PET–CT imaging (Nahrendorf et al. 2008b). Although the nanoparticle dose is reduced in nuclear imaging,

PET in turn does not offer high spatial resolution. Currently, the application of hybrid systems shows the greatest potential, which combine the high sensitivity of PET with the anatomic detail provided by CT or MRI. To enhance sensitivity, USPIO nanoparticles have been introduced. By attaching optical or nuclear reporters to the magnetic agents, the application of hybrid molecular techniques is made possible.

A dual-contrast agent detectable by MRI and fluorescence imaging was used in a recent study of atherosclerosis in mice (Nahrendorf et al. 2006b). Aortic plaques were identified through molecular imaging of raised concentrations of vascular cell-adhesion molecule 1 (VCAM-1). These cell-adhesion molecules participate in the inflammatory response of atherosclerotic lesions by stimulating the recruitment of leukocytes into the vessel wall.

In a new study of rodent models of atherosclerosis, a significant correlation between macrophage density and MRI signal was found using gadolinium-loaded immunomicelles. This contrast agent was targeted at the macrophage scavenger receptor (Amirbekian et al. 2007). Besides imaging macrophages, gadolinium-loaded immunomicelles have also made it possible to image oxidized low-density lipoproteins, which are present in large quantities in atheromatous plaques (Briley-Saebo et al. 2008).

Recent experiments demonstrate the ability to image macrophages with an iodine-containing contrast agent for CT (Hyafil et al. 2007). This approach may prove particularly useful in the clinical evaluation of atherosclerosis in carotid and coronary arteries because it provides high spatial and temporal resolutions. It would be a great enhancement to current methods of macrophage-associated atheroma imaging.

Trimodality systems are also emerging with a promise of improved sensitivity and higher specificity. A novel class of macrophage-targeted agents includes the nuclear tracer labeled with ^{64}Cu and a near-infrared fluorochrome, rendering a trimodality reporter for PET, MRI, and fluorescence imaging. A recent study compared the *in vivo* distribution of trireporter nanoparticle (^{64}Cu-TNP) to ^{18}FDG in atherosclerotic mice. PET signals were detected in plaque-laden territories in the aortic root for both ^{18}FDG and ^{64}Cu-TNP, but those received from atherosclerotic lesions persisted longer and were significantly higher for ^{64}Cu-TNP, validating the benefit of a trimodality system (Nahrendorf et al. 2008b). Novel methods for atherosclerotic imaging are continuously being researched, but their applicability in clinical settings is yet to be characterized.

29.3.2 Antitenascin-C Antibody

With currently available protocols, a definite diagnosis of myocarditis requires an invasive biopsy test, making a new diagnostic technique strongly desired. It has been noted that during the processes of wound healing and inflammation, an increased expression of an oligometric extracellular glycoprotein, called tenascin-C, occurs in the extracellular matrix and thus a monoclonal antibody against it has been identified as a potential imaging agent. These high levels of tenascin-C are caused by the disintegration of cell membranes during necrosis and the exposure of myosin heavy chains from inside the cells to the circulation, where the radiolabeled antimyosin antibody binds specifically to these exposed myosin molecules.

SPECT studies have been conducted and a recent analysis of rodent models with myocarditis showed a correlation between high concentrations of ^{111}In-labeled antitenescin-C and sites of myocardial inflammation. There was an almost eightfold increase in signal 6 h after injection and rapid clearance of radioactivity from the blood pool, which shows promise for radiolabeled antitenescin-C being useful in the imaging of myocarditis (Sato et al. 2002). In another study, patients at the acute and chronic stages were compared and positive results were obtained only for the former, suggesting that antimyosin imaging may be used for evaluating whether myocarditis is active or healed (Matsumori et al. 1996). Patients with dilated ventricles, in particular, obtained strongly positive results, suggesting a correlation between antimyosin uptake and poor prognosis. Radiolabeled antitenascin-C requires more research, but with current results it shows promise of being useful for the noninvasive diagnosis of myocarditis.

29.3.3   LTB4 Receptor

Active inflammatory cells produce a leukotriene B4 (LTB4) and secrete it as a potent chemotactic agent. In leukocytes, the leukotriene induces the adhesion and activation of the cells on the endothelium, while in neutrophils, LTB4 induces the formation of reactive oxygen species and the release of lysosome enzymes by these cells, contributing to the progression of inflammatory processes. Recently, a radiolabeled LTB4 receptor antagonist, ^{99m}Tc-RP517, has been developed for *in vivo* imaging of acute inflammation or infection. The tracer ^{99m}Tc-RP517 has a unique ability to label white blood cells *in vivo* after intravenous injection and has been noted to selectively localize to areas of inflammation (Serhan and Prescott 2000).

The feasibility of the application of ^{99m}Tc-RP517 as a targeted tracer for molecular imaging has been evaluated. The relative ^{99m}Tc-RP517 binding to human leukocyte subtypes was determined and ^{99m}Tc-RP517 uptake in inflamed canine myocardium was measured. It was concluded that ^{99m}Tc-RP517 binds to the neutrophil LTB4 receptor after intravenous injection and shows potential as an inflammation-imaging agent (Riou et al. 2002).

Due to the lipophilic nature of ^{99m}Tc-RP517 as a molecule, its application into the organism leads to high hepatobiliary clearance and thus large amounts of gastrointestinal uptake. Efforts have been made to eliminate this concern and thus alternative constructs of the LTB4 antagonist are currently being evaluated (van Eerd et al. 2003). The modified LTB4 antagonist (^{111}In-DPC11870-11) has shown specific receptor–ligand binding and infectious foci have been identified rapidly after injection with no significant accumulation of radioactivity in the gastrointestinal tract. This compound, having excellent characteristics for locating infectious and inflammatory foci, has the potential to replace ^{99m}Tc-RP517 and be used clinically.

29.4   IMAGING OF STENOSIS

Perfusion in the myocardium is a precisely controlled process, regulated by vessels and the endothelium. Any dysfunctions, such as vasoconstriction from acetylcholine and not enough nitric oxide release, are early signs of coronary artery disease (CAD). Coronary arteries have autoregulation that normalizes blood flow and oxygen supply by reducing resistance in the distal perfusion beds. However, when stenosis exceeds 85%–90% of the diameter of the lumen, the autoregulation becomes insufficient and resting blood flow is decreased, initiating the ischemic cascade (Salerno and Beller 2009).

Imaging of abnormal coronary flow has the potential to prevent more serious damage, such as ischemia, MI, or coronary aneurysms. It can help diagnose CAD with precise location and degree of development, as well as help identify wall motion abnormalities. Perfusion imaging is often carried out on patients with coronary artery bypass grafts and angioplasty to evaluate the success of ongoing treatment and evaluate for further procedures.

Myocardial perfusion imaging (MPI) is most effective in stress conditions that help visualize regional heterogeneity of coronary artery blood flow in the presence of CAD. Exercise or pharmacological stimulation with adenosine or dipyridamole induces a two- to fourfold increase in coronary vasodilation via both direct and endothelium-dependent flow-mediated processes. This makes it possible to image reduction in perfusion from flow-limiting stenosis, endothelial dysfunction, or adrenergic stimulation.

SPECT imaging is the most commonly applied technique to detect changes in perfusion. It can be done with radiotracers, such as thallium-201 (^{201}Tl), a potassium analog, which is taken up by viable myocytes in proportion to blood flow, or ^{99m}Tc-radiotracers, which show affinity to mitochondrial membranes. The uptake of these radiotracers is dependent on not only blood flow but also myocardial cellular integrity, which allows imaging to be done long after the first pass of the contrast agent, providing a significant advantage over other modalities. Additionally, imaging of the 3–4 h delayed redistribution permits ischemia to be clearly distinguished from myocardial scar.

There have been numerous studies on the efficacy of SPECT MPI. An analysis of 4480 patients with known or suspected CAD demonstrated mean sensitivity and specificity of 87% and 73%, respectively, for myocardial exercise-stress SPECT for detecting a >50% stenosis (Klocke et al. 2003). Another analysis of 2492 patients with known or suspected CAD demonstrated sensitivity and specificity of 89% and 75%, respectively, for vasodilator stress with dipyridamole or adenosine for the detection of a >50% stenosis (Klocke et al. 2003). SPECT MPI was validated as an important prognostic modality and is already in clinical use.

Attempts have also been made to image already existing ischemic areas based on myocardial blood flow. This technique is especially useful in deciphering infarct size. ^{99m}Tc-sestamibi, a lipophilic cation, can be injected intravenously into a patient and then imaged using SPECT technologies to successfully assess the size of ischemia (Gibbons et al. 2000). Measurements of infarct size with this technique have been applied in over 30 clinical trials of MI. The major advantage of these studies is the ability to compare different treatment strategies in acute MI using quantifiable assessment (Miller et al. 2010).

29.5 IMAGING OF LEFT VENTRICULAR REMODELING

Almost immediately after MI the affected ischemic tissue responds to the occurring changes. It begins a complex healing process and initiates an innate immune response. The heart attempts to regain mechanical stability and proper functionality, and as a result it initiates scar formation. With time the heart continues to change and undergoes left ventricular (LV) remodeling, which can involve hypertrophy and chamber dilation leading ultimately to chronic heart failure. During the development of postinfarction remodeling, the myocardium undergoes compensatory biochemical and structural alterations with specific changes in different regions of the heart depending on their location in respect to the source of MI. Different processes can be identified within three distinct anatomic regions: the MI itself, the surrounding border zone, and the remaining myocardium remote from the ischemic area.

Ventricular remodeling involves several key cell types and structural elements, including myocardial, endothelial, and inflammatory cells, as well as the extracellular matrix. A variety of regulatory systems become activated, such as MMPs, thrombin-dependent factor XIII (FXIII), and the renin–angiotensin system (RAS). The fiber architecture in the myocardium undergoes significant changes, as does the left ventricular capacity. The synapses in the muscle are affected and permanent alterations occur. Given the complexity of the processes involved in post-MI remodeling, imaging modalities must be equally diverse to effectively capture all of the featured changes in the remodeled myocardium.

29.5.1 AUTONOMIC INNERVATION

The autonomic nervous system is crucial for cardiac adaptation to the varying demands imposed upon the heart. It is directly involved in the physiological functionality of the cardiac muscle by regulating, for example, cardiac rhythm and LV ejection volume. The autonomic nervous system is also involved in the early response to pathologic stimuli and thus has been considered an attractive target for imaging in preventative care of individuals at the risk of cardiovascular disease (Bengel and Schwaiger 2004). Heart failure is a hyperadrenergic state, in which plasma norepinephrine is elevated and as a result cardiac β-adrenergic receptors become downregulated. These factors contribute to progressive impairment of postsynaptic signal transduction and degradation in LV systolic function. Therefore, there exists significant clinical interest in noninvasive approaches to determining altered sympathetic tones.

Several radiolabeled tracers, being either catecholamines or catecholamine analogs, have been identified (Figure 29.4). Among the most common is a norepinephrine analog ^{123}I-metaiodobenzylguanidine (^{123}I-MIBG) used in SPECT imaging, thanks to its unique resistance to monoamine oxidase and catechol-O-methyltransferase enzymes that normally interact with the

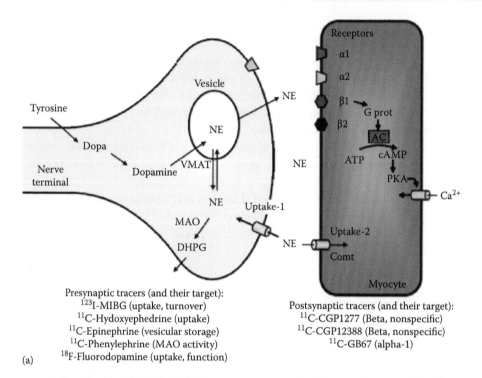

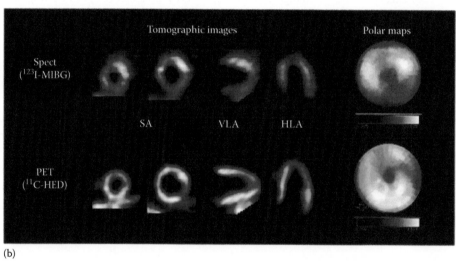

(b)

FIGURE 29.4 Radiotracers for presynaptic and postsynaptic neuroreceptor imaging. (a) Illustration of cardiac sympathetic neurotransmission and most commonly used radioligands for the assessment of cardiac presynaptic and postsynaptic function sympathetic innervation. (b) Images of the normal distribution of the presynaptic neuronal tracers (MIBG for SPECT; ^{11}C for PET) in healthy volunteers. *Abbreviations:* NE, norepinephrine; VMAT, vesicular monoamine transporter; MAO, monoaminooxidase; DHPG, dihydroxyphenylglycol; G prot, G-protein; AC, adenylcyclase; ATP, adenosinetriphosphate; cAMP, cyclic adenosinemonophosphate; PKA, protein kinase A; COMT, catechol-O-methyltransferase. (Reprinted from Sinusas, A.J. et al., *Circ. Cardiovasc. Imaging*, 1(3), 244, 2008. With permission.)

unaltered neurotransmitter norepinephrine. The imaging protocol is typically performed in two phases: at 15–30 min after ^{123}I-MIBG injection in the early phase and at 3–4 h in the delayed phase, allowing for determining both the uptake and washout of tracer signal (Carrio 2001). ^{123}I-MIBG is slightly different than other imaging markers as it can be used for imaging of ischemic myocardium at risk after acute MI. A persistent area of denervation within the LV has been determined by perfusion SPECT images taken during and ^{123}I-MIBG images taken 2 weeks after acute ischemic events (Sosnovik et al. 2005). Such presynaptic sympathetic denervation, as assessed by ^{123}I-MIBG, has significant impact on the pathophysiology of heart failure and leads to higher mortality (Chen et al. 2006). Recent findings have shown that ^{123}I-MIBG imaging can be used as a means of identifying HF patients that could benefit from β-adrenergic receptor blocker therapy (Nahrendorf et al. 2008b).

There are also several PET tracers, such as ^{11}C-meta-hydroxyephedrine (a synthetic norepinephrine analogue), that have been employed for sympathetic nerve imaging in human studies. PET offers higher resolution and sensitivity for interrogation of regional tracer distribution even in HF when the overall uptake is reduced (Bengel and Schwaiger 2004). PET also allows for the quantification and measurement of tracer kinetics. Various presynaptic tracers, such as ^{11}C-epinephrine and ^{11}C-phenylephrine, have been used together to allow for simultaneous evaluation of norepinephrine uptake, storage, and metabolism in presynaptic nerve terminals.

Postsynaptic receptors in turn have been targeted in PET imaging with compounds, such as ^{11}C-CGP12177, a hydrophilic nonselective β-adrenergic antagonist. *In vivo* experiments have already been done to verify the effectiveness of ^{11}C-labeled β1-adrenergic PET radioligands and ^{11}C-CGP12177 PET is being used to validate treatment therapies (Naya et al. 2009).

Imaging of the autonomic innervation in the myocardium has found application in clinical settings. It has been employed in oncology patients for the evaluation of chemotherapy-related LV dysfunction, in patients with chronic HF for assessment of risk of sudden death, as well as in diabetic individuals for determining diabetes-related cardiomyopathy. Such imaging has proven helpful in monitoring response to pharmacological treatment, including treatment with inhibitors of angiotensin or the β-adrenergic systems.

29.5.2 MATRIX METALLOPROTEINASES

The myocardial extracellular matrix is a medium for constant activity involving the synthesis and degradation of collagen. In a healthy heart, there is a dynamic balance between the two processes, but in a heart after MI, the formation of collagen begins to dominate over its degradation, allowing for LV remodeling and subsequently chamber enlargement.

MMPs are a family of zinc-dependent endopeptidases that play a pivotal role in ventricular remodeling through regulating the degradation of collagen (Visse and Nagase 2003). There are over 25 types of MMPs, differing in expression sites, substrate specificity, and molecular structure. Several MMP species identified within the myocardium may be deregulated in congestive heart failure. Those that are expressed in low levels in normal myocardial tissue tend to be upregulated in the post-infarction heart muscle. Using animal models with chronic heart failure, such a relationship has been demonstrated (Spinale 2002).

Also present in the myocardium are tissue inhibitors of the MMPs (TIMPs). These locally synthesized proteins bind to active MMPs and thereby regulate their proteolytic activity. However, there does not appear to be a concomitant increase in myocardial TIMP expression during LV remodeling and thus MMPs have been noted to be elevated in the progression of the remodeling process along with an abnormal increase in the degradation of the extracellular matrix. Pharmacological therapies aimed at a decrease in MMP activity are being tested. So far gene deletion of MMPs has been shown to have cardioprotective effects and attenuate LV enlargement (Ducharme et al. 2000).

MMPs and thus LV remodeling can be imaged post-MI *in vivo* by radiolabeling molecules that present high binding affinity to their catalytic domain (Su et al. 2005). MMP-targeted agents demonstrated a favorable biodistribution in a murine model of MI (Figure 29.5). One week after MI, ^{99m}Tc-labeled radiotracer (^{99m}Tc-RP805) and [201]TI were utilized for hybrid SPECT/CT imaging. A dual isotope protocol, involving parallel imaging of ^{99m}Tc-RP805 uptake and of ^{201}Tl perfusion, was performed. The study revealed corresponding ^{99m}Tc signal within the perfusion defect region, confirming MMP activation in that area. There was a noted fivefold increase in signal intensity at the site of interest compared to the rest of the body (Su et al. 2005). MMP-targeted imaging can be utilized to monitor therapeutic interventions directed at the inhibition of MMP activation.

Recently, MMP activity in MI was also evaluated using a near-infrared fluorescent probe (Chen et al. 2005). This molecular probe proved "smart" by becoming activated only after a peptide sequence was

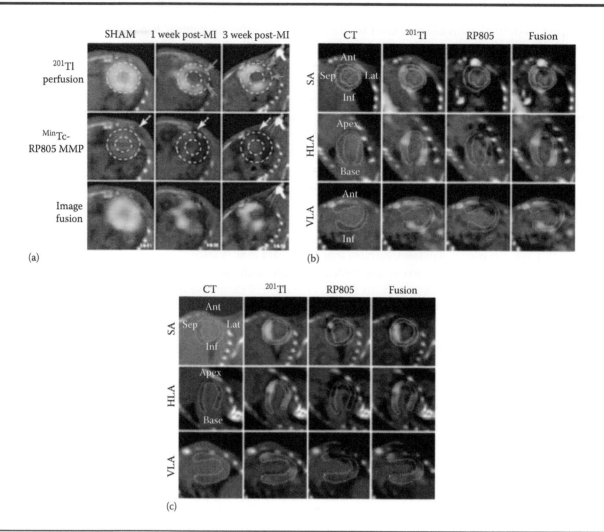

FIGURE 29.5   Hybrid micro-SPECT/CT reconstructed short-axis images were acquired without x-ray contrast (a) in control sham-operated mouse (left) and selected mice at 1 week (middle) and 3 weeks (right) after MI, after injection of ^{201}Tl (top row, green) and ^{99m}Tc-RP805 (middle row, red). A black-and-white (B&W) and multicolor fusion image is shown on bottom. Control heart demonstrates normal myocardial perfusion and no focal ^{99m}Tc-RP805 uptake within the heart, although some uptake is seen in chest wall at the thoracotomy site (dashed arrows). All post-MI mice have a large anterolateral ^{201}Tl perfusion defect (yellow arrows) and focal uptake of ^{99m}Tc-RP805 in defect area. A dashed circle is drawn around the heart to demonstrate the localization of ^{99m}Tc-RP805, the MMP radiotracer, within the infarcted area of the heart. Some activity is also seen in the peri-infarct border zone. Additional micro-SPECT/CT images were acquired by the use of a higher-resolution SPECT detector after the administration of x-ray contrast, at 1 week (b) and 3 weeks (c) after MI. The contrast agent permitted better definition of the LV myocardium, which is highlighted by white dotted line. Representative short-axis (SA), horizontal long-axis (HLA), and vertical long-axis (VLA) images are shown for two additional mice by use of the same format and color scheme. Focal uptake of ^{99m}Tc-RP805 is seen within the central infarct and peri-infarct regions, which again corresponds to ^{201}Tl perfusion defect. (Reprinted from Su, H. et al., *Circulation*, 112(20), 3157, 2005. With permission.)

recognized and cleaved by MMP-2 and MMP-9. Upon activation, the molecule changed its configuration and began releasing the fluorescent molecules. The use of this near-infrared fluorescence in an *in vivo* murine model made possible the determination of the time course of increased MMP activity after ischemic LV damage in the anterior wall of the heart. Other methods, such as gelatinase zymography and analysis of MMP-2 and MMP-9 mRNA levels, were utilized to confirm the obtained results. Through confocal microscopy, neutrophils were identified in addition to the MMPs within the infarct zone, which suggests MMPs can be used as biomarkers in future noninvasive imaging of protease activity.

29.5.3 FACTOR XIII

Another enzyme shown to have a significant role in post-infarction remodeling is factor XIII, a thrombin-activated substance involved in the final pathway of the coagulation cascade. The positive role of FXIII in infarct healing was suggested by clinical observation—lower FXIII protein content was detected in patients who had died due to MI than in those from the control cohort (Leuschner and Nahrendorf 2011). In studies involving preclinical animal models, mice with decreased levels of factor XIII demonstrated increased ventricular dilation and post-infarct rupture (Nahrendorf et al. 2006a). A further study suggested cardioprotective effects of a gene mutation associated with increased FXIII activity. Cardiac MRI of factor XIII-deficient mice demonstrated worse LV remodeling than in wild-type animals (Nahrendorf et al. 2006b). Numerous hypotheses have been made that the supplementation of factor XIII may have a beneficial role in post-infarct remodeling.

Imaging enhancers targeted at factor XIII have been employed. When active, factor XIII acts as a glutaminase that cross-links selected substrates to extracellular matrix proteins. A radiolabeled substrate analogue, ^{111}In-DOTA-FXIII, was applied, and after a short period of time, it gets accumulated in areas of increased factor XIII activity. A study in a murine model of MI has been conducted, where decreased levels of this ^{111}In-labeled peptide substrate were noted in animals treated with dalteparin, a direct thrombin inhibitor. After dalteparin treatment, the animals showed increased risk of infarct rupture, while mice with injected factor XIII presented increased factor XIII activity in the infarct zone and increased collagen synthesis as well as capillary density, suggesting improved post-MI healing (Nahrendorf et al. 2008a).

29.5.4 ACE Inhibitors and AT1 Antagonists

The activation of RAS has been proven to have quite unwanted effects in the failing heart. It shows increased activity in early HF and through its primary effector peptide, angiotensin II, has been linked to LV remodeling following the deterioration of myocardial function.

When the healing processes of the heart are not substantial, expression of prorenin, renin, and angiotensin-converting enzyme (ACE) is increased. All RAS components are locally produced in the heart and play important physiologic functions. RAS can be activated through signaling pathways mediated by the angiotensin II type I receptor (AT1). Its activity contributes to myocyte hypertrophy and apoptosis, as well as interstitial and perivascular collagen deposition (Aras et al. 2007). Inhibition of this pathway has been demonstrated to reverse the functional abnormalities caused by negative remodeling. Therefore, RAS-targeted therapies are continually put into practice.

Due to a need for noninvasive monitoring of those therapies, development in the synthesis of PET tracers targeting ACE inhibitors has been made and a number of ACE inhibitors and AT1 antagonists have been radiolabeled for molecular imaging (Shirani and Dilsizian 2008). So far, publications have focused on two ^{18}F-radiolabeled ACE inhibitors: [^{18}F]-captopril (FCAP) and [^{18}F]-fluorobenzoyl-lisinopril (FBL) (Hwang et al. 1991). FBL outperforms FCAP especially in the imaging of tissue-bound ACE, exhibiting higher affinity for tissue rather than plasma ACE. This property of FBL was confirmed in an *ex vivo* study of ischemic human hearts, where FBL was shown to specifically bind to tissue ACE with increased activity in the vicinity of the infarcted myocardium (Dilsizian et al. 2007). The study showed some degree of specificity for the binding of the radiolabeled ACE inhibitor in infarcted areas.

Recently, losartan, a drug interacting with the receptor of angiotensin II, was labeled with ^{99m}Tc and along with a fluoresceinated angiotensin peptide analogue used to image myocardial AT receptor density (Figure 29.6). The micro-SPECT images of an *in vivo* mouse model of acute MI showed the time course of activation of the local RAS and proved the feasibility of *in vivo* imaging of angiotensin II receptors (Verjans et al. 2008).

Early studies are promising in that the neurohormonal changes that take place within an infarction may help identify those at risk for developing significant heart failure after MI. More work is needed before these imaging agents can be introduced into clinical trials.

29.5.5 Fiber Architecture

The myocardium has a very structured, helix-like architecture, which has a prominent influence on ventricular function. The wall of a human ventricle is made of a helical spiral of fibers, which, when viewed from the apex, smoothly transitions from a left-handed orientation in the epicardium to a

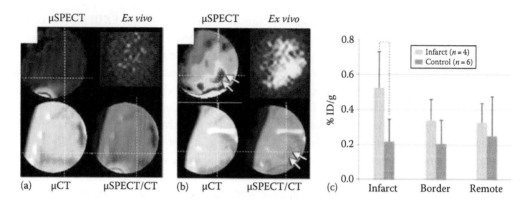

FIGURE 29.6 Noninvasive imaging of AT receptors with radiolabeled losartan. The micro-SPECT and micro-CT images are shown in a control mouse after technetium ^{99m}Tc losartan administration; no uptake in the heart can be seen (a) in the *in vivo* and *ex vivo* images. There is only some liver uptake on the bottom left of the SPECT image. (b) In the 3-week post-MI animal, significant radiolabeled losartan uptake is observed in the anterolateral wall (arrows). The infarct uptake on the *in vivo* image is confirmed in the *ex vivo* image. The histogram (c) demonstrates significantly (*) higher uptake in the infarcted region (0.524% ± 0.212% ID/g) as compared to control noninfarcted animals (0.215% ± 0.129% ID/g; $p < 0.05$). ID = injected dose. (Reprinted from *JACC Cardiovasc. Imaging*, 1(3), Verjans, J.W., Lovhaug, D., Narula, N., Petrov, A.D., Indrevoll, B., Bjurgert, E., Krasieva, T.B. et al., Noninvasive imaging of angiotensin receptors after myocardial infarction, 354–362. Copyright 2008, with permission from Elsevier.)

right-handed one in the subendocardium. The described structure is the main factor influencing ventricular torsion, stress, and strain, and subsequently a key factor in the formation of cardiomyopathies and LV remodeling.

As a consequence of the structural remodeling, regional and global myocardial functions are altered (Wickline et al. 1992). The mechanism, which links post-MI changes in structural components to progressive deterioration in cardiac function, remains unknown. Novel techniques, such as diffusion tensor MR imaging (DTI) show hope of providing insights into this relationship and replacing laborious and imprecise histological methods.

Recent studies have demonstrated that DTI may provide an alternative for rapid and nondestructive reconstruction of the three-dimensional fiber structure at high spatial resolution (Reese et al. 1995; Holmes et al. 2000; Geerts et al. 2002). To validate this approach, MR and histological angle measurements were compared and the results have so far been successful. A recent study investigated post-MI LV fiber structural alterations by *ex vivo* DTI in a porcine heart model (Wu et al. 2007). Six adult pigs underwent septal infarction and after 13 weeks were evaluated using *in vivo* cardiac MR to measure ventricular function. Subsequently, the six diseased hearts along with six controls were excised, submersed in formalin, and subjected to high-resolution DTI. Fractional anisotropy (FA), apparent diffusion coefficient (ADC), and transmural helix angle distribution were compared in the infarct, adjacent, and remote regions of the heart. Significant changes were found in FA and ADC values in the infarct, along with some less pronounced alterations in areas adjacent to the infarct. The myocardial double helix shifted its orientation to be more left-handed near the ischemic tissue. These findings of structural changes were confirmed by histological analysis. Another study evaluated microscopic structural changes caused by MI in rats 4 weeks after induced MI. DTI was performed on the formalin-fixed rat hearts, decreased diffusion anisotropy in the infarct was noted, and results were later confirmed through histological analysis. These findings suggest that DTI may become a useful tool for defining structural remodeling at the cellular and tissue level (Chen et al. 2003). However, studies have so far focused on *ex vivo* DTI, because post-acquisition strain correction was not available. Recently, progress has been made in data analysis and tools for *in vivo* cardiac DTI are becoming available. The next step for cardiac DTI is its verification in *in vivo* studies comparing normal and diseased myocardium.

29.6 IMAGING OF CELL DEATH

In every living multicellular organism, programmed cell death is a naturally occurring and vital process, commonly termed as apoptosis. It allows the maintenance of homeostasis and the constant disposal of cells that are no longer needed. However, certain circumstances such as local hypoxia in sites of ischemia may lead to abnormal, increased apoptotic activity, which is a sign associated with many cardiovascular pathologies. As an example, it has been estimated that up to 30% of the injured myocardial tissue undergoes pathological apoptotic activity, making cell death markers an attractive target for molecular imaging. Apart from happening within the myocardium during MI, apoptosis is also associated with atherosclerotic plaque instability, congestive heart failure, and allograft rejection of the transplanted heart.

Cell death can be initiated through two different mechanisms—the intrinsic and extrinsic pathways. The extrinsic pathway is started mainly through extracellular signals and targets cell membrane receptors like Fas, a common mammalian death receptor present in both humans and rodents. The intrinsic one is generated from within the cell through DNA damage, mitochondrial signals, and oncogene activation. Although initial triggers remain separate, upon the activation of a death-inducing signaling complex (DISC), the pathways have been proven to adopt the same effector caspase machinery, involving the activation of multiple caspase proteins (Blankenberg 2008). Molecular imaging takes advantage mainly of the extrinsic pathway as it involves cell-membrane components, such as phosphatidyl serine, which is present outside the cell during apoptosis and form suitable sites for the binding of imaging radiotracers. When cell disintegration progresses, cellular membranes lose potential, and intracellular structures dissipate, exposing proteins, such as histones and other organelle components, which have also been utilized for imaging techniques.

29.6.1 PHOSPHATIDYLSERINE AND ANNEXIN V

In the initial phase of apoptosis, after caspase-3 is activated, the integrity of the lipid bilayer of cell membranes is disturbed, and phospholipids, particularly phosphatidylserine (PS), show a tendency of flipping within the bilayer from the inner to the outer surface; low levels of ATP block the usually active translocase enzyme, which would otherwise inhibit the process. The resulting increased levels of PS on the outer shell serve as binding sites for annexin V and thus this protein has been explored as a potential labeling target (Figure 29.7).

Initial studies involved detecting apoptosis *in vitro* through fluorescence with (fluorescein isothiocyanate)-labeled annexin V (Koopman et al. 1994). Later studies reported labeling annexin V with biotin or with several radionuclides to facilitate various protocols for measuring apoptosis *in vivo* using animal models (Dumont et al. 2000, 2001). The first noninvasive studies involved [99mTc]-labeled annexin V and a standard gamma camera. To obtain radiolabeled annexin A5, the protein was derivatized with hydrazinonicotinamide (HYNIC), which allowed the binding of reduced [99mTc] (Blankenberg 2008). [99mTc]-labeled annexin V was then applied in post-MI patients to detect *in vivo* apoptosis (Hofstra et al. 2000). As a tracer, it has been tested in several clinical trials and shows great promise to be of benefit in clinical settings.

Patients with heart transplants have a high chance of taking advantage of the technology, since it is known that organ rejection is associated with myocyte death. A study was conducted on patients within a year after cardiac transplantation—the severity of rejection correlated directly with the intensity of annexin V binding (Narula et al. 2001). The results were confirmed with relevant biopsies and showed good dependencies—13 out of 18 patients had no [99mTc]-annexin V uptake and no rejection, while the other 5 had an uptake of [99mTc]-annexin V in the myocardial region and at least moderate transplant rejection. Under current medical guidelines, performing invasive endomyocardial biopsies is recommended 15–20 times during the first year after transplantation. The described findings open the possibility that serial annexin V imaging for apoptotic cells could be used as a noninvasive alternative to serial biopsies for detection of allograft rejection in patients with cardiac transplants.

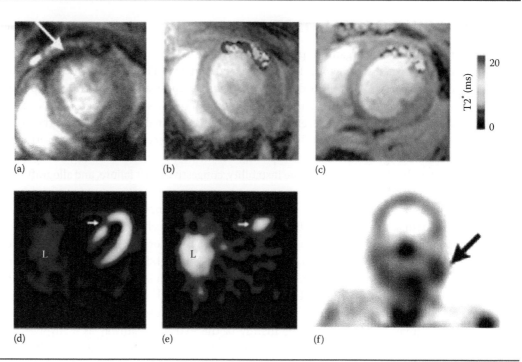

FIGURE 29.7 Apoptosis. (a–c) Molecular MRI of CM apoptosis *in vivo* in a mouse model of ischemia–reperfusion. (a) The apoptosis sensing magnetofluorescent nanoparticle, AnxCLIO-Cy5.5, accumulates in injured myocardium producing signal hypointensity (arrow) and a reduction in T2*. (b, c) *In vivo* T2* maps created in the regions of myocardium with equivalent degrees of hypokinesis in a mouse injected with AnxCLIO-Cy5.5 (b) and a mouse injected with the control probe (c). (d, e) Imaging of cell death with technetium-labeled annexin in a patient with an acute coronary syndrome, Perfusion defect (arrow) in the patient 6–8 weeks after the acute coronary syndrome. (e) Uptake of ^{99m}Tc annexin-V (arrow) at the time of the event correlates well with the perfusion defect. L, liver. F, Uptake of ^{99m}Tc annexin-V in a patient before carotid endarterectomy. A strong correlation was seen in this study between the uptake of the probe and macrophage content of the plaque. (Reprinted from Nahrendorf, M. et al., *Circ. Cardiovasc. Imaging*, 2, 56, 2009. With permission.)

Annexin V has also recently been applied to monitor novel treatment technologies. One of the studies tested a hypothesis that mesenchymal stem cells (MSCs) have cardioprotective effects, such as reducing myocyte apoptosis. Rat models with acute MIs and injected MSCs were utilized to detect the described effects of the stem cells. Twenty-four hours after MI, rats were injected with ^{99m}Tc-hydrazinonicotinamide (^{99m}Tc-HYNIC) annexin V and ^{201}Tl chloride and underwent dual-isotope SPECT/CT imaging. Only annexin V gave significant results but nonetheless they proved their hypothesis true (Godier-Furnemont et al. 2013).

The use of annexin V has several key advantages—it has very high affinity for apoptotic cells with low nanomolar to subnanomolar dissociation constant values, it can be easily reproduced with recombinant DNA technology, and the protein is not toxic *in vivo* (Boersma et al. 2005). However, other protein probes against PS have also been investigated. The C2A domain of a different membrane-trafficking protein, synpaptotagmin I, has a reasonably high affinity for PS and has been used with ^{99m}Tc-labeling in imaging of myocardial ischemia (Zhu et al. 2007). There are reports describing a novel PS-binding peptide (PS3–10), which shows selective binding to apoptotic cells and has potential as a probe for *in vivo* imaging (Shao et al. 2007). Many alternative approaches to cell death imaging other than annexin V are promising, but they still lack extensive research in human and animal studies that has been done with annexin V and its derivatives.

One of the noteworthy discoveries made on annexin V binding revealed that intermediate levels of PS exposure are present in cells that have been influenced by physiological stressors but do not yet have irreversible morphological features of apoptosis (Martin et al. 2000). Therefore, better understanding and more precise methods of imaging of annexin V may lead to the ability of early stress detection and the ability to halt or even reverse a variety of cell injuries, such as acute ischemia.

29.6.2 Antimyosin Antibodies

As cellular membranes lose their integrity, myosin heavy chains are released from inside the cell, making the exposed myosin in the extracellular matrix a possible target for molecular imaging. Monoclonal antibodies to myosin have been generated and initially labeled with technetium to assess the level of myosin exposure in patients with MI. A significant correlation with local necrosis was demonstrated (Khaw et al. 1986). Later studies utilized indium (^{111}In) in an attempt to visualize myocyte damage in MI. The indium-labeled antibodies were evaluated in rodents and advanced to clinical trials, showing satisfactory results (Johnson et al. 1989).

Apart from ischemia, myocarditis has also been a focus of myocyte-targeted imaging studies (Dec et al. 1990). A group of patients with suspected myocarditis was recruited to evaluate the reproducibility and diagnostic accuracy of antimyosin antibody uptake in disorders such as myocarditis and cardiac transplant rejection. The study showed positive results when images with ^{111}In labeling were compared with histological and clinical standards. The antimyosin scan was proven to give fewer false-negative results than the invasive endomyocardial biopsy, which misdiagnosed myocarditis 65% of the time. These initial studies gave promise for obviating biopsies in patients with suspected myocarditis (Narula et al. 1996).

29.6.3 Gadolinium-Enhanced MRI

Acute myocarditis still remains one of the most challenging cardiovascular diseases to diagnose and biopsies remain a common diagnostic procedure (Shauer et al. 2013). MRI has been recognized as a highly sensitive and specific diagnostic tool for myocarditis and guidance for endomyocardial biopsy. MR has a unique ability to image tissue change and to detect characteristic features of myocarditis, such as edema, which can be visualized using T2-weighted images, or hyperemia and capillary leak, which can be seen in gadolinium-enhanced T1-weighted MR images. In acute myocarditis, myocyte membranes disintegrate, allowing intravenously administered gadolinium to diffuse into the cells, and as a result contrast enhancement in the tissue can be seen.

Gadolinium enhancement is suitable for imaging any type of necrotic tissue, where irreversible damage had occurred. High diagnostic accuracy can be achieved by using MR T2-weighted imaging with early and late gadolinium enhancement, especially in patients with suspected acute myocarditis (Abdel-Aty et al. 2005). MRI has also been proven in the evaluation of causes for acute chest pain, as it provides means to noninvasively discriminate myocarditis from MI. Studies show that in MI the sub-endocardium is the first layer of the cardiac muscle to be affected, whereas in myocarditis, the infiltrates are characteristically located in the mid wall. This makes gadolinium-enhanced MRI a very appealing tool to be applied in clinical settings.

29.7 IMAGING OF THROMBOSIS

In every healthy organism, blood clotting is a constantly occurring process, which is usually effectively counterbalanced by fibrinolytic activity. The balance between thrombi formation and dissolution allows the body to be able to maintain an unobstructed flow of blood through the circulatory system. However, there are times when pathological changes take place and the balance between these two is distorted and leads to the occurrence of a phenomenon known as pathological thrombosis. If the changes are severe and repetitive, they may eventually lead to cardiovascular conditions, such as MI, stroke, atrial fibrillation, or venous thromboembolism. Thrombi have, therefore, become another target for imaging of cardiovascular diseases.

The current scope of the imaging of thrombosis allows physical detection and diagnosis of thrombi, as well as the characterization of the nature of detected thrombi by determining how prone they are to anticoagulation and thrombolysis. An ideal imaging modality is able to distinguish acute from chronic thrombi, which is becoming possible with new imaging approaches. Current strategies are targeted at the central events in thrombus formation including the generation of fibrin, platelet activation, and the cross-linking of fibrin strands by FXIII (see Section 29.5.3).

29.7.1 Fibrin

Both arterial and venous thrombi contain fibrin, which makes it an appealing target for thrombosis imaging. It has been demonstrated that fibrin is a stable imaging target in both newly formed and over-1-month-old thrombi (Sirol et al. 2005). To date, fibrin-specific research has been focused on techniques enhanced by EP2104R—a small gadolinium chelate commonly used as a contrast agent for MRI.

There have been numerous studies involving EP2104R-enhanced imaging of rodents, which document the ability of the agent to selectively visualize arterial emboli and coronary thrombi. With this contrast probe, a high local gadolinium concentration in the clots can be achieved, showing up to 18-times-higher signal intensity than in the blood pool (Uppal et al. 2010). It has been demonstrated that EP-2104R plays a critical role not only in the detection of arterial clots but also in the discrimination between thrombi of varying sizes and ages, especially in cases of chronic or organized thrombi, which can be difficult to characterize in unenhanced imaging (Sirol et al. 2005).

Fibrin-targeted PET probes for the detection of thrombi have also recently been investigated (Ciesienski et al. 2013). Three different peptides were chelated with $^{64}CuCl_2$ and evaluated in rat models of carotid artery thrombosis. The images showed a fourfold or higher signal to background ratio in the affected vessel parts for two of the three tested peptides, making them a potential PET enhancer.

However, the EP2104R approach currently still has numerous advantages. It involves MRI instead of nuclear techniques, so a tenfold improvement in spatial resolution can be observed, making it suitable to image small vessels. Typically MR tends to be less sensitive, but EP2104R capitalizes on three amplification strategies: it is a peptide, which allows for quick binding, it is highly selective to fibrin and not its precursor, fibrinogen, minimizing background signal from circulating fibrinogen, and it is multivalent, allowing four gadolinium molecules to conjugate to each peptide ligand, which further increases the thrombus target-to-background ratio and raises sensitivity. The current obstacle to overcome in EP2104R-enhanced imaging is to reduce the obligatory waiting period between the time of injection of the contrast agent and the time of imaging, which is set at a minimum of 2 h. To allow effective intervention in urgent situations, such as acute coronary syndromes or pulmonary embolism, quicker diagnosis would have to be possible. Unfortunately, the proposed PET methods do not overcome this problem either, requiring the same post-injection waiting time (Ciesienski et al. 2013).

29.7.2 Platelets: Receptor Integrin αIIbβ3 (Glycoprotein IIb/IIIa)

Another characteristic element of the clot that has been identified as a potential imaging factor of thrombosis is a glycoprotein complex, most often known as integrin αIIbβ3, which can be found on active thrombocytes in the organism. During platelet aggregation and endothelial adherence via calcium-dependent association, a complex of gpIIb and gpIIIa is formed on the surface of circulating platelets and acts as a receptor for fibrinogen.

In previous imaging studies, ligands to the glycoprotein IIbIIIa receptor have been conjugated to technetium for SPECT imaging. ^{99m}Tc-apcitide scintigraphy has been demonstrated to have the potential to be applied in cases of suspected recurrent DVT. The method in itself is very accurate, enabling the detection of most acute thrombi, and has few false-positive results in patients with previous DVT. ^{99m}Tc-apcitide was tested in phase III trials and shows potential for distinguishing acute from chronic DVT. In a subset of patients with acute DVT, ^{99m}Tc-apcitide achieved a sensitivity of 90% (Bates et al. 2003). However, there are still concerns regarding the method—its availability and cost, as well as the fact the interpreter is required to have significant skill and experience in the field. Accuracy has been proven to drop greatly when the interpreter is inexperienced or had not undergone adequate training (Bates et al. 2003).

29.8 CONCLUSIONS

29.8.1 POTENTIAL CLINICAL IMPACT

In the past 40 years, noninvasive imaging has undergone immense improvements, allowing scientists and clinicians to characterize pathophysiological cardiovascular structure, function, and perfusion. Molecular imaging has the potential to impact cardiovascular medicine in several aspects, including the assessment of risk, the early detection of disease, the development of personalized and targeted therapeutic regimens, and the monitoring of therapeutic efficacy and outcome. In addition to these direct implications, molecular imaging can affect clinical care indirectly—it will facilitate a more rapid development of novel pharmaceuticals and therapeutic strategies, as well as improve the basic understanding of cardiovascular pathophysiology.

From the biological perspective, research has focused on the development of contrast agents that provide sensitive and robust measures of thrombosis, myocardial ischemia, infarction, angiogenesis, tissue oxygenation, and presence of specific cell populations (neutrophils, monocyte/macrophages, fibroblasts, T cells, stem cells, etc.). Progress is being made and the library of available molecular imaging probes is rapidly expanding, making it possible to noninvasively assess nearly every key biologic process involved in cardiovascular pathologies.

From the engineering side, current modalities require increased sensitivity and better temporal and spatial resolution, as well as shorter time of acquisition. With unlimited scanning duration, a lot of these parameters can be optimized, but that creates unrealistic settings and limits *in vivo* applications. Hybrid systems provide a solution to some of these issues by combing high sensitivity and good spatial resolution with very comparable times of image acquisition, as in the case of PET/CT and SPECT/CT.

With molecular imaging, drug administration and its effects on the cardiovascular system can be precisely monitored, using targeted imaging technologies. More and more focus has been put on treatment development, such as observing changes in the heart structure and physiology after stem cell injection. With the absence of molecular imaging, it was difficult to assess the success of gene and stem cell therapy for the treatment of ischemic heart disease. However, recent studies have shown great promise in not only temporary treatment of ischemia, as done with stents or bypasses, but with restoring myocardial functionality to its original state by recreating new muscle tissue.

An important clinical application of molecular imaging in ischemic heart disease will be in providing means of quantitatively assessing treatment success in clinical trials, allowing the quantification of infarcted myocardium or ischemic area at risk. Such quantification might not have direct impact on the clinical management of individual patients, but it gives an invaluable tool to verify clinical trials, which for instance evaluate therapeutic agents intended to reduce infarct size by minimizing reperfusion injury.

PET, SPECT, and CT systems are currently extensively used in research facilities, but only those that are robust and provide clinically relevant endpoints will ultimately find application in clinical medicine. There exist barriers that make it challenging to advance novel approaches into clinical trials and subsequently into routine clinical practice. Most of the new imaging technologies and a majority of agents are still undergoing investigations and are not yet in the phase of approval for clinical use.

Costs of development play a significant role in the aforementioned process. Biotech companies are discouraged from investing into research on imaging agents, because only very few of the agents have potential of making it from bench to bedside. There have been instances in the past where investments in the range $100–$200 million were made from the private sector without any important discoveries being made. At the same time, a meaningful program is unlikely to cost less than that amount, so the risk has to be considered before a decision is made.

The U.S. Food and Drug Administration (FDA) has changed its requirements for imaging agent approval since the introduction of thallium-201 and technetium-99m to the hospitals over 20 years ago. The FDA now recommends that a medical imaging agent intended for diagnostic or therapeutic purposes in some way "be able to improve patient management decisions or improve patient outcomes." The newly stated requirement means that simple improvement of image quality does not guarantee agent approval and thus only a few new imaging agents have been added to the approved list in the past 10 years. Looking from the business perspective, a successful imaging agent should generate annual income roughly equivalent to the overall cost of development, including the cost of research on the other agents that never receive approval. It is difficult to not only acquire the necessary funding but also meaningful evidence for improvement in patient care outcomes in controlled clinical trials.

Moreover, although molecular imaging has a wide range of applications, there are some areas where conventional cardiovascular imaging has proven superior to the more novel technologies. Echocardiographic guidance, for instance, has long been used for valve replacement and repair. It has recently been extended to catheterization and electrophysiology lab procedures. Intraoperative echocardiography is used for procedures, such as closure of atrial and ventricular septal defects and paravalvular leaks, as well as pulmonary vein isolation. It has been proposed that a fusion of preoperative CT with transesophageal echocardiography could improve the safety margin for percutaneous aortic valve replacement. Unfortunately, this approach did not show promise of being used for these applications.

Overall, molecular imaging has contributed to great advancements in research done on various aspects of cardiovascular diseases, but in some cases, conventional methods of imaging proves to be invaluable, and most importantly, there still remains a great gap to be bridged before the existing imaging technologies are applied in clinical settings.

REFERENCES

Abdel-Aty, H., P. Boye et al. (2005). Diagnostic performance of cardiovascular magnetic resonance in patients with suspected acute myocarditis: Comparison of different approaches. *J. Am. Coll. Cardiol.* **45**(11): 1815–1822.

Amirbekian, V., M. J. Lipinski et al. (2007). Detecting and assessing macrophages in vivo to evaluate atherosclerosis noninvasively using molecular MRI. *Proc. Natl. Acad. Sci. USA* **104**(3): 961–966.

Aras, O., S. A. Messina et al. (2007). The role and regulation of cardiac angiotensin-converting enzyme for noninvasive molecular imaging in heart failure. *Curr. Cardiol. Rep.* **9**(2): 150–158.

Basu, S., H. Zhuang et al. (2009). Functional imaging of inflammatory diseases using nuclear medicine techniques. *Semin. Nucl. Med.* **39**(2): 124–145.

Bates, S. M., J. Lister-James et al. (2003). Imaging characteristics of a novel technetium Tc 99m-labeled platelet glycoprotein IIb/IIIa receptor antagonist in patients With acute deep vein thrombosis or a history of deep vein thrombosis. *Arch. Intern. Med.* **163**(4): 452–456.

Bengel, F. M. and M. Schwaiger. (2004). Assessment of cardiac sympathetic neuronal function using PET imaging. *J. Nucl. Cardiol.* **11**(5): 603–616.

Blankenberg, F. G. (2008). In vivo detection of apoptosis. *J. Nucl. Med.* **49**(Suppl. 2): 81S–95S.

Boersma, H. H., B. L. Kietselaer et al. (2005). Past, present, and future of annexin A5: From protein discovery to clinical applications. *J. Nucl. Med.* **46**(12): 2035–2050.

Briley-Saebo, K. C., P. X. Shaw et al. (2008). Targeted molecular probes for imaging atherosclerotic lesions with magnetic resonance using antibodies that recognize oxidation-specific epitopes. *Circulation* **117**(25): 3206–3215.

Carrio, I. (2001). Cardiac neurotransmission imaging. *J. Nucl. Med.* **42**(7): 1062–1076.

Chen, J., S. K. Song et al. (2003). Remodeling of cardiac fiber structure after infarction in rats quantified with diffusion tensor MRI. *Am. J. Physiol. Heart Circ. Physiol.* **285**(3): H946–H954.

Chen, J., C. H. Tung et al. (2005). Near-infrared fluorescent imaging of matrix metalloproteinase activity after myocardial infarction. *Circulation* **111**(14): 1800–1805.

Chen, J. W., M. Querol Sans et al. (2006). Imaging of myeloperoxidase in mice by using novel amplifiable paramagnetic substrates. *Radiology* **240**(2): 473–481.

Cheng, V. Y., P. J. Slomka et al. (2010). Impact of carbohydrate restriction with and without fatty acid loading on myocardial ^{18}F-FDG uptake during PET: A randomized controlled trial. *J. Nucl. Cardiol.* **17**(2): 286–291.

Ciesienski, K. L., Y. Yang et al. (2013). Fibrin-targeted PET probes for the detection of thrombi. *Mol. Pharm.* **10**(3): 1100–1110.

Dec, G. W., I. Palacios et al. (1990). Antimyosin antibody cardiac imaging: Its role in the diagnosis of myocarditis. *J. Am. Coll. Cardiol.* **16**(1): 97–104.

Dilsizian, V., W. C. Eckelman et al. (2007). Evidence for tissue angiotensin-converting enzyme in explanted hearts of ischemic cardiomyopathy using targeted radiotracer technique. *J. Nucl. Med.* **48**(2): 182–187.

Ducharme, A., S. Frantz et al. (2000). Targeted deletion of matrix metalloproteinase-9 attenuates left ventricular enlargement and collagen accumulation after experimental myocardial infarction. *J. Clin. Invest.* **106**(1): 55–62.

Dumont, E. A., L. Hofstra et al. (2000). Cardiomyocyte death induced by myocardial ischemia and reperfusion: Measurement with recombinant human annexin-V in a mouse model. *Circulation* **102**(13): 1564–1568.

Dumont, E. A., C. P. Reutelingsperger et al. (2001). Real-time imaging of apoptotic cell-membrane changes at the single-cell level in the beating murine heart. *Nat. Med.* **7**(12): 1352–1355.

Geerts, L., P. Bovendeerd et al. (2002). Characterization of the normal cardiac myofiber field in goat measured with MR-diffusion tensor imaging. *Am. J. Physiol. Heart Circ. Physiol.* **283**(1): H139–H145.

Gibbons, R. J., T. D. Miller et al. (2000). Infarct size measured by single photon emission computed tomographic imaging with (99m)Tc-sestamibi: A measure of the efficacy of therapy in acute myocardial infarction. *Circulation* **101**(1): 101–108.

Godier-Furnemont, A. F., Y. Tekabe et al. (2013). Noninvasive imaging of myocyte apoptosis following application of a stem cell-engineered delivery platform to acutely infarcted myocardium. *J. Nucl. Med.* **54**(6): 977–983.

Hansson, G. K. (2005). Inflammation, atherosclerosis, and coronary artery disease. *N. Engl. J. Med.* **352**(16): 1685–1695.

Hofstra, L., I. H. Liem et al. (2000). Visualisation of cell death in vivo in patients with acute myocardial infarction. *Lancet* **356**(9225): 209–212.

Holmes, A. A., D. F. Scollan et al. (2000). Direct histological validation of diffusion tensor MRI in formaldehyde-fixed myocardium. *Magn. Reson. Med.* **44**(1): 157–161.

Hwang, D. R., W. C. Eckelman et al. (1991). Positron-labeled angiotensin-converting enzyme (ACE) inhibitor: Fluorine-18-fluorocaptopril: Probing the ACE activity in vivo by positron emission tomography. *J. Nucl. Med.* **32**(9): 1730–1737.

Hyafil, F., J. C. Cornily et al. (2007). Noninvasive detection of macrophages using a nanoparticulate contrast agent for computed tomography. *Nat. Med.* **13**(5): 636–641.

Jaffer, F. A., P. Libby et al. (2007). Molecular imaging of cardiovascular disease. *Circulation* **116**(9): 1052–1061.

Johnson, L. L., D. W. Seldin et al. (1989). Antimyosin imaging in acute transmural myocardial infarctions: Results of a multicenter clinical trial. *J. Am. Coll. Cardiol.* **13**(1): 27–35.

Khaw, B. A., H. K. Gold et al. (1986). Scintigraphic quantification of myocardial necrosis in patients after intravenous injection of myosin-specific antibody. *Circulation* **74**(3): 501–508.

Klocke, F. J., M. G. Baird et al. (2003). ACC/AHA/ASNC guidelines for the clinical use of cardiac radionuclide imaging--executive summary: A report of the American College of Cardiology/American Heart Association Task Force on Practice Guidelines (ACC/AHA/ASNC Committee to Revise the 1995 Guidelines for the Clinical Use of Cardiac Radionuclide Imaging). *Circulation* **108**(11): 1404–1418.

Koopman, G., C. P. Reutelingsperger et al. (1994). Annexin V for flow cytometric detection of phosphatidylserine expression on B cells undergoing apoptosis. *Blood* **84**(5): 1415–1420.

Lee, W. W., B. Marinelli et al. (2012). PET/MRI of inflammation in myocardial infarction. *J. Am. Coll. Cardiol.* **59**(2): 153–163.

Leuschner, F. and M. Nahrendorf. (2011). Molecular imaging of coronary atherosclerosis and myocardial infarction: Considerations for the bench and perspectives for the clinic. *Circ. Res.* **108**(5): 593–606.

Martin, S., I. Pombo et al. (2000). Immunologic stimulation of mast cells leads to the reversible exposure of phosphatidylserine in the absence of apoptosis. *Int. Arch. Allergy Immunol.* **123**(3): 249–258.

Matsumori, A., T. Yamada et al. (1996). Antimyosin antibody imaging in clinical myocarditis and cardiomyopathy: Principle and application. *Int. J. Cardiol.* **54**(2): 183–190.

Miller, T. D., R. Sciagra et al. (2010). Application of technetium-99m sestamibi single photon emission computed tomography in acute myocardial infarction: Measuring the efficacy of therapy. *Q. J. Nucl. Med. Mol. Imaging* **54**(2): 213–229.

Murray, C. J. and A. D. Lopez. (1997). Alternative projections of mortality and disability by cause 1990–2020: Global Burden of Disease Study. *Lancet* **349**(9064): 1498–1504.

Nahrendorf, M., E. Aikawa et al. (2008a). Transglutaminase activity in acute infarcts predicts healing outcome and left ventricular remodelling: Implications for FXIII therapy and antithrombin use in myocardial infarction. *Eur. Heart J.* **29**(4): 445–454.

Nahrendorf, M., K. Hu et al. (2006a). Factor XIII deficiency causes cardiac rupture, impairs wound healing, and aggravates cardiac remodeling in mice with myocardial infarction. *Circulation* **113**(9): 1196–1202.

Nahrendorf, M., F. A. Jaffer et al. (2006b). Noninvasive vascular cell adhesion molecule-1 imaging identifies inflammatory activation of cells in atherosclerosis. *Circulation* **114**(14): 1504–1511.

Nahrendorf, M., D. E. Sosnovik et al. (2009). Multimodality cardiovascular molecular imaging: Part II. *Circ. Cardiovasc. Imaging* **2**: 56–70.

Nahrendorf, M., H. Zhang et al. (2008b). Nanoparticle PET-CT imaging of macrophages in inflammatory atherosclerosis. *Circulation* **117**(3): 379–387.

Narula, J., E. R. Acio et al. (2001). Annexin-V imaging for noninvasive detection of cardiac allograft rejection. *Nat. Med.* **7**(12): 1347–1352.

Narula, J., B. A. Khaw et al. (1996). Diagnostic accuracy of antimyosin scintigraphy in suspected myocarditis. *J. Nucl. Cardiol.* **3**(5): 371–381.

Naya, M., T. Tsukamoto et al. (2009). Myocardial beta-adrenergic receptor density assessed by 11C-CGP12177 PET predicts improvement of cardiac function after carvedilol treatment in patients with idiopathic dilated cardiomyopathy. *J. Nucl. Med.* **50**(2): 220–225.

Reese, T. G., R. M. Weisskoff et al. (1995). Imaging myocardial fiber architecture in vivo with magnetic resonance. *Magn. Reson. Med.* **34**(6): 786–791.

Riou, L. M., M. Ruiz et al. (2002). Assessment of myocardial inflammation produced by experimental coronary occlusion and reperfusion with ^{99m}Tc-RP517, a new leukotriene B4 receptor antagonist that preferentially labels neutrophils in vivo. *Circulation* **106**(5): 592–598.

Roger, V. L. (2011). Outcomes research and epidemiology: The synergy between public health and clinical practice. *Circ. Cardiovasc. Qual. Outcomes* **4**(3): 257–259.

Salerno, M. and G. A. Beller. (2009). Noninvasive assessment of myocardial perfusion. *Circ. Cardiovasc. Imaging* **2**(5): 412–424.

Sanz, J. and Z. A. Fayad. (2008). Imaging of atherosclerotic cardiovascular disease. *Nature* **451**(7181): 953–957.

Sato, M., T. Toyozaki et al. (2002). Detection of experimental autoimmune myocarditis in rats by 111In monoclonal antibody specific for tenascin-C. *Circulation* **106**(11): 1397–1402.

Serhan, C. N. and S. M. Prescott. (2000). The scent of a phagocyte: Advances on leukotriene b(4) receptors. *J. Exp. Med.* **192**(3): F5–F8.

Shao, R., C. Xiong et al. (2007). Targeting phosphatidylserine on apoptotic cells with phages and peptides selected from a bacteriophage display library. *Mol. Imaging* **6**(6): 417–426.

Shauer, A., I. Gotsman et al. (2013). Acute viral myocarditis: Current concepts in diagnosis and treatment. *Isr. Med. Assoc. J.* **15**(3): 180–185.

Shirani, J. and V. Dilsizian. (2008). Imaging left ventricular remodeling: Targeting the neurohumoral axis. *Nat. Clin. Pract. Cardiovasc. Med.* 5(Suppl. 2): S57–S62.

Sinusas, A. J., F. Bengel et al. (2008). Multimodality cardiovascular molecular imaging: Part I. *Circ. Cardiovasc. Imaging* **1**(3): 244–256.

Sinusas, A. J., J. D. Thomas et al. (2011). The future of molecular imaging. *JACC Cardiovasc. Imaging* **4**: 799–806.

Sirol, M., V. Fuster et al. (2005). Chronic thrombus detection with in vivo magnetic resonance imaging and a fibrin-targeted contrast agent. *Circulation* **112**(11): 1594–1600.

Sosnovik, D. E., E. A. Schellenberger et al. (2005). Magnetic resonance imaging of cardiomyocyte apoptosis with a novel magneto-optical nanoparticle. *Magn. Reson. Med.* **54**(3): 718–724.

Spinale, F. G. (2002). Matrix metalloproteinases: Regulation and dysregulation in the failing heart. *Circ. Res.* **90**(5): 520–530.

Su, H., F. G. Spinale et al. (2005). Noninvasive targeted imaging of matrix metalloproteinase activation in a murine model of postinfarction remodeling. *Circulation* **112**(20): 3157–3167.

Tang, T. Y., S. P. Howarth et al. (2009). The ATHEROMA (Atorvastatin Therapy: Effects on Reduction of Macrophage Activity) Study. Evaluation using ultrasmall superparamagnetic iron oxide-enhanced magnetic resonance imaging in carotid disease. *J. Am. Coll. Cardiol.* **53**(22): 2039–2050.

Tawakol, A., R. Q. Migrino et al. (2006). In vivo 18F-fluorodeoxyglucose positron emission tomography imaging provides a noninvasive measure of carotid plaque inflammation in patients. *J. Am. Coll. Cardiol.* **48**(9): 1818–1824.

Tegler, G., K. Ericson et al. (2012). Inflammation in the walls of asymptomatic abdominal aortic aneurysms is not associated with increased metabolic activity detectable by 18-fluorodeoxglucose positron-emission tomography. *J. Vasc. Surg.* **56**(3): 802–807.

Trivedi, R. A., C. Mallawarachi et al. (2006). Identifying inflamed carotid plaques using in vivo USPIO-enhanced MR imaging to label plaque macrophages. *Arterioscler. Thromb. Vasc. Biol.* **26**(7): 1601–1606.

Uppal, R., I. Ay et al. (2010). Molecular MRI of intracranial thrombus in a rat ischemic stroke model. *Stroke* **41**(6): 1271–1277.

van Eerd, J. E., W. J. Oyen et al. (2003). A bivalent leukotriene B(4) antagonist for scintigraphic imaging of infectious foci. *J. Nucl. Med.* **44**(7): 1087–1091.

Verjans, J. W., D. Lovhaug et al. (2008). Noninvasive imaging of angiotensin receptors after myocardial infarction. *JACC Cardiovasc. Imaging* **1**(3): 354–362.

Visse, R. and H. Nagase. (2003). Matrix metalloproteinases and tissue inhibitors of metalloproteinases: Structure, function, and biochemistry. *Circ. Res.* **92**(8): 827–839.

Wickline, S. A., E. D. Verdonk et al. (1992). Structural remodeling of human myocardial tissue after infarction: Quantification with ultrasonic backscatter. *Circulation* **85**(1): 259–268.

Wu, E. X., Y. Wu et al. (2007). MR diffusion tensor imaging study of postinfarct myocardium structural remodeling in a porcine model. *Magn. Reson. Med.* **58**(4): 687–695.

Zhu, X., Z. Li et al. (2007). Imaging acute cardiac cell death: Temporal and spatial distribution of ^{99m}Tc-labeled C2A in the area at risk after myocardial ischemia and reperfusion. *J. Nucl. Med.* **48**(6): 1031–1036.

30

Imaging Angiogenesis

Lawrence W. Dobrucki

30.1 INTRODUCTION

Molecular imaging of angiogenesis has been a rapidly growing field in the last decade, has become a part of routine clinical care of oncology patients, and steadily builds its importance for the cardiovascular community. Molecular imaging represents a very powerful tool for the noninvasive characterization, visualization, and quantification of various biological processes, including angiogenesis at the molecular and cellular level in living laboratory animals and humans. The successful application of this technology requires the availability of imaging probes that target molecules of interest, access to imaging instrumentation that enables the visualization of the probes' biodistribution, and widespread education of medical and lay communities and regulators to lower the reimbursement barriers of high medical care costs.

Recent developments in imaging technology and advances in genomic approaches to understand molecular events associated with pathophysiology of disease have resulted in widespread inclusion of molecular imaging in clinical trials, testing individually tailored pharmacological and cell-based therapeutic approaches to modulate angiogenic responses in cancer and cardiovascular research.

In contrast to traditional approaches to assess angiogenesis, which employ imaging anatomy and changes in hypoxia and perfusion, molecular imaging introduces molecular probes targeted at indicative biomarkers of physiological processes to provide not only early diagnostic and prognostic information but also a tool to assess the efficacy of therapeutic interventions. The overarching goal of molecular imaging is to provide a more global view of the disease process rather than focusing on isolated cellular events.

The advantages of such approaches are numerous; however, the imaging community currently faces a so-called "paradox of molecular imaging." The promise of imaging agents that would noninvasively identify various biological processes, including angiogenesis, and provide prognostic value is what attracts new researchers into this field. On the other hand, this excitement is tampered by the reality of a long and steep climb to translate these approaches to clinical applications.

This chapter provides a perspective on how molecular imaging with targeted probes can be successfully employed to study angiogenesis, which goes beyond providing only diagnostic and prognostic information but can also play a critical role for development of individual therapeutic interventions.

30.1.1 Molecular Physiology of Angiogenesis

Blood vessels were initially developed during the evolutionary process in animal species in which oxygen could not be carried to all cells via diffusion as in primitive organisms such as worms and fruit flies. Early discoveries of vascular networks and studies on the delivery of oxygen to distant organs using veins and arteries date back to seventeenth century. However, the last two decades have seen an explosion of interest in research focused on blood vessels, their development, and function, which has led to the understanding of key mechanisms implicated in the formation of vascular networks, as well as developing novel clinically approved therapeutic strategies to target vascular pathologies (Carmeliet 2005).

Formation of new vascular networks, in which new capillaries sprout from preexisting vessels, is termed angiogenesis and it represents a biological process that is orchestrated by a number of angiogenic factors and inhibitors (Carmeliet and Jain 2000). Angiogenesis is distinct from arteriogenesis and lymphangiogenesis, and is naturally implicated in numerous biological processes, such as reproduction (female reproductive cycle and embryogenesis), wound healing, and tissue remodeling. This process is usually focal and controlled by the actions of angiogenic inhibitors, thus is self-limiting in time (Folkman 2007). On the other hand, there are numerous macro and microvascular disorders characterized by either an excess (i.e., ocular and inflammatory disorders) or an insufficient number of blood vessels (i.e., ischemic heart disease or preeclampsia). These pathologies have recently become a focus of intense research efforts directed at developing (1) novel imaging techniques for early diagnosis and prognostication, and (2) novel individualized therapeutic strategies focused on the modulation of the angiogenic process (Sinusas 2004; Dobrucki et al. 2010a; Dobrucki and Sinusas 2010).

These efforts have been intensified in response to the outcomes of recent clinical trials designed to either stimulate new blood vessel growth and improve perfusion to hypoxic tissues or inhibit tumor-associated angiogenesis leading to apoptotic tumor growth reduction. Early preclinical studies in animal models had shown a potential benefit of targeted angiogenic therapy with growth factors and angiogenesis inhibitors and supported the transition to human trials (Fam et al. 2003).

The first angiogenic inhibitors were discovered by the Folkman group in 1980s, and by the mid-1990s, new drugs with antiangiogenic activity entered clinical trials (Folkman 2007). These trials proved the angiogenesis inhibitors to be highly effective in downregulation of blood vessel growth and, as a result, by April 2007 about 10 drugs with antiangiogenic activity have been approved by the Food and Drug Administration (FDA) in the United States for the treatment of cancer and age-related macular degeneration, whereas other currently FDA-approved drugs revealed antiangiogenic activity in addition to their activity directed against cancer.

Unfortunately, randomized clinical trials focused on promotion of angiogenesis in ischemic injury using locally administered growth factors such as vascular endothelial growth factor (VEGF) or fibroblast growth factor (FGF), which did not demonstrate a clear benefit in patients with peripheral artery disease or coronary artery disease (Lederman et al. 2002; Sneider et al. 2009; Mitsos et al. 2012).

Careful interpretation of these results revealed few potential factors found attributable to this unexpected failure, including suboptimal delivery strategies, use of a single growth factor at suboptimal dose, and duration of the therapy, which could lead to insufficient growth of new vessels or formation of nonfunctional vessels. Moreover, the evaluation of therapeutic angiogenesis in these trials was performed using rather insensitive techniques and focused on several clinical endpoints, including exercise tolerance, measures of quality of life and survival, peripheral pressure measurements, and imaging of tissue perfusion. Therefore, in recent years more focus was placed on the development of both novel therapeutic strategies, including genetic and cell-based approaches, and novel noninvasive imaging techniques to evaluate molecular events associated with angiogenesis (Dobrucki and Sinusas 2005a,b; Morrison and Sinusas 2010).

30.1.2 INDUCTION OF ANGIOGENESIS AND ROLE OF ANGIOGENIC FACTORS

Angiogenesis is stimulated by a number of external processes and orchestrated by a range of angiogenic factors and inhibitors. Initial stimuli such as ischemia, hypoxia, inflammation, genetic mutation, and shear stress lead to the interactions between various cell types including endothelial cells, smooth muscle cells, macrophages, and circulating stem cells, with each other and with the extracellular matrix (ECM).

The imbalance between oxygen delivery and demand or an insufficient nutrient supply in a given tissue or tumor has been found to be powerful stimulators of angiogenesis. Both the activation of preexisting vessels with quiescent endothelial cells and the upregulation of the transcriptional activator hypoxia-inducible factor-1 (HIF-1) lead to the transcription of key angiogenic modulators such as VEGF, the VEGF receptors (VEGFR-1 and VEGFR-2), and platelet-derived growth factor (PDGF). Additional regulating factors such as fibroblast growth factors (FGF) are locally secreted by endothelial, cancer, or stromal cells as well as the ECM (Folkman 2007; Haubner 2008).

VEGF, a key player in vessel growth and maturation, is initially activated by binding HIF-1 to the hypoxic response element. VEGF binds to its receptors on the surface of endothelial cells and is involved in downstream signal processing via the tyrosine kinase pathway, leading to a series of processes that result in endothelial cell proliferation, migration, and survival. Activated endothelial cells excrete proteolytic enzymes, such as matrix metalloproteinases (MMPs) and serine proteases implicated in the degradation of the basement membrane and the ECM surrounding the vessels. Besides their proteolytic functions, MMPs can also release proangiogenic matrix-bound factors and play an antiangiogenic role by cleaving matrix components into antiangiogenic factors (Rundhaug 2005).

The disruption of this balance between pro- and antiangiogenic roles of various factors implicated in angiogenesis contributes to the outcome of "the angiogenic switch," which results in either excessive angiogenesis found in malignant, ocular, and inflammatory disorders, or impaired angiogenesis manifested in ischemic vascular diseases, which prevents revascularization, healing, and regeneration.

Endothelial signaling through VEGF is not the only molecular pathway found in angiogenesis. Activated endothelial cells upregulate transmembrane cell adhesion molecules such as the integrins that aid in the cell–cell signaling, leading to the extracellular remodeling process. Specifically, the $\alpha v \beta 3$ integrin allows endothelial cells to interact with ECM, which aids in endothelial cell migration. Through signaling pathways, $\alpha v \beta 3$ integrin plays a critical role in the survival of cells undergoing angiogenesis (Brooks et al. 1994). Other numerous molecular pathways regulate angiogenesis by actions of several potent angiogenic factors such as angiopoietin (ANG-1), monocyte chemotactic protein-1 (MCP-1), granulocyte-macrophage colony stimulating factor (GM-CSF), hepatocyte growth factor (HGF), leptin, transforming growth factors (TGF-α and TGF-β), tumor necrosis factor alpha (TNF-α), brain-derived neurotrophic factor (BDNF), and many more. Finally, all these processes orchestrated by the actions of angiogenic factors result in the formation of endothelial tubes, which connect with the microcirculation forming operational new vasculature.

30.1.3 MOLECULAR TARGETS

In the cardiovascular system, the process of angiogenesis should result in improved perfusion and tissue oxygenation, resulting in reduced hypoxia and diminished ischemia. Effective therapeutic angiogenesis should restore perfusion (i.e., resting and stress myocardial perfusion) and improve functional parameters such as regional left ventricular function (Fam et al. 2003). Regarding malignant pathological processes, angiogenesis was found to be essential for the growth of solid tumors that require an adequate supply of oxygen and nutrients, which is possible only when new vessels are formed as a result of the "angiogenic switch" (Haubner 2008).

To assess the angiogenic activity in both cardiovascular system and tumors, a wide variety of methods have been developed, and historically many of them have aided in measuring physical parameters of the tissue as surrogates of angiogenesis such as regional perfusion, blood flow, blood volume, and permeability of the blood vessels. Angiogenesis-associated alterations in tissue oxygenation may be potentially more accurately measured by the analysis of regional pH, hypoxia, and metabolism. However, despite the fact that the imaging of physiological consequences of angiogenesis has proven to be extremely valuable, it is worth emphasizing that these methods assess only physiological changes related to angiogenesis but do not visualize the process of angiogenesis itself. Moreover, these traditional imaging approaches and the interpretation of measured parameters may be complicated by numerous factors, including changes in microvasculature structure, surface exchange area, vascular reactivity, and alterations in vascular permeability. Nevertheless, all these techniques have significantly contributed to our understanding of the angiogenic process and remain a critical step in the development and evaluation of therapeutic angiogenesis.

With the recent inventions in proteomics and genomics methodologies, our knowledge about molecular markers involved in angiogenesis has led to the conclusion that imaging of molecular signatures related to the angiogenic process might be a more specific and accurate method for the assessment of angiogenic activity in tumors and the cardiovascular system. Potential targets for imaging of angiogenesis were initially identified by the expert panel on angiogenesis at the Lake Tahoe Invitation Meeting of the American Society of Nuclear Cardiology held in July 2002 in Lake Tahoe, California (Cerqueira and Udelson 2003). Potential biological targets for imaging angiogenesis were believed to fall into three principal categories: (1) endothelial cell markers of angiogenesis including VEGF and $\alpha v \beta 3$ integrins, (2) nonendothelial cells involved with angiogenesis such as monocytes and stem cells, and (3) markers of the ECM including MMPs. Despite the wide array of potential targets for imaging angiogenesis, there are just three real active areas of research: imaging VEGF receptors using labeled VEGF; imaging the $\alpha v \beta 3$ integrin with ligand-like analogs including cyclic peptides and peptidomimetics; and imaging MMPs using probes modeled after MMP inhibitors.

As outlined in the following sections, the potential for targeted imaging of various endothelial and nonendothelial biological markers has been demonstrated in animal models of ischemia-induced and tumor-associated angiogenesis.

30.1.4 ENDOTHELIAL CELL TARGETS

As mentioned before, angiogenesis is generally defined as the growth of microvessel sprouts orchestrated by a range of angiogenic factors and involves endothelial cell proliferation and migration. In cancer, angiogenesis does not initiate malignancy but promotes tumor progression and growth. In contrast to cancer cells, endothelial cells are genomically stable, and therefore they were originally considered to be ideal targets for molecular imaging of angiogenesis. In addition, a potential target for imaging angiogenesis should relate to the molecular events associated with the initiation of the angiogenic process, which involves activation of various molecular markers localized to the endothelial cells. Therefore, evaluation of altered expression of integrins ($\alpha v\beta3$ and $\alpha v\beta5$), VEGF receptors, and FGF receptors led to the development of targeted imaging probes for imaging angiogenesis.

30.1.4.1 Vascular Endothelial Growth Factors

VEGF belongs to the group of fundamental mediators of angiogenesis and plays a pivotal role not only in angiogenesis but also in vascular development during embryogenesis (Ferrara and Davis-Smyth 1997). Besides its endothelial activity, VEGF interacts with bone-marrow-derived cells, monocytes, and tumor cell lines, and is also trophic for nerve cells, lung epithelial cells, and cardiac muscle fibers, further explaining that insufficient VEGF levels contribute not only to impaired angiogenic response but also to neurodegeneration, respiratory distress, and cardiac failure (Carmeliet 2005).

VEGF, as a central cytokine in the angiogenic process, mediates many cellular functions including release of other growth factors, endothelial cell proliferation, migration, and survival, which are essential in the development of new vasculature (i.e., during angiogenesis and tumorgenesis). There are at least five isoforms of human VEGF, which are either freely diffusible (i.e., VEGF165, VEGF145, and VEGF121) or cell-associated (i.e., VEGF206 and VEGF189). The VEGF receptor family consists of two principal receptors found on vascular endothelial cells, VEGFR-1 (or Flt-1) and VEGFR-2 (or Flk-1), and one receptor expressed on lymphatic endothelial cells (VEGFR-3) (Cross and Claesson-Welsh 2001).

The major VEGF isoform found in tumors is VEGF165, which can associate with cells via a heparin-like binding domain (Neufeld et al. 1996). In contrast, VEGF121 lacks affinity for heparin-like molecules and binds only to endothelial Flt-1 and Flk-1 receptors, which are responsive to hypoxia. Indeed, it was demonstrated that both acute and chronic lung hypoxia and myocardial infarction in rats resulted in significant levels of Flt-1 receptors. All this provided a basis for the development of tracers for monitoring both cardiovascular and tumor-associated angiogenesis structured after VEGF receptors and their ligands (Haubner 2008).

30.1.4.2 Integrins

Interesting molecular markers of the angiogenic process in both tumors and cardiovascular pathologies are integrins, which are heterodimeric membrane receptors comprised of α and β subunits that mediate interactions between cells, growth factors, and components of the ECM including collagen, laminin, fibronectin, and vitronectin. Despite the presence of over 24 different integrin receptors, only a few have been extensively studied. Of these, the most effort was put on developing the $\alpha v\beta3$ integrin antagonists to monitor angiogenesis. Indeed, $\alpha v\beta3$ integrin is highly expressed not only on the surface of angiogenic endothelial cells but is also on the surface of some tumor cells acting on both promoting and inhibiting angiogenesis (Eliceiri and Cheresh 2000). Considering this duality in actions of $\alpha v\beta3$ integrins, it should be emphasized that the interpretation of the signal obtained by integrin imaging is complex.

In addition to angiogenesis, αvβ3 integrin plays an important role in a number of pathological dysfunctions including osteoporosis, restenosis, inflammatory processes, and tumor metastasis. Therefore, αvβ3 integrin is a particularly interesting target for imaging probes and drug development.

The major signal transduction pathways associated with the cellular processes mediated by the integrins include processes involving endothelial cell adhesion, migration, proliferation, differentiation, survival, and response of the smooth muscle cells to mechanical stimuli. In addition to endothelial and cancer cells, αvβ3 integrin is expressed by smooth muscle cells, platelets, monocytes, T lymphocytes, and osteoclasts, which allows the interaction of these cells with a wide variety of ECM components. Fortunately, αvβ3 integrin expression in quiescent endothelial cells of established vessels is very low, whereas "angiogenic" cells demonstrate significant upregulation of αvβ3 expression. This unique biological behavior offers a tremendous advantage for targeted imaging of angiogenesis by providing a favorable target-to-background ratio.

30.1.5 Extracellular Matrix Proteins

Another important component of angiogenesis is signaling by molecules in the ECM, which include interactions between integrins (αvβ3 and αvβ5) with their ECM ligands. It was demonstrated that the natural ligand for several integrins is collagen. Therefore, collagen fragments conjugated with an imaging label have the potential for imaging angiogenesis. Similarly, because MMPs are capable of degrading proteins of the ECM including collagen, targeting MMPs may also prove useful for the imaging of angiogenesis.

MMPs are zinc-dependent endopeptidases consisting of more than 18 different members, which are divided into 5 classes: collagenases (MMP-1, MMP-8, and MMP-13), gelatinases (MMP-2 and MMP-9), stromolysins (MMP-3, MMP-7, MMP-10, MMP-11, and MMP-12), membrane-type MT-MMPs (MMP-14, MMP-15, MMP-16, and MMP-17), and nonclassified MMPs (MMP-18, MMP-19, MMP-20, MMP-23, and MMP-24) (Hidalgo and Eckhardt 2001).

Once activated, MMPs are controlled by the balance between synthesis of the proenzyme and expression of endogenous tissue-derived MMP inhibitors (TIMPs). An increased production of the proenzyme results in degradation of the basement membrane, a structural requirement for endothelial cell migration and the formation of new vasculature. Particularly, two MMP isoforms, namely gelatinases MMP-2 and MMP-9, were found to play important roles in the myocardial remodeling following myocardial infarction and in cancer where their expression levels are directly related to the tumor aggressiveness and metastatic potential (Haubner 2008).

Due to their important roles in ischemia-stimulated and tumor-associated angiogenesis, MMPs were proposed as potential targets in therapeutic interventions or as potential imaging agents. Over last two decades, significant effort was made to develop MMP inhibitors for both therapeutic and imaging applications; however, despite successes in utilizing MMP inhibitors in therapy, initial in vivo imaging data were somewhat disappointing to demonstrate the potential of this tracer class to monitor angiogenesis (Haubner 2008).

30.1.6 Nonendothelial Cell Targets

Several other potential avenues to assess angiogenic process in both the cardiovascular system and tumor have been recently proposed. These include the detection of HIF-1a activation, monitoring of the influx of blood-derived macrophages, monocytes, or circulating endothelial precursor cells, and expression of other markers involved in the maturation of developed neovasculature, such as ephrins, ephrin receptors, and semaphorins. Specifically, both monocytes and circulating endothelial precursor cells have been found to play pivotal roles in the stimulation of new vessel growth by releasing growth factors, proteases, and chemokines, which contribute not only to the angiogenesis but also to the collateral growth by recruiting other circulating cells such as T cells and bone-marrow-derived cells. Finally, tissue-derived stem cells and paracrine functions of other pluripotent progenitor cells also play an integral role in the cascade of events leading to both angiogenesis and arteriogenesis (Fam et al. 2003; Iagaru and Gambhir 2013).

30.2 UTILIZATION OF IMAGING MODALITIES

Over the last few decades, multiple interdisciplinary groups of engineers, chemists, and physiologists have developed various imaging modalities and technologies for mapping biological processes using specific biomarkers. These novel approaches have a great potential to mitigate the socioeconomic costs associated with the clinical management of diseases and to aid in developing individual treatment regimes that might result in more successful and efficient health care. The development of these technologies to noninvasively assess ischemia-stimulated cardiovascular angiogenesis and tumor-associated neovascularization is progressing currently in two directions. The first is the advancement of current imaging technology to develop high-resolution, high-sensitivity imaging systems and more recent efforts to develop dedicated hybrid (multimodal) imaging systems that combine two or more imaging modalities. The other direction is focused on the identification of unique biological signatures of angiogenesis and synthesis of targeted imaging probes using various constructs including antibodies, peptides, peptidomimetics, and nanoparticles.

Although many new modalities have been developed over the last decades, only a few are available for broad applications in both preclinical and clinical imaging research. The detailed description of operational parameters of various imaging modalities is beyond the scope of this chapter. However, it is worth emphasizing that the selection of an imaging technology to study angiogenesis should be based on the accessibility of the imaging system, the availability of targeted probes, and the properties of the biological system to be evaluated. Similarly, a number of unique markers of angiogenesis, including endothelial, nonendothelial, and ECM-associated biological targets, have been identified. However, among these potential targets only VEGF, VEGF receptors, and integrins in particular are the most promising biomarkers for the imaging of angiogenesis. Therefore, during recent years, attention has been directed to the development of molecular probes for these specific targets.

30.2.1 PET/SPECT IMAGING

Nuclear imaging approaches, including positron emission tomography (PET) and single-photon emission computed tomography (SPECT), have been the main modalities used for physiological imaging in patients (using perfusion tracers) and molecular imaging of biochemical processes. These modalities are characterized by their high sensitivity (in the picomolar range), reasonable spatial resolution (in the millimeter range), availability of instrumentation, and active development of targeted probes for imaging angiogenesis.

Early studies involving radiolabeled imaging probes were based on use of monoclonal antibodies specific for the epitopes expressed during tumor-associated angiogenesis. Initially, Collingridge and Li used human monoclonal anti-VEGF antibodies radiolabeled with [124]I and ^{123}I for PET and SPECT imaging, respectively (Li et al. 2001; Collingridge et al. 2002). However, suboptimal clearance rate of these imaging probes limited their use to preclinical use only.

To overcome this limitation, researchers employed in vivo screening of phage display peptide libraries, which resulted in the development of new probes with both the VEGF receptors and the corresponding ligands as the targets. Initial experiments involved radio-iodinated VEGF-121 and VEGF-165 isoforms, which were successfully used to assess tumor angiogenesis in humans (Li et al. 2001). Others used VEGF-121 labeled with ^{111}In for SPECT imaging of peripheral angiogenesis in rabbits or VEGF-121 labeled with ^{64}Cu for PET imaging of tumor and myocardial angiogenesis in laboratory animals (Cai et al. 2006; Rodriguez-Porcel et al. 2008). Another strategy involved using a fusion protein containing two fragments of human VEGF-121 cloned head to tail and labeled with the ^{99m}Tc chelator hydraziniumpyridine for SPECT (Backer et al. 2007) or ^{64}Cu for PET imaging of tumor-associated angiogenic vasculature (Levashova et al. 2009).

More recent studies suggest that a reporter gene imaging system can be developed to visualize and assess the efficacy of pro-angiogenic therapies involving VEGF including local administration of growth factors, gene vectors, and stem cells. Briefly, the system involves adenoviral delivery of mutated thymidine kinase (HSV1-sr39tk). Under the control of a cytomegalovirus promoter, the expression was localized to myocardial cells only. The reporter probe is ^{18}F-labeled fluoro-3-hydroxy-methyl-butylguanine (^{18}F-FHBG), which when phosphorylated by the HSV1-sr39tk is trapped in the myocardium for PET imaging. The feasibility of this reporter system linked to VEGF was demonstrated in preclinical animal models of myocardial angiogenesis in which in vivo imaging results correlated well with gamma well counting, thymidine kinase activity, and VEGF levels (Wu et al. 2002, 2004; Bengel et al. 2003; Inubushi et al. 2003).

As described in previous sections of this chapter, the ECM and integrins in particular are responsible for modulation of growth factor expression in response to mechanical strain and may play an integral role in initiation of angiogenesis. This biological behavior of αvβ3 integrins, characterized by a very low expression in quiescent endothelium and their upregulation in angiogenic cells, offers a tremendous advantage for targeted imaging of angiogenesis (Chen and Chen 2011; Cai and Conti 2013; Liu and Wang 2013).

Imaging angiogenic vessels through targeting of αvβ3 integrin was first proposed through a series of imaging studies using monoclonal antibodies against αvβ3 integrin. Unfortunately, poor blood clearance of these tracers motivated other investigators to use a number of αvβ3 antagonists based on the arginine-glycine-aspartate (RGD) binding sequence for targeted PET or SPECT imaging of angiogenesis (Sipkins et al. 1998).

A very high specificity of the RGD peptide for the αvβ3 integrin has been achieved by cyclization of a linear RGD pentapeptide, resulting in the cyclic pentapeptide (cRGDfV), which, when radiolabeled with ^{125}I, demonstrated a high affinity and good tumor targeting in preclinical models. However, strong liver, bile, and intestinal tract uptake of ^{125}I-cRGDfV was observed, which was responsible for relatively poor image quality. This limitation led to further chemical modifications of cRGD peptides to increase the hydrophilicity of cRGDyK by coupling sugar moieties, resulting in the development of Gluco-RGD (Haubner et al. 2001a) and galacto-RGD (Haubner et al. 2001b, 2004). In particular, ^{18}F-labeled galacto-RGD was the first integrin-specific PET radiotracer evaluated in humans for imaging tumor-associated angiogenesis and still is the most extensively studied αvβ3-integrin-specific PET tracer in the clinical setting. It demonstrated high affinity ($IC_{50} \sim 5$ nM), predominately renal excretion route, high stability in vivo, and fast blood clearance, and was well tolerated with no side effects. Unfortunately, the radiochemical synthesis of this tracer is complex and not easily automated, which hindered the widespread use of ^{18}F-galacto-RGD in the clinical routine.

Compared to ^{18}F-galacto-RGD, an advantage of ^{18}F-fluciclatide, developed by GE Healthcare, is the feasibility of its radiochemical synthesis. Its specificity toward integrins is also mediated by an RGD motif; however, by cyclization, introduction of disulfide bridges, and conjugation of polyethylene glycol (PEG) spacer, ^{18}F-fluciclatide was stable in vivo with ~74% intact tracer in the blood 1 h after injection. All this makes ^{18}F-fluciclatide feasible as a PET tracer in human subjects, which was demonstrated in studies where PET imaging was able to visualize all breast tumor lesions and liver metastases (Kenny et al. 2008).

Using RGD motif and click chemistry, Siemens Molecular Imaging developed a cyclic triazole-bearing RGD peptide (^{18}F-RGD-K5) for imaging angiogenesis in vivo. Its selectivity for αvβ3 integrin was 2.3 times higher than for other integrins and its affinity was determined as $K_d \sim 7.9$ nM. Initial studies in mice demonstrated predominantly renal clearance and exceptionally high in vivo stability (~98% of intact tracer in the blood at 1 h after injection) (Gaertner et al. 2010). Later, initial evaluation in 12 patients with breast cancer imaged with both ^{18}F-FDG and ^{18}F-RGD-K5 showed that, of a total of 157 FDG-positive lesions, 122 were also seen with ^{18}F-RGD-K5 and in most lesions FDG uptake was higher compared to ^{18}F-RGD-K5. However, no direct correlation was found between the uptake of these two tracers or between ^{18}F-RGD-K5 uptake and microvessel density assessed with histology.

Besides the above-mentioned RGD-based tracers, whose feasibility for imaging angiogenesis in humans was demonstrated in a series of preliminary studies and clinical trials, there is very active research focused on the development of novel peptide and nonpeptide radiolabeled agents for preclinical imaging of tumor-associated as well as both myocardial and peripheral ischemia-induced angiogenesis.

[111]Indium-labeled peptidomimetic quinolone (^{111}In-RP748) demonstrated nanomolar affinity and selectivity for activated αvβ3 integrins, which was confirmed in preclinical models of vascular injury (Harris et al. 2003; Sadeghi 2003). Focal uptake and blood clearance of ^{111}In-RP748 was evaluated in rodents and canines at early and late times after myocardial infarction and correlated with perfusion (assessed with ^{99m}Tc-sestamibi), tissue hypoxia (assessed with nitroimidazole ^{99m}Tc-BRU5921), and histochemical analysis of angiogenic markers (Meoli et al. 2004; Kalinowski et al. 2008).

More recently, a technetium-99m-labeled SPECT tracer (^{99m}Tc-NC100692, maraciclatide) was introduced for the imaging of αvβ3 integrin expression in various preclinical animal models of tumor- and ischemia-induced angiogenesis. This tracer, characterized by a very high affinity (~1 nM) and renal clearance route, was demonstrated to successfully evaluate αvβ3 integrin expression in subcutaneous and orthotopic tumors (Dearling et al. 2013), peripheral angiogenesis in murine model of hindlimb ischemia (Hua et al. 2005; Dobrucki et al. 2009), and myocardial angiogenesis in mice lacking the MMP-9 gene (Lindsey et al. 2006) and in rats subjected to the local adenoviral delivery of the insulin-like growth factor-1 (IGF-1) injected into the peri-infarct region (Figure 30.1) (Dobrucki et al. 2010b).

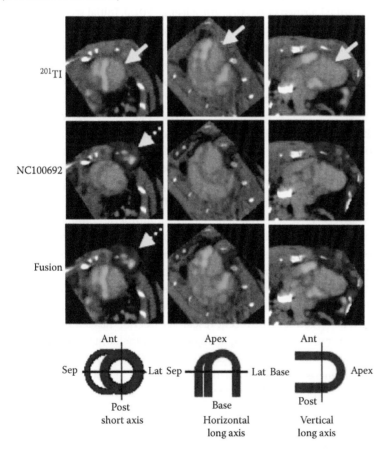

FIGURE 30.1 In vivo micro-SPECT-CT images of ^{201}Tl perfusion (top row, green) and ^{99m}Tc-NC100692 (middle row, red) in IGF-1 rat at 4 weeks post MI, reconstructed in short and horizontal and vertical long axes and fused (bottom row) with a reference contrast CT image (grayscale). All post-MI rats had an anterolateral ^{201}Tl perfusion defect (yellow solid arrows) and focal uptake of ^{99m}Tc-NC100692 in the defect area. The contrast agent permitted better definition of myocardium, allowing differentiation of focal myocardial uptake of targeted radiotracer from uptake within chest wall at the thoracotomy site (dashed yellow arrows). (Reprinted from *J. Mol. Cell Cardiol.*, 48(6), Dobrucki, L.W., Tsutsumi, Y., Kalinowski, L., Dean, J., Gavin, M., Sen, S., Mendizabal, M., Sinusas, A.J., and Aikawa, R., Analysis of angiogenesis induced by local IGF-1 expression after myocardial infarction using microSPECT-CT imaging, 1071–1079. Copyright 2010, with permission from Elsevier.)

Other groups have developed different constructs based on the RGD motif, including coupling 1,4,7,10-etraazacyclododecane-1,4,7,10-tetraacetic acid (DOTA) to cRGDyK (for ^{64}Cu PET imaging) or 1,4,7-triazacyclononane-1,4,7-triacetic acid (NOTA) to cRGDyK (for ^{68}Ga PET imaging) (Chen et al. 2004; Breeman and Verbruggen 2007; Eo et al. 2013). Both tracers have nanomolar affinity for αvβ3 integrin and are predominantly excreted via the renal pathway, and their in vivo feasibility was successfully evaluated in xenograft models of breast (MDA-MB-435) and colorectal (SNU-C4) cancer. Overall, it seems that ^{68}Ga-labeled RGD peptides have higher potential for clinical translation compared to ^{64}Cu-cRGD tracers. In contrast to ^{18}F- and ^{64}Cu-labeled compounds, ^{68}Ga can be eluted from commercially available ^{68}Ge/^{68}Ga generator systems, and therefore it does not require the access to an on-site cyclotron. In addition, the quick (~10 min) and relatively easy radiolabeling procedure and the short half-life of ^{68}Ga (68 min) are ideally suited for preparation of small peptidic tracers with optimal in vivo pharmacokinetics.

Despite the fact that chemical modifications of cyclic RGD peptide, such as by coupling sugar moieties or PEGylation, resulted in high-affinity, high-specificity targeted tracers with optimal pharmacokinetic properties for imaging angiogenesis, multiple groups have recently worked to further improve the targeted uptake, especially for detection of αvβ3 expression in areas with high physiological background. This approach resulted in the development of multimeric tracers, which are particularly useful in areas with multivalent binding sites and clustering of integrins (Liu 2009; Beer et al. 2011; Mittra et al. 2011; Shi et al. 2011; Laitinen et al. 2012, 2013; Ji et al. 2013). Initial studies involved the synthesis of monomeric, dimeric, and tetrameric RGD tracers by connecting multiple cRGDfE peptides via PEG linkers. In vitro binding affinity increased according to the number of cRGD peptides in the compound, which was further confirmed in biodistribution studies. In other studies, ^{64}Cu-labeled RGD dimers (DOTA-E[cRGDyK]2 and DOTA-E[cRGDfK]2) showed better tumor retention compared to their monomeric analogs (Chen et al. 2004) but nearly threefold lower integrin avidity when compared to the tetramer ^{64}Cu-DOTA-E[e[cRGDfK]2]2 (Wu et al. 2005). Other comparisons performed for ^{68}Ga-labeled dimeric cRGD probes led to similar conclusions (Oxboel et al. 2013).

Recently, Notni et al. developed a promising cRGDfK trimer that uses tri-azacyclononane-phosphinic acid (TRAP) as ^{68}Ga chelator. Compared to ^{18}F-galacto-cRGD, ^{68}Ga-TRAP(RGD)3 demonstrated over 7-fold higher affinity for αvβ3 integrin and 3.9-fold higher tumor uptake in a preclinical melanoma xenograft model (Notni et al. 2011).

Despite the relatively active research in developing multimeric compounds based on well-characterized cRGD pentapeptide backbone, it is worth emphasizing that a higher absolute uptake of the multimeric compound does not directly lead to improved imaging characteristics.

Currently, there is great interest in the utilization of radiolabeled, targeted nanoparticles and macromolecules for the molecular imaging of angiogenesis. Through careful optimization of passive uptake, nanoparticle structure, and composition, it is possible to achieve controlled in vivo circulation and tissue-specific targeting. This recent research direction has led to the development of targeted nanoparticles for molecular imaging of angiogenesis, including RGD nanoprobes containing a dendritic core, functionalized with a polyethylene oxide (PEO), and labeled with ^{76}Br or ^{64}Cu for PET imaging of angiogenesis (Lesniak et al. 2007; Almutairi et al. 2009).

As outlined in this section, radiolabeled tracers targeted at angiogenesis are currently evaluated in several preclinical animal models. These novel approaches in combination with the development of novel instrumentation and image analysis techniques may provide a new strategy to assess angiogenesis in the onset of novel genetic or cell therapies.

30.2.2 Ultrasound Imaging

Ultrasound is a relatively inexpensive, clinically available imaging modality, which is characterized by high temporal and spatial resolution providing real-time imaging. It has been employed to image angiogenesis by the acoustic ultrasonic detection of stable or inertial cavitations of microbubbles or gas-filled nanoparticle contrast agents. This technique named "contrast-enhanced ultrasound" (CEU) can be used to image molecular processes within the vascular compartment only. US microbubbles are pure intravascular probes, and their cellular targeting is based on chemical modification

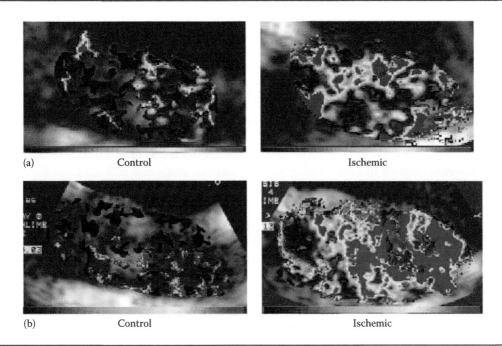

FIGURE 30.2   Examples of color-coded CEU images reflecting retention fraction of integrin-targeted microbubbles in control and ischemic proximal hindlimb adductor muscles from an (a) untreated and (b) FGF-2–treated rat 4 days after iliac artery ligation. Color scales appear at bottom. (Reprinted from Leong-Poi, H. et al., *Circulation*, 111(24), 3248, 2005. With permission.)

of the bubble shell to facilitate attachment to cells involved in angiogenesis (i.e., leukocytes) or by the conjugation of specific ligands such as antibodies, peptides, or peptidomimetics to the microbubble surface (Kaufmann and Lindner 2007; Lindner 2009; Piedra et al. 2009).

Although successful targeting depends on many factors, including ligand density on the microbubble surface, receptor density on the endothelium, and shear stress in the vasculature, it has been demonstrated that CEU is sensitive enough to detect a single microbubble within the vascular compartment. Several studies have targeted either VEGF receptors or $\alpha v \beta 3$ integrins for CEU imaging of angiogenesis in preclinical animal models including tumor and ischemic vascular remodeling. A feasibility study of a microbubble functionalized with echistatin to detect αv integrin expression was initially evaluated in vivo in Matrigel and in situ in a glioma model of tumor angiogenesis (Leong-Poi et al. 2003). Other studies have used similar approaches to noninvasively assess peripheral angiogenesis in a rodent model of hindlimb ischemia. An elegant study by Leong-Poi demonstrated that targeted microbubbles could successfully evaluate therapeutic angiogenesis stimulated in an ischemic rat hindlimb by local intramuscular sustained release of FGF-2 (see Figure 30.2) (Leong-Poi et al. 2005). In this study, signals obtained from targeted microbubbles were intense, localized to the ischemic muscle, and peaked before changes in peripheral perfusion, indicating that early angiogenic response can be detected with CEU before major changes in perfusion were observed.

Other groups have focused on using different peptides and peptidomimetic agents; however, all these targets were restricted to the intravascular compartment, including endothelial cell-specific sequence containing an RRL (arginine-arginine-leucine) motif for CEU imaging of tumor angiogenesis (Weller et al. 2005).

More recently, microbubbles functionalized with vascular growth factors, including VEGF antibodies targeted to their receptors such as VEGFR-2, were tested for the CEU imaging of peripheral and tumor-associated angiogenesis (Willmann et al. 2008).

Although the feasibility of CEU to noninvasively assess biological events including angiogenesis have been firmly established in preclinical animal models, the potential clinical role of this technology is still unclear and depends on further refinement of targeted microbubble conjugation chemistry, availability of novel vascular targets, and development of new instrumentation that will utilize the unique diagnostic information obtained from molecular CEU imaging.

30.2.3 MAGNETIC RESONANCE IMAGING

In contrast-enhanced MRI studies, the strong magnetization of magnetic particles is used to improve the signals originating from the intrinsic magnetization of the nuclei in the body. Typical contrast agents utilize gadolinium (Gd), which brightens regions retaining the contrast agent in T1-weighted images, or iron oxides that change the magnetic susceptibility, resulting in darkening of the affected voxels in T_2^*-sensitive images (Dobrucki et al. 2010a).

Investigators first proposed use of paramagnetic contrast agents containing Gd^{3+} paramagnetic liposomes targeted to endothelial αvβ3 integrin via the LM609 monoclonal antibody (Sipkins et al. 1998). However, targeted in vivo imaging with monoclonal antibodies suffered from slow blood clearance, which resulted in unsatisfactory image quality. Therefore, other groups proposed use of peptidomimetic αvβ3 integrin antagonists conjugated to magnetic Gd nanoparticles for imaging tumors using a clinical 1.5 T MRI scanner (Winter et al. 2003). More recently, liposomal nanoparticles functionalized with two angiogenesis-specific targeting ligands (cRGD and galectin-1 specific peptides) were used for bimodal (MRI and fluorescence) imaging of the angiogenesis process (Kluza et al. 2009). This multitargeting approach significantly improved nanoparticle retention as compared with single-ligand targeting. The major disadvantage of Gd-based contrast agents for MR imaging of angiogenesis is their relatively low sensitivity in the millimolar range. To overcome this limitation, other approaches were proposed, including use of super paramagnetic iron oxide (SPIO) constructs that provide much higher sensitivity and are less toxic than Gd conjugates. As a result of this strategy, ultrasmall RGD-conjugated SPIO nanoparticles were developed and evaluated for αvβ3 integrin-targeted T_2^*-weighted imaging of the tumor vasculature (Zhang et al. 2007). These RGD-SPIO nanoparticles specifically bound to tumor vasculature and were responsible for irregular decrease in signal density within αvβ3 integrin positive tumor vessels. Using a similar approach, Wolters et al. evaluated cyclic NGR (cNGR) tripeptide-labeled iron oxide particles that homed to activated endothelial cells of angiogenic blood vessels in tumor-bearing mice (Figure 30.3) (Wolters et al. 2012). Another strategy to take advantage of the high resolution of MRI and high sensitivity of other functional modalities such as nuclear or optical techniques involved the development of multimodal nanoparticles for targeted imaging of angiogenesis. Montet et al. proposed the use of RGD-SPIO-Cy5.5 nanoparticles for MRI-fluorescence imaging of angiogenesis (Montet et al. 2006). A similar concept involving RGD-SPIO-DOTA was proposed by Dijkgraaf for MRI/PET imaging of tumor with a high degree of accuracy (Dijkgraaf et al. 2009). Recent data suggest that dual-modality (PET/MRI) imaging probes should be of particular interest for angiogenesis imaging, especially considering recent development of clinical PET/MRI integrated systems (Liu et al. 2013).

Despite the progress in the quantification of MR signals originating from molecular targeted probes, there are still many sources of potential error, which makes MRI techniques less quantitative than other molecular imaging modalities. For example, SPIO conjugates have better toxicology profile than Gd conjugates, but the quantification of SPIO signals strongly depends on various imaging parameters, pulse sequences, field strengths, and voxel volumes. Other factors that affect signal quantification from molecular MRI probes are blood flow and pressure within target tissue. All this suggests that, despite potential advantages of MRI regarding spatial and temporal resolution and integration with metabolic or physiological indices, MR molecular imaging will require clinical validation.

30.2.4 OPTICAL IMAGING

Optical imaging, despite its limited clinical applicability due to the finite and small penetration depth of optical signal, belongs to a very versatile and highly sensitive set of tools for molecular imaging in small animals. It is based on the detection of visible spectrum photons emitted from living cells, tissues, or animals, and is categorized into two modalities: bioluminescence (BLI) utilizing luciferase and its derivatives, and fluorescence imaging (FLI) utilizing photons originating from fluorescent compounds (fluorophores) excited by photons generated by an external light source. While there are not many targeted probes for BLI, BLI reporter systems are suitable for biomedical

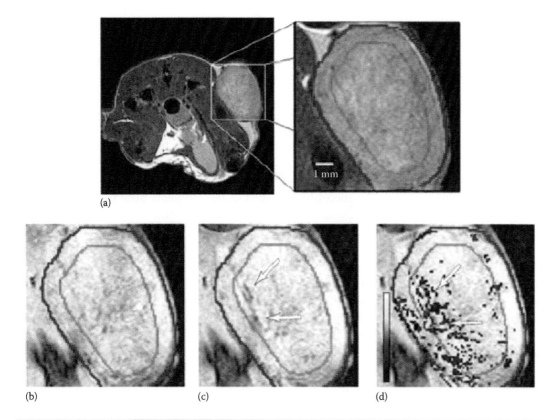

FIGURE 30.3 Axial sections through the tumor, where the red line represents the tumor contour and the purple line divides the rim and core section (1 mm-thick rim). (a) A spin echo image that shows the location of the tumor in a mouse of the cNGR-SPIO group. (b, c) Pre- and post-contrast GE images, respectively. (d) ΔR_2^* mapping (post and pre-contrast). Color bar range is 0 (black) to 100 (white). (From Wolters, M., Oostendorp, M., Coolen, B.F., Post, M.J., Janssen, J.M., Strijkers, G.J., Kooi, M.E., Nicolay, K., and Backes, W.H.: Efficacy of positive contrast imaging techniques for molecular MRI of tumor angiogenesis. *Contrast Media Mol. Imaging*. 2012. 7(2). 130–139. Copyright Wiley-VCH Verlag GmbH & Co. KGaA. Reproduced with permission.)

research purposes including small animal imaging due to the extremely low background signal, high signal-to-noise ratio, high sensitivity, short acquisition time (minutes), and high throughput, which allow imaging multiple animals during a single session (Snoeks et al. 2010). In contrast to BLI, there is a wide range of available fluorophores and targeted optical probes for FLI, spanning from fluorescent dyes to nanoparticles, such as quantum dots, and to fluorescent proteins, which can be expressed in transgenic animals. However, similar to BLI, in vivo FLI suffers from limited penetration depth due to tissue absorbance, scattering, and autofluorescence.

The last decade has seen a rapid growth of both optical imaging applications and development of BLI and FLI systems to assess biological processes in small animal models. The majority of imaging systems for BLI and FLI are still limited to two-dimensional planar information, which is nonquantitative and with poor spatial resolution. To overcome this limitation, much effort has been put recently to the development of a whole-body small animal imaging strategy that pushes further the limits of optical imaging in terms of sensitivity, quantification, and spatial resolution.

30.2.4.1 Gene Reporter Systems

VEGF and VEGF receptors that are involved in the angiogenic process are attractive targets for molecular imaging of angiogenesis, and, as a result, several approaches to image and quantify VEGF receptor and VEGF expression have been proposed.

VEGF actions in angiogenesis are mediated through the VEGF receptors (VEGFR1 and VEGFR2) expressed on endothelial cells. To study these interactions, several transgenic animal models have been recently developed. A prime example of such a model is a VEGFR2-luciferase transgenic mouse model, in which luciferase expression is controlled by the VEGFR2 promoter region causing endothelial cells within angiogenic vasculature to express luciferase. This model has been validated in numerous studies including a cutaneous wound-healing model, in which the increase in luciferase signal (and thus VEGFR2 gene activation) peaked at 7–10 days after the wound was inflicted (Zhang et al. 2004).

To further expand the proposed model, the same group utilized co-registration of the BLI signal originating from VEGFR2-luciferase system and FLI signal produced by the red fluorescent SMF-mCherry breast cancer cells inoculated orthotopically. This approach allowed the simultaneous quantification of both tumor growth (with FLI) and pro-angiogenic signaling (with BLI), the ratio of which was used to characterize the angiogenic properties of the tumors (Zhang et al. 2004).

VEGF signaling has been also successfully studied using transgenic mice that express green fluorescent protein (GFP) or bioluminescent luciferase under control of the VEGF promoter (Kishimoto et al. 2000; Wang et al. 2006). The latter model, named pVEGF-TSTA-fl, utilizes a two-step transcriptional amplification (TSTA) system, which resulted in 50-fold increase in luciferase expression compared to the direct one-step system (Iyer et al. 2001). This transgenic model successfully demonstrated the correlation between VEGF expression and BLI signal in vitro and in in vivo wound-healing assays and tumor xenograft mouse models.

Other transgenic animal models have been recently developed to image not only angiogenic VEGF signaling but also existing vasculature. These models include transgenic mice with endothelial specific expression of GFP driven by the promoters of the endothelial specific receptor tyrosine kinase (Tie2) or endothelial nitric oxide synthase (eNOS) (Motoike et al. 2000; Hillen et al. 2008)

Both Tie2-GFP and eNOS-GFP mouse models are well suited to study angiogenesis and interactions between cancer cells (i.e., xenografts) and tumor-associated vasculature, and allow serial in vivo fluorescence imaging over time (Yang et al. 2001; van Haperen et al. 2003).

Tie2-Biotag mouse belongs to the most recent animal models that allow studying angiogenic signaling. This model is characterized by the expression of biotin ligase (BirA) under the control of the endothelial-specific Tie2 promoter. In this model, a cluster of BirA substrate sequences (Biotag) fused to a transmembrane domain can be targeted in vivo with highly specific avidinated imaging probes. As a result, biotinylated endothelial cells involved in angiogenesis can be imaged with a modality of choice using avidinated multimodal agents.

30.2.4.2 Fluorescent Probes

In addition to the bioluminescent (luciferase) and fluorescent (GFP) gene reporter systems described in the previous section, there are imaging approaches that utilize fluorescent imaging probes targeted at various epitopes involved in the process of angiogenesis. These targeted probes consist of a fluorophore conjugated to a highly specific antibody, peptide sequence, or nanoparticle. Recent development of these fluorescent probes has been focused on two vascular targets: VEGF and VEGF receptors, and $\alpha v \beta 3$ integrins.

As shown earlier, VEGFR-2 was often used to image both tumor and cardiovascular angiogenesis due to its local and temporal upregulation during angiogenesis. Initial work involved using single-chain VEGF (scVEGF) with an N-terminal cysteine-containing tag (Cys-tag) (Backer et al. 2006).

The Cys tag was then used for site-specific attachment of various agents for targeted imaging of the VEGF receptor in in vitro and in vivo tumor angiogenesis (Blankenberg et al. 2006; Levashova et al. 2008). More recently, scVEGF labeled with Cy5.5 has been used for near-infrared fluorescence imaging of VEGF receptor in angiogenic tumor-associated vasculature (Backer et al. 2007; Levashova et al. 2009).

In addition to VEGF and VEGF receptors, cRGD peptide sequence, which specifically targets $\alpha v \beta 3$ and $\alpha v \beta 5$ integrins, has been a focus of intensive research to develop targeted probes for fluorescence imaging of angiogenesis. Based on this specific interaction between cRGD and $\alpha v \beta 3$ integrin, many fluorescent cRGD peptides have been developed for fluorescence imaging of angiogenesis. These targeted agents include near-infrared probes such as Cy7-cRGD, Cy5.5-cRGD, QD705-cRGD, or cRGD coupled to IRDye 800CW to take advantage of the ideal fluorescent characteristics for in vivo imaging. Others have successfully developed nanoparticle constructs, which allowed multimodality molecular imaging of angiogenesis. Mulder et al. described a green fluorescent quantum dot core paramagnetic nanoparticle functionalized with cRGD peptide that was suitable for both fluorescence and MRI of the expression of $\alpha v \beta 3$ integrin (Mulder et al. 2008). More recently, Li et al. developed lipid-encapsulated infrared quantum dots with cRGD presented at the outside of the albumin microsphere and radiolabeled with ^{64}Cu for PET, fluorescence, and Cerenkov luminescence imaging of angiogenesis (Li et al. 2015).

Despite these recent efforts to improve probe specificity and binding affinity, it was demonstrated that although many tumor cells are positive for $\alpha v \beta 3$ integrin, many cRGD peptide-based imaging probes localize mainly at the tumor endothelium and not in the tumor itself (Schraa et al. 2002). To overcome this limitation, small molecule nonpeptide $\alpha v \beta 3$ antagonist fused to the near-infrared fluorophore (commercially available as IntegriSense) was synthesized and validated in in vitro and in vivo models. Unlike cRGD peptide-based imaging agents, extravasation of IntegriSense does occur, and the probe localizes in endothelium and $\alpha v \beta 3$ integrin-positive tumor cells. Both IntegriSense and ProSense (a cathepsin-based protease-activated fluorescent probe) were recently used to visualize both angiogenic blood vessels and tumor, and to quantify the effect of angiogenic treatment (see Figure 30.4) (Kossodo et al. 2010).

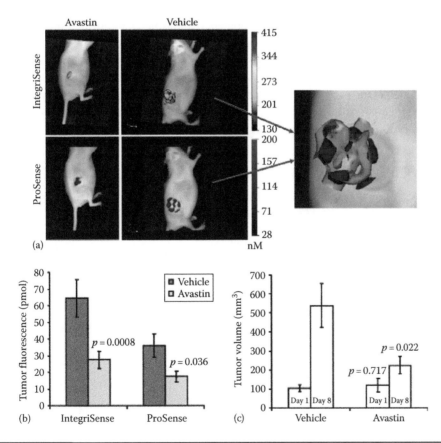

FIGURE 30.4 Effect of antiangiogenic avastin treatment on integrin and cathepsin B signals in A673 tumor-bearing mice. (a) Representative volume renderings of IntegriSense and ProSense. Note the differential localization of fluorescent signal with IntegriSense and ProSense in the same animals (IntegriSense in blue and ProSense in red). (b) Quantification of tumor fluorescence shows that IntegriSense and Prosense signal in tumors is decreased following avastin treatment ($p = 0.0008$ and $p = 0.013$, respectively). (c) Avastin also significantly inhibits tumor growth as determined by tumor volume caliper measurements (by 58.4%, $p = 0.022$). (With kind permission from Springer Science+Business Media: *Mol. Imaging Biol.*, Dual in vivo quantification of integrin-targeted and protease-activated agents in cancer using fluorescence molecular tomography (FMT), 12(5), 2010, 488–499, Kossodo, S., Pickarski, M., Lin, S.-A., Gleason, A., Gaspar, R., Buono, C., Ho, G. et al.)

The development of small animal dedicated fluorescence-mediated tomography (FMT), algorithms to unmix optical signals of various wavelengths, and long circulating fluorescent targeted macromolecules has shifted the imaging focus from intravital microscopy to 3D whole body imaging, which has provided a powerful tool to study different processes and interactions such as angiogenesis and the tumor growth in the same animal. Furthermore, optical imaging can be combined with other imaging modalities, which offers the opportunity for high-sensitivity imaging of angiogenesis and anatomical co-localization of optical signals with high spatial resolution.

30.3 FUTURE DIRECTIONS

The future of the noninvasive imaging of angiogenesis may rest on the development of new imaging technologies, including new instrumentation and discovery of new biological markers of angiogenic process, which will lead to the synthesis of novel targeted imaging agents. Several imaging agents targeted at VEGF or $\alpha v\beta 3$ integrin have already proved successful in monitoring angiogenesis in various preclinical animal models, and some of them have already progressed to clinical trials. Also, quantitative and highly sensitive imaging modalities such as nuclear and optical technologies are being combined with higher resolution imaging methods such as CT and MRI to facilitate accurate image quantification and anatomical localization. This combination of anatomical and functional imaging modalities will also require the development of multimodal multifunctional imaging agents (typically based on nanoparticle constructs), which can utilize the strengths of this multimodal approach. Such application of nanotechnology to imaging angiogenesis is an exciting new area of research, but further studies are needed to fully evaluate the feasibility of applying this technology to the clinical arena. The overarching goal of these new targeted imaging technologies is to provide new tools for prognostication as well as improvement in more individualized care of patients with oncologic and cardiovascular pathologies.

REFERENCES

Almutairi, A., R. Rossin et al. (2009). Biodegradable dendritic positron-emitting nanoprobes for the noninvasive imaging of angiogenesis. *Proc. Natl. Acad. Sci. USA* **106**(3): 685–690.

Backer, M., Z. Levashova et al. (2007). Molecular imaging of VEGF receptors in angiogenic vasculature with single-chain VEGF-based probes. *Nat. Med.* **13**(4): 504–509.

Backer, M. V., V. Patel et al. (2006). Surface immobilization of active vascular endothelial growth factor via a cysteine-containing tag. *Biomaterials* **27**(31): 5452–5458.

Beer, A. J., H. Kessler et al. (2011). PET imaging of integrin alphaVbeta3 expression. *Theranostics* **1**: 48–57.

Bengel, F. M., M. Anton et al. (2003). Noninvasive imaging of transgene expression by use of positron emission tomography in a pig model of myocardial gene transfer. *Circulation* **108**(17): 2127–2133.

Blankenberg, F. G., M. V. Backer et al. (2006). In vivo tumor angiogenesis imaging with site-specific labeled (99m)Tc-HYNIC-VEGF. *Eur. J. Nucl. Med. Mol. Imaging* **33**(7): 841–848.

Breeman, W. A. and A. M. Verbruggen. (2007). The 68Ge/68Ga generator has high potential, but when can we use 68Ga-labelled tracers in clinical routine? *Eur. J. Nucl. Med. Mol. Imaging* **34**(7): 978–981.

Brooks, P. C., R. A. Clark et al. (1994). Requirement of vascular integrin alphavbeta3 for angiogenesis. *Science* **264**(5158): 569–571.

Cai, H. and P. S. Conti. (2013). RGD-based PET tracers for imaging receptor integrin alphavbeta3 expression. *J. Label. Comp. Radiopharm.* **56**(5): 264–279.

Cai, W., K. Chen et al. (2006). PET of vascular endothelial growth factor receptor expression. *J. Nucl. Med.* **47**(12): 2048–2056.

Carmeliet, P. (2005). Angiogenesis in life, disease and medicine. *Nature* **438**(7070): 932–936.

Carmeliet, P. and R. K. Jain. (2000). Angiogenesis in cancer and other diseases. *Nature* **407**(6801): 249–257.

Cerqueira, M. and J. Udelson. (2003). Lake Tahoe invitation meeting 2002. *J. Nucl. Cardiol.* **10**(2): 223–257.

Chen, K. and X. Chen. (2011). Integrin targeted delivery of chemotherapeutics. *Theranostics* **1**: 189–200.

Chen, X., R. Park et al. (2004). MicroPET and autoradiographic imaging of breast cancer alpha v-integrin expression using 18F- and 64Cu-labeled RGD peptide. *Bioconjug. Chem.* **15**(1): 41–49.

Collingridge, D., V. Carroll et al. (2002). The development of [(124)I]iodinated-VG76e: A novel tracer for imaging vascular endothelial growth factor in vivo using positron emission tomography. *Cancer Res.* **62**(20): 5912–5919.

Cross, M. J. and L. Claesson-Welsh. (2001). FGF and VEGF function in angiogenesis: Signalling pathways, biological responses and therapeutic inhibition. *Trends Pharmacol. Sci.* **22**(4): 201–207.

Dearling, J. L., J. W. Barnes et al. (2013). Specific uptake of 99mTc-NC100692, an alphavbeta3-targeted imaging probe, in subcutaneous and orthotopic tumors. *Nucl. Med. Biol.* **40**(6): 788–794.

Dijkgraaf, I., A. Beer et al. (2009). Application of RGD-containing peptides as imaging probes for alphavbeta3 expression. *Front. Biosci.* **14**: 887–899.

Dobrucki, L. W., E. D. de Muinck et al. (2010a). Approaches to multimodality imaging of angiogenesis. *J. Nucl. Med.* **51**(Suppl. 1): 66S–79S.

Dobrucki, L. W., D. P. Dione et al. (2009). Serial noninvasive targeted imaging of peripheral angiogenesis: Validation and application of a semiautomated quantitative approach. *J. Nucl. Med.* **50**(8): 1356–1363.

Dobrucki, L. W. and A. J. Sinusas. (2005a). Cardiovascular molecular imaging. *Semin. Nucl. Med.* **35**(1): 73–81.

Dobrucki, L. W. and A. J. Sinusas. (2005b). Molecular imaging: A new approach to nuclear cardiology. *Q. J. Nucl. Med. Mol. Imaging* **49**(1): 106–115.

Dobrucki, L. W. and A. J. Sinusas. (2010). PET and SPECT in cardiovascular molecular imaging. *Nat. Rev. Cardiol.* **7**(1): 38–47.

Dobrucki, L. W., Y. Tsutsumi et al. (2010b). Analysis of angiogenesis induced by local IGF-1 expression after myocardial infarction using microSPECT-CT imaging. *J. Mol. Cell Cardiol.* **48**(6): 1071–1079.

Eliceiri, B. P. and D. A. Cheresh. (2000). Role of alpha v integrins during angiogenesis. *Cancer J.* **6**(Suppl. 3): S245–S249.

Eo, J. S., J. C. Paeng et al. (2013). Angiogenesis imaging in myocardial infarction using 68Ga-NOTA-RGD PET: Characterization and application to therapeutic efficacy monitoring in rats. *Coron. Artery Dis.* **24**(4): 303–311.

Fam, N. P., S. Verma et al. (2003). Clinician guide to angiogenesis. *Circulation* **108**(21): 2613–2618.

Ferrara, N. and T. Davis-Smyth. (1997). The biology of vascular endothelial growth factor. *Endocr. Rev.* **18**(1): 4–25.

Folkman, J. (2007). Angiogenesis: An organizing principle for drug discovery? *Nat. Rev. Drug Discov.* **6**(4): 273–286.

Gaertner, F. C., M. Schwaiger et al. (2010). Molecular imaging of alfavbeta3 expression in cancer patients. *Q. J. Nucl. Med. Mol. Imaging* **54**(3): 309–326.

Harris, T., S. Kalogeropoulos et al. (2003). Design, synthesis, and evaluation of radiolabeled integrin alpha v beta 3 receptor antagonists for tumor imaging and radiotherapy. *Cancer Biother. Radiopharm.* **18**(4): 627–641.

Haubner, R. (2008). Noninvasive tracer techniques to characterize angiogenesis. *Handb. Exp. Pharmacol.* (185 Pt 2): 323–339.

Haubner, R., B. Kuhnast et al. (2004). [18F]Galacto-RGD: Synthesis, radiolabeling, metabolic stability, and radiation dose estimates. *Bioconjug. Chem.* **15**(1): 61–69.

Haubner, R., H. J. Wester et al. (2001a). Glycosylated RGD-containing peptides: Tracer for tumor targeting and angiogenesis imaging with improved biokinetics. *J. Nucl. Med.* **42**(2): 326–336.

Haubner, R., H. J. Wester et al. (2001b). Noninvasive imaging of alpha(v)beta3 integrin expression using 18F-labeled RGD-containing glycopeptide and positron emission tomography. *Cancer Res.* **61**(5): 1781–1785.

Hidalgo, M. and S. G. Eckhardt. (2001). Development of matrix metalloproteinase inhibitors in cancer therapy. *J. Natl. Cancer Inst.* **93**(3): 178–193.

Hillen, F., E. L. Kaijzel et al. (2008). A transgenic Tie2-GFP athymic mouse model; a tool for vascular biology in xenograft tumors. *Biochem. Biophys. Res. Commun.* **368**(2): 364–367.

Hua, J., L. W. Dobrucki et al. (2005). Noninvasive imaging of angiogenesis with a Tc-99m-labeled peptide targeted at alpha(v)beta(3) integrin after murine hindlimb ischemia. *Circulation* **111**(24): 3255–3260.

Iagaru, A. and S. S. Gambhir. (2013). Imaging tumor angiogenesis: The road to clinical utility. *AJR Am. J. Roentgenol.* **201**(2): W183–W191.

Inubushi, M., J. C. Wu et al. (2003). Positron-emission tomography reporter gene expression imaging in rat myocardium. *Circulation* **107**(2): 326–332.

Iyer, M., L. Wu et al. (2001). Two-step transcriptional amplification as a method for imaging reporter gene expression using weak promoters. *Proc. Natl. Acad. Sci. USA* **98**(25): 14595–14600.

Ji, S., Y. Zhou et al. (2013). Monitoring tumor response to linifanib therapy with SPECT/CT using the integrin alphavbeta3-targeted radiotracer 99mTc-3P-RGD2. *J. Pharmacol. Exp. Ther.* **346**(2): 251–258.

Kalinowski, L., L. W. Dobrucki et al. (2008). Targeted imaging of hypoxia-induced integrin activation in myocardium early after infarction. *J. Appl. Physiol.* **104**(5): 1504–1512.

Kaufmann, B. and J. Lindner. (2007). Molecular imaging with targeted contrast ultrasound. *Curr. Opin. Biotechnol.* **18**(1): 11–16.

Kenny, L. M., R. C. Coombes et al. (2008). Phase I trial of the positron-emitting Arg-Gly-Asp (RGD) peptide radioligand 18F-AH111585 in breast cancer patients. *J. Nucl. Med.* **49**(6): 879–886.

Kishimoto, J., R. Ehama et al. (2000). In vivo detection of human vascular endothelial growth factor promoter activity in transgenic mouse skin. *Am. J. Pathol.* **157**(1): 103–110.

Kluza, E., D. van der Schaft et al. (2009). Synergistic targeting of alpha(v)beta(3) integrin and Galectin-1 with heteromultivalent paramagnetic liposomes for combined MR imaging and treatment of angiogenesis. *Nano Lett.*

Kossodo, S., M. Pickarski et al. (2010). Dual in vivo quantification of integrin-targeted and protease-activated agents in cancer using fluorescence molecular tomography (FMT). *Mol. Imaging Biol.* **12**(5): 488–499.

Laitinen, I., J. Notni et al. (2013). Comparison of cyclic RGD peptides for alphavbeta3 integrin detection in a rat model of myocardial infarction. *EJNMMI Res.* **3**(1): 38.

Laitinen, I., K. Pohle et al. (2012). Evaluation of 68Ga-labeled RGD peptides for αvβ3 integrin detection in myocardial infarct. *J. Nucl. Med.* **53**(Suppl. 1): 460.

Lederman, R. J., F. O. Mendelsohn et al. (2002). Therapeutic angiogenesis with recombinant fibroblast growth factor-2 for intermittent claudication (the TRAFFIC study): A randomised trial. *Lancet* **359**(9323): 2053–2058.

Leong-Poi, H., J. Christiansen et al. (2003). Noninvasive assessment of angiogenesis by ultrasound and microbubbles targeted to alpha(v)-integrins. *Circulation* **107**(3): 455–460.

Leong-Poi, H., J. Christiansen et al. (2005). Assessment of endogenous and therapeutic arteriogenesis by contrast ultrasound molecular imaging of integrin expression. *Circulation* **111**(24): 3248–3254.

Lesniak, W. G., M. S. Kariapper et al. (2007). Synthesis and characterization of PAMAM dendrimer-based multifunctional nanodevices for targeting alphavbeta3 integrins. *Bioconjug. Chem.* **18**(4): 1148–1154.

Levashova, Z., M. Backer et al. (2009). Imaging vascular endothelial growth factor (VEGF) receptors in turpentine-induced sterile thigh abscesses with radiolabeled single-chain VEGF. *J. Nucl. Med.* **50**(12): 2058–2063.

Levashova, Z., M. Backer et al. (2008). Direct site-specific labeling of the Cys-tag moiety in scVEGF with technetium 99m. *Bioconjug. Chem.* **19**(5): 1049–1054.

Li, J., L. W. Dobrucki et al. (2015). Enhancement and wavelength-shifted emission of Cerenkov luminescence using multifunctional microspheres. *Phys. Med. Biol.* **60**(2): 727–739.

Li, S., M. Peck-Radosavljevic et al. (2001). Characterization of (123)I-vascular endothelial growth factor-binding sites expressed on human tumour cells: Possible implication for tumour scintigraphy. *Int. J. Cancer* **91**(6): 789–796.

Lindner, J. (2009). Contrast ultrasound molecular imaging of inflammation in cardiovascular disease. *Cardiovasc. Res.* **84**(2): 182–189.

Lindsey, M. L., G. P. Escobar et al. (2006). Matrix metalloproteinase-9 gene deletion facilitates angiogenesis after myocardial infarction. *Am. J. Physiol. Heart Circ. Physiol.* **290**(1): H232–H239.

Liu, S. (2009). Radiolabeled cyclic RGD peptides as integrin alpha(v)beta(3)-targeted radiotracers: Maximizing binding affinity via bivalency. *Bioconjug. Chem.* **20**(12): 2199–2213.

Liu, Y., Y. Yang et al. (2013). A concise review of magnetic resonance molecular imaging of tumor angiogenesis by targeting integrin alphavbeta3 with magnetic probes. *Int. J. Nanomed.* **8**: 1083–1093.

Liu, Z. and F. Wang. (2013). Development of RGD-based radiotracers for tumor imaging and therapy: Translating from bench to bedside. *Curr. Mol. Med.* **13**(10): 1487–1505.

Meoli, D. F., M. M. Sadeghi et al. (2004). Noninvasive imaging of myocardial angiogenesis following experimental myocardial infarction. *J. Clin. Invest.* **113**(12): 1684–1691.

Mitsos, S., K. Katsanos et al. (2012). Therapeutic angiogenesis for myocardial ischemia revisited: Basic biological concepts and focus on latest clinical trials. *Angiogenesis* **15**(1): 1–22.

Mittra, E. S., M. L. Goris et al. (2011). Pilot pharmacokinetic and dosimetric studies of (18)F-FPPRGD2: A PET radiopharmaceutical agent for imaging alpha(v)beta(3) integrin levels. *Radiology* **260**(1): 182–191.

Montet, X., K. Montet-Abou et al. (2006). Nanoparticle imaging of integrins on tumor cells. *Neoplasia* **8**(3): 214–222.

Morrison, M. and A. Sinusas. (2010). Molecular imaging approaches for evaluation of myocardial pathophysiology: Angiogenesis, ventricular remodeling, inflammation, and cell death. In B. L. Zaret and G. A. Beller (eds.), *Clinical Nuclear Cardiology: State of the Art and Future Directions.* Elsevier, Philadelphia, PA, pp. 691–712.

Motoike, T., S. Loughna et al. (2000). Universal GFP reporter for the study of vascular development. *Genesis* **28**(2): 75–81.

Mulder, W. J., D. P. Cormode et al. (2008). Multimodality nanotracers for cardiovascular applications. *Nat. Clin. Pract. Cardiovasc. Med.* **5**(Suppl. 2): S103–S111.

Neufeld, G., T. Cohen et al. (1996). Similarities and differences between the vascular endothelial growth factor (VEGF) splice variants. *Cancer Metastasis Rev.* **15**(2): 153–158.

Notni, J., J. Simecek et al. (2011). TRAP, a powerful and versatile framework for gallium-68 radiopharmaceuticals. *Chemistry* **17**(52): 14718–14722.

Oxboel, J., M. Brandt-Larsen et al. (2013). Comparison of two new angiogenesis PET tracers Ga-NODAGA-E[c(RGDyK)] and Cu-NODAGA-E[c(RGDyK)]; in vivo imaging studies in human xenograft tumors. *Nucl. Med. Biol.*

Piedra, M., A. Allroggen et al. (2009). Molecular imaging with targeted contrast ultrasound. *Cerebrovasc. Dis.* **27**(Suppl. 2): 66–74.

Rodriguez-Porcel, M., W. Cai et al. (2008). Imaging of VEGF receptor in a rat myocardial infarction model using PET. *J. Nucl. Med.* **49**(4): 667–673.

Rundhaug, J. E. (2005). Matrix metalloproteinases and angiogenesis. *J. Cell Mol. Med.* **9**(2): 267–285.

Sadeghi, M. M. (2003). Imaging avb3 integrin in vascular injury: Does this reflect increased integrin expression or activation? *Circulation* **108**(Suppl. 17): 1868.

Schraa, A. J., R. J. Kok et al. (2002). Targeting of RGD-modified proteins to tumor vasculature: A pharmacokinetic and cellular distribution study. *Int. J. Cancer* **102**(5): 469–475.

Shi, J., Y. Zhou et al. (2011). Evaluation of in-labeled cyclic RGD peptides: Effects of peptide and linker multiplicity on their tumor uptake, excretion kinetics and metabolic stability. *Theranostics* **1**: 322–340.

Sinusas, A. (2004). Imaging of angiogenesis. *J. Nucl. Cardiol.* **11**(5): 617–633.

Sipkins, D., D. Cheresh et al. (1998). Detection of tumor angiogenesis in vivo by alphaVbeta3-targeted magnetic resonance imaging. *Nat. Med.* **4**(5): 623–626.

Sneider, E. B., P. T. Nowicki et al. (2009). Regenerative medicine in the treatment of peripheral arterial disease. *J. Cell. Biochem.* **108**(4): 753–761.

Snoeks, T. J., C. W. Lowik et al. (2010). 'In vivo' optical approaches to angiogenesis imaging. *Angiogenesis* **13**(2): 135–147.

van Haperen, R., C. Cheng et al. (2003). Functional expression of endothelial nitric oxide synthase fused to green fluorescent protein in transgenic mice. *Am. J. Pathol.* **163**(4): 1677–1686.

Wang, Y., M. Iyer et al. (2006). Noninvasive indirect imaging of vascular endothelial growth factor gene expression using bioluminescence imaging in living transgenic mice. *Physiol. Genomics* **24**(2): 173–180.

Weller, G., M. Wong et al. (2005). Ultrasonic imaging of tumor angiogenesis using contrast microbubbles targeted via the tumor-binding peptide arginine-arginine-leucine. *Cancer Res.* **65**(2): 533–539.

Willmann, J., R. Paulmurugan et al. (2008). US imaging of tumor angiogenesis with microbubbles targeted to vascular endothelial growth factor receptor type 2 in mice. *Radiology* **246**(2): 508–518.

Winter, P., S. Caruthers et al. (2003). Molecular imaging of angiogenesis in nascent Vx-2 rabbit tumors using a novel alpha(nu)beta3-targeted nanoparticle and 1.5 tesla magnetic resonance imaging. *Cancer Res.* **63**(18): 5838–5843.

Wolters, M., M. Oostendorp et al. (2012). Efficacy of positive contrast imaging techniques for molecular MRI of tumor angiogenesis. *Contrast Media Mol. Imaging* **7**(2): 130–139.

Wu, J. C., I. Y. Chen et al. (2004). Molecular imaging of the kinetics of vascular endothelial growth factor gene expression in ischemic myocardium. *Circulation* **110**(6): 685–691.

Wu, J. C., M. Inubushi et al. (2002). Positron emission tomography imaging of cardiac reporter gene expression in living rats. *Circulation* **106**(2): 180–183.

Wu, Y., X. Zhang et al. (2005). microPET imaging of glioma integrin {alpha}v{beta}3 expression using (64) Cu-labeled tetrameric RGD peptide. *J. Nucl. Med.* **46**(10): 1707–1718.

Yang, M., E. Baranov et al. (2001). Whole-body and intravital optical imaging of angiogenesis in orthotopically implanted tumors. *Proc. Natl. Acad. Sci. USA* **98**(5): 2616–2621.

Zhang, C., M. Jugold et al. (2007). Specific targeting of tumor angiogenesis by RGD-conjugated ultrasmall superparamagnetic iron oxide particles using a clinical 1.5-T magnetic resonance scanner. *Cancer Res.* **67**(4): 1555–1562.

Zhang, N., Z. Fang et al. (2004). Tracking angiogenesis induced by skin wounding and contact hypersensitivity using a Vegfr2-luciferase transgenic mouse. *Blood* **103**(2): 617–626.

31

Imaging of Hypoxia, Apoptosis, and Inflammation

Stavros Spiliopoulos and Athanasios Diamantopoulos

31.1 INTRODUCTION

Innovative molecular imaging technologies have radically increased the quality and efficiency of preclinical studies by enabling the in vivo monitoring of disease progression and real-time investigation of therapeutic protocols. Today, molecular in vivo imaging of cell apoptosis, hypoxia, and inflammation is included in many up-to-date experimental protocols, mainly in the field of oncological research, as an integral component for the comprehensive understanding of the pathophysiology of various pathologies such as cancer and neurodegenerative and cardiovascular diseases, thus contributing to the development and testing of novel pharmaceutical agents.

The main radiological modalities used for the in vivo imaging of hypoxia, apoptosis, and inflammation in small animal models include micromagnetic resonance imaging (micro-MRI), microcomputed tomography (micro-CT), micropositron emission tomography (micro-PET), microsingle photon emission computed tomography (micro-SPET), micro-high-frequency ultrasound (micro-HFUS), optical coherence tomography (OCT), bioluminescence, fluorescence, and Fourier transform infrared spectroscopy imaging. This chapter will focus on the fundamentals of hypoxia, apoptosis, and inflammation in vivo imaging in small animals using the aforementioned modalities, as well as the possible clinical implications deriving from these research protocols.

It should also be taken under consideration that imaging of small animals requires anesthesia, and anesthetic agents have been reported to generate an intense effect on the animal's physiology distorting the image data acquired. As a result, the appropriate anesthetic regime and monitoring systems during image acquisition are of foremost importance (Tremoleda et al. 2012).

31.2 IMAGING OF APOPTOSIS

Apoptosis is the phenomenon of programmed cell death in which cells that are no longer required to the organism are removed from the organism without triggering inflammation. Apoptosis is known to occur not only during physiologic homeostatic processes but also during the development of various pathologies including cancer (Thompson 1995; Fink and Cookson 2005).

The term apoptosis (from the Greek *apo* [off] and *ptosis* [fall], meaning "dropping off") was chosen in order to describe the paraphysiologic result of the composite series of cellular and molecular changes such as nuclear condensation, cytoplasmic shrinkage, and cytoplasmic membrane alteration (Fink and Cookson 2005). Imaging of the apoptotic event could provide a much faster way to predict the effectiveness of cancer chemotherapy than the currently used morphologic assessment. In cancer, apoptosis state-of-the-art imaging is utilized for the assessment response to therapy, while in cardiovascular disease, it is extremely useful for the management of myocardial infarction, unstable atherosclerotic plaques, and cardiac allograft rejection. Moreover, in vivo apoptosis imaging is imperative for the development of up-to-date personalized therapy. Today, apoptosis can be detected in vivo using real-time molecular imaging modalities, as apoptotic cells undergo particular biochemical alterations that can be exploited as targets for various molecular imaging agents. At present, there are four main categories of apoptosis imaging agents, based on the cellular processes that they are able to detect: (1) phosphatidylserine-binding agents, (2) caspase-activation binding agents, (3) altered membrane permeability binding agents, and (4) mitochondrial membrane potential collapse binding agents (Tait 2008).

31.2.1 PHOSPHATIDYLSERINE-BINDING AGENTS

During apoptosis, phosphatidylserine becomes externalized on the cell's surface. The most widely investigated phosphatidylserine-binding agent is radiolabeled with ^{99m}Tc annexin V, a global intracellular human protein presenting affinity for membrane phosphatidylserine (Munoz et al. 2007). Alternative forms of radiolabeled annexin V for radionuclide imaging with PET, SPECT, or fluorescence include the self-chelating annexin V mutants V-117 and V-128, proteins presenting an endogenous site for ^{99m}Tc chelation (Tait et al. 2000; Jin et al. 2004). These new-generation annexins have been investigated in small animal models and demonstrated superior results regarding renal uptake (50%–75% radiotracer reduction) and specific localization to the apoptotic sites compared to hydrazinonicotinamide–annexin V (Tait et al. 2006).

Other agents developed to recognize membrane-bound phosphatidylserine for the detection of apoptotic cells include zinc-dipicolylamine coordination complex, cationic liposomes, 12-residue phosphatidylserine-binding peptide, and C2A domain of synaptotagmin I (Bose et al. 2004; Hanshaw et al. 2005; Shao et al. 2007). Of interest, the latter develops a calcium-dependent bond with the negatively charged membrane phospholipids such as phosphatidylserine and, therefore,

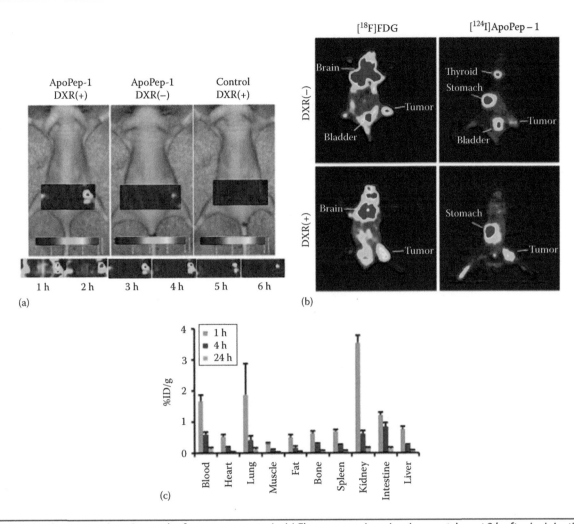

FIGURE 31.1 ApoPep-1 as an imaging probe for tumor apoptosis. (a) Fluorescence imaging. Images taken at 2 h after i.v. injection of FITC-labeled peptides into A549 tumor-bearing mice untreated (DXR(−)) or treated with doxorubicin (DXR(+)). Images at different time points after the injection of FITC-ApoPep-1 are shown in lower panels. (b) Micro PET imaging. Images were taken to assess the uptake of [^{18}F]FDG by untreated or treated H460 tumor at 1 h after injection. Subsequently, [^{124}I]ApoPep-1 was injected into the mice and images were taken at 5 h after injection. Note the physiological uptake of [^{18}F]FDG by the brain and of free iodide (^{124}I) by the thyroid and stomach. Representative images of three separate experiments are shown. (c) Biodistribution analysis. The uptake of [^{131}I]ApoPep-1 into blood and organs were counted at 1, 4, and 24 h after injection into rats. Data represent percent injected dose per gram (%ID/g) ± S.D. (n = 4). (Reprinted from Wang, K. et al., *J. Control. Release*, 148(3), 283, 2010. With permission.)

C2A and its mutants can be used for both SPECT and MRI imaging (^{99m}Tc labeling of a fusion protein for SPECT and with superparamagnetic iron oxide particles and Gd^{3+} for MRI imaging) (Jung et al. 2004; Zhao et al. 2006; Krishnan et al. 2008; Thapa et al. 2008). C2A complexes demonstrated significantly higher disassociation constant compared with different types of annexin V (Alam et al. 2010). Finally, the results using novel molecular imaging probes such as ApoPep-1 (apoptosis-targeting peptide-1), which targeted apoptotic cells in tumor tissue using fluorescent in vivo imaging by binding to the 1H histone exposed on the apoptotic cells' surface, have been recently reported. The authors concluded that ApoPep-1 holds great promise as a in vivo apoptosis imaging probe and that histone H1 constitutes a unique molecular signature for apoptotic imaging (Figure 31.1) (Wang et al. 2010).

31.2.2 CASPASES ACTIVATION-BINDING AGENTS

Caspases are a family of inactive proteases activated under apoptotic conditions and involved in the apoptotic signal transduction (Danial and Korsmeyer 2004). Various optical imaging approaches have been developed to detect caspase activity in vivo mainly using fluorescence or bioluminescence, and appropriate probes are produced in an inactive form that is activated only by specific enzymatic activity (Laxman et al. 2002; Gross and Piwnica-Worms 2005).

The 5-pyrrolidinylsulfonylisatin class of nonpeptidyl caspase inhibitors, which binds to the cysteine residue of the active site of a given caspase, has been used for apoptosis imaging. However, caspase 3 activation does not selectively involve only apoptosis (Spires-Jones et al. 2008). As a result, a 2′-fluoroethyl-1,2,3-triazole with a subnanomolar affinity for caspase 3 has been identified. This agent demonstrates high labeling efficiencies and high in vivo stability, as well as rapid uptake and elimination. At present, PET imaging using 18F-labeled isatin analogues, such as the 18F-labeled isatin 18F-isoprenylcysteine carboxyl methyltransferase 11 (ICMT-11) labeled by click radiochemistry, is under investigation as to determine their role as apoptotic radiotracers (Chen et al. 2009; Nguyen et al. 2009).

31.2.3 ALTERED MEMBRANE PERMEABILITY-BINDING AGENTS

Another characteristic of apoptotic cells is the phenomenon of membrane permeability alteration, which provides a valid target for membrane-based probes of in vivo apoptosis imaging. Two novel radiopharmaceutical agents, $N,N9$-didansylcystine and 5-dimethylamino-1-naphthalenesulfonyl-α-ethyl-fluoroalanine (NST-732) with the trade name Apo-Sense, have been recently developed. It has been proposed that these ^{18}F containing compounds are excluded from normal cells but accumulate in apoptotic cells when membrane phospholipids are altered early in apoptosis enabling in vivo PET imaging (Aloya et al. 2006). The ability of this molecule to rapidly concentrate within damaged cells potentially provides superior imaging compared to annexin V (Korngold et al. 2008).

31.2.4 MITOCHONDRIAL MEMBRANE POTENTIAL COLLAPSE-TARGET AGENTS

Collapse of the mitochondrial membrane electrochemical potential occurring during the apoptotic process enabled the development of another category of in vivo apoptosis-detecting agents—lipophilic radiolabeled probes based on phosphonium cations that accumulate in healthy mitochondria due to the negative mitochondrial transmembrane potential (Green and Kroemer 2004). These probes decrease in cellular uptake in vitro as mitochondrial potential is decreased, and in vivo, these probes show highest uptake in the heart and kidneys during PET imaging (Madar et al. 2007). These probes, in contrast to the previously discussed agents, show a decreased uptake in apoptotic cells and, therefore, are indicated for the imaging of apoptosis in organs with numerous mitochondria and increased metabolic activity, such as the heart and kidneys, but might not produce adequate signal in organs with lower metabolic activity (Tait 2008).

31.2.5 OTHER APOPTOSIS IMAGING MODALITIES

Noninvasive in vivo apoptosis imaging modalities include MR spectroscopy (MRS), HFUS and mid-infrared region spectroscopy (MIRS), and synchrotron radiation–based Fourier transform midinfrared spectromicroscopy (SF-MIS), while invasive intravascular apoptosis imaging can be obtained using OCT (Blankenberg and Norfray 2011). Assessment of mobile lipids and choline using proton MRS has been described for the in vivo imaging of apoptosis. Apoptosis-specific changes detected with MRS include selective increase in methylene relative to methyl mobile lipid proton signal intensities, as the increase in –CH_2– resonance, occurred following the administration of various apoptotic drugs and growth factor deprivation, while a different MRS profile necrosis was detected in necrosis. Authors reported that the of CH_2 to CH_3 signal intensity ratio was correlated with other markers of apoptosis, such as fluorescent annexin V cytometry and DNA ladder formation (Blankenberg et al. 1996). HDUS (≥10 MHz) has also been reported to depict apoptotic cells in vivo mainly because of its ability to detect products that result from the condensation and fragmentation of the cellular nucleus that forms during the late phases of apoptosis. Transducers operating at 10–60 MHz generate ultrasound wavelengths similar to the size of individual cells and nuclei (10–20 μm) and, therefore, are sensitive to their modifications in size and morphology, occurring during apoptosis. However, soft tissue remains the main limitation of HFUS due to the poor beam penetration (2–5 cm depth) (Czarnota et al. 1999). MIRS (~2.5–15.5 μm wavelength) constitutes a micro-imaging technique demonstrating pronounced real-time sensitivity to the hydrogen bond structure linking cellular water molecules to ions and other small molecules (radicals and small organic acids) present during apoptosis. Another midinfrared spectromicroscopic technique, SF-MIS, enables the in vivo study of single cells through an intense midinfrared light spectrum obtainable from a cyclotron, as spectral changes correspond to changes in proteins, DNA, and RNA, as the cell enters from the G1 to the S phase of the cellular circle and into mitosis (Holman et al. 2008, 2010).

Finally, invasive intravascular imaging of apoptosis is made possible using OCT laser light–based technology, which emits a central wavelength of 1325 nm resulting in an axial resolution of 9 μm. Similar to the HFUS, OCT is able to detect several scattering and reflective interfaces that arise within a cell during the apoptotic process, as apoptotic cells demonstrate marked increases in integrated backscatter, while necrotic cells demonstrate a decrease in backscatter (Farhat et al. 2011).

31.3 IMAGING OF HYPOXIA

Hypoxia derives from the Greek words *hypo*, meaning "under/below," and "oxygen" and is defined as the reduction of oxygen supply to a tissue below physiological levels, despite adequate blood perfusion of the tissue. Hypoxia has been widely investigated in the ambit of stroke and it is also considered as a common feature of a variety of cancer types, associated with neo-angiogenesis, augmented cancer aggressiveness, and resistance to therapy (Koukourakis et al. 2002; Takasawa et al. 2008). As a result, various imaging modalities for the detection of hypoxia and for the preclinical study of hypoxia-targeted drugs, which constitute promising new treatment regimens are currently under investigation (Rapisarda and Melillo 2012). PET, SPET, and MRI are the three most frequently used modalities for the development of hypoxia in vivo imaging protocols (Rajendran et al. 2006; Serganova et al. 2006; Yuan et al. 2006).

The ideal hypoxia PET tracer is yet to be established and its effectiveness will depend both on tracer sensitivity and on biodistribution. Nearly 15 years ago, Yamamoto et al. utilized the inhalation of ^{15}O-labeled molecular oxygen gas ($^{15}O_2$) in a greyhound model to achieve the quantification of regional myocardial oxygen consumption ($MMRO_2$) and oxygen extraction fraction, obtaining excellent results of even superior accuracy compared to invasive oxygen tension measurements, establishing this method as the gold standard of noninvasive hypoxia imaging (Yamamoto et al. 1996). Newly developed PET tracers used for tumoral hypoxia imaging include nitroimidazole-based hypoxia-avid PET tracers, such as fluorine 18 (^{18}F) fluoromisonidazole (^{18}F-FMISO) and copper 64 (^{64}Cu) diacetyl-bis-N4 methylthiosemicarbazone (^{64}Cu-ATSM). The two nitoinidazoles-based tracers such as ^{18}F-FMISO are reduced due to hypoxia and bind to extra- and intracellular substances, while ^{64}Cu-ATSM imaging depends on the selective deposition of ^{64}Cu on the hypoxic tissue. ^{64}Cu-ATSM has been reported to demonstrate various binding ability in different cancer types, while a lower hypoxia-specific fraction is achieved compared to ^{18}F-FMISO. On the other hand, radiolabeled glucose analogue [^{16}F]-2-fluorodeoxyglucose (^{16}F-FDG) that has been used successfully for two decades is based on the increased glycolytic activity of tumor cells, resulting in increased glucose transport and hexokinase function (Serganova et al. 2006; Reischl et al. 2007; Kurihara et al. 2012). Both ^{64}Cu-ATSM and FDG imaging resulted in satisfactory in vivo imaging of hypoxic tumor cell regions in a Japanese white rabbit VX2 tumor model (Obata et al. 2003). More recent 2-nitroimidazole-based agents include iodine 124 (^{124}I)-labeled iodoazomycin arabinoside used for SPET imaging of hypoxia, while other authors applied Iodine 124 (^{124}I)-labeled iodoazomycin galactopyranoside (IAZGP) in rodent models using micro-PET achieving excellent consistency with independent measures of hypoxia such as direct oxygen tension probe measurements and immune-histochemical hypoxia markers (pimonidazole and EF5), while the optimal imaging result of tumor hypoxia was detected 6 h after the administration of the tracer (Riedl et al. 2008). In conclusion, besides tumoral hypoxia, 2-nitroimidazoles are also considered promising racers for the in vivo brain hypoxia mapping following stroke. Although cerebral blood flow thresholds defining the penumbra have been determined using multi-tracer ^{15}O PET, the application of nitroimidazole-based tracers yielded very satisfactory results in both animal models and clinical protocols (Takasawa et al. 2008).

MRI-based protocols for hypoxia include nuclear magnetic resonance spectroscopy, which detects abnormally increased tissue lactate levels and decreased ATP levels, as well as tissue pH. However, this modality demonstrates low sensitivity and spatial resolution (Gaertner et al. 2012). Blood oxygen level detection (BOLD) is a functional MRI application widely used for the detection of changes in tissue perfusion by the amount of oxygenated blood. However, the main limitation of BOLD is the fact that it cannot determine the level of oxygen in tissues or describe tumoral cellular modifications at a molecular level (Krishna et al. 2001). The most promising developing MRI technology in the field of hypoxia imaging is electron paramagnetic resonance imaging as it can detect cellular oxygen levels (Elas et al. 2003). Accordingly, PET imaging is still considered superior to the various MRI-based protocols for hypoxia imaging, while other in vivo imaging techniques include electron spin resonance and optical near-infrared techniques, which are presently under investigation (Brun et al. 1997).

31.4   IMAGING OF INFLAMMATION

Inflammation is known to have a significant role in various pathological conditions including arthritis, pulmonary disease, as well as cardiovascular disease and cancer. A critical step in this process is the activation of leukocytes and their migration into the extravascular space (Kaufmann and Lindner 2007). In vivo imaging is believed to be of great importance to track involved cells and anti-inflammatory drugs and to study the activators responsible for the specific pathophysiological pathways. In this way, in vivo effectiveness of specific targeted drugs may be evaluated be applying modern methods of molecular imaging.

Imaging techniques of inflammation may be performed in a clinical setting using ultrasound, MRI, CT, PET, planar scintigraphy, SPECT, and endoscopic techniques (Gotthardt et al. 2010). In preclinical studies and in the in vivo imaging of inflammation processes, the most widely used imaging techniques are fluorescence techniques, micro-PET, SPECT, ultrasound, and micro-MRI.

More specifically, imaging of inflammation using ultrasound is performed by taking advantage of the specific interactions of targeted ultrasound microbubbles and leukocytes and or leukocyte mediators (Lindner et al. 2000a–c; Christiansen et al. 2002; Kaufmann and Lindner 2007). This technique has already been used in mouse kidney to study ischemia–reperfusion-associated inflammation by targeting P-selectin with monoclonal antibodies (Lindner et al. 2001), the detection of allograft-transplant rejection by targeting intercellular adhesion molecule 1 (ICAM-1) (Weller et al. 2003), and the detection of Crohn's disease using the mucosal addressin cell adhesion molecule-1 (Bachmann et al. 2006).

PET techniques have also been used to image inflammation processes in preclinical small animal imaging studies in several interesting protocols. The main interest regarding this technology is the development of specific labeled compounds that are able to bind to cyclooxygenase and in this way observe both the levels and the location of this inflammation-related molecules in the subject's body (Herschman 2003). Another interesting concept is the use of protein kinases as targets to study inflammation (Ortu et al. 2002; Herschman 2003). In this respect, a recent study has shown successful monitoring and quantitative assessment of experimental joint inflammation in mice using ^{18}F-FDG PET-CT techniques (Irmler et al. 2010). The main advantage of PET and SPECT techniques, whenever used to visualize specific processes in a small animal experimental setting, is the ability to accurately perform quantitative measurements independent of the part of the subject's body under investigation and enable direct application of similar techniques in a clinical setting (Chatziioannou 2005).

Another imaging modality to visualize inflammation in vivo is micro-MRI. Several studies investigated the use of specific paramagnetic micro- or nanoparticles that can target inflamed activated endothelium in experimentally induced stroke models and in such way use MRI to visualize neuro-inflammation (Barber et al. 2004; Deddens et al. 2012). These particles are usually functionalized using a ligand for cell adhesion molecules such as P- and E-selectin and vascular cell adhesion molecule 1 (VCAM-1) (Barber et al. 2004; Jin et al. 2009; van Kasteren et al. 2009).

The most frequently used agents for molecular MRI techniques are superparamagnetic iron oxide particles, followed by low-molecular-weight paramagnetic gadolinium polyaminocarboxylate chelates such as Gd-DTPA and Gd-DOTA (Deddens et al. 2012). Finally, MRI molecular techniques have also been used in order to visualize the presence of abscesses in a rat model by intravenous injection of a Gd-DTPA complex (Brasch et al. 1984).

Fluorescence techniques and, more specifically, near-infrared fluorescence imaging have also been applied to image inflammation in in vivo experimental models (Frangioni 2003). It has been reported that the use of special fluorescent probes activated by the interaction with the cathepsin B shows high activity of the aforementioned protease in inflamed atherosclerotic lesions. In order to demonstrated that, researchers used an optical tomography system that successfully demonstrated the high activity of cathepsin B in these vulnerable plaques (Haller et al. 2008). Similar techniques have also been applied to noninvasively image physiological, cellular, and subcellular processes in lung inflammation disease (Haller et al. 2008). In conclusion, there is a growing interest in developing compounds that target cysteine proteases in order to visualize inflammatory processes in a wide range of experimental models (Turk et al. 2000, 2001; Yasuda et al. 2005; Vasiljeva et al. 2007; Weidauer et al. 2007).

REFERENCES

Alam, I. S., A. A. Neves et al. (2010). Comparison of the C2A domain of synaptotagmin-I and annexin-V as probes for detecting cell death. *Bioconjug. Chem.* **21**(5): 884–891.

Aloya, R., A. Shirvan et al. (2006). Molecular imaging of cell death in vivo by a novel small molecule probe. *Apoptosis* **11**(12): 2089–2101.

Bachmann, C., A. L. Klibanov et al. (2006). Targeting mucosal addressin cellular adhesion molecule (MAdCAM)-1 to noninvasively image experimental Crohn's disease. *Gastroenterology* **130**(1): 8–16.

Barber, P. A., T. Foniok et al. (2004). MR molecular imaging of early endothelial activation in focal ischemia. *Ann. Neurol.* **56**(1): 116–120.

Blankenberg, F. G. and J. F. Norfray. (2011). Multimodality molecular imaging of apoptosis in oncology. *AJR Am. J. Roentgenol.* **197**(2): 308–317.

Blankenberg, F. G., R. W. Storrs et al. (1996). Detection of apoptotic cell death by proton nuclear magnetic resonance spectroscopy. *Blood* **87**(5): 1951–1956.

Bose, S., I. Tuunainen et al. (2004). Binding of cationic liposomes to apoptotic cells. *Anal. Biochem.* **331**(2): 385–394.

Brasch, R. C., H. J. Weinmann et al. (1984). Contrast-enhanced NMR imaging: Animal studies using gadolinium-DTPA complex. *AJR Am. J. Roentgenol.* **142**(3): 625–630.

Brun, N. C., A. Moen et al. (1997). Near-infrared monitoring of cerebral tissue oxygen saturation and blood volume in newborn piglets. *Am. J. Physiol.* **273**(2 Pt 2): H682–H686.

Chatziioannou, A. F. (2005). Instrumentation for molecular imaging in preclinical research: Micro-PET and Micro-SPECT. *Proc. Am. Thorac. Soc.* **2**(6): 533–536, 510–511.

Chen, D. L., D. Zhou et al. (2009). Comparison of radiolabeled isatin analogs for imaging apoptosis with positron emission tomography. *Nucl. Med. Biol.* **36**(6): 651–658.

Christiansen, J. P., H. Leong-Poi et al. (2002). Noninvasive imaging of myocardial reperfusion injury using leukocyte-targeted contrast echocardiography. *Circulation* **105**(15): 1764–1767.

Czarnota, G. J., M. C. Kolios et al. (1999). Ultrasound imaging of apoptosis: High-resolution non-invasive monitoring of programmed cell death in vitro, in situ and in vivo. *Br. J. Cancer* **81**(3): 520–527.

Danial, N. N. and S. J. Korsmeyer. (2004). Cell death: Critical control points. *Cell* **116**(2): 205–219.

Deddens, L. H., G. A. Van Tilborg et al. (2012). Imaging neuroinflammation after stroke: Current status of cellular and molecular MRI strategies. *Cerebrovasc. Dis.* **33**(4): 392–402.

Elas, M., B. B. Williams et al. (2003). Quantitative tumor oxymetric images from 4D electron paramagnetic resonance imaging (EPRI): Methodology and comparison with blood oxygen level-dependent (BOLD) MRI. *Magn. Reson. Med.* **49**(4): 682–691.

Farhat, G., V. X. Yang et al. (2011). Detecting cell death with optical coherence tomography and envelope statistics. *J. Biomed. Opt.* **16**(2): 026017.

Fink, S. L. and B. T. Cookson. (2005). Apoptosis, pyroptosis, and necrosis: Mechanistic description of dead and dying eukaryotic cells. *Infect. Immun.* **73**(4): 1907–1916.

Frangioni, J. V. (2003). In vivo near-infrared fluorescence imaging. *Curr. Opin. Chem. Biol.* **7**(5): 626–634.

Gaertner, F. C., M. Souvatzoglou et al. (2012). Imaging of hypoxia using PET and MRI. *Curr. Pharm. Biotechnol.* **13**(4): 552–570.

Gotthardt, M., C. P. Bleeker-Rovers et al. (2010). Imaging of inflammation by PET, conventional scintigraphy, and other imaging techniques. *J. Nucl. Med.* **51**(12): 1937–1949.

Green, D. R. and G. Kroemer. (2004). The pathophysiology of mitochondrial cell death. *Science* **305**(5684): 626–629.

Gross, S. and D. Piwnica-Worms. (2005). Spying on cancer: Molecular imaging in vivo with genetically encoded reporters. *Cancer Cell* **7**(1): 5–15.

Haller, J., D. Hyde et al. (2008). Visualization of pulmonary inflammation using noninvasive fluorescence molecular imaging. *J. Appl. Physiol.* **104**(3): 795–802.

Hanshaw, R. G., C. Lakshmi et al. (2005). Fluorescent detection of apoptotic cells by using zinc coordination complexes with a selective affinity for membrane surfaces enriched with phosphatidylserine. *ChemBioChem* **6**(12): 2214–2220.

Herschman, H. R. (2003). Micro-PET imaging and small animal models of disease. *Curr. Opin. Immunol.* **15**(4): 378–384.

Holman, H. Y., H. A. Bechtel et al. (2010). Synchrotron IR spectromicroscopy: Chemistry of living cells. *Anal. Chem.* **82**(21): 8757–8765.

Holman, H. Y., K. A. Bjornstad et al. (2008). Mid-infrared reflectivity of experimental atheromas. *J. Biomed. Opt.* **13**(3): 030503.

Irmler, I. M., T. Opfermann et al. (2010). In vivo molecular imaging of experimental joint inflammation by combined (18)F-FDG positron emission tomography and computed tomography. *Arthritis Res. Ther.* **12**(6): R203.

Jin, A. Y., U. I. Tuor et al. (2009). Magnetic resonance molecular imaging of post-stroke neuroinflammation with a P-selectin targeted iron oxide nanoparticle. *Contrast Media Mol. Imaging* **4**(6): 305–311.

Jin, M., C. Smith et al. (2004). Essential role of B-helix calcium binding sites in annexin V-membrane binding. *J. Biol. Chem.* **279**(39): 40351–40357.

Jung, H. I., M. I. Kettunen et al. (2004). Detection of apoptosis using the C2A domain of synaptotagmin I. *Bioconjug. Chem.* **15**(5): 983–987.

Kaufmann, B. A. and J. R. Lindner. (2007). Molecular imaging with targeted contrast ultrasound. *Curr. Opin. Biotechnol.* **18**(1): 11–16.

Korngold, E. C., F. A. Jaffer et al. (2008). Noninvasive imaging of apoptosis in cardiovascular disease. *Heart Fail. Rev.* **13**(2): 163–173.

Koukourakis, M. I., A. Giatromanolaki et al. (2002). Hypoxia-inducible factor (HIF1A and HIF2A), angiogenesis, and chemoradiotherapy outcome of squamous cell head-and-neck cancer. *Int. J. Radiat. Oncol. Biol. Phys.* **53**(5): 1192–1202.

Krishna, M. C., S. Subramanian et al. (2001). Magnetic resonance imaging for in vivo assessment of tissue oxygen concentration. *Semin. Radiat. Oncol.* **11**(1): 58–69.

Krishnan, A. S., A. A. Neves et al. (2008). Detection of cell death in tumors by using MR imaging and a gadolinium-based targeted contrast agent. *Radiology* **246**(3): 854–862.

Kurihara, H., N. Honda et al. (2012). Radiolabelled agents for PET imaging of tumor hypoxia. *Curr. Med. Chem.* **19**(20): 3282–3289.

Laxman, B., D. E. Hall et al. (2002). Noninvasive real-time imaging of apoptosis. *Proc. Natl. Acad. Sci. USA* **99**(26): 16551–16555.

Lindner, J. R., M. P. Coggins et al. (2000a). Microbubble persistence in the microcirculation during ischemia/reperfusion and inflammation is caused by integrin- and complement-mediated adherence to activated leukocytes. *Circulation* **101**(6): 668–675.

Lindner, J. R., P. A. Dayton et al. (2000b). Noninvasive imaging of inflammation by ultrasound detection of phagocytosed microbubbles. *Circulation* **102**(5): 531–538.

Lindner, J. R., J. Song et al. (2000c). Noninvasive ultrasound imaging of inflammation using microbubbles targeted to activated leukocytes. *Circulation* **102**(22): 2745–2750.

Lindner, J. R., J. Song et al. (2001). Ultrasound assessment of inflammation and renal tissue injury with microbubbles targeted to P-selectin. *Circulation* **104**(17): 2107–2112.

Madar, I., H. Ravert et al. (2007). Characterization of membrane potential-dependent uptake of the novel PET tracer 18F-fluorobenzyl triphenylphosphonium cation. *Eur. J. Nucl. Med. Mol. Imaging* **34**(12): 2057–2065.

Munoz, L. E., B. Frey et al. (2007). The role of annexin A5 in the modulation of the immune response against dying and dead cells. *Curr. Med. Chem.* **14**(3): 271–277.

Nguyen, Q. D., G. Smith et al. (2009). Positron emission tomography imaging of drug-induced tumor apoptosis with a caspase-3/7 specific [18F]-labeled isatin sulfonamide. *Proc. Natl. Acad. Sci. USA* **106**(38): 16375–16380.

Obata, A., M. Yoshimoto et al. (2003). Intra-tumoral distribution of (64)Cu-ATSM: A comparison study with FDG. *Nucl. Med. Biol.* **30**(5): 529–534.

Ortu, G., I. Ben-David et al. (2002). Labeled EGFr-TK irreversible inhibitor (ML03): In vitro and in vivo properties, potential as PET biomarker for cancer and feasibility as anticancer drug. *Int. J. Cancer* **101**(4): 360–370.

Rajendran, J. G., D. L. Schwartz et al. (2006). Tumor hypoxia imaging with [F-18] fluoromisonidazole positron emission tomography in head and neck cancer. *Clin. Cancer Res.* **12**(18): 5435–5441.

Rapisarda, A. and G. Melillo. (2012). Overcoming disappointing results with antiangiogenic therapy by targeting hypoxia. *Nat. Rev. Clin. Oncol.* **9**(7): 378–390.

Reischl, G., D. S. Dorow et al. (2007). Imaging of tumor hypoxia with [^{124}I]IAZA in comparison with [18F] FMISO and [^{18}F]FAZA—First small animal PET results. *J. Pharm. Pharm. Sci.* **10**(2): 203–211.

Riedl, C. C., P. Brader et al. (2008). Imaging hypoxia in orthotopic rat liver tumors with iodine 124-labeled iodoazomycin galactopyranoside PET. *Radiology* **248**(2): 561–570.

Serganova, I., J. Humm et al. (2006). Tumor hypoxia imaging. *Clin. Cancer Res.* **12**(18): 5260–5264.

Shao, R., C. Xiong et al. (2007). Targeting phosphatidylserine on apoptotic cells with phages and peptides selected from a bacteriophage display library. *Mol. Imaging* **6**(6): 417–426.

Spires-Jones, T. L., A. de Calignon et al. (2008). In vivo imaging reveals dissociation between caspase activation and acute neuronal death in tangle-bearing neurons. *J. Neurosci.* **28**(4): 862–867.

Tait, J. F. (2008). Imaging of apoptosis. *J. Nucl. Med.* **49**(10): 1573–1576.

Tait, J. F., D. S. Brown et al. (2000). Development and characterization of annexin V mutants with endogenous chelation sites for (99m)Tc. *Bioconjug. Chem.* **11**(6): 918–925.

Tait, J. F., C. Smith et al. (2006). Improved detection of cell death in vivo with annexin V radiolabeled by site-specific methods. *J. Nucl. Med.* **47**(9): 1546–1553.

Takasawa, M., R. R. Moustafa et al. (2008). Applications of nitroimidazole in vivo hypoxia imaging in ischemic stroke. *Stroke* **39**(5): 1629–1637.

Thapa, N., S. Kim et al. (2008). Discovery of a phosphatidylserine-recognizing peptide and its utility in molecular imaging of tumour apoptosis. *J. Cell. Mol. Med.* **12**(5A): 1649–1660.

Thompson, C. B. (1995). Apoptosis in the pathogenesis and treatment of disease. *Science* **267**(5203): 1456–1462.

Tremoleda, J. L., A. Kerton et al. (2012). Anaesthesia and physiological monitoring during in vivo imaging of laboratory rodents: Considerations on experimental outcomes and animal welfare. *EJNMMI Res.* **2**(1): 44.

Turk, B., D. Turk et al. (2000). Lysosomal cysteine proteases: More than scavengers. *Biochim. Biophys. Acta* **1477**(1–2): 98–111.

Turk, V., B. Turk et al. (2001). Lysosomal cysteine proteases: Facts and opportunities. *EMBO J.* **20**(17): 4629–4633.

van Kasteren, S. I., S. J. Campbell et al. (2009). Glyconanoparticles allow pre-symptomatic in vivo imaging of brain disease. *Proc. Natl. Acad. Sci. USA* **106**(1): 18–23.

Vasiljeva, O., T. Reinheckel et al. (2007). Emerging roles of cysteine cathepsins in disease and their potential as drug targets. *Curr. Pharm. Des.* **13**(4): 387–403.

Wang, K., S. Purushotham et al. (2010). In vivo imaging of tumor apoptosis using histone H1-targeting peptide. *J. Control. Release* **148**(3): 283–291.

Weidauer, E., Y. Yasuda et al. (2007). Effects of disease-modifying anti-rheumatic drugs (DMARDs) on the activities of rheumatoid arthritis-associated cathepsins K and S. *Biol. Chem.* **388**(3): 331–336.

Weller, G. E., E. Lu et al. (2003). Ultrasound imaging of acute cardiac transplant rejection with microbubbles targeted to intercellular adhesion molecule-1. *Circulation* **108**(2): 218–224.

Yamamoto, Y., R. de Silva et al. (1996). Noninvasive quantification of regional myocardial metabolic rate of oxygen by 15O2 inhalation and positron emission tomography: Experimental validation. *Circulation* **94**(4): 808–816.

Yasuda, Y., J. Kaleta et al. (2005). The role of cathepsins in osteoporosis and arthritis: Rationale for the design of new therapeutics. *Adv. Drug Deliv. Rev.* **57**(7): 973–993.

Yuan, H., T. Schroeder et al. (2006). Intertumoral differences in hypoxia selectivity of the PET imaging agent 64Cu(II)-diacetyl-bis(N4-methylthiosemicarbazone). *J. Nucl. Med.* **47**(6): 989–998.

Zhao, M., X. Zhu et al. (2006). 99mTc-labeled C2A domain of synaptotagmin I as a target-specific molecular probe for noninvasive imaging of acute myocardial infarction. *J. Nucl. Med.* **47**(8): 1367–1374.

32

Vessel Wall Imaging

Stavros Spiliopoulos

32.1 INTRODUCTION

Cardiovascular diseases are considered among the leading causes of morbidity and mortality worldwide (Rooke et al. 2013). Although the vast majority of vascular pathologies, including atherosclerosis, involves the vascular wall, in clinical practice current standard clinical imaging methods are mainly limited to the evaluation of luminal stenosis, which presents numerous limitations in the evaluation of vessel wall and in the in vivo characterization of arterial pathologies such as atherosclerotic plaque and inflammation (Rengier et al. 2013). Imaging varies from plain vessel lumen evaluation and indirect assessment of vessel wall pathology to vessel wall visualization and direct evaluation using both invasive and noninvasive imaging modalities. Modern imaging modalities allow the visualization not only of the general and indirect features of layer pathology but also the direct depiction of vascular anatomy and histopathology in a variety of diseases involving the vessel wall, such as atherosclerotic disease, vasculitis, and cancer.

Although invasive modalities such as intravascular ultrasound (IVUS) and optical coherence tomography (OCT) can be used to visualize vessel wall in large animals such as the rabbit and the dog (Yin et al. 2011), small animal imaging does not allow in vivo intravascular investigation because of the small vessel caliber. This chapter describes the current imaging modalities employed for vessel wall imaging and its pathology with special focus on atherosclerosis in experimental in vivo small animal models.

32.2 SMALL ANIMAL VESSEL WALL ANATOMY

The size of the mouse arterial wall thickness is <50 mm. The arterial wall of small animals such as rats and mice consists of three tunics (from inside to outside): the tunica intima, the tunica media, and the tunica adventitia. Although the main anatomic and histologic structure of the vasculature in humans and mice are similar, differences such as thinner walls in mouse arteries and the prominent presence of cardiomyocytes around the pulmonary veins of mice have been noticed (Laflamme et al. 2012). Moreover, in a mice coronary artery study, the authors reported that the microstructure of the three arterial layers slightly differs between species. In the coronary arteries of mice, the tunica intima consists of an endothelial layer, an inconspicuous subendothelial connective tissue, and an undulating internal elastic lamina. The tunica media consists of circularly arranged smooth-muscle fibers, and the tunica adventitia consists of collagen and elastic fibers (Khan et al. 2006). Of note, histologically altered arterial wall presenting different biomechanical and microstructural properties can be achieved using various transgenic mice models such as the n fibulin-5 null mice, enabling the investigation of a variety of cardiovascular diseases (Stone et al. 2004; Wan et al. 2010; Wan and Gleason 2013).

32.3 CONVENTIONAL AND MOLECULAR MAGNETIC RESONANCE IMAGING (MRI)

Micro-MRI is a well-established noninvasive, high-resolution imaging modality for the vascular system of small animal models due to its capability to offer a variety of structural and functional measurements of the vessel wall down to the cellular and molecular level. The combination of high spatial resolution and excellent soft-tissue contrast makes MRI an appropriate modality for the visualization and characterization of thin vessel walls of small animals. The use of a high-field magnet and dedicated radio frequency coils provides high-resolution (10–100 μm) images, offering the possibility of direct visualization of the majority of vascular pathologies including lumen abnormalities such as stenosis or occlusion, vessel wall thrombus formation, and dissection or rupture, as well as the detection, characterization, and monitoring of atherosclerosis (Weissleder 2002). In general, atherosclerotic plaque components may be differentiated and characterized by using unenhanced vessel wall MRI, while additional data regarding the vessel wall can be acquired by contrast-enhanced MIR and molecular MRI with the use of targeted contrast agents that enhance particular cells or even molecules (Makowski and Botnar 2013). Fayad et al. used serial MRI with a 9.4 T, 89 mm vertical bore system (Bruker Instruments, Billerica, MA) at the 400 MHz proton frequency with a 30 mm internal diameter gradient insert to in vivo detect, quantify, and characterize atherosclerotic lesions in apolipoprotein E-deficient (APOE$^{-/-}$) mice, achieving strong correlation with the subsequent histopathological analysis (Fayad et al. 1998). Manka et al. using electrocardiography (EKG) and respiratory-gated MRI, in order to suppress motion artifacts, managed to accurately quantify injury-induced atherosclerotic aortic stenosis in APOE$^{-/-}$ mice, again achieving a satisfactory correlation with histopathology (Manka et al. 2000). Of note, serial MRI is a suitable in vivo imaging modality to detect and quantify with accuracy vessel wall positive remodeling, and thus the dynamic process of the outward expansion of the vessel without concurrent luminal narrowing occurring at the early stages of plaque evolution (20 weeks in APOE$^{-/-}$ mice) (Weinreb et al. 2007). Serial MR microscopy (MRM) is another imaging modality that can be employed for the in vivo detection and quantification of arterial wall atherosclerosis and its progression, as well as vessel wall remodeling. In an experimental APOE$^{-/-}$ mice model, Choudhury et al. visualized both progressive atherosclerotic lesions within the aortic wall accompanied by positive remodeling using 9.4 T MRM. Proton density-weighted (PDW) images were acquired (TR = 2000, TE = 9 ms) using four signal averages, while the image resolution was 109 × 109 × 500 μm^3 (Choudhury et al. 2003). Moreover, MRI using reduction of respiratory and cardiac motion-related artifacts techniques has been used to precisely evaluate plaque evolution and to characterize plaque vulnerability by enabling the differentiation between the fibrous, lipid, and cellular atherosclerotic lesion components, even in

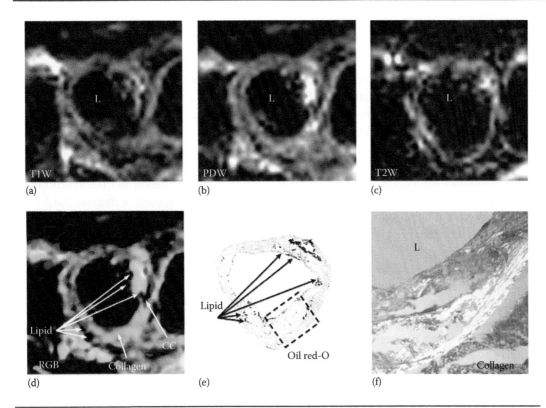

FIGURE 32.1 In vivo MRI analysis of lesion composition. T1W (a), PDW (b), and T2W (c) images obtained by in vivo MRI of the ascending aorta of an apoE$^{-/-}$ mouse fed Western diet for 32 weeks. (d) A red-green-blue (RGB) color composite of the three image modalities denotes the lipid areas (red), collagen (green), and cholesterol crystals (black, void space). Lipid (e) and collagen (f) content, as assessed by oil red O and picrosirius red staining, respectively, in a histopathological cross-section of the aorta reveals a close match between histopathology and MRI and highlights the ability to noninvasively determine plaque composition in mouse atherosclerotic plaque. (*Note*: Dotted box in (e) corresponds to area shown in (f) under higher magnification; ×250; L denotes lumen.) (Reprinted from Trogan, E. et al., *Arterioscler. Thromb. Vasc. Biol.*, 24(9), 1714, 2004. With permission.)

small-caliber vessels as those of in APOE$^{-/-}$ mice. According to Trogan et al., collagen appears bright on the T1-weighted (T1W) and PDW images, and lipid-rich regions appears as hyperintense signals on the T1W and hypointense signals on the PDW and T2-weighted (T2W) images (Figure 32.1) (Trogan et al. 2004).

Recently, serial molecular micro-MRI was performed in a Kunming mice model of injured common carotid artery using transfused endothelial progenitor cells (EPCs) labeled with Fe_2O_3-poly-L-lysine (Fe_2O_3-PLL) in order to explore the temporal and spatial migration of EPCs homing to the damaged endothelium. The authors reported that significantly more frequent larger MR signal voids of vessel wall on T2WI was revealed in the labeled EPC transfusion group 15 days after carotid artery injury. Histopathology confirmed the accumulation and distribution of transfused EPCs at the injured vessel wall, suggesting that MRI could be used as an in vivo tracking modality of EPCs homing to the endothelium (Chen et al. 2011).

The development of molecular micro-MRI with targeted contrast agents permitted the measurement of the plaque molecular composition and the detection pathological changes of the vessel wall at the molecular level (Corti and Fuster 2011). Molecular MRI of atherosclerotic biomarkers such as fibrin and fibronectin has been described for the detection of both atherogenesis and atherosclerotic plaque progression in an apoE$^{-/-}$ mouse model using a CLT1 peptide targeted macrocyclic Gd(III) chelate. The authors performed high-spatial-resolution MRI before and up to 35 min after i.v. injection of 0.1 mmol Gd/kg of CLT1-dL-(DOTA-Gd)4 or a nonspecific control agent, and reported that the former produced stronger enhancement in the atherosclerotic lesions of the aortic wall compared to control, while cross-sectional MRI detected progressive thickening of the atherosclerotic aortic wall, which correlated well with histological staining. The authors concluded that molecular MRI with CLT1-dL-(DOTA-Gd)4 could be used as a noninvasive method for the detection and monitoring of the progression of atherosclerosis (Wu et al. 2013).

Vessel wall inflammation can be detected using molecular micro-MRI using specific paramagnetic nanoparticles functionalized by P- and E-selectin and vascular cell adhesion molecule 1 (VCAM-1) that can target inflamed activated endothelium in experimental induced stroke models (Barber et al. 2004; Jin et al. 2009).

Finally, both conventional and molecular MRI methods have been used for the in vivo monitoring of abdominal aortic aneurysms (AAA) in an aortic aneurysm mouse model. Specifically, Kink et al. developed an AAA C57BL/6 mouse model using angiotensin II infusion and TGF b neutralization in which conventional high-resolution multisequence MRI was applied to monitor AAA temporal progression and molecular MRI following i.v. infusion of paramagnetic/fluorescence micellar nanoparticles functionalized with the CNA-35 collagen binding protein in order to enable collagen imaging. The authors demonstrated that, in the specific animal model, the temporal longitudinal monitoring of the AAA was achievable with high-resolution conventional, multisequence, black-blood MRI, which successfully detected the primary aortic wall fibrotic response, and the presence of a secondary channel of dissection and progressive aortic enlargement, while molecular MRI using collagen-targeted CAN-35 nanoparticles resulted in satisfactory aortic wall collagen visualization with a good histological correlation, a fact associated with AAA degeneration and risk of rupture (Klink et al. 2011).

32.4   MOLECULAR IMAGING

32.4.1   MICRO-PET AND MICRO-SPECT

Micro-PET is a molecular imaging modality that can be used for the investigation of cellular and molecular pathophysiology of the vessel wall. Although it demonstrates limited spatial and temporal resolution and poor anatomic context capabilities compared with MRI or CT, micro-PET enables highly sensitive and quantitative in vivo measurements of biologic or biochemical processes using positron emitters such as ^{11}C and ^{18}F (for OH^-). Recent developments of micro-PET technology allow accurate attenuation correction with isotope transmission or CT, while list-mode acquisition enables the simultaneous reconstruction of sub-second-resolution first-pass dynamic respiratory and ECG-gated images in mice with a volumetric imaging FOV (field of view) >10 cm and a reconstructed resolution of ~1 µL. Moreover, next generation of small-animal PET systems using submillimeter detectors are approaching the 0.2–0.3 mm limit of the positron range for ^{18}F and ^{11}C isotopes (de Kemp et al. 2010). Laitinen et al. proposed the ^{18}F-galacto-RGD tracer uptake for the in vivo imaging of vascular inflammation in atherosclerotic lesions of hypercholesterolemic LDLR$^{-/-}$ apoB$^{100/100}$ mice. ^{18}F-galacto-RGD is a PET tracer that binds to the $\alpha_v\beta_3$ integrin expressed by macrophages and endothelial cells in atherosclerotic lesions. The authors demonstrated that the biodistribution of the tracer was higher in the atherosclerotic than in the normal aorta and that its uptake was associated with macrophage density, marking ^{18}F-galacto-RGD a potential tracer for inflammation imaging of aortic atherosclerotic lesions (Laitinen et al. 2009). Specifically, FDG-PET has been reported to be able to distinguish inflamed from noninflamed plaque in atherosclerosis animal models. Inflammatory cell (in particular macrophage) accumulation in the intimal layer of the arterial wall is indicative of atherosclerotic plaque vulnerability. According to Davies et al. (2010), the depiction and quantification of aortic wall macrophage content was made possible using FDG-PET imaging.

Hybrid PET-CT and PET-MRI imaging can be used to outline the anatomic distribution of the tracer, further improving spatial resolution and contrast recovery in submillimeter structures, as demonstrated recently in mouse cardiac imaging (Spinelli et al. 2008). Micro-SPECT and micro-SPECT-CT allow the study of the coronary artery functions and related diseases, such as ischemia, infarction, and atherosclerosis (Liu et al. 2002; Tsui et al. 2005). Diepenbrock et al. recently reported their results from in vivo PT/CT imaging of the vessel wall in a carotid artery of apoE knockout mice model of atherosclerosis, using a high-resolution (0.7 mm full width at half-maximum) quadHIDAC small animal PET scanner (Oxford Positron Systems, Weston-on-the-Green, United Kingdom). Interestingly, the uptake in the cuff region was increased over the control. Vessel-shape plot demonstrated that the lumen diameter did not sharply change with start and end of the implanted cuff, but continuously decreased toward the cuff and gradually increased downstream from the cuff. This corresponded to plaque formation and was attributed to the low and laminar shear stress upstream and low and oscillatory shear stress downstream from the cuff. The authors concluded that the specific PET/CT in vivo protocol atherosclerotic lesions analysis in the carotid cuff model was much faster than conventional analysis, due to the direct visual comparison to the control side, and the immediate visual feedback of possibly false-positive outcomes. The results were also confirmed by histology (Diepenbrock et al. 2013).

Both ^{99m}Tc-annexin V, (125)I-MCP-1 and [(18)F]-fluoro-2-deoxyglucose are effective in identifying apoptotic cell death, macrophage infiltration, and metabolic activity in atheromatic lesions of animal models. Additionally, as the expression of $\alpha_v\beta_3$ integrin increased in activated endothelial cells and vascular smooth muscle cells after a vascular injury, various radiolabeled high-affinity peptides can be used to target and therefore visualize areas of vascular damage. These advances in micro-SPECT technology could overcome the weaknesses of conventional CT and MRI modalities in vessel wall imaging (Blankenberg et al. 2002).

32.4.2 IMMUNOFLUORESCENCE

Real-time confocal wide-field microscopy imaging was recently employed to visualize thrombus formation in the microcirculation of an in vivo mouse model of laser-induced vascular injury. Using high-speed, near-simultaneous acquisition of images of multiple fluorescent probes and of a bright-field channel, the investigators managed to visualize platelet deposition, tissue factor accumulation, and fibrin generation following endothelial injury (Falati et al. 2002). Quantitative fluorescence microscopy has been used for the visualization and quantification of very low density lipoprotein (VLDL) molecules within the arterial wall of rats. As the primary event in atherogenesis is lipid accumulation within the vessel wall, Rutledge et al. sought to visualize the interactions of VLDL surface and core lipids with the artery wall. In this study, the surface lipid in VLDL was traced using the phospholipid-like fluorescent probe 1,1'-dioctadecyl-3,3,3',3'-tetramethyl-indocarbocyanine (DiI) and the core of VLDL particles was traced by fluorescently labeling apolipoprotein B with TRITC (tetramethylrhodamine isothiocyanate). Subsequently, the fluorescently labeled VLDLs were perfused through the carotid arteries of rats, and their arterial wall accumulation was evaluated by quantitative fluorescence microscopy. Interestingly, both DiI and TRITC were primarily visualized in the endothelial layer, while DiI also depicted deeper in a subendothelial position (Rutledge et al. 2000).

Using intravital confocal fluorescent microscopy, the in vivo imaging of interactions between unusually large von Willebrand factor multimers and platelets on the surface of the vascular wall was possible using a GFP ADAMTS13$^{-/-}$ mouse model (green fluorescent protein-expressing transgenic mice, deficient in the gene encoding the protease that cleaves UL-VWF, adisintegrin-like and metalloprotease with thrombospondin type 1 motif 13) (Rybaltowski et al. 2011). Conventional and confocal fluorescence microscopy for the direct visualization of blood vessels in small animals has also been performed using a specially formulated aqueous solution containing DiI, a lipophilic carbocyanine dye, which integrates within the membranes of the endothelial cell. The specific technique using confocal microscopy enables the acquisition of high-quality serial optical sections from thick tissue sections and their three-dimensional reconstruction, especially at low magnification. Compared to the previously described methods of vessel visualization such as fluorescent dye-conjugated dextran, the corrosion casting technique, endothelial cell-specific markers, and lectins, the authors conclude that confocal microscopy following the application of the lipophilic carbocyanine dye DiI is a more simplified method for the visualization of small diameter blood vessels such as those of the retinal vasculature including angiogenic sprouts and pseudopodial processes (Li et al. 2008).

Finally, molecularly targeted nanoparticles (quantum dots) linked with arginine-glycine-aspartic acid (RGD) peptides to target newly formed and/or forming blood vessels expressing αvβ3 integrin were used for intravital microscopy (IV-100, Olympus, Center Valley, PA) with subcellular ~0.5 μm resolution in order to directly observe the binding of nanoparticle conjugates to tumor blood vessels in living mice (Smith et al. 2008). Quantum dots (qdots) are single, nanocrystalline, colloidal semiconductors with exceptional fluorescent properties that have been used to target tissue-specific vascular biomarkers (Akerman et al. 2002). Using this methodology, the authors managed to directly visualize the binding of RGD qdots to tumor vessel endothelium.

32.5 MICROCOMPUTED TOMOGRAPHY (CT) AND HIGH-FREQUENCY ULTRASOUND

Micro-CT is another well-established small animal imaging methodology that can generate 3-D image datasets and provide image resolution scaled to small animals (50–100 mm^3 of voxels) and equivalent contrast to that of clinical whole-body CT (300–600 mm^3 of voxels). ECG and respiratory gated micro-CT can also be employed to avoid motion artefacts (de Kemp et al. 2010). In order to visualize the coronary arterial wall and its atherosclerotic pathology ex vivo, high spatial resolution contrast-enhanced micro-CT and 3-D reconstruction post-processing can be employed. Conventional contrast-enhanced micro-CT using iodinated contrast media will provide information regarding vessel wall integrity such as dissection, rupture, and the presence of atherosclerotic steno-occlusive lesions. However, in order to provide more detailed imaging of the vessel wall, more sophisticated techniques using novel alternative x-ray contrast agents are required. Zhu et al. (2007) have demonstrated that early morphological changes in coronary arteries of hypercholesterolemic pigs can be detected using the x-ray contrast agent osmium tetroxide (OsO$_4$) and 3-D CT. Pai et al. reported the feasibility of micro-CT in small animals using OsO$_4$ solution in heart specimens of C57BL/6 wild-type mice and apoE$^{-/-}$ mice at 10 μm resolution (Figure 32.2). Osmium tetroxide is a tissue-staining contrast agent, visible in CT, which is retained in the vessel wall and the surrounding tissues during

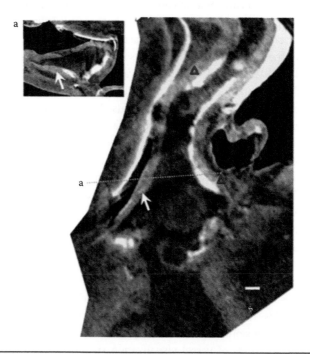

FIGURE 32.2 Longitudinal and transverse (inset panel a) cross-sectional views of the aorta of a 13-week-old female apoE mouse on a Western diet. Plaque deposits in the aortic wall are indicated by hollow triangles. The dotted "a" line marks the location of the transverse cross-section shown in inset (a). The solid arrows indicate the wall of the aorta. Hyperintense areas outside the aorta are osmium-stained pericardial fatty tissue. Scale bar: 100 μm. (Reprinted from Pai, V.M. et al., *J. Anat.*, 220(5), 514, 2012. With permission.)

the fixation process and is then cleared from the vessel lumen. As OsO_4 favorably binds to lipids, it outlines lipid deposition within the vessel wall. With this method, the arterial walls of coronary arteries as small as 45 μm in diameter were visible with the same clarity using both a table-top micro-CT scanner and a synchrotron CT scanner (Pai et al. 2012).

High-frequency ultrasound imaging is another promising imaging modality enabling the in vivo assessment of blood flow and vessel wall pathophysiology. Wang et al. (2001) first used ultrasound technology to measure aneurysms in mice. New high-frequency transducer systems with frequencies ranging from 20 to 60 MHz result in high imaging resolution and offer the ability to reconstruct dynamic signals from cylindrical volumes on the order of 43 μm diameter and 66 μm depth over an FOV of 10 mm width and 5 mm depth, and therefore depict micro vessels from 30 to 100 μm in diameter. Consequently, this technology enables the dynamic imaging of vessel wall morphology and cap thickness and plaque morphology in small animal experimental models (McVeigh 2006). The quantification of aortic diameter and wall thickness in mice with of AngII-induced AAAs with high-frequency ultrasound has been also successfully reported. Finally, the dynamic properties of vessels can also be measured, obtaining additional information about aneurysm distensibility, while tissue Doppler imaging can be utilized to measure in vivo arterial wall motion in small animals (Martin-McNulty et al. 2005; Ramaswamy et al. 2013).

32.6 PLAIN X-RAY AND DIGITAL SUBTRACTION ANGIOGRAPHY

Fluoroscopy and plain x-ray films are capable of detecting moderate and marked vessel calcifications. However, their efficacy decreases for smaller calcifications. The sensitivity of fluoroscopy has been reported to be 52% compared with electron-beam CT (EBCT), although it demonstrated high specificity rates. Therefore, more sophisticated modalities such as micro-CT should be used in order to depict small foci of calcifications (Johnson et al. 2006). In an experimental rat model, Price et al. used warfarin for 2 weeks to produce focal calcification of the elastic lamellae of the aorta and the aortic valve. Warfarin is a vitamin K antagonist and promotes calcification by averting γ-carboxylation of the MGP, an inhibitor of calcification. After 5 weeks, calcifications were apparent on plain radiographs (Price et al. 1998).

Intra-arterial digital subtraction angiography (DSA) is performed through the acquisition of images before and after the intra-arterial injection of an extracellular contrast agent. Although DSA is the "gold standard" imaging modality for the evaluation of small animal vessel lumen stenosis, only limited information on the vessel wall morphology and pathology, such as the presence of vessel wall dissection or rupture and calcifications, could be demonstrated (Badea et al. 2008). DSA enables the distinction between non-flow-limiting and flow-limiting dissections because of its real-time flow-evaluation capability. Such flow-based characterizations are not accurate using conventional contrast-enhanced micro-CT.

REFERENCES

Akerman, M. E., W. C. Chan et al. (2002). Nanocrystal targeting in vivo. *Proc. Natl. Acad. Sci. USA* **99**(20): 12617–12621.

Badea, C. T., M. Drangova et al. (2008). In vivo small-animal imaging using micro-CT and digital subtraction angiography. *Phys. Med. Biol.* **53**(19): R319–R350.

Barber, P. A., T. Foniok et al. (2004). MR molecular imaging of early endothelial activation in focal ischemia. *Ann. Neurol.* **56**(1): 116–120.

Blankenberg, F. G., C. Mari et al. (2002). Development of radiocontrast agents for vascular imaging: Progress to date. *Am. J. Cardiovasc. Drugs* **2**(6): 357–365.

Chen, J., Z. Y. Jia et al. (2011). In vivo serial MR imaging of magnetically labeled endothelial progenitor cells homing to the endothelium injured artery in mice. *PLoS One* **6**(6): e20790.

Choudhury, R. P., Z. A. Fayad et al. (2003). Serial, noninvasive, in vivo magnetic resonance microscopy detects the development of atherosclerosis in apolipoprotein E-deficient mice and its progression by arterial wall remodeling. *J. Magn. Reson. Imaging* **17**(2): 184–189.

Corti, R. and V. Fuster. (2011). Imaging of atherosclerosis: Magnetic resonance imaging. *Eur. Heart J.* **32**(14): 1709–1719.

Davies, J. R., D. Izquierdo-Garcia et al. (2010). FDG-PET can distinguish inflamed from non-inflamed plaque in an animal model of atherosclerosis. *Int. J. Cardiovasc. Imaging* **26**(1): 41–48.

de Kemp, R. A., F. H. Epstein et al. (2010). Small-animal molecular imaging methods. *J. Nucl. Med.* **51**(Suppl. 1): 18S–32S.

Diepenbrock, S., S. Hermann et al. (2013). Comparative visualization of tracer uptake in in vivo small animal PET/CT imaging of the carotid arteries. In *Eurographics Conference on Visualization (EuroVis)*, Leipzig, Germany.

Falati, S., P. Gross et al. (2002). Real-time in vivo imaging of platelets, tissue factor and fibrin during arterial thrombus formation in the mouse. *Nat. Med.* **8**(10): 1175–1181.

Fayad, Z. A., J. T. Fallon et al. (1998). Noninvasive In vivo high-resolution magnetic resonance imaging of atherosclerotic lesions in genetically engineered mice. *Circulation* **98**(15): 1541–1547.

Jin, A. Y., U. I. Tuor et al. (2009). Magnetic resonance molecular imaging of post-stroke neuroinflammation with a P-selectin targeted iron oxide nanoparticle. *Contrast Media Mol. Imaging* **4**(6): 305–311.

Johnson, R. C., J. A. Leopold et al. (2006). Vascular calcification: Pathobiological mechanisms and clinical implications. *Circ. Res.* **99**(10): 1044–1059.

Khan, H., A. A. Khan et al. (2006). Comparative histology of coronary arteries in mammals. *J. Anat. Soc. India* **55**(1): 33–36.

Klink, A., J. Heynens et al. (2011). In vivo characterization of a new abdominal aortic aneurysm mouse model with conventional and molecular magnetic resonance imaging. *J. Am. Coll. Cardiol.* **58**(24): 2522–2530.

Laflamme, M. A., M. M. Sebastian et al. (2012). Cardiovascular. In P. Treuting, S. Dintzis, D. Liggitt, and C. W. Frevent (eds.), *Comparative Anatomy and Histology. A Mouse and Human Atlas*. Academic Press, Oxford, U.K.

Laitinen, I., A. Saraste et al. (2009). Evaluation of alphavbeta3 integrin-targeted positron emission tomography tracer 18F-galacto-RGD for imaging of vascular inflammation in atherosclerotic mice. *Circ. Cardiovasc. Imaging* **2**(4): 331–338.

Li, Y., Y. Song et al. (2008). Direct labeling and visualization of blood vessels with lipophilic carbocyanine dye DiI. *Nat. Protoc.* **3**(11): 1703–1708.

Liu, Z., G. A. Kastis et al. (2002). Quantitative analysis of acute myocardial infarct in rat hearts with ischemia-reperfusion using a high-resolution stationary SPECT system. *J. Nucl. Med.* **43**(7): 933–939.

Makowski, M. R. and R. M. Botnar. (2013). MR imaging of the arterial vessel wall: Molecular imaging from bench to bedside. *Radiology* **269**(1): 34–51.

Manka, D. R., W. Gilson et al. (2000). Noninvasive in vivo magnetic resonance imaging of injury-induced neointima formation in the carotid artery of the apolipoprotein-E null mouse. *J. Magn. Reson. Imaging* **12**(5): 790–794.

Martin-McNulty, B., J. Vincelette et al. (2005). Noninvasive measurement of abdominal aortic aneurysms in intact mice by a high-frequency ultrasound imaging system. *Ultrasound Med. Biol.* **31**(6): 745–749.

McVeigh, E. R. (2006). Emerging imaging techniques. *Circ. Res.* **98**(7): 879–886.

Pai, V. M., M. Kozlowski et al. (2012). Coronary artery wall imaging in mice using osmium tetroxide and micro-computed tomography (micro-CT). *J. Anat.* 220(5): 514–524.

Price, P. A., S. A. Faus et al. (1998). Warfarin causes rapid calcification of the elastic lamellae in rat arteries and heart valves. *Arterioscler. Thromb. Vasc. Biol.* **18**(9): 1400–1407.

Ramaswamy, A. K., M. Hamilton, 2nd et al. (2013). Molecular imaging of experimental abdominal aortic aneurysms. *Sci. World J.* **2013**: 973150.

Rengier, F., P. Geisbusch et al. (2013). State-of-the-art aortic imaging: Part I. Fundamentals and perspectives of CT and MRI. *VASA* **42**(6): 395–412.

Rooke, T. W., A. T. Hirsch et al. (2013). Management of patients with peripheral artery disease (compilation of 2005 and 2011 ACCF/AHA Guideline Recommendations): A report of the American College of Cardiology Foundation/American Heart Association Task Force on Practice Guidelines. *J. Am. Coll. Cardiol.* **61**(14): 1555–1570.

Rutledge, J. C., A. E. Mullick et al. (2000). Direct visualization of lipid deposition and reverse lipid transport in a perfused artery: Roles of VLDL and HDL. *Circ. Res.* **86**(7): 768–773.

Rybaltowski, M., Y. Suzuki et al. (2011). In vivo imaging analysis of the interaction between unusually large von Willebrand factor multimers and platelets on the surface of vascular wall. *Pflugers Arch.* **461**(6): 623–633.

Smith, B. R., Z. Cheng et al. (2008). Real-time intravital imaging of RGD-quantum dot binding to luminal endothelium in mouse tumor neovasculature. *Nano Lett.* **8**(9): 2599–2606.

Spinelli, A. E., G. Fiacchi et al. (2008). Iterative EM reconstruction of cardiac small animal PET images using system point spread function modelling and MAP with anatomical priors. In *Nuclear Science Symposium*, Institute of Electrical and Electronics Engineers, Piscataway, NJ, pp. 5147–5152.

Stone, E. M., T. A. Braun et al. (2004). Missense variations in the fibulin 5 gene and age-related macular degeneration. *N. Engl. J. Med.* **351**(4): 346–353.

Trogan, E., Z. A. Fayad et al. (2004). Serial studies of mouse atherosclerosis by in vivo magnetic resonance imaging detect lesion regression after correction of dyslipidemia. *Arterioscler. Thromb. Vasc. Biol.* 24(9): 1714–1719.

Tsui, B. M. W., Y. C. Wang et al. (2005). Feasibility of micro-SPECT/CT imaging of atherosclerotic plaques in a transgenic mouse model. In M. A. Kupinski and H. H. Barrett (eds.), *Small-Animal SPECT Imaging.* Springer, New York, pp. 215–224.

Wan, W. and R. L. Gleason, Jr. (2013). Dysfunction in elastic fiber formation in fibulin-5 null mice abrogates the evolution in mechanical response of carotid arteries during maturation. *Am. J. Physiol. Heart Circ. Physiol.* **304**(5): H674–H686.

Wan, W., H. Yanagisawa et al. (2010). Biomechanical and microstructural properties of common carotid arteries from fibulin-5 null mice. *Ann. Biomed. Eng.* **38**(12): 3605–3617.

Wang, Y. X., B. Martin-McNulty et al. (2001). Angiotensin II increases urokinase-type plasminogen activator expression and induces aneurysm in the abdominal aorta of apolipoprotein E-deficient mice. *Am. J. Pathol.* **159**(4): 1455–1464.

Weinreb, D. B., J. G. Aguinaldo et al. (2007). Non-invasive MRI of mouse models of atherosclerosis. *NMR Biomed.* **20**(3): 256–264.

Weissleder, R. (2002). Scaling down imaging: Molecular mapping of cancer in mice. *Nat. Rev. Cancer* **2**(1): 11–18.

Wu, X., N. Balu et al. (2013). Molecular MRI of atherosclerotic plaque progression in an ApoE(−/−) mouse model with a CLT1 peptide targeted macrocyclic Gd(III) chelate. *Am. J. Nucl. Med. Mol. Imaging* **3**(5): 446–455.

Yin, J., X. Li et al. (2011). Novel combined miniature optical coherence tomography ultrasound probe for in vivo intravascular imaging. *J. Biomed. Opt.* **16**(6): 060505.

Zhu, X. Y., M. D. Bentley et al. (2007). Early changes in coronary artery wall structure detected by microcomputed tomography in experimental hypercholesterolemia. *Am. J. Physiol. Heart Circ. Physiol.* **293**(3): H1997–H2003.

Index

Printed and bound by CPI Group (UK) Ltd, Croydon, CR0 4YY

01/11/2024

01782604-0020